Surimi and Surimi Seafood

Second Edition

FOOD SCIENCE AND TECHNOLOGY

A Series of Monographs, Textbooks, and Reference Books

Editorial Advisory Board

Gustavo V. Barbosa-Cánovas Washington State University–Pullman
P. Michael Davidson University of Tennessee–Knoxville
Mark Dreher McNeil Nutritionals, New Brunswick, NJ
Owen R. Fennema University of Wisconsin–Madison
Richard W. Hartel University of Wisconsin–Madison
Y. H. Hui Science Technology System, West Sacramento, CA
Lekh R. Juneja Taiyo Kagaku Company, Japan
Marcus Karel Massachusetts Institute of Technology
Daryl B. Lund University of Wisconsin–Madison
Sanford Miller Virginia Polytechnic Institute & State University
David B. Min The Ohio State University
Seppo Salminen University of Turku, Finland
James L. Steele University of Wisconsin–Madison
John H. Thorngate III University of California–Davis
Pieter Walstra Wageningen University, The Netherlands
John R. Whitaker University of California–Davis
Rickey Y. Yada University of Guelph, Canada

76. Food Chemistry: Third Edition, *edited by Owen R. Fennema*
77. Handbook of Food Analysis: Volumes 1 and 2, *edited by Leo M. L. Nollet*
78. Computerized Control Systems in the Food Industry, *edited by Gauri S. Mittal*
79. Techniques for Analyzing Food Aroma, *edited by Ray Marsili*
80. Food Proteins and Their Applications, *edited by Srinivasan Damodaran and Alain Paraf*
81. Food Emulsions: Third Edition, Revised and Expanded, *edited by Stig E. Friberg and Kåre Larsson*
82. Nonthermal Preservation of Foods, *Gustavo V. Barbosa-Cánovas, Usha R. Pothakamury, Enrique Palou, and Barry G. Swanson*
83. Milk and Dairy Product Technology, *Edgar Spreer*
84. Applied Dairy Microbiology, *edited by Elmer H. Marth and James L. Steele*

85. Lactic Acid Bacteria: Microbiology and Functional Aspects, Second Edition, Revised and Expanded, *edited by Seppo Salminen and Atte von Wright*
86. Handbook of Vegetable Science and Technology: Production, Composition, Storage, and Processing, *edited by D. K. Salunkhe and S. S. Kadam*
87. Polysaccharide Association Structures in Food, *edited by Reginald H. Walter*
88. Food Lipids: Chemistry, Nutrition, and Biotechnology, *edited by Casimir C. Akoh and David B. Min*
89. Spice Science and Technology, *Kenji Hirasa and Mitsuo Takemasa*
90. Dairy Technology: Principles of Milk Properties and Processes, *P. Walstra, T. J. Geurts, A. Noomen, A. Jellema, and M. A. J. S. van Boekel*
91. Coloring of Food, Drugs, and Cosmetics, *Gisbert Otterstätter*
92. *Listeria*, Listeriosis, and Food Safety: Second Edition, Revised and Expanded, *edited by Elliot T. Ryser and Elmer H. Marth*
93. Complex Carbohydrates in Foods, *edited by Susan Sungsoo Cho, Leon Prosky, and Mark Dreher*
94. Handbook of Food Preservation, *edited by M. Shafiur Rahman*
95. International Food Safety Handbook: Science, International Regulation, and Control, *edited by Kees van der Heijden, Maged Younes, Lawrence Fishbein, and Sanford Miller*
96. Fatty Acids in Foods and Their Health Implications: Second Edition, Revised and Expanded, *edited by Ching Kuang Chow*
97. Seafood Enzymes: Utilization and Influence on Postharvest Seafood Quality, *edited by Norman F. Haard and Benjamin K. Simpson*
98. Safe Handling of Foods, *edited by Jeffrey M. Farber and Ewen C. D. Todd*
99. Handbook of Cereal Science and Technology: Second Edition, Revised and Expanded, *edited by Karel Kulp and Joseph G. Ponte, Jr.*
100. Food Analysis by HPLC: Second Edition, Revised and Expanded, *edited by Leo M. L. Nollet*
101. Surimi and Surimi Seafood, *edited by Jae W. Park*
102. Drug Residues in Foods: Pharmacology, Food Safety, and Analysis, *Nickos A. Botsoglou and Dimitrios J. Fletouris*
103. Seafood and Freshwater Toxins: Pharmacology, Physiology, and Detection, *edited by Luis M. Botana*
104. Handbook of Nutrition and Diet, *Babasaheb B. Desai*
105. Nondestructive Food Evaluation: Techniques to Analyze Properties and Quality, *edited by Sundaram Gunasekaran*
106. Green Tea: Health Benefits and Applications, *Yukihiko Hara*

107. Food Processing Operations Modeling: Design and Analysis, *edited by Joseph Irudayaraj*
108. Wine Microbiology: Science and Technology, *Claudio Delfini and Joseph V. Formica*
109. Handbook of Microwave Technology for Food Applications, *edited by Ashim K. Datta and Ramaswamy C. Anantheswaran*
110. Applied Dairy Microbiology: Second Edition, Revised and Expanded, *edited by Elmer H. Marth and James L. Steele*
111. Transport Properties of Foods, *George D. Saravacos and Zacharias B. Maroulis*
112. Alternative Sweeteners: Third Edition, Revised and Expanded, *edited by Lyn O'Brien Nabors*
113. Handbook of Dietary Fiber, *edited by Susan Sungsoo Cho and Mark L. Dreher*
114. Control of Foodborne Microorganisms, *edited by Vijay K. Juneja and John N. Sofos*
115. Flavor, Fragrance, and Odor Analysis, *edited by Ray Marsili*
116. Food Additives: Second Edition, Revised and Expanded, *edited by A. Larry Branen, P. Michael Davidson, Seppo Salminen, and John H. Thorngate, III*
117. Food Lipids: Chemistry, Nutrition, and Biotechnology: Second Edition, Revised and Expanded, *edited by Casimir C. Akoh and David B. Min*
118. Food Protein Analysis: Quantitative Effects on Processing, *R. K. Owusu-Apenten*
119. Handbook of Food Toxicology, *S. S. Deshpande*
120. Food Plant Sanitation, *edited by Y. H. Hui, Bernard L. Bruinsma, J. Richard Gorham, Wai-Kit Nip, Phillip S. Tong, and Phil Ventresca*
121. Physical Chemistry of Foods, *Pieter Walstra*
122. Handbook of Food Enzymology, *edited by John R. Whitaker, Alphons G. J. Voragen, and Dominic W. S. Wong*
123. Postharvest Physiology and Pathology of Vegetables: Second Edition, Revised and Expanded, *edited by Jerry A. Bartz and Jeffrey K. Brecht*
124. Characterization of Cereals and Flours: Properties, Analysis, and Applications, *edited by Gönül Kaletunç and Kenneth J. Breslauer*
125. International Handbook of Foodborne Pathogens, *edited by Marianne D. Miliotis and Jeffrey W. Bier*
126. Food Process Design, *Zacharias B. Maroulis and George D. Saravacos*
127. Handbook of Dough Fermentations, *edited by Karel Kulp and Klaus Lorenz*
128. Extraction Optimization in Food Engineering, *edited by Constantina Tzia and George Liadakis*

129. Physical Properties of Food Preservation: Second Edition, Revised and Expanded, *Marcus Karel and Daryl B. Lund*
130. Handbook of Vegetable Preservation and Processing, *edited by Y. H. Hui, Sue Ghazala, Dee M. Graham, K. D. Murrell, and Wai-Kit Nip*
131. Handbook of Flavor Characterization: Sensory Analysis, Chemistry, and Physiology, *edited by Kathryn Deibler and Jeannine Delwiche*
132. Food Emulsions: Fourth Edition, Revised and Expanded, *edited by Stig E. Friberg, Kare Larsson, and Johan Sjoblom*
133. Handbook of Frozen Foods, *edited by Y. H. Hui, Paul Cornillon, Isabel Guerrero Legarret, Miang H. Lim, K. D. Murrell, and Wai-Kit Nip*
134. Handbook of Food and Beverage Fermentation Technology, *edited by Y. H. Hui, Lisbeth Meunier-Goddik, Ase Solvejg Hansen, Jytte Josephsen, Wai-Kit Nip, Peggy S. Stanfield, and Fidel Toldrá*
135. Genetic Variation in Taste Sensitivity, *edited by John Prescott and Beverly J. Tepper*
136. Industrialization of Indigenous Fermented Foods: Second Edition, Revised and Expanded, *edited by Keith H. Steinkraus*
137. Vitamin E: Food Chemistry, Composition, and Analysis, *Ronald Eitenmiller and Junsoo Lee*
138. Handbook of Food Analysis: Second Edition, Revised and Expanded, Volumes 1, 2, and 3, *edited by Leo M. L. Nollet*
139. Lactic Acid Bacteria: Microbiological and Functional Aspects: Third Edition, Revised and Expanded, *edited by Seppo Salminen, Atte von Wright, and Arthur Ouwehand*
140. Fat Crystal Networks, *Alejandro G. Marangoni*
141. Novel Food Processing Technologies, *edited by Gustavo V. Barbosa-Cánovas, M. Soledad Tapia, and M. Pilar Cano*
142. Surimi and Surimi Seafood: Second Edition, *edited by Jae W. Park*

Surimi and Surimi Seafood

Second Edition

Edited by
Jae W. Park

Taylor & Francis
Taylor & Francis Group

Boca Raton   London   New York   Singapore

A CRC title, part of the Taylor & Francis imprint, a member of the
Taylor & Francis Group, the academic division of T&F Informa plc.

Published in 2005 by
CRC Press
Taylor & Francis Group
6000 Broken Sound Parkway NW, Suite 300
Boca Raton, FL 33487-2742

© 2005 by Taylor & Francis Group, LLC
CRC Press is an imprint of Taylor & Francis Group

No claim to original U.S. Government works
Printed in the United States of America on acid-free paper
10 9 8 7 6 5 4 3 2 1

International Standard Book Number-10: 0-8247-2649-9 (Hardcover)
International Standard Book Number-13: 978-0-8247-2649-2 (Hardcover)
Library of Congress Card Number 2004059362

This book contains information obtained from authentic and highly regarded sources. Reprinted material is quoted with permission, and sources are indicated. A wide variety of references are listed. Reasonable efforts have been made to publish reliable data and information, but the author and the publisher cannot assume responsibility for the validity of all materials or for the consequences of their use.

No part of this book may be reprinted, reproduced, transmitted, or utilized in any form by any electronic, mechanical, or other means, now known or hereafter invented, including photocopying, microfilming, and recording, or in any information storage or retrieval system, without written permission from the publishers.

For permission to photocopy or use material electronically from this work, please access www.copyright.com (http://www.copyright.com/) or contact the Copyright Clearance Center, Inc. (CCC) 222 Rosewood Drive, Danvers, MA 01923, 978-750-8400. CCC is a not-for-profit organization that provides licenses and registration for a variety of users. For organizations that have been granted a photocopy license by the CCC, a separate system of payment has been arranged.

Trademark Notice: Product or corporate names may be trademarks or registered trademarks, and are used only for identification and explanation without intent to infringe.

Library of Congress Cataloging-in-Publication Data

Surimi and surimi seafood / edited by Jae W. Park. — 2nd ed.
    p. cm. (Food science and technology ; 142)
Includes bibliographical references and index.
ISBN 0-8247-2649-9 (alk. paper)
1. Surimi. 2. Fishery processing. I. Park, Jae Won, 1953- II. Series.

SH336.S94S96 2004
664'.94—dc22   2004059362

Taylor & Francis Group
is the Academic Division of T&F Informa plc.

Visit the Taylor & Francis Web site at
http://www.taylorandfrancis.com

and the CRC Press Web site at
http://www.crcpress.com

Preface

To accommodate the fast-paced surimi and surimi seafood industry, this second edition is edited 5 years subsequent to the publication of the first edition. This revised book has been expanded with new chapters (Isolation of Functional Fish Proteins, New Developments in Japan; Sensory; Sanitation and HACCP; Microbiology and Pasteurization). Most of the remaining chapters have also been extensively revised and expanded with newly updated information.

The production of surimi using various resources has continued to grow little by little every year. The world consumption of surimi seafood, however, has not extensively grown although active developments are noticeable in Eastern Europe and South America. The future of surimi remains promising, regardless of a possible crisis in surimi resources. New technology in isolating functional fish proteins, which provides a yield of theoretically achievable recovery, is likely to play a major role in the future of surimi. This process is also likely to utilize small pelagic fish as a functional food ingredient.

Including the recent establishment of the Surimi School South America (January 2004), we now have four formal surimi schools in the world. Since its first establishment in Astoria, Oregon, in 1993, the Surimi School has been offered every year in Oregon, where a one-day Surimi Industry Forum has also been available since 2001. The Forum covers a wide range of industrial problems and issues with completely different agendas each year. Surimi School overseas has been offered every other year since 1996 (in Bangkok, Thailand), 1999 (in Massy, France), and 2004 (in Lima, Peru, alternatively with Santiago, Chile). With the success of the OSU Surimi School, I hope this book will continue to serve as a major informational handbook for surimi and surimi seafood research.

Jae W. Park
Oregon State University
July 1, 2004

Acknowledgments

I would like to thank all the authors who have joined me in this second edition for their dedication and tolerance of my impatience. Special thanks go to my research and editorial assistant Angela Hunt for her creative and consistent work. In addition, I would like to recognize all of my former students and visiting professors/scientists for their dedicated work, which has enabled me to write my chapters. Finally, I would like to thank my kids, Duke and Caroline, for their understanding of my life and never-ending workload.

Editor

Jae W. Park, Ph.D., is Professor of Food Science and Technology at the Oregon State University (OSU) Seafood Lab, Astoria. He is the author, co-author, or editor of more than 100 journal articles, book chapters, patents, and booklets. Almost all of his publications have dealt with surimi, surimi seafood, or related technologies. He is a professional member of the Institute of Food Technologists and has served as its Division Chair for Aquatic Food Products (2002–2003). His professional membership also extends to the Society of Japanese Fisheries Science, the Korean Fisheries Society, and National Fisheries Institute. Before becoming a professor in 1992, he was a Director of Technical Services at SeaFest/JAC Creative Foods, Motley, Minnesota, which later merged under Louis Kemp/ConAgra Foods. He founded the OSU Surimi Technology School (www.surimischool.org) in Astoria, Oregon; Bangkok, Thailand; Massey, France; Lima, Peru/Santiago, Chile, in 1993, 1996, 1999, and 2004, respectively. Park received his B.S. degree (1980) in animal science from Kon-Kuk University, Seoul, Korea; his M.S. degree (1982) in meat science from the Ohio State University; and his Ph.D. degree (1985) in food science from North Carolina State University, Raleigh.

Contributors

Patricio Carvajal, Ph.D., Research Associate, Dept. of Food Science, North Carolina State University, has actively studied the cryoprotection and cryostabilization of fish proteins.

Yeung Joon Choi, Ph.D., Professor, Dept. of Seafood Science, Gyeong Sang National University, Tong Yeung, South Korea. He has actively studied seafood biochemistry and enzymes, and is recognized as a leader in surimi research in Korea.

Mark Daeschel, Ph.D., Professor, Dept. of Food Science and Technology, Oregon State University, Corvallis. He has actively studied food safety and microbiology and has taught microbiology at the Surimi School in Bangkok.

Véronique Dubosc, Flavor Chemist, Activ International (Mitry, France), has presented flavor lectures at the OSU Surimi School in Paris and Oregon.

Joseph Frazier, Processing Specialist, National Food Processors Association, Seattle, Washington. He is involved in the thermal processing of various seafood products and led the U.S. surimi seafood industry research on pasteurization.

Pascal Guenneuges, Ph.D., President, International Seafood Consulting. He has been actively involved in surimi trading for the past 10 years. He has given lectures on surimi supply and demand at the OSU Surimi School in Oregon and Bangkok.

Herbert O. Hultin, Ph.D., Professor at Gloucester Marine Station, University of Massachusetts, Amherst, Massachusetts. He has developed new protein recovery processes using a pH shift and patented the process. He is recognized as a world leader in muscle chemistry.

Osamu Inami, Senior Chemist, T. Hasegawa, Japan, actively developed various colors in surimi seafood applications. He participated in the OSU Surimi School as a speaker in 1993.

Alan Ismond, P.Eng., has a degree in chemical engineering and 30 years' experience in the food industry. He is a partner in Aqua-Terra Consultants, the leading consulting firm in environmental and process engineering products, serving the seafood processing and rendering industries on the U.S. west coast.

Craig Johnson, Assistant General Manager, Konica Minolta Instrument Systems Division, has taught color measurement to various segments of food industry in the United States.

Jacek Jaczynski, Ph.D., Assistant Professor at West Virginia University, has recently published five publications dealing with electron beam and surimi seafood for microbiological and physicochemical changes.

Ik-Soon Kang, Ph.D., Oscar Meyer/Kraft Foods, Madison, Wisconsin.

Byung Y. Kim, Ph.D., Professor Emeritus, Dept. of Food Science, Kyung Hee University, Suwon, South Korea, has participated in the OSU Surimi Technology School as a rheology speaker. He is recognized as a leader in seafood rheology in Korea.

Edward Kolbe, Ph.D., Professor, Dept. of Bioresource Engineering, Oregon State University, has regularly participated in the Annual OSU Surimi Technology School as a speaker.

Kunihiko Konno, Ph.D., Professor, Hokkaido University, is recognized as a leader in muscle biochemistry and is the editor of *Fisheries Science*.

Contributors

Hurdor Kristensen, Ph.D., Assistant Professor, University of Florida, is recognized as a leading seafood chemist in warm-water species.

Tyre C. Lanier, Ph.D., Professor, Dept. of Food Science, North Carolina State University. Previously edited the book *Surimi Technology* for Marcel Dekker in 1991. He has been involved in surimi research and teaching for more than 25 years. He has participated in the OSU Surimi Technology School in Oregon and Paris as a speaker, and is recognized as a leader in surimi science and technology.

Gabriel J. Lauro, Ph.D., Director of Natural Color Research Center, California State University – Pomona, has been involved in research, teaching, and manufacturing of natural colorants for more than 35 years.

T.M. John Lin, Ph.D., General Manager for Pacific Surimi has participated regularly in the OSU Surimi School in Oregon as a speaker, and is recognized as an industry leader in surimi technology.

Grant A. MacDonald, Director, MacDonald and Associates Ltd., Nelson, New Zealand. Has been actively researching surimi and seafood quality for 20 years. He is recognized as a leader in freezing of seafoods, seafood quality and processing of Southern Hemisphere cold-water fish species. Consulting Food Scientist and teaches Seafood Science at Massey University and Auckland University.

Amby Mankoo, Vice President — Creative, Takasago International, actively created various seafood flavors and works with key industry accounts for product development.

Charles Manley, Ph.D, Vice President — Science and Technology of Takasago International Corp. (USA). He is recognized and respected as one of the leading flavor scientists in the industry. He is a former president of the Institute of Food Technology (IFT), the most highly regarded society for food science and technology in the world, and has served as president of the Flavor and Extract Manufacturers' Association (FEMA), the leading trade association, representing over 110 companies.

Michael T. Morrissey, Ph.D., Professor, Dept. of Food Science, Director of OSU Seafood Lab, Oregon State University, has participated in the OSU Surimi Technology School in Oregon and Bangkok as a speaker.

Jean-Marc Sieffermann, Ph.D., Professor, ENSIA, Massy, France, has presented lectures on sensory science at the OSU Surimi School in France and Oregon. He is actively involved in consumer behaviors in sensory science, and has published numerous articles on the subject. He is recognized as a leader in sensory science and has given invited lecturers to various research institutes and private industries.

Yi-Cheng Su, Ph.D., Assistant Professor, Dept. of Food Science & OSU Seafood Lab, Oregon State University. He has taught microbiology at the OSU Surimi School in Oregon. He is actively involved in seafood safety research.

Jirawat Yongsawatdigul, Ph.D., Assistant Professor, School of Food Science, Suranaree University of Technology, Nakhon Rachasima, Thailand. He has participated in the OSU Surimi School in Bangkok as a speaker, and is recognized as a leader in surimi research in Southeast Asia.

Won B. Yoon, Ph.D., Group Leader at Seafood Technology, CJ Foods Research Center, Seoul, Korea. He has recently published more than ten papers on seafood and hydrocolloid rheology and is recognized as a leading rheologist in the Korean food industry.

Contents

PART I Surimi

Chapter 1 Surimi Resources 3
1.1 Introduction 4
1.2 Cold-Water Whitefish Used for Surimi 5
    1.2.1 Alaska Pollock 5
        1.2.1.1 History of the Pollock Industry 5
        1.2.1.2 Current Trends in Pollock Harvests 8
        1.2.1.3 Management 11
    1.2.2 Pacific Whiting 13
    1.2.3 Arrowtooth Flounder 16
    1.2.4 Southern Blue Whiting and Hoki 17
    1.2.5 Northern Blue Whiting 19
    1.2.6 Other Whitefish Resources (South America) 19
1.3 Tropical Fish Used for Surimi 19
    1.3.1 Threadfin Bream (*Nemipterus* spp.) 20
    1.3.2 Lizardfish (*Saurida* spp.) 21
    1.3.3 Bigeye Snapper (*Priacanthus* spp.) 22
    1.3.4 Croaker (*Sciaenidae*) 23
    1.3.5 Other Species 23

1.4 Pelagic Fish Used for Surimi 25
1.5 Conclusions: Changes in Surimi Supply and Demand 26
References 29

Chapter 2   Surimi: Manufacturing and Evaluation 33

2.1 Introduction 35
2.2 Processing Technology and Sequence 37
    2.2.1 Heading, Gutting, and Deboning 37
    2.2.2 Mincing 38
    2.2.3 Washing and Dewatering 39
    2.2.4 Refining 40
    2.2.5 Screw Press 41
    2.2.6 Stabilizing Surimi with Cryoprotectants 42
    2.2.7 Freezing 44
    2.2.8 Metal Detection 47
2.3 Biological (Intrinsic) Factors Affecting Surimi Quality 48
    2.3.1 Effects of Species 48
    2.3.2 Effects of Seasonality and Sexual Maturity 49
    2.3.3 Effects of Freshness or Rigor 52
2.4 Processing (Extrinsic) Factors Affecting Surimi Quality 53
    2.4.1 Harvesting 53
    2.4.2 On-Board Handling 55
    2.4.3 Water 58
    2.4.4 Time/Temperature of Processing 60
    2.4.5 Solubilization of Myofibrillar Proteins during Processing 62
    2.4.6 Washing Cycle and Wash Water Ratio 66
    2.4.7 Salinity and pH 69
2.5 Processing Technologies that Enhance Efficiency and Profitability 76
    2.5.1 Neural Network 76
    2.5.2 Processing Automation: On-Line Sensors 80
    2.5.3 Digital Image Analysis for Impurity Measurement 81
    2.5.4 Innovative Technology for Wastewater 82
    2.5.5 Fresh Surimi 84
2.6 Decanter Technology 86
2.7 Surimi Gel Preparation for Better Quality Control 91
    2.7.1 Chopping 91
        2.7.1.1 2% and 3% Salt 92

		2.7.1.2	Moisture Adjustment	92
		2.7.1.3	Chopping Temperature	92
		2.7.1.4	Effect of Vacuum	93
	2.7.2	Cooking		93
		2.7.2.1	Plastic Casing and Stainless Steel Tube	93
		2.7.2.2	Cooking Method Resembling Commercial Production of Crabstick	93
2.8	Summary			95
Acknowledgments				97
References				98

Chapter 3 Process for Recovery of Functional Proteins by pH Shifts 107

3.1	Introduction	108
3.2	Characteristics of Dark Muscle Fish Crucial to Surimi Processing	109
	3.2.1 Dark Muscle	109
	3.2.2 Lipids	113
	3.2.3 Muscle Proteins	115
	3.2.4 Processing of Pelagic Fish	117
	3.2.4.1 Alkaline Processing	119
	3.2.4.2 Problems with Processing Dark-Muscled Species	120
3.3	A New Approach for Obtaining Functional Protein Isolates from Dark-Muscled Fish	123
3.4	Summary	132
References		133

Chapter 4 Sanitation and HACCP 141

4.1	Introduction	142
4.2	Sanitation	143
4.3	Good Manufacturing Practices (GMPs)	144
4.4	Hazard Analysis Critical Control Point (HACCP)	145
4.5	Principles of the HACCP System	146
4.6	HACCP for Surimi Production	148
4.7	HACCP for Surimi Seafood Production	149
4.8	Microbiological Standards and Specifications for Surimi Seafood	151
4.9	Sanitation Standard Operating Procedures (SSOPs)	153

4.10	Cleaners and Sanitizers	154
4.11	Verification	158
References		161

Chapter 5 Stabilization of Proteins in Surimi 163

5.1	Introduction		164
5.2	Myosin and Fish Proteins		165
5.3	Stability of Myosin		166
	5.3.1	Stabilization of the Globular Head Region	166
	5.3.2	Helix-Coil Transition in the Myosin Rod	168
5.4	Intrinsic Stability of Fish Muscle Proteins		170
	5.4.1	Influence of Animal Body Temperature	170
	5.4.2	Naturally Occurring (Protecting and Nonprotecting) Osmolytes	172
	5.4.3	Antifreeze Proteins	175
5.5	Stability of Frozen Surimi Proteins		176
	5.5.1	Cold Destabilization	177
	5.5.2	Ice Crystallization	177
	5.5.3	Hydration and Hydration Forces	179
	5.5.4	Other Destabilizing Factors during Frozen Storage	181
5.6	Mechanisms for Cryoprotection and Cryostabilization		184
	5.6.1	Solute Exclusion from Protein Surfaces	184
	5.6.2	Ligand Binding	188
	5.6.3	Antioxidants	189
	5.6.4	Freezing-Point Depression	189
	5.6.5	Cryostabilization by High Molecular Weight Additives	191
	5.6.6	Vitrification	195
5.7	Processing Effects on Surimi Stability		196
	5.7.1	Fish Freshness	196
	5.7.2	Leaching	197
	5.7.3	Freezing Rate	197
5.8	Stabilized Fish Mince		198
5.9	Stabilization of Fish Proteins to Drying		201
	5.9.1	Processes for Drying of Surimi	202
		5.9.1.1 Freeze-Drying	202
		5.9.1.2 Spray-Drying of Surimi	205
	5.9.2	Potential Additives and Mechanisms of Lyoprotection	206

| Contents | xvii |

5.10 Future Developments in Fish Protein Stabilization 210
References 213

Chapter 6 Proteolytic Enzymes and Control in Surimi 227

6.1 Introduction 228
6.2 Classification of Proteolytic Enzymes 231
    6.2.1 Acid Proteases (Lysosomal Cathepsins) 232
        6.2.1.1 Cathepsin A 233
        6.2.1.2 Cathepsin B 234
        6.2.1.3 Cathepsin C (Dipeptidyl Transferase) 235
        6.2.1.4 Cathepsin D 236
        6.2.1.5 Cathepsin E 237
        6.2.1.6 Cathepsin H 237
        6.2.1.7 Cathepsin L 238
    6.2.2 Neutral and Ca^{2+}-Activated Proteinases 241
    6.2.3 Alkaline Proteinases 245
6.3 Sarcoplasmic vs. Myofibrillar Proteinases 249
6.4 Control of Heat-Stable Fish Proteinases 250
    6.4.1 Proteinase Inhibitors 250
        6.4.1.1 Inhibitors of Serine Proteinases 251
        6.4.1.2 Inhibitors of Cysteine Proteinases 252
        6.4.1.3 Inhibitors of Metalloproteinases 253
    6.4.2 Food-Grade Proteinase Inhibitors 253
    6.4.3 Minimization of Proteolysis by Process Control 259
6.5 Summary 260
References 260

Chapter 7 Waste Management and By-Product Utilization 279

7.1 Introduction 281
7.2 Surimi Waste Management and Compliance 283
    7.2.1 Measurements Needed for Compliance 284
        7.2.1.1 Accurate Wastewater Flow Meters 284
        7.2.1.2 Correct Measurement of Solid Concentration 285
        7.2.1.3 Correct Reporting of Tonnage 286

	7.2.2	How to Implement a Waste Management Program	286
		7.2.2.1 Plant Audit and Mass Balance	286
		7.2.2.2 Water Reduction/Reuse	287
		7.2.2.3 Waste Solids Recovery	287
7.3	Solid Waste		288
	7.3.1	Fish Meal and Fish Protein Hydrolysates	288
	7.3.2	Fish Oil Recovery	290
	7.3.3	Specialty Products	292
7.4	Surimi Wastewater		293
	7.4.1	Chemical Methods	295
	7.4.2	Biological Methods	297
		7.4.2.1 Aerobic Process	297
		7.4.2.2 Anaerobic Process	298
	7.4.3	Physical Methods	299
		7.4.3.1 Dissolved Air Flotation	299
		7.4.3.2 Heat Coagulation	300
		7.4.3.3 Electrocoagulation	301
		7.4.3.4 Centrifugation	301
		7.4.3.5 Membrane Filtration	303
7.5	Recovery of Bioactive Components and Neutraceuticals		306
	7.5.1	Bioactive Compounds	306
		7.5.1.1 Enzymes	307
		7.5.1.2 Other Waste Compounds	309
	7.5.2	Recovery of Bioactive Compounds	310
7.6	Opportunities and Challenges		311
	7.6.1	The Limitations of Fish Solids Recovery	311
		7.6.1.1 Quality Impediments	311
		7.6.1.2 Environmental Limitations	312
		7.6.1.3 Marketing Impediments	312
		7.6.1.4 Proximity to Market	312
		7.6.1.5 Labor and Maintenance Considerations	313
	7.6.2	Current and Future Potential	313
7.7	Summary		315
References			316

Chapter 8 Freezing Technology 325

8.1 Introduction 326
8.2 Horizontal Plate Freezers 327

8.3	Airflow Freezers		331
	8.3.1	Spiral Freezer	331
	8.3.2	Tunnel Freezer	332
	8.3.3	Blast Freezer	334
8.4	Brine Freezers		336
	8.4.1	Sodium Chloride (NaCl)	337
	8.4.2	Calcium Chloride ($CaCl_2$)	337
	8.4.3	Glycols	338
	8.4.4	Other	338
8.5	Cryogenic Freezers		338
	8.5.1	Liquid Nitrogen (LN)	339
	8.5.2	Carbon Dioxide (CO_2)	340
8.6	Freezing the Product		342
8.7	Freezing Capacity		344
8.8	Freezing Time		347
8.9	Some "What-If" Effects on Freezing Time		356
	8.9.1	Block Thickness	357
	8.9.2	Cold Temperature Sink, T_a	357
	8.9.3	Heat Transfer Coefficient, U	359
8.10	Energy Conservation		362
	8.10.1	Freezer Design and Operation	363
	8.10.2	Refrigeration Machinery Options	366
	8.10.3	A Blast Freezer Case	367
8.11	Conclusions		368
Acknowledgments			368
References			369

PART II Surimi Seafood

Chapter 9 Surimi Seafood: Products, Market, and Manufacturing 375

9.1	Introduction		376
	9.1.1	Surimi-Based Products in Japan and the United States	377
		9.1.1.1 Japanese Market	377
		9.1.1.2 The U.S. Market	380
	9.1.2	Market Developments in France	383
	9.1.3	Surimi Seafood Products in Other Countries	385

9.2 Manufacture of Surimi-Based Products 388
    9.2.1 Kamaboko 388
    9.2.2 Chikuwa 390
    9.2.3 Satsuma-age/Tenpura 390
    9.2.4 Hanpen 392
    9.2.5 Fish Ball 393
    9.2.6 Surimi Seafood 395
        9.2.6.1 Filament Meat Style 396
        9.2.6.2 Solid Meat Style 417
9.3 Other Processing Technology 419
    9.3.1 Ohmic Heating 419
    9.3.2 High Hydrostatic Pressure 424
    9.3.3 Least-Cost Linear Programming 425
Acknowledgments 428
References 430

Chapter 10 Surimi Gelation Chemistry 435

10.1 Introduction 436
10.2 Protein Components of Surimi 437
    10.2.1 Myofibrillar Proteins 437
        10.2.1.1 Myosin 439
        10.2.1.2 Actin 441
        10.2.1.3 Other Myofibrillar Proteins 441
        10.2.1.4 Thick Filament Assembly 442
    10.2.2 Stroma Proteins 444
    10.2.3 Sarcoplasmic Proteins 444
        10.2.3.1 Heme Proteins 446
        10.2.3.2 Enzymes 447
10.3 Lipid Components of Fish Muscle 450
10.4 Bonding Mechanisms during Heat-Induced Gelation
    of Fish Myofibrillar Proteins 451
    10.4.1 Hydrogen Bonds 451
    10.4.2 Ionic Linkages (Salt Bridges) 452
    10.4.3 Hydrophobic Interactions 455
    10.4.4 Covalent Bonds 456
        10.4.4.1 Disulfide Bonds 456
        10.4.4.2 Rheological Behavior of Cross-Linked
                  Protein Gels 459
        10.4.4.3 Role of Disulfide Bonding in
                  Myosin/Actomyosin Gelation 459

		10.4.4.4	Covalent Cross-Linking during Setting.. 461
		10.4.4.5	Protein Stability Effects on Setting.. 467
		10.4.4.6	Endogenous Transglutaminase (TGase).. 468
		10.4.4.7	Exogenous TGase Addition................. 469

10.5 Factors Affecting Fish Protein Denaturation and Aggregation.. 471
    10.5.1 The Importance of Muscle pH (Acidity)............. 475
    10.5.2 The Frozen Storage Stability of Surimi............ 476
10.6 Summary: Factors Affecting Heat-Induced Gelling Properties of Surimi.. 476
References... 477

Chapter 11 Rheology and Texture Properties of Surimi Gels.. 491

11.1 Introduction 493
11.2 Fundamental Test 496
    11.2.1 Force and Stress... 496
    11.2.2 Deformation and Strain...................................... 498
    11.2.3 Flow and Rate of Strain 499
    11.2.4 Rheological Tests Using Small Strain (Deformation).. 501
        11.2.4.1 Compressive Test for Surimi Gel........ 501
        11.2.4.2 Shear Type Test for Surimi Paste and Gels... 501
        11.2.4.3 Stress Relaxation Test........................ 503
        11.2.4.4 Oscillatory Dynamic Test 508
    11.2.5 Rheological Testing Using Large Strain (Failure Test) 518
        11.2.5.1 Axial Compression for Cylinder Type Gels... 519
        11.2.5.2 Compressive Test for Convex Shape Samples 520
        11.2.5.3 Compressive Test for Rod-Type Samples... 521
        11.2.5.4 Torsion Test 523
11.3 Empirical Tests.. 528
    11.3.1 Punch (Penetration) Test..................................... 529

	11.3.2	Texture Profile Analysis (TPA)	534
	11.3.3	Relationship between Torsion and Punch Test Data	537
11.4	Effects of Processing Parameters on Rheological Properties of Surimi Gels		540
	11.4.1	Effects of Fish Freshness/Rigor Condition	540
	11.4.2	Effect of Refrigerated Storage of Gels	540
	11.4.3	Effect of Sample Temperature at Measurement	541
	11.4.4	Effect of Moisture Content	543
	11.4.5	Effect of Low Temperature Setting	544
	11.4.6	Effect of Freeze-Thaw Abuse	546
	11.4.7	Effect of Functional Additives	546
	11.4.8	Texture Map	548
11.5	Viscosity Measurements		549
	11.5.1	Measurement of Dilute Extract	550
	11.5.2	Measurement of Surimi Seafood Pastes	550
	11.5.3	Rheological Behavior of Surimi Paste	556
11.6	Practical Application of Dynamic Rheological Measurements		558
	11.6.1	Gelation Kinetics of Surimi Gels	559
		11.6.1.1 Thermorheological Properties and Kinetic Model	559
		11.6.1.2 Nonisothermal Kinetic Model	561
		11.6.1.3 Rubber Elastic Theory	562
		11.6.1.4 Dependence of Gelation Temperature on Moisture Content	563
		11.6.1.5 Activation Energy during Gelation	565
	11.6.2	Estimation of Steady Shear Viscosity of Fish Muscle Protein Paste	567
		11.6.2.1 Steady Shear Viscosity of Surimi Paste and the Cox-Merz Rule	567
		11.6.2.2 Superimposed Viscosity Using the Cox-Merz Rule	568
		11.6.2.3 Concentration Dependence of the Viscosity of Surimi Paste	569
11.7	Summary		574
Acknowledgements			575
References			576

Chapter 12 Microbiology and Pasteurization of Surimi
Seafood 583

12.1 Introduction 585
12.2 Growth of Microorganisms in Foods 585
12.3 Surimi Microbiology 587
12.4 Microbial Safety of Surimi Seafood 590
    12.4.1 *Listeria Monocytogenes* 591
    12.4.2 *Clostridium Botulinum* 593
12.5 Pasteurization of Surimi Seafood 596
12.6 Process Considerations and Pasteurization
Verification for Surimi Seafood 600
    12.6.1 Principles of Thermal Processing to
Surimi Seafood Pasteurization 601
    12.6.2 D-Value 602
    12.6.3 z-Value 602
    12.6.4 F-Value (Lethality Value) 605
    12.6.5 General Considerations for Heat Process
Establishment or Verification 606
    12.6.6 Study Design and Factors Affecting
Pasteurization Process 607
    12.6.7 Temperature Distribution Test Design 608
    12.6.8 Heat Penetration Test Design 610
    12.6.9 Initial Temperature (IT) and Product Size 611
    12.6.10 Product Preparation/Formulation 611
    12.6.11 Heat Resistance of Selected "Target"
Microorganism 614
    12.6.12 Analyzing the Pasteurization
Penetration Data 617
12.7 Temperature Prediction Model for Thermal Processing
of Surimi Seafood 624
12.8 Predictive Model for Microbial Inactivation during
Thermal Processing of Surimi Seafood 626
12.9 New Technologies for Pasteurization: High-Pressure
Processing and Electron Beam 628
    12.9.1 High-Pressure Processing 628
    12.9.2 Food Irradiation 629
    12.9.3 Electron Beam 630
    12.9.4 Electron Penetration in Surimi Seafood 631
    12.9.5 Microbial Inactivation in Surimi Seafood 633
    12.9.6 Effect of E-Beam on Other Functional
Properties of Surimi Seafood 634

12.10 Packaging Considerations.. 637
References.. 638

Chapter 13 Ingredient Technology for Surimi and
Surimi Seafood 649

13.1 Introduction 650
13.2 Ingredient Technology... 653
    13.2.1 Water... 653
    13.2.2 Starch... 657
        13.2.2.1 What Is Starch?................................. 657
        13.2.2.2 Modification of Starch...................... 657
        13.2.2.3 Starch as a Functional Ingredient
                for Surimi Seafood 659
    13.2.3 Protein Additives... 668
        13.2.3.1 Whey Proteins.................................... 669
        13.2.3.2 Egg White Proteins........................... 673
        13.2.3.3 Plasma Proteins................................. 676
        13.2.3.4 Soy Proteins....................................... 677
        13.2.3.5 Wheat Gluten and Wheat
                Flour.. 681
    13.2.4 Hydrocolloids 681
        13.2.4.1 Carrageenan....................................... 682
        13.2.4.2 Konjac.. 683
        13.2.4.3 Curdlan... 684
        13.2.4.4 Alginate... 685
    13.2.5 Cellulose... 686
    13.2.6 Vegetable Oil and Fat Replacer 686
    13.2.7 Food-Grade Chemical Compounds.................... 689
        13.2.7.1 Oxidizing Agents............................... 689
        13.2.7.2 Calcium Compounds 689
        13.2.7.3. Transglutaminase (TGase).................. 692
        13.2.7.4 Phosphate 695
        13.2.7.5 Coloring Agents................................. 696
13.3 Evaluation of Functional Ingredients............................... 699
    13.3.1 Texture 699
    13.3.2 Color 700
    13.3.3 Formulation Development and
        Optimization.. 700
Acknowledgements... 701
References.. 702

Chapter 14 Surimi Seafood Flavors ... 709

14.1 Introduction ... 710
14.2 What Is Flavor? ... 712
    14.2.1 Creation of a Flavor ... 712
    14.2.2 Natural Product Chemistry ... 713
        14.2.2.1 Solvent Extraction ... 715
        14.2.2.2 Gas Chromatography–Olfactometry (GCO) ... 716
        14.2.2.3 Headspace Analysis ... 716
    14.2.3 Building a Flavor ... 717
14.3 Basic Seafood Flavor Chemistry ... 720
    14.3.1 Sources of Flavor Ingredients ... 720
        14.3.1.1 Natural Extracts ... 720
        14.3.1.2 Synthetic Components ... 721
    14.3.2 The Importance of Lipids in Fish Flavors ... 721
    14.3.3 Important Components Found in Seafood Extracts ... 724
        14.3.3.1 Volatile Compounds ... 725
        14.3.3.2 Nonvolatile Compounds ... 726
14.4 Additives and Ingredients Used in Flavors ... 727
    14.4.1 Glutamate ... 728
    14.4.2 Ribonucleotides ... 728
    14.4.3 Hydrolyzed Proteins ... 728
    14.4.4 Yeast Extracts ... 729
14.5 The "Off Flavors" of Seafood ... 730
14.6 Effects of Processing on Seafood ... 731
14.7 Flavor Release and Interactions ... 731
14.8 Effects of Ingredients on Flavor ... 733
    14.8.1 Sorbitol and Sugar ... 734
    14.8.2 Starch ... 734
    14.8.3 Surimi (Raw material) ... 735
    14.8.4 Egg Whites and Soy Proteins ... 736
    14.8.5 Vegetable Oil ... 736
    14.8.6 Salt ... 736
14.9 Processing Factors Affecting Flavors ... 737
    14.9.1 Adding Additional Flavor/Flavor Components ... 737
    14.9.2 Addition Points ... 737
    14.9.3 Encapsulation ... 738
    14.9.4 Storage Conditions and Shelf Life ... 738

14.10 Flavor Regulations and Labeling 739
    14.10.1 United States 739
    14.10.2 European Union (EU) 740
    14.10.3 Japan 741
    14.10.4 A Potential World List — The United Nations 742
    14.10.5 Religious Certification Issues 743
    14.10.6 Worldwide Issues 743
14.11 Summary 744
References 745

Chapter 15 Color Measurement and Colorants for Surimi Seafood 749

15.1 Introduction 751
15.2 Understanding Color and Measurement 753
    15.2.1 Development of Color Language 753
    15.2.2 Color Space 754
    15.2.3 Instrument Development 756
    15.2.4 Tristimulus Values 756
    15.2.5 L*a*b* Color Space 757
    15.2.6 Indices 759
    15.2.7 Measuring Color 759
        15.2.7.1 Tristimulus Measurement 759
        15.2.7.2 Spectrophotometric Measurement 762
15.3 Coloring Surimi Seafood 765
    15.3.1 Preparation of Surimi Paste for Crabsticks 766
    15.3.2 Color Application to Crabsticks 766
    15.3.3 General Principles 766
15.4 Colorants 769
    15.4.1 Colorants Requiring Certification 769
    15.4.2 Colorants Not Requiring Certification 770
        15.4.2.1 Carmine (21 CFR 73.100, EEC No. 120, CI No. 75470, CI Natural Red 4) 771
        15.4.2.2 Cochineal Extract (21CFR 73.100, EEC No. E120, CI Number 75470, Natural Red) 773
        15.4.2.3 Paprika (21CFR 73.345, EEC No. E 160c) 776

| | | 15.4.2.4 | Annatto (21 CFR 73.30, EEC No. E 160b, CI 75120, CI Natural Orange 4) | 780 |

            15.4.2.5 Turmeric (21CFR 73.600, EEC No. E 100, CI No. 75300, CI Natural Yellow 3) 783
            15.4.2.6 Grape Color (21CFR 73.169, EEC No, E163) and Other Anthocyanins 784
            15.4.2.7 Beet Juice Concentrate (21CFR 73.260) 785
            15.4.2.8 Caramel (21 CFR73.85, EEC No. 150) 786
        15.4.3 Monascus Colorants 787
        15.4.4 Nature Identical Colorants 787
            15.4.4.1 Canthaxanthin (21 CFR 73.73.75, EEC No. E161g) 787
            15.4.4.2 β-Carotene (21 CFR 73.95) 788
        15.4.5 Other Naturally Derived Colorants 788
            15.4.5.1 Titanium Dioxide (E171, CI No. 77891, CI Pigment White 6) 788
            15.4.5.2 Calcium Carbonate (EE 170, CI No. 77220, CI pigment White 18) 788
            15.4.5.3 Vegetable Oil 789
    15.5 Color Quality 789
        15.5.1 Final Product Color 789
        15.5.2 Colorant Quality 789
        15.5.3 Acceptance Criteria 789
        15.5.4 Acceptance Tolerance 792
    15.6 Labeling 794
        15.6.1 Requirements in the United States 794
        15.6.2 Religious Requirements 795
    15.7 Summary 796
    References 796
    Additional Reading 798

Chapter 16 Application of Sensory Science to Surimi Seafood 803

    16.1 Introduction 805
        16.1.1 What Is Sensory Evaluation? 805
        16.1.2 Why Should We Care about Sensory Evaluation? 805

	16.1.3	Brief History 806
	16.1.4	Fundamentals of Sensory Evaluation 807
		16.1.4.1 Complex Nature of Sensory Measurement 807
		16.1.4.2 Sensory Perception Is More than Product Measurement 807
		16.1.4.3 Multidimensional Sensory Answer 809
16.2	Who Is Sensory Evaluation Working For? 811
	16.2.1	Research and Development (R&D) 811
	16.2.2	Production 812
	16.2.3	Marketing 812
16.3	Developing a Sensory Approach 813
	16.3.1	What Is the Problem? 813
	16.3.2	Sensory Human Resources 817
		16.3.2.1 Experimenter 817
		16.3.2.2 Panel 818
	16.3.3	Sensory Laboratory and Sensory Test Conditions 820
		16.3.3.1 Analytical Studies 820
		16.3.3.2 Hedonic Studies 822
	16.3.4	Sensory Tests 823
		16.3.4.1 Difference Tests 823
		16.3.4.2 Threshold Tests 823
		16.3.4.3 Descriptive Tests 824
	16.3.5	Statistics for Descriptive Analysis 826
		16.3.5.1 Basic Statistical Analysis 828
		16.3.5.2 Variance Analysis 828
		16.3.5.3 Multivariate Analysis 828
	16.3.6	Consumer Tests 828
		16.3.6.1 Declarative Methodologies 829
		16.3.6.2 Behavioral Methodologies 829
	16.3.7	Summary: Which Tests for Which Panelists? 831
16.4	Correlating Sensory Evaluation with Instrumental and Consumer Measures 832
	16.4.1	Why Is Instrumental Not Enough? 832
	16.4.2	Linking Consumer Data with Analytical Data: Preference Mapping Techniques 835
16.5	Conclusion: Sensory Evaluation from the Lab to the Consumers 837
References 844

Contents

Chapter 17 New Developments and Trends in Kamaboko and Related Research in Japan 847

17.1 History of Kamaboko 848
17.2 Variations in Kamaboko Products in Japan 849
    17.2.1 Steamed Kamaboko on a Wooden Board: Itatsuke Kamaboko 850
    17.2.2 Grilled Kamaboko on Wooden Board: Yakinuki Kamaboko 851
    17.2.3 Grilled Kamaboko on Bamboo Stick: Chikuwa 851
    17.2.4 Deep-Fried Kamaboko: Age-Kamaboko 851
    17.2.5 Boiled Kamaboko: Hanpen and Tsumire 852
    17.2.6 Crab Leg Meat Analog, Crabstick: Kani-ashi Kamaboko or Kanikama 853
    17.2.7 Fish Sausage and Ham 853
    17.2.8 Other Kamaboko 853
17.3 Change in Fish Species Used for Kamaboko Production 855
17.4 Trends of Kamaboko Products: Quality, Variety, and Nutrition 857
17.5 Scientific and Technological Enhancement in Kamaboko in Japan during the Past 10 to 15 Years 859
    17.5.1 Recent Progress in Gelation Mechanism 859
    17.5.2 Recent Progress in the Understanding of Setting 860
    17.5.3 Recent Progress in the Myosin Denaturation Study in Relation to Gelation 862
    17.5.4 Myosin Rod Aggregation at High Temperature in Relation to Gelation 863
    17.5.5 Biochemical Index for the Quality Evaluation of Frozen Surimi 865
References 866

Appendix Code of Practice for Frozen Surimi 869

Index 887

Part I
Surimi

1

Surimi Resources

PASCAL GUENNEUGUES, PH.D.

International Seafood Consulting,
Mill Creek, Washington

MICHAEL T. MORRISSEY, PH.D.

Oregon State University Seafood Laboratory,
Astoria, Oregon

CONTENTS

1.1 Introduction 4
1.2 Cold-Water Whitefish Used for Surimi 5
    1.2.1 Alaska Pollock 5
        1.2.1.1 History of the Pollock Industry 5
        1.2.1.2 Current Trends in Pollock Harvests 8
        1.2.1.3 Management 11

  1.2.2 Pacific Whiting 13
  1.2.3 Arrowtooth Flounder 16
  1.2.4 Southern Blue Whiting and Hoki 17
  1.2.5 Northern Blue Whiting 19
  1.2.6 Other Whitefish Resources (South America) 19
1.3 Tropical Fish Used for Surimi 19
  1.3.1 Threadfin Bream (*Nemipterus* spp.) 20
  1.3.2 Lizardfish (*Saurida* spp.) 21
  1.3.3 Bigeye Snapper (*Priacanthus* spp.) 22
  1.3.4 Croaker (*Sciaenidae*) 23
  1.3.5 Other Species 23
1.4 Pelagic Fish Used for Surimi 25
1.5 Conclusions: Changes in Surimi Supply
  and Demand 26
References 29

1.1 INTRODUCTION

The surimi industry has changed dramatically over the past decade. A decrease in Alaska pollock harvests, from over 6.5 million metric tons in the late 1980s to less than 3 million metric tons since the year 2000, has opened the door for the utilization of new species in the surimi industry. Southeast Asia initiated the expansion by utilizing threadfin bream to make surimi (itoyori), which now represents 25% of the total volume of surimi production. New technologies have opened the way for new resources to be used as raw material for surimi. The use of protease inhibitors, for example, has made it possible to use Pacific whiting and other species for surimi production. Decanter technology and new washing techniques have allowed the processing of surimi from fatty fish such as mackerel. In addition, several advances have been made to increase the yield of surimi production, thereby greatly improving the economics of the surimi process.

  The global decrease in whitefish supply has strengthened the demand for other product forms (e.g., fillets and blocks) made from Alaska pollock, while the surimi seafood industry has learned to use lower-quality surimi (i.e., lower gel functionality and darker color) to process surimi products, thus favoring the production of surimi from other species. The international marketplace for surimi has changed as well. Japanese consumption of surimi seafood continuously declined during the 1990s and early 2000s as the younger generations are gradually shifting to a more Western and meat diet.[1] This shift has been partly offset by an increasing market for surimi-based products in Europe, Russia, and Southeast Asia, and these new markets are more open to different sources of raw materials and new methods of processing.

The key terms for the 21st century are "fisheries management" and "sustainable fisheries." There will be increased emphasis on stock assessment, fisheries by-catch, conservation, and maximum utilization of what is harvested, as well as an emphasis on ecological issues.

The surimi supply fluctuated between 450,000 and 550,000 metric tons (t) during the 1990s, but has increased slightly in recent years and is now close to 600,000 t. This increase is due to the production of non-pollock surimi, particularly tropical fish in Southeast Asia and more recently pelagic fish in South America. The production of pollock surimi has stabilized after a significant decrease during the 1990s. This is primarily due to the increased yield in production from traditional surimi processing operations. Surimi is now produced from resources that span the globe, as shown in Figure 1.1. This chapter discusses surimi production and fisheries resources for surimi.

1.2 COLD-WATER WHITEFISH USED FOR SURIMI

1.2.1 Alaska Pollock

1.2.1.1 History of the Pollock Industry

Alaska pollock (*Theragra chalcogramma*) still represents the largest fishery biomass used for surimi production. The fish has white flesh, tends to be uniform in size, can be captured in large volumes by both coastal vessels and factory trawlers,

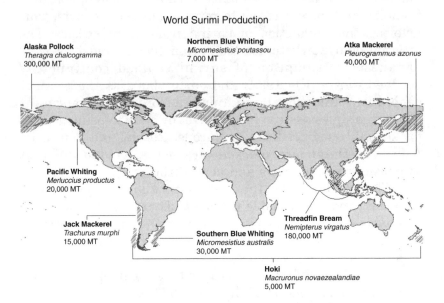

Figure 1.1 World map showing major surimi resources and surimi production.

and makes a high-quality surimi product. At about 3 million t, Alaska pollock represents the largest global whitefish catch and ranks second only to the Peruvian anchovy for total catch.[2]

The pollock fisheries for surimi began in the mid-1960s after Japanese scientists discovered the function of cryoprotectants in preserving protein functionality for frozen, washed fish mince in 1960. This allowed Japanese vessels to fish at-sea for extended periods of time and to harvest, process, and store frozen surimi in the vessels. In addition, the domestic surimi-based seafood industry in Japan was revolutionized as manufacturers could now receive product year-round.[3] Alaska pollock soon became the resource of choice for the rapidly growing surimi industry. It also became the largest resource for industrial fish fillet blocks that are used by the industry for processing breaded fish and prepared food.

The next major change occurred in the mid-1980s with the "Americanization" of the fishery. In 1977, the Exclusive

Economic Zone (EEZ) was implemented for U.S. waters. However, there were neither sufficient U.S. vessels nor U.S. markets to take advantage of the expanding fishing grounds. During the earlier years of the EEZ, licensed foreign vessels (e.g., Japanese or Korean) or joint-venture (JV) operations between U.S. catcher vessels and foreign processing ships harvested most of the Alaska pollock. Today in U.S. waters, U.S. flagged vessels harvest and process all Alaska pollock. In the Russian EEZ, the fishery is a combination of domestic ships and foreign fisheries (Japanese and Korean), as well as JV operations.

The Alaska pollock industry is a combination of shore-based processors, catcher/processors, and motherships (smaller trawlers transferring the trawl bag at-sea to processing vessels). Shore-side operations occur primarily in Dutch Harbor, and factories are also located in Akutan, Kodiak, and, Sandpoint. By 1999, there were seven companies producing surimi at these locations, which were close enough to the fishing grounds to allow shore-based vessels to capture pollock and return to offload at plants, usually within 48 hr.

At-sea, operations by catcher/processors permit fishing throughout the Bering Sea. These are large-capacity vessels that can process 500 t of fish per day. During the 1980s, these vessels were critical to the production of surimi for world markets. Since the beginning of the 2000s, however, these vessels have gradually shifted over 50% of their production to fillet blocks.

The law allocates the fishing quota (TAC: total allowable catch) between the shore plants, factory trawlers, motherships, and CDQ (community development quota). The Pollock Conservation Cooperative was formed for at-sea catcher/processors.[4] This represents a private contractual agreement among the nine processing companies that own catcher/processors and harvest Alaska pollock. They work within the framework of the federal fishery management structure and are allocated a percentage of the quota. This has changed the fishery from a "race to fish" Olympic fishery to one where it can be run more as a business operation, thereby improving the efficiency and economic return of the fishery.

1.2.1.2 Current Trends in Pollock Harvests

The Alaska pollock fishing areas are divided into several regions in the Northern Pacific. The Eastern Bering Sea off Alaska and the Okhotsk Sea region in Russian waters are traditionally the most productive areas. In 1997, the total catch of Alaska pollock was 3.56 million t, of which the majority came from the Russian sector. Due to overfishing and poor management practices, the total Russian catch declined to less than 1.0 million t. For example, the Western Bering Sea had a harvest high of 1.25 million t in 1989 and decreased to a low of 0.40 million t by 1996. There was also a decline in the other Asian regions during the 1990s due to overfishing by Korea and Japan in their coastal waters.

The U.S. pollock fishery has fared better. Since incorporation of the EEZ in 1977, the U.S. capture has been consistent, averaging approximately 1.2 million t per year. The lowest year was 1987, at 0.9 million t, while the highest capture was in 1991, at 1.6 million t. In 2003, the recorded harvest was 1.4 million t, which was considered sustainable.[5]

Although there are several different species currently being used for surimi production, Alaska pollock is still the most utilized. Pollock is also used for several other seafood products, including fillets, minced blocks, and headed and gutted (H&G) products.[6] It is important to understand the trends in the world catch of whitefish and how this might affect the use of pollock for products other than surimi. Figure 1.2 shows the trend of whitefish harvests over the past 10 years in different fishing areas.

There was a steady decline from 1995 to 2002, from approximately 9 million t down to 6 million t.[2] The main reasons for the decline were decreases in the cod fisheries in the North Atlantic, declines in the South American hake fisheries, and a collapse of the Russian Alaska pollock industry in the North Pacific. By 2004, the worldwide harvest of whitefish had stabilized at 6.5 million t, with Alaska pollock at around 2.9 million t. The Russian pollock fishery decreased from a high of 4.0 million t to approximately 1.0 million t, but there is general consensus that the bottom has been

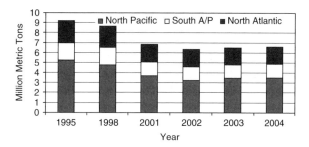

Figure 1.2 Total harvests of whitefish in main fishing areas. A/P: Atlantic/Pacific.

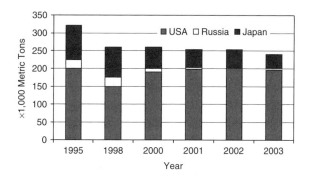

Figure 1.3 Production of Alaska pollock surimi by main producers.

reached and that the stocks can recover. Whether management of the fisheries is strong enough to help this recovery in Russian waters remains to be seen.

The global surimi production from Alaska pollock from all countries is approximately 250,000 t per year as shown in Figure 1.3. Russian surimi production has become insignificant as the majority of its harvest is directed to the whitefish market of fillets and H&G. The Japanese production of pollock surimi has also continued to decline and is less than 40,000 t.

Because the U.S. production of Alaska pollock is the largest harvest of whitefish in the world, it is important to know the processing trends and which products are being produced. During the first half of the 1990s, the majority of the U.S.

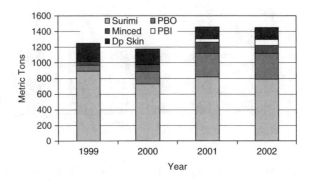

Figure 1.4 U.S. Alaska pollock utilization. Dp skin, deep skinned fillet; PBI, pin bone in fillet; PBO, pin bone out fillet.

harvest was directed toward surimi production. However, due to the reduced supply of pollock H&G (some of which is reprocessed as fillets in China) and reduced production of fillet blocks from Russia, by the end of the 1990s, there was an increased demand for U.S. pollock fillet blocks in the world market.[5] For this reason, the utilization of U.S. Alaska pollock for surimi has decreased over the past years, down to about 50% of the U.S. pollock capture, while there is increased production of alternative products made from pollock (Figure 1.4). There has been a significant increase in the production of pinbone out (PBO) fillets, which are becoming more popular in the world marketplace. The increasing demand for whitefish products and the lower demand for Alaska pollock surimi should drive, over time, more pollock to alternative markets, particularly fillet production.

The U.S. surimi industry had been criticized for low yields and a high percentage of surimi waste. U.S. surimi production has stabilized in recent years at around 200,000 t although the utilization of pollock for surimi production has decreased. The main reason for this stems from the increase in production recovery. The reauthorization of the Magnuson Fisheries Act stressed the need for increased utilization of what was harvested. As shown in Figure 1.5, the recovery of surimi from raw fish has doubled from the early 1990s until today. Many

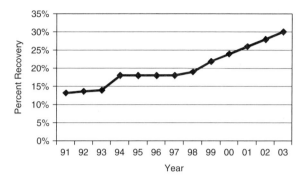

Figure 1.5 Increased recovery of surimi in Alaska pollock processing operations.

surimi operations now recover more than 25% of the fish into surimi. A dramatic increase in recovery occurred from 1998 onward, particularly as a result of better cutting of the fish and implementation of the recovery of meat from the frames and washwater. World demand for lower-quality surimi has also allowed processors to market recovery grade or to blend it with primary grades to produce medium/low-quality surimi.

A better understanding of surimi processing through several research efforts has allowed surimi manufacturers to improve their process operations and yields (see Chapter 2). New improvements in technology and a better understanding of processing techniques have helped increase recovered protein at both ends of the surimi processing line. Additional cuts (e.g., collar cut) at the beginning of the surimi process provided higher mince recovery, as did the use of meat separators on the frames. New designs of high-speed decanters that can separate washed muscle protein have also allowed recovery of small meat particles from the washwater effluent of surimi operations.

1.2.1.3 Management

The most dramatic change in the management of the U.S. Alaska pollock fishery has been the formation of the Pollock

Conservation Cooperative (PCC). The PCC was formed in December 1998 to "promote the rational and orderly harvest of pollock by the catcher/processor sector of the Bering Sea and Aleutian Islands off Alaska." The PCC consists of eight companies that own 19 catcher/processors and are allowed to harvest Alaska pollock under the American Fisheries Act (AFA).[7] Each member company of the PCC receives a percentage of the quota as specified by the AFA. The North Pacific Fisheries Management Council determines the quotas. For the 2003 season, this translated to approximately 480,000 t Alaska pollock for the PCC members.

The advantages of forming the cooperative are several — from both a fisheries/processing standpoint as well as improving returns to the company. In the past, the pollock season was divided into two or three seasons and specific quotas within those seasons. It was essentially an Olympic fishery as the boats raced each other to capture the quota within the specified dates. This caused inefficiencies in fishing and processing operations as the boats had to focus on the speed of fishing and processing throughput rather than recovery. With a specific quota allocated to each company, they can now use several strategies in capturing and processing the fish in an effort to maximize the recovery and the return on this fixed amount of fish. For example, boats can move to schools of different-sized fish, thereby optimizing the size for an automatic filleting operation onboard the vessel. There has also been a noticeable decrease in by-catch as unwanted species can be avoided over time. Moreover, companies can focus on fish quality and processing the products that will bring the best return. This has led to higher yields in surimi processing, as previously mentioned. The formation of the PCC has also spurred investment in processing equipment to improve utilization of the fish that are harvested. Finally, the PCC system may, in the long run, favor the production of fillets vs. surimi by factory trawler because the return on fillets is higher, on average, over time. Previously, the Olympic system favored faster throughput and surimi processing while the PCC system allows at-sea processors and their companies to be more selective in their fishery, products, and eventual marketplace.

The formation of the PCC has also addressed the overcapitalization issue that plagued U.S. fisheries during the 1990s. Several vessels were retired from the fishery; and under the present structure, there is less need to build new vessels to speed up the capture of pollock. Although controversial when it first began, most observers agree that the PCC is a rational organization that works to the benefit of the fishery from both an economical and biological standpoint. It has stabilized the U.S. pollock fishery and helped make it sustainable.

The U.S. Alaska pollock fishery has always been considered a shining example of how federal management of a fishery can work well. Others have taken notice. The Marine Stewardship Council, an environmental organization that certifies sustainable fisheries, is currently evaluating the U.S. Alaska pollock fishery for certification.[8]

1.2.2 Pacific Whiting

Utilization of Pacific whiting (*Merluccius productus*) for surimi production showed rapid growth in the 1990s. At one time, Pacific whiting surimi represented approximately 20% of surimi production in the United States but decreased significantly during the 2002–2003 seasons as quotas were low and poor prices also kept production low.

Pacific whiting is a gadoid fish related to pollock and is one of many whitings harvested throughout the world. The majority of whitings — Argentine (*M. hubbsi*), Chilean (*M. gayi*), Peruvian (*M. gayi peruanus*), South African/Namibian (*M. capensis*), and others — are fished primarily for products such as fillets or H&G.[9] All whiting/hakes have white flesh and a delicate taste and texture, and some are highly prized as a premier eating fish. Of all the fish in the whiting/hake family, none has a greater combination of quality problems associated with it than the Pacific whiting off the West Coast of the United States and Canada. This stock has a number of biological and intrinsic quality characteristics that make controlling product quality difficult. These characteristics include a high variability in recruitment, complex migration patterns, and a relatively soft flesh that is often made even softer through the presence of high levels of protease enzymes in the muscle tissue.[10]

The presence of protease in whiting flesh is associated with a myxosporidean parasite that is endemic throughout the Pacific whiting stocks.[11] This parasite is microscopic and is not associated with any human health hazard. However, it appears to induce an immune response in the fish and an increased amount of lysosomal proteases called cathepsins.[12,13]

During the late 1980s and early 1990s, scientists at several laboratories investigated the use of different food-grade protease inhibitors in whiting surimi production. The most active inhibitor was found to be beef plasma protein, which is commonly used at about 1% concentration.[14] Using beef plasma protein, high-quality surimi could be manufactured from either at-sea or shore-side operations. However, with the outbreak of mad cow disease, the use of beef plasma has become problematic, and Pacific whiting surimi with beef plasma has been banned by most buyers or even by regulation first in Europe, then in Japan and Korea, and finally in the United States.

Several other protease inhibitors such as egg white and whey protein have been identified and are also used in the production of whiting surimi. Research is continuing in this area of inhibitor development from both natural sources and using recombinant DNA for a soy cystatin inhibitor that has shown promise.[15] A study by Yongsawatigul and Park[16] showed that rapid heating of Pacific whiting surimi is also sufficient to make good-quality surimi seafood products without the use of protease inhibitors, but the protease activity in Pacific whiting surimi continues to remain an obstacle to the marketing of this product.

Pacific whiting has been captured off the West Coast of the United States since the 1960s, primarily through foreign fisheries that could operate to within 12 miles of the U.S. coastline. With the implementation of the EEZ in 1977, a combination of foreign fisheries and JV-mothership operations occurred. Because of the increasing demand for surimi worldwide, the Alaska at-sea fishing industry took an interest in utilizing whiting for surimi production.

Prior to 1990, Russia, Korea, Poland, and others either froze the majority of Pacific whiting whole or used it as an H&G product. Part of the reason for this was the presence of a muscle protease in the flesh that would rapidly break down myofibrillar proteins and prevent the formation of a surimi gel. Once this problem was solved by the use of protease inhibitors, U.S. whiting surimi production began. In 1991, 217,505 t of Pacific whiting was harvested, mainly by at-sea operations, and the majority was made into surimi. In 1992, the first shore-based processing plants were built in Astoria and Newport, Oregon. By 1994, 250,000 t Pacific whiting were harvested, of which 72,000 t were landed on the shore-side, the majority in Oregon where there are three surimi plants. Pacific whiting now represents more than 50% of the volume of all landings in Oregon, which is a dramatic change from its traditional fisheries.

Pacific whiting spawns in the winter off Baja California, Mexico, and then migrates northward along the West Coast of the United States during the spring. The fishery begins in May for the at-sea fisheries, and in June for the shore-side operations. Because of the enzymatic activity in Pacific whiting, it is critical that fish are landed and processed within 24 hr of capture. For the most part, this occurs in both the at-sea fishery and shore-side operations. Whiting boats in Oregon leave their homeports and are usually in the whiting grounds within 6 to 8 hr. A 2-hr tow is often successful in bringing aboard between 50,000 and 70,000 kg fish, which are then chilled by refrigerated seawater. Coastal boats normally offload their catch within 12 to 16 hr post-capture. Longer periods cause softening and degradation of the tissue for surimi production.[17]

The ex-vessel price for whiting is one of the lowest for any white-fleshed fish. Over the 1995 to 2003 seasons, fishermen received U.S.$0.025 to 0.05 per pound for whiting. They are still able to make a profit at these low prices as trips could be made on a daily basis and volumes were often greater than 120,000 lb per trip. Pacific whiting surimi has received a lower price than pollock in the global marketplace and the majority was exported to Korea.

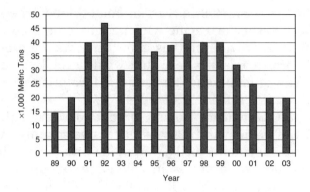

Figure 1.6 Pacific whiting surimi production.

Originally, the Pacific whiting stocks were considered healthy, as strict management quotas have been applied since the late 1970s. The harvests averaged over 200,000 t through the 1990s, and the season length was close to 100 days for fishing. Poor recruitment into the fishery, however, found managers having to reduce the allowable quota. The fishery dropped from a level of 232,000 t in 2000 to 129,600 t by 2002, and was declared overfished and required a rebuilding plan. However, additional data from 2004 showed that the stocks were healthy and the quota could be significantly increased to about 250,000 t once again.

The at-sea whiting sector has also formed a Whiting Conservation Cooperative that has allowed a more planned harvest of the fish and helped maximize returns.[18] While this bodes well for the fishery, record low prices for Pacific whiting surimi and a reluctance to use beef plasma protein in the product have created marketing problems in the industry. With a weak demand for Pacific whiting surimi, the production can be expected to decrease or remain at low levels despite the increased quotas (Figure 1.6).

1.2.3 Arrowtooth Flounder

Arrowtooth flounder (*Atheresthes stomias*) is a relatively large flounder found off the West Coast of the United States and northward to the Bering Sea. The allowable biological catch (ABC) was estimated to be over 100,00 t for 2001 in the Eastern Bering Sea and more than twice that in the Gulf of Alaska.[19] There was minimal harvest of arrowtooth flounder from these regions for several reasons. Arrowtooth flounder suffers from the same quality problems as Pacific whiting — that is, a heat-stable protease enzyme in the muscle tissue.[20] This gel softening can be overcome by food-grade protease inhibitors such as beef plasma protein, and early commercial runs of arrowtooth surimi have proven successful with these inhibitors. Babbitt and co-workers[21] have also developed a new surimi processing technique using a decanter in place of washing operations. This has worked well with arrowtooth and preliminary trials have been successful.

Another major concern for arrowtooth flounder utilization in the high-volume surimi industry is the lack of automated machinery for cutting the flesh (J.K. Babbitt, personal communication, 1998). Most filleting operations or mincers used in the surimi industry are designed for round fish, such as pollock and whiting. Most flatfish filleting machines are geared for fish less than 20 inches long, although they could be reconfigured for larger fish if the demand was sufficient. In addition, there are concerns about size uniformity in arrowtooth flounder. Sizes ranging from 500 to 5000 g are common and it is usually more economical to hand-fillet the fish. The greatest impediment to arrowtooth flounder utilization, however, is the by-catch issue. A directed arrowtooth flounder fishery will usually take a significant by-catch of halibut and other, more valued species. The arrowtooth flounder fishery will close down once the by-catch limits are reached and, often, less than 10% of the allowable arrowtooth catch is landed. Until the fishing technology/management issue is resolved, the use of arrowtooth flounder for surimi production will be on a limited basis.

1.2.4 Southern Blue Whiting and Hoki

Southern blue whiting (*Micromesistius australis*) and hoki (*Macuronus novaezelandiae* and *Macruronus magellanicus*) are restricted in distribution to the sub-Antarctic waters.[22] Hoki is a major New Zealand fishery, of which a small percentage is used for surimi production. During the latter half of the 1980s, hoki surimi potential reached its peak and in 1988, approximately 23,000 t were produced, representing 3% of the world's surimi supply. Hoki surimi proved to be of excellent quality and high gel strength, and most of the supply went to Japan and Korea. During the 1990s, there was a shift away from surimi production, and now the majority of the hoki harvest is used for fillets and frozen blocks. The quotas in New Zealand have averaged around 250,000 t and only a small percentage is now produced as surimi: however, there are reports of increasing hoki surimi production off the Chilean coast.

Several research groups have investigated hoki as a surimi source and have found it to be of excellent quality.[23,24] MacDonald et al.[25] compared hoki and Southern blue whiting surimi to pollock surimi and found similar gel strength results over different heating regimes. MacDonald et al.[26] showed changes in proximate composition during the season, with the highest protein and lowest moisture content occurring in the spring months in the Southern Hemisphere (October). Lipid content varied throughout the year, ranging from 1.6 to 3.7%, with the highest average lipid occurring in the months prior to winter spawning (July).

There are two distinct populations of Southern blue whiting that are actively fished, one of which (i.e., *Micromesistius australis australis*) is found near the Falkland Islands and Argentine Patagonia in the western South Atlantic and also off South Georgia, South Shetland and South Orkney Islands, and in the southeastern Pacific off Chile. The other population (*Micromesistius australis pallidus*) lives on the various banks and rises around the South Island of New Zealand. The majority of Southern blue whiting surimi is now produced in Argentina and Chile. There is also some Southern blue whiting surimi produced by Japanese vessels fishing in New Zealand waters. After peaking in 1998, surimi production has remained stable at around 30,000 t. Southern blue whiting is often contaminated with parasites that make it unsuitable for fillet production and therefore most of the catch is directed to surimi production.

1.2.5 Northern Blue Whiting

Since the early 1990s, a French factory trawler from the North Atlantic has produced Northern blue whiting (*Micromesistius Poutassou*) surimi. Northern blue whiting gives a high-quality surimi similar to Southern blue whiting. Production volume increased since the early 2000s when a second vessel went into operation in the Faroe Islands, to reach around 7,000 t in 2003. In 2003, shore-based surimi plants were also built in Russia and the Faroe Islands to process this fish from either fresh or frozen material but the low market price of the surimi has stalled these operations. However, surimi produced from frozen fish showed extremely low quality, simply due to the nature of low thermal stability of cold-water species (see Chapter 2).

1.2.6 Other Whitefish Resources (South America)

Chilean whiting (*Merluccius gayii*) has been used in limited quantities by surimi shore plants in Chile, yielding about 1000 t per year. This fish is more commonly made into fillets and has had a steady market for fillet products over the past decade.

1.3 TROPICAL FISH USED FOR SURIMI

The fish species used in Southeast Asia for the production of surimi are mainly threadfin bream (*Nemipterus* spp.), bigeye snapper (*Priacanthus* spp.), croakers (*Pennahia* and *Johnius* spp.), lizardfish (*Saurida* spp.), and goat fish/red mullet (*Upeneus* spp, *Parupeneus* spp.). These species are also commonly used in southern subtropical Japan and are known as itoyori, kinmedia, guchi, eso, and himeji, respectively.[27] Other species are also used, depending on availability (seasonality) and price. These include conger eel (*Congresoxs* spp.), barracuda (*Sphyraena* spp.), hairtail (*Trichiurus* spp.), and leather jacket

(*Stephanoleptis Cirrhifer, Navodon Modestus*). In 2002, it is estimated that about 200,000 t of surimi were made from tropical fish.

The largest surimi-producing country was Thailand at 140,000 t (50% itoyori, 15% kinmedai, 15% eso, and 10% himeji), while India produced 40,000 t (70% itoyori), Vietnam 20,000 t, and China around 10,000 t. Indonesia, Myanmar, Pakistan, and Malaysia are also currently developing their surimi industries using tropical species. There are continuing questions, however, in the fisheries management and sustainability of the fisheries and industry in these areas. However, there is no question that tropical fish surimi has become a major player in the marketplace and will remain so for the time being.

1.3.1 Threadfin Bream (*Nemipterus* spp.)

The use of threadfin bream for surimi production has increased dramatically over the past decade and will continue to play a major role in surimi markets. It has been shown to make high-quality surimi with good gel strength.[28] Threadfin bream belongs to the family Nemipteridae and about ten species are commonly found in the Indo-West Pacific region in tropical and subtropical waters. Threadfin bream forms an important part of the trawl catch, with the greater catches being landed in Thailand, India, Indonesia, the Philippines, and Malaysia.

These fish are benthic, inhabiting marine waters on sandy or muddy bottoms usually in depths of 20 to 50 m, feeding on small benthic invertebrates and small fish. Males are usually larger and some species may be protogynous hermaphrodites. Two prolonged spawning seasons occur from November to February and another from May to June. Catches of threadfin bream are usually not identified by species; but in the Southeast Asian region, the main species caught include *Nemipterus peronii, N. marginatus, N. mesoprion, N. nematophorus,* and *N. japonicus.*[30] Most of these species are 10 to 15 cm in length. The larger-sized individuals are usually sorted out for sale as whole fish for direct consumption. The size of

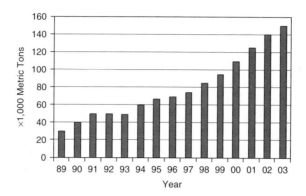

Figure 1.7 Production of surimi from threadfin bream and other tropical fish in Southeast Asia.

fish typically used for surimi processing is about 30 g per fish and they are manually headed and gutted before being subjecting to deboning.

The supply of "fresh-chilled" threadfin bream for the surimi factories in Thailand now comes from southern Thailand, with significant landings from the Andaman Sea area and also from the Myanmar waters. Most factories are also using "frozen on-board" threadfin bream caught in Indonesian waters. India has increased production considerably over the past 5 years and is now the second major producer of threadfin bream surimi.

Itoyori surimi is now well accepted in Japan and commands high prices. This has resulted in increased production of surimi from threadfin bream, as shown in Figure 1.7. Because of the white color, smooth texture, strong gel-forming ability of the fish meat, and easy processing, threadfin bream surimi is widely used as a raw material for Japanese "kamaboko" and surimi-based crabstick (*kani-kama*).

1.3.2 Lizardfish (*Saurida* spp.)

In south Japan, lizardfish (*Synodantidae*) have long been considered a high-grade raw material for surimi and kamaboko, with high meat yield, white color, good flavor, and high gel-forming ability.[30] However, the freshness and gel-forming ability decreases very quickly over time even in ice, and only very fresh raw materials are used in Japan. In the Asean region, lizardfish are considered a low-market value fish and are landed in large quantities. It gives a white surimi with low gel-forming ability. Lizardfish (*Saurida waniese*) are commonly used in Thailand for dried fish products and their fresh minced meat is used for fish cake products and surimi. Lizardfish surimi is also used for fish cakes in Japan and Korea. The main species of lizardfish in these countries are *S. tumbil* and *S. undosquamis* (usual size is 10 to 15 cm).

With a better understanding of the quality changes of lizardfish, there is the potential to improve the quality and quantity of lizardfish surimi from the region. Studies by Nozaki et al.[31] and Yasui et al.[32] indicate that the gel-forming ability of lizardfish falls rapidly during ice storage, due to the formation of formaldehyde and dimethyamine. The improvement of the gel-forming ability of lizardfish by washing the minced meat with sodium pyrophosphate solution has been reported. Ng et al.[33] showed that sodium pyrophosphate leaching of lizardfish (*S. tumbil*) was effective when formaldehyde levels did not exceed 50 ppm.

1.3.3 Bigeye Snapper (*Priacanthus* spp.)

Bigeye snapper (kinme-dai) belongs to the family *Priacanthidae* and, in the South China Sea area, is represented by two species: (1) *Priacanthus tayenus*, which is more abundant, and (2) *P. maracanthus*. Both species have a bright crimson color with a thick, tough skin. Unlike threadfin bream, which often have burst-belly when kept in ice for too long, bigeye snapper have a longer shelf life in ice. The minced meat is usually darker than threadfin bream due to the presence of a strip of dark meat on the caudal area, but has a higher gel-forming ability. Bigeye snapper is abundant in trawl catch, and often landed in substantial quantities. Due to its appearance and thick skin, it is not consumed directly and is therefore a suitable raw material for surimi manufacture. It can reach a size of 30 cm; however, the average commercial harvest for surimi is between 10 and 15 cm.

1.3.4 Croaker (*Sciaenidae*)

Croaker has long been a preferred species for the traditional kamaboko industry in Japan, especially in the Odawara area and southern Japan. Kamaboko made from croaker or "guchi" surimi has a distinctive taste and texture, distinguishing it from the relatively blander taste of kamaboko made from Alaska pollock. In the subtropical areas of Japan, China, and Taiwan, the species of croaker commonly used include the blackmouth croaker (*Atrobucca nibe*), white croaker (*Argyrosomus argenteus*), and yellow croaker (*Pseudoscianena polytis*).

In the Asean region, the croaker species used for surimi are generally smaller, comprised mainly of *Pennahia* and *Johnius* spp. These species are abundant in the trawl fishery, especially from the coastal, muddy waters off Sarawak in the South China Sea and off Myanmar in the North Andaman Sea, near large river mouths. The usual size of croaker ranges between 10 and 15 cm, and is usually sorted according to size rather than species. Croaker surimi from these species is generally darker in color than threadfin bream surimi but in Japan can fetch a high price.

1.3.5 Other Species

Goatfish (*Upeneus* spp.) and red mullet (*Parupeneus* spp.) are landed fresh from Thai waters and frozen from Indonesia. The bigger sized fish (100 to 200 g) are sold whole round or processed into skin-on fillets for Europe and other markets. The smaller-sized fish are processed into surimi called *himeji surimi*. It is slightly pinkish in color due to its skin color and has low gel strength. Several other fish species are used for surimi manufacture although they are not as abundant as the above species. These include the pike-conger eel (*hamo*), the hairtail (*Tachiuo*), barracuda (*kamasu*), and leather jacket (*klathi*).

The pike-conger eel (*Muraenesocidae*) has a cylindrical, eel-shaped body without scales, with three main species caught in the Southeast Asia region: *Congresox talabonoides*,

Congresox talabon, and *Muraenosox cinereus.* The first two species are generally larger (usually about 150 to 200 cm in total length) and are the preferred species for fish balls and fish cakes. The meat is extremely white and has a high meat yield of about 68%. The pike-conger eels are found off the coast of India eastward to Celebes, the Philippines, and the South China Sea. The fish live over soft bottom, down to 100 m, feeding mainly on bottom-living fishes, and are caught mainly by hand lines and trawl.

The hairtail (*Trichiurus lepturus*) has a very elongated and compressed body, is steel blue in color with a metallic sheen, and is silvery gray when dead. It is common in the Indo-Pacific region, up to Japan in the north and southward to Queensland, Australia. It is commonly caught in coastal waters and trawling grounds, feeding on crustaceans and fish, with a typical size of 70 to 90 cm and a maximum of 110 cm. Although it has a low gel-forming ability and the surimi is generally darker in color, it is used in Japan for surimi because of its good flavor.

Barracuda (*Sphyraenidae*) is sometimes used as raw material for surimi when in season, although its gel-forming ability is generally low. The main species used include the smaller species *Sphyraena langsar* and *S. obtusata.* The larger specie *S. jello* is usually used as salted-dried fish. Leather jacket (*Stephanoleptis cirrhifer, Navodon modestus*) has a very hard and thick skin and lots of bones. It is processed into surimi in the southern region of India and gives a low-quality grayish surimi.

The Chinese communities in Southeast Asia consider fish balls a premium fish jelly product and have used selected species based on their strong gel-forming ability and protein stability to the relatively high temperatures in the region. The two fish species used for making premium-quality fish balls include coral fish (*Caesio erythrogaster*) and wolf herring (*Chirocentrus dorab*). These two species form very stable gels that are "slow setting," and "fresh" fish balls are often formed manually in the markets in front of the customers. In Singapore, there are also specialty fish ball-noodle stalls well known for their "coral fish" fish balls.

Leather jacket (*Stephanolepsis cirhifer, Navodon modestus*) has a hard skin and its meat includes lots of bones. It is processed as dried fish for the Asian market and has also been used for surimi production. It gives a low-grade surimi that can be used for fish-ball and fried surimi products.

Peru has made some forays into making surimi from Peruvian whiting (*Merluccius gayi peruanus*) but that was quickly overfished and operations have stopped. Bereche (*Larimus pacificus*) and lumptail sea robin (*Prionotus stephanophyrs*) and other species have also been used in Peru to produce approximately 1000 t of surimi per year. Some other efforts have also been made to make surimi from Peruvian anchovies (*Engraulis ringens*).

1.4 PELAGIC FISH USED FOR SURIMI

Pelagic fish have traditionally been used in Japan for surimi production, particularly atka mackerel (*Pleurogrammus monopterygius*). This surimi has a dark color, relatively low gel strength, and a strong "fishy" taste. It is used for traditional, low-priced surimi products in Japan (fried fish cakes, etc.). The production of atka mackerel surimi in Japan was complemented in the 1990s by the import of jack mackerel (*Trachurus murphyi*) surimi produced in Chile. The resource for this fish in Chile is large (around 1.5 million t per year) and healthy; and because the fish can be caught in large volumes close to the coast, it is landed inexpensively. There has been an active research program in Japan using similar species to show the potential use of these species for surimi products.[34]

The market decrease in Japan of these surimi products, however, prompted a fall in atka mackerel surimi prices and the subsequent decrease in production of atka mackerel surimi (from 60,000 to 40,000 t) by the end of the 1990s while most of the jack mackerel surimi operations in Chile closed down. At the same time, however, the use of new technology (decanter technology for jack mackerel surimi production), resulting in better-quality surimi (higher gel, better color, less fishy taste and smell), opened new markets for jack mackerel surimi in Europe where it is now widely used for crabstick

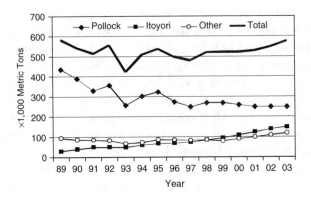

Figure 1.8 Trends in world surimi production.

production by mixing with whitefish surimi. With this new market, production increased in Chile up to 15,000 t by 2003.

1.5 CONCLUSIONS: CHANGES IN SURIMI SUPPLY AND DEMAND

Since the turbulent 1990s, surimi production has stabilized and even increased slightly (2 to 3%) between 2000 and 2003 with a total volume of slightly below 600,000 t (Figure 1.8). The decline of the Russian and Japanese pollock fisheries led to a large decrease in pollock surimi production in these countries during the early 1990s. A concomitant decrease of fillet production in Russia caused an increase in demand for whitefish fillet products from other regions in the world, with a subsequent increase in fillet production from Alaska. This led to a decrease in the production of high-quality primary grade U.S. pollock surimi (see Alaskan pollock surimi section in this chapter) and opened the door for other resources to be made into surimi. Alaska pollock now represents less than 50% of the world's surimi production.

The decrease in pollock harvest has been partially offset by the significant increase in yields in pollock surimi processing. These increased yields are due to several reasons, including:

1. Introduction of new cutting machines so that more muscle meat is brought into the processing operation
2. Improved processing with closer attention paid to the washing ratios, times, temperature, and recovery
3. Use of centrifuges in recovering small meat particles from the wash water
4. Recovery of meat from the frames and off-cuts (belly flaps, etc.) for surimi production
5. Improved fishery management by the U.S. Government.

This also resulted in a shift away from producing "high-grade" surimi to producing large quantities of lower-grade pollock surimi. This low-grade production now represents around 40% of the total produced. A lower grade product will allow the use of protein that was formerly lost in surimi processing waste and used for fish meal production. Because of this increase in recovery, pollock surimi production in the United States has stabilized in the 2000s to around 200,000 t.

Over this same time period, the use of threadfin bream (itoyori) has shown a steady increase, as have other species. While pollock surimi production was decreasing, tropical fish surimi production was dramatically increasing in Southeast Asia during the 1990s and early 2000s. Thailand threadfin bream (itoyori) production has grown over 80,000 t, followed by India (20,000 t), Vietnam, Indonesia, China, Burma, and Pakistan. Recently, surimi production from other tropical fish surimi also rapidly increased in Thailand, India, and Vietnam, particularly lizardfish (eso), bigeye (kinmedai), croaker (guchi), and red mullet (himeji). More and more tropical species have also been tested and used in surimi production. Today, production of tropical fish surimi other than itoyori represents over 50,000 t in Thailand and, including India and Vietnam, more than 80,000 t.

In the late 1990s and early 2000s, South America also developed production of surimi from pelagic species, particularly jack mackerel in Chile and gunnard fish in Peru. These fisheries are expected to exceed 20,000 t of surimi in 2004. On the other hand, the production of atka mackerel surimi in Japan has significantly decreased.

Threadfin bream surimi can be as high in quality as Alaska pollock when processed properly from fresh fish. It is the main species used in Southeast Asia for surimi seafood production and has found a good market niche in Japan, where it has replaced approximately one third of the pollock surimi supply. Unlike Alaska pollock, threadfin bream surimi is processed from manually headed and gutted fish instead of fillets, as their small size prohibits the use of automated fillet processing equipment. This surimi tends to position itself as a less expensive alternative to Alaska pollock. However, based on the nature of using frozen fish (as much as 40%) in surimi processing in Thailand, overall gel quality of threadfin bream surimi is similar to medium-grade Alaska pollock surimi. It should be noted that decent-grade surimi can be made from frozen warm-water species while extremely poor-quality surimi is made from frozen cold-water fish (see Chapter 2).

While the production of recovery-grade pollock, tropical species, and pelagic fish surimi is rapidly increasing, there is also more demand for lower-grade surimi as the markets have undergone changes during the past decade. For better or worse, surimi products have found their niche as an inexpensive fish product in many countries, thus requiring cheaper raw materials. Technological advances in surimi processing have allowed the use of lower-grade surimi in surimi-based seafood. This is especially true for the European market, but also Southeast Asia, Korea, and China tend to use more of these lower grades of surimi for manufacturing crabsticks, fish balls, fish cakes, and other surimi products. The continuing decline of the Japanese market for surimi seafood since the early 1990s and the general recession of the Japanese economy have also driven Japanese surimi seafood producers into using lower qualities of surimi instead of high-grade Alaskan pollock. This lower-quality product trend has raised some concerns in the marketplace, especially with regard to introducing new consumers to surimi-based seafood products. On the other hand, inexpensive surimi seafood products are opening opportunities for new mass markets of fish products while high-quality fish remains relatively expensive.

New processing methods such as the pH-shift method (see Chapter 3) have potential for increasing the use of small pelagic species and improving the recovery-grade surimi made from traditional species. Overall, the surimi industry is not suffering from a lack of resources and it can be expected that a number of new species will be utilized in surimi production during the coming years through new technology and the development of new industrial fisheries. The major concern in 2004 is the lack of growth in markets for surimi-based seafood, which will continue to affect surimi demand and price.

REFERENCES

1. S.C. Sonu. Surimi Supply, Demand, and Market of Japan. NOAA Technical Report NMFS. Southwest Region, National Marine Fisheries Service, NOAA, Long Beach, CA, 2002.
2. FAO. State of the World Fisheries and Aquaculture. FAO Fisheries web site (www.fao.org/sof/sofia/index_en.htm), 2004.
3. J.W. Park, T.M. Lin, and J. Yongsawatdigul. New Developments in Manufacturing of Surimi and Surimi Seafood. *Food Rev. Int.,* 13(4), 577–610, 1997.
4. PCC. Joint Report of the Pollock Conservation Cooperative (PCC) and High Seas Catchers' Cooperative. Pollock Conservation Cooperative Research Center (www.sfos.uaf.edu/pcc), University of Alaska, Fairbanks, AK, 2003.
5. AFSC. Pollock research at Alaska Fisheries Science Center (www.afsc.noaa.gov/species/pollock.htm), Alaska Fisheries Science Center, Seattle, WA, 2004.
6. H. Johnson. *Annual Report on the United States Seafood Industry, 6th ed.* Jacksonville, OR: H.M. Johnson & Associates, 2003.
7. APA. At-Sea Processors Association (APA) (www.atsea.org), Seattle, WA, 2004.
8. MSC. Marine Stewardship Council (MSC) Assessment Report for the United States Bering Sea and Aleutian Islands Pollock Fishery. MSC web site at msc.org, 2003.
9. J. Alheit and T.J. Pitcher. *Hake: Fisheries, Ecology and Markets.* New York: Chapman & Hall, 1995.

10. G. Sylvia, and M.T. Morrissey. *Pacific Whiting: Harvesting, Processing, Marketing and Quality Assurance.* Corvallis, OR: Oregon Sea Grant, 1992.

11. M. Patashnik, H.S. Groninger, H. Barnett, G. Kudo, and B. Koury. Pacific whiting (*Merlucius productus*). I. Abnormal muscle texture caused by Myxosporidean induced proteolysis. *Mar. Fish Rev.,* 44(5), 1–12, 1982.

12. H. An, M.Y. Peters, and T.A. Seymour. Roles of endogenous enzymes in surimi gelation. *Trends in Food Sci. Technol.,* 7, 321–327, 1996.

13. R. Porter, B. Koury, and G. Kudo. Inhibition of protease activity in muscle extracts and surimi from Pacific whiting (*Merlucius productus*), arrowtooth flounder (*Atheresthes stomias*). *Mar. Fish Rev.,* 55(3), 10–15, 1993.

14. M.T. Morrissey, J.W. Wu, D. Lin, and H. An. Protease inhibitor effects on torsion measurements and autolysis of Pacific whiting surimi. *J. Food Sci.,* 58, 1050–1054, 1993.

15. I.S. Kang and T.C. Lanier. Bovine plasma protein functions in surimi gelation compared with cysteine protease inhibitors. *J. Food Sci.,* 64, 842–884, 1999.

16. J. Yongsawatdigul and J.W. Park. Thermal aggregation and dynamic rheological properties of Pacific whiting and cod myosins as affected by heating rate. *J. Food Sci.,* 64, 679–683, 1999.

17. G. Peters, M.T. Morrissey, G. Sylvia, and J. Bolte. Linear regression, neural network and induction analysis to determine harvesting and processing effects on surimi quality. *J. Food Sci.,* 61, 876–880, 1996.

18. G. Sylvia and H. Munro. The Achievements of the Pacific "Whiting Conservation Cooperative": Rational Collaboration in a Sea of Irrational Competition. Report to Coastal Oregon Marine Experiment Station, Newport, OR, 2004.

19. NMFS. National Marine Fisheries Services in Alaska (www.fakr.noaa.gov), Juneau, AK, 2004.

20. D. Wasson, K. Reppond, J.K. Babbitt, and J.S. French. Effect of additives on functional properties of arrowtooth flounder surimi. *J. Aqua Food Prod. Technol.,* 1(3/4), 147–165, 1992.

21. J.K. Babbitt, K. Reppond, G. Kamath, A.C. Hardy, C.J. Pook, N. Kokubu, and S. Berntsen. Evaluation of processes for producing arrowtooth flounder surimi. *J. Aqua Food Prod. Technol.*, 2(4), 89–95, 1993.

22. NZ Fisheries. Summary of assessments of sustainability and current TACC and recent catch levels and the status of stocks for the 2000–2001 fishing year — Southern blue whiting fishery. New Zealand Ministry of Fisheries (www.fish.govt.nz), 2004.

23. M. Okada. The gel-forming capacity of some hake species from South America. In Technical Consultation on the Latin America Hake Industry. *FAO Fisheries Report*, 203(1), 153, 1978.

24. G.A. MacDonald, J. Lelievre, and N.D.C. Wilson. Strengths of gels prepared from washed and unwashed of hoki (*Macuronus novaezelandiae*) stored in ice. *J. Food Sci.*, (55)4, 976-978, 1990.

25. G.A. Macdonald, J. Stevens, and T.C. Lanier. Characterization of hoki and Southern blue whiting compared to Alaska pollock surimi. *J. Aqua Food Prod. Technol.*, 3(1), 19–38, 1994.

26. G.A. MacDonald, B.I. Hall, and P. Vlieg. Seasonal changes in hoki (*Macruonus novaezelandaie*). *J. Aqua Food Prod. Technol.*, 11(2), 35–51, 2002.

27. S.M. Tan, K. Miwa, and S. Kongaya. Advances in the fish processing industry in Southeast Asia. In *Proceedings of the International Seafood Research Meeting*, Mie University, Japan, 1994.

28. J. Yongsawatdigul, A. Worratao, and J.W. Park. Effect of endogenous transglutaminase on threadfin bream surimi gelation. *J. Food Sci.*, 67, 3258–3263, 2002.

29. B.C. Russell. Nemipterid fishes of the world. *FAO Species Catalogue*, Vol. 12. Rome, Italy: United Nations Food and Agricultural Association, 1990.

30. Y. Itoh, T. Maekawa, P. Suwansakornkul, and A. Obatake. Seasonal variation in gel-forming characteristics of three lizard species. *Fish Sci.*, 61, 942–947, 1995.

31. Y. Nozaki, R. Kanazu, and Y. Tabata. Freezing Storage of lizardfish for kamaboko preparation. *Refrigeration*, 53, 473–480, 1978.

32. A. Yasui and P.Y. Lim. Changes in chemical and physical properties of lizardfish meat during ice and frozen storage. *Nippon Shokuhin Kyogo Gakkaishi,* 34(1), 54-60, 1987.

33. M.C. Ng, H.K. Lee, K. Sophomphang, S. Rungjratanam, O. Kongpun, W. Suwannarak, and L.K. Low. Utilization of lizardfish *Saurida tumbil* for surimi production. In *Advances in Fish Processing Technology in SE Asia in Relation to Quality Management,* SEAFDEC, Singapore, 1997.

34. H.H. Chen, S.N. Lou, T.Y. Chen, Y.W. Chen, and C.H. Lee. Gelation properties of horse mackerel (*Trachurus japaonicus*) surimi as affected by operation condition, *Food Sci.,* 23, 45–55, 1996.

2

Surimi: Manufacturing and Evaluation

JAE W. PARK
Oregon State University Seafood Lab,
Astoria, Oregon

T.M. JOHN LIN
Pacific Surimi, Warrenton, Oregon

CONTENTS

2.1 Introduction 35
2.2 Processing Technology and Sequence 37
    2.2.1 Heading, Gutting, and Deboning 37
    2.2.2 Mincing 38
    2.2.3 Washing and Dewatering 39
    2.2.4 Refining 40
    2.2.5 Screw Press 41

		2.2.6	Stabilizing Surimi with Cryoprotectants	42
		2.2.7	Freezing	44
		2.2.8	Metal Detection	47
	2.3	Biological (Intrinsic) Factors Affecting Surimi Quality		48
		2.3.1	Effects of Species	48
		2.3.2	Effects of Seasonality and Sexual Maturity	49
		2.3.3	Effects of Freshness or Rigor	52
	2.4	Processing (Extrinsic) Factors Affecting Surimi Quality		53
		2.4.1	Harvesting	53
		2.4.2	On-Board Handling	55
		2.4.3	Water	58
		2.4.4	Time/Temperature of Processing	60
		2.4.5	Solubilization of Myofibrillar Proteins during Processing	62
		2.4.6	Washing Cycle and Wash Water Ratio	66
		2.4.7	Salinity and pH	69
	2.5	Processing Technologies that Enhance Efficiency and Profitability		76
		2.5.1	Neural Network	76
		2.5.2	Processing Automation: On-Line Sensors	80
		2.5.3	Digital Image Analysis for Impurity Measurement	81
		2.5.4	Innovative Technology for Wastewater	82
		2.5.5	Fresh Surimi	84
	2.6	Decanter Technology		86
	2.7	Surimi Gel Preparation for Better Quality Control		91
		2.7.1	Chopping	91
			2.7.1.1 2% and 3% Salt	92
			2.7.1.2 Moisture Adjustment	92
			2.7.1.3 Chopping Temperature	92
			2.7.1.4 Effect of Vacuum	93
		2.7.2	Cooking	93
			2.7.2.1 Plastic Casing and Stainless Steel Tube	93

2.7.2.2 Cooking Method Resembling
Commercial Production
of Crabstick 93
2.8 Summary 95
Acknowledgments 97
References 98

2.1 INTRODUCTION

Surimi is stabilized myofibrillar proteins obtained from mechanically deboned fish flesh that is washed with water and blended with cryoprotectants. Surimi is an intermediate product used in a variety of products ranging from the traditional "kamaboko" products of Japan to surimi seafood, otherwise known as shellfish substitutes. Before 1960, surimi was manufactured and used within a few days as a refrigerated raw material because freezing commonly deteriorated muscle proteins and induced protein denaturation, which resulted in poor functionality. However, with the discovery of cryoprotectants, the surimi industry was able to tap into previously unexploitable resources.

Nishiya et al.,[1] at the Hokkaido Fisheries Research Station of Japan, discovered a technique that prevented freeze denaturation of proteins in Alaska pollock (*Theragra chalcogramma*) muscle. This technique required the addition of low molecular weight carbohydrates, such as sucrose and sorbitol, in the dewatered myofibrillar proteins prior to freezing. The carbohydrates worked to stabilize the actomyosin, which is highly unstable during frozen storage.[2]

The discovery that the functional properties of the myofibrillar proteins were protected during frozen storage when carbohydrates were incorporated revolutionized the industry. Prior to 1960, Alaska pollock from the North Pacific Ocean and the Bering Sea was a largely unexploited resource. During the 1960s and 1970s, however, by using carbohydrates, Alaska pollock could be utilized to meet the increased demands of the Japanese kamaboko industry.[3,4] Consequently, production and sales in Japan, as well as other surimi-producing countries, increased because the industry was no longer limited by the availability of fresh fish or the proximity of the fishery resource.

To a large extent, the history of world surimi production started with the Japanese fish processing industry. Early developments, however, have expanded the industry into the United States, Korea, and Southeast Asia. With increased surimi production in the United States, the involvement of Japan in world surimi production decreased. Since 1989, the annual U.S. production of surimi has reached about 150,000 to 220,000 metric tons (t). The world surimi production over the past 15 years, which ranged between 420,000 to 580,000 t, is also shown in Chapter 1.

The surimi industry mainly utilizes Alaska pollock for surimi production, which covers 50 to 70% of total surimi, but its proportion has been continuously reduced. Since 1991, efforts to use other species have also been successful, through technical and marketing advances in Japan. Currently, a number of different species are being utilized in commercial surimi production. The most suitable species for surimi processing are those with white flesh and low fat content, including Pacific whiting (*Merluccius productus*) from the Pacific coast of the United States and Canada; hoki (*Macruronus navaezelandiae*) from New Zealand and Chile; Southern blue whiting (*Micromesistius australis*) from Chile and Argentina; Northern blue whiting (*Micromesistius poutassou*) from EEC waters; threadfin bream (*Nemipterus japonicus*) from Thailand, Malaysia, and India; yellow croaker (*Pseudosciaena manchurica*) from the south of Japan; and from Peru, bereche (*Larimus pacificus*), lumptail sea robin (*Prionotus stephanophyrys*), and giant squid (*Dosidiscus gigas*). In addition, underexploited species, which have a higher portion of red or dark muscle and/or a higher fat content, such as pink salmon (*Oncorhynchus gorbuscha*), atka mackerel (*Pleurogrammus azonus*), Japanese sardine (*Sardinops melanostrictus*), Chilean jack mackerel (*Trachurus murphyi*), Peruvian anchovy (*Engraulis ringens*), and Pacific herring (*Culpea harengus*), can be used for the production of low-grade surimi. Surimi resources and their biological availability are discussed in Chapter 1.

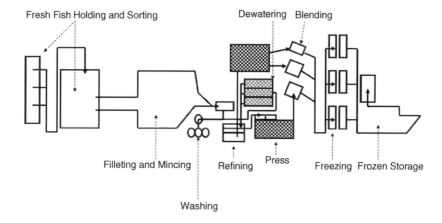

Figure 2.1 Flow chart of surimi manufacturing. (Adapted from Reference 5.)

This chapter reviews the current practice of surimi manufacturing from light muscle fish based on a sequential process and discusses technologies and approaches that enhance production efficiency and profitability.

2.2 PROCESSING TECHNOLOGY AND SEQUENCE

The surimi manufacturing process is outlined in the processing flowchart[5] (Figure 2.1). It starts from holding fish and sorting by size and ends with freezing and frozen storage. Detailed processing steps are discussed.

2.2.1 Heading, Gutting, and Deboning

Mechanical fish meat separators, developed by companies in Japan, Germany, Korea, and the United States, are modern sanitary machines that remove virtually all the flesh from the frame of a properly prepared fish. According to Pigott,[6] there are several methods to prepare fish for deboning. One is to remove head, gut, and thoroughly clean the belly walls prior to deboning the carcass. The other is to fillet the fish and then debone the fillet. The former method can better retain meat recovery but caution is needed to ensure that all viscera are completely removed. The inclusion of liver or other intestinal components in the mince can cause a severe shelf-life problem. During the gutting and filleting steps, plenty of water, along with a wheel brush, is used to separate the fillets from the undesirable parts of the fish. Factory trawlers, where desalinization is costly, commonly use refrigerated seawater up to the point of deboning. Shoreside operations, however, use refrigerated fresh water.

In Pacific whiting surimi manufacturing, the removal of fillets with a high concentration of black spotting associated with myxosporidean parasites is indispensable in preventing the devaluation of the product. Morrissey et al.[7] reported that up to 4 or 5% of Pacific whiting can have these defects, which are visible as black hair-like striations. Although they present no health hazard, they are easy to see and are unacceptable for aesthetic reasons.

2.2.2 Mincing

It is most common to use a roll-type meat separation technique for the mincing/deboning operation. The dressed fish is pressed between a traveling rubber belt and a steel drum with numerous orifices of 3 to 5 mm in diameter. The fish meat is pressed through the orifices into the interior of the drum, while separating skin, bone, hard cartilage, and other impurities into the exterior of the drum. The medium orifice size of 3 to 4 mm appears to be optimal for retaining quality and yield.[8,9]

A mechanical deboner, with a relatively large orifice size (>5 mm), consequently yields larger meat particles. As a result, it makes it more difficult to remove the sacroplasmic proteins and other impurities during the subsequent washing process. Although using a larger orifice size could improve recovery yield, the quality of surimi would be compromised as it diminishes the washing efficiency. On the other hand, mincing fish with a relatively smaller orifice size of 1 to 2 mm would enhance washing efficiency, but a significant portion of fine meat particles would be lost during the washing process, thus resulting in lower recovery.

The size and texture of fish are also factors in selecting the right mincing machine for optimal recovery and quality. Fish of smaller size or firmer texture would benefit from a smaller orifice diameter. In mincing smaller fish, as in surimi processing from warmwater fish (threadfin bream, lizardfish, and others), the use of a large orifice would generate more bone fragments and/or broken skin in mince.

Skinned and deboned fillets give cleaner minced meat because blood, membrane, and other contaminants have been removed. A headed and gutted carcass, on the other hand, results in a higher final yield of minced flesh, but the quality is relatively low. Another method is skin-on fillet (butterfly shape), which increases yield and keeps quality loss to a minimum. More recently, it is not uncommon to see H&G (headed and gutted) fish directly subjected to the deboning machine, and resulting in higher recovery.

2.2.3 Washing and Dewatering

Washing is an essential step in removing water-soluble proteins, primarily sarcroplasmic proteins, which is thought to impede the gel-forming ability of surimi, and other impurities that also reduce product quality. Sacroplasmic proteins exist in the fluids within and between muscle fibers, and include many metabolic enzymes that diminish the stability of functional proteins during storage. Myofibrillar proteins, the primary components that possess the ability to form a three-dimensional gel network, constitute approximately 70% of the total proteins in minced fish meat. A reduction in water-soluble proteins in turn concentrates the myofibrillar proteins, thus enhancing the functional property of surimi.

A proper washing process, therefore, is vital to achieve high-quality surimi with high recovery. An insufficient washing process could result in a substantial loss of gel quality during frozen storage. On the other hand, over-washing could cause a substantial loss of fine particles and excessive moisture content. Maintaining water temperature near or below

5°C is also crucial for cold-water species (Alaska pollock) and temperate-water species (Pacific whiting) that contain a significant amount of texture-softening proteolytic enzymes.

The number of washing cycles and the volume of water vary with fish species, freshness of fish, structure of the washing unit, and the desired quality of the surimi.[10] In the early 1990s, it was common to have water/mince ratios of 5:1 to 10:1 with three to four washing cycles. Lin et al.[11] reported that for a shore-side operation, 29.1 liters (L) of wastewater were generated to produce 1 kilogram (kg) of surimi. As the cost of using fresh water and treating wastewater continues to increase, substantial efforts have been made by the industry to reduce water usage and achieve better washing efficiency. An effective washing process can now be accomplished with two washing cycles at water/meat ratios of less than 2:1. In comparison, at-sea processors can achieve the same washing effect with less water than shore-based processors due to the difference in freshness of the fish. The fish used in the shore-based processor tend to be older (20 to 100 hr). As fish age in the storage tank (even in the ice storage), heme pigments (hemoglobin and myoglobin) become easily denatured, making the washing process less efficient.

Washing efficiency is often affected by various factors. In addition to the water/meat ratio and age of fish, there is the shape of the washing tank (round vs. square), the speed of the agitator, the shape of the agitator (vertical vs. horizontal), and water temperature. Square-shaped tanks seem to work better than round-shaped tanks because the former can generate a counter-current washing effect. When the agitating paddle is placed horizontally rather than vertically, the washing efficiency is higher. When the agitator is operated too fast, it might also result in a temperature rise as well as difficulty in dewatering by the screw press. The optimum speed (rpm) for agitation must be determined based on specific operations (i.e., 20 to 40 rpm).

2.2.4 Refining

Before the final dewatering under a screw press, impurities (such as skin, fine bones, scales, and connective tissues) are removed by the refiner. According to Kim and Park,[12] the approximate composition of refiner discharge was 81.4% moisture, 1.9% lipid, 15.4% protein, and 1.0% ash. They also found that the majority of protein was stroma proteins derived from connective tissue. This clearly indicates that the refining process is used to separate connective tissues from washed mince. Running the refiner at a slower speed with a smaller screen size will result in cleaner surimi with less recovery. On the other hand, running the refiner at a faster speed with a larger screen size will enhance recovery but with a risk of higher impurities. Screen sizes of 1.5 to 1.7 mm are commonly used in commercial applications. Normally, 15 to 20% of the meat is rejected from the primary refiner and goes to the secondary refiner for secondary surimi production. The secondary surimi, as compared to the primary surimi, has higher impurities, lower whiteness, and lower gel strength.

Conventionally, all manufacturers used a Fukoku refiner until the recent introduction of the Brown refiner (Covina, California). According to the U.S. industry experts, the Brown refiner is designed as a more user-friendly structure in cleaning and adjusting the paddle height. However, the selection should be made after careful evaluation based on individual operation.

2.2.5 Screw Press

The moisture content of meat increases from 82 to 85% to 90 to 92% after repeated washing. It is essential, therefore, to remove the excess water prior to blending with cryoprotectants and freezing. The desirable moisture content of the meat, prior to blending, ranges between 80 and 82%. The length and speed of the screw, the volume reduction ratio, and the perforation of the screens determine the effectiveness of water removal. For example, a screw press with a larger volume reduction ratio and longer screw can achieve the same dewatering effect at higher speed compared to the screw press with a small volume reduction operated at slower speed.

Screens with 0.5 to 1.5-mm perforations are commonly used in industry. In addition, screens with smaller perforations are usually placed at the end section to preserve/recovery. It is not uncommon to use a 0.1 to 0.3% salt mixture of NaCl and $CaCl_2$ to facilitate the removal of water from the screw press. The use of salt often results in increased gel values when testing is done immediately. Added salt positively contributes to the unfolding of protein structure, resulting in better gel strength when testing is done within a few days after manufacturing. However, this added salt enhances protein denaturation during frozen storage and, consequently, shortens the frozen shelf life of surimi. Therefore, mechanical dewatering without salt is best to maintain the frozen stability of surimi.

2.2.6 Stabilizing Surimi with Cryoprotectants

The addition of cryoprotectants is important to ensure maximum functionality of frozen surimi because freezing induces protein denaturation and aggregation. Sucrose and sorbitol, alone or mixed at approximately 9% w/w to dewatered fish meat, serve as the primary cryoprotectants in the manufacture of surimi. However, 6% sucrose is typically used in surimi manufactured from warm-water species perhaps due to higher thermal stability. Further study must be conducted to compare frozen stability of surimi made from cold- and warm-water species using 6% sucrose. In addition, a mixture (1:1) of sodium tripolyphosphate and tetrasodium pyrophosphate at 0.2 to 0.3% is commonly used as both a chelating agent, which makes metal ions in surimi inactive, and as a pH adjusting agent.

Cryoprotectants were originally incorporated into the dewatered meat using a kneader. Currently, silent cutters are often used because they uniformly distribute cryoprotectants faster and the temperature increases less during chopping. Commercial practices for mixing cryoprotectants (100 kg per batch) using a kneader and a silent cutter are 6 min and 2.5 min, respectively. The temperature of the mix must not exceed 10°C because at temperatures greater than 10°C, protein functionality could be damaged, particularly for cold-water species.

Since 1991, with the commercial surimi processing of Pacific whiting, enzyme inhibitors, such as beef plasma protein, egg whites, or potato extracts, have been used in conjunction with cryoprotectants, gel enhancers, and color enhancers. Enzyme inhibitors are commonly formulated with sucrose, sorbitol, sodium tripolyphosphate, tetrasodium pyrophosphate, calcium carriers (calcium lactate, calcium sulfate, calcium citrate, or calcium caseinate), sodium bicarbonate, mono- or diglyceride, and partially hydrogenated conola oil.[13] The formulation of these ingredients varies, depending on the company. Therefore, there are slight differences from one company to another. The addition of enzyme inhibitors or calcium compounds, however, before freezing surimi is not necessary, especially because added calcium compounds can actually enhance protein denaturation during frozen storage. Instead, these calcium compounds can be added when the surimi paste is prepared to make slow-cooked gels.

Due to the recent outbreaks of BSE (bovine serum encephalopathy, or mad-cow disease) in the EU, Japan, Canada, and the United States, the use of beef plasma as an enzyme inhibitor has been prohibited. However, Park and co-workers[14] demonstrated that fast heating (i.e., conventional way of crabstick manufacturing or ohmic heating) is a suitable alternative for surimi paste containing Pacific whiting surimi or other surimi with proteolytic enzyme problems. Consequently, a majority of Pacific whiting surimi is now processed with only cryoprotectants (sucrose, sorbitol, and phosphate).

Other efforts have been made to introduce new cryoprotectants. Roquette Corporation (Keokuk, IA) introduced a short-chain glucose polymer (LD and SD) as an effective cryoprotectant (Figure 2.2).[15] They performed well during 8-mo frozen storage. In addition, Cargill Corporation (Minnetonka, MN) introduced trehalose, which is a disaccharide that is 45% the sweetness of sucrose. As shown in Figure 2.3,[16] trehalose effectively replaced sorbitol and/or sucrose (see Figure 2.3B, F, G, A, D, and H) during 12-mo frozen storage. Trehalose treatments without phosphate (see Figure

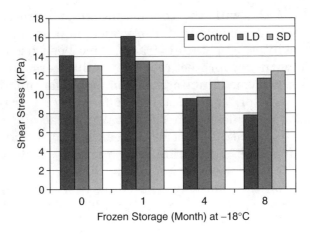

Figure 2.2 Shear stress of gels as affected by two glucose polymers as potential cryoprotectants during 8-month frozen storage.

2.3F and G) also showed improved frozen stability. However, the positive role of trehalose in the absence of phosphate must be further studied.

2.2.7 Freezing

In commercial applications, surimi is formed in a standard 10-kg block in a plastic bag (3 to 7 mil), which is then placed on a stainless steel tray. The trays are then placed in a contact plate freezer and held for approximately 2.5 hr or until the core temperature reaches –25°C. After inspecting the frozen surimi blocks with a metal detector, two 10-kg frozen surimi blocks are packed into a cardboard box. Drum freezing of surimi, on the other hand, offers the prospect of rapid freezing,[17] which enhances surimi quality and results in frozen surimi chips, which provide a more convenient product form. However, drum freezing may not be preferred by at-sea processors where storage space is limited. Further details on surimi freezing are discussed in Chapter 8.

Productivity-driven operations typically bottleneck at the freezing step because of the time involved. The effects of the freezing rate on the gelation properties of surimi are often

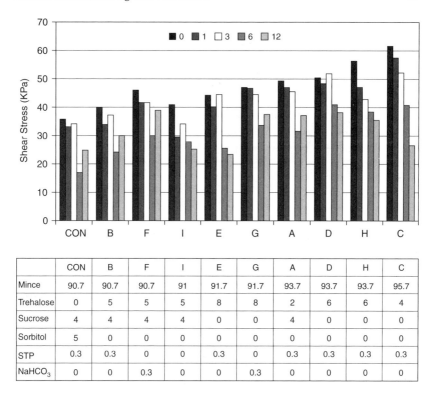

Figure 2.3 Shear stress of gels as affected by trehalose as potential cryoprotectants during 12-month frozen storage.

questioned. Reynolds et al.[18] examined the effects of various freezing methods on the biochemical and physical properties of surimi. As shown in Figure 2.4, fresh surimi blocks containing cryoprotectants were frozen (1) using a conventional plate freezer, (2) by placing a fresh block on the bottom floor of −18°C freezer (slow freezing), and (3) by flake freezing in liquid nitrogen spray after extruding 2- to 3-mm-thick surimi (fast freezing). The time to reach −18°C was 132, 1436, and 17 min, respectively.

By comparing the results of the conventionally frozen block samples, surprisingly no significant differences (p < 0.05) resulted in the rheological properties between samples up to 9 mo. However, there were striking visual differences

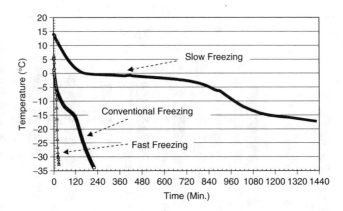

Figure 2.4 Freezing rate of three different freezing methods for surimi.

between the two block types. The conventionally frozen samples had a smooth white appearance with no visible ice crystals. The slow-frozen blocks, on the other hand, were more translucent (glossy), exhibiting a darker (grayish) appearance and showed visible ice crystals. In addition, when the block was tempered, it had a crystalline look. It also tempered faster than the conventionally frozen block. From their appearance, it seemed that the slow-frozen samples would have resulted in lower gel strengths but this was not the case up to 9-mo storage.[18] However, after 18 mo, a significantly lower texture value was obtained for slow-frozen surimi compared to the other samples (Figure 2.5). Even at 18 mo the shear strain value of freeze-dried surimi stored at –18°C did not change compared to 0 mo.[18] The effect of various freezing rates on gel texture did not appear to be significant up to 9-mo storage. However, continuous long-term frozen storage (>18 mo) definitely affected gel texture.

Freeze-dried surimi kept at –18°C showed no changes up to 9 mo, but significantly reduced shear stress after 18 mo (Figure 2.5). Shear strain, denoting the cohesiveness of gels, did not show any changes at 18 mo.[18]

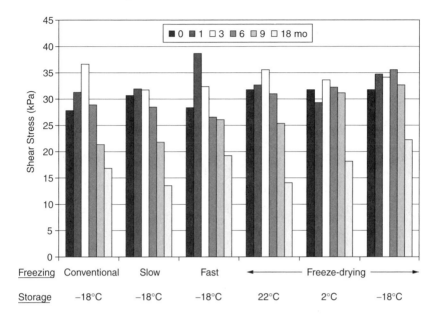

Figure 2.5 Shear stress of gels as affected by various freezing rates and storage conditions during 18-mo frozen storage. (From Reference 18 with permission.)

2.2.8 Metal Detection

Metal detection is a critical control point for the surimi HACCP program. The FDA's Health Hazard Evaluation Board has supported regulatory action against products with metal fragments of 7 to 25 mm in length.[19] Corrective actions shall be taken if metal inclusion occurs. Processors must make sure that the unsafe product does not reach the consumer and must take corrective actions to address the cause of the deviation.

The most common types of metallic contamination include ferrous, copper, aluminum, lead, and various types of stainless steel. Of these, ferrous metals are the easiest to detect. In surimi manufacturing equipment, stainless steel alloys are most commonly used and are the most difficult to detect, especially the nonmagnetic grades such as 316 and 304L. Other factors also affect the sensitivity of metal detection, including the shape of the metal, orientation of the metal, aperture dimension, position of the metal in the aperture, environmental conditions, condition of the product (frozen vs. chilled), operation frequency, and throughput speed.[20]

As for the limit of calibration, there is a different setting between shore-side operation and on-board operation. Due to the continuous motion, on-board calibration is very difficult. In most U.S. operations, the calibration metal for shore-side operations is 2 to 3 mm for ferrous or non-ferrous and 3 to 4 mm for stainless steel, while on-board operations use 3 to 4.5 mm for ferrous or non-ferrous and 4.5 to 5 mm for stainless steel. This limit is far below the FDA's action limit of 7 mm.

2.3 BIOLOGICAL (INTRINSIC) FACTORS AFFECTING SURIMI QUALITY

2.3.1 Effects of Species

In addition to Alaska pollock, there are a number of species that are utilized as raw material for commercial surimi processing. Depending on the species used, however, the functional and compositional properties of the surimi vary. The functional properties of surimi depend on composition, but cannot generally be predicted from compositional analysis. It is, therefore, important for processors to understand the relationships between the physico-chemical functions of fish and the functional and compositional properties of surimi.

With the development of Pacific whiting surimi, the importance of understanding the intrinsic enzymes in the fish has been highlighted. An et al.[21] identified the enzymes in Pacific whiting as cathepsins B, H, and L. They behave differently with different environmental conditions, such as pH, temperature, and ionic strength. Cathepsin B and H are easily washed off during surimi processing, while cathepsin L remains in the muscle tissue. Cathepsin L has an optimum temperature of 55°C and causes textural deterioration when the surimi paste is slowly heated. Therefore, enzyme inhibitors are required unless the surimi is cooked rapidly using either an ohmic heater[14] or microwave, or is thinly extruded and cooked rapidly, as in crabstick processing.

Arrowtooth flounder (*Atheresthes stomias*) is another fish that requires enzyme inhibitors to minimize textural deterioration due to a heat-stable enzyme.[22] Gel weakening at around 55 to 60°C has also been reported in threadfin bream (*Nemipterus bathybius*),[23] Atlantic menhaden (*Brevoorti tyrannus*),[24] white croaker (*Micropogon opercularis*), oval filefish (*Navodon modestus*),[25] and lizardfish (*Saurida* spp.).[26]

Alaska pollock (*Theragra chalgogramma*) has been known as a fish that gives no proteolytic enzymes. However, recent studies[27,28] indicate that Alaska pollock is infested with microsporia, which induce gel softening. Kimura and colleagues[27] revealed the presence of cysts of unidentified microsporian and multinucleate bodies of *Ichithyophonus hoferi* in Alaska pollock. The infested muscle contains a protease that degrades myofibrillar proteins at around 50 to 60°C. They also reported that cycteine protease inhibitors were able to reduce the enzyme activity, whereas specific inhibitors of serine proteases and aspartic proteases were ineffective.

To make surimi from oily/dark or red-fleshed fish, such as mackerel, sardine, and salmon, certain steps must be taken to negate the effects of the oil and heme proteins. Heme proteins, such as myoglobin and hemoglobin, account for the red color of dark muscle. In addition, fat oxidation in dark muscle is promoted by heme proteins, which causes an offensive, rancid odor to develop.[29] It is therefore suggested that 0.1 to 0.5% $NaHCO_3$ in the first washing solution and a decanter be used to remove the extra oil. Addition of 0.05 to 0.1% sodium pyrophosphate and the use of a vacuum during washing are also recommended to remove heme proteins. Details of discoloration are discussed later in this chapter (see Section 2.4.3).

2.3.2 Effects of Seasonality and Sexual Maturity

Compositional properties of fish vary as the fishing season changes. Compositional properties of Alaska pollock were reported by AFDF.[30] The results showed that protein content was highest (19.0%) in November and lowest (16.5%) in May, while moisture content was highest (82.3%) in July and lowest (80.2%) in November. In addition, Morrissey and coworkers[31] conducted a seasonal analysis of the protein, moisture, fat, and ash contents for Pacific whiting over a 3-year period (Figure 2.6). The highest moisture reading (84.5%) for Pacific whiting was recorded in April, while the lowest reading (80 to 82%) was recorded at the end of October. Protein content was at its lowest (14 to 15%) in April, and then increased and held relatively steady (15.5 to 16.5%) after June. Fat held fairly steady (0.5 to 1.5%) until August, and then started to increase (1.5 to 2.5%) in October. Therefore, for Pacific whiting surimi, both yield and quality increase during the summer months.[32]

These results were of significant interest to the shoreside industry. Consequently, in 1996, the opening date for Pacific whiting harvests changed from April 15 to May 15 for at-sea processors and to June 15 for shore-side processors. In addition, Sylvia et al.[33] incorporated seasonality effects and quality factors into bio-economic models for Pacific whiting. Results also showed that by delaying the harvest, economic gains could be made by the industry and conservation of the resource could be enhanced.

Alaska pollock also has very similar trends of compositional properties of flesh as the season changes. Spring-season pollock showed higher moisture content at 82.7%, and lower fat and protein contents at 0.2% and 15.6%, respectively (Table 2.1). Ash content was not affected by the season (C. Crapo, personal communication, 2004).

Tokunaga and Nishioka[29] found seasonal changes in the fat content of sardine harvested in the middle Pacific Ocean. In August, the fat content was as high as 33% and was the lowest in April at 3%. Consequently, to manufacture surimi from sardines in summer, due to the higher fat content, special technologies using $NaHCO_3$ and a centrifuging decanter must be applied.

In general, fish harvested during the feeding period produce the highest-quality surimi. During this period, fish muscle has the lowest moisture content and pH, as well as

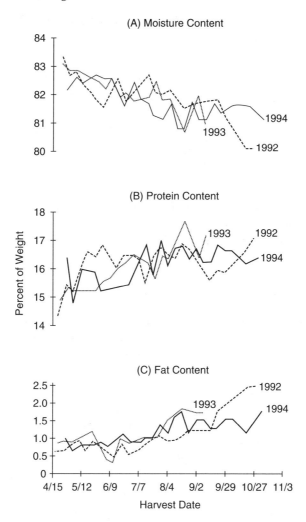

Figure 2.6 Compositional properties of Pacific whiting from April to October.

the highest total protein.[34] Therefore, fish harvested during and after the spawning season produce the lowest-quality surimi. It has been established that spawning fish have a relatively higher pH and tend to retain more water. Consequently, it is difficult to remove the extra water from the

TABLE 2.1 Compositional Properties of Alaska Pollock Flesh as Affected by Season

	Moisture (%)	Fat (%)	Protein (%)	Ash (%)
Spring (spawning)	82.7 (±1.6)	0.2 (±0.1)	15.6 (±1.8)	1.1 (±0.3)
Fall	80.9 (±1.2)	0.6 (±0.12)	16.9 (±1.4)	1.1 (±0.3)

Source: From C. Crapo, personal communication, 2004

washed meat. To easily remove the extra water, muscle tissue characteristics must be altered by either lowering the pH or increasing the salinity of the final washwater.[35] However, this alteration can lead to significantly reduced quality, particularly after frozen storage.

2.3.3 Effects of Freshness or Rigor

The freshness of fish is primarily time/temperature dependent. On at-sea vessels, processing occurs within 12 hr, while at shore-side operations processing occurs within 24 to 100 hr for pollock, depending on the location of the fishing grounds. Due to endogenous enzymes activated by rising temperatures, Pacific whiting is processed within a shorter time period: immediately on at-sea vessels and within 20 hr after harvest at shore-side plants.

The biochemical and biophysical changes during the development of *rigor mortis* induce significant changes in the functional properties of muscle proteins. Fish should be processed as soon as possible after going through *rigor*. Prior to passing through this stage, about 5 hr in the case of pollock surimi, it is difficult to remove the "fishy" odor, various membranes, and other contaminants that affect product quality.[6] However, Park et al.[36] reported that significantly higher protein content and yield, reduced cooking loss, and enhanced gel-forming ability are associated with surimi processed from manually filleted pre-*rigor* tilapia fish.

The length of time that fish can be held in ice or refrigeration before processing varies, depending on the species.

Surimi: Manufacturing and Evaluation

The effect of time is especially prominent in fish, such as Pacific whiting, that have intrinsic enzyme problems. Peters and Morrissey[37] investigated time/temperature effects on the compositional and functional quality of Pacific whiting surimi processed at Oregon shore-side operations. Their study suggested that if kept refrigerated, Pacific whiting should be processed within approximately 24 hr of capture. Otherwise, the quality begins to decline. If Pacific whiting are not cooled quickly, processing must occur within 8 to 10 hr.

Current practice of surimi from threadfin bream or other warm-water species in Thailand utilizes 40% frozen fish mixed with 60% fresh fish. Manufacturing of surimi from frozen cold-water species is virtually impossible and results in no or extremely poor gels. It is possible, however, to make decent surimi from frozen warm-water species simply because of its higher thermal stability. However, better surimi can be manufactured from warm-water species if fresh fish is used.

2.4 PROCESSING (EXTRINSIC) FACTORS AFFECTING SURIMI QUALITY

2.4.1 Harvesting

Surimi quality is affected by the harvesting conditions and methods used for capture, as well as the on-board handling methods and vessel storage conditions. Intrinsic quality factors (e.g., protein content, post-spawning condition) are important considerations even before the nets are put in the water. The geographic location of the fishing grounds may also affect quality and determine factors, such as the size of the fish or the amount of time required to deliver the fish to the processing plant.

Several factors in the actual capture of fish can also affect final product quality. These include at-sea weather conditions, capture methods, size of tow, length of tow, salt uptake, and the temperature of the fish post-capture. Most of these factors are interactive and, at times, it is difficult to weigh the importance of each factor separately. In this section, the types of vessels commonly used in the surimi industry, the on-board handling options, and the effect they might have on surimi quality are discussed.

In the Alaska pollock and Pacific whiting fisheries, there are several methods of harvesting and transporting fish. Since the discovery of cryoprotectants in 1960, large factory trawlers, 70 to 150 m in length, have been engaged in the pollock fisheries. These large boats spend several months off-shore capturing and processing pollock into at-sea frozen surimi blocks.[9] Because the fish are processed on board the vessel, the time between capture and final product production is usually short. In addition, factory trawlers are able to fish in distant offshore waters that smaller trawlers cannot reach and generally produce a high-quality product. One disadvantage, however, is the size of capture, as a single trawl might hold more than 100 metric tons (t). Crushing and bruising of fish in the cod-end of the trawl can, therefore, cause some quality problems when compared to fish at the front end of the net. Nonetheless, factory trawlers have proven to be an efficient method of fish capture for surimi production.

Another fishing method that combines large factory ships and shore-based trawlers is the use of motherships. These are large processing vessels that receive the "trawl-bag" from smaller trawlers. In general, a smaller tow size (20 to 50 t) is expected because of the smaller catcher boats involved. The bags are transferred at sea, hauled aboard the mothership, and then processed into surimi. It is important to coordinate the activities of the catcher vessels with the motherships to allow efficient transfer of fish. Consequently, this method allows smaller vessels to work farther offshore and provide raw material for at-sea processing.

A third method of capture is the use of smaller coastal trawlers (25 to 50 m in length) that offload at shore-based processing plants. An important factor that can affect final surimi quality is the steaming time from the fishing grounds to the processing plants. For these reasons, successful shore-side operations are located near the fishing grounds (i.e., Dutch Harbor and Kodiak, Alaska, for the pollock fishery; Astoria and Newport, Oregon, and Ucluelet and Port Alberni, BC, Canada, for the Pacific whiting fishery). Coastal boats

Surimi: Manufacturing and Evaluation

that are very conscious of time/temperature factors between fish capture and processing produce high-quality raw material that will be processed into higher-grade surimi. Coastal boats are at a disadvantage, however, when schools of fish move from nearby fishing grounds, thus requiring longer travel times. Other problems exist in very rough weather, which causes bruising of fish held in the seawater tanks aboard the vessels.

Outside the larger fishing nations, there are a variety of operations, including iceboats and coastal trawlers, that freeze whole fish on-board the vessel for later onshore surimi processing. An example of this fishing technique occurs in the Gulf of Thailand where threadfin bream are harvested. The average size of threadfin bream is extremely small (30 to 50 g) and requires manual labor to remove the head and intestinal organs. Consequently, boats that travel substantial distances to harvest fish find it easier to freeze fish on-board the vessel and offload for processing at specific ports. Although freezing of gadoid fish, such as pollock and whiting, severely decreases protein functionality and gelation, threadfin bream appear to withstand protein denaturation during frozen storage and produce an acceptable product.

2.4.2 On-Board Handling

Time and temperature of the fish between capture and processing can be considered two of the most important factors that affect final surimi quality. Factory trawlers have the advantage of processing at sea and usually produce a final product within 12 hr after harvesting the fish. However, few factory trawlers have systems that allow them to chill the fish while waiting to be processed. This may not be a major concern while fishing for Alaska pollock in January (the "A" season) in the Bering Sea, but it may be an important factor while the same vessel is fishing for pollock during September (the "B" season) when temperatures can be higher. During certain years, shore-side pollock trawlers may have to fish 48 to 100 hr away from their homeport. Holding temperatures at 4 to 6°C can make a significant difference in surimi quality compared to fish held close to 0°C (G. Peters, personal communication, 1998).

The Pacific whiting fishery is a summer fishery off the West Coast of the United States, and fish may be landed in water temperatures of 15°C and deck temperatures as high as 24°C. In the whiting fishery, it is recommended that fish be cooled down rapidly and landed for processing within 24 hr post-harvest.[38] This is due, in part, to the delicate nature of the whiting flesh and its propensity to undergo proteolytic hydrolysis after capture. Proteolysis is an enzymatic reaction that is time and temperature dependent, and faster chilling of the fish, as well as low-temperature storage, will help offset longer storage time effects.[39]

There are different types of refrigeration systems that can be used to reduce the temperature. Several reviews describe the advantages and disadvantages of each system.[40–43] The most common systems in the United States, however, are refrigerated seawater (RSW), slush ice (SI), and champagne seawater systems (CSW).

On small-scale fishing vessels, slush ice is the method of choice, as it is simple and relatively inexpensive. RSW systems were introduced into fishing operations during the 1960s. This system uses on-board refrigeration to lower the temperature of the catch. Water pumps within the hold circulate the water and eliminate potential hot spots. SI and CSW systems, on the other hand, use a varied ratio of ice to seawater to chill the systems, depending on the length of the fishing trip. The main difference, however, between CSW and SI is that CSW uses a system of forced air to create bubbles that agitate the ice:water mixture.

The major concerns with these on-board storage systems include maintaining adequate circulation and potential overloading of the system. Although pumps and compressed air facilitate mixing, temperature differences can develop in the fishing hold.[44] Overloading the system with fish also decreases proper circulation. This will allow some fish temperatures to decrease to 0°C within a short period of time, while other fish will maintain temperatures closer to the water temperature in which they were harvested.

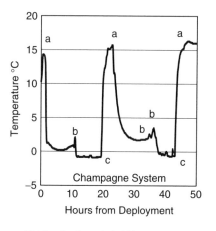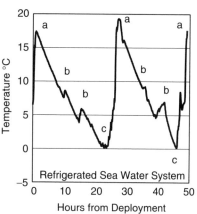

a: Finish unloading, start of trip
b: Addition of tow into fish hold
c: Begin unloading at shore-side

Figure 2.7 Temperature profiles of tank on fishing vessels using two chilling systems (Champagne ice system and Refrigerated sea water system). (Adapted from Reference 32.)

Ample time is also needed to reduce the temperature to 0°C before introducing fish into the hold. Figure 2.7 describes the temperature profile of both a CSW and RSW system on two shore-based whiting vessels. In the CSW system, the temperature is below 0°C (the effect of ice and seawater). When fish are introduced into the hold, there is a momentary spike as the ice melts to take heat from the fish. The CSW, however, is sufficiently fortified with ice so the temperature quickly returns to 0°C. In the RSW system, the crew introduced fish into the hold before the system reached 0°C. A second haul was made a few hours later. The graph shows that the fish reached 0°C approximately 14 hr after the initial harvest, which was considerably slower than the CSW example. However, if the RSW system had been brought down to temperature, before harvest, a temperature profile similar to that of the CSW system would have occurred.

The tow size and tow time length can also influence final product quality. Larger factory trawlers will tow for several hours with large nets to obtain harvests of 100 to 150 t per haul. Smaller coastal vessels, on the other hand, tend to have shorter tows and hauls of 20 to 70 t. Despite limited research on the effects of tow size on surimi quality, there is an increasing belief in the whitefish industry that shorter tows and smaller captures produce a better product. Consequently, in the New Zealand hoki fishery (primarily for fillets), most boat captains are under instructions to limit their tows to less than 20 t (G. MacDonalds, personal communication, 1998).

A delicate balance exists in fishing operations between volume and quality. The economics of surimi, however, are such that fish must be harvested and transported in large volumes to be profitable. There is a tendency, even in smaller shore-side vessels, to take one long tow with a capture of 50 to 70 t to fill the tanks. This may cause crushing of the fish or overload the refrigeration system. Fishermen might argue, however, that they need to fish large volumes to make fishing profitable. A potential solution would be to make contractual agreements between fishermen and processors for tow size, as well as time and temperature parameters.

2.4.3 Water

The important quality factors associated with water are temperature, hardness or mineral content, pH, and salinity. The level of chlorination in the water should also be considered because of its bleaching and deodorizing effects.[35] The water must be refrigerated to a temperature below which the fish muscle proteins can retain their maximum functional properties. The temperature of the water can vary, based on the thermostability of the fish proteins. Warm-water fish can therefore tolerate higher water temperatures than cold-water fish without reducing protein functionality.[45] Misima et al.[46] also reported that Ca^{2+}-activated myofibrillar Mg^{2+}-ATPase was highest at the habitat temperature (25.5°C) of carp. However, considering the changes in air temperature during processing, the recommended water temperature for obtaining maximum quality is 5°C or less.

Theoretically, soft water with minimum levels of minerals such as Ca^{2+}, Mg^{2+}, Fe^{2+}, Mn^{2+} is recommended for washing. Hard water causes deterioration of texture and color quality during frozen storage.[47] In addition, Ca^{2+} and Mg^{2+} are responsible for the texture change, particularly when surimi is frozen, while Fe^{2+} and Mn^{2+} are responsible for the color change.[48] Furthermore, the pH of the water must be maintained at approximately that of pre-*rigor* fish muscle tissue (6.8 to 7.0) to obtain higher water retention of the gels. All factory trawlers, where fresh water is generated using a steam distillator, have extremely low mineral contents in the washwater. Most shore-side operations, on the other hand, use either municipal water or well water purified using a softner and/or reverse osmosis membrane.

Well water, depending on geographical locations, often shows signs of high hardness and typically contains cations (Ca^{2+} and Mg^{2+}). For large-scale operations, a lime-soda process can be used to soften hard water by removing Ca^{2+} and Mg^{2+}. Hard water is treated with a combination of slake lime, $Ca(OH)_2$ and soda ash, Na_2CO_3. Calcium precipitates as $CaCO_3$ and magnesium precipitates as $Mg(OH)_2$. The residual content of sodium (Na^+) in the processing water does not directly affect the quality of the surimi. However, the intake of sodium from surimi could negatively affect the nutritional profile of the subsequent surimi seafood. Therefore, unless reverse osmosis (RO) is used to remove sodium ions, the use of a different softening method that employs a magnet to remove Ca^{2+} and Mg^{2+} would be ideal. According to U.S. surimi industry experts, various minerals found in the processing water, either from city water or well water, range between less than 1 and 6 ppm (sodium, copper, calcium, magnesium).

Before washing, the salinity of fish mince is approximately 0.7%. Moisture removal gradually increases when the percent salt concentration of the washwater is increased.[48,49] It is common, therefore, to use a mixture of NaCl and $CaCl_2$ at 0.1 to 0.3% in the final washwater. However, special care must be given. Residual salt in surimi (0.2 to 0.4%) would minimize the loss of soluble proteins during washing and give slightly improved gel strength when surimi testing is conducted soon after the surimi is frozen. However, this residual salt content will accelerate the denaturation of fish proteins during frozen storage (beyond 2 mo), resulting in a much shorter frozen shelf life.

Processing water has also been treated with either ozone (O_3) or UV light to disinfect bacteria and other pollutants. Because these chemicals are dissipated easily during processing, they are used as a processing aid without labeling. Ozone is approved as a sanitizer for contact with food and food equipment because it is effective against many microbes and leaves no residue after it reacts and decomposes. Unlike another strong oxidizer (i.e., chlorine), ozone does not react with organic materials to produce undesirable compounds and does not leave an unpleasant taste. Therefore, ozone has begun to replace chlorine in treating drinking water and processing water for reuse.

According to the industrial trials with ozone water in surimi manufacturing, it reduces bacterial counts as well as increases whiteness. Haraguchi et al.[50] demonstrated the antimicrobial effect of ozone on jack mackerel. After exposing 0.6 ppm for 30 to 60 min, reduced bacterial counts (2 to 3 logs) and extended shelf life (by 1.2 to 1.6 times) were observed. The effect of ozone on surimi color can be explained based on its bleaching effect. Several reports were made on the discoloration effects of ozone for mackerel meal and swine powder[51,52] and mackerel surimi.[53] According to Buckley et al.,[54] the porphyrin structure of the heme pigment is destroyed during ozonation.

2.4.4 Time/Temperature of Processing

In commercial shore-side surimi operations, Pacific whiting are usually delivered to processing plants within 6 to 12 hr after harvest. Due to the limited capacity of processing facilities, the fish are often kept in holding tanks with ice water (~0°C) for up to another 6 to 14 hr. Thus, fish are usually 6 to 24 hr post-harvest before they are subjected to surimi processing. During this holding period, if the fish are not handled properly, the temperature can rise. Prolonged holding time and elevated temperatures can cause severe proteolysis of myofibrillar proteins. Consequently, proteolysis before and during processing causes more myofibrillar proteins to be dissolved as water-soluble waste.[55-57]

The effects of various post-harvest storage temperatures and times on the proteolysis of Pacific whiting and its relationship to changes in protein solubility were evaluated by Lin and Park.[58] Degradation of MHC (myosin heavy chain) increased rapidly during the post-harvest storage period (Figure 2.8). This degradation occurred even when the temperature was maintained at 0°C. In commercial operations, Pacific whiting is commonly kept in holding tanks (~0°C) up to 14 hr before fish are subjected to processing. Results showed that 23.5% of MHC degradation occurred for this storage time (0°C, 14 hr). Greater degradation of MHC was observed with prolonged storage. More than 70% of the MHC was degraded when fish were stored at 0°C for 72 hr.

MHC degradation was also affected by storage temperatures. Fish kept at 5°C showed higher degradation than those stored at 0°C, suggesting that ice water was more efficient than refrigeration in controlling proteolysis. When temperatures increased further, degradation occurred more rapidly. Within a 14-hr storage period, degradation nearly doubled when temperatures increased from 0 to 10°C. After holding at 20°C for 2 hr, 31.6% MHC was degraded, which was equivalent to the degradation that occurred when fish were stored at 0°C for 24 hr. These data, therefore, indicated that both time and temperature were critical to MHC degradation.

Low temperatures also slowed proteolysis. With prolonged storage time, however, severe degradation occurred although the storage temperature was maintained at 0°C.[58] According to An et al.,[21] in the temperature range of 0 to 5°C, the activity of cathepsin L was insignificant, while cathepsin B exhibited half its maximal activity, and cathepsin H retained about a fifth of its maximal activity. Therefore, cathepsins B and H might contribute to the degradation occurring at low-temperature storage. Consequently, to minimize proteolysis, fish should be processed promptly upon landing or kept at 0°C, if holding is necessary.

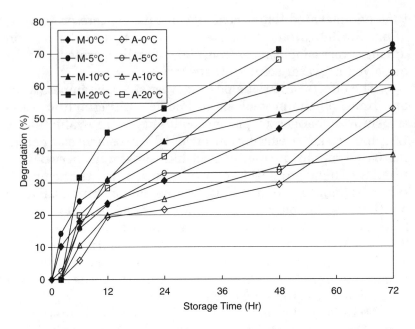

Figure 2.8 Degradation of myosin heavy chain and actin of Pacific whiting at various post-harvest storage conditions. M = myosin heavy chain; A = actin. Numbers followed by M or A indicate storage temperatures.

The trend of actin degradation was similar to that of myosin heavy chain (MHC), but to a lesser extent.[58] Actin degradation also increased as storage time and temperature increased (Figure 2.8). At 0°C, degradation of actin was not significant in the first 3 to 6 hr. When prolonged storage time was allowed, however, degradation increased substantially. Consequently, actin was degraded by about 20% after 12 hr at 0°C. Similar to MHC degradation, actin degraded more rapidly at 5°C than at 0°C. As temperatures increased further, the degradation of actin occurred more rapidly.

2.4.5 Solubilization of Myofibrillar Proteins during Processing

Efficient washing is the most important step in surimi processing to ensure maximum gelling as well as colorless and odorless surimi. Minced fish meat contains approximately two thirds myofibrillar proteins. The remaining one third consists of blood, myoglobin, fat, and sarcoplasmic proteins, which impede the final quality of surimi gels. Washing increases the quality of surimi and extends the frozen shelf life by removing the undesirable one third, thus concentrating the functional myofibrillar proteins.

Lin et al.[49] reported that a considerable amount of myofibrillar proteins, myosin and actin, were lost in wastewater during washing and dewatering. The loss of myosin was small in the first wastewater, increased substantially in the second, and then remained nearly constant throughout the rest of processing. Generally, the washing of minced flesh with water removes sarcoplasmic proteins, which in turn concentrates the myofibrillar proteins. However, the higher losses of myofibrillar proteins, particularly with MHC, during washing and dewatering, have become of particular interest (Figure 2.9).

Babbitt[59] suggested the use of continuous washing cycles to remove all undesirable compounds and to reduce the loss of solids. Extensive washing, however, might cause myofibrillar proteins to dissolve in water. The mechanical force during dewatering, which results in the loss of myofibrillar proteins as particulates, could also be responsible for the loss of functional myofibrillar proteins in surimi processing. This confirms the study of Lin and Park,[58] in which the solubility of myofibrillar proteins increased as the number of washing cycles increased. Consequently, because sarcoplasmic proteins and undesirable non-protein nitrogen compounds (i.e., trimethylamine and dimethylamine) are fairly soluble in water, the removal of these components could be accomplished in the first washing step using enough water to solubilize and remove them.

As the minced fish muscle is washed, it increases in weight due to an increase in the water uptake ability of the muscle proteins. At the same time, there is a decrease in the ionic strength, which is commonly expressed as equivalent millimolar sodium chloride. The minced fish tissue becomes swollen and is difficult to dewater when low volume washes are successively conducted. Therefore, before excessive swell-

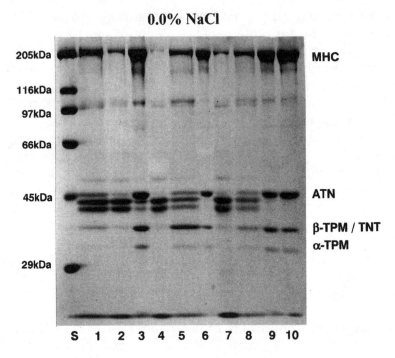

Figure 2.9 Electrophoretic patterns of proteins soluble in water (0% NaCl). S: high molecular weight standard; lane 1: 1st washing step (WS) of the first washing cycle (WC); lane 2–3: the first and second WS of the 2 WC, respectively; lane 4–6: the 1, 2, 3rd WS of the 3 WC, respectively; lane 7–10: the 1, 2, 3, and 4th WS of the 4 WC, respectively. MHC: myosin heavy chain; ATN: actin; β-TPM/TNT: β-tropomyosin/troponin-T; α-TPM: α-tropomyosin. (From Reference 61 with permission.)

ing occurs, thereby preventing dispersion, it is necessary to extract sarcoplasmic proteins in a large volume of water.[60]

Lin and Park[58] reported that, by using fresh water, most of the sarcoplasmic proteins were easily solubilized and removed in the first washing step of four successive washing cycles. In the second washing step, the residual sarcoplasmic proteins were continuously removed by sacrificing a relatively small amount of myosin heavy chain, actin, troponin, and tropomyosin. In subsequent washing steps, no significant

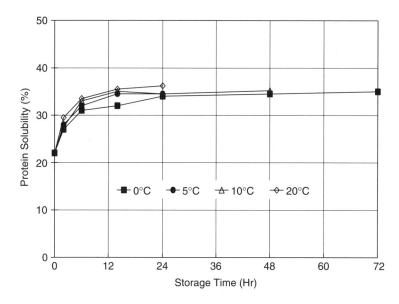

Figure 2.10 Solubility of Pacific whiting proteins during washing as affected by various storage conditions.

amounts of sarcoplasmic proteins were found, while the density of myofibrillar proteins dramatically increased. This observation suggests that most of the sarcoplasmic proteins are readily soluble in water and removed during the first washing step. To remove the residual sarcoplasmic proteins, the loss of a small amount of myofibrillar proteins is inevitable in the second washing step. Once the sarcoplasmic proteins were completely removed, further washing caused a severe loss of myofibrillar proteins.

The solubility of proteins during washing increased as well when fish were held for a longer period and/or at higher temperature (Figure 2.10). When fish were kept at a condition similar to commercial operations (0°C, 14 hr), before washing, total protein loss increased from 22.8% (0°C, 0 hr) to 33.8% (0°C, 14 hr). Furthermore, the loss of proteins increased slightly after 14 hr and reached a maximum loss of 35% at 72 hr.

In addition, as storage temperatures increased, more proteins were solubilized during washing. The temperature effect, however, was significant only for storage times up to 14 hr. After 24-hr storage, no difference in total protein loss was observed among the various temperatures.[58] Consequently, the solubility of proteins was maximized and maintained at approximately 35%, regardless of storage time and temperature. In addition, protein solubility appeared limited to a certain level when the water/meat ratio, washing cycle, and washing time were constant.

2.4.6 Washing Cycle and Wash Water Ratio

Washing is one of the most critical steps in surimi manufacturing. Large amounts of water are used to remove the sarcoplasmic proteins, blood, fat, and other nitrogenous compounds from the minced fish flesh. Texture, color, and odor of the final product are greatly improved when these impurities are removed by washing. However, rising utility costs, limited water sources, and pollution problems have prompted surimi manufacturers to consider minimizing water usage and reduce wastewater disposal.

The degree of washing required to produce good-quality surimi depends on the type, composition, and freshness of the fish. The number of washing cycles and water/meat ratios employed for washing vary among surimi processors. A water:meat ratio ranging from 4:1 to 8:1 is often employed by on-shore processors. This washing process was often repeated three to four times to ensure sufficient removal of sarcoplasmic proteins in the earlier days of surimi manufacturing. On the other hand, at-sea processors use a lower water/meat ratio (1:1 to 3:1) with only one or two washing cycles, due to their limited access to fresh water. Generally, the overall water/meat ratio used for washing ranges from 1:1 to 10:1. Increased water usage for washing usually results in more protein loss and increased wastewater disposal.[61] It was estimated that approximately 50% of total proteins were lost during washing.[62–64] In addition, for on-shore processing plants, 30 L of wastewater were generated per kilogram surimi produced.[58] However, with efforts made by the surimi industry, it is currently estimated that less than 10 to 15 L of water for shore-side operations and less than 5 to 7 L for at-sea operations are used for the production of 1 kg surimi.

The question has been raised as to whether more water usage guarantees better surimi quality or if it is unnecessary and wasteful. In early studies, it was shown that the gel strength of surimi continued to increase as the number of washing cycles increased, which resulted in more concentrated myofibrillar proteins.[65] However, Lin and Park[61] indicated that the majority of sarcoplasmic proteins are fairly soluble and removed during the initial washing steps. Subsequent washing removes the residual sarcoplasmic proteins along with a small amount of myofibrillar proteins. Consequently, after the sarcoplasmic proteins are completely removed, further washing causes a severe loss of myofibrillar proteins. Therefore, excessive washing not only increases the cost of water usage and wastewater treatment, but also results in a loss of myofibrillar proteins. Thus, for surimi processors to maximize quality and yield, as well as minimize water usage and wastewater disposal, careful control of the water/meat ratio, washing cycles, and washing time is important.

A logical approach to achieve the same washing effect with less water would be to increase the washing time and the number of washing cycles with a lower water/meat ratio. Theoretically, when a longer washing time is allowed and a constant water/meat ratio is maintained, more extractable proteins will dissolve in water until the equilibrium stage is reached. In addition, an increased number of washing cycles, at a constant overall water/meat ratio, have a higher dilution factor, and more extractable proteins will be removed per unit of water used.

Lin and Park[66] investigated minimizing water usage for leaching by reducing the water/meat (W/M) ratio and increasing the wash cycles (WC) and wash time (WT). Increased WT did not enhance the removal of sarcoplasmic proteins once equilibrium was reached. However, increased WC continuously removed residual sarcoplasmic proteins from the mince. At the low W/M ratio (2:1 or 1:1), regardless of WC and WT,

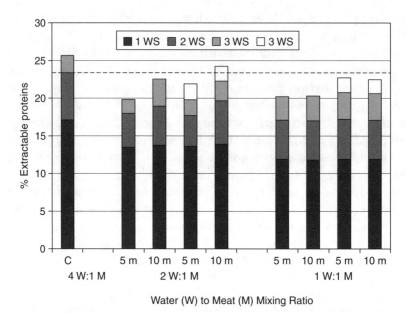

Figure 2.11 Effects of various washing conditions on total extractable proteins. 5m and 10m = 5 min and 10 min of washing time; 4W:1M, 2W:1M, and 1W:1M = water to meat mixing ratio; 1–4WS = 1–4 washing steps; C = control with 4:1 water to meat with 3 washing steps for 5 min each. Dotted line indicates 23.5%.

no significant loss of myofibrillar proteins occurred. Myosin heavy chain content, water retention, and whiteness of the washed mince, however, decreased when the W/M ratio was reduced. Increasing WC and/or WT, however, enhanced these properties but also resulted in a higher moisture content.

Based on the estimated content of sarcoplasmic proteins in whiting flesh at about 23.5% of total proteins,[61] Lin and Park[66] suggested that four washing cycles with a water/meat (W/M) ratio of 2:1 and 10-min washing time for each washing step appeared to be sufficient for the leaching process. Three washing cycles with a W/M ratio of 2:1 and 10-min washing time could be also applied (Figure 2.11).

Lin and Park[61] also showed that the relative amount of MHC increased as washing was repeated. However, this trend

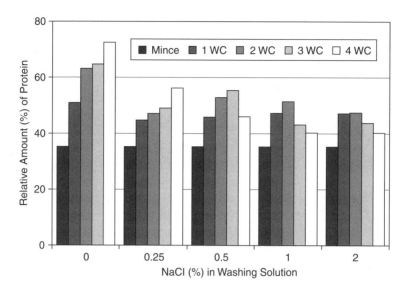

Figure 2.12 Effect of washing cycles and salt concentrations on the relative amount (%) of myosin heavy chain in washed whiting meat. (Adapted from Reference 61.)

was only effective at lower salt concentrations. Consequently, three and four washing cycles negatively affected the relative amount of MHC in the washed meat at 0.5% or higher salt concentration (Figure 2.12).

2.4.7 Salinity and pH

Compositional properties of fish vary when the fishing season changes. Generally, fish harvested during the feeding period produce the highest quality surimi. During this period, fish muscle has the lowest moisture content and pH, as well as the highest total protein content. Consequently, because the muscle of spawning fish has a higher pH and tends to retain more water, fish harvested during and after the spawning season produce the lowest quality surimi. Water removal from the washed meat of these fish, therefore, is difficult, but can be improved by either lowering the pH or increasing the salinity of the final washwater.

Myofibrillar proteins are generally classified as salt-soluble proteins. However, a previous study[58] revealed that a considerable amount of myofibrillar proteins was lost in surimi waste streams. The loss of myofibrillar proteins during surimi processing could be due, in part, to the nature of their water solubility. Stefansson and Hultin[60] and Wu et al.[67] reported that myofibrillar proteins solubilize in water and low ionic strength solutions. Hennigar et al.[68] and Choi et al.[69] reported that gels could be prepared using fish muscle without NaCl, which also suggests that myofibrillar proteins from fish muscle could be soluble in water or very low ionic strength solutions. In addition, extensive washing of muscle might cause decreased salt concentration in the tissue and allow myofibrillar proteins to become highly associated with water. Matsumoto[70] reported that when protein extracted by salt solution was repeatedly washed with water, most of the protein was solubilized. Also, increased washing time[36] or washing cycles using the same amount of water[64] resulted in more proteins, possibly myofibrillar proteins, being solubilized.

Presumably, washing time and washing cycles are crucial in changing the ionic strength of fish muscle during washing. The muscle of saltwater fish contains a variety of salts that are approximately equivalent to a solution with an ionic strength (μ) of 0.11 to 0.15.[60,61,67] Protein solubilization might be inhibited by keeping an ionic strength around 0.145 μ, which corresponds to 0.85% (0.145 M) NaCl,.[67,71,72] When the ionic strength of the extracting (washing) solution was decreased sufficiently, almost all the proteins of washed cod muscle were soluble (Figure 2.13). This data clearly indicates that fish myofibrillar proteins are water soluble if they are washed continuously.

Park and Lanier[73] showed that the addition of salt shifted the denaturation transitions to lower temperatures and decreased the enthalpies of heat denaturation. These results suggest that the addition of salt might cause a partial unfolding of proteins and increase sensitivity to denaturation.

In addition, Lin and Park[61] evaluated the effects of salt concentrations and washing cycles on the extraction of proteins. Sarcoplasmic proteins were readily soluble in water (0%

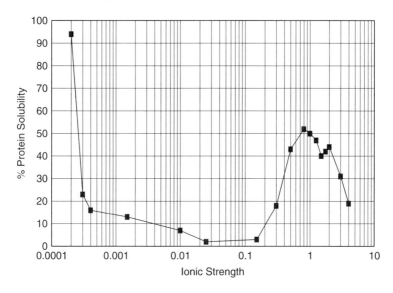

Figure 2.13 Effects of ionic strength on solubility of cod myofibrillar proteins. (Adapted from Reference 60.)

NaCl) and easily removed in the initial washing steps (Figure 2.9). Myofibrillar proteins became relatively soluble and were lost during extensive washing. Control of the water/meat ratio, washing time, and washing cycles were critical in reducing the loss of myofibrillar proteins. Washing with 0.25%, 0.5%, and 1.0% NaCl solutions also reduced the loss of myofibrillar proteins (Figure 2.14). However, these solutions were not effective in removing the sarcoplasmic proteins, even with increased washing cycles (Figure 2.15). Washing with 2.0% NaCl also resulted in low removal of sarcoplasmic proteins and, in addition, myofibrillar proteins were severely lost (Figure 2.16).

Another factor that affects myofibrillar protein solubility is pH.[63,74–80] As the pH approaches the isoelectric point, the negative and positive charges among the protein molecules are nearly equal. Therefore, protein molecules are strongly associated with each other through ionic linkages.[81] Proteins have reduced solubility at the isoelectric point because protein–water interactions are replaced by protein–protein interactions. Consequently, at pH above or below the isoelectric

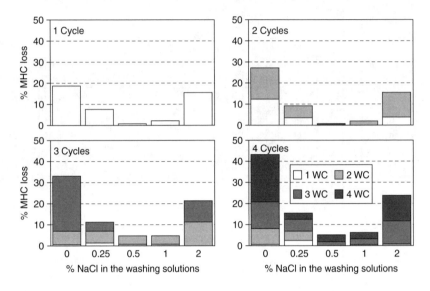

Figure 2.14 Loss of myosin heavy chains at various washing conditions. (Adapted from Reference 61.)

point, the protein acquires an increasing net negative or positive charge. These net charges provide more binding sites for water and cause repulsion among proteins, thus increasing protein solubility.[82]

Muscle proteins are highly soluble at either extremely acidic or alkaline pH.[80] Solubility increased rapidly as the pH shifted either from 5 to 4, or from 10 to 11. Protein solubility reached a maximum value at pH 2 and 12. Similar results were also reported for other species, including cod myofibrillar proteins,[60] salmon myosin,[83] rockfish proteins,[84] and Pacific whiting proteins.[78] At 10 mM NaCl, between pH 5 and 10, solubility was low but increased dramatically as the pH was shifted to either acidic or alkaline pH. At 600 mM NaCl, the isoelectric point was shifted to the acidic direction by about two pH units, resulting in aggregation of proteins at low pH, although the solubility of myosin heavy chain (MHC) improved between pH 6 and 10 (Figure 2.17).

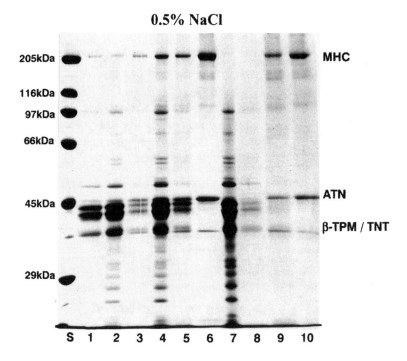

Figure 2.15 Electrophoretic patterns of proteins soluble in water (0.5% NaCl). S: high molecular weight standard; lane 1: 1st washing step (WS) of the first washing cycle (WC); lane 2–3: the first and second WS of the 2 WC, respectively; lane 4–6: the 1, 2, 3rd WS of the 3 WC, respectively; lane 7–10: the 1, 2, 3, and 4th WS of the 4 WC, respectively. MHC: myosin heavy chain; ATN: actin; β-TPM/TNT: β-tropomyosin/troponin-T; α-TPM: α-tropomyosin. (From Reference 61 with permission.)

The effects of several different chloride salts, and one iodide salt, on the solubility of protein from washed cod mince were determined by Stefansson and Hultin[60] (Figure 2.18). The general patterns of solubility were very similar. The major pH effect of the different cations was observed with magnesium and calcium chlorides, which at higher concentrations reduced the solubility of the muscle proteins.

In the unbuffered systems of the final washed meat, small changes in proton concentrations could cause large

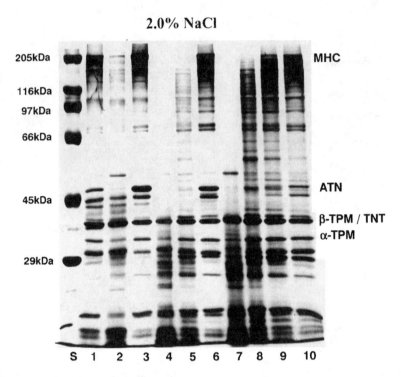

Figure 2.16 Electrophoretic patterns of proteins soluble in water (2.0% NaCl). S: high molecular weight standard; lane 1: 1st washing step (WS) of the first washing cycle (WC); lane 2–3: the first and second WS of the 2 WC, respectively; lane 4–6: the 1, 2, 3rd WS of the 3 WC, respectively; lane 7–10: the 1, 2, 3, and 4th WS of the 4 WC, respectively. MHC: myosin heavy chain; ATN: actin; β-TPM/TNT: β-tropomyosin/troponin-T; α-TPM: α-tropomyosin. (From Reference 61 with permission.)

changes in pH. When added at an ionic strength of 0.01, magnesium and calcium chlorides reduced the pH by about 0.2 units, approximately 6.5 compared to 6.7. This difference in pH is possibly caused by the strong interactions between the muscle protein side groups, as well as the divalent cations releasing more protons than the monovalent cations. This decrease in pH would probably be sufficient to cause the differences in protein solubility (Figure 2.18).

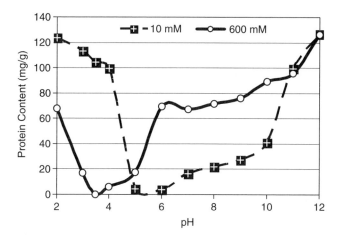

Figure 2.17 Solubility of Pacific whiting muscle proteins prepared at various pH and two IS levels (10 mM and 600 mM NaCl). (From Reference 80 with permission.)

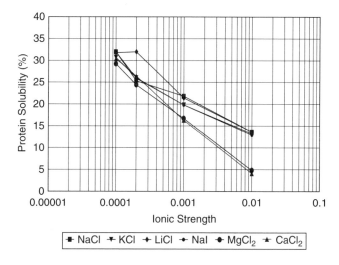

Figure 2.18 Effects of different salts on protein solubility of washed cod mince. (Adapted from Reference 60.)

2.5 PROCESSING TECHNOLOGIES THAT ENHANCE EFFICIENCY AND PROFITABILITY

2.5.1 Neural Network

A multitude of factors affect surimi quality. These can be broken down into three main categories: fish characteristics, site/trip characteristics, and processing operation factors (Figure 2.19). These, in turn, can be further broken down into numerous subcategories. For example, under fish characteristics, there are intrinsic factors such as weight, length, maturity, seasonality, water, protein, lipid content of the flesh, condition factor, etc. Extrinsic factors, on the other hand, include on-board handling and necessary factors that can be controlled in surimi production.

There is an increasing awareness that intrinsic and extrinsic factors affect final surimi quality characteristics, such as gel strength (breaking force), gel cohesiveness (defor-

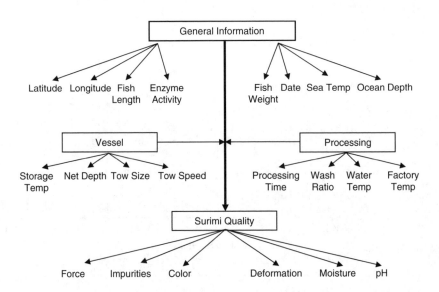

Figure 2.19 Interaction of intrinsic and extrinsic factors affecting final surimi quality.

mation), color, pH, moisture, and impurities. There has also been an increased interest in collecting as much data as possible on surimi operations, from fishing to production. Problems arise, however, in the data analysis, such as determining the most critical factors, knowing the factors that can be controlled, and choosing factors that should be studied more indepth.

Like other food processing systems, surimi processing is becoming more automated and computer driven. Therefore, information such as water/meat ratios, pH, and temperature can be easily collected automatically. Furthermore, most surimi production companies have quality control systems that allow technicians to take samples throughout the processing operations and record information on an hourly basis. Fishing operations themselves are highly regulated and there are complete databases on captured fish and their intrinsic characteristics, which include length, weight, sexual maturity, etc. The problems arise when a head technician notices surimi quality factors that are out of the tolerance range and then must determine the cause.

Food scientists in the past often looked at the cause–effect relationship in isolation. That is, if there is low gel strength, there is a tendency to look for one cause (e.g., a temperature aberration or poor washing conditions). However, there may be a host of factors that contribute to poor gel strength and many of these can be interactive. New computer systems, such as fuzzy logic and neural networks, allow scientists and technicians to look at a wide range of variables to determine their effect, in an interactive manner, on final product quality.

The artificial neural network (ANN) is a relatively new computer technology that allows the technician to correlate a large number of variables with desired quality outputs.[85] ANN is a method of nonparametric modeling that is ideally suited for the generalization of patterns from large data sets. The key word is "patterns," as ANN does not give results represented by equations or specific relationships. Instead, ANN attempts to mimic the "software of the mind," in that it

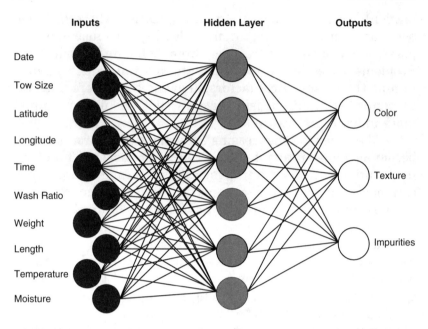

Figure 2.20 Neural network mechanism.

recognizes patterns and imitates the human reasoning process about a particular problem or operation.

The basic architecture of ANN is the concept of connecting a large number of very simple processing elements (neurons) that are highly connected (Figure 2.20). The value of the hidden layer neurons is determined by summing the value of each input neuron multiplied by its connection strength. The output value (surimi quality) is calculated in the same manner by summing the value of the hidden layer neurons multiplied by the connection strengths. In this way, the ANN system "learns" in a similar manner to the human brain, that is, by a method of back-propagation (experience feedback). The outputs are influenced by the inputs (variables in the surmise operation) and more input increases the accuracy of the ANN system.[32] Therefore, ANN is a very useful tool as an analysis system for large data sets.

Surimi: Manufacturing and Evaluation

The usefulness of the program lies in the interconnectedness of the different variables. Some inputs, such as time, temperature, and wash ratios, may show stronger connections and will have a greater influence on the final output, such as gel strength, color, etc. In the real world of surimi processing, factors such as salt, pH, and moisture influence gel strength, but also interact with each other. ANN allows process engineers and technicians to study these interactions and help determine factors that are important for final product quality.

At the Oregon State University Seafood Laboratory, data was collected on the characteristic, harvesting, and surimi processing variables of Pacific whiting for the years 1992 to 1994. More than 80 different input variables that might influence surimi quality were collected into a computer database for ANN analysis.[86] Through back-propagation and automatic learning by ANN, relationships were determined for all the variables and their interactions simultaneously. Consequently, several factors that affect quality in the Pacific whiting surimi industry were described. The important variables include time from harvest to process, storage temperature, date, salinity of flesh, pH of flesh, meat/water ratios, geographic location, and the length and weight of fish. The three major factors influencing the quality of Pacific whiting are (1) the time it takes to process the product, (2) the temperature at which the fish are stored until processing, and (3) the time of year the fish are processed.

Time and temperature from harvest to final product vs. gel strength were determined (Figure 2.21). It is obvious from the graph that time and temperature are interrelated. Data showed that maximum quality is obtained if the fish are processed within 4 to 10 hr post-capture and kept at temperatures below 4°C. The optimum time to process also depends on the storage temperature of the fish. If the fish are stored at refrigerated temperatures, processing the fish 6 to 10 hr post-capture should allow the surimi manufacturer to maximize quality. For nonrefrigerated whiting, surimi quality begins to drop rapidly after 7 hr; whereas for refrigerated fish (<4°C), surimi quality declines 15 hr after capture.

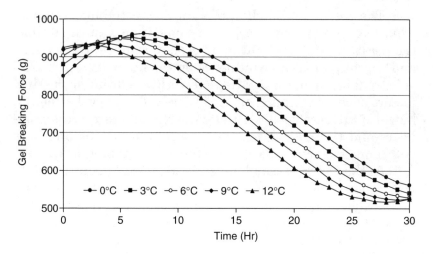

Figure 2.21 Effect of storage time and temperature on gel strength.

2.5.2 Processing Automation: On-Line Sensors

On-line sensors that measure the various physical and chemical characteristics of food can be powerful tools for processing and quality control. Physical attributes such as pressure, weight, temperature, and flow rate, along with chemical attributes such as moisture, fat, and protein content, can be monitored to control and ensure product quality.[87]

Maintaining target moisture levels during surimi production is of utmost importance to any processor. The moisture content directly governs the overall quality and consistency of surimi. Variations in moisture content during processing can unknowingly alter the finished surimi properties, including gel strength (breaking force), gel cohesiveness (deformation), whiteness, impurities, and, most importantly, overall grade.

A new technology using near-infrared (NIR) energy in the transmission mode (i.e., passing NIR energy completely through the surimi) can monitor moisture or protein contents in surimi within the processing pipeline. As surimi flows through the flow cell, different wavelengths of NIR energy are passed completely through the surimi at a rate of 60 individual scans per second. The resulting NIR absorbance is then measured by the sensor module and, through the calibration process, the moisture content is instantly calculated.[88] This value is then sent to a display so that manual adjustments can be made. The system could also be equipped with a series of calibrated analog output signals that could then be sent to an existing control system or directly to the screw press. In the past several years, there have been few attempts made in the U.S. surimi industry using NIR technology, but the outcome was quite limited. Therefore, further research must be done to utilize this new technology.

There are other areas where sensors can be successfully used in surimi processing. For example, the automatic injection of a cryoprotective mixture into a silent cutter or a ribbon blender can be used to guarantee the consistency of the product. In addition, automatic rejection of parasite-contaminated whiting fillets, before mincing, could eliminate the problem of impurities in surimi, and thus increase the overall grade.

2.5.3 Digital Image Analysis for Impurity Measurement

Impurities, which often consist of broken tissues from exterior skins and/or internal belly flap, also determine the quality of surimi. Although the impurity is not a health or safety issue, a large amount of impurity adversely affects the finished product and reduces consumer satisfaction. In measuring impurity counts, it is true that trained and experienced technicians cannot agree on impurity counts for the same sample due to various factors associated with measurement: color, size, number or level of impurity, consistency of sample presentation or lighting, and technician fatigue.

A digital image analysis technique for counting impurities in surimi has consequently been commercially developed by a Canadian manufacturer and successfully applied by the leading U.S. surimi manufacturers. According to an industry expert (P. Knight, personal communication, 2003), the SPX Speck Expert program (Figure 2.22) delivers objective, accurate, and

Figure 2.22 Digital image analyzer for surimi impurity. (Courtesy of Nutech Analytical, Inc.)

repeatable counts for surimi impurity. The linear array captures a sample image and then divides the image into pixels. Pixel values form a grey-scale histogram from 0 (black) to 255 (white). Based on company standards for impurity counts, the size and number of impurities are identified and counted. Results are displayed as colored overlays on the impurities and recorded in *.txt files that can be saved, along with the image, to an MS Excel spreadsheet. The archive function allows counts and images to be reviewed at any time, as well as shared among users and clients. The application of an objective standard using digital counting eliminates failures due to human limitations and also improves the finished product.

2.5.4 Innovative Technology for Wastewater

Surimi processing requires large amounts of chilled fresh water during the washing process. According to Lin et al.,[49] the majority (~75%) of wastewater was discharged from screeners, whereas a relatively minor amount of wastewater was released from dehydrators and screw presses. The protein content of wastewater streams was between 0.46 and 2.34%. The wastewater in successive discharge points revealed a decrease in protein, non-protein nitrogen, fat, and ash. Because sarcoplasmic proteins and undesirable non-protein nitrogen compounds (i.e., trimethylamine and

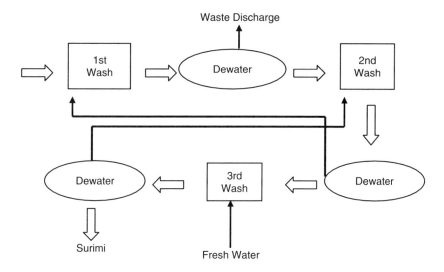

Figure 2.23 Water recycle with countercurrent washing.

dimethylamine) are fairly soluble in water, the removal of these components could be accomplished in the first wash step using enough water to solubilize them. Minced fish solids (~40 to 50%) are lost during the washing and dewatering processes as well.[62,63] Myofibrillar proteins of white fleshed fish, previously considered insoluble in water, are also lost in the washwater.[61,66,67]

Washwater can be recycled using the principle of countercurrent washing (Figure 2.23), as proposed by Lee.[48] Theoretically, recycling of the washwater, except the first washwater, could reduce water usage by two thirds. The first washwater, however, must be discarded due to high levels of undesirable impurities. Recycling would relieve the burden of expensive waste management and also save energy costs related to fresh water production and water refrigeration. For some reason, however, this process has been neglected by surimi manufacturers. With increased requirements by environmental agencies, water recycling must be thoroughly investigated on a commercial production scale for economic feasibility.

Innovative processing technologies applied to surimi processing over the past decade have enhanced process efficiency, resulting in product quality improvements. Decanter centrifuges are now commonly used to recover fine particles lost through the dewatering screens and screw presses.[89] In addition, recovered meat is either recycled in a main processing line or marketed as low-grade surimi.

Lin et al.[49] demonstrated the use of microfiltration to recover insoluble particulates as a low-cost alternative to centrifuging. The particulates lost in the final dewatering process (screw press) were successfully recovered and concentrated using a perforated stainless steel rotary screen and microfiltration. According to Lin et al.,[49] replacing 10% surimi with meat recovered using microfiltration had the same gel quality as regular surimi with respect to gel hardness, elasticity, water retention, and color. The surimi production rate could also be increased by 1.7% (based on finished product) by adding recovered meat.

Lin et al.[49] also used ultrafiltration to recover soluble protein and for water purification. The aerobic plate count, chemical oxygen demand value, turbidity, and protease levels of wastewater were substantially reduced by ultrafiltration. In addition, use of ultrafiltration equipment provides the possibility of recycling processing water in the leaching system. However, proteins concentrated by ultrafiltration had considerable dark color and strong odors and therefore were suggested for use in a high-protein animal feed.

Details on issues of surimi processing waste and by-products are discussed in Chapter 7.

2.5.5 Fresh Surimi

Before 1960, all surimi used in the Japanese kamaboko industry was fresh surimi.[4] There was no method available to control freeze denaturation until the discovery of cryoprotectants. Pipatsattayanuwong et al.[90] hypothesized the following benefits of fresh surimi over frozen surimi:

1. Fresh surimi can be produced at a lower cost because additives and freezing are not necessary.
2. Fresh surimi can be used to develop no-sugar products and also increase the market value of certain products (e.g., "never-frozen" surimi seafood).
3. Fresh surimi can exhibit better gel functionality and be used at a reduced level.

In addition, Park and Pipatsattayanuwong[91] determined that the use of fresh surimi over commercial frozen surimi could contribute a cost savings of $0.172 per kilogram of finished product to users.

When 50% of frozen and fresh surimi were used respectively in surimi seafood, significantly higher stress (by approximately 25%) and higher strain values (by approximately 5%) were obtained (Figure 2.24). Furthermore, Figure 2.24 illustrates that 42.2% fresh surimi is as good as 50% frozen surimi. The calculation was made based on increased quantity and higher quality of available proteins obtained from fresh surimi as compared with those from frozen surimi. The shelf life, however, of fresh Pacific whiting surimi was a maximum of 5 days with respect to microbiological and functional properties.[90]

In Japan and Korea, where shore-side surimi processing plants are located in closer proximity to users, a small portion of fresh surimi is still commercially produced, on a small scale, using locally available fish species. Until the establishment of shore-side operations on the Oregon coast, the feasibility of using fresh surimi in the United States was extremely low due to the geographical distance of Alaska pollock surimi plants (i.e., Dutch Harbor, Akutan, Sandpoint, and Kodiak, AK) to surimi users. Currently, in Thailand and India, surimi seafood manufacturers are in close proximity to the surimi manufacturers. If the production of fresh surimi and its utilization are arranged properly, there will be a significant cost saving to the surimi manufacturers and surimi seafood manufacturers.

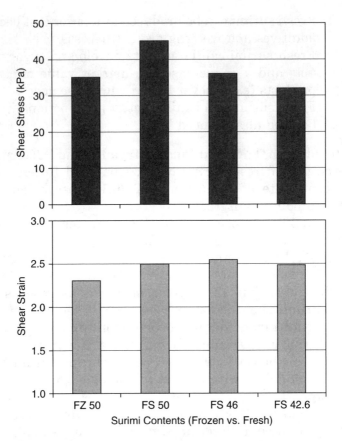

Figure 2.24 Comparison of fresh and frozen surimi based on tectural values of gels. Both frozen and fresh surimi samples were from the same batch and evaluated based on the equal moisture contents (75%). FZ = frozen surimi; FS = fresh surimi. Numbers frolled by FZ or FS indicate percent surimi content used in surimi seafood gels.

2.6 DECANTER TECHNOLOGY

One innovative processing technology applied to surimi processing over the past decade is the decanter centrifuge. It has enhanced efficiency as well as improved product yield. The decanter has been applied to surimi processing with two different objectives: (1) to recover fine insoluble particles from washwater, and (2) to replace the conventional screw press.

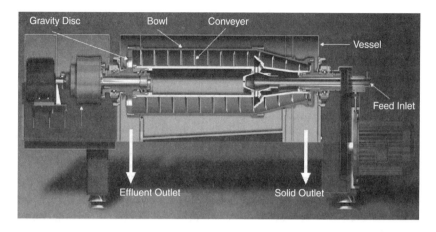

Figure 2.25 Cut-away view of a decanter centrifuge. (Courtesy of Alfa Laval.)

The recovery strategy of meat particles using decanter centrifuges is to get fine particles lost through conventional screens and screw presses used for the dewatering process.[89,92] Consequently, the recovered meat is either recycled in a main processing line or marketed as low-grade surimi. When a decanter centrifuge is used, losses in functional proteins can be reduced by 50% compared to using rotary screens and screw presses alone.[92]

The main component of the decanter centrifuge (Figure 2.25 and Figure 2.26) is the rotating bowl, which consists of a cylindrical part and a conical part. Inside the bowl, a conveyor, which rotates at a slightly different speed than the bowl, conveys the solids toward the solid discharge port. Surimi washwater is fed by an inlet tube to the conical–cylindrical junction of the decanter bowl through the hollow decanter conveyor shaft. When leaving the inlet pipe, the surimi washwater is smoothly accelerated in the bowl interior to full rotational speed. As a consequence of the high centrifugal force applied, the fine protein particles settle out as a deposit on the inner bowl wall. The screw conveyor continuously transports the settled solids toward the conical end. The conveyor speed is the differential speed that is determined by

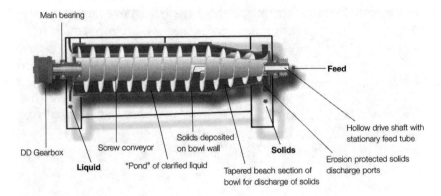

Figure 2.26 Cross-sectional view of a decanter centrifuge bowl. (Courtesy of Alfa Laval.)

the rotational speed difference between the main bowl and the screw conveyor.

Three main factors that influence the separation of mince and water in a decanter centrifuge are:

4. *The design of the decanter* (geometrical configuration, the bowl diameter, length, and speed, the differential speed of the conveyor relative to the bowl, and the conveyor type):
    A. Bowl diameter: a large bowl diameter increases the solids-handling capacity and centrifugal forces, but also dictates a lower main speed in order to not exceed the mechanical limitations.
    B. Bowl length: increasing the length of the bowl generally improves the liquid clarification as the residence time in the gravitational field increases.
    C. Beach length is the length of the conical beach near the solids discharge. A longer beach length gives a drier cake at the expense of solids recovery due to the decreased residence time. At low differential speeds (3 to 4 rpm), beach length does not significantly affect the moisture content of the cake. At high differential speeds (>15 rpm), however, a longer beach length is required for a dried cake.

Surimi: Manufacturing and Evaluation

   D. Pool depth: deeper pool results in longer residence time, higher capacity, lower solid content in the bowl, and less shear force on the proteins. However, increasing pool depth will shorten beach length, causing high moisture content in the cake.
   E. Bowl speed: at a higher centrifugal force, a higher solid setting rate, lower moisture content in the recovered solid, and higher recovery are obtained. However, when the shear force increases, the amount of fine particles increases and longer residence time is required.
   F. Differential speed: when differential speed is increased, the recovery increases at the expense of cake dryness. However, at a high flow rate, increasing the differential speed results in a drier cake.
5. Composition of the liquid and the particles to be separated (density, viscosity, size, distribution, configuration, and concentration of the particles)
6. Temperature and feed rate
   A. Temperature: at higher temperatures, lower viscosity, higher yield, and lower solids are expected. High temperature also deteriorates quality and induces higher bacteria counts
   B. Flow rate: as the flow rate increases, the sludge spends less time under the centrifugal force and solid recovery is therefore expected to decrease. If the flow capacity is exceeded, an overflow will occur.

Surimi produced through the decanter is assigned as recovery-grade surimi. Recovery-grade surimi possesses fairly good color and low amounts of impurities (Table 2.2). The gel deformation values of recovery-grade surimi are also comparable to those of the primary surimi produced through a screw press. The gel-breaking force values, however, vary with the solid content of the recovered meat: the higher the solid content, the higher the gel-breaking force values. In typical commercial operations, the solids content of recovered meat can vary from 10.5 to 18.1%.

TABLE 2.2 Comparison of Recovery-Grade and Primary-Grade Surimi

	Primary-Grade Surimi	Recovery-Grade Surimi
Moisture (%)	75.6	78.6
Impurities[a]	13	15
Gel deformation (cm)	1.51	1.50
Gel breaking force (g)	575	231
Color L* value	81.5	80.9
Color a* value	4.5	4.4

[a] Impurity count was made based on 40 g surimi using a Codex Code method (Appendix).

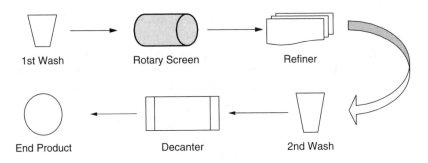

Figure 2.27 Flow diagram for the use of decanter centrifuge to replace a conventional screw press.

In addition to the use of a decanter to recover fine particles from washwater, Babbitt[93] has demonstrated the function of a decanter to replace a conventional screw press in the surimi lines. Functional and compositional properties of surimi manufactured using a decanter were not lower than those of conventional surimi. However, the decanter process gives a much higher processing yield. As indicated previously, the decanter surimi process (Figure 2.27) is a simplified operation, and it increases yield and results in consistent product quality.[94] Consequently, the decanter centrifuge will play an important role in surimi processing in order to maximize recovery and productivity.

However, it is not uncommon to use NaCl and/or $CaCl_2$ as a processing aid for easy water removal with a decanter. Once the residual content of salt is not properly monitored (>0.2%), the shelf life of surimi during frozen storage is significantly shortened. Special attention must be given during decanter surimi processing, particularly for the control of residual salt concentration; otherwise, the gel quality of frozen surimi will deteriorate at an accelerated rate.

2.7 SURIMI GEL PREPARATION FOR BETTER QUALITY CONTROL

Better quality assessment is the key to time savings and better quality control. The quality of surimi is determined based on a number of characteristics, some being more important than others. These include gel strength, color, moisture content, impurities, and microbiological counts. Other properties affecting the final quality are pH, protein content, fat content, cryoprotectants, and other food-grade additives. Among all the properties related to surimi quality, there is no doubt that gel properties— namely, gel strength — are of primary interest in surimi production and trade.[95]

A Codex Code for frozen surimi, under the guidelines of the FAO/WHO, was developed by the governments of Japan and the United States. For details, please refer to the appendix in this book. This code was developed particularly to assess the gel properties of Alaska pollock surimi, with an emphasis on two different methods for gel analysis: the Japanese punch (penetration) test and the U.S. torsion test. It also suggested that the buyer and seller decide the test that would be appropriate, depending on their particular interests.

Because the pros and cons for these two methods for gel analysis are described in Chapter 11, this section discusses advanced gel preparation methods for better quality control.

2.7.1 Chopping

Surimi, partially thawed, is subjected to comminution (chopping) with 3% salt (punch test) or 2% salt (torsion test).

Regardless of moisture content of surimi, the moisture of the surimi batch is adjusted to 78% for torsion, while no adjustment is made for punch test.

2.7.1.1 2% and 3% Salt

The higher the salt concentration (up to 5 to 7%), the higher the gel strength. Based on the processing nature of surimi seafood (i.e., kamaboko or crabstick), the level of salt for commercial practices is between 1.2 and 2%. The use of 3% in gel testing, as opposed to a maximum level of 2% in the surimi seafood, will apparently show a higher gel value than the value that would be obtained at 1.2 to 2.0% salt in commercial products.[95] The use of 2% salt in gel testing would be more practical, while the use of 3% salt could overestimate the gel quality of the surimi.

2.7.1.2 Moisture Adjustment

The punch test prepares gels without considering the effect of moisture on the gel value. Moisture contents of commercial surimi vary between 72 and 77%, while the majority falls between 73 and 76%. There is a significant effect of moisture on shear stress (similar to breaking force measured from the punch test).[96] In addition, moisture affects the protein content of surimi. This approach leaves us to question: do we want to measure the quality of protein or the quantity of protein (inversely, moisture content)? Surimi must be analyzed for its quality as affected by the protein quality while keeping moisture content (inversely, protein content) equal for all surimi.[95]

2.7.1.3 Chopping Temperature

The effect of chopping temperature on gel values is explained in depth in Chapter 9. Due to the dominance of Alaska pollock as a raw material for surimi seafood, it was always suggested to keep the chopping temperature of surimi below 13 to 15°C. This is the chopping temperature most commonly used in gel preparation, regardless of species. However, a recent study conducted by Park and co-workers[97] clearly stated that the

Surimi: Manufacturing and Evaluation

optimum chopping temperature depends significantly on the thermal sensitivity of the fish, which is related to its habitat temperatures: optimum chopping temperatures for cold-water species (Alaska pollock), temperate-water species (Pacific whiting), and warm-water species (threadfin bream, bigeye snapper, and lizardfish) are 0 to 5°C, 10 to 15°C, and 20 to 25°C, respectively. By comminuting surimi at the optimum temperature, the gel value can increase by 20 to 30%. Detailed information can be found in Figure 9.15 of Chapter 9.

2.7.1.4 Effect of Vacuum

Mechanical chopping creates numerous air pockets in surimi paste. Therefore, the removal of these air pockets is essential to accurately evaluate the quality of surimi. The conventional chopping process uses an open, silent cutter or stone mill that applies smashing (chopped paste, after scooping handful, is strongly smashed onto the chopping bowl). Depending on the manufacturers, it is repeated 5 to 20 times. However, most manufacturers in the United States use a vacuum, silent cutter (Figure 2.28) to overcome the problem of air pockets. When air pockets are not properly removed, it is common to see a broken casing while cooking. Surimi gel texture could also be downgraded if the punch plunger hits air pockets hidden in the gel matrix.

2.7.2 Cooking

2.7.2.1 Plastic Casing and Stainless Steel Tube

Plastic casings with a 3-cm diameter have been used as a standard. However, it is difficult to maintain the same compactness for every sample. It is also time consuming. If a stainless steel tube is used instead, it will give the same consistency in density and it is also reusable.

2.7.2.2 Cooking Method Resembling Commercial Production of Crabstick

Gel cooking methods currently used, as indicated in the Codex Code (see appendix), are based on waterbath heating. Using

Figure 2.28 Vacuum silent cutter for gel preparation. (Courtesy of Stephan Machinery.)

stainless steel tubes (2 cm in diameter) in a 90°C waterbath, it takes 12 to 13 min to obtain an equivalent temperature at the geometric center as a plastic casing (Figure 2.29). In the case of plastic casings (3.0 cm diameter), it might take 20 to 25 min to reach 90°C at the center. This results in extremely slow cooking that could cause activation of proteolytic proteases, resulting in significantly lower gel values than what would be obtained during commercial crabstick production, which takes 45 to 50 sec to reach the internal temperature of a thin sheet (<2 mm).

The use of rheological values of surimi gels slowly cooked in a waterbath may not resemble the formation of gel texture under commercial crabstick production where the heat is conducted very fast.[95] Fast cooking can result in very good gels from low- to mid-grade surimi and/or enzyme-laden Pacific whiting surimi without enzyme inhibitors.[98] The best method for gel cooking must be determined based on the nature of commercial production. It is highly suggested that the best

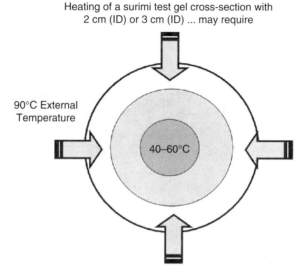

Figure 2.29 Heating pattern of a surimi test gel cross-section. (Adapted from Reference 99.)

method for gel cooking resembling crabstick processing is either ohmic or microwave heating. Both methods provide rapid, linear heating. The former gel cooking method was developed and investigated at the Oregon State University (OSU) Seafood Lab (Astoria, OR), while the latter was studied at North Carolina State University (Raleigh, NC).

2.8 SUMMARY

Various efforts were made for surimi manufacturing technology during the past two decades. The most impressive achievements were the development of surimi from species other than pollock and the increase in yield. The yield increase was achieved through changes in fishery management (United States), improved cutting machines, and decanter technology.

Multidisciplinary approaches must be taken to become better surimi processors. First, it is necessary to collect large

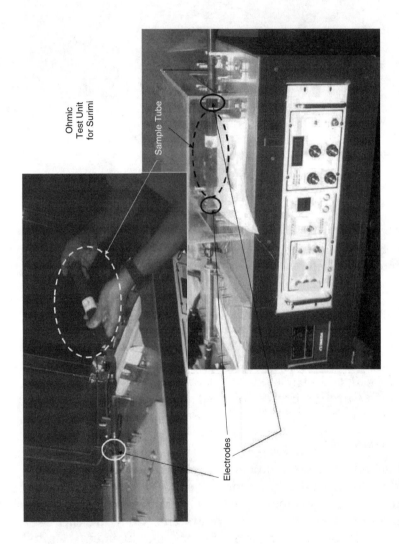

Figure 2.30 Ohmic test unit for surimi. (Courtesy of OSU Seafood Lab.)

Bench-top **Microwave** Testing Unit for Surimi

Figure 2.31 Bench-top microwave testing unit for surimi. (Courtesy of IMS, Cary, NC.)

amounts of data and understand how to analyze it. Second, look for obvious trends between the quality of surimi and extrinsic and intrinsic factors associated with surimi processing. Furthermore, it would be wise to run experiments and look for advice from the literature and experts. When scientific evidence is provided, do not hesitate to get out of old tradition.

ACKNOWLEDGMENTS

The authors would like to thank the leading experts in the U.S. surimi industry, including Phil Knight, Richard Draves, Don Graves, Pete Maloney, Mike Burns, Tom Junod, and Victor Covone for their invaluable information related to surimi manufacturing.

REFERENCES

1. K. Nishiya, F. Takeda, K. Tamoto, O. Tanaka, and T. Kubo. Studies on freezing of surimi (fish paste) and its application. III. Influence of salts on quality of fish meat. Monthly Report of Hokkaido Fisheries Research Laboratory, Fisheries Agency, Japan, 17, 373–383, 1960.

2. D.N. Scott, R.W. Porter, G. Kudo, R. Miller, and B. Koury. Effect of freezing and frozen storage of Alaska pollock on the chemical and gel-forming properties of surimi. *J. Food Sci.*, 58, 353–358, 1988.

3. AFDF. *Surimi — It's American Now*. Anchorage, AK: Alaska Fisheries Development Foundation, 1987.

4. M. Okada. The history of surimi and surimi based product in Japan. In R.E. Martin and R.L. Collette, Eds. *Engineered Seafood Including Surimi*. Park Ridge, NJ: Noyes Data Corp., 1990.

5. R. Draves. Surimi processing. Presented at the *11th OSU Surimi School*, Astoria, OR, 2003.

6. G.M. Pigott. Surimi: the "high tech" raw materials from minced fish flesh. *Food Reviews Int.*, 2, 213–246, 1986.

7. M.T. Morrissey. Unpublished data. Oregon State University Seafood Lab, Astoria, OR, 1995.

8. F. Takeda. Technological history of frozen surimi industry. *New Food Ind.*, 13, 27–31, 1971.

9. C.M. Lee. Surimi process technology. *Food Technol.*, 38(11), 69–80, 1984.

10. G.A. Adu, J.K. Babbitt, and D.L. Crawford. Effect of washing on the nutritional and quality characteristics of dried minced rockfish flesh. *J. Food Sci.*, 48(4), 1053–1060, 1983.

11. T.M. Lin and J.W. Park. Study of myofibrillar protein solubility during surimi processing: effects of washing cycles and ionic strength. Presented at the *PFT Annual Meeting*, Mazatlan, Mexico, 1995.

12. J.S. Kim and J.W. Park. Gelatin from solid by-products of Pacific whiting surimi processing. Abstract #76A-10. Presented at the *IFT Annual Meeting*, Chicago, IL. July 12–16, 2003.

13. J.W. Park and M.T. Morrissey. The need for developing surimi standard. In G. Sylvia and M.T. Morrissey, Eds. *Quality Assurance and Quality Control for Seafood*. Corvallis, OR: Oregon Sea Grant, 1994.

14. J. Yongsawatdigul, J.W. Park, E. Kolbe, Y. AbuDagga, and M.T. Morrissey. Ohmic heating maximizes gel functionality of Pacific whiting surimi. *J. Food Sci.*, 60, 10–14. 1995.

15. A. Hunt, J.W. Park, and C. Jaundoo. Cryoprotection of Pacific whiting surimi using a non-sweet glucose polymer. Abstract #100-4. Presented at the *IFT Annual Meeting*, New Orleans, LA. June 23–27, 2001.

16. A. Hunt, J.W. Park, and H. Zoerb. Trehalose as functional cryoprotectant for fish proteins. Abstract #99-6. Presented at the *IFT Annual Meeting*, Anaheim, CA. June 16–19, 2002.

17. T.C. Lanier. Measurement of surimi composition and functional properties. In T.C. Lanier and C.M. Lee, Eds. *Surimi Technology*. New York: Marcel Dekker, 1992.

18. J. Reynolds, J.W. Park, and Y.J. Choi. Physicochemical properties of Pacific whiting surimi as affected by various freezing and storage conditions. *J. Food Sci.*, 67(6), 2072–2078. 2002.

19. FDA. *Fish and Fisheries Products Hazards and Control Guidance, 3rd ed.* Washington, D.C.: U.S. Food and Drug Administration. 2001.

20. A. Lock. *The Guide to Reducing Metal Contamination in the Food Processing Industry.* Tampa, FL: Safeline Ltd., 1990.

21. H. An, V. Weerasinghe, T.A. Seymour, and M.T. Morrissey. Cathepsin degradation of Pacific whiting surimi proteins. *J. Food Sci.*, 59, 1013–1017 and 1033, 1994.

22. D.H. Greene and J. Babbitt. Control of muscle softening and protease parasite interactions in arrowtooth flounder (*Atheresthes stomias*). *J. Food Sci.*, 55, 579–580, 1990.

23. H. Toyohara and Y. Shimizu. Relation between the modori phenomenon and myosin heavy chain breakdown in the threadfin-bream gel. *Agric. Biol. Chem.*, 52, 255–257, 1988.

24. S.M. Boye and T.C. Lanier. Effect of heat stable alkaline protease activity of Atlantic menhaden (*Brevoorti tyrannus*) on surimi gels. *J. Food Sci.*, 53, 1340–1342 and1398, 1988.

25. H. Toyohara, T. Sakata, K. Yamashita, M. Kinishita, and Y. Shimizu. Degradation of oval-filefish meat gel caused by myofibrillar proteinase(s). *J. Food Sci.*, 55, 364–368, 1990.

26. O. Esturk and J.W. Park. Thermal stability of fish proteins from various species. Abstract #102-2. Presented at the *IFT Annual Meeting*, Chicago, IL. July 12–16, 2003.

27. I. Kimura, M. Nasu, G. Ito, M. Satake, D. Graves, Y. Kogeichi, K. Ogawa, and S. Watabe. Protease activity in walleye pollock muscle which has been infested by parasites. *Fisheries Sci.*, 68, 1549–1552, 2002.

28. S. Noguchi. Effect of parasites on the quality of frozen surimi. Presented at the *11th World Congress of Food Science and Technology*, Seoul, Korea. April 22–27, 2001.

29. T. Tokunaga and F. Nishioka. The improvement of technology for surimi production from fatty Japanese sardine. In proceedings of a national technical conference, *Fatty Fish Utilization: Upgrading from Feed to Food*, Raleigh, NC. 1988. UNC Sea Grant Publication 88-04.

30. AFDF. Ground fish quality chart. Anchorage, AK: Alaska Fisheries Development Foundation, 1992.

31. G. Sylvia, S. Larkin, and M.T. Morrissey. Quality and resource management: bioeconomic analysis of the Pacific whiting industry. In O. Bellwood, H. Choat, and N. Saxena, Eds. *Recent Advances in Marine Science and Technology*. Townsville, QS, Australia: James Cook University, 1994.

32. G. Peters. Determination of Quality Parameters for the Pacific Whiting Fishery Using Neural Networks and Induction Modeling. Ph.D. thesis, Oregon State University, Corvallis, OR, 1995.

33. G. Sylvia, S. Larkin, and M.T. Morrissey. Intrinsic quality and fisheries management: bioeconomic analysis of the Pacific whiting industry. In D.A. Hancock, D.C. Smith, A. Grant, and J.P. Beumer, Eds. *Developing and Sustaining World Fisheries Resources: The State of Science and Management*. Collingsworth, Australia: CSIRO Publishing, 1997.

34. Anonymous. Primary processing of surimi, secondary processing from surimi and its marketing. Presented at *Surimi Workshop by Overseas Fishery Cooperation Foundation and Japan Deep Sea Trawlers Assoc.,* Seattle, WA, 1984.

35. C.M. Lee. Surimi manufacturing and fabrication of surimi-based products. *Food Technol.,* 40(3), 115–124, 1986.

36. J.W. Park, R.W. Korhonen, and T.C. Lanier. Effects of rigor mortis on gel-forming properties of surimi and unwashed mince prepared from tilapia. *J. Food Sci.,* 55, 353–355,360, 1990.

37. G. Peters and M.T. Morrissey. Processing parameters affecting the quality of Pacific whiting. Presented at the *Meeting of Oregon Trawl Commission,* Astoria, OR, 1994.

38. M.T. Morrissey, J.W. Park, and J. Yongsawatdigul, Innovative processing in the seafood industry: the potential for ohmic heating and high hydrostatic pressure. Presented at the *First International Symposium of Biochemical Engineering and Food Technology at ITESM* — Campus Queretaro, Queretaro, Mexico, 1994.

39. M.T. Morrissey, G. Peters, J. Bolte, and G. Sylvia. Predicting seafood quality: use of new computer systems. In A. Bremner, C. Davis, and B. Austin, Eds. *Making the Most of the Catch.* Hamilton, Australia: AUSEAS, National Seafood Centre, 1997.

40. G.A. Gibbard and S.W. Roach. Standard for an RSW System. Tech. Rep. 676, Fish Mar. Serv., Vancouver, BC, 1976.

41. E. Kolbe, C. Crapo, and K. Hilderbrand. Ice requirements for chilled seawater systems. *Mar. Fish Rev.,* 47(4), 33–42, 1985.

42. M. Kraus. RSW-treatment of herring and mackerel for human consumption. In J.R. Burt, Ed. *Pelagic Fish: The Resource and Its Exploitation.* Oxford, England: Fishing News Books, 1992.

43. H.H. Huss. Quality and Quality Changes in Fresh Fish. FAO Fisheries Technical Paper 348. Food and Agricultural Organization of the United Nations. Rome, Italy, 1995.

44. P. Dion and M.T. Morrissey. The effects of superchilling on Oregon trawl caught fish. Report to the Oregon Trawl Commission, Astoria, OR, 1998.

45. K. Arai, K. Kawamura, and C. Hayashi. The relative thermostabilities of the actomyosin-ATPase from the dorsal muscles of various fish species. *Bull. Jap. Soc. Fish.*, 39, 1077–1082, 1973.

46. T. Misima, H. Mukai, Z. Wu, K. Tachibana, and M. Tsuchimoto. Resting metabolism and myofibrillar Mg^{++}-ATPase activity of carp acclimated to different temperatures. *Nippon Suisan Gakkaishi*, 59, 1213–1218, 1993.

47. FAO. Codex Code for Fish and Fishery Products. Food and Agriculture Organization of the United Nations. Rome, Italy: FAO, 1997.

48. C.M. Lee. Countercurrent and continuous washing systems. In R.E. Martin and R.L. Collette, Eds. *Engineered Seafood Including Surimi*. Park Ridge, NJ: Noyes Data Corp, 1990.

49. T.M. Lin, J.W. Park, and M.T. Morrissey. Recovered protein and reconditioned water from surimi processing waste. *J. Food Sci.*, 60, 4–9, 1995.

50. T. Haraguchi, U. Shimidu, and K. Aiso. Preserving effect of ozone to fish. *Bull. Jap. Soc. Sci. Fish.*, 35(9), 915–919, 1969.

51. H.S. Chang, C.C. Chen, and H.M. Chung. The discoloration effect of ozone on swine hemoglobin. *Food Science (Taiwan)*, 22, 766–775, 1995.

52. K.P. Lin and H.M. Chang. Effect of ozonation on color and functional properties of mackerel meal. *Food Science (Taiwan)*, 22, 218–226, 1995.

53. S.T. Jiang, M.L. Ho, S.H. Jiang, L. Lo, and H.C. Chen. Color and quality of mackerel surimi as affected by alkaline washing and ozonation. *J. Food Sci.*, 63(4), 652–655, 1998.

54. R.D. Buckley, J.D. Hackney, and K. Clark. Ozone and human blood. *Arch. Environ. Health*, 30, 40–43, 1975.

55. T. Suzuki. Frozen minced meat (surimi). In *Fish and Krill Protein: Processing Technology*. London: Applied Sci Publishers Ltd., 1981, 115–147.

56. M. Patashnik, H.S. Groninger Jr., H. Barnett, G. Kudo, and B. Koury. Pacific whiting, miraculous productus. I. Abnormal muscle texture caused by myxosporidian-induced proteolysis. *Marine Fish Rev.*, 44(5), 1–12, 1982.

57. Y.L. Xiong and C.J. Brekke. Changes in protein solubility and gelation properties of chicken myofibrils during storage. *J. Food Sci.*, 54, 1141–1146, 1989.

58. T.M. Lin and J.W. Park. Protein solubility in Pacific whiting afffected by proteolysis during storage. *J. Food Sci.*, 61, 536–539, 1996.

59. J. Babbitt. The use of a decanter centrifuge to prepare Alaska pollock surimi. In J.S. French and J. Babbitt, Eds. *Evaluation of Factors Affecting the Consistency, Functionality, Quality, and Utilization of Surimi*. Kodiak, AK: Alaska Fisheries Development Foundation. 1990, 26.

60. G. Stefansson and H.O. Hultin. On the solubility of cod muscle in water. *J. Agric. Food Chem.*, 42, 2656–2664, 1994.

61. T.M. Lin and J.W. Park. Extraction of proteins from Pacific whiting mince at various washing conditions. *J. Food Sci.*, 61, 432–438, 1996.

62. G.A. Adu, J.K. Babbitt, and D.L. Crawford. Effect of washing on the nutritional and quality characteristics of dried minced rockfish flesh. *J. Food Sci.*, 48, 1053–1060, 1983.

63. R. Pacheco-Aguilar, D.L. Crawford, and L.E. Lampila. Procedures for the efficient washing of minced whiting (*Miraculous productus*) flesh for surimi production. *J. Food Sci.*, 54, 248–252, 1989.

64. T.S. Yang and G.W. Froning. Selected washing processes affect thermal gelation properties and microstructure of mechanically deboned chicken meat. *J. Food Sci.*, 57, 325–329, 1992.

65. F. Nishioka. Leaching treatment. In H. Shimizu, Ed. *Science and Technology of Fish Paste Products*. Tokyo: Koseisha-Koseikaku Publishing Co. 1984, 62–73.

66. T.M. Lin and J.W. Park. Effective washing conditions reduce water usage for surimi processing. *J. Aquat. Food Product Technol.*, 6(2), 65–79, 1997.

67. Y.J. Wu, M.T. Atallah, and H.O. Hultin. The proteins of washed, minced fish muscle have significant solubility in water. *J. Food Biochem.*, 15, 209–218, 1991.

68. C.J. Hennigar, E.M. Buck, H.O. Hultin, M. Peleg, and K. Vareltzis. The effect of washing and sodium chloride on mechanical properties of fish muscle gels. *J. Food Sci.*, 53, 963–964, 1988.

69. M.R. Choi, J.W. Park, Y. Feng, and H.O. Hultin. Reduction of ionic strength in fish proteins and its effect on gel characteristics. Abstract #76E-12. Presented at the *IFT Annual Meeting*, Anaheim, CA. June 16–19, 2002.

70. J.J. Matsumoto. Identity of M-actomyosin from aqueous extract of the squid muscle with the actomyosin-like protein from salt extract. *Bull. Jap. Soc. Sci. Fish.*, 25, 38–43, 1959.

71. S.C. Sonu. Surimi. NOAA Technical Memorandum NMFS No. 13. Southwest Region, National Marine Fisheries Service, National Oceanic and Atmospheric Administration, Terminal Island, CA. 1986.

72. B. Trevino, V. Moreno, and M. Morrissey. Functional properties of sardine surimi related to pH, ionic strength, and temperature. In M.N. Voigt and J.R. Botta, Eds. *Advances in Fisheries Technology and Biotechnology for Increased Profitability*. Lancaster, PA: Technomic Publishing Co. 1990, 413–422.

73. J.W. Park and T.C. Lanier. Scanning calorimetric behavior of tilapia myosin and actin due to processing of muscle and protein denaturation. *J. Food Sci.*, 54, 49–51, 1989.

74. E.A. Foegeding. Functional properties of turkey salt-soluble proteins. *J. Food Sci.*, 52, 1495–1499, 1987.

75. T.S. Yang and G.W. Froning. Study of protein solubility during the washing of mechanically deboned chicken meat. *Poultry Sci.*, 69(1), 147, 1990.

76. S.L. Turgeon, S.F. Gauthier, and P. Paquin. Emulsifying property of whey peptide fractions as a function of pH and ionic strength. *J. Food Sci.*, 57, 601–604,634, 1992.

77. F.J. Monahan, J.B. German, and J.E. Kinsella. Effect of pH and temperature on protein unfolding and thiol/disulfide interchange reactions during heat-induced gelation of whey proteins. *J. Agric. Food Chem.*, 43, 46–52, 1995.

78. Y.J. Choi and J.W. Park. Acid-aided protein recovery from enzyme-rich Pacific whiting. *J. Food Sci.*, 67(8), 2962–2967, 2002.

79. Y.S. Kim, J.W. Park, and Y.J. Choi. New approaches for the effective recovery of fish proteins and their physicochemical characteristics. *J. Fisheries Sci.*, 69, 1231–1239, 2003.

80. S. Thawornchinsombut and J.W. Park. Roles of pH in solubility and conformational changes of Pacific whiting muscle proteins. *J. Food Biochem.*, 28,135–154, 2004.

81. J.E. Kinsella. Relationship between structure and functional properties of food proteins. In P.F. Fox and J.J. Condon, Eds. *Food Proteins.* New York: Applied Science Publisher, 1981, 51–103.

82. R. Hamm. Biochemistry of meat hydration. *Adv. Food Res.*, 10, 355–423, 1960.

83. T.M. Lin and J.W Park. Solubility of salmon myosin as affected by conformational changes at various ionic strengths and pH. *J. Food Sci.*, 62, 215–218, 1998.

84. J. Yongsawatdigul and J.W. Park. Effects of alkali and acid solubilization on gelation characteristics of rockfish muscle proteins. *J. Food Sci.*, 69(7), C499–505, 2004.

85. H. Ni and S. Gunasekaren. Food quality prediction with neural networks. *Food Technol.*, 52(10), 60–65, 1998.

86. G. Peters, M.T. Morrissey, G. Sylvia, and J. Bolte. Linear regression, neural network, and induction analyses to determine harvesting and processing effects on surimi quality. *J. Food Sci.*, 61, 876–880, 1996.

87. J. Giese. On-line sensors for food processing. *Food Technol.*, 47(5), 88–95, 1993.

88. D. Brown. On-line moisture measurements during production. Presented at the *Surimi Conference,* University of Washington, Seattle, WA, 1993.

89. J. Babbitt. The use of a decanter centrifuge to prepare Alaska pollock surimi. In J.S. French and J. Babbitt, Eds. *Evaluation of Factors Affecting the Consistency, Functionality, Quality, and Utilization of Surimi.* Anchorage, AK: Alaska Fisheries Development Foundation, 1990.

90. S. Pipatsattayanuwong, J.W. Park, and M.T. Morrissey. Functional properties and shelf life of fresh surimi from Pacific whiting. *J. Food Sci.*, 60, 1241–1244, 1995.

91. J.W. Park and S. Pipatsattayanuwong. Unpublished data. Oregon State University Seafood Lab, Astoria, OR, 1994.

92. T.C. Lanier, P.K. Manning, T. Zettering, and G.A. MacDonald. Process innovations in surimi manufacture. In T.C. Lanier and C.M. Lee, Eds. *Surimi Technology.* New York: Marcel Dekker, 1992, 167–180.

93. J. Babbitt. Demonstration of a new decanter surimi process. Progress Report (Grant No. 96-1-015) to Alaska Science & Technology Foundation, Anchorage, AK, 1997.

94. M. Burns. Payback in less than 30 production days. In AlfaPlus technical bulletin. Alfa Laval, Denmark: Alfa Laval. 1997.

95. J.W. Park. Surimi gel preparation and texture analysis for better quality control. In M. Sakaguchi, Ed. *More Efficient Utilization of Fish and Fisheries Products.* Amsterdam, Netherlands: Elsevier, 2004, 333–341.

96. W.B. Yoon, J.W. Park, and B.Y. Kim. Linear programming in blending various components in surimi seafood. *J. Food Sci.,* 62, 561–564, 567, 1997.

97. O. Esturk, J.W. Park, and S. Thawornchinsombut. Thermal sensitivity of fish proteins from various species on rheological properties. *J. Food Sci.,* 69(7), E412–416, 2004.

98. J. Yongsawatdigul and J.W. Park. Linear heating rate affects gel formation of Alaska pollock and Pacific whiting. *J. Food Sci.,* 61, 149–153, 1996.

99. T. Lanier. Surimi Chemistry. Presented at the *12th OSU Surimi School,* Astoria, OR. April 12–15, 2004.

3

Process for Recovery of Functional Proteins by pH Shifts

HERBERT O. HULTIN, PH.D.
University of Massachusetts, Amherst, Massachusetts

HORDUR G. KRISTINSSON, PH.D.
University of Florida, Gainesville, Florida

TYRE C. LANIER, PH.D.
North Carolina State University, Raleigh, North Carolina

JAE W. PARK, PH.D.
Oregon State University, Astoria, Oregon

CONTENTS

3.1 Introduction 108
3.2 Characteristics of Dark Muscle Fish Crucial
    to Surimi Processing 109
    3.2.1 Dark Muscle 109
    3.2.2 Lipids 113
    3.2.3 Muscle Proteins 115
    3.2.4 Processing of Pelagic Fish 117
        3.2.4.1 Alkaline Processing 119
        3.2.4.2 Problems with Processing
                Dark-Muscled Species 120
3.3 A New Approach for Obtaining Functional Protein
    Isolates from Dark-Muscled Fish 123
3.4 Summary 132
References 133

3.1 INTRODUCTION

The demand for fish protein throughout the world is increasing faster than can be met with traditional resources. This demand has, in fact, led to overfishing of many of the more traditional species and has required governmental intervention to prevent eradication of these species. Despite the current bleak situation and the economic disruption that the loss of fish stocks has caused in many parts of the world, there are still abundant fish that are underutilized in the sense that they are not primarily directed toward human food.

Dark-muscled fish species currently make up about 40% of the total fish catch worldwide. There is great interest in utilizing the large quantities of currently available, low value, fatty, pelagic fish for human food. However, cost-effective ways to improve the quality of the materials are required.[1–3] Likewise, much fish flesh remaining on the skeletons after filleting operations is shunted into low-value uses. Due to contamination with other tissue components such as skin, backbone, blood, etc., many of the problems associated with recovery of the proteins of these by-products are similar to those encountered with the dark-muscled species.

The advantages of a process for isolating a high-quality human food from these raw materials are obvious. To upgrade products made from the pelagic species and by-products would not only add economic value, but would also be a more responsible use of an important resource. The purpose of this chapter is to give an overview of problems encountered in dealing with the processing of small pelagic species; discuss the technological problems; indicate, where possible, some potential solutions to processing small pelagic species; discuss new insights on the basis of recent research; and present a new process for producing surimi that might overcome many of the problems associated with the processing of small pelagic species and other low-value fish raw materials into human foods. Fillets of white muscle species can also be processed by these techniques; the problems encountered with these materials, however, are much reduced compared with those containing significant amounts of other tissues.

3.2 CHARACTERISTICS OF DARK MUSCLE FISH CRUCIAL TO SURIMI PROCESSING

3.2.1 Dark Muscle

Fish muscle is unique in that the light muscle fibers are generally clearly separated from the red (dark) muscle fibers, as opposed to most homeotherms, in which the muscle cell types are mixed. The greatest portion of the problems associated with producing surimi from small pelagic species is the content of red muscle. Suzuki and Watabe[4] have reported that surimi prepared from fresh sardine ordinary muscle has a "quality ... equal to the high grade of Alaska pollock surimi." In addition, mackerel light muscle surimi had a higher and more constant sensory score, lower lipid oxidation, and less change in true strain values than comparable surimi made from whole mackerel muscle.[5] The light muscle of pelagic species is, however, darker than the same muscle of white-fleshed fish and produces a surimi with a lower "L" value.[6] In addition, mackerel surimi prepared with dark muscle contents from 0 to 100% gave decreasing percentages of water in cooked gels and decreasing strain values and fold test scores with increasing amounts of dark muscle (unpublished observations). Consequently, the higher lipid contents, less stable proteins, greater concentrations of heme proteins, lower ultimate pH values, higher proteolytic activities, and higher concentrations of sarcoplasmic proteins are all characteristics of dark muscle that have been suggested to contribute to the difficulties in making high-quality surimi from raw material with high contents of dark muscle.[1,2,4,7–10]

The differences between the dark and light (ordinary) muscles of dark muscle fish relate to their functions. As with most animal species, ordinary muscle is considered an anaerobic organ whose function is to provide energy quickly and intensively. Ordinary muscle tires easily and primarily uses glycogen as its energy source. The dark (or red) muscle, on the other hand, is designed for long-term exercise and is used by migrating species that travel great distances.[11] Dark muscle relies on oxidative metabolism of lipid as its principal source of energy. This is the reason for the high content of oil in the muscle, which in lean, bottom-dwelling fish is carried mostly in the liver.

The other major reactant required for producing energy for fish red muscle is oxygen, which is transported throughout muscle by the blood. Red muscle fibers (cells) are smaller in diameter than light muscle fibers but have the same number of capillary vessels surrounding each one. This gives dark muscle up to 10 times more capillaries than light muscle.[12] Thus, the blood supply to the red fibers is correspondingly higher than to the light fibers. With the higher amount of blood, there is also a higher amount of the pro-oxidant hemoglobin. High concentrations of myoglobin are required to bind and transport oxygen within the muscle cell; thus, this pigment is also present in high concentrations in dark muscle.

In one series of experiments,[13] unbled mackerel light muscle was found to have 6.1 µmol hemoglobin per kilogram tissue with no detectable myoglobin. Unbled mackerel dark muscle contained 159 µmol of hemoglobin per kilogram tissue and 342 µmol of myoglobin per kilogram. Because hemoglobin is a tetramer and thus has a molecular mass four times that of myoglobin, almost two thirds of the heme protein by weight in mackerel dark muscle is hemoglobin. This very high concentration of heme proteins in the dark muscle of mackerel makes it difficult to control the pro-oxidative activity of these proteins. Bleeding freshly caught mackerel reduced the hemoglobin content in light muscle by 45%, but removed only 24% of the hemoglobin from the dark muscle. This low removal of blood from the dark muscle may be related to the very large numbers of minute capillaries that surround the small dark muscle cells.

Mitochondria are few in ordinary muscle but make up 15 to 25% or more of the total volume of the red muscle cell.[14] The red muscle of fish from colder waters has more mitochondria than those from warmer environments.[11] The mitochondrion is the organelle responsible for the final stages of the oxidation of energy-yielding substrates to produce ATP with reduction of molecular oxygen to water.

Although molecular oxygen is critical and necessary for biological oxidations in all aerobic organisms, by-products of the reactions of molecular oxygen can be destructive to cellular components. Good evidence for this is the large number and types of cellular components that are designed to counteract the destructive effects of these reactive oxygen species. These protective agents include enzymes, reducing agents, metal chelating agents, and both water-soluble and lipid-soluble free radical scavengers.

Molecular oxygen can undergo four separate one-electron reductions (Figure 3.1). Although it is a very strong oxidizing agent, molecular oxygen does not normally react with most cellular components because it has both kinetic and thermodynamic restrictions on its activity. Molecular oxygen is a double free radical because it has two unpaired electrons. The interaction of molecular oxygen with ground-state molecules, such as unsaturated fatty acids, is spin forbidden. However, once the superoxide radical is formed, this spin restriction is removed. This one-electron reduction of molecular oxygen to superoxide is thermodynamically unfavorable, although all of

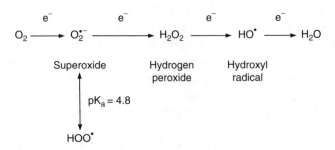

Figure 3.1 Changes of molecular oxygen with electron reduction.

the other single-electron reductions and the two- and four-electron reductions of molecular oxygen to hydrogen peroxide and water are thermodynamically favorable.[15,16]

Superoxide and the protonated neutral hydroperoxyl radical (HOO•), hydrogen peroxide, and the highly reactive hydroxyl radical can cause damage to cellular components. These reactive oxygen species are normally prevented from forming in biological systems by reducing molecular oxygen with enzymes that bind the reactive intermediates such that only relatively stable products are released, for example, water in the case of mitochondrial respiration.

The redox potentials of certain cellular components make the thermodynamic barriers for the first reduction of molecular oxygen to superoxide possible under some conditions. These components include reduced flavins, some non-heme iron proteins, quinols, and semiquinones. These compounds are found in membrane systems, especially mitochondria.

Approximately 80% of the oxygen used in aerobic cells is processed by mitochondria, and some 1 to 4% of this "leaks" out in the form of reactive oxygen species.[17] Another source of superoxide is the autoxidation of oxyhemoglobin[18] or oxymyoglobin.[19] The superoxide produced can be readily converted to other reactive forms by interacting with transition metals, such as iron. In a storage study of mackerel light and dark muscle, data was obtained that led to the hypothesis that the mitochondria of dark, but not light, mackerel muscle contribute to the development of rancidity.[20]

One of the principal reasons why dark muscle tissue is difficult to process into good surimi is its propensity to undergo oxidations. Oxygen activation is the first and most crucial step in the process and occurs in the living animal. Good handling practices may not be able to eliminate this problem entirely because peroxides are present in the living animal.[21]

3.2.2 Lipids

Dark muscle fish are often referred to as fatty fish. This is a reflection of their high lipid content. In addition, there are strong seasonal fluctuations in composition in many of these fatty species.[22–24] For example, Okada[1] reported that the lipid content in northern Pacific sardine varied from a low of around 6% in March to a high of around 28% in late summer. Likewise, Licciardello (personal communication, 1987) observed that the light muscle of Atlantic mackerel varied from 3.8% in April to 13.4% in November, while the low value for mackerel dark muscle was 7.5% in April and the high was 19.1% in September. The presence of a high lipid content has important implications in the storage, processing, stability, and nutritional value of fish muscle. It is also one reason for the rapid sensory quality losses observed.

The polar lipids of the membrane systems in muscle contain a higher percentage of the highly polyunsaturated fatty acids than do the neutral triacylglycerols. Whereas the high neutral lipid content of the fatty species, and especially the dark muscle, is related to the need of these species for a sustainable energy source, the highly unsaturated nature of the membrane lipids is necessary for the metabolic functional requirements of the membrane.[25] It is also necessary for the membrane phospholipids to stay sufficiently fluid for the membranes to carry out their metabolic functions. Membrane phospholipids from cold-water species contain up to 50% of their fatty acids as the 5- and 6-double bonded eicosapentaenoic and docosahexaenoic acids, respectively[26]

Fatty acid composition is not the only difference between the neutral oils and the polar membrane lipids. Because the polar phospholipids of the membrane exist primarily as a bilayer, they have a very large surface area exposed to the aqueous phase of the cell. The effective concentration of a component in the non-water-soluble phase of the cell is the surface area that is exposed to that aqueous phase. It can be estimated that on an equal weight basis, the area of the polar phospholipids is approximately two orders of magnitude greater than that of the neutral triacylglycerols.[27] In fish muscle that contains 10% neutral lipid and 1% phospholipid, the phospholipid fraction would have ten times more exposure to pro-oxidants in the aqueous phase than the triacylglycerols at the surface of the oil droplets.

In addition to their greater surface area, membrane lipids are found in association with components that can accelerate their oxidation. It was mentioned above that the mitochondrial inner membrane processes most of the molecular oxygen of the cell and that reactive oxygen species may escape from its electron transport chain. Other membrane systems also have electron transport systems that, although they may not be as active as the mitochondrial inner membrane, can still produce reactive oxygen species. In addition, membrane components, such as cytochromes or non-heme iron proteins, can convert species like superoxyl radicals (or the protonated HOO•) into more reactive species, such as the hydroxyl radical. The juxtaposition of these membrane components and the highly unsaturated fatty acids would encourage oxidation of the fatty acids.

Another characteristic of the polar membrane lipids that might affect their rate of oxidation is the pH of their immediate environment. When the phospholipid bilayer forms in the membrane, the charged heads of the polar lipids are exposed to the aqueous phase. This produces a net negative charge at the surface of the membrane at neutral pH. The negatively charged surface could attract hydrogen ions, thus producing a lower pH.[28] The net effect of low pH on lipid oxidation, however, is not well understood.

The conversion of superoxide ($O_2^{•-}$) to the thermodynamically more reactive HOO• would favor oxidative reactions. The HOO• also has the ability to penetrate into the hydrophobic lipid region of the bilayer, which would make it an even more effective pro-oxidant. On the other hand, some pro-oxidative processes are less favored at low pH. An example of the latter is the lower activity of the sarcoplasmic reticulum for reducing ferric iron to the reactive ferrous iron in the presence of NADH at pH values less than 6.8.[29]

3.2.3 Muscle Proteins

The current theory of gel formation from washed, minced fish muscle proteins holds that a high concentration of salt (NaCl) is required to solubilize the myofibrillar proteins, particularly myosin and actomyosin, which can then gel upon heating as the proteins denature, interact, and aggregate. Typically, a final salt concentration of 0.4 to 0.6 M at a neutral or slightly alkaline or acid pH is used. Good protein gels, however, can be obtained under conditions in which there is little or no solubility of the myofibrillar proteins, including myosin.[30,31] Consequently, it can be assumed that more than one mechanism contributes to gel formation.

It has recently been demonstrated that all the muscle fiber proteins in cod[32] and several other white-fleshed species, as well as mackerel light muscle, can be almost completely extracted in solutions of physiological ionic strength, 150 mM, or less. There is, however, a difference in the treatment required to achieve this solubility. With white-fleshed species, it is only necessary to wash the samples and extract in a sufficient volume of water to reduce the ionic strength to less than 0.3 mM. In the case of mackerel light muscle, however, muscle tissue has to be first washed with a neutral solution of moderate ionic strength. This removes some proteins, which then allows the remaining myofibrillar proteins, including myosin, to become extractable in the low ionic strength solution.

The possibility must therefore be considered that treatment of muscle proteins to prepare them to become soluble in water (very low ionic strength solutions) leads to their ability to produce a good gel. The difference in the requirements for muscle protein solubility in water between several white-fleshed species and dark muscled fish correlated with the initial pH of the muscle tissue.[33] The white-fleshed species had initial pH values above 6.6 while the pH values of the dark-fleshed fish were less than 6.6. When the pH of one of the white-fleshed fish, cod, was incubated at pH values below 6.6, the solubility of its proteins took on the characteristics of the dark-fleshed fish. This phenomenon in mackerel light muscle has been related to the effect of low pH on proteins that maintain the structure of the thick filaments (M protein) and Z-disk (α-actinin and desmin).[34]

The proteins of muscle tissue are usually grouped into three categories based on their solubility characteristics. One group is the sarcoplasmic proteins, which are soluble in water or solutions of dilute salt. Salt-soluble proteins are another group and are generally equated with the myofibrillar proteins. These are defined as those proteins soluble in salt concentrations greater than 0.3 M, with or without pH adjustment or the presence of components such as magnesium ion and ATP. Proteins not soluble in either of these extracting solutions comprise the final group and are called stromal proteins. These proteins are primarily connective tissue proteins but also include denatured myofibrillar proteins and membrane proteins.

Sarcoplasmic proteins are defined rather nebulously as being those soluble in low salt concentrations or even water. It is not possible, however, to really extract in "water" because the proteins are already in a salt solution in the tissue. The addition of water simply reduces the salt concentration. It is also important to remember that the "extractability" of proteins, such as the sarcoplasmic proteins, cannot necessarily be directly equated with their "solubility." In addition, some sarcoplasmic proteins, although soluble in a solvent, may not be extractable from muscle tissue in the same solvent. Two factors may prevent extraction of soluble proteins: (1) either the proteins bind to muscle subcellular structures; or (2) they may be located in cellular compartments not reached by the solvent, such as the mitochondrial matrix.[35]

The dark muscle of pelagic species has been reported to contain higher concentrations of sarcoplasmic proteins than the light muscle. Furthermore, the extractability of the sarcoplasmic proteins of dark muscle depends more on the extracting medium.[4] The presence of sarcoplasmic proteins has often been cited as one of the reasons for the poorer gelation characteristics of fish dark muscle compared to light muscle. The theory is that the sarcoplasmic proteins bind to the myofibrillar proteins and thus interfere with the formation of gels.[1,9,36]

Several recent studies, however, have brought into question whether the sarcoplasmic proteins, in fact, interfere with gel formation. These studies have either demonstrated no effect on gelation or an actual gel enhancement by the sarcoplasmic proteins.[37,38] The resolution of these different hypotheses could be that there is a single cause that leads to both greater solubilization of proteins and an improvement in the gel-forming ability of the cytoskeletal and myofibrillar proteins. Some research has led to the conclusion that the presence or absence of sarcoplasmic proteins contributes, at most, only a minor effect on the gelation ability and that good solubilization of the sarcoplasmic proteins occurs under the same conditions that allow the water solubility of the myofibrillar proteins to be expressed. In addition, Nishioka et al.[39] concluded that the function of washing was to improve the quality of the myofibrils, and this improved gel formation, not removal of the sarcoplasmic proteins. Furthermore, Morioka et al.[40] showed that the concentrations of 94, 64, and 40 kDa components in the sarcoplasmic fraction had a positive correlation with the strength of the gel formed.

Dark muscle also has higher proteolytic activity than white muscle.[9] This can cause modori, the weakening of the gel that occurs if the gel is held too long at a temperature around 50 to 60°C. Some differences occur in the myosin isoforms of the two muscle types as well. Light muscles have three different light chains, whereas dark muscles have only two; there are three or four myosin isoforms in light muscles and two in dark muscles.[41]

3.2.4 Processing of Pelagic Fish

Due to the potential importance of pelagic fish as a human food source, much research has been directed toward this end.

It has been suggested that the processing of pelagic species into surimi would be an appropriate way to utilize these fish.[1] The washing of the fish that takes place during surimi processing removes many of the components that cause low quality and poor stability of products prepared from these species. In addition, because the muscle is minced, components can be added that would aid in improving quality and/or stability.

Many components in the muscle tissues of dark-muscled species are present in higher concentrations than in white-fleshed fish, which contributes to their low economic value. Acid content is higher (i.e., lower pH), which can cause more rapid denaturation of the proteins. For this reason, the gelation characteristics of these species are often inferior to those of white-fleshed fish. In addition, dark-muscled fish contain more pro-oxidants and pigments. These cause rancid, fishy odors to develop and may produce highly colored products. Pelagic species also have high contents of histidine. Histidine may be converted to histamine, considered an important contributor to scombroid poisoning.

Factors that make pelagic species more difficult to successfully process into surimi include their high content of dark muscle (and thus blood); the seasonality of the catch and its composition; the small size of many species; and the heavy contamination of the water used to process the tissue, which leads to pollution problems. Attempts have been made, however, to modify the processing procedures to overcome some of these problems. Often, an adjustment in one processing parameter to deal with one problem exacerbates another problem.

There seems to be little doubt that a major problem with pelagic fish species is their high content of dark muscle. For this reason, much research has focused on removing the dark muscle before processing into surimi. It has been reported that surimi prepared from very fresh sardine with the dark muscle removed has a quality equal to that of high-grade Alaska pollock surimi.[4] Ishikawa et al.[42] describe experiments where surimi equal in quality to super A class Alaska pollock surimi was prepared from sardine and mackerel light muscle.

Several techniques have been suggested to remove dark muscle.[9,43] The location of the dark muscle along the lateral line near the skin, for example, makes deep skinning a feasible process for some species. Another method freezes the skin of a fillet onto a drum. The skin is then cut away, removing some dark meat and subcutaneous fat. As more dark muscle is removed, however, so is some light muscle, and yields subsequently decrease.

Dark muscle also has greater mechanical strength than light muscle due to its greater content of connective tissue. Therefore, much dark muscle can be removed during the refining process because it remains intact and is retained when the softer light muscle is pushed through the orifices. Water under pressure has been used to remove the softer light meat while leaving the dark meat attached to the skin. This process, however, requires much water.

In addition, because of its high fat content, the red muscle is less dense than the white muscle. By controlling the density of a solution in which the minced meat is suspended, much of the dark muscle will float while the light muscle settles. The separation, however, as in the case of refining, is not very efficient.

A major problem with all of these techniques, however, is that they substantially reduce yield, not only by taking out the red muscle but also by loss of some light muscle. The small size and low economic value of many dark-muscled fish also prevents consideration of a more efficient hand separation process to remove the dark muscle.

3.2.4.1 Alkaline Processing

The addition of alkali in the surimi washwater produces a higher-quality product than just using water.[9,44,45] Various concentrations of sodium bicarbonate can be added in one or more of the wash steps to increase the pH. Sodium chloride is also sometimes added. It has been suggested that gelation is improved after this type of washing process because the "solubility of the sarcoplasmic proteins" is increased and there is a "decreased rate of denaturation as the muscle pH is increased."[9] This process also releases more of the fat than washing without alkali. The addition of sodium chloride of 30 mM or greater to the washwater aids in the removal of heme proteins from minced or homogenized muscle tissue from Atlantic herring (*Clupea harengus*), especially at low pH (e.g., pH 6), but there is a concomitant loss of 10 to 16% of the muscle proteins. At pH 7, however, there is little effect of salt on heme protein extraction.

In an improved process, a mixture of sodium pyrophosphate and sodium bicarbonate was used to wash fish muscle that had been homogenized into fine fragments for 20 to 30 min, under vacuum.[44] Pyrophosphate was added to dissociate the actomyosin. The vacuum treatment and small particle size favor removal of impurities, that is, fat, colored, and odorous substances. Due to the small size of the fragments, however, the protein must be recovered with a decanter centrifuge rather than by a rotary sieve or screw press.

Alkaline processes, in addition to better gelation, produce surimi with better color, less fat, and better flavor than the standard process. A major drawback, however, is that the protein yields are lower. Typical yields of surimi from fillets range from 55 to 65% in the standard process. When the alkaline process for mackerel light muscle was used, less than 40% of the fillet protein was recovered.

A process using an underwater mincing in alkaline solution gave further improvement in gel strength (64%) compared to the process where tissue disruption occurs in air.[46] Disrupting fish muscle tissue under water can reduce development of rancid odors as well. This improvement is probably due to a rapid dilution of hemoglobin in the tissue.[47]

3.2.4.2 Problems with Processing Dark-Muscled Species

The seasonally fluctuating lipid content of the muscle tissue of dark-muscled species also causes problems.[48] Surimi produced at high fat concentrations is often of lower quality than surimi prepared from the same species at a time when the fat content is lower. It has been recommended, therefore, when the fat is unusually high that a preliminary treatment to remove some of the fat may be warranted.

It is generally accepted that it is "impossible to make surimi from small pelagic species that are not fresh."[4] Langmyhr et al.[7] echoed that sentiment when they stated that "to find ways of making good surimi from stored fatty fish is a great challenge." Production of good-quality surimi from frozen fish has been difficult, although it can be done successfully if the fish has not been stored too long and is homogenized into small fragments.[9,44] The color of surimi from frozen pelagics, however, is not good. It is possible to make good gels from dark-muscled fish if the fish is fresh, washed with alkaline brine, and if a setting procedure is used.[1] The color is not, however, as good as from a white-fleshed fish.

Development of fishy and rancid odors from lipid oxidation is a serious problem in processing dark-muscled fish into surimi,[9] but these odors can be reduced with early antioxidative intervention. Surimi from light mackerel muscle lost no sensory quality in one year of storage at −20°C and only about 10% of its true strain value when there was early treatment with antioxidants.[5] Comparable surimi prepared from whole muscle lost about 20% of its true strain value and had only slight declines in sensory scores over the 52-week period. In addition to inhibiting lipid oxidation, an antioxidant mixture added to washed bluefish or Atlantic mackerel mince produced significantly higher strain values.[49]

The amount of hemoglobin present in washed trout fillets and Atlantic mackerel are sufficient to account for all the lipid oxidation that occurs in raw fillets.[13,47] It has also been reported that heme pigments are a major catalyst in the development of rancidity in fish after cooking.[50] In preparing mackerel fillets for storage at −20°C, as little as 1-min exposure to the hemoglobin normally present in the tissue before washing in an antioxidant solution was sufficient to give a significantly shorter shelf life.[47] If, however, minced fish tissue can be washed under conditions that prevent lysing of the red blood cells, the effect of the hemoglobin is greatly reduced.[13] Presumably, some of the red blood cells are ruptured during mincing. The contribution of hemoglobin as a pro-oxidant, though, is expected to be highly species-dependent since the hemoglobins from different species have very different characteristics.[51,52]

The major factor that controls the pro-oxidative activity of fish hemoglobin is pH.[53] Lag phases of lipid oxidation of approximately 12 hr, 4 days, and 12 days were obtained in a model system of washed cod muscle in the presence of trout hemolysate at pH 6.0, 7.2, and 7.6, respectively. Important changes in hemoglobin structure occur with decreasing pH. These include dissociation of the tetrameric holoprotein to dimers and monomers, as well as the dissociation of heme from both the dimeric and monomeric species.[54] These dissociation processes are strongly affected by pH, increasing with decreasing pH. They are also dependent on temperature and the presence of organic phosphates, such as ATP. ATP favors dissociation and may be one of the reasons why blood is most effective as a pro-oxidant in freshly killed fish. The dimeric and monomeric species of hemoglobin are far more powerful pro-oxidants than the tetrameric species.[55,56] Dissociated heme pigment is not only an extremely reactive pro-oxidant, but is also very hydrophobic and would tend to associate with the hydrophobic components of muscle tissue, which include lipids.

Myoglobin is present at high concentrations in dark muscle but there is very little present in fish light muscle. Although it is widely reported that hemoglobin is less susceptible to oxidation than myoglobin, this is only true of the tetrameric form of hemoglobin. The constituent chains of hemoglobin, once dissociated, oxidize much more rapidly than do the oxymyoglobins.[55] The high concentration of total heme monomers contributed by hemoglobin and myoglobin in fish dark muscle[13] is a major reason why the dark muscle of fish is so unstable and why it is difficult to process dark-muscled fish.

Interestingly, the post-mortem age of the fish had almost no bearing on the sensory odor quality of surimi prepared from Atlantic mackerel muscle. There was also evidence that the freshest mackerel made surimi with the lowest quality based on its odor characteristics. In addition, the evidence showed that poor-quality mackerel made satisfactory surimi from the point of view of sensory odor. These results were obtained in the absence of antioxidants during processing.[57]

The difficulties in dealing with the dark (and light) muscle tissue of fatty pelagic fish species, however, are only part of the problem. It is often difficult to obtain, in an economical way, only the muscle tissue. Even if the samples can be headed and gutted, some attendant tissues such as the black layer on the belly flaps and the kidney tissue along the backbone make processing whole or headed and gutted fish even more difficult than processing just the muscle tissue.

Until now, there has been no entirely satisfactory way of mechanically removing the head and abdominal cavity contents. Some interesting approaches to this problem were discussed by Langmyhr et al.[7] These techniques have included putting fish on conveyer belts under pressure to squeeze out the intestines and roe. This, however, would not remove the head. Another method involves washing pieces of capelin for 45 min in weak acid at 20°C or at neutral pH at 40°C.[58] This allows separation of the skin and belly lining from the flesh.

In addition, using a water jet to remove the bones also eliminated dark pigments along the backbone. Unfortunately, the resulting mince had poor functional properties. In addition, there is a process called "nobbing." Here, a slant cut is made behind the head of the fish slanting toward the ventral and caudal regions. This removes the head plus part of the abdominal contents. However, an efficient, economical way to obtain muscle tissue with minimal contamination has not been achieved.

3.3 A NEW APPROACH FOR OBTAINING FUNCTIONAL PROTEIN ISOLATES FROM DARK-MUSCLED FISH

Recently, an approach has been developed to overcome some of the problems that are caused by the nature of the pelagic species and the processing that is consequently required.[59,60] The overall process concept is simple. The proteins of the muscle tissue are first solubilized. The solubilization can be accomplished in 5 to 10 volumes of water with alkali added to obtain approximately pH 10.5 or higher, or with acid added to about pH 3.5 or lower. It is usually necessary to choose the pH at which the consistency of the solution decreases to a value that allows the removal of undesirable materials. If it is desired to remove cellular membranes, it is generally necessary to go to a consistency of about 50 mPa·sec or lower. The mixture is then centrifuged. This allows the light oil fraction to rise to the top of the suspension. At the same time, the lipids of the membrane are removed in the sediment. Thus, both lipid fractions are removed due to density differences compared to the main protein solution. Other insoluble impurities, such as bone or skin, are also sedimented at this stage. The muscle proteins are then precipitated and collected by a process such as centrifugation.

The easiest way to precipitate proteins is by adjusting the pH to a value near the isoelectric point of the majority of the proteins, that is, about 5.2 to 5.5. Strangely, almost all the muscle proteins become insoluble under these conditions. This includes the sarcoplasmic proteins, which are mostly washed away during standard surimi manufacture. The nonprotein-soluble materials from the muscle tissue remain in the supernatant fraction after centrifugation and can subsequently be removed. The water remaining in the collected protein contains the same concentration of impurities found in the supernatant fraction. Additional washes of the sedimented protein at the same pH can be used to decrease the concentration of these soluble impurities if necessary. The overall process is illustrated diagrammatically in Figure 3.2.

It was previously shown that the major contractile proteins of muscle tissue, myosin, and actin could be solubilized in water at neutral pH at very low ionic strength (essentially in water).[32] A possible explanation for this solubilization that is consistent with the data is that the myofibrillar proteins have a negative charge at neutral pH. In water or solutions of very low ionic strength, the repulsive forces from these negatively charged side chains are sufficient to drive the individual protein molecules apart when sufficient water is made available. Conditions that reduce the repulsive forces, such as shielding of the charges by salts, allow the proteins to associate, most probably driven by hydrophobic interactions.

The requirement of very low salt concentrations for solubilization imposed serious limitations on how this knowledge might be used in a practical situation. However, the above

Process for Recovery of Functional Proteins by pH Shifts

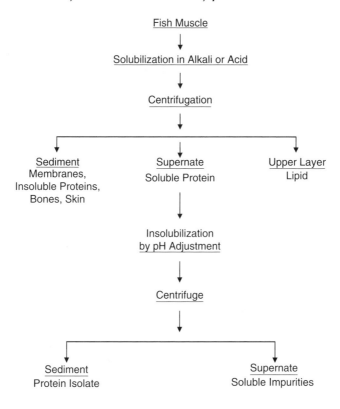

Figure 3.2 Scheme of process using pH shifts for producing surimi from fish muscle.

theory predicted that proteins could also be solubilized when the net charge, positive or negative, on the proteins was sufficiently high to overcome the effects of moderate salt concentrations. This turned out to be the case as the muscle protein could be solubilized at pH values of 3.5 and below or 10.5 and above in the presence of physiological concentrations (about 150 mM) of salt. The increase in net positive charge on the proteins at low pH comes primarily from neutralization of the negative charges of the carboxylate side chains of aspartic and glutamic acids, which have pKa values of about 4.7. The increase in net negative charge comes from deprotonation of basic groups such as the imidizole side chains of histidine, the guanidyl side chains of arginine, and the lysyl side chain, and from deprotonation of the phenolic side chains (e.g., tyrosyl).

A typical process would be carried out in the following way. Muscle tissue is homogenized at a tissue concentration of one part muscle to five to nine parts water. The pH is then adjusted to the appropriate pH value to solubilize the protein at either alkaline or acid pH. Although the concentration of the muscle tissue in the homogenate is important because the consistency of the material depends in part on the amount of protein present, factors other than protein concentration are also important. Consistency can also be reduced by the use of sodium chloride or by holding the homogenized tissue on ice for several minutes.[61] In addition, it has been determined that viscosity is probably not as important in membrane separation as is the detachment of the membrane from the various cytoskeletal proteins that connect the myofibrillar structures with the various membrane tissues of the cell.[62]

Addition of divalent cations such as calcium and magnesium together with an organic acid such as citric acid, which contains more than one carboxylate group and has a hydroxyl group, allows much easier separation of the membranes from the proteins. With this treatment, commercially available decanter centrifuges should be effective in sedimenting cellular membranes. Removal of membrane at this stage, however, is not always necessary to achieve the desired stability to oxidative rancidity.

It is necessary to reduce the muscle tissue to a fine particle size. If the tissue is simply ground, as in normal surimi processing, solubilization is incomplete. Homogenization of the tissue to produce fine particles also has the advantage that it allows very rapid mixing of the soluble cellular components with the added washwater. A drawback of standard surimi processing is the length of time it takes to wash the tissue because extraction rates depend on the diffusion times of the cellular components from the broken cells. We found that a 20-min extraction time of ground mackerel muscle produced somewhat less than 80% extraction of soluble protein (unpublished observations). On the other hand, extraction of homogenized tissue appears to be almost instantaneous.

After the first centrifugation in a laboratory-style batch centrifuge, there may be a fourth layer formed in addition to the original three: oil on top (only with fatty fish), protein solution in the center, and sediment on the bottom. The sediment contains cellular membranes, connective tissue, and other contaminants such as bone and skin, depending on the material processed. The fourth layer is a very weak gelatinous material that forms on top of the sediment. It is greater than 98% water, which means that it contains very little insoluble protein. However, it is reasonable to assume that the protein content of the aqueous portion of this very weak gel contains essentially the same soluble proteins at the same concentration as does the main aqueous supernatant fraction. Thus, it is important to reduce this layer to a minimum, to collect it when the soluble protein solution is separated, or to recycle it back to undergo the separation process again. The amount of this gel-like material that forms depends on protein concentration, pH, species, and most likely other factors as well. The gel is so weak that if the soluble protein fraction is siphoned off, within a few minutes the gel will collapse on itself and turn to liquid and can then be poured off.

Almost all the muscle tissue proteins are soluble at a pH of 3.5 or less or 10.5 and above. The actual solubility, however, may vary somewhat, depending on the species and the muscle type. Greater than 98% of cod and mackerel light muscle proteins are solubilized. The solubility of the proteins of mackerel red muscle varied from 75 to almost 100% under the same conditions. The variability may be related to post-mortem age and time of exposure to pH values below 6.6.

After solubilization, the protein is made insoluble and the proteins are subsequently recovered. A number of techniques can be used to achieve this. One efficient way is to change the pH into a range near the isoelectric point(s) of the muscle proteins. The proteins then precipitate while the soluble muscle tissue components, such as salts, nucleotides, sugar phosphates, amino acids, small peptides, etc., remain in solution. A straightforward way to collect the precipitated proteins is by centrifugation. Other techniques such as filtration are also possible.

With this method, two phases would be formed after centrifugation. The supernatant fraction contains the nonprotein, water-soluble components and a small amount of soluble protein. Most of the protein is therefore in the sediment. The moisture content of the protein sediment formed will depend on pH, ionic strength, and the centrifugal force used.

Because the water-holding capacity of the proteins at their isoelectric point is poor, further moisture can be removed by applying pressure. Any of the soluble components present in the supernatant fraction would be expected to be present in the free water associated with the sedimented protein. If it was required to further reduce the contamination of the precipitated proteins by soluble muscle components, additional washings could be carried out at the same pH and ionic strength at which the proteins were precipitated. To avoid the use of excessive amounts of water, water from any subsequent washes of the sedimented protein could be reused for suspension of new raw material.

The protein isolate can be adjusted to a neutral pH with bicarbonate or other base along with cryoprotectants before freezing the material. This surimi can be made into gels having good functional properties. Sample results obtained with proteins isolated from five species of fish and mechanically separated turkey are given in Table 3.1. The data represent a range of values obtained at different laboratories by different people with limitations on the knowledge of the raw material used. Thus, only comparisons made in the horizontal rows are truly valid. The data does, however, give a range of results, and these data may have some interest.

Studies to determine the effect of extreme pH values on the conformational changes of the muscle proteins have concentrated on what are arguably the two most important proteins in the fish tissue related to the functionality and stability of the protein isolates, myosin and hemoglobin.[68] Results of conformational and structural changes at pH 2.5 and 11 suggest that at the acid pH, myosin may fully dissociate into its six subunits while it does not dissociate completely at pH 11. Both the acid and alkaline pH values led to significant conformational changes in the globular head fraction of the myosin

Process for Recovery of Functional Proteins by pH Shifts

TABLE 3.1 Textural Properties of Three Different Forms of Fish Proteins Prepared using a pH Shift and Conventional Washing (Surimi)

Raw Material	Alkali Torsion Stress (kPa)	Strain	Alkali Punch Force (g)	Defor. (mm)	Fold Test	Acid Torsion Stress (kPa)	Strain	Acid Punch Force (g)	Defor. (mm)	Fold Test	Surimi[1] Torsion Stress (kPa)	Strain	Surimi Punch Force (g)	Defor. (mm)	Fold Test
Farmed catfish[2]	29	1.8	850	9.6	5	26	1.8	464	9	4.5	14	1.5	665	8.5	3
Herring															
Fresh[3]	56	1.6	871	9.2	5	58	1.8	566	9.2	5					
Aged 6 days[3]	—	—	464	6.2	3	—	—	498	7.3	3					
Frozen[4]	95	2.2	552	12.1	5	77	2	478	11.6	5					
Cod															
Fillets[5]	68	2.3	478	14.5	5										
Deboned mince[5,6]	34	0.9	304	7.7	2										
Cod frame[7]	—	—	277	10.1	5	—	—	225	9.1	5			290	7	
Atlantic croaker[8]	—	—	490	6.6	—	—	—	196	5.1	—			33	4.3	
Pacific whiting[9]	—	—	284	8.8	—	—	—	196	9.2	—			—	—	
Turkey, mechnically separated[10]	—	—	992	9	—	—	—	—	—	—			—	—	

Fold test score: 1–5; 5 is best (folds twice without cracking).

[1] Surimi-prepared with three washes of 1 part minced fish to 3 parts water.
[2] Farmed catfish – Kristinsson, University of Florida.
[3] Reference 63.
[4] Frozen Icelandic herring, (Y. Feng, personal communication).
[5] Reference 64.
[6] The mechanically deboned mince was considered to be of borderline quality for sale by the fish processor who donated it.
[7] Cod frames were of indeterminate quality, (Y. Liang, personal communication).
[8] Gel samples were small (1.4 cm long, 1.4 cm diameter) and probed with a 3.175-mm diameter ball. These punch test results cannot be compared directly with others in the table because of the size difference. Regular gels 3 cm high by 3 cm deep would give higher values.[65]
[9] Protein isolates adjusted to 78% moisture before cooking.[66]
[10] Reference 67.

heavy chains. These changes suggest a conversion to a molten globular configuration. Most of the myosin light chains were also lost at both pH treatments. This would mean that on pH readjustment to neutrality, heavy chains would not show calcium ATPase activity although they do take on a structural form similar to the native state. However, the unfolding and folding processes, which occur under various conditions during the alkaline and acid solubilization processes, may improve some functional properties of the proteins.

Because hemoglobin is most probably the prime catalyst of lipid oxidation in fish, and particularly so in pelagic species and many by-products such as racks, the effect of acid and alkali treatment on the ability of hemoglobin to carry out oxidations is significant. As described above, acid pH causes a rapid breakdown of the hemoglobin into monomers and dimers and the release of free heme. Thus, the low pH required in the acid process can lead to rapid oxidation of the lipid fraction of the tissue, and therefore it is important that the exposure time of the homogenized tissue to low pH be kept to a minimum. Hemoglobin that has been adjusted to a low pH and then returned to neutral pH has an increased propensity to act as a pro-oxidant and this propensity increases with lower pH and increased length of time.[69] Hemoglobin is much more stable at alkaline pH than it is at acid or even neutral pH. This is reflected in its much lower pro-oxidative activity at high pH than it has at acid pH.

There are, however, conformational changes at both acid and alkaline pH that indicate a greater exposure of hydrophobic groups. Although these conformational changes to greater hydrophobicity do not have a great effect on solubility with isolated cod hemoglobin, they allow for greater binding of the treated cod heme pigment to the myofibrillar protein fraction of the muscle tissue than non-pH-challenged heme proteins. It is clear that rapid handling and pH control are therefore necessary to keep the hemoglobin sufficiently soluble so it can be removed in the soluble fraction in this new process. This is particularly a problem when the acid solubilization process is used.

The pH shift process, however, offers several advantages over the traditional one. The new process gives improved processing yield. Greater than 85% yields are generally obtained from fillets compared to the 55 to 70% from the standard process. It is clear that this process recovers most of the sarcoplasmic proteins as well. Furthermore, there was no indication that these sarcoplasmic proteins interfered with gel formation. As discussed above, there is evidence that they may, in fact, enhance gelation. Recovery of the sarcoplasmic proteins would make the process advantageous even for white-fleshed species because yields would be improved compared to standard surimi processing methods.

The pH shift process is also faster than the standard surimi processing because it does not depend on diffusion processes to extract the water-soluble materials from broken muscle cells. Impurities such as skin and bone or contaminating metals, etc., would also be removed in the sediment of the first centrifugation. In addition to the ability of the proteins to form high-quality gels, they have other excellent functional properties, such as emulsification, foaming, and water-holding capacity. For example, protein isolates prepared by either the alkaline or acid solubilization process have superior emulsifying properties compared to untreated myofibrillar proteins.[70] Furthermore, the pH shift process produces a functional protein isolate that is low in lipid and has a high protein recovery. Greater stability to lipid oxidation can also be expected due to the removal of most of the lipid from the product. Most rancid odors and substances contributing to high thiobarbituric acid-reactive substances are removed during this process.

There is improved safety with the pH shift process. Lipid-soluble toxins such as polychlorinated biphenyls are removed and cholesterol levels are also reduced. Heavy metals such as mercury would be removed if they were in a lipid-soluble form (e.g., methyl mercury). In addition, all substances used are GRAS (generally recognized as safe) and the process reduces the number of bacteria by about one order of magnitude. A further reduction of 28% was achieved by the acid solubilization process and a reduction of 47% by the alkali process.

Low-cost, abundant protein sources that are not now used directly for human food can be used in this process. Headed and gutted fish can be used directly because skin and bones are easily removed in the process. Furthermore, it has been found that the protein isolates of some species retain their functional properties even after frozen storage of the whole fish. Consequently, some pelagic species need not be extremely fresh to produce good-quality gels.

There is also a lower cost of pollution control with this process. The processing water has a low biological oxygen demand because of its low protein content. It also has a relatively low salt content and its natural water-soluble components have been greatly diluted. Thus, the washwater is capable of multiple reuse. Furthermore, its low protein content should make it amenable to recycling by ultrafiltration and similar techniques. In addition, no components are added during the process that would contribute to pollution.

3.4 SUMMARY

A new procedure has been developed to isolate muscle proteins from raw muscle tissue by a process of shifting pH values. The first step requires a solubilization of muscle homogenized in water at a ratio of roughly 1:5 to 1:9. The protein can be solubilized by alkalinization to approximately 10.5 or above, or by acidification to approximately 3.5 or below. The insoluble material less dense than the aqueous medium (oil or fat), can then be removed by flotation, and insoluble impurities more dense than water can be removed by sedimentation, centrifugation, or filtration. The more dense impurities include cellular membranes, as well as skin and bone. The major aqueous fraction is then adjusted to a pH in the range of the isoelectric points of the muscle proteins (i.e., 5.2 to 5.5) to precipitate the proteins. The insoluble proteins are collected by centrifugation or sieving. The pH of the isolated protein can then be adjusted to any value desired with the addition of cryoprotectants and subsequently frozen to produce surimi.

The use of the pH shift processes for producing protein isolates is especially suited to low-value raw materials that are difficult to process by the standard surimi process of simply washing minced muscle tissue. Some cautionary points are discussed, such as the presence of heme proteins. The procedure offers several advantages, including higher yields, higher-quality protein, improvement of functional properties, reduction of pollutants, removal of most lipids, and efficient removal of insoluble impurities.

REFERENCES

1. M. Okada. Utilization of small pelagic species for food. In R.E. Martin, Ed. *Proceedings of the Third National Technical Seminar on Mechanical Recovery & Utilization of Fish Flesh.* Washington, D.C.: National Fisheries Institute, 1980, 265–282.

2. J. Opstvedt. A national program for studies on the value for "surimi" production of industrial fish species in Norway. In R.E. Martin and R.L. Collette, Eds. *Proceedings of the International Symposium on Engineered Seafood Including Surimi.* Washington, D.C.: National Fisheries Institute, 1985, 218–224.

3. G. Gunning. Focus: record fish catch. *Iceland Business,* 4, 3, 1997.

4. T. Suzuki and S. Watabe. New processing technology of small pelagic fish protein. *Food Rev. Int.,* 2, 271–307, 1986/1987.

5. S.D. Kelleher, H.O. Hultin, and K.A. Wilhelm. Stability of mackerel surimi prepared under lipid-stabilizing processing conditions. *J. Food Sci.,* 59, 269–271, 1994.

6. S.-H. Jiang, M.-L. Ho, S.-H. Jiang, L. Lo, and H.-C. Chen. Color and quality of mackerel surimi as affected by alkaline washing and ozonation. *J. Food Sci.,* 63, 652–655, 1998.

7. E. Langmyhr, J. Opstvedt, R. Ofstad, and N.K. Sørensen. Potential conversion of North Atlantic fatty species into surimi and surimi-derived products. In N. Davis, Ed. *Fatty Fish Utilization: Upgrading from Feed to Food.* Raleigh, NC: UNC Sea Grant College Program Publication, 1988, 79–117.

8. G.M. Hall and N.H. Ahmad. Surimi and fish mince products. In G.M. Hall, Ed. *Fish Processing Technology.* London: Blackie Academic and Professional, 1997, 74–91.

9. Y. Shimizu, H. Toyohara, and T.C. Lanier. Surimi production from fatty and dark-fleshed fish species. In T.C. Lanier and C.M. Lee, Eds. *Surimi Technology.* New York: Marcel Dekker, 1992, 181–207.

10. N.K. Sørensen and A. Mjelde. Preservation of pelagic fish quality for further processing on board and ashore. In J.R. Burt, R. Hardy, and K.J. Whittle, Eds. *Pelagic Fish: The Resource and Its Exploitation.* Oxford: Fishing News Books, Blackwell Scientific Publications Ltd., 1992, 38–54.

11. S.M. Kisia. Structure of fish locomotory muscle. In J.S. Datta-Munshi and H.M. Gutta, Eds. *Fish Morphology: Horizon of New Research.* Rotterdam: A.A. Balkema, 1996, 169–178.

12. Q. Bone. Locomotor muscle. In W.S. Hoar and D.J. Randall, Eds. *Fish Physiology.* New York: Academic Press, 1978, 361–424.

13. M.P. Richards and H.O Hultin. Contributions of blood and blood components to lipid oxidation in fish muscle. *J. Agric. Food Chem.,* 50, 555–564, 2002.

14. R.M. Love. *The Chemical Biology of Fishes, Advances 1968–1977.* Vol. 2, London: Academic Press, 1980.

15. A. Naqui and B. Chance. Reactive oxygen intermediates in biochemistry. *Annu. Rev. Biochem.,* 55, 137–166, 1986.

16. V.P. Skulachev. Role of uncoupled and non-coupled oxidations in maintenance of safely low levels of oxygen and its one-electron reductants. *Quart. Rev. Biophys.,* 29, 169–202, 1996.

17. J.Z. Byczkowski and T. Gessner. Biological role of superoxide ion-radical. *Int. J. Biochem.,* 20, 569–580, 1988.

18. J. Everse and N. Hsia. The toxicities of native and modified hemoglobins. *Free Radical Biol. Med.,* 22, 1075–1099, 1997.

19. M. Krüger-Ohlsen and L.H. Skibsted. Kinetics and mechanism of reduction of ferrylmyoglobin by ascorbate and D-isoascorbate. *J. Agric. Food Chem.,* 45, 668–676, 1997.

20. D. Petillo, H.O. Hultin, J. Krzynowek, and W.R. Autio. Kinetics of antioxidant loss in mackerel light and dark muscle. *J. Agric. Food Chem.,* 46, 4128–4137, 1998.

21. T. Nakamura, R. Tanaka, Y. Higo, K. Taira, and T. Takeda. Lipid peroxide levels in tissues of live fish. *Fisheries Sci.,* 64, 617–620, 1998.

22. R.G. Ackman and C.A. Eaton. Mackerel lipids and fatty acids. *Can. Inst. Food Technol. J.,* 4, 169–174, 1971.

23. T. Ohshima, S. Wada, and C. Koizumi. Lipid contents and compositions of various parts of sardine caught in different seasons. *J. Tokyo Univ. Fish.,* 75, 169–188, 1988.

24. N.M. Bandarra, M.L. Batista, M.L. Nunes, J.M. Empis, and W.W. Christie. Seasonal changes in lipid composition of sardine (*Sardina pilchardus*). *J. Food. Sci.,* 62, 40–42, 1997.

25. M.V. Bell, R.J. Henderson, and J.R. Sargent. The role of polyunsaturated fatty acids in fish. *Comp. Biochem. Physiol.,* 83B, 711–719, 1986.

26. R.L. Shewfelt. Fish muscle lipolysis — a review. *J. Food Biochem.,* 5, 79–100, 1981.

27. H.O. Hultin. Role of membranes in fish quality. In F. Jessen, Ed. *Fish Quality — Role of Biological Membranes.* Copenhagen: Nordic Council of Ministers, 1995, 13–35.

28. P. Fromherz and B. Masters. Interfacial pH at electrically charged lipid monolayers investigated by the lipoid pH-indicator method. *Biochim. Biophys. Acta,* 356, 270–275, 1974.

29. R.E. McDonald and H.O. Hultin. Some characteristics of the enzymic lipid peroxidation system in the microsomal fraction of flounder skeletal muscle. *J. Food Sci.,* 52, 15–21, 27, 1987.

30. H.-S. Chang, Y. Feng, and H.O. Hultin. Role of pH in gel formation of washed chicken muscle at low ionic strength. *J. Food Biochem.,* 25, 439–457, 2001.

31. Y. Feng and H.O. Hultin. Effect of pH on the rheological and structural properties of gels of water-washed chicken breast muscle at physiological ionic strength. *J. Agric. Food Chem.,* 49, 3927–3935, 2001.

32. G. Stefansson and H.O. Hultin. On the solubility of cod muscle proteins in water. *J. Agric. Food Chem.,* 42, 2656–2664, 1994.

33. S.D. Kelleher, Y. Feng, H.O. Hultin, and M.B. Livingston. Role of initial muscle pH on the solubility of fish muscle proteins in water. *J. Food Biochem.*, 28, 279–292, 2004.

34. Y. Feng and H.O. Hultin. Solubility of proteins of mackerel light muscle at low ionic strength. *J. Food Biochem.*, 21, 479–496, 1997.

35. H.O. Hultin, Y. Feng, and D.W. Stanley. A re-examination of muscle protein solubility. *J. Muscle Foods*, 6, 91–107, 1995.

36. J.W. Park, T.M. Lin, and J. Yongsawatdigul. New developments in manufacturing of surimi and surimi seafood. *Food Rev. Int.*, 13, 577–610, 1997.

37. K. Morioka and Y. Shimizu. Contribution of sarcoplasmic proteins to gel formation of fish meat. *Nippon Suisan Gakkaishi*, 56, 929–933,1990.

38. W.-C. Ko and M.-S. Hwang. Contribution of milkfish sarcoplasmic protein to the thermal gelation of myofibrillar protein. *Fisheries Sci.*, 61, 75–78, 1995.

39. F. Nishioka, T. Tokunaga, T. Fujiwara, and S. Yoshioka. Development of a new leaching technology and a system to manufacture high quality frozen surimi. In *Proceedings of the Meetings of Commission C2, Chilling and Freezing of New Fish Products*, Sept. 18–20. Paris: International Institute of Refrigeration, 1990, 123–130.

40. K. Morioka, T. Nishimura, A. Obatake, and Y. Shimizu. Relationship between the myofibrillar protein gel strengthening effect and the composition of sarcoplasmic proteins from Pacific mackerel. *Fisheries Sci.*, 63, 111–114, 1997.

41. I. Martinez, R. Ofstadt, and R.L. Olsen. Myosin isoforms in red and white muscles of some marine teleost fishes. *J. Muscle Res. Cell Motil.*, 11, 489–495, 1990.

42. S. Ishikawa, K. Nakamura, and Y. Fujii. Test program to manufacture sardine-based products and frozen surimi. I. Effects of freshness of material and fish dressing methods. *Tokai Fisheries Res. Agency Japan*, 20,59–66, 1977.

43. K.E. Spencer and M.A. Tung. Surimi processing from fatty fish. In F. Shahidi and J.R. Botta, Eds. *Seafoods: Chemistry, Processing Technology and Quality*. London: Blackie Academic & Professional, 1994, 288–319.

44. T. Tokunaga and F. Nishioka. The improvement of technology for surimi production from fatty Japanese sardines. In N. Davis, Ed. *Fatty Fish Utilization: Upgrading from Feed to Food, Proceedings of a National Technical Conference.* Raleigh, NC: UNC Sea Grant College Program Publication, 1988, 143–159.

45. M. Nonaka, F. Hirata, H. Saeki, and Y. Sasamoto. Manufacture of highly nutritional fish meat for food stuff from sardines. *Nippon Suisan Gakkaishi,* 55, 1575–1581, 1989.

46. N. Katoh, A. Hashimoto, N. Nakagawa, and K.-I. Arai. A new attempt to improve the quality of frozen surimi from Pacific mackerel and sardine by introducing underwater mincing of raw materials. *Nippon Suisan Gakkaishi,* 55, 507–513, 1989.

47. M.P. Richards, S.D. Kelleher, and H.O. Hultin. Effect of washing with or without antioxidants on quality retention of mackerel fillets during refrigerated and frozen storage. *J. Agric. Food Chem.,* 46, 4363–4371, 1998.

48. T. Ohshima, T. Suzuki, and C. Koizumi. New developments in surimi technology. *Trends Food Sci. Technol.,* 4, 157–163, 1993.

49. H.M. Bakir, H.O. Hultin, and S.D. Kelleher. Gelation properties of fatty fish processed with or without added NaCl, cryoprotectants and antioxidants. *Food Res. Int.,* 27, 443–449, 1994.

50. C. Koizumi, S. Wada, and T. Ohshima. Factors affecting development of rancid off odor in cooked fish meats during storage at 5 degrees C. *Nippon Suisan Gakkaishi,* 53, 2003–2009, 1987.

51. F.B. Jensen, A. Fago, and R.E. Weber. Hemoglobin structure and function. In S.F. Perry and B.L. Tufts, Eds. *Fish Physiology, Vol 17: Fish Respiration.* Academic Press, San Diego, 1998, 1–40.

52. M.P. Richards and H.O. Hultin. Effects of added hemolysates from mackerel, herring and rainbow trout on lipid oxidation of washed cod muscle. *Fisheries Sci.,* 69, 1298–1300, 2003.

53. M.P. Richards and H.O. Hultin. Effect of pH on lipid oxidation using trout hemolysate as a catalyst: a possible role for deoxyhemoglobin. *J. Agric. Food Chem.,* 48, 3141–3147, 2000.

54. W.P. Griffith and I.A. Kaltashov. Highly asymmetric interactions between globin chains during hemoglobin assembly revealed by electrospray ionization mass spectrometry. *Biochemistry,* 42, 10024–10033, 2003.

55. K. Shikama and A. Matsuoka. Human hemoglobin. A new paradigm for oxygen binding involving two types of AB contacts. *Eur. J. Biochem.*, 270, 4041–4051, 2003.

56. N. Griffon, V. Baudin, W. Dieryck, A. Dumoulin, J. Pagnier, C. Poyart, and M.C. Marden. Tetramer-dimer equilibrium of oxyhemoglobin mutants determined from auto-oxidaton rates. *Protein Sci.*, 7, 673–680, 1998.

57. S.D. Kelleher, L.A. Silva, H.O. Hultin, and K.A. Wilhelm. Inhibition of lipid oxidation during processing of washed, minced Atlantic mackerel. *J. Food Sci.*, 57,1103–1108, 1119, 1992.

58. O. Eide, T. Børresen, and T. Strøm. Minced fish production from capelin (*Mallotus villosus*). A new method for gutting, skinning and removal of fat from small fatty fish species. *J. Food Sci.*, 47, 347–349, 354, 1982.

59. H.O. Hultin and S.D. Kelleher. Process for Isolating a Protein Composition from a Muscle Source and Protein Composition. U.S. Patent No. 6,005,073. December 21, 1999.

60. H.O. Hultin and S.D. Kelleher. High Efficiency Alkaline Protein Extraction. U.S. Patent No. 6,136,959. October 24, 2000.

61. I. Undeland, S.D. Kelleher, H.O. Hultin, J. McClements, and C. Thongraung. Consistency and solubility changes in herring (*Culpea harengus*) light muscle homogenates as a function of pH. *J. Agric. Food Chem.*, 51, 3992–3998, 2003.

62. Y. Liang. Improved Techniques for Separating Muscle Cell Membranes from Solubilized Muscle Proteins. Ph.D. dissertation, University of Massachusetts, Amherst, MA, 2003.

63. I. Undeland, S.D. Kelleher, and H.O. Hultin. Recovery of functional proteins from herring (*Culpea harengus*) light muscle by an acid or alkaline solubilization process. *J. Agric. Food Chem.*, 50, 7371–7379, 2002.

64. M.C. Chang. Thermal Gelation Properties of Muscle Protein Gels Made from Different Meat Sources Using Alkaline or Acid Solubilization Process. MS thesis, University of Massachusetts, Amherst, MA, 2003.

65. M. Perez-Mateos, P.M. Amato, and T.C. Lanier. Gelling properties of Atlantic croaker surimi processed by acid or alkaline solubilization. *J. Food Sci.*, 69, FCT328–FCT333, 2004.

66. Y.S. Kim, J.W. Park, and Y.J. Choi. New approaches for the effective recovery of fish proteins and their physicochemical characteristics. *Fisheries Sci.,* 69, 1231–1239, 2003.

67. Y. Liang and H.O. Hultin. Functional protein isolates from mechanically deboned turkey by alkaline solubilization with isoelectric precipitation. *J. Muscle Foods,* 14, 195–205, 2003.

68. H.G. Kristinsson. Conformational and Functional Changes of Hemoglobin and Myosin Induced by pH: Functional Role in Fish Quality. Ph.D. dissertation, University of Massachusetts, Amherst, MA, 2002.

69. H.G. Kristinsson and H.O. Hultin. Changes in trout hemoglobin conformations and solubility after exposure to acid and alkali pH. *J. Agric. Food Chem.,* 52, 3633–3643, 2004.

70. H.G. Kristinsson and H.O. Hultin. Effect of low and high pH treatment on the functional properties of cod muscle proteins. *J. Agric. Food Chem.,* 51, 5103–5110, 2003.

4

Sanitation and HACCP

YI-CHENG SU, PH.D.
Oregon State University, Astoria, Oregon

MARK A. DAESCHEL, PH.D.
Oregon State University, Corvallis, Oregon

CONTENTS

4.1 Introduction 142
4.2 Sanitation 143
4.3 Good Manufacturing Practices (GMPs) 144
4.4 Hazard Analysis Critical Control Point (HACCP) 145
4.5 Principles of the HACCP System 146
4.6 HACCP for Surimi Production 148
4.7 HACCP for Surimi Seafood Production 149

4.8 Microbiological Standards and Specifications
     for Surimi Seafood.. 151
4.9 Sanitation Standard Operating Procedures
     (SSOPs) 153
4.10 Cleaners and Sanitizers... 154
4.11 Verification... 158
References.. 161

4.1 INTRODUCTION

The most important prerequisite for a quality food is that it be safe. The consequences of not ensuring that the consumer receives food free of hazardous materials are enormous and far-reaching. It has been estimated by the Centers for Disease Control and Prevention that food-borne diseases cause more than 75 million illnesses; 325,000 hospitalizations; and 5000 deaths annually in the United States.[1] In addition to the impact on public health, food-borne illness costs billions of dollars from multiple negative outcomes that include healthcare costs, lawsuits, increased insurance costs, business closings, and the erosion of trust that consumers have in the safety of the food supply. One illness outbreak associated with a particular food product can cause catastrophic damage to that industry, from which recovery can take many years. Moreover, regulatory agencies may increase surveillance and initiate new regulatory requirements.

The old axiom that an ounce of prevention is worth a pound of cure has never been more applicable than it is with food safety. Ensuring food safety is not rocket science but rather a systematic discipline requiring attention to detail, especially in adhering to processing and sanitation protocols. Record keeping is of paramount importance. As far as regulators and lawyers are concerned, if you did not record it, you did not do it. Knowing how to keep food safe is not the issue. The challenge lies in how to effectively apply information in a systematic, continuous, and economical way. Regulatory agencies may provide oversight by having laws, guidelines, and surveillance programs. However, the responsibility of food safety lies primarily with the food processors and it is in their best interest to embrace any opportunity to learn, practice, and improve upon it.

The purpose of this chapter is not to provide a comprehensive examination of seafood safety and step-by-step protocols to achieve it. There are many such documents available, both as hard-copy text and web site contributions. Rather, the intent here is to provide a ready reference of essential food safety and sanitation information for surimi manufacturers within the context of this text on surimi manufacture. We present the essentials of the cornerstone regulations of sanitation, the Good Manufacturing Practices (GMPs), and the complementary Standard Sanitation Operating Procedures (SSOPs), which together provide the foundation of the mandated Hazard Analysis Critical Control Point (HACCP) programs. Information is also provided regarding how to develop optimum cleaning and sanitizing procedures for surimi-based processing operations.

4.2 SANITATION

What is sanitation? The motto of the National Sanitation Foundation is:

> Sanitation is a way of life that is expressed in the clean home, the clean farm, the clean business and industry, the clean neighborhood, the clean community. Being a way of life it must come from within the people; it is nourished by knowledge and grows as an obligation and an ideal in human relations.

More pragmatically for those in the food processing business, it is defined as:

> That condition of cleanliness that must prevail continuously in the food processing environment to prevent adulteration and assure the production of clean, safe, and wholesome foods.

This working definition is derived from the Food, Drug, and Cosmetic (FD&C) Act of 1938, which mandates that foods are safe, wholesome, unadulterated, and produced under sanitary conditions. Although the act provided enforcement authority, it was not until the establishment of the Good Manufacturing Practices (GMPs) with their final form in 1986 that regulations were in place and formed a framework for agency surveillance and inspection. The GMPs are part of the Code of Federal Regulations (CFRs) and are identified as Title 21, Part 110. These are easily accessible in their entirety from Federal Government websites (www.gpoaccess.gov).

4.3 GOOD MANUFACTURING PRACTICES (GMPS)

The GMPs regulate the production and distribution of domestic and imported food in the United States. They are comprehensive in scope and mandate that food not be adulterated or produced under unsanitary conditions. Moreover, it is management that is responsible for adherence to the regulations. They are divided into subparts with specific requirements:

1. Personnel
2. Plants and grounds
3. Sanitary operations
4. Sanitary facilities and operations
5. Equipment and utensils
6. Processing and controls
7. Warehousing and distribution

An additional part addresses "defect action levels," which sets tolerance levels for certain unavoidable contaminants in some food as long as they do not present a health hazard. This is decided on a case-by-case basis and is based on scientific information. For example, in blue fin and other freshwater herring, the product is unacceptable if the number of parasitic cysts exceeds the defect action level of 60 parasitic cysts per 100 fish (for fish averaging 1 lb or less). The cysts at low levels are considered an aesthetic issue rather than one of public health significance.

The GMPs are also used as a checklist for inspection, with regulators looking at certain critical factors. These include but are not limited to temperature control, time and temperature control, separation of raw foods from finished foods, sanitation procedures, sanitary food handling, personal hygiene, record keeping, and corrective action procedures in place. The most practical approach to maintain compliance with the GMPs is to have in place a set of Sanitation Standard Operating Procedures (SSOPs). These are not regulations, but rather a specific plan for meeting the GMP regulations. Moreover, they are required for any food processor that falls under the FDA's seafood HACCP requirement.

The best and most comprehensive source of information on seafood processing SSOPs can be obtained from several University Extension websites, including the Seafood Information Network at the University of California at Davis (seafood.ucdavis.edu). Working examples of seafood SSOPs are provided at these sites, as well as a calendar of training programs on seafood safety.

4.4 HAZARD ANALYSIS CRITICAL CONTROL POINT (HACCP)

Hazard Analysis Critical Control Point (HACCP) is a systematic, scientific approach to improve product safety through controlled processes. It is a simple, logical, but highly specialized control system developed through thorough understanding of a commodity to which it is to be applied. HACCP was first deliberated by the First National Food Protection Conference in 1972[2] and has been recommended by several subcommittees of the National Academy of Sciences[3-5] to be employed in foods as the inspectional technique of choice. The evolution of HACCP in the food industry resulted in a mandatory requirement for the seafood industry in 1997.

The HACCP system can be considered a two-step approach to ensure the quality and safety of food products.[6] The first step in developing the system is to conduct a comprehensive analysis of hazards that are likely to happen to a specific product. These might include the consideration of raw materials, ingredients, processes for controlling hazards, consumer populations at risk, and other potential safety concerns related to the product. The second step of HACCP development is to determine the control measures to eliminate the hazards associated with the products. These might include the identification of critical control points at which hazards can be controlled, determination of monitoring procedures, establishment of verification systems, and record keeping.

The HACCP program is a long-term commitment to ensuring product safety and its development requires thorough knowledge in materials handling, processing, production, and storage of a particular food product. Each plant should have an individual HACCP program specifically designed to fit its need due to the variations among processing procedures.

4.5 PRINCIPLES OF THE HACCP SYSTEM

There are seven principles adopted by the National Advisory Committee on Microbiological Criteria of Foods[7] for developing an HACCP plan:

1. Conduct a hazard analysis:
   - HACCP is a total system approach and the hazard analysis is the first critical step in the development of an HACCP program. The hazard analysis should address all kinds of significant hazards that are likely to occur, whether a microbial, chemical, physical, or even an economical concern (Table 4.1). Hazards that are not significant or not likely to occur will not require further consideration in the HACCP plan. Results of the analysis provide a basis for the determination of Critical Control Points (CCPs).

TABLE 4.1 Potential Hazards for Surimi Seafood

Physical Hazards	Chemical Hazards	Biological Hazards
Glass	Allergens	Spoilage bacteria
Metal	Naturally occurring toxins	Bacterial pathogens
Other foreign subjects	Animal drug residues	Parasites
	Pesticide residues	
	Unapproved additives	

2. Identify the critical control points (CCPs) in the process:
   - A Critical Control Point (CCP) is any point or step during the production at which a control can be applied to prevent, eliminate, or reduce a hazard of concern to an acceptable level. The determined CCPs should allow the system to control all kinds of hazards identified from the analysis and be plant-specific. Different facilities preparing the same products can have different CCPs due to differences in facility layout, equipment, raw materials and ingredients, or processing procedures.
3. Establish critical limits for CCPs:
   - A critical limit is a criterion that must be met for each CCP. The critical limits must be established based on the best information available for producing safe products as well as being attainable. One thing to keep in mind when establishing a limit is that any product that fails to meet the critical limit must have a corrective action taken.
4. Establish CCP monitoring procedures:
   - Monitoring CCPs is an important observation or measurement to determine if there is a loss of control during the production. A deviation that occurs at a CCP will result in taking corrective action. Monitoring is essential for tracking the operation and providing documentation to verify the HACCP program. All records associated with monitoring should be signed and dated by the individuals conducting the activity.
5. Establish corrective actions:
   - The corrective actions are procedures to be followed when a deviation at a CCP occurs. A corrective action may be as simple as reprocessing or repackaging, or may require total product destruction. However, the correction must be specific for each CCP, and the action taken must demonstrate that the CCP has been brought under control. It is important that the employees be given the authority to stop production

if the process becomes out of control. This empowerment is critical to a successful HACCP program.
6. Establish effective record-keeping:
   • Good record keeping is important for a successful HACCP plan. Records should be accurate and reflect the process, deviation, and corrective action taken. The HACCP plan should be placed on file and the records must be reviewed daily by an individual who did not produce the records.
7. Establish verification procedures:
   • Once an HACCP is established, it needs to be continually evaluated, modified, and upgraded. Verifications may involve the use of a scientific or technical process to verify the critical limits for each CCP, a review of the program to ensure it is working properly, and conducting an independent audit to examine the performance of the HACCP program.

4.6 HACCP FOR SURIMI PRODUCTION

Surimi production requires the use of various machinery for cutting, chopping, mixing, and washing fish flesh. The more machinery involved in production, the greater the chance that contamination or cross-contamination will occur. The degree of contamination depends largely on the initial microbial load of the raw materials. Therefore, it is critical to use fish that contain low numbers of microorganisms because one single fish with a high microbial load can increase the contamination of the entire surimi batch. A clean water rinse of fish fillets before mincing generally removes some microbial load on the fillet and subsequently reduces contamination in the surimi.

Raw surimi is used as the major ingredient of surimi seafood and will be cooked and/or pasteurized during the production of surimi seafood. Therefore, the contamination of microorganisms in raw surimi can be considered as a quality rather than a safety concern. The only hazard that might occur during raw surimi production is the potential inclusion of metal fragments. Surimi block should therefore pass through a metal detector as the CCP before frozen storage.

Current industrial settings for the metal detector are discussed in Chapter 2.

4.7 HACCP FOR SURIMI SEAFOOD PRODUCTION

Surimi seafoods are often vacuumed-packed and sold under refrigerated storage. Although variations among procedures exist for producing different types of products, a common flowchart for surimi seafood production can be used for CCP determination (Figure 4.1). For detailed processing sequences, refer to Chapter 9. The potential hazards for surimi seafood can include the inclusion of metal fragments and the existence of human pathogens, such as *Listeria monocytogenes* and *Clostridium botulinum*. Therefore, CCPs for eliminating or reducing these hazards from surimi seafood include the pasteurization process, rapid cooling and low-temperature storage, and metal detection.

1. Pasteurization:
   - Pasteurization is a heat process designed to eliminate targeted bacterial pathogens and reduce total populations of spoilage bacteria in products. Although bacterial spores usually survive the heat process, a properly pasteurized product should contain a minimal amount of spoilage bacteria and be free of pathogens. Critical considerations for pasteurization of surimi seafood, including target pathogens, the heat resistance of organisms, heat penetration during the process, time and temperature determination, and process validation, are discussed in Chapter 12.
2. Rapid cooling and low-temperature storage:
   - Rapid cooling of pasteurized products will prevent the germination of bacterial spores and the growth of spore-forming bacteria such as *Bacillus* and *Clostridium* species. The rapid cooling procedure also prevents heat-injured bacterial cells from recovering and regaining their ability to grow at refriger-

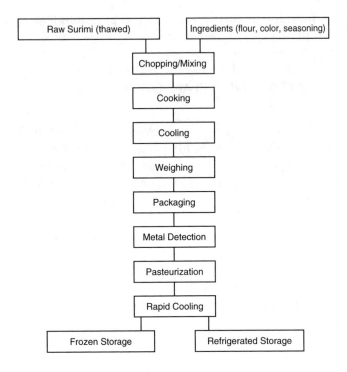

Figure 4.1 Common flow chart for surimi seafood production.

ation temperatures. Pasteurized surimi seafood should be cooled from 60°C (140°F) to less than 21.1°C (70°F) within 2 hr and to less than 4.4°C (40°F) within another 4 hr to prevent spore germination as well as retard the growth of spoilage bacteria.[8] Vacuum-packed surimi seafoods that are sold under refrigerated storage should be kept at temperatures below 3°C (37.4°F) to prevent the growth and toxin production of non-proteolytic types of *Clostridium botulinum*.

3. Metal detection:
   - Foreign objects such as metal fragments can cause injury to consumers and should be considered possible hazards associated with surimi seafood production. Metal fragments can be produced through

metal-to-metal contact, especially during mechanical cutting or blending operations during surimi and surimi seafood production. The FDA's Health Hazard Evaluation Board has supported regulatory action against products with metal fragments of 0.3 in. (7 mm) to 1.0 in. (25 mm) in length.[9] Current settings for the industry are discussed in Chapter 9. The possible inclusion of metal fragments in surimi and surimi seafood can be reduced by frequent examination of cutting and blending equipment for damage or missing parts, or be eliminated by passing finished products through a metal detector.

4.8 MICROBIOLOGICAL STANDARDS AND SPECIFICATIONS FOR SURIMI SEAFOOD

Sampling is a process to examine randomly collected units from materials or products for compliance with defined criteria. Details on developing a sampling plan can be obtained from the International Committee on Microbiological Specifications for Food[10] and will therefore not be discussed in detail here. An example of recommended microbiological standards and sampling plans for ready-to-eat crabmeat is shown in Table 4.2. In this case, a two-class attribute plan is used for testing *Staphylococcus aureus,* while a three-class attribute plan is used for testing aerobic plate count, coliform, *Escherichia coli,* and *Vibrio parahaemolyticus.* For *S. aureus,* five sample units are tested and none of them can contain populations greater than 10^3/g. For aerobic plate count, five sample units are tested and none can exceed 10^6 (M: the margin separates acceptable and defective quality), and no more than two of the five samples tested can exceed 10^5 (m: margin separates good and acceptable quality).

The microbiological criteria and sampling plans should be applied as part of the HACCP program. Surimi seafood should meet the microbiological safety criteria assigned to ready-to-eat fishery products (Table 4.3) to ensure quality and safety. The integration of microbial criteria into an HACCP-based system provides higher protection for food consumption.

TABLE 4.2 Recommended Microbiological Criteria and Sampling Plans for Ready-to-Eat Crabmeat

Microorganism	n	c	m	M
Aerobic plate count	5	2	10^5	10^6
Coliform	5	2	500	5000
Escherichia coli	5	1	11	500
Staphylococcus aureus	5	0	10^3	—
Vibrio parahaemolyticus	10	1	100	1000

Note: n: number of samples analyzed based on health hazard. c: maximum number allowed for unsatisfactory but acceptable results. m: margin separates good and acceptable quality. M: margin separates acceptable and defective quality.

Source: Adapted from Reference 10.

TABLE 4.3 Microbiological Safety Levels for Ready-to-Eat Fishery Products

Bacterial Pathogen	Action Level
Clostridium botulinum	Presence of toxin or viable spores or cells in products that will support their growth
Enterotoxigenic *Escherichia coli* (ETEC)	10^3/g or positive for toxins (LT or ST)
Listeria monocytogenes	Presence of cell
Salmonella species	Presence of cell
Staphylococcus aureus	10^4/g or positive for enterotoxin
Vibrio cholerae	Presence of toxigenic O1 or non-O1 cell
Vibrio parahaemolyticus	10^4/g
Vibrio vulnificus	Presence of cell

Source: Adapted from Reference 9.

Surimi seafoods that are properly pasteurized should be free of *Salmonella, Listeria monocytogenes, Vibrio cholerae,* and *Vibrio vulnificus.*

An effective HACCP system must be developed by each food processor and tailored specifically to its product, processing, and distribution. The system should include sanitation consideration because safe products cannot be produced in a facility with sanitation problems. Therefore, an adequate sanitation program (sanitation standard operating procedures — SSOPs) should be in place before the HACCP system is developed. A plant can have a well-designed HACCP program but fail to make it function properly because of poor sanitation or personal hygiene practices.

4.9 SANITATION STANDARD OPERATING PROCEDURES (SSOPS)

The SSOPs are plans for carrying out an effective plant program to reduce the likelihood of microbial contamination from occurring before, during, and after food processing. The SSOPs are essentially the map to success. Work by them and they will work for you. Although fish processing may appear to be an independent operation, it is just one link in the chain of food safety custody that extends from the harvest catch to the consumer or food service operation that prepares and serves the final product. However, it is the processor that shoulders most of the burden and responsibility of ensuring a safe product, whereas the raw product usually harbors many of the microorganisms that contaminate processing equipment and finished products.

It is therefore prudent to closely monitor the microbial quality of incoming raw product, reject all raw product that has unacceptable levels of microorganism by specifying limits in acceptance specifications or COAs (Certificates of Acceptance), and systematically test the finished product. By reducing the microbial input into the facility via raw product, the likelihood of microbial contamination in the finished product and the burden on the sanitation program will be reduced. Although cleaners and sanitizers can be used to reduce bacterial contamination, they have their limits, especially in grossly contaminated environments. Systematic testing of the finished product validates if the SSOPs and HACCP program are effective. More importantly, it confirms that the responsibility in the chain of custody has been met.

4.10 CLEANERS AND SANITIZERS

Much has been written about cleaners and sanitizers and their relative merits in different applications. Practically speaking, the best source of information is often the sanitation chemical technical representative. These individuals have enormous practical experience in knowing which applications/products are best for the operation. They are backed by large corporations that are constantly involved in developing new and more effective cleaning and sanitizing products and application systems. Take advantage of them — as they are the experts!

Cleaning and sanitizing are two separate operations with different functions, yet they are intimately linked by the common objective of achieving a "sanitary" environment. Cleaning is the physical removal of dirt, microorganisms, and food debris from plant processing equipment and structural surfaces. Sanitizing is the destruction of microorganisms through chemical treatment or a heat process. Cleaning alone will not ensure a sanitary state. Therefore, cleaning and sanitizing are always conducted sequentially, with cleaning followed immediately by sanitizing.

A water rinse step is often integrated before, after, and between the cleaning and sanitizing steps. This will facilitate the removal of loosely adherent dirt and debris, and allow for the use of reduced amounts of cleaning chemicals. An intermediate rinse will remove any residual cleaner that could interfere with the action of the sanitizer. A last rinse is applied to remove any sanitizer; however, many sanitizers are allowed to remain on food contact surfaces if they do not exceed certain levels.

In general, cleaning consists of four steps:

1. Solubilized cleaner is applied to objects or surfaces.
2. Dirt and debris are loosened from objects or surfaces.
3. Dirt and debris are solubilized or suspended in the cleaning water.
4. The cleaned surfaces are rinsed to remove residual cleaner, dirt, and debris.

Cleaning can be achieved either by soaking (small objects) or by spray application. A clean–in-place (CIP) system is often used to clean tanks and other holding vessels where the equipment is hard-plumbed with devices designed to deliver cleaner/rinse water under high velocity. Mechanical energy is often utilized to remove soils, either with high-velocity sprays or abrasives integrated into the cleaning solution. Oftentimes, there is no substitute for old-fashioned elbow grease with a scrub brush, especially in hard-to-reach areas.

Many factors can affect the efficacy of cleaners. These include water source, temperature and velocity of application, type of cleaner, concentration, and contact time. Hard water makes it much more difficult for detergents to function and often the water must be softened to optimize cleaning. An ideal temperature for cleaner/detergent action is in the range of 130 to 160°F. Excessive temperature is often counter-productive because it can "cook" organic debris onto surfaces and make removal difficult. The type of soil and/or food debris is also an important criterion for cleaner selection. In fish plants where there is significant organic matter, alkaline cleaners are preferred because of their protein and lipid emulsifying properties. Acid detergents are occasionally used to remove hard water scale build-up from certain equipment. Degreasing detergents are beneficial when lipid build-up on equipment exceeds the capability of alkaline cleaners.

Microbial biofilms tend to form in plant locations that are continually moist and out of sight and, thus, out of mind. Areas such as floor drains and ceiling condensates are common locations for biofim formation. Biofilms consist of a matrix of organic matter, soil, and microorganisms. The matrix can adhere tightly to metal, concrete, or synthetic surfaces and is resistant to removal by cleaning. Areas of the fish processing plant that might be conducive to the formation of microbial biofilms therefore need special cleaning attention and should be frequently examined, cleaned, and sanitized.

The food safety threat lies in the recognition that biofilms can serve as a reccurring source of microbial contamination if not completely removed. Moreover, the matrix serves as a protective shelter for bacteria by neutralizing the killing

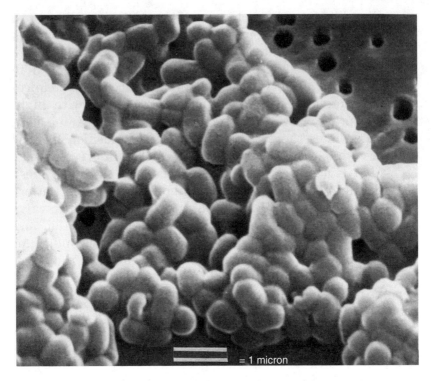

Figure 4.2 Scanning electron micrograph depicting cells of *Listeria monocytogenes* adhering to a surface. Note the cells stick to each other and to the surface with polysaccharide material the cells produce.

power of sanitizers (Figure 4.2 and Figure 4.3). *Listeria monocytogenes*, an environmental microbial contaminant, can easily form biofilms and is a serous threat to food safety in all types of food processing plants.[11,12]

All sanitizers have the same simple objective of killing any remaining microorganisms after an object or equipment surface is cleaned. The keyword in the previous statement is "after" cleaning. Sanitizers are designed to inactivate the small number of microorganisms that might be present after cleaning. A surface that has dirt or debris on it cannot be effectively sanitized because the sanitizer will be inactivated through interaction with such materials. Therefore, sanita-

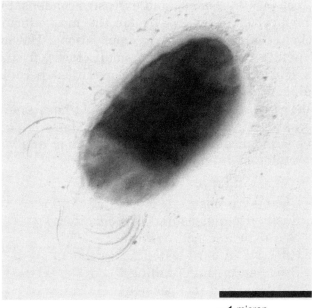

= 1 micron

Figure 4.3 Scanning electron micrograph of a single cell of *Escherichia coli* O157:H7 adering to a surface via its polysaccharide slime layer.

tion is not considered a sterilization process; rather, its purpose is to destroy microorganisms of public health significance. Although some bacterial spores might survive the sanitizing process, they are generally not considered a food safety threat.

Effective sanitizing can be accomplished by the application of either heat or chemicals. Heat sanitizing has advantages and disadvantages when compared to chemical sanitizers. Heat can be applied either dry or moist. The conductivity of heat is far greater in an aqueous environment and thus much more efficient as a sanitizer. In addition, heat is non-selective to microorganisms, can penetrate into surfaces, leaves no chemical residue, and can be easily monitored (temperature). However, there are a number of factors that limit the use of thermal sanitizing. Among these are high energy costs, worker safety, and excessive condensate formation. Chemical sanitizing is by far the most efficient, and offers choices and flexibility in applications. However, the limited ability to penetrate into small cracks, fissures, and biofilms makes thorough cleaning a critical perquisite for chemical sanitation.

A desirable chemical sanitizer should have many of the following properties. In reality, none will have all, but most will have economical and effective application in a variety of food processing environments:

1. Water solubility
2. Cleaning properties
3. Easily used, measured, and monitored
4. Non-toxic and non-irritating
5. Effective in hard water
6. Broad spectrum of killing
7. Available and inexpensive
8. Stable in storage and when diluted
9. Active over a range of temperatures
10. Active over a range of pH values

Listed in Table 4.4 are the main categories of commonly used sanitizers, along with their comparative advantages and disadvantages. Other sanitizers, including bromine, acid sanitizers, acid anionic sanitizers, UV light, hydrogen peroxide, ozone, and several others, can be used for more specific applications. Refer to Marriott[13] and Troller[14] for more details.

4.11 VERIFICATION

Cleaning and disinfection can be accomplished effectively by following a five-step procedure: rinse, clean, rinse, sanitize, and rinse. However, it should be verified with appropriate analyses to determine if the process is working properly. Commonly used techniques for detecting microorganisms after the disinfecting process include swabbing, direct surface contact, final rinse water, and air quality.

TABLE 4.4 Commonly Used Sanitizers with Comparative Advantages and Disadvantages

Sanitizers	Pros	Cons	Application Limit w/o Rinse
Chlorine	Very inexpensive Broad spectrum of killing Many available forms Easy to monitor	Irritating to skin Corrosive to metal pH dependent Activity decreases with organics	200 ppm
Idophors	Low concentrations effective Nonirratating to skin Less corrosive Unaffected by hard water	Will stain plastics Relatively more expensive Heat unstable pH sensitive	25 ppm
Quats	Provides residual activity Cleaning properties Odorless, colorless Temperature stable	Expensive Some bacteria resistant Film forming Difficult to monitor	25 ppm
Heat (steam)	Nonselective to microorganism Can penetrate into surfaces Leaves no chemical residue Ease to monitor (temperature)	High energy costs Worker safety Excessive condensate formation	No limit

Source: Adapted from References 13, 14, 15, and 16.

1. Swabbing:
   - The swab test is the most commonly used technique to determine microbial contamination on surfaces. This test is usually done by swabbing a fixed surface area (usually 4 × 4 in) with a sterile swab. Microorganisms, if present on the surface, are transferred to a dilution bottle and enumerated as colony-forming units in a suitable growth medium. Swabbing is especially useful in checking areas such as joints and valves that are difficult to clean and disinfect.
2. Direct surface contact:
   - Direct surface contact is an easier procedure than the swab test to determine the existence of microorganisms on surfaces. It uses contact Petri dishes or slides containing either selective or general medium to pick up microorganisms. Through the direct contact of surface and agar on the dishes or slides, microorganisms can be enumerated after the required incubation. This technique is simple to use and requires no medium preparation. However, it can only be applied to smooth surfaces, which limits its application.
3. Final rinse water:
   - It is sometimes difficult to use the swab or direct surface contact technique to verify the cleaning and disinfecting process, especially for a CIP system. In such cases, analysis of the final rinse water using a membrane filtration technique is an alternative to determine the existence of microorganisms through agar incubation.
4. Air quality:
   - Air may be a source of microbial contamination, especially for mold spores in a wet and moldy environment. Therefore, air should be filtered to remove microorganisms and spores before being introduced into the production area. In addition, maintaining a slightly positive pressure in the production area will help keep a clean air environment and eliminate possible contamination as a result of poor air quality.

The main purpose of "verification" is to take action of control before a loss occurs. Although corrective actions can be taken to improve the cleaning and disinfecting process according to results of these analyses, all these techniques require at least overnight incubation before results are available. Therefore, it is often too late to correct a critical problem using these techniques. Other techniques, which give a "real-time" result of analysis, are highly desirable.

While no real-time analyses were available for verification of the cleaning and disinfecting process at the time this chapter was prepared, ATP (adenosine triphosphate) bioluminometric assay can be used for a "near-real-time" analysis. This technique detects bacterial cell and food residue via the reaction of ATP, which is found in all cells, with luciferin and luciferase to create light output. The presence of bacterial cells or food residue on the surface is detected with the production of light measurable by a luminometer, and the intensity of the light output is converted into a result of either a clean or dirty zone. This technique is easy to use and provides a quick check of the cleaning and disinfecting process in just a few minutes. However, this technology cannot distinguish between bacterial cells and food residues. Microbiological tests are therefore still needed to determine if the disinfecting process fails to eliminate microbial contamination.

REFERENCES

1. P.S. Mead, L. Slutsker, V. Dietz, L. McCaig, J. Bresee, G. Shapiro, P. Griffin, and R. Tauxe. *Emerging Infectious Diseases.* Atlanta, GA: Centers for Disease Control and Prevention, 1999, Vol. 5, No 5.

2. APHA. *Proceedings of the 1971 National Conference on Food Protection.* Washington, D.C.: Food and Drug Administration, 1972.

3. NAS. *An Evaluation of the Role of Microbiological Criteria for Foods and Food Ingredients.* Washington, D.C.: National Academy of Sciences, National Academy Press, 1985.

4. NAS. *Meat and Poultry Inspection.* The Scientific Basis of the Nation's Program. Washington, D.C.: National Academy of Sciences, National Academy Press, 1985.

5. NAS. *Poultry Inspection, The Basis for a Risk-Assessment Approach.* Washington, D.C.: National Academy of Sciences, National Academy Press, 1987.

6. M. Hudak-Roos and E.S. Garrett. Model seafood surveillance project: an update. In *Proceedings of the 13th Annual Conference of the Tropical and Subtropical Fisheries Technological Society of the Americas,* Gulf Shores, AL, 1988. SGR-94:6-13.

7. NACMCF. HACCP: principles and applications, In M.D. Pierson, D.A. Corlett Jr., Eds. *Hazard Analysis and Critical Control Point System.* New York: Nostrand Reinhold, 1992.

8. B.H. Himelbloom, J.S. Lee, and R.J. Price. Microbiology and HACCP in surimi seafood. In J.W. Park, Ed. *Surimi and Surimi Seafood.* New York: Marcel Dekker, 2000, 325–341.

9. FDA. *Fish and Fisheries Products Hazards and Controls Guidance, 3rd ed.* Center for Food Safety and Applied Nutrition, United States Food and Drug Administration. Washington, D.C., 2001, http://www.cfsan.fda.gov/~comm/haccp4.html.

10. ICMSF. *Microorganisms in Foods. 2. Sampling for Microbiological Analysis: Principles and Specific Applications, 2nd ed.* International Committee on Microbiological Specifications for Food. Toronto: University of Toronto Press, 1986.

11. C.K. Bower, J. McGuire, and M.A. Daeschel. The adhesion and detachment of bacteria and spores on food contact surfaces. *Trends Food Sci. Technol.,* 7, 152–157, 1996.

12. M.A. Daeschel and J. McGuire. Interrelationships between protein surface adsorption and bacterial adhesion. *Biotechnol. Genetic Eng. Rev.,* 15, 413–438, 1998.

13. N.G. Marriott. *Principles of Food Sanitation.* Gaithersburg, MD: Aspen Publications, 1999.

14. J.A. Troller. *Sanitation in Food Processing.* Orlando, FL: Academic Press, 1983.

15. D. McSwane, N. Rue, and R. Linton. *Essentials of Food Safety and Sanitation. 3rd ed.* Upper Saddle River, NJ: Prentice Hall, 2003.

16. S. Knochel. Cleaning and sanitation in seafood processing. In H.H Huss, Ed. *Assurance of Seafood Quality.* FAO Fisheries Technical Paper. No. 334. Rome, FAO. 1993, http://www.fao.org/DOCREP/003/T1768E/T1768E00.htm#TOC.

5

Stabilization of Proteins in Surimi

PATRICIO A. CARVAJAL and TYRE C. LANIER
North Carolina State University, Raleigh, North Carolina

GRANT A. MACDONALD
MacDonald & Associates, Nelson, New Zealand

CONTENTS

5.1 Introduction 164
5.2 Myosin and Fish Proteins.. 165
5.3 Stability of Myosin 166
    5.3.1 Stabilization of the Globular Head Region 166
    5.3.2 Helix-Coil Transition in the Myosin Rod 168
5.4 Intrinsic Stability of Fish Muscle Proteins................ 170
    5.4.1 Influence of Animal Body Temperature.......... 170

    5.4.2 Naturally Occurring (Protecting and
          Nonprotecting) Osmolytes 172
    5.4.3 Antifreeze Proteins 175
5.5 Stability of Frozen Surimi Proteins 176
    5.5.1 Cold Destabilization 177
    5.5.2 Ice Crystallization 177
    5.5.3 Hydration and Hydration Forces 179
    5.5.4 Other Destabilizing Factors during
          Frozen Storage 181
5.6 Mechanisms for Cryoprotection and
    Cryostabilization 184
    5.6.1 Solute Exclusion from Protein Surfaces 184
    5.6.2 Ligand Binding 188
    5.6.3 Antioxidants 189
    5.6.4 Freezing-Point Depression 189
    5.6.5 Cryostabilization by High Molecular Weight
          Additives 191
    5.6.6 Vitrification 195
5.7 Processing Effects on Surimi Stability 196
    5.7.1 Fish Freshness 196
    5.7.2 Leaching 197
    5.7.3 Freezing Rate 197
5.8 Stabilized Fish Mince 198
5.9 Stabilization of Fish Proteins to Drying 201
    5.9.1 Processes for Drying of Surimi 202
          5.9.1.1 Freeze-Drying 202
          5.9.1.2 Spray-Drying of Surimi 205
    5.9.2 Potential Additives and Mechanisms of
          Lyoprotection 206
5.10 Future Developments in Fish Protein
    Stabilization 210
References 213

5.1 INTRODUCTION

Aggregation of surimi proteins at any time prior to thermal processing will reduce the subsequent ability of these proteins to properly gel. Premature aggregation can occur routinely during leaching, dewatering, freezing, and especially frozen storage of surimi. Surimi proteins will denature and aggregate during extended frozen storage unless mixed intimately with cryoprotective additives before freezing. Interestingly, cryoprotective compounds have also been found to help stabilize fish protein during drying and high-pressure treatment.

The chemical and physical stability of surimi proteins, especially myosin, the major protein constituent of fish muscle, provides the basis for understanding surimi cryoprotection. While many different chemical compounds (e.g., food acids, amino acids, sugars, sugar alcohols, antifreeze proteins, and phosphates) have been shown to have a cryoprotective effect on surimi proteins, only sucrose (table sugar), sorbitol (a sugar alcohol), and polyphosphates are currently used commercially for this purpose. Several mechanisms whereby added compounds can exert a cryoprotective effect on surimi proteins are discussed.

5.2 MYOSIN AND FISH PROTEINS

Striated fish muscle is composed of fibers, and bundled within these are smaller myofibrils, which themselves are composed of contractile proteins arranged in repeating end-on-end units called sarcomeres (see Chapter 10). Within each sarcomere, the thick and thin filaments play a central role in contraction. The thick filament is primarily composed of myosin, the most abundant myofibrillar protein. Natural thick filaments consist of several hundred self-associated myosin molecules. Myosin is largely responsible for the functional properties of surimi, including its gel-forming ability.

Myosin can be extracted from fresh muscle by treatment with solutions of high ionic strength (i.e., high salt content) because this causes the thick filaments to depolymerize. If the salt concentration is rapidly reduced back to physiological levels after extraction, then it is possible to reform myosin filaments that resemble the natural thick filaments. This effect implies that the "glue" that holds the thick filaments together is largely electrostatic. The increase in ionic strength allows counter-ions to neutralize each charge, cancelling the attractive force and allowing the myosin molecules to dissociate from one another (see Figure 10.3 in Chapter 10).

Each myosin molecule is composed of two 200-kDa heavy chains. The N-terminal end of each heavy chain is folded into a globular head called a heavy meromyosin subfragment-1 (HMM-S1 or S1), which forms an elongated pear-shaped head, typical of a globular protein, being 60% α-helix and 15% β-structure. It has ATPase activity and binds to actin (thin filaments). On each of the globular heads, two small light chains of approximately 20 kDa are noncovalently attached. The remainder of the heavy chains participate in a long, fibrous structure, the myosin rod. Refer to Chapter 10 for a more complete explanation of the myosin molecule assembly.

5.3 STABILITY OF MYOSIN

The stability of the myosin molecule is defined by the degree to which either of the two distinctive conformational regions — the globular head and the coiled-coil rod — is maintained in the native "folded" conformations.

5.3.1 Stabilization of the Globular Head Region

The amino acid sequence of the myosin head is a mixture of polar and nonpolar residues distributed along the chain with no discernible pattern. In aqueous solution, it cannot remain as a fully extended polypeptide because, while the polar and charged side chains and the carbonyl and amide groups of the peptide backbone would be able to form hydrogen bonds with water, the nonpolar side chains cannot. Their physical presence disrupts the hydrogen-bonded structure of water without making any compensating hydrogen bond with the solvent. To minimize this water structuring effect, these side chains tend to clump together in a similar manner to oil droplets dispersed in water.

The clustering of hydrophobic side chains from diverse parts of the polypeptide chain also causes the entire polypeptide to become more compact. When the nonpolar side chains come together to form a hydrophobic core, they simulta-

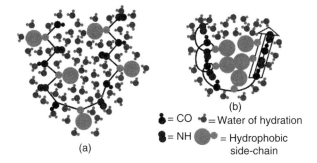

Figure 5.1 (a) Unfolded and fully hydrated polypeptide chain; (b) folded chain, right. Note that, in the unfolded state, both the backbone (CO and NH) and side chains atoms of protein interact with water molecules and in the folded state they are secluded from water by forming hydrophobic packing and internal hydrogen bonds. (From Reference 1 with permission.)

neously drag the polar backbone carbonyl (CO) and amide (NH) groups into this "oily" interior of the protein (Figure 5.1).[1] While these polar groups previously formed hydrogen bonds to water when the chain was extended, they are now unable to do so. The result is that most peptide NH and CO groups of the folded protein hydrogen bond to each other. It is this clustering of the hydrophobic groups and the concurrent formation of backbone amide-carbonyl hydrogen bonds that gives rise to the highly folded secondary structure of any globular protein or protein region. Although most hydrophobic side chains and polar backbone groups are buried by this conformational change, a few remain on the surface of the folded polypeptide along with most of the polar side chains.

This globular structure is only marginally stable and is highly susceptible to chemical or physical change. Chemical change is defined as any modification involving covalent bonds. Physical changes include protein unfolding and/or undesirable adsorption to surfaces or other proteins (aggregation).

Recent research has led to the hypothesis that partially unfolded globular proteins readily aggregate. This partially unfolded state (also called molten globular state) generally

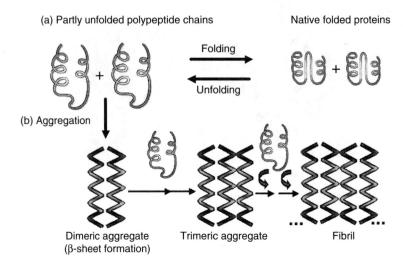

Figure 5.2 (a) Partly unfolded polypeptide chains formed from native globular proteins can briefly expose hydrophobic regions on their surfaces, and (b) these regions may bind similar surfaces in nearby folding molecules instead of becoming buried inside the structure leading to aggregation. The problem is made much worse by concentrating the globular protein (like freezing) since this aggregate may continue associating with other partly unfolded chains and so on. Aggregates vary in size from soluble dimers and trimers up to insoluble fibrillar structures. (From Reference 2 with permission.)

adopts a collapsed conformation that is more compact than the completely unfolded conformation and has substantial secondary structure but little tertiary structure. They have large patches of contiguous surface hydrophobicity and are more prone to aggregation than the native state or the completely unfolded state. Aggregation is often irreversible, and aggregates often contain high levels of nonnative, intermolecular β-sheet structures (Figure 5.2).[2]

5.3.2 Helix-Coil Transition in the Myosin Rod

An isolated myosin chain is generally unstable in an aqueous environment and must associate with other chains to gain stability. This is true not only of the heavy chains of the

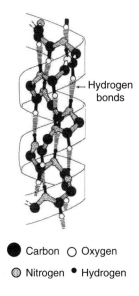

● Carbon ○ Oxygen
◉ Nitrogen • Hydrogen

Figure 5.3 Structure of the helix. Stability is conferred by hydrogen bonds, oriented axially. A single turn of the helix can be disrupted only if three successive bonds are broken. (From Reference 3 with permission.)

myosin rod, but also of proteins in other fibrous structures such as collagen and keratin.

Hydrogen bonds are the critical factor in maintaining the native helical conformation of the myosin rod. Hydrophobic interactions play a secondary role because hydrophobic residues make up less than 25% of the protein rod. The opposite is true for the globular head, in which hydrophobic residues make up more than 50% of the amino acids. In the helical state, axially oriented hydrogen bonds form between neighboring points along the helix (Figure 5.3),[3] bolstered by the "knob into hole" arrangement of the hydrophobic residues (see Chapter 10). A more neutral pH is needed; otherwise, repulsive forces between these charged side groups would outweigh the stabilizing force of the hydrogen bonds.

In addition to the balance that must be maintained between intermolecular hydrogen bonding and side chain charge, a third and equally critical force arises from the kinetic motion of molecules, as determined largely by temperature. The large degree of random motion resulting from a rise in temperature tends to diminish the hydrogen bond formation needed to stabilize the coiled-coil helices of the myosin rod. As a result, the helix will collapse, thereby hydrating the CO and NH backbone groups by exposing them to the solvent.

It is currently well accepted that the heat capacity changes observed during protein unfolding are largely the result of changes that occur in the hydration of groups that were previously buried from the solvent in the native (folded) state. Nonpolar atoms make a positive contribution to the heat capacity when they become solvent exposed, while polar atoms make a negative contribution when they become hydrated. Thus, the positive heat capacity change that can be measured upon unfolding of a globular protein, such as the myosin heads, can be rationalized in terms of the great exposure of nonpolar rather than polar surfaces. Likewise, the negative heat capacity changes that occur upon unfolding of the rod coiled-coil indicate that exposure of the polar peptide backbone (CO and NH groups) is the largest contributor to changes in the surface of this portion of the protein.

5.4 INTRINSIC STABILITY OF FISH MUSCLE PROTEINS

5.4.1 Influence of Animal Body Temperature

It is generally accepted that fish myosin is less stable than mammalian myosin. Moreover, the stability of fish myosin differs considerably among fish species, being closely associated with the temperature at which the fish lives; that is, the colder the environmental temperature, the more labile (less stable) the myosin.[4-6] The stability of myosin from warm-water species may even approach that of warm-blooded animals. For some stenothermal species (e.g., salmon and trout), the fish live in a relatively narrow thermal niche, whereas, other eurythermal species can adapt to large changes in environmental temperature on a seasonal basis and their muscle proteins can undergo extensive remodeling to accommodate this variation in environmental temperature. For example, it has been reported that carp, which can inhabit a wide range of environmental temperatures, will express three types of myosin isoforms with different thermal stabilities in relation to environmental temperatures.[7] The stability of the proteins may also be a function of harvest time and/or location in certain species.

Johnston and Goldspink[8] determined the thermodynamic *activation* parameters of the Mg^{2+}-activated myofibrillar ATPase from the white muscle of teleost fish, which inhabit a wide range of thermal environments from Antarctic to tropical waters. They plotted $\log_{10} V_{max}$ against the reciprocal of the absolute temperature (K) over the range 0 to 18°C and then calculated the activation energy. There was a strong positive correlation between the mean annual habitat temperature of the species and the activation energy, with species falling in the order Antarctic < North Sea < Mediterranean < Indian Ocean < African equatorial lakes (Figure 5.4).[8] The half-life of inactivation of the ATPase varied by 350 times between the two temperature extremes.

Hashimoto et al.[6] studied the thermostability of 40 species of fish and rabbit using Arrhenius plots of the first-order rate constants (k_d) for *inactivation* of myosin Ca^{2+}ATPase. Similar to the results obtained by Johnston and Goldspink,[7] there was a strong relationship between the thermostability of fish myofibrils and the environmental temperature at which the species lives (Figure 5.5).[6] Most of these and similar studies are characterized by a loss of ATPase activity and thus have correlated strongly with the aggregation of myosin heads.

The stability of the myosin rod regions has also been reported to differ among species. For example, Rodgers et al.[9] reported that myosin rod purified from Antarctic fish was less heat-stable than rabbit myosin rod. It has been argued that for a cold-adapted protein to operate efficiently, it requires a relatively greater conformational freedom at that temperature that results from a more flexible molecule. It is therefore

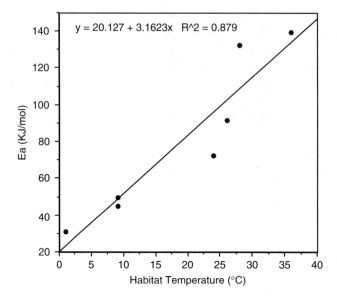

Figure 5.4 Increasing activation energy for activation of fish myofibrillar ATPase with habitat temperature for seven species of fish. Habitats for the different species ranged from Antarctica, North Sea, Indian Ocean, to African equatorial hot springs. (Adapted from Reference 8.)

likely that the greater efficiency of cold-adapted proteins, conferred by enhanced protein flexibility, also results in less thermostability.[10]

5.4.2 Naturally Occurring (Protecting and Nonprotecting) Osmolytes

It has been recognized for some time that many plants, animals, and microorganisms that have adapted to environmental stresses such as extreme temperatures, anhydrobiosis, perturbing solutes, and high hydrostatic pressure accumulate significant intracellular concentrations of small organic molecules. From this observation came the hypothesis that these so-called osmolytes have the ability to protect the cellular components against denaturing environmental stresses. The disaccharide trehalose is the principal osmolyte present and

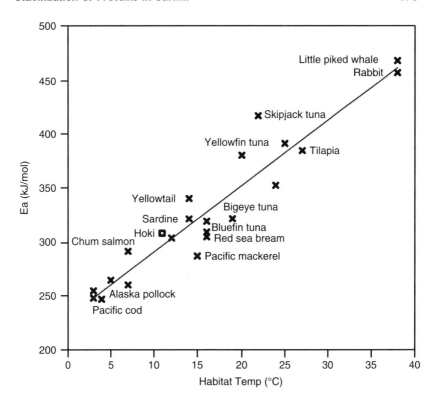

Figure 5.5 Increasing activation energy for deactivation of fish myofibrillar Ca^{++} ATPase with habitat temperature for 20 species of fish, little piked whale, and rabbit. (Adapted from Reference 6.) Hoki data from MacDonald and Hall (unpublished data) using a mean habitat temperature at 11°C.

protects the resurrection plant against desiccation; baker's yeast is a similar example.

Methylamine compounds, particularly trimethylamine oxide (TMAO), are compatible osmolytes that commonly occur in the tissues of marine organisms. High TMAO levels in polar fish are thought to increase the osmotic concentration, thus depressing the freezing point of the body fluids. TMAO is also known to be a counteracting solute that protects proteins against various destabilizing forces. In elasmobranches, for example, TMAO counteracts the effects of high urea content

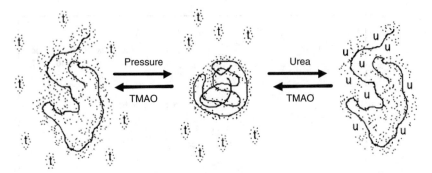

Figure 5.6 Model of solute and pressure effects on protein folding. Small dots represent water molecules. Addition of TMAO (t), excluded from the protein hydration layer, favors folding since that reduces the amount of excluded ordered water (middle). Addition of urea (u) enhances unfolding since it maximizes favorable binding sites (right). Pressure can favor unfolding (left) but only if there is a net expansion as water is released during folding. (From Reference 11 with permission.)

on proteins. Although urea is highly perturbing to enzyme systems, sharks and rays retain large quantities of urea in their fluids as an osmolyte. TMAO has demonstrated the ability to counteract the negative effects of urea on enzyme activity when accumulated in a 2:1 ratio with urea.

TMAO may also counteract the effects of hydrostatic pressure on enzyme function in deep-sea animals. Testing has found that TMAO *in vitro* is able to counteract the loss of activity in some enzymes that results from high hydrostatic pressure (Figure 5.6).[11] In shallow marine animals, TMAO is usually found at less than 100 mmol/kg. However, deep-sea teleosts and other animals have a TMAO content up to 300 mmol/kg. This content increases with the depth at which the creature is found and perhaps is related to pressure.

Unfortunately, TMAO can be demethylated to dimethylamine (DMA) and formaldehyde during post-mortem handling or frozen storage, as shown in the following reaction:

$$(CH_3)_3NO \rightarrow (CH_3)_2NH + HCOH$$

It has been shown that the amount of formaldehyde-induced aggregation is enhanced by physical abuse to the catch prior to freezing and by temperature fluctuations during frozen storage. The occurrence of formaldehyde markedly deteriorates the quality of the fish protein by inducing crosslinking (aggregation). As a result, fish fillets become sponge-like in texture and are unsuitable for surimi products because of the decrease in myofibril solubility at high salt concentration. In hake muscle held frozen for a prolonged time and/or at high temperatures, a nonextractable residue was obtained that consisted of a network of myosin in the form of interconnected and/or aggregated thick filaments rich in β-sheet structure, suggesting that myosin solubility decreased.[12]

An endogenous enzyme, TMAO demethylase (TMAOase), is responsible for the formaldehyde synthesis that occurs during frozen storage of fish meat or surimi. Elevated levels of this enzyme have been detected in kidney, spleen, pyloric caecum, and liver of fishes, and may occur also in both dark and ordinary muscle of gadoid and nongadoid fish.[13] Recently, TMAOase was purified from the myofibrillar fraction of walleye pollock.[14] This particular enzyme was characterized as a single acidic protein with an apparent molecular mass of 25 kDa, required Fe^{2+} for its activity, and exhibited optimum activity at pH 7.0. The enzyme activity was retained after heating at 80°C for 30 min, suggesting an extremely high thermal stability.

5.4.3 Antifreeze Proteins

Biological antifreeze molecules constitute a diverse class of proteins found in Arctic and Antarctic fish, as well as amphibians, trees, plants, and insects. These compounds are unique in that they have the ability to inhibit the growth of ice by lowering the freezing point of solutions noncolligatively. Consequently, they are essential for the survival of organisms inhabiting environments where sub-zero temperatures are routinely encountered. The proteins, which are found in northern cod and Antarctic fish, are up to 500 times more effective at lowering the freezing temperature than any other known solute molecule because of the unique aspects of their tertiary structures. These proteins act by specifically adsorbing to the surface of ice crystals as they form, thereby preventing ice crystal growth.[15]

There are two types of biological antifreeze molecules: (1) the antifreeze proteins (AFPs) and (2) the antifreeze glycoproteins (AFGPs). AFPs are further divided into four types, each possessing a very different primary, secondary, and tertiary structure. In contrast, AFGPs are subject to considerably less structural variation. A typical AFGP is composed of a repeating tripeptide unit (threonyl–alanyl–alanyl) in which the threonine residue is glycosylated.

The exact mechanisms whereby these molecules inhibit ice crystal growth at the molecular level remain a source of intense debate. Studies on the ice–water interface of AFPs suggested that this interface itself is not an abrupt transition as typically represented in static models. Most recent evidence shows that the loss of organized ice structure at the interface is fairly gradual, occurring over approximately 10 Å.

A theory that is gaining support, however, is that these proteins may owe their antifreeze properties to traits similar to those of collagen, which involves an extended repeating peptide chain and the consequential long-range polarization of water in the ice–protein interface. Considerable evidence that the antifreeze glycoproteins exist in an unfolded state has already been collected. Furthermore, the antifreeze protein binding sites are relatively flat and engage a substantial proportion of the protein surface in ice binding. The antifreeze protein from sculpin and northern flounders appears to have high helical content; however, the high helical content may not be the state these proteins assume in their natural functional state.

5.5 STABILITY OF FROZEN SURIMI PROTEINS

Two physical changes — low temperature and ice crystal formation — distinguish frozen systems from ambient. Low temperature slows most deteriorative reactions by decreasing molecular mobility. In addition, intramolecular hydrophobic interactions, which stabilize many protein native conformations, become weaker as the temperature decreases, just as hydrogen bonding becomes more important to stabilization at lower temperature. More dramatic is the change in the state of water, which in most foods makes up the bulk of the volume, serving to separate and suspend the remaining constituents. In intact animal and plant tissues, dehydration of cells as a result of water freezing may cause disruption of membranes or cell walls and distortion of tissue structure, adversely affecting thawed texture and water-holding properties. Large ice crystals will, of course, aggravate this tissue damage, particularly when they form intracellularly. Surimi already exists as a paste with little intact cellular structure. Concentration of solutes, as water crystallizes out of solution, can cause serious imbalances in its colloidal chemistry, which leads to damage of the proteins, aggregation, and loss of functionality.

5.5.1 Cold Destabilization

It is conceivable that cold destabilization of myofibrillar proteins, resulting from a weakening of the intramolecular hydrophobic interactions that stabilize the native protein structure,[16] would be a major factor in the instability of fish proteins. However, temperature reduction alone rarely, if ever, produces serious damage to living systems.[17] Studies on the insolubilization of protein in cod[18] have indicated that the quality of meat was impaired less by supercooling than by freezing. Thus, the undesirable side effects that often accompany preservation by freezing usually cannot be attributed to sub-zero reductions in temperature, and must therefore be caused primarily by liquid–solid transformation.[19,20] Recently, Strambini and Gabellieri[19] showed that several proteins exhibited long lifetimes for phosphorescence emission of tryptophan residues in supercooled solution, indicating that low temperatures are not a destabilizing factor.

5.5.2 Ice Crystallization

When ice forms, its crystalline structure excludes almost all solutes, so that ice is a nearly pure, single-component phase. Consequently, the structure of ice is almost unchanged by the presence of solutes in an ice-solution sample.[21]

A further consideration with regard to the presence of ice crystals is that physical changes may occur on rewarming, especially during frozen storage. Water molecules can diffuse from smaller ice crystals to larger ones, which grow and increase the probability of strain in the biological material due directly to the presence of ice. This is known as migratory recrystallization[22] and takes place over a wide range of temperatures. It is fastest when small crystals initiated at low temperatures are stored at higher temperatures; but under all circumstances, the rate of recrystallization increases at higher storage temperatures, which justifies, in part, the ultra-low temperature used in cryopreservation. It has been estimated that ice crystals initiated in bovine muscle with a mean diameter of 60 μm would double in size after 21 days storage at –10°C or 50 days storage at –20°C.[23]

Frozen meat products suffer quality losses during storage that can be higher than those taking place in any other stage of the process (i.e., freezing and thawing).[24,25] Love[26] commented that almost all of the deterioration occurs as a result of frozen storage, especially at temperatures near the melting point, and almost none is the direct result of freezing itself.

Temperature gradients in products can produce moisture migration because crystals in areas of low temperatures grow at the expense of those in areas of higher temperature. Formation and modification of ice crystals lead to a redistribution of water, which affects its reentering the original sites (protein rehydration, retention capacity of water).[27] The hydration sphere that surrounds proteins imparts stability to the three-dimensional structure, which is strongly dependent on the network of hydrogen bonds. Thus, any modification of the medium that promotes dehydration (migration of water associated with proteins to form ice crystals) or rearrangement (recrystallization) will result in an interruption of the hydrogen bonding system and the exposure of hydrophobic or hydrophilic zones, thus leaving unprotected and vulnerable regions. This favors intramolecular interactions, leading to alterations of the three-dimensional, or intermolecular structures, which induce protein–protein interactions and finally aggregation.[28,29] As solutes concentrate, denaturation and aggregation are favored because the proteins destabilize.

Strambini and Gabellieri[19] demonstrated, by intrinsic phosphorescence emission, that the solidification of water alters the native fold of globular proteins in the inner core of the macromolecule. They found a generalized increase in flexibility, reflecting an extensive unfolding of the polypeptide. The results suggest that the dominant perturbation originates from a direct interaction of the macromolecule with ice. These authors concluded that the strain on the native protein fold in frozen solutions can be attributed to the adsorption of the macromolecule onto the surface of ice.

5.5.3 Hydration and Hydration Forces

It is possible for modest volumes of water to remain unfrozen, in the presence of ice crystals, at tens of degrees Celsius below the equilibrium freezing temperature. The existence of unfrozen water at freezing temperatures, in the presence of ice or other nucleators, can be essentially attributed to the effect of the solute or macromolecule to alter the structure of the bulk water at the immediate surface. This is called the water of hydration, or unfrozen water (Figure 5.7).[30]

Upon freezing, the removal of water as ice causes protein surfaces to be brought into close proximity. If this distance becomes close enough (about 1 nm), a very large force, called the hydration force, can be measured. The origin of this force is believed due to the ordering of water at the surface, which propagates from the surface with decreasing strength. Experiments have shown that the structure of the water between surfaces is different than that of bulk water, and that such a difference can give rise to long-range interactions between proteins in solution (Figure 5.8).[31] The strength of the interaction appears to be proportional to the surface charge, and the force arises because work must be done to remove additional water from the system. The myosin rod is highly charged and, thus, hydration forces would be stronger at its surface. By contrast, the heads of myosin are less charged and exert weaker hydration forces.

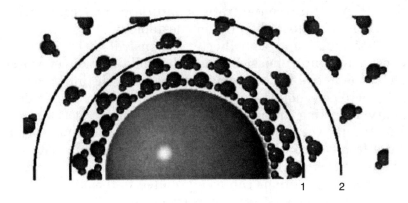

Figure 5.7 Layers of hydration around a surface. The density of water molecules around a surface is highest in the proximity of the surface and decays with a distance to the density observed for water molecules in the bulk solution. This results in several layers of hydration (1, 2, ..., n). (From Reference 30 with permission.)

Hydration forces have also been observed on hydrophobic surfaces. However, water molecules facing hydrophobic surfaces do not adsorb or form hydrogen bonds with the surface. Instead, they form a self-assembled structure with its own hydrogen-bonded network. Water molecules forming this self-assembled structure are claimed to be more ordered than in bulk water, although their hydrogen bonds are not stronger than in pure water. This restructuring of water on hydrophobic surfaces is entropically unfavorable because it disrupts the existing water structure and imposes a new, more ordered structure on the adjacent water molecules.

The stability of the frozen protein system is essentially determined by the existence and strength of hydration forces at the protein surfaces. Hydration forces between hydrophobic and hydrophilic surfaces, respectively, have been identified to decay exponentially with a decay length about the diameter of a water molecule. The freezing out of water into ice will tend to remove hydration water from surfaces where the

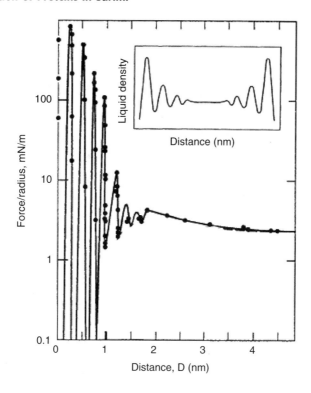

Figure 5.8 Effect of separation on force between closely spaced charged surfaces. Below 2 nm (about 8 water-molecule diameters) the force becomes increasingly oscillatory, with a periodicity equal to the diameter of water molecules (2.5 Å), indicative of the existence of diffuse water layers between the surfaces. Inset: water density profile between two surfaces. Note that at the middle separation of the two surfaces the oxcillation becomes null, or water assumes bulk behavior. (From Reference 31 with permission.)

hydration forces are less strong, leading to aggregation of neighboring protein surfaces (Figure 5.9).[32]

5.5.4 Other Destabilizing Factors during Frozen Storage

Muscle pH affects not only the denaturation rate at high temperature, but also the denaturation rate during frozen

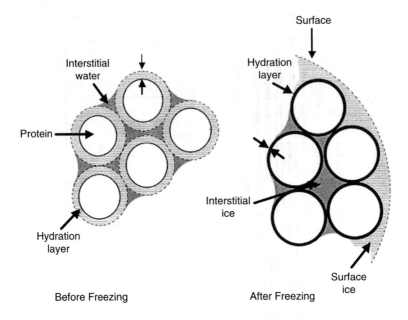

Figure 5.9 Effect of freezing on surface water. The most important effect of freezing is the removal of bulk water (interstitial water) into ice. This dehydration causes protein surfaces to come into close contact, at which point a new bond forms. Note that the hydration layer become smaller but stronger. The persistence of this hydration layer over time will determinate the stability of the system. (From Reference 32 with permission.)

storage. The extent of denaturation of mackerel actomyosin as measured by ATPase activity at various pH values during frozen storage is shown in Figure 5.10.[33] It is apparent that below pH 6.5, the myofibrillar proteins are unstable and rapidly lose their ATPase activity, which is an indicator of gel-forming ability.

The gel-forming ability of fresh fish muscle is optimal at neutral pH, decreasing with decreases in pH.[34] Thus, pH control during surimi manufacture is important for maintaining the gel-forming ability of surimi.[33] Fortunately, the meat pH of white-fleshed species such as Alaska pollock does not change much after death and is easily kept neutral during

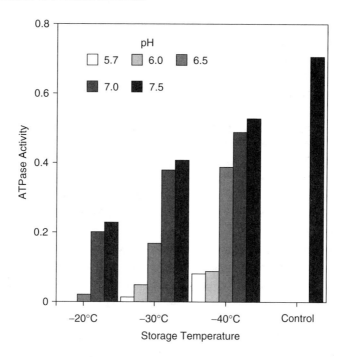

Figure 5.10 Effect of pH on the ATPase activity of mackerel actomyosin denaturation rate after 1 month storage at –20°C, –30°C, and –40°C. ATPase activity was measured to determine extent of denaturation. (Adapted from Reference 33.)

surimi processing and subsequent frozen storage. However, in dark-meat fish, such as sardine and mackerel, rapid glycolysis occurs after death. The generation and accumulation of lactic acid causes the pH of such fish meat to drop sharply, reaching pH 5.6.

The presence of calcium and metal ions can also destabilize surimi during frozen storage. These may be present in surimi through use of unusually hard water in the leaching stage of the manufacturing process, or from uncoated metal piping in the factory. In the past, there have been attempts by surimi manufacturers to boost the gelling ability of surimi by the addition of calcium salts, as this enhances the activity of the endogenous transglutaminase cross-linking enzyme and can additionally add salt bridges, which strengthen the gel (see Chapter 10). However, these effects are needed later in the process, at the time of cooking the prepared, salted paste to make surimi seafoods, not when the raw surimi is kept in frozen storage. The presence of elevated calcium levels in the raw surimi during frozen storage contributes to premature aggregation of the proteins, which reduces the gelling ability of the surimi.[35]

Should TMAOase still be active in surimi, and sufficient TMAO still present after the surimi leaching process during manufacture, then formaldehyde production is also a possibility during frozen storage of the surimi. This would, of course, lead to a loss of gelling ability in the surimi due to the aggregation induced by the presence of formaldehyde. Generally, however, the water leaching process of surimi manufacture removes critical co-factors to the activity of TMAO demethylase, such that formaldehyde production is greatly reduced.[36] Addition of sugars as a cryoprotectant also diminishes the activity of the TMAOase.[37] Another practical means of preventing the autolytic production of formaldehyde is to store fish at temperatures below −30°C and minimize temperature fluctuations in the cold store.

5.6 MECHANISMS FOR CRYOPROTECTION AND CRYOSTABILIZATION

5.6.1 Solute Exclusion from Protein Surfaces

Biochemists have long known that certain substances, such as sucrose and glycerol, when present at high concentration (>0.5 M), stabilize the activity of enzymes in solution and during freeze-thawing. Natural production of glucose by the liver is thought partially responsible for freeze tolerance in certain sub-Arctic frog species[38] and other species. Based on the results of freeze-thaw experiments on labile enzymes[39] and myofibrillar proteins[40–42], and a review of the literature on protein freezing, it has been determined that cryopreservation can be explained by the same general mechanism that Timasheff, Arakawa, and others have defined for solute-induced protein stabilization in nonfrozen, aqueous solution.[43–50]

Extensive studies on a variety of co-solvents that lower the solubility of globular proteins or stabilize them (e.g., glycerol, sugars, amino acids, and salts) have shown that these cosolvents are preferentially excluded from the domain of the protein; that is, in their presence, the protein is preferentially hydrated. Such exclusion is thermodynamically unfavorable because it increases the chemical potential (activity) of both the protein and the co-solvent. In the protein unfolding (denaturation) equilibrium, the effect of the solvent is defined by the relative affinities (exclusion or binding) of the solvent for the protein in its two end states, native (N) and denatured (D).

$$N \underset{\text{Stabilizer}}{\overset{K(S)}{\rightleftharpoons}} D, \quad K = \frac{[D]}{[N]} \quad (5.1)$$

The effect of the stabilizer on the equilibrium constant of the reaction, K, is a function of the stabilizer (S) and can be expressed in terms of the equilibrium constant by the Wyman linkage.[51]

$$-\left(\frac{\delta \Delta G}{\delta \mu}\right) = \Delta \upsilon_S = \upsilon^D - \upsilon^N \quad (5.2)$$

where $\Delta \upsilon_S$ is the difference in the number of co-solvent molecules bound by the denatured and native molecules, m is the chemical potential (partial molar free energy), and ΔG is the Gibbs free energy. Because ΔG = RT lnK, K being the equilibrium constant for the reaction, and μ = RT lna_s, where as is the thermodynamic activity coefficient for additive S, then

$$\left(\frac{\delta \ln K}{\delta \ln a_S}\right) = \Delta \upsilon_S = \upsilon^D - \upsilon^N \quad (5.3)$$

Hence, if an additive is a stabilizer, the reaction is shifted to the left and from Equation 5.3 $\Delta \upsilon_S$ would be negative. This means there must be less binding of the stabilizer to the denatured state.

Because the surface of contact between protein and solvent constitutes an interface, there must be in this surface an interfacial (surface) tension. Low molecular weight carbohydrate cryoprotective additives such as sucrose and sorbitol perturb the cohesive force of water and hence its surface tension. The excess or deficiency of the additive in the surface layer can be calculated by Gibbs' adsorption isotherm[52]:

$$d\gamma = -\Gamma d\mu \qquad (5.4)$$

where γ is surface tension and Γ is the adsorption (excess concentration) of co-solvent at the surface. Thus, if a substance increases the surface tension of water, its excess in the surface layer will be negative, that is, preferentially excluded from the interface.

Timasheff and Arakawa[50] compared the ratio of the preferential interactions measured experimentally by dialysis equilibrium to that calculated from the surface tension effect for a range of proteins and stabilizers, and concluded that this mechanism appears to give rise to the observed preferential exclusion of sugars, amino acids, and structure-stabilizing salts. Stabilization by increased surface tension of the medium appears to be the dominant factor leading to the thermal stability of proteins, particularly globular proteins. In addition, as observed earlier, the preferential hydration of proteins in the presence of sugars is due to the ability of the sugar to increase the surface tension of water. Increased surface tension of water in the presence of polyols and sugars has been attributed to stronger or more extensive hydrogen bonding between solute hydroxyl groups and water molecules, as suggested from various spectroscopic and thermodynamic studies.[53]

An alternative explanation is that stabilization is due to specific effects of these solutes on the hydration forces at the protein surface. Solutes could affect hydration forces if they were either adsorbed onto the protein–water interface, in which case they would produce an interface with an altered capacity to polarize water and an altered surface mobility, or excluded from the interface and thus create a barrier between macromolecules.

A primary factor contributing to the instability of proteins at subambient temperatures is the decreasing strength of intramolecular hydrophobic interactions, which stabilize the native protein structure, with decreasing temperature.[16] Because hydrophobic interactions are a thermodynamic response to the structuring of water by apolar groups on the protein surface, agents that affect water structure can affect protein stability. While the protein intramolecular hydrophobic interactions are weakened by lowering the temperature, there is a counteracting effect of increasing solute concentration and thus increasing solute surface tension resulting from the freezing of pure water into ice. A more in-depth discussion of the solute exclusion theory is given in MacDonald and Lanier.[54]

The validity of the solute exclusion principle in surimi was illustrated in studies[41,42] that investigated the cryoprotective properties of sodium lactate. Overall, both freeze-thaw and heat denaturation studies showed that sodium lactate was a more effective stabilizer than sucrose. However, in each case, above a certain concentration, lactate actually destabilized the actomyosin.

Figure 5.11 shows the results of freeze-thaw experiments on the denaturation of tilapia actomyosin.[41] A loss of Ca^{2+}ATPase activity showed the extent of denaturation during freeze-thawing. The cryoprotective effect increased to a maximum at about 6% (w/v) sodium lactate concentration with 80% of the activity recovered. In contrast, for sucrose, the level of cryoprotection increased monotonically with sugar concentration. On a percent basis, sodium lactate appeared to be about four times more effective than sucrose because 25% sucrose would be needed to give an equivalent degree of cryoprotection to the optimum for sodium lactate.

It was concluded that sodium lactate, like sucrose, stabilized the actomyosin by the mechanism of increasing solution surface tension. Surface tension measurements showed that the concentrations of sodium lactate, which corresponded to stabilization and destabilization of heated actomyosin, closely correlated with concentrations where surface tension either increased or decreased, respectively. The decrease in surface tension was ascribed to dimer formation at higher

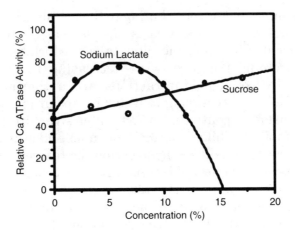

Figure 5.11 Cryoprotection of tilapia actomyosin by sodium lactate and sucrose. Actomyosin (2.5 mg/mL, pH 7.0 with 25 mM Tris-maleate in 0.6 M KCl) was frozen in liquid nitrogen for 3 min then held in a bath at –5°C for 60 min and thawed for 5 min at 25°C. Unfrozen control was taken as 100% Ca^{2+} ATPase activity. Results are the average of replicate experiments. (Adapted from Reference 41.)

concentrations, creating an amphiphilic molecule.[41,55] In comparison, the surface tension of sucrose solutions increased with increasing concentration. This correlates with previous studies that established a monotonic relationship between protein stabilization and sucrose concentration. The surface tensions of sucrose solutions were also lower than the corresponding sodium lactate solutions up to 20% sodium lactate.

Sodium lactate is a particularly useful example of an additive stabilizing by the preferential exclusion mechanism, because higher concentrations of sodium lactate actually destabilized actomyosin, and the transition from stabilization to destabilization was associated with either an increase or decrease in solution surface tension with sodium lactate concentration.

5.6.2 Ligand Binding

An exception to the general destabilization of proteins by cosolvents that bind to the protein is the case when oligomeric proteins are stabilized by binding with specific ligands. For example, the nucleotides ATP, ADP, and IMP exerted a protective effect on fish actomyosin stored at −20°C while the nucleotide catabolites, inosine and hypoxanthine, destabilized these proteins.[56] Myosin is a hexamer that consists of two identical heavy chains and two pairs of essential and regulatory light chains with a specific nucleotide binding site. Binding ATP induces a conformational change in the myosin structure, which leads to increased stability of the protein.[57]

The importance of the specificity of the ligand–protein interaction is shown by the contrary findings of Carpenter and colleagues,[58–61] who reported that low concentrations of Zn^{2+} ions acted synergistically with sugars to stabilize the tetrameric ATPase, phosphofructokinase, during freeze-thawing and freeze-drying, whereas MacDonald et al.[62] demonstrated that even at very low levels, Zn^{2+} destabilized fish actomyosin during freeze-thawing.

5.6.3 Antioxidants

Polyphosphates at 0.2 to 0.3% are commonly added as synergists to the cryoprotective effect of carbohydrate additives in the manufacture of surimi.[63–65] Phosphates may serve to chelate calcium ions that can induce protein aggregation.[66] As antioxidants of lipids, they may also protect proteins from denaturation induced by hydrolysis or autooxidation of phospholipids.[67] Other antioxidants may also serve a useful function in protecting unsaturated lipids and proteins during frozen storage. Erickson[68] has recently reviewed a wide range of antioxidants and their application to frozen foods, including seafoods.

5.6.4 Freezing-Point Depression

Ice formation concentrates reactants, particularly salts and hydrogen ions (pH depression), which thermodynamically destabilize proteins.[29,69] The volume of ice formed increases with decreasing temperature below the freezing point. Thus, by depressing the freezing point with low molecular weight solutes, the deleterious effects of freeze concentration on

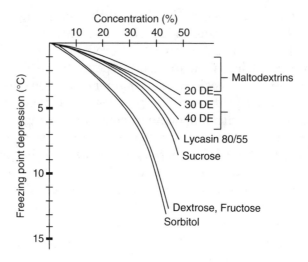

Figure 5.12 Freezing point depression of solutions by addition of various carbohydrates. (Adapted from Reference 70.)

proteins at any sub-freezing storage temperature (above the operative glass transition temperature; see below) should be lessened.

The presence of any solute will lower the freezing point of water in a colligative fashion. Ideally, its performance is related to its concentration according to Raoult's law, in which the mole fraction of solute is equal to the proportional reduction of solution vapor pressure compared to the pure solvent. Figure 5.12 illustrates the freezing point depression effected by a number of food-grade carbohydrates as a function of concentration.[70]

The freezing point of water is defined as the temperature at which the vapor pressure of the solid phase (ice) is equal to the vapor pressure of the liquid phase.[71,72] Thus, on an equal mass basis, lower molecular weight solutes will have the greater molarity and therefore the greater effect on freezing point depression. Interactions between solute and water molecules that affect the vapor pressure of the water may cause deviations from ideal behavior.[52,73] The most general cause for non-ideality in solutions of macromolecules arises from the size and shape, as expressed by their molecular excluded volume.[52] This excluded volume, which accounts for the volume occupied by the solute, is therefore not available to other molecules in the solution, and can be unexpectedly large for asymmetric rods or random coils.

5.6.5 Cryostabilization by High Molecular Weight Additives

Carpenter and Crowe[39] theorized that certain high molecular weight polymers (e.g., polyvinylpyrrolidone, polyethylene glycol, and dextran) are good cryoprotectants because they are stearically excluded from the protein surface by their size.[39,74] However, an additional mechanism has been postulated by other workers to explain the cryoprotective effects of many high molecular weight polyols[39,74,75] and glucose polymers (starch hydrolysis products),[76] a mechanism that takes into account the dramatic increase in viscosity that occurs as these polymers are freeze concentrated and form glasses.

With continued lowering of the temperature below the freezing point, the rates of diffusion-limited deteriorative reactions decrease steadily as ice freezes out and solution viscosity increases, which lowers the diffusivity of the dissolved reactants (i.e., molecular mobility).[77,78] Solutes that most interact with water to affect solution viscosity and water mobility would have the greater effect in reducing reaction rates at any subfreezing temperature.[73] Solution viscosity reaches a maximum at the glass transition temperature (Tg) of a solution.[76,79]

Figure 5.13 illustrates the glass transition temperature of a simple solution as a function of solute concentration. At higher concentrations of solute, the Tg occurs at temperatures above freezing. Thus, the mixture cools to form a "candy" glass directly from the liquid state. At solute concentrations below the point Cg', when the temperature falls below the freezing curve, the solution will exist either as a viscous supersaturated solution in the liquid state or, more commonly, as a mixture of ice crystals and supersaturated solution. Under these conditions, the system is termed a "rubber," exhibiting

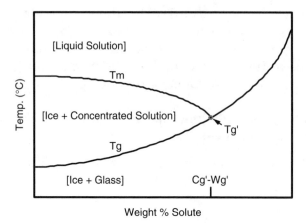

Figure 5.13 Schematic state diagram for a binary system in which the solute component does not crystallize.

a high viscosity due to the presence of ice crystals and/or strong intermolecular solute interactions/entanglements. The rubber state is characterized by a departure from Arrhenius kinetics of chemical and enzymatic reactions, following instead Williams-Landel-Ferry kinetics. The rate of freezing and the constancy of storage temperature will influence ice crystal sizes and numbers, and likewise the solute concentration.[17,80] Thus, Tg can be altered by time factors not shown in the two-dimensional phase diagram in Figure 5.13.

As the glass transition temperature is approached, reactions become diffusion limited due to immobilization of the water within a solute structure.[78,81] The glassy state is attained when solute–solute interactions supersede solute–water attractions and occurs at the Tg for the given solute and concentration. Viscosities near 10^{14} Pa.s are characteristic.[17] The structure is amorphous, like that of liquid, with no crystallinity other than that of the enmeshed ice crystals. If complete vitrification (glass formation) is achieved, no further ice formation will accompany decreases in temperature.

According to the cryostabilization theory of Levine and Slade, deteriorative processes cease in the practical timeframe once the glass state is achieved. Mobility or diffusion

Stabilization of Proteins in Surimi

in a glassy system is claimed to be virtually nonexistent.[79,82–84] However, Fennema[77] argues that no food system is so simple that a single Tg can be ascribed because microenvironments within the food will have successively lower Tg values. For this reason, he asserts the concept of a glass transition zone of temperature, below which ultimate stability is achieved due to water immobilization.

The requirement in cryostabilization for use of polymers exhibiting relatively large molecular weights arises from their ability to form glasses at higher freeze temperatures. This is due to their propensity to entangle as well as form hydrogen and other bonds, imparting a greater viscosity at any given concentration. The glass transition of a carbohydrate is weakly dependent on structure, with a much stronger dependence on molecular weight. Levine and Slade[76,82] have published tables and graphs that indicate a generally direct relationship between molecular weight (or, inversely, dextrose equivalent, DE, for starch hydrolysis products) and Tg′. Roos and Karel[85] and Roos[86] also found a relationship between increasing Tg′ and increasing molecular weight for a series of glucose polymers, although they found that the "effective molecular weight" was a better predictor of a polymer's ability to change Tg′ than the molecular weight average. Branching of the molecule will affect the solution properties of polymers, such that a strictly direct relationship between MW (molecular weight) and Tg′ will hold only for a homologous series.

It is possible to increase the overall glass transition temperature (or zone) of a food, and therefore theoretically improve the food's stability, by adding polymers that form glasses at high temperatures. The simplest equation that has been used to estimate the Tg′ is the Gordon and Taylor equation[87]:

$$Tg = \frac{w_1 Tg_1 + kw_2 Tg_2}{w_1 + kw_2} \quad (5.5)$$

where *Tg1* and *Tg2* are the glass transition values (in K) of components 1 (water) and 2 (solute) of the sample, respectively; *w1* and *w2* are the weight fractions of components 1 and 2 of the sample, respectively; and k is an empirical constant.

TABLE 5.1 Relationship between Maltodextrin DE, Average Molecular Weight (MW), and Glass Transition Temperature (Tg′)

Maltodextrin	DE	Mean MW	T Tg′ (°C)
M050	5	3600	−5
M100	10	1800	−11
M150	15	1200	−13
M180	18	1000	−15
M200	20	900	−16
M250	25	720	−19
Sucrose	N/A	350	−40

Source: Courtesy of Grain Processors Corporation, Muscatine, Iowa.

Although Roos and Karel[85] found the Gordon and Taylor equation (which was originally developed for binary polymer systems) useful for predicting the Tg′ in amorphous carbohydrates, care should be taken when using it to estimate the Tg′ of complex foods. Further and more complex models are discussed in Roos.[86] Thus far, only maltodextrins and some other sugar polymers such as polydextrose have been investigated for their cryoprotective properties as agents for raising Tg′ in foods. Table 5.1 illustrates the Tg′ values of selected maltodextrins as compared with sucrose. Carvajal et al.[88] investigated the role of maltodextrins in the cryoprotection of surimi. They found that, with increasing molecular weight, the ability of the maltodextrin to cryoprotect surimi during freezing/thawing diminished greatly, but its ability to cryoprotect during isothermal storage at −20°C remained high. This was taken as evidence that maltodextrins do primarily cryoprotect by raising the Tg′, although the lower molecular weight maltodextrins may also cryoprotect to some extent by a solute exclusion mechanism.

Apart from the deleterious changes associated with phase changes as muscle or minces are frozen and thawed, such as enzymatic processes, it is possible that cold destabilization of myofibrillar proteins[16] could be a factor in the instability of fish proteins. Some of the principal mechanisms for myosin denaturation, aggregation, and the subsequent loss of functionality or quality may not be diffusion controlled and hence may not be greatly affected by glass formation. Likewise, very small molecules such as oxygen may be able to diffuse within the glass formed at or below the Tg of the system, such that oxidation of lipids and proteins may go unhindered.[89] Thus, total reliance on the restriction of molecular mobility by high molecular weight components to achieve cryoprotection may be inadvisable. For these and other practical reasons, developing cryoprotectant "cocktails" that are designed specifically for each food and its processing/storage environment is suggested.[40]

5.6.6 Vitrification

A vitrified biological system is one in which no (or practically no) ice is present and all the fluid is immobilized in a glass. Aspects of this state apparently contribute to the ability of certain insects to survive freezing[90] and freeze resistance in plants.[91] Research on this topic is seriously being pursued in cryobiology with the purpose of eventually being able to successfully freeze, without loss of viability, entire human organs.[91–94]

To attain such a physical state in a biological system will likely require a combined approach of adding glass-forming cryoprotectants to raise Tg, high pressure during freezing and/or thawing to lower the freezing/melting point (see later discussion) and raise Tg, and extremely fast freezing rates. Successful vitrification of small cells and embryos has been obtained in the laboratory,[95] but there are equal problems of thawing the specimen without massive recrystallization and damage to the organism. For example, Kresin et al.[96] determined that even if cooling rates high enough to avoid substantial crystallization are achieved, the critical warming rates needed to avoid devitrification and ice recrystallization are 2 to 3 orders of magnitude higher. For practical reasons, it is unlikely that total vitrification (i.e., without ice formation) will be a feasible approach to food preservation, particularly surimi.

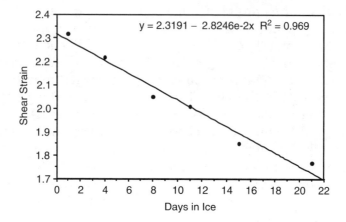

Figure 5.14 Loss of cohesiveness (shear strain) of gels made from hoki stored in ice. Gels were made from mince comminuted with 2.5% salt, 0.2% phosphate and cooked at 60°C for 40 min. (Adapted from Reference 97.)

5.7 PROCESSING EFFECTS ON SURIMI STABILITY

5.7.1 Fish Freshness

It is axiomatic that the quality of frozen fish material will depend on the quality of fish initially frozen. That is, the final quality will be a function of both the rate of quality loss and the initial value. For hoki, the functional property of gelation is lost in a linear manner with time stored in ice (Figure 5.14).[97] However, within the first few hours post-harvest, there may actually be an increase in gel-forming ability due to changes in functional properties associated with the development of rigor.

The nucleotides ATP, ADP, and IMP have been shown to exert a protective effect on fish actomyosin stored at –20°C while the nucleotide catabolites inosine and hypoxanthine destabilized these proteins.[56] This finding may help explain why fresh fish, with consequent higher concentrations of ATP, ADP, and IMP are more stable during frozen storage than less fresh fish.[98,99]

5.7.2 Leaching

Leaching fish mince with fresh water was initially used in Japan in the 1910s as a means of removing fats, oils, and fishy odors as well as removing color. It soon became clear that such washing also resulted in an increase in the product's gel strength. It is now recognized that the leaching process performs several functions. It removes sarcoplasmic proteins, which interfere with the gelation of myofibrillar proteins in surimi-based products. Leaching also improves frozen storage stability[33] by reducing the activity of trimethylamine oxidase, removing lipids, undesirable blood, pigments and odorous substances, and by concentrating the myofibrillar proteins.

5.7.3 Freezing Rate

It is generally believed that storage temperature is more important than freezing rate in determining the quality of frozen fish. However, the rate at which muscle is frozen determines ice crystal size, locality of ice crystals, and sites of highly concentrated salt solutions, and some of these changes in frozen muscle may be linked to protein alterations.

According to Love and Ironside,[100] protein solubility of cod muscle was not greatly affected at different freezing rates if the muscles were thawed without storage. However, in further studies,[101] protein solubility dropped to a minimum of 68% when cod muscle was frozen in 70 min (time for temperature to drop from 0 to $-5°C$) and stored for one year at $-29°C$. Protein solubility of cod muscle frozen faster and slower than 70 min and then stored for 1 year at $-29°C$ was about 85%. It is believed that with a freezing time of 70 min, a single ice crystal was formed in each fiber within cod muscle and that this may have contributed to protein denaturation by localized salt concentration.

Generally, it is advisable to aim for a freezing time that is faster than 70 min. Certainly, for red meat, it has been shown, using ATPase activity and differential scanning calorimeter measurements, that the lower the freezing rate, the greater the myosin denaturation.[102] It is therefore worthwhile to analyse current freezing practices to check that the freezing rate is sufficiently fast to ensure that stability is not compromised during subsequent storage.

5.8 STABILIZED FISH MINCE

The process for making stabilized mince is much simpler, and therefore cheaper, than making surimi, which may be a viable option for manufacture in conjunction with fillet processing on factory ships at-sea, with surimi processing being conducted on shore where water is more available. MacDonald et al.[37] conceived the development of stabilized (unleached) fish mince as a material that could be substituted to some extent for surimi in some manufactured products, or be stored frozen as a raw material for surimi manufacture at a later date and distant location. To test this concept, fish (New Zealand hoki, a lean gadoid species) were mechanically deboned and minced and some of the mince was stabilized by mixing in 12% sucrose and 0.2% phosphates (Figure 5.15).[37] Samples of headed and gutted (H&G) fish, minced fish, and minced and stabilized fish were frozen at −20°C and −50°C, respectively. Surimi was manufactured from the same fresh fish and stored at −30°C for comparison.

To test the stability of surimi made from these samples, smaller samples were taken each month for 6 months and made into surimi. These were tested for gel-forming ability. After 6 months' storage, there was a clear separation of treatments into four groups on the basis of the deformability (shear strain at failure) when gels were prepared by two-stage cooking (25°/90°C) (Figure 5.16).[37] There was no significant difference ($P < 0.01$) between freshly prepared surimi, surimi prepared from stabilized mince, and surimi prepared from stabilized mince stored at −20 and −50°C. These samples produced very cohesive gels. Less cohesive ($P < 0.05$), but still acceptable gels, were obtained from surimi made from H&G fish stored at −50°C and unstabilized mince stored at −50°C. Weaker gels were obtained from surimi made from H&G fish stored at −20°C, and very weak gels were obtained from unstabilized mince stored at −20°C.

Stabilization of Proteins in Surimi

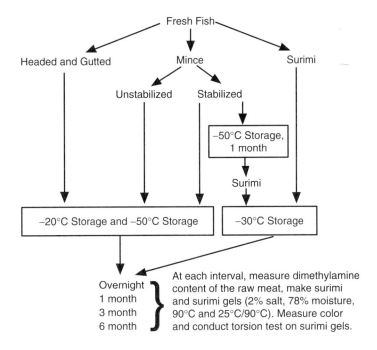

Figure 5.15 Experimental design to evaluate the stabilized mince concept.

Gadoid fish possess an enzyme-substrate system that generates dimethylamine (DMA) and formaldehyde in equal molar quantities during frozen storage. Formaldehyde is a potent denaturant of proteins.[103] It is widely known that enzyme reactions proceed much more slowly as the temperature decreases. Additionally, enzymes and substrates within animal tissue are compartmentalized, such that the reactants are physically separated and do not readily react until the tissue is physically broken down. Thus, formaldehyde generation in fish proceeds much faster at warmer frozen storage temperatures, and is also more rapid in minced fish than in intact muscle.

After 6 months' storage, dramatic differences in the DMA content could be seen among the treatments (Figure 5.17).[37] H&G fish showed only low levels of DMA generation, with

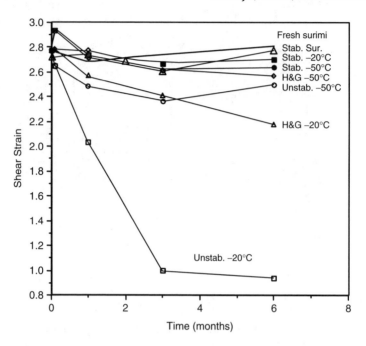

Figure 5.16 Change in cohesiveness of gels as measured by true strain at failure during storage of samples up to 6 months. Gels were heated in two stages, 25°C (2 hr) and then 90°C (15 min). (Adapted from Reference 37.) FRESH SURIMI: surimi made from fresh fish at time of stabilized mince manufacture; STAB. SUR.: surimi made from stabilized mince following its frozen storage at −50°C for 30 days and then stored at −30°C; STAB. −20, −50°C: stabilized mince made from fresh fish and stored at −20, −50°C; UNSTAB. −20, −50°C: mince stored at −20 and −50°C; H&G −20, −50°C: headed and gutted trunks stored at −20 and −50°C.

less at −50°C than at −20° C ($P < 0.05$). Minces showed greatly elevated DMA production. Low-temperature (−50°C) storage was very effective in reducing the activity of the enzyme, as DMA levels were much less in treatments stored at −50°C ($P < 0.05$). Remarkably, the addition of cryoprotectants also decreased DMA production by an additional 60% in minced samples stored at −20°C ($P < 0.01$). This was an unexpected finding, which may have other applications in stabilizing the

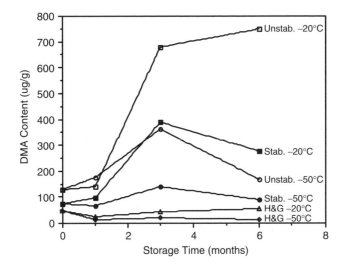

Figure 5.17 Dimethylamine levels in the various treatments during frozen storage. See Figures 5.15 and 5.16 for treatment details. (Adapted from Reference 37.)

cut surfaces of frozen fillets. Possibly, the addition of 12% sucrose to the mince sufficiently reduced water activity to inhibit TMAOase activity.

By combining low-temperature storage with cryoprotection, it was possible to stabilize the gel-forming properties of the fish mince. More recent studies showed that mince (Pacific whiting) could be stabilized with 6% sucrose rather than the 12% used in the hoki study.[104,105]

5.9 STABILIZATION OF FISH PROTEINS TO DRYING

One drawback of frozen surimi as a functional food ingredient is the necessity to freeze, transport, and thaw a prodigious amount of water along with the functional protein. Whereas most food protein ingredients are traded in dry form, frozen surimi contains in excess of 70% water. Drying of such a sensitive protein material as fish muscle is difficult because both heating and drying are involved in the process, which can lead to protein denaturation.

5.9.1 Processes for Drying of Surimi

5.9.1.1 Freeze-Drying

Freeze-drying is the gentlest means of drying a protein, but is not a practical option for surimi as a food ingredient commodity due to the expense of such a process. Freeze-dried surimi has been commercially produced in Japan for use as a specialized binder in preparing fabricated herring roe, fillet blocks, and restructured products from beef, pork, and chicken. The process is depicted in Figure 5.18. A sufficient concentration of sucrose added along with a small amount of polyphosphate protects the proteins from denaturation during drying and storage; however, above 5% by weight on a wet basis, there is little improvement in stability. For vacuum drying on an industrial scale, the surimi is placed on trays or belts in a sealed chamber. The temperature and pressure are reduced and the product is frozen. After a sufficient vacuum is reached, hot water coils or electric resistance heating carefully heat the frozen fish so that the ice is sublimed directly to vapor without melting within the fish. Unit operation of a typical vacuum freeze-drier is shown in Figure 5.19. The dehydrated material is hygroscopic and must be packed in sealed containers; vacuum or inert gas packing is common.

Freeze-dried materials have 80 to 160 times more surface area than air-dried materials.[106] Accordingly, they are more susceptible to influences of the storage environment, particularly humidity. Dried proteins are more stable at low relative humidity.[107] The chemical stability of solid-state proteins decreases with increasing moisture in the solid due to changes in either dynamic activity or conformation stability of the protein, or due to water serving as a reactant and/or a medium for mobilization of reactant.[108,109]

However, the storage stability does not improve significantly below a relative humidity (RH) of 15% at 20°C or below 10% at 35°C. Low-temperature storage is important for retaining good functionality in freeze-dried surimi.[107] When freeze-

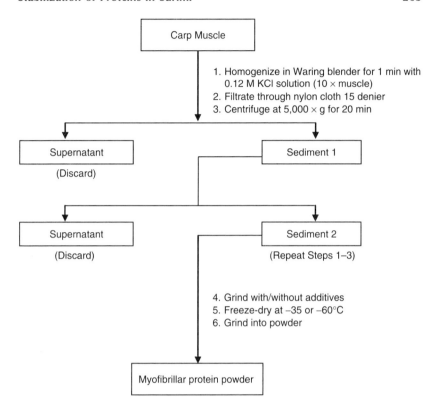

Figure 5.18 Preparation of lyophilized myofibrillar powder from carp muscle.

drying was conducted with platen temperature at 60°C and an absolute pressure of 0.18 to –0.03 torr, the total ATPase activity of the surimi powder did not significantly change during 5 months' storage at 15°C in a desiccator (RH = 15%).

Yoo and Lee[110] evaluated the thermoprotective properties of sorbitol on freeze-dried red hake fish mince. As shown in Table 5.2, the gel-forming ability of freeze-dried fish mince protein was improved with an increase in sorbitol concentration. This indicated that sorbitol protected the fish mince protein from losing functional properties during freeze-drying, as evidenced by 96.5% greater gel-forming ability at 4% sorbitol as compared to that of the control.

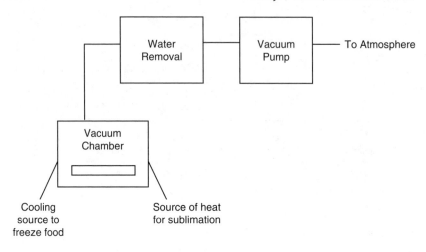

Figure 5.19 Unit operation in a freeze-drying unit.

TABLE 5.2 Effect of Sorbitol on Gel-Forming Ability of Freeze-Dried/Refined Fish Mince

Sorbitol Level (%)	Compressive Force kg (% increase)	Penetration Force g (% increase)
0.0	2.0	209
2.8	3.7 (83.5%)	218 (4.3%)
4.0	3.9 (96.5%)	220 (5.3%)

Kanna et al.[111] found that both freeze-drying and drying by absorption to silica gel at 0°C caused little denaturation of muscle proteins and less decrease in salt solution solubility than air drying (Figure 5.20).[111] The freeze-dried muscle was light and porous, while both air-dried and silica gel-dried muscle were hard and cohesive. When immersed in water, both the silica gel- and vacuum-dried muscle reconstituted quickly to give excellent texture and appearance, but the air-dried muscle did not. These findings indicate that protein denaturation and rehydratability of dried fish muscle depend on the temperature of the drying process.

Recently, a study compared the quality of freeze-dried vs. conventionally frozen whiting surimi.[112] Low-temperature

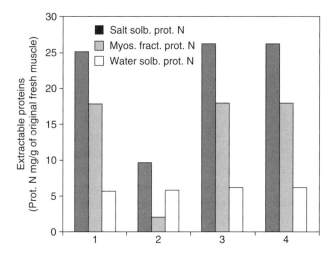

Figure 5.20 Changes in proteins extracted from muscle by drying. 1: Fresh muscle; 2: Dehydrated muscle by air-drying; 3: Dehydrated muscle by silica gel; 4: Dehydrated muscle by freeze drying.

storage (−18°C) was found to be important in retaining excellent functionality in freeze-dried surimi for up to 9 months' storage. Huda et al.[113] also reported similar superior quality for freeze-dried surimi powders from various species, although some proteolytic degradation was observed at 6 and 9 months of storage.

5.9.1.2 Spray-Drying of Surimi

The more common method of drying food proteins is by spray-drying. Conventional (atmospheric) spray-drying necessitates higher temperatures than freeze-drying or drying on silica gel, and is therefore more detrimental to protein quality although new methods of vacuum spray-drying are now being developed. To be spray-dried, the fish proteins must also be in a dispersed, low-viscosity state.

Apparently, conventional spray-drying of whole fish protein has been successfully accomplished by only one group,[114,115] who prepared a functionally active fish protein powder from Alaska pollock. The dressed fish was minced and milled in a colloidal mill along with sucrose, which was added to prevent protein denaturation. The slurry was spray-dried to obtain a product containing 65% protein, 4% fat, and 24% sucrose. The authors overcame the problem of high feed viscosity by adding carbonic acid to reduce the pH to the isoelectric point of the proteins. Upon drying, carbonic acid is decomposed to carbon dioxide and water, thus restoring the pH of the dried powder to neutrality.

Conventional (non-vacuum) spray-drying may impose significant stresses on the protein structure and stability, and may lead to reductions in enzymatic activity, alterations of the native fold, α-helix $\rightarrow$ β-sheet transitions, as well as rearrangement of the native β-sheet elements.[116]

5.9.2 Potential Additives and Mechanisms of Lyoprotection

Lyoprotection is widespread in nature; many plants and animals from several phyla have the ability to survive complete dehydration, a condition known as anhydrobiosis. A common theme in these organisms is the accumulation of large amounts of sugars, especially the disaccharides sucrose and/or trehalose.[117–121] There are several hypotheses to explain why trehalose is particularly effective, none of which completely accounts for experimental observations. Some theories are based on the interaction of trehalose with biological structures, while others are based solely on the interaction of trehalose with water and the thermophysical properties of its aqueous solutions.

Many additives, when present along with protein in aqueous solutions prior to drying, markedly stabilize the protein against denaturation. Disaccharides, such as sucrose, lactose, and, more recently, the totally non-reducing sugar trehalose, have been used widely as protectants during air-drying or freeze-drying of a variety of materials. Tzannis and Prestrelski[122] utilized moisture to probe the interaction between a model protein, trypsinogen, and sucrose in the solid state, following spray-drying. Both dynamic and equilibrium moisture uptake studies indicated the presence of an optimal protein–sucrose hydrogen bonding network. At low sucrose content, a preferential protein–sucrose hydrogen bonding interaction was dominant, resulting in protein stabilization. However, at high carbohydrate concentrations, preferential sugar–sugar interactions prevailed, resulting in a phase separation within the formulation matrix. The preferential incorporation of the sucrose molecules in a sugar-rich phase reduced the actual amount of the carbohydrate available to interact with the protein, and thereby decreased the number of effective protein–sucrose contacts. Consequently, the protein could not be effectively protected during spray-drying. They hypothesized that the observed phase separation at this sucrose concentration, originating from its exclusion from the protein in solution before spray-drying, was further accompanied by preferential clustering of the sucrose molecules.

Miller et al.[123] found similar behavior during freezing and vacuum-drying LDH in the presence of trehalose. Following freeze-thawing, the recovered activity increased with increasing trehalose concentration. The activity began to level off at about 80% recovery at a trehalose concentration of 300 mM. The vacuum-drying/rehydration data exhibited an interesting peak in recovered activity (76%) at an initial trehalose concentration of approximately 100 mM. The addition of more trehalose diminished the recovery activity. They attributed this to partial crystallization of trehalose. Presumably, crystalline trehalose would not be able to interact with the enzyme in a protective capacity. Using differential scanning calorimetry, they found that partial trehalose crystallization occurred in mixtures with an initial trehalose concentration greater than 200 mM. Such samples required longer drying times due to slow diffusion of water through the sample. However, from a kinetic point of view, a long drying time increases the probability and degree of trehalose crystallization.

Recent work indicates that the stresses that arise during freezing and drying are fundamentally different, thereby necessitating separate mechanisms to explain protein stabilization during each of these processes.[124] That is, many agents that preserve proteins in solution and during freezing/thawing fail to preserve them in the dried state. Some disaccharides are known to stabilize proteins in the dried state, and trehalose, has received special attention in this regard because of its greater stability to inversion as compared to sucrose.[123,125,126]

Crowe[127] suggested that disaccharides lyoprotect proteins by a "water replacement" mechanism. That is, he proposed that certain physiological solutes "replace" the lost water around the polar residues of macromolecules. Carpenter and Crowe[128] agreed that this is likely the mechanism by which certain solutes maintain dried proteins in their native conformation. These authors showed that leakage from model membrane systems (liposomes, vesicles) is prevented as a result of the interaction of disaccharide with the polar head groups of the membrane phospholipids.[129–131]

Some researchers have suggested that the only requirement for preservation of structure and function in freeze-dried membranes, liposomes, and proteins is the ability of the additive (e.g., sugar or polysaccharide) to form a glass at all water contents and temperatures during the freeze-drying process.[132,133] Green and Angell[134] suggested that the high efficacy of trehalose is connected with its glass-forming characteristics. On the basis of mole percent glucose ring, trehalose had the highest glass transition temperature (Tg) among the tested sugars, followed by maltose, sucrose, and glucose. This order of Tg was also the same as the order of efficacy of the carbohydrates in biopreservation. Intracellular processes in trehalose/water solutions can be brought to a halt by viscous slow-down (i.e., suspending life), while the water content remains quite high relative to other saccharide/water systems of the same Tg.[135] In seeds that commonly accumulate sucrose and oligosaccharides, the formation of intracellular glasses is associated with the survival of desiccation and the maintenance of seed viability during storage.[136,137]

Miller at al.[123] investigated a new class of additives that combines synergistically with trehalose. Based on the results of several preliminary trials with various compounds, they found that certain mixtures of trehalose and borate ions dramatically increase the Tg of trehalose. The Tg of pure, dry trehalose is approximately 115°C, while it increases to 150°C for a freeze-dried trehalose/sodium tetraborate mixture with a boron:trehalose mole ratio of 0.33. Using ^{11}B NMR, they verified that the borate ions form cross-links with the hydroxyl groups of the trehalose molecule, thereby forming trehalose–borate complexes. This cross-linking raises the solution viscosity and, at sufficiently low temperatures, promotes the formation of a glass. Furthermore, they found that the degree of trehalose–borate complexation could be controlled by adjusting the pH of the aqueous medium (unpublished).

Freeze-thawing and vacuum-drying/rehydration of LDH in the presence of 300 mM trehalose resulted in the recovery of 80 and 65% of the original activity, respectively. For vacuum-dried mixtures, boron concentrations below 1.2 mole boron/mole trehalose had no effect on the recovered LDH. After several weeks of storage in either humid (100% relative humidity) or warm (45°C) environments, vacuum-dried formulations that included trehalose and borate showed greater enzymatic activities than those prepared with trehalose alone. They attributed this stability to the formation of a chemical complex between trehalose and borate.

More recently, Aldous and co-workers[132] suggested that the ability to crystallize a stoichiometric hydrate from the amorphous phase lends exceptional stabilization properties to certain carbohydrates often used in lyophilization. This stabilization occurs during long-term storage, where the presence of low concentrations of water is inevitable. In this case, they suggest that the amorphous saccharide is stabilized by incorporation of trace amounts of water into the crystalline dihydrate rather than into the amorphous phase. This prevents water from acting as a plasticizer of the amorphous phase, which would lead to a decrease in Tg. However, the rates of both the crystallization of the hydrate and the migration of water in such concentrated solutions have not been examined in detail. Although such processes are thermodynamically favorable, they are kinetically inhibited due to low molecular mobility in the amorphous "phase."

Sola-Penna and Meyer-Fernandes[138] showed that trehalose has a larger hydrated volume than other related sugars (e.g., sucrose maltose, glucose, and fructose). According to their results, trehalose occupies at least a 2.5 times larger volume than sucrose, maltose, glucose, and fructose. They correlated this property with the ability to protect the structure and function of pyrophosphatase and glucose 6-phosphate dehydrogenase against thermal inactivation. When the concentrations of all sugars were corrected by the percentage of the occupied volume, they presented the same effectiveness. Their results suggest that because of this large hydrated volume, trehalose can substitute more water molecules in the solution, and this property is very close to its effectiveness. They concluded that the higher size exclusion effect is responsible for the difference in efficiency of protection against thermal inactivation of enzymes.

This effect can be correlated with Xie and Timasheff's[139] study on the thermodynamic mechanism of protein stabilization by trehalose. They carried out preferential interaction measurement of ribonuclease A by 0.5 M trehalose at 52°C at pH 5.5 and 2.8, where the protein is native and unfolded respectively, and at 20°C where the protein is native at both pH values. At the low temperature, the interaction was preferential exclusion. At 52°C, the interaction was that of preferential binding, greater to the native than the unfolded protein. The temperature dependence of the preferential interactions of 0.5 M trehalose with ribonuclease "A" showed that it is the smaller preferential binding to the unfolded protein compared to the native one that gives rise to protein stabilization.

Recent frozen storage trials (T.C. Lanier and J.W. Park, personal communication, 2004) have indicated that trehalose functions about equally in cryoprotective effectiveness as sucrose or sorbitol when incorporated into surimi before freezing. Apparently, the advantages of trehalose as a protein stabilizer may be more evident in drying proteins than in freezing.

5.10 FUTURE DEVELOPMENTS IN FISH PROTEIN STABILIZATION

Using cryoprotectants to extend the shelf life of leached fish minces (i.e., the development of surimi) has been an important technological development in the seafood industry. Surimi has allowed the commercial exploitation of otherwise underutilized fish species. Currently, the most commonly used cryoprotectants for cold-water species are sucrose and sorbitol, typically added in a blend of 4%+4% with 0.2% phosphates. This level of addition varies, depending on species. For example, only 5% sucrose is added to surimi made from Southeast Asian warm-water species in which the muscle proteins are more stable.

These cryoprotectants have been chosen for their effectiveness, low cost, availability, and low tendency to cause Maillard browning. Sorbitol is often substituted for all or a portion of the sucrose because of its lesser sweetness. There is however, interest in identifying alternative cryoprotectants and blends with improved ability to stabilize myofibrillar proteins and with reduced sweetness for specific applications.

A wide variety of compounds will cryoprotect labile proteins during freeze-thawing (as have been previously reviewed.[40,52,140] These include sugars, amino acids, polyls, methylamines, carbohydrate polymers, synthetic polymers (e.g., polyethylene glycol [PEG]), other proteins (e.g., bovine serum albumin [BSA]), and even inorganic salts (e.g., potassium phosphate and ammonium sulfate). For most proteins, the cryoprotectant must be at a relatively high concentration to confer maximum protection. Exceptions are polymers such as PEG, BSA, and polyvinylpyrrolidone (PVP), which at concentrations of even less than 1% (wt./vol.) fully protect sensitive enzymes such as lactate dehydrogenase or phosphofructokinase. Antioxidants and metal chelators such as phosphate compounds may also act to extend the shelf life of surimi and other food proteins, serving as cryoprotective adjuncts.

In the cryostabilization of the viability of freeze-tolerant organisms, specific antifreeze proteins (also called thermal hysteresis proteins [THPs]), stress proteins (cold and heat), and ice nucleating agents may also be important. These may one day find use in the stabilization of food proteins such as surimi.[141–145]

Leistner[146] introduced the "hurdle" concept to stabilization of foods against microbial attack; individual additives or process steps that inhibit or destroy microorganisms can be viewed as "hurdles" to the growth of spoilage or harmful microorganisms in foods. Each such "hurdle," while alone is not able to ensure stability, contributes additively, or sometimes synergistically, to the overall microbial stability of the food. In the same way, the chemical and physical stability of frozen foods are advanced by individual factors, which can include cryoprotectant additives, alone or in a mixture. Because the class of additives known as "cryoprotectants" is so diverse, each mechanistic type can be viewed as a separate "hurdle" to the chemical or physical instability caused by freezing. To date there has been little research investigating the potential for synergistic interactions between different cryoprotectants, particularly those that can act via different mechanisms.

Perhaps a better mechanistic understanding of how cryoprotectants exert their effect on proteins will also lead to new developments. A better understanding of the role of cryoprotectants on water structure, and how this in turn affects protein stability, is also fundamental.

In summary, we can expect future developments to come from the following areas:

- Improved understanding of the changes in intrinsic stability of the myofibrillar proteins from the raw material fish, between different species, with season, and the need to cryoprotect these proteins
- New technologies for harvesting and handling fish that will impact the stability of the muscle proteins during subsequent processing and storage will influence which cryoprotectant additives and mechanisms will be most effective for extending shelf life
- Identifying new cryoprotectants (both intrinsic and added)
- Increased availability of hitherto expensive cryoprotectants and lyoprotectants such as trehalose
- Identifying synergistic blends of cryoprotectants
- New process developments, such as the new acid-aided process, and developments in drying technology

REFERENCES

1. Z. Weng and C. DeLisi. Amino acid substitutions: effects on protein stability. In *Encyclopedia of Life Sciences.* London: Nature Publishing Group (www.els.net), 2001.

2. R.J. Ellis and J.T. Pinheiro. Danger-misfolding proteins. *Nature,* 416, 483–484, 2002.

3. G. Pollack. Helix-coil transition in the myosin rod. In *Muscle & Molecules: Uncovering the Principles of Biological Motion.* Seattle, WA: Ebner & Sons Publishers, 1990, 106.

4. J.J. Connell. The relative stabilities of the skeletal muscle myosins of some animals. *Biochem. J.,* 80, 503–509, 1961.

5. I.A. Johnston, W. Davison, and G. Goldspink. Adaptions in Mg^{2+}-activated ATPase activity induced by temperature acclimation. *FEBS Lett.,* 50, 293–295, 1975.

6. A. Hashimoto, A. Kobayashi, and K. Arai. Thermostability of fish myofibrillar Ca-ATPase and adaption to environmental temperature. *Bull. Japan Soc. Sci. Fish.,* 48, 671–684, 1982.

7. K. Konno, K. Arai, and S. Watanabe. Regulatory proteins from dorsal muscle of the carp. *J. Biochem. (Tokyo),* 82, 931–938, 1977.

8. I.A. Johnston and G. Goldspink. Thermodynamic activation parameters of fish myofibrillar ATPase enzyme and evolutionary adaptations to temperature. *Nature,* 257, 620–622, 1975.

9. M.E. Rodgers, T. Karr, K. Biedermann, H. Ueno, and W.F. Harrington. Thermal stability of myosin rod from various species. *Biochemistry,* 26(26), 8703–8708, 1987.

10. I.A. Johnston and N.J. Walesby. Molecular mechanisms of temperature adaption in fish myofibrillar adenosine triphosphatases. *J. Comp. Physiol.,* 119, 195–206, 1977.

11. P.H. Yancey. Water stress, osmolytes and proteins. *Am. Zool.,* 41, 699–709, 2001.

12. P. Torrejon, M.L. del Mazo, M. Tejada, and M. Careche. Aggregation of minced hake during frozen storage. *Eur. Food Res. Technol.,* 209, 209–214, 1999.

13. C.G. Sotelo and H. Rehbein. In N.F. Haard, and B.K. Simpson, Eds. *Seafood Enzymes.* New York: Marcel Dekker, 2000, 167–190.

14. M. Kimura, N. Seki, and I. Kimura. Occurrence and some properties of trimethylamine-N-oxide demethylase in myofibrillar fraction from walleye pollack muscle. *Fish Sci.,* 66, 725–729, 2002.

15. G.L. Fletcher, C.L. Hew, P.L. Davies. Antifreeze proteins of teleost fishes. *Annu. Rev. Physiol.,* 63, 359–390, 2001.

16. P.L. Privalov. Cold denaturation of proteins. *Crit. Rev. Biochem. Mol. Biol.,* 25, 281– 305, 1990.

17. F. Franks. *Biophysics and Biochemistry at Low Temperatures,* Cambridge, U.K.: Cambridge University Press, 1985, 206.

18. R.M. Love. New factors involved in the denaturation of frozen cod muscle protein. *J. Food. Sci.,* 27, 544–550, 1962.

19. G.B. Strambini and E. Gabellieri. Proteins in frozen solutions: Evidence of ice-induced partial unfolding. *Biophysics,* J 70, 971–976, 1996.

20. O. Fennema. Activity of enzymes in partially frozen aqueous systems, In R.B. Duckworth, Ed. *Water Relations of Foods.* London: Academic Press, 1975, 397–413.

21. J. Wolfe and G. Bryant. Freezing, drying, and/or vitrification of membrane solute–water systems. *Cryobiology,* 39, 103–129, 1999.

22. A. Calvelo. Recent studies on meat freezing. In R. Lawrie, Ed. *Developments in Meat Science, Vol. 2.* London, New Jersey: Applied Science Publishers, 1981, 125.

23. B.W. Grout. The effect of ice formation during cryopreservation of clinical systems. In B.J. Fuller and B.W.W. Grout, Eds. *Clinical Applications of Cryobiology.* Boca Raton, FL: CRC Press, 1991, 82–91.

24. M. Jul. *The Quality of Frozen Food.* London: Academic Press, 1984, 1.

25. R. Hamm. Functional properties of the myofibrillar system and their measurements. In P.J. Bechtel, Ed. *Muscle as Food.* Orlando, FL: Academic Press, 1986, 135–199.

26. R.M. Love. The freezing of animal tissue. In H.T. Meryman, Ed. *Cryobiology.* New York: Academic Press, 1966, 317–399.

27. J.J. Matsumoto. Denaturation of fish muscle proteins during frozen storage. In O. Fennema, Ed. *Proteins at Low Temperatures.* Washington, D.C.: American Chemical Society, 1979, 206–224.

28. G. Tarborsky. Protein alterations at low temperatures. An overview. In O. Fennema, Ed. *Proteins at Low Temperatures.* Washington, D.C.: American Chemical Society, 1979, 1.

29. S.Y. Shenouda. Theories of protein denaturation during frozen storage of fish flesh. *Adv. in Food Res.,* 26, 275–311, 1980.

30. D. Grasso, K. Strevett, M. Butkus, and K. Subramanian. A review of non-DLVO interactions in environmental colloidal systems. *Rev. Environ. Sci. Biotechnol.,* 1, 17–38, 2002.

31. J. Israelashvili. Solvation, structural and hydration forces. In *Intermolecular & Surface Forces.* San Diego, CA: Academic Press, 1992 260–287.

32. P.J. Parker and A.G. Collins. Dehydration of flocs by freezing. *Environ. Sci. Technol.,* 33, 482–488, 1999.

33. J.J. Matsumoto and S.F. Noguchi. Cryostabilization of protein in surimi, In T.C. Lanier and C.M. Lee, Eds. *Surimi Technology.* New York: Marcel Dekker, 1992, 357–388.

34. Y. Shimizu, W. Simidu, and T. Ikeuchi. Studies on jelly strength of kamaboko. III. Influence of pH on jelly strength. *Bull. Japan Soc. Sci. Fish.,* 20, 209–212, 1954.

35. N.G. Lee and J.W. Park. Calcium compounds to improve gel functionality of Pacific whiting and Alaska pollock surimi. *J. Food Sci.,* 63, 969–974, 1998.

36. J.F. Holmquist, E.M. Buck, and H.O. Hultin. Properties of kamaboko made from red hake (*Urophycis chuss*) fillets, mince, or surimi. *J. Food. Sci.,* 49, 192–196, 1984.

37. G.A. MacDonald, N.D.C. Wilson, and T.C. Lanier. Stabilised mince: an alternative to the traditional surimi process. In *Proc. IIR Conf. on Chilling and Freezing of New Fish Products.* Aberdeen, U.K.: Torry Research Station, 1990.

38. K.B. Storey. Life in a frozen state: adaptive strategies for natural freeze tolerance in amphibians and reptiles. *Am. J. Physiol.*, 258, R559–R568, 1990.

39. J.F. Carpenter and J.H. Crowe. The mechanism of cryoprotection of proteins by solutes. *Cryobiology*, 25, 244–255, 1988.

40. G.A. MacDonald and T.C. Lanier. The role of carbohydrates as cryoprotectants in meats and surimi. *Food Technol.*, 45, 150–159, 1991.

41. G.A. MacDonald and T.C. Lanier. Actomyosin stabilization to freeze-thaw and heat denaturation by lactate salts. *J. Food Sci.*, 59, 101–105, 1994.

42. G.A. MacDonald, T.C. Lanier, H.E. Swaisgood, and D.D. Hamann. Mechanism for stabilization of fish actomyosin by sodium lactate. *J. Agric. Food Chem.*, 44, 106–112, 1996.

43. K. Gekko and S.N. Timasheff. Mechanism of protein stabilization by glycerol: preferential hydration in glycerol–water mixtures. *Biochemistry*, 20, 4667–4676, 1981.

44. T. Arakawa and S.N. Timasheff. Stabilization of protein structure by sugars. *Biochemistry*, 21, 6536–6544, 1982.

45. T. Arakawa and S.N. Timasheff. Preferential interactions of proteins with salts in concentrated solutions. *Biochemistry*, 21, 6545–6552, 1982.

46. T. Arakawa and S.N. Timasheff. Mechanism of protein salting in and salting out by divalent cation salts: balance between hydration and salt binding. *Biochemistry*, 23, 5912–5923, 1984.

47. T. Arakawa and S.N. Timasheff. Protein stabilization and destabilization by guanidinium salts. *Biochemistry*, 23, 5924–5929, 1984.

48. T. Arakawa and S.N. Timasheff. The stabilization of proteins by osmolytes. *Biophys. J.*, J 47, 411–414, 1985.

49. J.C. Lee and S.N. Timasheff. The stabilization of proteins by sucrose. *J. Biol. Chem.*, 256, 7193–7201, 1981.

50. S.N. Timasheff and T. Arakawa. Stabilization of protein structure by solvents, In T.E. Creighton, Ed. *Protein Structure — A Practical Approach*. Oxford: IRL Press, 1989, 331–346.

51. J. Wyman. Linked functions and reciprocal effects in hemoglobin: a second look. *Adv. Protein Chem.*, 19, 223–286, 1964.

52. K.E. Holde. *Physical Biochemistry.* Englewood Cliffs, NJ: Prentice Hall, 1985, 287.

53. C. Branca, S. Maagazu, G. Maisano, P. Migliardo, V. Villari, and A.P. Sokolov. The fragile character and structure-breaker role of alpha, alpha-trehalose: viscosity and Raman scattering findings. *J. Phys. Condens. Matter,* 11, 3823–3832, 1999.

54. G.A. MacDonald and T.C. Lanier. Cryoprotectants for improving frozen food quality, In M.C. Erickson and Y.C. Hung, Eds. *Quality in Frozen Foods.* New York: Chapman & Hall, 1997, 197–232.

55. C.H. Holten, A. Muller, and D. Rehbinder. *Lactic Acid. Properties and Chemistry of Lactic Acid and its Derivatives.* Weinheim: Verlag Chemie GmbH, 1971, 566.

56. S.T. Jiang, B.O. Hwang, and C.T. Tsao. Effect of adenosinenucleotides and their derivatives on denaturation of myofibrillar proteins *in vitro* during frozen storage at −20°C. *J. Food Sci.,* 52, 117–123, 1987.

57. D. Mornet, A. Bonet, E. Audemard, and J. Bonicel. Functional sequences of the myosin head. *J. Musc. Res. Cell Motility,* 10, 10–24, 1989.

58. J.F. Carpenter, S.C. Hand, L.M. Crowe, and J.H. Crowe. Cryoprotection of phosphofructokinase with organic solutes: characterisation of enhanced protection in the presence of divalent cations. *Arch. Biochem. Biophys.,* 250, 505–512, 1986.

59. J.F. Carpenter, B. Martin, L.M Crowe, and J.H. Crowe. Stabilization of phosphofructokinase during air-drying with sugars and sugar/transition metal mixtures. *Cryobiology,* 24, 455–464, 1987.

60. J.F. Carpenter, L.M. Crowe, and J.H. Crowe. Stabilization of phosphofructokinase with sugars during freeze-drying: characterization of enhanced protection in the presence of divalent cations. *Biochim. Biophys. Acta,* 923, 109–115, 1987.

61. K.C. Hazen, L.D. Bourgeois, and J.F. Carpenter. Cryoprotection of antibody by organic solutes and organic solute/divalent cation mixtures. *Arch. Biochem. Biophys.,* 267, 363–371, 1988.

62. G.A. MacDonald, T.C. Lanier, and F.G. Giesbrecht. Interaction of sucrose and zinc for cryoprotection of surimi. *J. Agric. Food Chem.*, 44, 113–118, 1996.

63. J.W. Park and T.C. Lanier. Combined effects of phosphates and sugar or polyol on protein stabilization of fish myofibrils. *J. Food Sci.*, 52, 1509–1513, 1987.

64. J.W. Park, T.C. Lanier, and D.P. Green. Cryoprotective effects of sugar, polyols, and/or phosphates on Alaska pollock surimi. *J. Food Sci.*, 53, 1–3, 1988.

65. Y. Kumazawa, Y. Oozaki, S. Iwami, I. Matsumoto, and K. Arai. Combined protective effect of inorganic pyrophosphate and sugar on freeze-denaturation of carp myofibrillar protein. *Nippon Suisan Gaikkaishi*, 56, 105–113, 1990.

66. H. Saeki. Gel-forming ability and cryostability of frozen surimi processed with $CaCl_2$-washing. *Fish Sci.*, 62, 252–256, 1996.

67. J.P.H. Wessels, C.K. Simmonds, P.D. Seaman, and L.W.J. Avery. The Effect of Storage Temperature and Certain Chemical and Physical Pretreatments on the Storage Life of Frozen Hake Mince Blocks. Fishing Industry Research Institute, Capetown, South Africa, 1981.

68. M.C. Erickson. Antioxidants and their appliction in frozen foods, In M.C. Erickson and Y.C. Hung, Eds. *Quality in Frozen Food*. New York: Chapman & Hall, 1997, 233–263.

69. D.B. Volkin and A.M. Klibanov. Minimizing protein inactivation, In T.E. Creighton, Ed. *Protein Function a Practical Approach*. New York: IRL Press, 1989, 1–23.

70. M. Serpelloni. The food applications of sorbitol, mannitol, and hydrogenated glucose syrups. Presented at *International Symposium of Polyols and Polydextrose*. Paris, France, 1985.

71. L. Pauling. *General Chemistry*. 2nd ed. San Francisco: Freeman, 1953, 344–347.

72. A.L. DeVries. The role of antifreeze glycopeptides and peptides in the freezing avoidance of Antarctic fishes. *Comp. Biochem. Physiol.*, 90B, 611–621, 1988.

73. H.T. Meryman. Cryoprotective agents. *Cryobiology*, 8, 173–183, 1971.

74. J.F. Carpenter, S.J. Petrelski, T.J. Anchordoguy, and T. Arakawa. Interactions of stabilizers with proteins during freezing and drying, In J.L. Cleland and R. Langer, Eds. *Formulation and Delivery of Proteins and Peptides.* Washington, D.C.: American Chemical Society, 1994, 134–147.

75. T.C. Lanier and T. Akahane. Method of Retarding Denaturation of Meat Products. US. Patent No. 4,572,838, 1986.

76. H. Levine and L. Slade. A polymer physico-chemical approach to the study of commercial starch hydrolysis products (SHPs). *Carbohydr. Polym.,* 6, 213–244, 1986.

77. O. Fennema. Water and ice, In O. Fennema, Ed. *Food Chemistry.* New York: Marcel Dekker, 1996, 1–94.

78. R. Parker and S.G. Ring. A theoretical analysis of diffusion-controlled reactions in frozen foods. *Cryo-Lett.,* 16, 197–208, 1995.

79. H. Levine and L. Slade. Principles of cryostabilization technology from structure/property relationships of carbohydrate/water systems. A review. *Cryo-Lett.,* 9, 21–63, 1988.

80. F. Franks. Complex aqueous systems at subzero temperatures, In D. Simantos and J.L. Multon, Eds. *Properties of Water in Foods.* Dordrecht, Netherlands: Martinus Nijhoff Pub, 1985, 497–509.

81. L.R.G. Treloar. *Introduction to Polymer Science.* New York, Springer-Verlag, 1970, 61.

82. H. Levine and L. Slade. A food polymer science approach to the practice of cryostabilization technology. Comments. *Agric. Food Chem.,* 1, 315–396, 1989.

83. L. Slade, H. Levine, J. Ievolella, and M. Wang. The glassy state phenomenon in applications for the food industry: Application of the food polymer science approach to structure–function relationships of sucrose in cookie and cracker systems. *J. Sci. Food Agric.,* 63, 133–176, 1993.

84. L. Slade and H. Levine. Glass transitions and water–food structure interactions. *Adv. Food Nutr. Res.,* 38, 103–269, 1995.

85. Y. Roos and M. Karel. Water and molecular weight effects on glass transitions in amorphous carbohydrates and carbohydrate solutions. *J. Food Sci.,* 56, 1676–1681, 1991.

86. Y.H. Roos. *Phase Transitions in Foods*. San Diego, CA: Academic Press, 1995, 360.

87. M. Gorden and J.S. Taylor. Ideal copolymers and second-order transitions of synthetic rubers. I. Non-crystalline copolymers. *J. Appl. Chem.*, 2, 493–500, 1952.

88. P.A. Carvajal, G.A. MacDonald, and T.C. Lanier. Cryostabilization mechanism of fish muscle proteins by maltodextrins. *Cryobiology*, 38, 16–26, 1999.

89. N.C. Brake and O.R. Fennema. Glass transition values of muscle tissue. *J. Food Sci.*, 64, 10–15, 1999.

90. J.M. Wasylyk, A.R. Tice, and J.G. Baust. Partial glass formation: a novel mechanism of insect cryoprotection. *Cryobiology*, 25, 451–458, 1988.

91. A.G. Hirsh. Vitrification in plants as a natural form of cryoprotection. *Cryobiology*, 24, 214–228, 1987.

92. G.M. Fahy, D.R. MacFarlane, C.A. Angell, and H.T. Meryman. Vitrification as an approach to cryopreservation. *Cryobiology*, 21, 407–426, 1984.

93. G.M. Fahy, D.I. Levy, and S.E. Ali. Some emerging principles underlying the physical properties, biological actions, and utility of vitrification solutions. *Cryobiology*, 24, 196–213, 1987.

94. D.R. MacFarlane. Physical aspects of vitrification in aqueous solutions. *Cryobiology*, 24, 181–195, 1987.

95. H. Ishimori. Vitrification of mouse and bovine embryos using a mixture of ethylene glycol and dimethyl sulfoxide. *Snow Brand R&D Report*, 106, 115–152, 1996.

96. M. Kresin, I. Herschel, and G. Rau. Expectations and limitations of vitrification as a method for cryopreservation. *Cryobiology*, 30, 573–, 1994.

97. G.A. MacDonald, J. Lelievre, and N.D. Wilson. Strength of gels prepared from washed and unwashed minces of hoki (*Macruronus novaezelandiae*) stored in ice. *J. Food Sci.*, 55, 976–978 & 982, 1990.

98. W.J. Dyer and J. Peters. Factors influencing quality changes during frozen storage and distribution of frozen products, including glazing, coating and packaging, In R. Kreuzer, Ed. *Freezing and Irradiation of Fish,* London, U.K.: Fishing News (Books) Ltd., 1969, 317–322.

99. Y. Fukuda, Z. Tarakita, and K. Arai. Effect of freshness of chub mackerel on the freeze denaturation of myofibrillar protein. *Bull. Japan Soc. Sci. Fish,* 50, 845–852, 1984.

100. R.M. Love and J.I.M. Ironside. Studies on protein denaturation in frozen fish. II. Preliminary freezing experiments. *J. Sci. Food Agric.,* 9, 604–608, 1958.

101. R.M. Love. Studies on protein denaturation in frozen fish. III. The mechanism and site of denaturation at low temperatures. *J. Sci. Food Agric.,* 9, 609–617, 1958.

102. J.R. Wagner and M.C. Anon. Effect of freezing rate on the denaturation of myofibrillar proteins. *J. Food Technol.,* 20, 735–744, 1985.

103. T.A. Gill, R.A. Keith, and B. Smith-Lall. Textural deterioration of red hake and haddock muscle in frozen storage as related to chemical parameters and changes in myofibrillar proteins. *J. Food Sci.,* 44, 661, 1979.

104. R.E. Simpson, E. Kolbe, G.A. MacDonald, T.C. Lanier, and M.T. Morrissey. Frozen stabilized mince as a source of Pacific whiting surimi. In *Developments in Food Engineering. Proceedings of the 6th International Conference on Engineering and Food.* Tokyo: Blackie Academic & Professional, 1994, 1041–1043.

105. R.E. Simpson, M.T. Morrissey, E. Kolbe, T.C. Lanier, and G.A. MacDonald. Effects of varying sucrose concentrations in Pacific whiting (*Merluccius productus*) stabilized mince used for surimi production. *J. Aquat. Food Prod. Technol.,* 3, 41–52, 1994.

106. S. Ohata, A. Hiyoshi, A. Wakaizumi, T. Asari, and K. Suminoe. Studies of lipid oxidation of freeze dried food during storage. I. Effect of porosity on lipid oxidation of freeze dried food during storage. *Nippon Shokuhin Kogyo Gakkaishi,* 14, 241–243, 1967.

107. Y. Matsuda. The methods of preparation and preservation of fish meat powder having kamaboko-forming ability. *Bull. Japan Soc. Sci. Fish,*, 49, 1293–1295, 1983.

108. M.J. Hageman. The role of moisture in protein stability. *Drug Dev. Ind. Pharm.*, 14, 2047–2070, 1988.

109. L.N. Bell, M.J. Hageman, and J.M. Bauer. Impact of moisture on thermally induced denaturation and decomposition of lyophilized bovine somatotropin. *Biopolymers*, 35, 201–209, 1995.

110. B. Yoo and C.M. Lee. Thermoprotective effect of sorbitol on proteins during dehydration. *J. Agric. Food Chem.*, 41, 190–192, 1993.

111. K. Kanna, T. Tanaka, K. Kakuda, and T. Shimizu. Denaturation of fish proteins by drying, III. Protein denaturation and histological changes in dehydrated fish muscle. *Bull. Tokai Reg. Fish Res. Lab.*, 68, 51–60, 1971.

112. J. Reynolds, J.W. Park, and Y.J. Choi. Physicochemical properties of Pacific whiting surimi as affected by various freezing and storage conditions. *J. Food Sci.*, 67, 2072–2078, 2002.

113. N. Huda, A. Abdullah, and A.S. Babji. Functional properties of surimi powder from three Malaysian marine fish. *Int. J. Food Sci. Technol.*, 36, 401–406, 2001.

114. H. Niki and S. Igarashi. Some factors in the production of active fish-protein powder. *Bull. Japan Soc. Sci. Fish*, 48, 1133–1137, 1982.

115. H. Niki, Y. Matsuda, and T. Suzuki. Dried forms of surimi, In T.C. Lanier and C.M. Lee, Eds. *Surimi Technology*. New York: Marcel Dekker, 1992, 209–243.

116. H. Constantino, K. Griebenow, P. Mishra, R. Langer, and A. Klibanov. Fourier transform infrared spectroscopic investigation of protein stability in the lyophilized form. *Biochim. Biophys. Acta*, 1253, 69–74, 1995.

117. F.A. Hoekstra, L.M. Crowe, and J.H. Crowe. Differential desiccation sensitivity of corn and pennisetum pollen linked to their sucrose contents. *Plant Cell Environ.*, 12, 83–91, 1989.

118. R. Suau, A. Cuevas, V. Valpuesta, and M. Reid. Arbutin and sucrose in the leaves of the resurrection plant *myrothamnusflabellifolia*. *Phytochemistry*, 30, 2555–2556, 1991.

119. K. Madin and J.H. Crowe. Anhydrobiosis in nematodes — Carbohydrate and lipid metabolism during dehydration. *J. Exp. Zool.*, 193, 335–342, 1975.

120. J.S. Clegg. The origin of trehalose and its significance during the formation of encysted dormant embryos of Artemia salina. *Comp. Biochem. Physiol.*, 14, 135–143, 1965.

121. K.L. Koster and A.C. Leopold. Sugars and dessication tolerance in seeds. *Plant Physiol.*, 88, 829–832, 1988.

122. S. Tzannis and S. Prestrelski. Activity-stability considerations of trypsinogen during spray drying: Effects of sucrose. *J. Pharm. Sci.*, 88, 351–359, 1999.

123. D.P. Miller, R.E. Anderson, and J.J. dePablo. Stabilization of lactate dehydrogenase following freeze-thawing and vacuum-drying in the presence of trehalose and borate. *Pharmaceut. Res.*, 15, 1215–1221, 1998.

124. J.H. Crowe, J.F. Carpenter, and L.M. Crowe. Are freezing and dehydration similar stress vectors — A comparison of modes of interaction of stabilizing solutes with biomolecules. *Cryobiology*, 27, 219–231, 1990.

125. L.M. Crowe, D.S. Reid, and J.H. Crowe. Is trehalose special for preserving dry biomaterials? *Biophys. J.*, 71, 2087–2093, 1996.

126. S.B. Leslie, E. Iraeli, B. Lighthart, J.H. Crowe, and L.M. Crowe. Trehalose and sucrose protect both membranes and proteins in intact bacteria during drying. *Apl. Environ. Microbiol.*, 61, 3592–3597, 1995.

127. J.H. Crowe. Anhydrobiosis: an unsolved problem. *Am. Naturalist*, 105, 563–574, 1971.

128. J.F. Carpenter and J.H. Crowe. Modes of stabilization of a protein by organic solutes during desiccation. *Cryobiology*, 25, 459–470, 1988.

129. J.H. Crowe and L.M. Crowe. Role of vitrification in stabilization of dry liposomes. *Biophys. J.*, 64, 219–231, 1993.

130. J.H. Crowe, L.M. Crowe, and J.F. Carpenter. Interactions of sugars with membranes. *Biochim. Biophys. Acta*, 947, 367–384, 1988.

131. J.H. Crowe and L.M. Crowe, and D. Chapman. Infrared spectroscopic studies on interactions of water and carbohydrates with a biological membrane. *Arch. Biochem. Biophys.*, 232, 400–407, 1984.

132. B. Aldous, A.D. Auffret, and F. Franks. The crystallization of hydrates from amorphous carbohydrates. *Cryo-Lett.*, 16, 181–186, 1995.

133. H. Levine and L. Slade. Another view of trehalose for drying and stabilizing biological materials. *BioPharm*, 5, 36–40, 1992.

134. J. Green and C. Angell. Phase relations and vitrification in saccharide-water solutions and the trehalose anomaly. *J. Phys. Chem.*, 93, 2880–2882, 1989.

135. S. Ding, J. Fan, J. Green, Q. Lu, E. Sanchez, and C. Angell. Vitrification of trehalose by water loss from its crystalline dehydrate. *J. Therm. Anal.*, 47, 1391–1405, 1996.

136. W.Q. Sun, T.C. Irving, and A.C. Leopold. The role of sugar, vitrification and membrane phase-transition in seed desiccation tolerance. *Physiol. Plantarum*, 90, 621–628, 1994.

137. W.Q. Sun. Glassy state, seed storage stability: the WLF kinetics of seed viability loss at T>Tg and the plasticization effect of water on storage stability. *Ann. Bot. London*, 79, 291–297, 1997.

138. M. Sola-Penna and J.R. Meyer-Fernandes. Stabilization against thermal inactivation promoted by sugars on enzyme structure and function: why is trehalose more effective than other sugars? *Arch. Biochem. Biophys.*, 360, 10–14, 1998.

139. G.F. Xie and S.N. Timasheff. The thermodynamic mechanism of protein stabilization by trehalose. *Biophys. Chem.*, 64, 25–43, 1997.

140. I. Matsumoto, T. Nakakuki, Y. Ito, and K. Arai. Preventative effect of various sugars against denaturation of carp myofibrillar protein caused by freeze-drying. *Nippon Suisan Gakkaishi*, 58, 1913–1918, 1992.

141. R.E. Feeney and Y. Yeh. Antifreeze proteins: Current status and possible food uses. *Trends Food Sci. Technol.,* 9, 102–106, 1998.

142. F. Franks, J. Darlington, T. Schenz, S.F. Mathias, L. Slade, and H. Levine. Antifreeze activity of Antarctic fish glycoprotein and a synthetic polymer. *Nature,* 325, 146–147, 1987.

143. A. Mizuno, M. Mitsuiki, S. Toba, and M. Motoki. Antifreeze activities of various food components. *J. Agric. Food Chem.,* 45, 14–18, 1997.

144. S.R. Payne and O.A. Young. Effects of pre-slaughter administration of antifreeze proteins on frozen meat quality. *Meat Sci.,* 41, 147–155, 1995.

145. H. Chao, P.L. Davies, and J.F. Carpenter. Effects of antifreeze proteins on red blood cell survival during cryopreservation. *J. Exp. Biol.,* 199, 2071—2076, 1996.

146. L. Leistner. Hurdle effect and energy saving. In W.K. Downey, Ed. *Food Quality and Nutrition.* Essex, U.K.: Elsevier, 1978, 553–557.

6

Proteolytic Enzymes and Control in Surimi

YEUNG JOON CHOI, PH.D.
Gyeongsang National University, Tongyeong, Korea

IK-SOON KANG, PH.D.
Oscar Meyer/Kraft Foods, Madison, Wisconsin

TYRE C. LANIER, PH.D.
North Carolina State University, Raleigh, North Carolina

CONTENTS

6.1 Introduction 228
6.2 Classification of Proteolytic Enzymes 231
    6.2.1 Acid Proteases (Lysosomal Cathepsins) 232
        6.2.1.1 Cathepsin A 233

        6.2.1.2 Cathepsin B...234
        6.2.1.3 Cathepsin C (Dipeptidyl
                 Transferase) 235
        6.2.1.4 Cathepsin D 236
        6.2.1.5 Cathepsin E..237
        6.2.1.6 Cathepsin H 237
        6.2.1.7 Cathepsin L.......................................238
     6.2.2 Neutral and Ca^{2+}-Activated Proteinases 241
     6.2.3 Alkaline Proteinases 245
6.3 Sarcoplasmic vs. Myofibrillar Proteinases................. 249
6.4 Control of Heat-Stable Fish Proteinases................... 250
     6.4.1 Proteinase Inhibitors.. 250
             6.4.1.1 Inhibitors of Serine Proteinases........ 251
             6.4.1.2 Inhibitors of Cysteine Proteinases 252
             6.4.1.3 Inhibitors of Metalloproteinases........ 253
     6.4.2 Food-Grade Proteinase Inhibitors 253
     6.4.3 Minimization of Proteolysis by Process
             Control... 259
6.5 Summary.. 260
References 260

6.1 INTRODUCTION

The freshness of the fish from which surimi is made affects the textural quality of the surimi gel (surimi seafood). The freshness factor primarily affects the folded state of the long-chain protein polymer. As the saying "garbage in, garbage out" suggests, the inherent gelling quality of the raw fish muscle proteins cannot be improved by surimi processing. Therefore, denaturation (unfolding) of the protein prior to the time of surimi seafood processing impairs the gel-forming ability of the proteins during subsequent heating of salted surimi paste (see Chapter 10).

Another important factor in gelling quality, which may or may not be directly linked with freshness, is the integrity (intactness) of the muscle protein polymers, particularly of myosin or actomyosin. Siebert[1] pointed out that fish muscle typically exhibits protein autolysis (cleavage by endogenous proteolytic enzymes, "proteinases") at a rate 10 times that of mammalian muscle. In addition, various types of proteinases are found among fish species, which are optimally active from acid to alkaline conditions.[2]

In the living animal, muscle proteins are degraded by endogenous proteinases even as new proteins are being synthesized.[3,4] Post-mortem protein degradation in fish muscle is due to the continued activity of the inherent proteolytic enzymes, without the counter-action of protein synthesis. The physiological changes occurring upon fish death also promote muscle autolysis because the muscle can no longer maintain homeostasis of temperature, pH, salt concentrations, and cellular integrity required for enzyme compartmentalization. It is, therefore, desirable to reduce or restrict the proteolytic activity in muscle as early as possible after harvest to preserve the muscle protein integrity.

Softening of fish muscle can be induced by cooking, and in such cases is characterized by rapid and severe degradation of myosin. Several heat-stable proteinases have been shown to contribute to the thermal softening of fish muscle, that is, lysosomal proteinases (primarily cathepsins), alkaline proteinase, calpain, and collagenase.[5] Likewise, extremely high proteolytic activity is detected during the relatively slow heating processes used in the manufacture of certain large-diameter or relatively thick surimi gels or surimi seafoods,[6–10] most notably those manufactured from Pacific whiting, arrowtooth flounder, Atlantic menhaden, and white croaker.

Shimizu et al.[11] classified 49 different fish species, including saltwater, freshwater, and cartilaginous fish, plus squids, prawns, and land animal meats (chicken and rabbit) into four groups based on the effects of the heat-induced proteolysis of the muscle, which results in gel softening, and endogenous transglutaminase activity, which strengthens gels by cross-linking proteins during low-temperature "setting" (see Chapter 10).

1. Those that exhibit little gel strengthening during "setting" (30 to 50°C) and little gel softening during heating (50 to 70°C); this group included sharks,

needlefish, marlin, chicken, and rabbit (all land animal meats).
2. Those that exhibit little gel strengthening during "setting" (30 to 50°C), but considerable gel softening during heating (50 to 70°C); this group consisted primarily of certain red-meat (pelagic) fish.
3. Those that exhibit considerable gel strengthening during "setting" (30 to 50°C) and also considerable gel softening during heating (50 to 70°C); this group included sardines, croakers, and some warm-water fish such as threadfin bream.
4. Those that exhibit considerable gel strengthening during "setting" (30 to 50°C) and little gel softening during heating (50 to 70°C); this group included flying fish, barracuda, and grub fish.

It is apparent that the protein gel/polymer building forces of heat denaturation and/or enzymic-mediated crosslinking by transglutaminase would be thwarted to some degree by the presence of heat-stable proteinases, especially if heat processing involves significant dwell time (often only minutes are needed) of the product in the critical 50 to 70°C range wherein heat-stable proteinases exert the greater part of their degrading action.

The price of surimi is intimately linked with its gelling ability. Therefore, the presence of endogenous heat-stable or heat-activated proteinases, which can degrade myosin and thus impair protein gelation, is an important economic issue in the industry. Equally important to surimi gelling ability is avoiding denaturation of myosin during surimi processing. This has mandated the use of very fresh fish and cold handling and processing temperatures for fish intended for surimi processing. These conditions minimize the post-mortem proteolysis that often occurs immediately post-harvest in whole fish muscle handled less stringently.[12] Thus, it is the heat-stable/heat-activated proteinases (those most active at 50 to 70°C) that pose the greatest practical problem in the surimi industry.

These heat-stable proteinases may potentially arise in muscle from a number of sources: microbial contamination, leaching of digestive juices into the muscle during extended holding of whole fish, contamination of muscle by bits of organ tissues due to improper cleaning techniques, muscle reaction to or contamination by parasitic organisms, and abnormally high levels of such enzymes naturally occurring in the muscle. The rapid chilling and handling/processing typical to surimi processing, however, minimize the involvement of microbial enzymes, such that only tissue enzymes are of practical concern to surimi gelling properties.

The remainder of this chapter reviews the types of proteinases implicated in the impairment of heat-induced surimi gelation and possible means of controlling the action of these enzymes to ensure the gel quality of surimi seafood.

6.2 CLASSIFICATION OF PROTEOLYTIC ENZYMES

Approximately 144 proteolytic enzymes of animal tissues have been classified.[13] These can be divided into two main families: (1) exopeptidases (proteinases), which are restricted to terminal peptide linkages, and (2) endopeptidases, which are not restricted. In surimi, the endopeptidases have the more serious effect because internal cleavage reduces the size of the protein polymer, and thus its gel-forming ability decreases dramatically (Figure 6.1). The International Union of Biochemistry (IUB) Nomenclature Committee[14] further subdivided these two families into classes according to the difference in the catalytic mechanism within endopeptidases and their similar activity on substrates in exopeptidases. Thus, fish endopeptidases can also be classified into four main subgroups of serine, cysteine, aspartic, and metalloproteinases, according to the chemical group of their active site (although histidine-linked proteinases have also been reported).[15] Only serine and cysteine (also termed "thiol" and "sulfhydryl") proteinases appear to be involved in surimi gel degradation during heating.

From a practical standpoint, fish muscle proteinases can also be grouped according to the optimum pH of their activity on muscle proteins, such as acid, neutral, and alkaline pro-

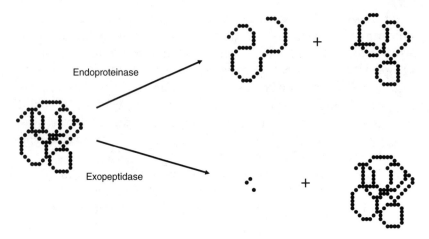

Figure 6.1 Action of endoproteinases and exopeptidases on protein structure. (Adapted from Reference 184.)

teases. The following discusses the proteinases most important in fish muscle within each of these three groups.

6.2.1 Acid Proteases (Lysosomal Cathepsins)

Lysosomal cathepsins are responsible for the intracellular protein degradation that is physiologically and pathologically important in living tissue.[16,17] Wasson[5] noted abundant literature citations attesting to the presence of acid proteinases in fish, but concluded there was little work supporting the involvement of acid proteinases in surimi gel degradation. In the time since that review was published, however, considerable evidence has amassed supporting the role of particular heat-stable cathepsins in species important to surimi production, such as Pacific whiting.[18]

Of the 13 lysosomal proteinases known to exist, 8 have thus far been isolated in skeletal muscle cells, these being A, B_1, B_2 (lysosomal carboxypeptidase B), C, D, E, H, and L (Table 6.1). Of these, cathepsin L is implicated in surimi gel degradation during cooking.[18]

Siebert[1] reported the catheptic activity of muscles from various fishes, including cod, herring, sole, flounder, trout, and

TABLE 6.1 Properties of Some Lysosomal Proteinases, Cathepsins A–L, Found in Muscle

Enzymes	Molecular Weight	Functional Group	Optimal pH	Target Proteins (Ref.)
A (endo-, exo-)	100 kDa	-OH	5.0–5.2	Less effect on intact proteins (53, 181, 182)
B1 (endo-, exo-)	25 kDa	-SH	5.0	Myosin, actin, collagen
B2 (exo-)	47–52 kDa	-SH	5.5–6.0	Broad specificity (33, 35, 53)
C (exo-)	200 kDa	-SH	5–6	Less effect on intact protein (39, 42, 183)
D (endo-)	42 kDa	-COOH	3.0–4.5	Myosin, actin, titin, nebulin, M- and C-proteins (45, 46, 47)
E	90–100 kDa	-COOH	2–3.5	Less effect on intact protein (55–57)
H (endo-, amino-)	28 kDa	-SH	5.0	Actin, myosin (16, 20, 60)
L (endo-)	24 kDa	-SH	3.0–6.5	Actin, myosin, collagen, α-actinin, troponin-T, -I (16, 66, 68)

carp. In carp muscle, three cathepsins were subsequently isolated and denoted as A, B, and C.[19] During spawning migration, salmon exhibited 3 to 7 times higher activity of cathepsins B, D, H, and L as compared to those measured during feeding migration.[20] Activities of cathepsins L, B, H, G, and D have been detected in the muscles of both dark-fleshed fatty fishes (anchovy and gizzard-shad) and white-fleshed lean fishes (sea bass and sole).[21]

6.2.1.1 Cathepsin A

Various types of proteinases act synergistically for the purpose of physiological digestion *in vivo*. This synergistic action might be expected to continue in the muscle after death. Carp

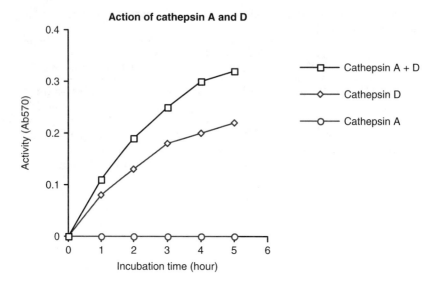

Figure 6.2 Action of cathepsin A, D and A+D on hemoglobin at pH 4. (Adapted from Reference 22.)

cathepsin A, when added simultaneously with cathepsin D, increased the hydrolytic activity of the latter enzyme (Figure 6.2). Makinodan et al.[22] indicated that cathepsin A further hydrolyzed the peptide products released by cathepsin D.

6.2.1.2 Cathepsin B

Cathepsin B is the best known and most thoroughly investigated lysosomal thiol proteinase. It was originally isolated from liver lysosome[23] and was later recognized as being comprised of two molecular components: B′ (or cathepsin B_1) and cathepsin B_2.[24–26] Cathepsin B_1 is a thiol endopeptidase[27] having a molecular weight between 24 and 28 kDa with a pI of 5.0 to 5.2.[28,29] It has maximum activity at pH 6.0 and is unstable above pH 7.0.[30] In contrast, cathepsin B_2 has a molecular weight of 47 to 52 kDa,[30] hydrolyzes Bz-Gly-Arg at pH 5.5 to 6.0,[31] and shows amidase activity at pH 5.6.[32] In addition, cathepsin B_1 was reported to hydrolyze four different synthetic substrates (BAA, BAPA, BANA, and BAEE), while cathepsin B_2 hydrolyzed only N-α-benzoly-L-argininamide (BAA).[26]

Cathepsin B purified from carp,[19] grey mullet,[33] tilapia,[34] and mackerel[35] evidenced molecular weights of 23 to 29 kDa and an optimal pH range of 5.5 to 6.0. For activation, the enzyme required thiol compounds, such as 2-mercaptoethanol, cysteine, dithiothreitol, and glutathione, as well as metal chelating reagents (EDTA, EGTA, or citric acid). Among these, cysteine was the most effective, while 2-mercaptoethanol was the least effective. Cathepsin B was greatly inhibited by iodoacetic acid, iodoacetamide, TLCK, and TPCK, and moderately inhibited by pCMB and NEM. Cathepsin B prepared from different sources showed different proteolytic character on various substrates. Carp cathepsin B degraded muscle myosin heavy chain, actin, and troponin-T(TN-T) but was inactive on tropomyosin.[36]

Most of the autolytic activity occurring at pH 6.5 was attributed to cathepsin B and/or L in salmon caught during spawning migration.[20] Cathepsin B activity in Pacific whiting fillets was 170% that of cathepsin L at 55°C.[18] However, whiting cathepsin B was substantially removed during the surimi washing process, probably because the enzyme exists in the sarcoplasmic fluid of the fish.[37] Furthermore, cathepsins B along with D, H, and L have been known to be major proteinases in post-mortem muscle degradation.[16,38]

6.2.1.3 Cathepsin C (Dipeptidyl Transferase)

Cathepsin C is unlikely to act on intact proteins directly, but has the highest specific activity among all the lysosomal peptidases, suggesting that it further digests the resulting peptide fragments from the action of cathepsin D.[39,40] Substrates most susceptible to attack by this enzyme are dipeptidyl amides or esters, bearing a free α-amino (or α-imino) group in the NH_3-terminal position. Most of the compounds found to possess substrate activity have glycyl, L-alanyl, or L-seryl residues as the NH_2-terminus. If either leucine or lysine is present in this position, then the substrate activity is lost.[41]

Carp cathepsin C was reported to be stable at 60°C for 20 min, while carp cathepsin B was inactivated completely.[19] Hameed and Haard[42] found that cathepsin C (25 kDa) from Atlantic squid was Cl- and sulfhydryl dependent, being inhibited by sulfhydryl enzyme inhibitors such as iodoacetate, PCMB, and $HgCl_2$, but not by EDTA (which inhibits metalloproteases), PMSF (inhibits serine proteases), pepstatin A (inhibits acid proteases), or puromycin (inhibits amidopeptidases). From the sarcoplasmic fluid in Pacific whiting, cathepsin C showed optimum activity at pH 7.0, which was greater than the optimum activity at pH 6.0 in true cod.[37]

6.2.1.4 Cathepsin D

Cathepsin D was first detected in fish skeletal muscle by Siebert[1] and later identified as cathepsin D by Mekinodan and Ikeda.[43] It is believed to play a significant role in texture degradation during chilled storage in post-mortem muscle.[44,45] Huang and Tappel[39] reported that cathepsin D is one of the most important cathepsins in post-mortem tenderization because it directly attacks intact muscle proteins to produce peptides that can be further broken down by other cathepsins. This enzyme is an aspartyl-type proteinase, having several isomers with pIs 5.7 to 6.8, optimum pH 3.0 to 4.5, and a broad range of activity on muscle substrates.[46,47]

Cathepsin D is strongly inhibited by pepstatin, a specific inhibitor of carboxyl proteinases, while thiol proteinase inhibitors have negligible effects on enzyme activity.[33] Cathepsin D rapidly degrades titin, connectin, C-protein, M-protein, and myosin (both heavy and light chains),[47–49] and slowly breaks down actin, troponinT/I, and tropomyosin.[46,50]

Despite the fact that cathepsin D has strong activity as well as synergistic action (Figure 6.2) with cathepsins A, B, or C in muscle protein degradation,[51] its lower activity at near-neutral pH and generally low heat stability raise doubts as to its ultimate importance in the heat-activated proteolysis of surimi gels during cooking.[5] Makinodan et al.[52] reported that carp cathepsin D hydrolyzed myofibrils optimally at pH 3 to 4, but not above pH 6.0. Cathepsin D from grey mullet had maximum activity at pH 4.0 and was stable up to 45°C, with progressive activity loss by 34 and 100% at 50°C and 70°C, respectively.[33] Chicken cathepsin D was also stable when stored at −14°C for months or heated at 60°C, but was completely inactivated at 70°C. Jiang et al.[53] suggested the participation of cathepsin D in the post-mortem degradation of fish muscle, based on increased myofibril fragmentation when incubated with tilapia cathepsin D at pH 6.0 and 6.5.

6.2.1.5 Cathepsin E

Lapresle and Webb[54] purified an aspartic proteinase in a direct staining method from rabbit spleen that was distinct from cathepsin D and named it cathepsin E. No cathepsin E from skeletal muscle was reported until 1978 when Venugopal and Bailey[55] found it in bovine and porcine diaphragm muscle. Cathepsin E from salmon also has the peculiar property of exhibiting a very acidic optimum at pH 2.8.[56] Like cathepsin D, cathepsin E is a COOH-group-dependent endopeptidase. They are both inhibited by pepstatin and diazoacetylnorleucine methyl ester with Cu^{2+},[57,58] but cathepsin E is distinguished by the inhibitor from *Ascaris lumbricoides*, which does not inhibit cathepsin D.[59] Bovine cathepsin E activity was diminished considerably when frozen for 2 weeks.[55] This low stability and relatively acid optimum suggest that it is not important to surimi gelation.

6.2.1.6 Cathepsin H

Cathepsin H was named by Kirschke,[60,61] who isolated it from rat liver. The enzyme shows both thiol endopeptidase and aminopeptidase activities with a molecular weight of 28 kDa. It has two multiple forms showing a maximum activity at pH 7.[62] Stauber and Ong[63] demonstrated its presence in skeletal muscle. Cathepsin H was reported to break down myosin 2 to 3 times faster than cathepsin B.[44,47] In white muscle of spawning migration salmon, cathepsin H activity was 3 times higher than fish in feeding migration.[20] In Pacific whiting fillets, cathepsin H along with B and L were easily detectable cathepsin proteinases.[18] Cathepsin H showed maximum activity at 20°C, which was 75% of cathepsin L at 55°C. However, its activity was not detected after extensive washing during surimi production.

6.2.1.7 Cathepsin L

Okitani et al.[64] purified skeletal muscle cathepsin L (24 kDa) first from rabbit, demonstrating that it was optimally active on myosin at pH 4.1. Taylor et al.[65] identified cathepsin L in skeletal muscle by an immunohistochemical method. It degrades myosin heavy chain, actin, α-actinin, troponin T, and I.[66] Cathepsin L has been estimated to have 10 times greater activity per protease molecule against myosin than cathepsin B.[44,47] The enzyme has several multiple forms (pI 5.8 to 6.1) and a wide range of pH (3.0 to 6.5) for activity. It is strongly inhibited by iodoacetate, leupeptin, and antipain, but not by pepstatin or phenylmethane sulfonyl fluoride.[67] From mackerel white muscle, Ueno et al.[68] purified a pepstatin insensitive cathepsin L-like protease that was inhibited by leupeptin and antipain, and reactivated with the addition of sulfhydryl reagents. Yamashita and Konagaya[69] reported that the enzyme most responsible for post-mortem softening of salmon meat during spawning migration was cathepsin L. In continued research,[56] they found that about 80% of the autolytic activity at low pH from 3 to 5 was due to cathepsin L and the remaining was attributed to cathepsins D and E.

At pH 6.0 and 50°C, a cathepsin L-like enzyme from anchovy exhibited its maximal activity on casein and N-benzoyl-D, L-arginine-β-naphthylamide, and hydrolyzed at the position of Phe_1, Asn_1, Val_{13}, Glu_{14}, Val_{19}, and Gly_{24} of the insulin β-chain.[70] Cathepsin L was also identified as the predominant proteinase responsible for autolysis of arrowtooth flounder muscle at elevated temperature.[71] About 77% of the total cathepsins B+L+L-like activities in minced, leached, and NaCl-ground from mackerel was maintained even after 8 weeks' storage at –20°C.[72]

Cathepsin L from both fillets and surimi of Pacific whiting has the highest activity at 55°C, and thus can degrade gel texture during conventional cooking of surimi seafood.[18]

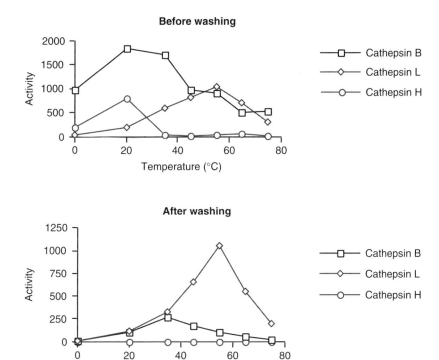

Figure 6.3 Activities of cathepsins (B, H, L) in fish mince before and after washing of the meat. Activities are expressed as relative fluorescence intensity. (Adapted from Reference 184.)

Among the highly active lysosomal cathepsins (B, H, and L) in whiting fillets, L was the predominant cathepsin in surimi, while B was highest in the intact fillets. This indicates that cathepsin L remained after the extended washing process used for surimi manufacture, while the others were leached away from the myofibrillar proteins (Figure 6.3). Cathepsin L purified from Pacific whiting consists of a single peptide with a molecular weight of 28.8 kDa and a pH optimum near 5.5, although it readily degrades the myofibrillar proteins at near-neutral pH.[73]

Cathepsin activity showed considerable difference, depending on processing. The specific activity of cathepsin B

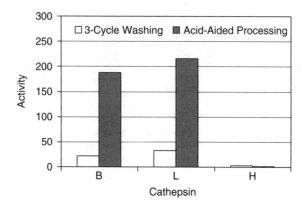

Figure 6.4 Cathepsin activities of recovered protein from Pacific whiting after 3-cycle washing and acid-aided processing. (Adapted from Reference 74.)

and L from recovery protein after three-cycle washing was lower than the activity of those from acid-aided (a new process using protein charges and isolation) recovery protein (Figure 6.4). This study indicates that acid-aided solubilization might have enhanced the cathepsin activity, particularly for Pacific whiting. Cathepsin H was completely removed after three-cycle washing and with acid-aided processing.[74] Fish proteins treated with acid and alkaline pH showed the highest activities of cathepsin L-like enzymes. Dramatic reduction in activity was distinctively observed at pH 11 and 12. Cathepsin B-like enzymes appeared to be highly activated at the acid treatment. However, the alkaline process removed them dramatically.[75]

The high levels of cathepsin L, thought to be responsible for the great propensity of surimi from Pacific whiting to autolytically degrade during heating, are linked to infection of the fish muscle by *Kudoa*, a species of the *Myxosporidean* parasite.[5] Similar infestation has been noted in other species that exhibit textural softening during cooking, such as barracuda, tuna, flatfish, salmon, and herring.[76] Wasson[5] discussed the relationship between parasite infestation and acidic protease activity in fish and concluded that, while many studies show a correspondence between infestation and proteolytic activity in the muscle, the issue is complex because other studies do not show this correlation. She concluded that there may be a connection to the complex and largely uncharted life cycles of the *Myxosporidean* species that makes their presence in affected tissues an indicator of high proteolytic activity, but not necessarily a measure of the individual activity levels.

6.2.2 Neutral and Ca^{2+}-Activated Proteinases

The complete disappearance of the Z-disk and weakening of the actin-myosin interaction induced by neutral protease termed "calpain" was initially detected in muscle left in a Ca^{2+} (1 mM)-containing solution.[77] Since this recognition, considerable research has been conducted into the role of calpain in the textural degradation (tenderizing) of mammalian muscle during post-mortem storage. The enzyme is a ubiquitous protease and exists in the cell cytoplasm or is associated with myofibrils, primarily in the region of the Z-disk (66% on Z-disk, 20% in I-band, and 14% in A-band.[78] Although the enzyme does not degrade actin or myosin, it causes degradation of the Z-disk, titin, troponin-T, and desmin.[28,79]

The Z-disk in pre-rigor muscle is a highly ordered structure linking thick and thin filaments mainly by titin and nebulin on one side, and connecting myofibrils to the sarcolemma by the protein structure called costameres on the other side (Figure 6.5). The constitutional proteins of costameres and those in the Z-line, such as desmin, nebulin, vinculin, and titin, are all excellent substrates for calpain.[80] As the Z-line structure degrades following death of the fish, the two major zipper actions between thick and thin filaments, and myofibril and sarcolemma, disintegrate so that the actin and myosin can be highly exposed to other readily available proteinases. While the main effect on intact fillets is unsightly gaping of the myotomes, this exposure of myofibrillar proteins to other proteinases may be the most important role of calpains to the quality of surimi.

Compared to the extensive studies on the role of mammalian calpain in mammalian muscle tenderization, research

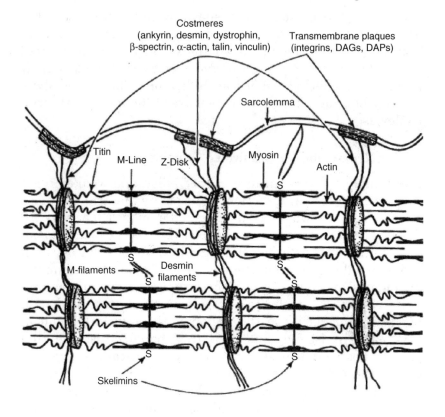

Figure 6.5 Schematic diagram showing the structure and protein compositions of costameres in striated muscle relatives to Z-disks and the myofibrillar lattice. (From Reference 80 with permission.)

on fish calpain is very limited. Makinodan and Ikeda[81] first detected neutral protease in fish muscle from carp and Red Sea bream. The activity of carp neutral protease, however, was neither activated by Ca^{2+} nor affected by 2-mercaptoethanol or iodoacetic acid, indicating that this enzyme was quite different from mammalian calpain. Thus, these authors regarded it as being a new neutral endopeptidase.[82]

True fish calpain (80 kDa), activated by Ca^{2+} and inhibited by iodoacetic acid, was purified from carp muscle by Taneda et al.[83] Calpains, thus far, have also been isolated from bass, sea-trout,[84] tilapia,[85] carp,[86] and Chinook salmon.[87]

The carp enzyme demonstrated degradation of myofibrils with the release of α-actinin, troponin-T, and troponin-I. Tsuchiya and Seki[88] reported that the major contribution of calpain to post-mortem Z-line degradation is from its ability to digest native α-actinin from myofibrils, which is involved in the anchoring of F-actin filaments to the Z-line. Tsuchiya et al.[89] demonstrated initial degradation of α-actinin in carp muscle stored at 20°C, while other myofibrillar protein bands (myosin, actin, tropomyosin, troponin-T, and troponin-I) remained unchanged. Calpain also destroyed titin, a giant protein anchored in the Z-line connected to α-actitin. Thus, the loss of the two components could collaboratively affect Z-disk degradation.[90,91]

In ordinary cells, two calpain isozymes have been isolated on the basis of their calcium requirements, calpain I (μ-calpain) and calpain II (m-calpain), which were maximally active at 50 to 70 μM and 1 to 5 mM Ca^{2+}, respectively. Both calpains from vertebrate muscles are composed of a catalytic subunit (80 kDa) and a regulatory subunit (30 kDa). Type m-calpain would seem unlikely to be involved in post-mortem changes due to its high Ca^{2+} requirement, which exceeds the levels in post-mortem muscle. Koohmaraie[92] reported that 24 to 28% of μ-calpain activity remained in the purified myofibrils incubated with calpain at pH 5.5 to 5.8 at 5°C. This level was sufficient to reproduce the changes in the myofibrils associated with post-mortem storage.

From an immunoblotting study with specific polyclonal antibodies to fish α-actinin, Papa et al.[84] detected rapid degradation of α-actinin in bass muscle stored at 4°C, indicating that calpain may be involved in muscle degradation during the early stages of fish storage. Wang et al.[93,94] reported that about 50% of calpain activity at pH 5.5 remained when fish muscle was incubated at 25°C for 58 min, also indicating initial involvement of the enzyme in fish texture degradation. Calpastain (300 kDa), a specific inhibitor to calpain, was identified from carp muscle that inhibited caseinolytic activity of the carp calpain II.[95] The calpastatin showed a unique property of heat stability (Figure 6.6) by maintaining its full activity against calpain II after heating at 100°C for

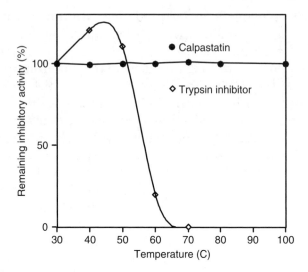

Figure 6.6 Heat stability of calpastatin and the trypsin inhibitor from carp muscle. (Adapted from Reference 95.)

5 min. Similar calpastatin activity was also detected in mammalian muscle.[96]

Calpain, however, was not detected in fish muscle until recently. Therefore, the possible role of calpain in surimi has not been researched sufficiently. Zinc, however, is an inhibitor of calpain activity,[97] and Koohmaraie[98] reported that infusion of ovine carcasses with 10% of their live weight by $ZnCl_2$ (50 mM) blocked post-mortem texture degradation up to 14 days, while 10% infusion with $CaCl_2$ (0.3 M) accelerated the process to within 24 hr as opposed to 3 to 7 days.[99] Similar studies might be repeated on fish muscle to ascertain the role of calpain in surimi texture degradation.

Although calpain does not hydrolyze the major myofibrillar components of actin and myosin directly in surimi and thus would not be expected to degrade surimi gels alone, a number of studies have reported that Z-disk degradation by calpain is a major factor contributing to post-mortem degradation of myofibrillar proteins due to their further digestion by other proteinases.

6.2.3 Alkaline Proteinases

For proteolytic enzymes to be actively involved in myofibrillar proteins and gel degradation, they should meet the following three criteria suggested by Koohmaraie[92]:

1. The proteinase must be located inside skeletal muscle cells.
2. The enzyme must have the capability to induce changes in myofibrils in an *in vitro* system, which reproduces those changes normally observed postmortem.
3. The enzyme must have access to the substrate(s).

The characteristics of alkaline proteinases in skeletal muscle appear to more closely meet these requirements than do cathepsin enzymes, which are compartmentalized inside lysosomes and thus must be released in order to contact the myofibrillar substrates. Based on immunohistochemical localization studies, alkaline proteinase was detected in mast cells of skeletal muscle tissue.[100,101] Therefore, alkaline proteinases can easily interact with skeletal proteins, such as actin and myosin. More importantly, for surimi processing, many of the alkaline proteinases are heat stable and also become active at the neutral pH of fish meat paste.

Alkaline proteinase in fish muscle was reported when Makinodan et al.[102] recognized weaker elastic gel formation at 60°C than during incubation at 30 to 40°C or above 70°C (Figure 6.7). They also found that crudely purified proteinase prepared from fish muscle functioned optimally at 60°C and pH 8.0. Less firm and gummier textures were detected in gels of Atlantic croaker with concurrent degradation of myosin heavy chain and tropomyosin when slowly heated to 70°C for 2 hr.[103] In this case, the disappearance of myosin heavy chain increased with extended heating time.[10]

A unique property of alkaline proteinase from fish muscle is that its activity is not detectable below 50°C, while considerable activity is observed around 60°C.[104] Maximum activity of alkaline proteinase was found at 60 to 65°C and pH 7.7 to 8.1 with few exceptions for the light muscle of 21 species of marine and 4 species of freshwater fish.[105]

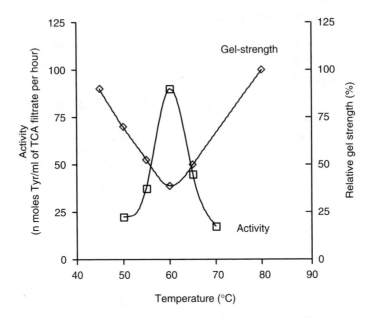

Figure 6.7 Autolytic activity of muscle homogenate at pH 6.8 with respect to gel strength curve of fish gels as affected by temperature. (Adapted from Reference 10.)

After various studies of poor gelation in croaker paste, Makinodan et al.[10] reported that, of the proteinolytic fractions studied (cathepsin D, neutral proteinase, calpain, and alkaline proteinase), only alkaline proteinase could act at the pH (6.8) and process temperature (60°C) of the meat paste. The conclusion that enzymes other than alkaline proteinase likely contribute little to heat-induced gel degradation was because cathepsin D was not active near neutral pH, while both neutral protease and calpain are heat labile.

Two different alkaline proteinases were isolated from white croaker skeletal muscle.[106–108] Both enzymes exhibited optimum activity at 60°C, but differed in pH optima (8.5 and 9.1, respectively). The former proteinase was characterized as being a trypsin-like serine enzyme, which showed a great capacity for degrading intact myofibrils *in vitro*,[109] while the latter shared several similar properties to other alkaline proteinases previously reported in Atlantic salmon,[110] cod, herring,[111] white croaker,[107] rainbow trout, sardine, mackerel,[2] Atlantic croaker,[112] carp,[113] and bigeye snapper.[114] The similar properties of these alkaline proteinases are their narrow range of pH (7.5 to 8.5), temperature optima at about 60°C with no detectable activity below 50°C or above 70°C, and relatively high molecular weight (560 to 920 kDa).

The specific activity of the carp alkaline proteinase increased about 300-fold with homogeneous isolation from the crude extract, exhibiting a molecular weight of 780 kDa.[113] This carp proteinase degraded most of the intracellular proteins, such as myofibrillar, mitchondrial, lysosomal, microsomal, and sarcoplasmic proteins.[115] Homogeneous alkaline proteinase (2000-fold) was also purified from white croaker skeletal muscle, showing an optimum pH of 8.0 with four different subunit complexes $(\alpha\beta\gamma_2\delta_2)_4$ ranging in molecular weight from 45 to 57 kDa.[104] The enzyme was stable at 50°C for 1 hr, which was attributed to this subunit complex. Hase et al.[116] reported that each subunit in fish alkaline proteinase had a particular role and only a certain subunit carried the catalytic activity. Studying the dissociation of the carp muscle alkaline protease $(\alpha\beta\gamma_2\delta_2)_4$ with 2-mercaptoethanol, pH, heat, and urea, they described the individual functions of the subunits: (1) α-subunit — for binding between subunits or between enzymes and some functional proteins; (2) β-subunit — as an inhibitor in the quaternary structure of the enzyme; (3) γ-subunit — for retaining the δ-subunit in its active conformation; and (4) δ-subunit — as the catalytic unit in proteolysis.

Boye and Lanier reported that high activity of an alkaline proteinase occurred near 60°C in Atlantic menhaden muscle. This activity was presumably endogenous as fish muscle was obtained directly from live animals with no gut contamination or parasite infestation. Presumably, the enzyme binds to the myofibrillar fraction as surimi processed from this species also displays a high heat-stable proteolytic activity.

Choi et al.[117] later purified and characterized the alkaline proteinases from Atlantic menhaden. Two fractions were purified 62.9- and 986.5-fold compared to the crude muscle extract, which exhibited molecular weights of 707 and 450 kDa, respectively. Both are probably serine-type proteinases and optimum caseinolytic activity was shown at pH 8.0 and 55°C. Because both were heat stable and active in the presence of NaCl, it is likely that these are the active proteinases involved in the degradation of surimi gels from Atlantic menhaden, although other serine proteinases were also isolated from this species.[118]

The content of alkaline proteinase in fish muscle can also be influenced by contamination from organ tissues of the fish, such as kidney and liver, due to poor cleaning techniques. These organs contain proteinase activity several-fold higher than is present in skeletal muscle.[119] Webb et al.[120] reported that hand-separated meat rather than mechanically separated fish meat resulted in a stronger heat-induced gel product when cooked. It can be assumed that hand separation involved more careful cleaning of organ tissues from the fillets. When kidney and liver tissue contaminated the paste of Atlantic cod and pollock, a rapid loss of salt-extractable protein during storage at $-5°C$ (due to inherent trimethylamine oxide demethylase activity) and degradation of fish tissue upon comminuting and cooking occurred.[112,121]

A third source of alkaline proteolytic activity may be contamination by tryptic or chymotryptic enzymes originating in the gut. Such contamination of the muscle is more often due to holding of whole fish too long prior to processing and/or the use of fish that are actively feeding at harvest.[122]

In threadfin bream, Toyohara et al.[123] reported that a heat-stable serine proteinase, termed the gel-degradation inducing factor (GIF), caused myosin heavy chain degradation. This GIF activity degraded myosin heavy chain at pH 7.0 in the presence of NaCl, but not in the absence of NaCl, as opposed to another heat-stable alkaline proteinase isolated from the same material that was inactivated by NaCl addition. Considering that addition of NaCl is essential in the manufacture of surimi seafood, and that the normal pH for surimi is near 7.0, the reported GIF proteinase is likely to be actively involved in surimi gel degradation during processing and cooking. However, this GIF was detected in the sarcoplasmic fraction of muscle, which should be easily eliminated with proper washing during surimi processing.

6.3 SARCOPLASMIC VS. MYOFIBRILLAR PROTEINASES

The localization of the enzyme in fish muscle and its solubility in water can affect its role in surimi gelation. Proteinases in the sarcoplasm are easily extractable, while those in myofibrils are tightly associated and not easily leached during surimi manufacture. For this reason, the gelling properties of the minced muscle of certain species can be similar to that of surimi made from it, while for others it is not.[124,125] For example, the heat-stable proteinase in oval filefish was found to be tightly associated with the myofibrils.[126] This association was so strong that the enzyme was not eliminated even in the presence of detergents such as Triton X-100, Tween 20, Brij 35, cholic acid, and deoxycholic acid. From the sarcoplasmic fraction of the fish, the authors identified the existence of an inhibitor to this myofibrillar proteinase; evident is the fact that more extensive breakdown occurred in a washed myofibril (similar to surimi) gel compared to a gel made from minced muscle. Similarly, Shimizu et al.[127] reported that the factor responsible for texture degradation at 60°C in croaker muscle was evident in the myofibrillar fraction but not in the sarcoplasmic fraction. Conversely, the heat-stable serine proteinase in threadfin bream responsible for degrading its gels was demonstrated to be present only in the sarcoplasm.[123]

Kinoshita et al.[128] classified twelve fish species based on the extractability characteristics of the proteinase from the muscle (i.e., sarcoplasmic or myofibrillar associated), the optimum temperature (50 or 60°C) for myosin heavy chain degradation, and the sensitivity of the proteinase activity to n-butanol. Among the twelve species, six had proteinases that were easily extractable in the sarcoplasma, three possessed proteinase tightly associated with the myofibrils, and three others possessed proteinases of both types.

Pacific whiting is a good example of a species that contains proteolytic enzymes of both types. While cathepsins B, H, and L present in this species show comparable activities in fish mince, a large portion of cathepsin B and almost all of cathepsin H are removed during the washing process of surimi manufacture, while cathepsin L is not washed out, indicating that it alone is tightly associated with the myofibrils.[18]

Cao et al.[129] reported that a myofibril-bound serine proteinase (MBSP) from carp readily decomposed myosin heavy chain in myofibrils and surimi gels at 55°C and 60°C, respectively. This MBSP was thus regarded as the proteinase most probably involved in the gel weakening effect. The molecular weight of an MBSP purified from lizardfish muscle was estimated at 50 kDa and involved subunits of 28 kDa. The N-terminal amino acid sequence showed high homology to fish trypsin (64 to 77%). When the purified MBSP was stored at –35°C in the presence of 50% ethylene glycol (v/v), its activity was entirely preserved over 6 months, being stable against freezing and thawing.[130]

MBSP was also implicated in the proteolysis of tilapia surimi, attaining the highest activity at 65°C. Soybean trypsin inhibitor and leupeptin significantly inhibited its activity. The myosin heavy chain completely disappeared when incubated at 65°C for 4 hr.[131]

6.4 CONTROL OF HEAT-STABLE FISH PROTEINASES

6.4.1 Proteinase Inhibitors

Inhibitors of proteinases are often grouped on the basis of their reaction mechanism, origin, or structural similarity. An enzyme inhibitor is any substance that reduces the rate of an enzyme-catalyzed reaction. Protease inhibitors mimic the usual protein substrate by binding to the active site of the proteinase. Proteinase inhibitors can be broadly grouped into three classes based on their specificity: (1) those that react with more than one class of proteinases (active site classification), (2) those that are ostensibly specific for all proteinases within one of the classes, and (3) those that show high selectivity for a single proteinase within one of the classes.[132]

Enzyme inhibition can be either reversible or irreversible. In irreversible inhibition, the enzyme activity cannot be regained by physical means; whereas in reversible inhibition, the enzyme activity is regenerated upon removal or displacement of the inhibitory molecule.[133] Most of the known irreversible inhibitors are synthetic substances that are primarily used to analytically determine the active site class of a proteinase. Reversible inhibitors are naturally occurring proteins and usually have a measurable kinetic binding constant.[134]

Specific inhibitors are active-site-directed substances and combine with the catalytic or substrate-binding sites of the enzyme to form a stable complex. Specific protease inhibitors of natural origin are either proteins or peptides with all the typical characteristics of the substrate, but nonspecific inhibitors are rare in nature except for α-2 macroglobulin, which inhibits proteinases of all classes.[133]

6.4.1.1 Inhibitors of Serine Proteinases

The protein inhibitors of serine proteinases fall into several groups according to various schemes.[132] Ovoinhibitor is an inhibitor of serine proteinases and is similar to ovomucoid in its properties.[135] Ovoinhibitors from egg white of chicken inhibited the proteolytic enzymes bovine trypsin, bovine α-chymotrypsin, subtilisin, and *Asperigillus oryzae* alkaline proteinase. Sites for trypsin and α-chymotrypsin (or subtilisin) were independent and noncompetitive. Arginine residues were found essential for the trypsin-inhibitory activity of ovoinhibitors.[136] Guinea pig plasma contains four major trypsin inhibitors (i.e., contrapsin, α-1-antiproteinase, α-2 macroglobulin, and murinoglobulin). Contrapsin inactivated trypsin but did not significantly affect chymotrypsin, pancreatic elastase, or pancreatic kalikrein.[137]

Proteinaceous proteinase inhibitors are widely distributed in the plant kingdom. Trypsin and chymotrypsin inhibitors were identified in legume seeds, soybeans, lima beans, garden beans, and ground nuts.[138] Two prevalent trypsin inhibitors in soybeans are the Kunitz soybean inhibitor and the Bowman-Birk inhibitor. They differ markedly from each other in size, amino acid composition, structure, and biochemical properties. A large number of proteinase inhibitors have been separated and isolated from potatoes, which contain chymotrypsin inhibitor I.

6.4.1.2 Inhibitors of Cysteine Proteinases

Chicken egg white contains a cysteine proteinase inhibitor, or cystatin (MW 12.7 kDa). There is reportedly very little effect of pH (between pH 4 and 9) or ionic strength (up to 0.64) on its inhibitory activity against ficin.[139,140] This cystatin has two major forms with pI values of 6.5 and 5.6, respectively, and are referred to as A and B.[141] The two major forms are immunologically identical and neither contains any carbohydrate. Chicken egg white cystatin inhibits a number of cysteine proteinases, including ficin, papain, cathepsin B, cathepsin H, cathepsin L, and dipeptidyl peptidase I, but not clostipain or streptococcal proteinase, and it only weakly inhibits bromelain.[140]

Proteinaceous kininogens also exhibit inhibitory activities against many cysteine proteinases. L-kininogen from pig plasma (MW 55 kDa) inhibited μ- and m-calpains, B, L, and L-like cathepsins, and papain, and was stable at pH 3.0 to 10.5. It could inhibit the proteolysis of mackerel myosin heavy chain caused by a purified L-like cathepsin at 55°C.[142] There are three types of kininogens designated as high molecular weight kininogen (H-kininogen, 120 kDa), low molecular weight kininogen (L-kininogen, 68 kDa), and T-kininogen in mammalian blood plasma.[143] Both L- and H-kininogens are strong inhibitors for cathepsin B and L.[144]

Commercial pineapple stem acetone powder is a rich source of an unusually complex mixture of proteases (bromelains) and their polypeptide inhibitors. The inhibitors appear to be remarkably stable even under extreme conditions; no loss of activity was observed when they were kept at pH 3.0 for several weeks, or when kept at 90°C for 10 min at pH 7.0. These competitively inhibit papain and ficin.[145]

6.4.1.3 Inhibitors of Metalloproteinases

Ovostatin (ovomacroglobulin) is a large molecule having a tetrameric structure with a molecular weight of 780 kDa. It inhibits a wide range of endoproteinases, including thermolysin (a metal-ion-requiring proteinase) and collagenase. Its structure and mechanism of action is like that of the serum proteinase inhibitor α-2 macroglobulin.[146] Ovostatin from duck inhibits both metalloproteinases and serine proteinases, whereas the one from fowl inhibits metalloproteinases only.[135]

Calpastatin, a specific calpain inhibitor from grass prawn (*Penaeus monodon*), revealed four beef μ-calpain and two beef m-calpain binding domains, respectively. It was stable during 1 hr of incubation at 30°C between pH 4.5 and 10.0, and was shown to be a highly specific inhibitor of calpain.[147] White potato contains a carboxypeptidase inhibitor. This inhibitor is stable upon exposure to 80°C for 5 min. It is also stable after 50 min of incubation at pH 7.8 with either trypsin or chymotrypsin at room temperature.[138]

6.4.2 Food-Grade Proteinase Inhibitors

The discovery that beef plasma addition to Pacific whiting surimi would prevent the heat-induced degradation of gels during cooking[8,148] opened this fishery to the production of an acceptable surimi. In addition, this discovery initiated considerable interest in the topic of food-grade inhibitors for fish proteinases.

In nature, proteinase inhibitors are ubiquitous, being present in numerous tissues of animals, plants, and microorganisms to prevent undesired proteolysis during life. Among these, beef plasma protein (BPP), egg white, milk whey proteins, and white potato extracts have been used as food-grade inhibitors in surimi.[8,149–151] Both BPP and egg white were reported to be effective on gels from Pacific whiting,[152] arrowtooth flounder,[153] Atlantic menhaden,[8] and Alaska pollock.[153] Potato extracts[154] were first reported as useful against fish muscle (Atlantic croaker) proteinases by Lanier et al.,[151] and later were used in surimi from Pacific whiting[155] and arrowtooth flounder.[7] Whey protein concentrate, a by-product from

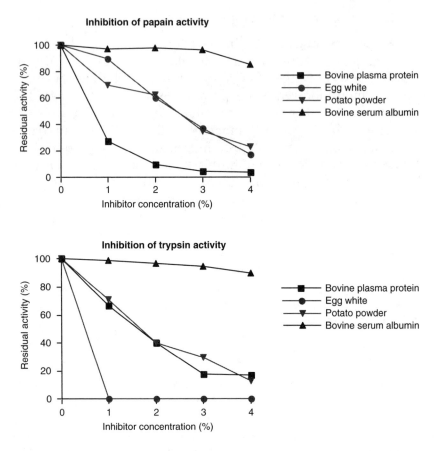

Figure 6.8 Inhibition of papain and trypsin activity by various proteinase inhibitors with BAPNA as the substrate. (Adapted from Reference 149.)

cheese manufacturing, diminished gel degradation of Pacific whiting surimi, although it was not as effective as BPP.[156] For detailed information on these protein additives, refer to Chapter 13.

From results of inhibitor activity staining (Figure 6.8), BPP showed the highest papain (cysteine proteinase) inhibitory activity while egg white was best in trypsin (serine proteinase) inhibition.[149] However, Hamann et al.[8] found BPP superior to egg white in inhibiting the heat-stable proteinase of Atlantic menhaden, which was later shown to consist of serine proteinases.[117] Generally, cysteine proteinases appear to be the most significant type responsible for gel degradation in surimi, such as Pacific whiting,[157] arrowtooth flounder,[158] white croaker,[104] Atlantic croaker,[112] and chum salmon[69] surimi. Serine proteinases, in contrast, are reported to be important in Atlantic and Gulf menhaden[117,118] and possibly in white croaker.[159]

In addition to potato extracts, other plants have been explored as possible sources for food-grade inhibitors of fish proteinases, including rice bran,[160] tomato leaves,[161] cone peppers, and others. None of these, however, including potato, have proven commercially viable. Currently, no economical food-grade inhibitor material has been found superior to BPP in the inhibition of surimi proteinases in fish species (Figure 6.9).[152,155]

BPP contains both α_2-macroglobulin, a unique, broad-spectrum inhibitor (of all four classes of proteinases) by virtue of its "entrapment" mechanism of inhibition,[162] and kininogen, a cysteine proteinase-specific inhibitor.[163] Seymour et al.[164] reported that the gel-enhancing ability of Pacific whiting surimi by BPP was due to the combined effects of proteolytic inhibition, covalent cross-linking by transglutaminase and/or α_2-macroglobulin, and gel formation by bovine serum albumin or fibrinogen.[163] In support of this combined effect by BPP, Kang and Lanier[165] found that when using a recombinant-derived cystatin (cysteine proteinase inhibitor) in Pacific whiting surimi, an inhibitory activity (using azocasein as substrate) 10 times higher than that of the 1% BPP typically used in industry was required to produce an equivalent improvement in gel-strength (Figure 6.10).

Commonly used edible inhibitor additives, however, may have unavoidable negative effects on the quality of surimi seafood, such as off-flavor and coloration from BPP,[155] sulfurous odors from egg white,[7] and off-color from potato extract.[155] Moreover, many people can have allergic reactions to the non-inhibitor components of these materials. BPP has the added disadvantage that several religions and cultures object to its use as a food ingredient, despite the fact that the

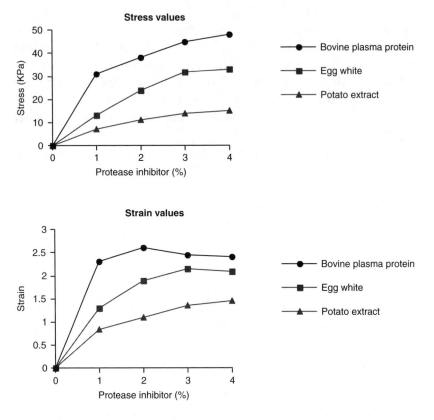

Figure 6.9 Strain and stress values of Pacific whiting surimi gels, heat-set at 60°C for 30 min followed by 90°C for 15 min, as affected by various concentrations of protease inhibitors. (Adapted from Reference 152.)

levels added to surimi are comparable or even less than those naturally present in fresh meats.

An even more serious objection has arisen in recent years involving the use of BPP in surimi. There has been a great degree of uncertainty surrounding just how the active protein ("prion") that is thought to cause bovine spongiform encephalopathy (BSE, or mad cow disease), and an associated uncertainty regarding the ability of the bovine prion to be transmitted to, and result in a similar manifestation of brain

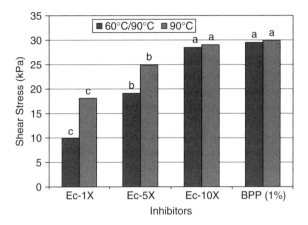

Figure 6.10 Comparison of mean stress values of Pacific whiting surimi gels containing BPP or various concentrations of recombinant (*E. coli*) soybean cystatin and subjected to one of two heating conditions. Ec: *E. coli* cystatin (crude solution of *E. coli* lysate). [a,b,c] means with different superscript differ significantly (P<0.05) within the same cooking regime. 1X: 30,000 inhibitor unit (IU); 5X: 150,000 (IU); 10X: 300,000 (IU). Cooking conditions: 60°C/90°C: 30 min incubation at 60°C plus 15 min cooking at 90°C; 90°C: 15 min cooking at 90°C. (Adapted from Reference 165.)

disease, in humans. Thus, almost any tissues of cattle other than the skeletal meat have come under scrutiny as being possible agents for transmission of a BSE-like disease to human consumers. This large uncertainty has led to an almost universal abandonment of the use of BPP as a proteinase inhibitor in surimi.

Recombinant (gene cloning) technology has been utilized recently to develop food-grade proteinase inhibitors that might overcome many of the limitations of the naturally available inhibitor materials. Using approved microorganisms as a major source of enzyme (inhibitor) production, the new system presents several advantages, such as controlled and uniform composition (and thus predictable safety) of the inhibitor, enhanced purity for greater effectiveness at lower usage levels (thus negligible organoleptic effects), lowered allergenicity, and possibly cleaner labeling. In addition, the production system can easily employ different types of inhibitor sources (gene) and/or microbial hosts for continued yield improvement, as well as improved safety.

Abrahamson et al.[166] reported that human cystatin C was expressed from *Escherichia coli* with relatively high yield and fully active inhibition against papain or cathepsin B. The proteinase inhibitor from rice grain was also cloned into *E. coli*.[167,168] Various sources of plants are reported to possess cysteine proteinase inhibitors, such as potato tubers,[169] soybeans,[170] corn,[171] rice,[172] and tomato leaves.[161]

With regard to soy proteinase inhibitors, three inhibitor genes (L_1, R_1. and N_2) were inserted into *E. coli* to express these target proteins. Among the three, both R_1 and N_2 showed more inhibitory activity than E-64 against Western corn rootworm gut proteinases.[173] The inhibitors (12 kDa) were also reported to lose activity after heating at 100°C for 15 min.[174] This suggested that such inhibitors might be able to neutralize acid proteinase activity in surimi during cooking.

Kang and Lanier[165] reported that the crude supernatant obtained from sonication of the soy-gene cloned *E. coli* produced 120 times higher inhibitory activity (per gram of protein) than that of BPP against protease purified from Pacific whiting fillets. The crude soy recombinant inhibitor was able to produce Pacific whiting surimi gels of the same strength (stress value) using 10 times less quantity than BPP with no off-color or off-flavor problems. This inhibitor also has the potential to be infused into intact fish fillets, which have been undervalued thus far due to their propensity to soften during oven (slow) cooking.[175] Recent efforts to clone this gene into a food-grade organism that expresses the inhibitor exogenous to the cell have met with some success.[176] Other workers have reported success in recombinant expression of a chicken cystatin gene for production of a food-grade cystatin that could be used in surimi.[177,178] Food-grade inhibitors used in seafood is summarized in Table 6.2.

TABLE 6.2 Food-Grade Proteinase Inhibitors Available for Seafood

Inhibitors	Target Proteinase Type	Ref.
Plasma proteins:		
Bovine	Serine, cysteine	117, 149, 164, 165
Pig	Cysteine	140, 185
Chicken	Serine	186
Egg white	Serine, cysteine	149
Whey protein	Cysteine	156
Legume	Serine	187
Potato extract	Metallo	151, 180
Green tea polyphenols	Metallo	188
Spinach extract	Serine	189

6.4.3 Minimization of Proteolysis by Process Control

Because most heat-stable proteinase activity is rapidly inactivated once the temperature exceeds about 70°C, and is relatively inactive below 50°C, it is possible to minimize the influence of even relatively high levels of endogenous heat-stable proteinase if heating is rapid through the 50 to 70°C temperature range. Practically, this is difficult or impossible with conventional (radiant or steam) heating methods for many of the steamed or baked kamaboko products of Japan because the dimensions of these products are too large to rapidly heat the center to avoid proteolysis. An early method of minimizing the problem was to cook kamaboko in layers, cooking the inside first and applying successive layers with a cook cycle between each application to allow rapid heat transfer to newly applied layers. In the manufacture of crab-like surimi seafood by the sheet method (see Chapter 9), the layer of surimi paste is thin enough to allow rapid heating, which obviates much of the problem that might arise from proteinase content.

New rapid heating methods, such as radio frequency (RF) heating, ohmic heating,[179] and microwave heating,[180] to counteract heat-induced texture degradation show promise to decrease reliance upon the addition of food-grade proteinase inhibitors into surimi. Ohmic heating has already been commercialized for some types of surimi seafood processing, and new microwave techniques also show promise for commercial application.

6.5 SUMMARY

Heat-induced gelation of surimi varies widely according to various factors, such as fish species (determining protein stability, content, etc.), fish freshness or degree of decomposition, proteolytic enzyme content and type(s), endogenous transglutaminase content, and cooking times/temperatures. Among these, proteinase activity is a primary factor in certain fish species, and a variable factor in almost all fish species, depending on feeding conditions at harvest, holding conditions prior to processing, cleaning efficiency of carcasses, and/or parasitization. Both acidic and alkaline proteinases can be involved, these being primarily of the cysteine or serine active site types. Their effect on surimi seafood gel texture, however, can be controlled by the addition of food-grade inhibitors and/or rapid heating. At present, with the elimination of the most effective food-grade inhibitor, BPP, due to the occurrence of BSE, there are only two food-grade inhibitors used in commercial surimi production: egg whites and whey protein concentrates. While these efficiently protect surimi seafood from degradative enzymatic activity during heating, their addition can have adverse effects on product acceptability. Through recombinant technology, however, there is the promise of developing new food-grade inhibitors that may be more acceptable. Likewise, new heating technologies may also obviate the need for the addition of any inhibitor.

REFERENCES

1. G. Siebert. Protein-splitting enzyme activity of fish flesh. *Experientia*, 14, 65–66, 1958.
2. Y. Makinodan, H. Toyohara, and S. Ikeda. Comparison of muscle proteinase activity among fish species. *Comp. Biochem. Physiol.*, 79B, 129–134, 1984.

3. J.C. Waterlow and M.L. Stephen. Lysine turnover in man measured by intravenous infusion of L-[^{14}C]-lysine. *Clin. Sci.*, 33, 507, 1967.

4. E. Salkowaski. Ueber autodigestion der Organe. *Z. Klin. Med.*, 17(Suppl.), 77, 1890.

5. D.H. Wasson. Fish muscle proteases and heat-induced myofibrillar degradation: A review. *J. Aquat. Food Prod. Tech.*, 1, 23–41, 1992.

6. M.T. Morrissey, P.S Hartley, and H. An. Proteolytic activity in Pacific whiting and effect of surimi processing. *J. Aquat. Food Prod. Tech.*, 4, 5–18, 1995.

7. R.W. Porter, B. Koury, and G. Kudo. Inhibition of protease activity in muscle extracts and surimi form Pacific whiting, *Meruccius productus*, and arrowtooth flounder, *Atheresthes stomias*. *Marine Fish Rev.*, 55, 10–15, 1993.

8. D.D. Hamann, P.M. Amato, M.C. Wu, and E.A. Foegeding. Inhibition of modori (gel weakening) in surimi by plasma hydrolysate and egg white. *J. Food Sci.*, 55, 665–669, 795, 1990.

9. S.W. Boye and T.C. Lanier. Effects of heat-stable alkaline protease activity of Atlantic menhaden (*Brevoorti tyrannus*) on surimi gels. *J. Food Sci.*, 53, 1340–1342, 1398, 1988.

10. Y. Makinodan, H. Toyohara, and E. Niwa. Implication of muscle alkaline proteinase in the textural degradation of fish meat gel. *J. Food Sci.*, 50, 1351–1355, 1985.

11. Y. Shimizu, R. Machida, and S. Takenami. Species variations in the gel-forming characteristics of fish meat paste. *Bull. Jap. Soc. Sci. Fish,* 47, 95–104, 1981.

12. G.A. MacDonald, S.P. Davies, and B.I. Hall. Kinetics of early post mortem texture deterioration of hoki (*Macruronus novaezelandiae*) determined by tensile properties and myofibril fragmentation index. Paper #10-7, Presented at the *Annual Meeting of IFT,* Orlando, FL, 1997.

13. A.J. Barrett and J.K. McDonald. *Mammalian Proteinases: A Glossary and Bibliography. Vol. 1: Endopeptidases.* London, U.K.: Academic Press, 1980.

14. IUB Nomenclature Committee. Enzyme Nomenclature. New York: Academic Press, 1984.

15. K. Musch, G. Siebert, J.V. Davies, and M. Ebert. Differenzierung von zweikathepsinen aus dorschmuskel. *Hoppe-Seyler's Z. Physiol. Chem.*, 352, 878–882, 1971.

16. H. An, V. Weerasinghe, T.A. Seymour, and M.T. Morrissey. Cathepsin degradation of Pacific whiting surimi proteins. *J. Food Sci.*, 59, 1013–1017, 1994.

17. H. Kirschke and A.J. Barrett. Chemistry of lysosomal proteases. In H. Glaumann and F.J. Ballard, Eds. *Lysosomes: Their Role in Protein Breakdown.* London: Academic Press, 1987, 193–238.

18. J.W. Coffey and C.D. Duve. Digestive activity of lysosomes. *J. Biol. Chem.*, 243, 3255–3263, 1968.

19. Y. Makinodan and S. Ikeda. Studies on fish muscle protease V. On the existence of cathepsins A, B and C. *Bull. Jap. Soc. Sci. Fish*, 37, 1002–1006, 1971.

20. M. Yamashita and S. Konagaya. High activities of cathepsins B, D, H and L in the white muscle of chum salmon in spawning migration. *Comp. Biochem. Physiol.*, 95B, 149–152, 1990.

21. J.-H. Pyeun, D.-S. Lee, D.-S. Kim, and M.-S. Heu. Activity screening the proteolytic enzyme responsible for post-mortem degradation of fish tissues. *J. Korean Fish Soc.*, 29, 296–308, 1996.

22. Y. Makinodan, H. Toyohara, and S. Ikeda. Combined action of carp muscle cathepsins A and D on proteins. *Bull. Jap. Soc. Sci. Fish*, 49, 1153, 1983.

23. L.M. Greenbaum and J.S. Fruton. Purification and properties of beef spleen cathepsin B. *J. Biol. Chem.*, 226, 173, 1957.

24. K. Otto and S. Bhaki. Studies on cathepsin B1: specificity and properties. *Hoppe-Seyler's Z. Physiol. Chem.*, 350, 1577, 1969.

25. H. Keilova and B. Keil. Isolation and specificity of cathepsin B. *FEBS Lett.*, 4, 295–298, 1969.

26. P. Distelmaier, H. Hubner, and K. Otto. Cathepsin B1 and B2 in various organs of the rat. *Enzymologia*, 42, 363–375, 1972.

27. S.S. Husain and J. Baqai. Evidence for two types of sulfhydryl groups in cathepsins B-1 and B-1A. *Fed. Proc. Fed. Am. Soc. Exp. Biol.*, 35, 1460–1460, 1976.

28. D.E. Goll, Y. Otsuka, P.A. Nagainis, J.D. Shannon, S.K. Sathe, and M. Muguruma. Role of muscle proteinases in maintenance of muscle integrity and mass. *J. Food Biochem.*, 7, 137, 1983.

29. S.G. Franklin and R.M. Metrione. Chromatographic evidences for the existence of multiple forms of cathepsin B1. *Biochem. J.*, 127, 207, 1972.

30. A.J. Barrett. A new assay for cathepsin B1 and other thiol proteinases. *Anal. Biochem.*, 47, 280, 1972.

31. V. Ninjoor, S.L Taylor, and A.L. Tappel. Purification and characterization of rat liver lysosomal cathepsin B2. *Biochim. Biophys. Acta*, 370, 308, 1974.

32. K. Otto. Cathepsin B1 and B2. In A.J. Barrett and J.T Dingle, Eds. *Tissue Proteinase*. Amsterdam, Netherlands: North-Holland, 1971, 1–28.

33. M.J. Bonete, A. Manjon, R. Llorca, and J.L. Iborra. Acid proteinase activity in fish. II. Purification and characterization of cathepsins B and D from Mujil Auratus muscle. *Comp. Biochem. Physiol.*, 78B, 207–213, 1984.

34. S.V. Sherekar, M.S. Gore, and V. Ninjoor. Purification and characterization of cathepsin B from the skeletal muscle of fresh water fish, *Tilapia mossambica*. *J. Food Sci.*, 53, 1018–1023, 1988.

35. M. Matsumiya, A. Mochizuki, and S. Otake. Purification and characterization of cathepsin B from ordinary muscle of common mackerel *Scomber japonicus*. *Nippon Suisan Gakkaishi*, 55, 2185–2190, 1989.

36. K. Hara, A. Suzumatsu, and T. Ishihara. Purification and characterization of cathepsin B from carp ordinary muscle. *Nippon Suisan Gakkaishi*, 54, 1243–1252, 1988.

37. M.C. Erickson, D.T. Gordon, and A.F. Anglemier. Proteolytic activity in the sarcoplasmic fluids of parasitized Pacific whiting (*Merluccius productus*) and unparasitized true cod (*Gadus macrocephalus*). *J. Food Sci.*, 48, 1315–1319, 1983.

38. A. Quali, A. Garrel, C. Obled, C. Deval, and C. Valin. Comparative action of cathepsins D, B, H, L and of a new lysosomal cysteine proteinase on rabbit myofibrils. *Meat Sci.*, 19, 83–100, 1987.

39. F.L. Huang and A.L. Tappel. Action of cathepsins C and D in protein hydrolysis. *Biochim. Biophys. Acta,* 236, 739–748, 1971.

40. R.M. Metrione, A.G. Neves, and J.S. Fruton. Purification and properties of dipeptidyl transferase (cathepsin C). *Biochemistry,* 5, 1597–1604, 1966.

41. M.J. Mycek. Cathepsins. In G.E. Perlmann and L. Lorand, Eds. *Methods in Enzymology, Vol XIX.* New York: Academic Press, 1970, 285–315.

42. K.S. Hameed and N.F. Haard. Isolation and characterization of cathepsin C from Atlantic short finned squid *Illex illecebrosus. Comp. Biochem. Physiol.,* 82B, 241–246, 1985.

43. Y.. Makinodan and S. Ikeda. Studies on fish muscle proteases. III. Purification and properties of a protease active in acid pH range. *Bull. Jap. Soc. Sci. Fish,* 35, 758–766, 1969.

44. J.W.C. Bird and J.H. Carter. Proteolytic enzymes in striated and nonstriated muscle. In K. Wildenthal, Ed. *Degradative Processes in Heart and Skeletal Muscle.* New York: Elsevier/North Holland Biomedical Press, 1980, 51–85.

45. F.M. Robbins, J.E. Walker, S.H. Colten, and S. Chatterjee. Action of proteolytic enzymes on bovine myofibrils. *J. Food Sci.,* 44, 1672–1677, 1979.

46. E.A. Ogunro, R.B. Lanmam, J.R. Spencer, A.G. Ferguson, and M. Lesch. Degradation of canine cardiac myosin and actin by cathepsin D isolated from homologous tissue. *Cardiovasc. Res.,* 11, 621–629, 1979.

47. W.N. Schwartz and J.W.C. Bird. Degradation of myofibrillar proteins by cathepsins B and D. *Biochem. J.,* 167, 811–820, 1977.

48. M.G. Zeece, K. Katoh, R.M. Robson, and F.C. Parrish, Jr. Effect of cathepsin D on bovine myofibrils under different conditions of pH and temperature. *J. Food Sci.,* 51, 769–773, 780, 1986.

49. O.A. Young, A.E. Graafhuis, and L.C. Davey. Post-mortem changes in cytoskeletal proteins of muscle. *Meat Sci.,* 5, 41–55, 1980.

50. T. Matsumoto, A. Okitani, Y. Kitamura, and H. Kato. Mode of degradation of myofibrillar proteins by rabbit muscle cathepsin D. *Biochim. Biophys. Acta,* 755, 76–80, 1983.

51. Y. Makinodan and S. Ikeda. Studies on fish muscle protease VI. Separation of carp muscle cathepsins A and D, and some properties of carp muscle cathepsin A. *Bull. Jap. Soc. Sci. Fish.*, 42, 239–247, 1976.

52. Y. Makinodan, T. Akasaka, H. Toyohara, and S. Ikeda. Purification and properties of carp muscle cathepsin D. *J. Food Sci.*, 47, 647–652, 1982.

53. S.T. Jiang, Y.T. Wang, and C.S. Chen. Lysosomal enzyme effects on the postmortem changes in tilapia (*Tilapia nilotica* × *T. aurea*) muscle myofibrils. *J. Food Sci.*, 57, 277–279, 282, 1992.

54. C. Lapresle and T. Webb. The purification and properties of a proteolytic enzyme, rabbit cathepsin E, and further studies on rabbit cathepsin D. *Biochem J.*, 84, 455–462, 1962.

55. B. Venugopal and M.E. Bailey. Lysosomal proteinases in muscle tissue and leukocytes of mcat animals. *Meat Sci.*, 2, 228–239, 1978.

56. M. Yamashita and S. Konagaya. Differentiation and localization of catheptic proteinases responsible for extensive autolysis of mature chum salmon muscle (*Oncorhynchus keta*). *Comp. Biochem. Physiol.*, 103B, 999–1003, 1992.

57. A.J. Barrett and J.T. Dingle. Terminology of the protease. In A.J. Barrett and J.T. Dingle, Eds. *Tissue Proteinases*. Amsterdam, Netherlands: North-Holland Publishing Co., 1971, 9–10.

58. H. Keilova and C. Lapresle. Inhibition of cathepsin E by diazoacetyl-methyl ester. *FEBS Lett.*, 9, 348–350, 1970.

59. H. Keilova and V. Tomasek. The specificity of a bovine fibroblast cathepsin D. I. Action on the s-sulfo derivatives of the insulin A and B chains and of porcine glucagon. *Biochim. Biophys. Acta*, 284, 461–464, 1972.

60. H. Kirschke, J. Langner, B. Wiederanders, S. Ansorge, P. Bohley, and H. Hansen. Cathepsin H: an endoaminopeptidase from rat liver lysosomes. *Acta Biol. Med. (Ger.)*, 36, 185, 1977.

61. H. Kirschke, J. Langner, B. Wiederanders, S. Ansorge, P. Bohley, and U. Broghammer. Intrazellularer Proteinabbau. VII. Kathepsin L und H: Zwei neue Proteinasen aus Rattenleberlysosomen. *Acta Biol. Med. (Ger.)*, 35, 285, 1976.

62. P. Locnikar, T. Popovic, T. Lah, I. Kregar, J. Babnik, M. Kopitar, and V. Turk. The bovine cysteine proteinases, cathepsins B, H, and S. In V. Truk and L.J. Vitale, Eds. *Proteinases and Their Inhibitors: Structure, Function and Applied Aspects.* Oxford: Pergamon Press, 1981.

63. W.T. Stauber and S. Ong. Fluorescence demonstration of a cathepsin muscles. *Histochem. J.,* 14, 585, 1982.

64. A. Okitani, U. Matsukura, H. Kato, and M. Fujimaki. Purification and some properties of a myofibrillar protein-degrading protease, cathepsin L, from rabbit skeletal muscle. *J. Biochem.,* 87, 1113–1143, 1980.

65. M.A.J. Taylor, R.E. Almond, and D.J. Etherington. The immunohistochemical location of cathepsin L in rabbit skeletal muscle: evidence for a fiber type dependence distribution. *Histochemistry,* 86, 379–383, 1987.

66. U. Matsukura, A. Okitani, T. Nishimuro, and H. Kato. Mode of degradation of myofibrillar proteins by an endogenous protease, cathepsin L. *Biochim. Biophys. Acta,* 662, 41, 1981.

67. H. Kirschke, H.J. Kargel, S. Riemann, and P. Bohley. Cathepsin L. In V. Turk and L.J. Vitale, Eds. *Proteinases and Their Inhibitors: Structure, Function and Applied Aspect.* Oxford: Pergamon Press, 1981, 93.

68. R. Ueno, K. Sakanaka, S. Ikeda, and Y. Horguchi. Characterization of pepstatin insensitive protease in mackerel white muscle. *Nippon Suisan Gakkaishi,* 54, 699–707, 1988.

69. M. Yamashita and S. Konagaya. Participation of cathepsin L into extensive softening of the muscle of chum salmon caught during spawning migration. *Nippon Suisan Gakkaishi,* 56, 1271–1277, 1990.

70. M.S. Heu, H.R. Kim, D.M. Cho, J.S. Godber, and J.H. Pyeun. Purification and characterization of cathepsin L-like enzyme from the muscle of anchovy, *Engraulis japonica. Comp. Biochem. Physiol.,* 18B, 523–529, 1997.

71. W. Visessanguan, A.R. Menino, S.M. Kim, and H. An. Cathepsin L: a predominant heat-activated proteinase in arrowtooth flounder muscle. *J. Agric. Food Chem.,* 49, 2633–2640, 2001.

72. M.-L. Ho, C.-H. Chen, and S.-T. Jiang. Effect of mackerel cathepsins L and L-like, and calpain on the degradation of mackerel surimi. *Fisheries Science,* 66, 558–568, 2000.

73. T.A. Seymour, M.T. Morrissey, M.Y. Gustin, and H. An. Purification and characterization of Pacific whiting protease. *J. Agric. Food Chem.,* 42, 2421–2427, 1994.

74. Y.J. Choi and J.W. Park. Acid-aided protein recovery from enzyme-rich Pacific whiting. *J. Food Sci.,* 67, 2962–2967, 2002.

75. Y.S. Kim, J.W. Park, and Y.J. Choi. New approaches for the effective recovery of fish proteins and their physicochemical characteristics. *Fisheries Science,* 69, 1231–1239, 2003.

76. J. Lom and J.R. Arthur. A guideline for the preparation of species descriptions in Myxosporea. *J. Fish Dis.,* 12, 151–159, 1989.

77. C.L. Davey and K.V. Gilbert. Studies in meat tenderness. 7. Changes in the fine structure of meat during aging. *J. Food Sci.,* 34, 69, 1969.

78. M. Koohmaraie. Muscle proteinases and meat aging. *Meat Sci.,* 36, 93–104, 1994.

79. M. Koohmaraie, J.E. Schollmeyer, and T.R. Dutson. Effect of low calcium-requiring calcium activated factor on myofibrils under varying pH and temperature conditions. *J. Food Sci.,* 51, 28, 1986.

80. R.G. Taylor, G.H. Geesink, V.F. Thompson, M. Koohmaraie, and D.E. Goll. Is Z-disk degradation responsible for postmortem tenderization? *J. Anim. Sci.,* 73, 1351–1367, 1995.

81. Y. Makinodan and S. Ikeda. Studies on fish muscle protease. VIII. On the existence of protease active in neutral pH range. *Bull. Jap. Soc. Sci. Fish,* 42, 665–670, 1976.

82. Y. Makinodan, M. Hirotsuka, and S. Ikeda. Neutral proteinase of carp muscle. *J. Food Sci.,* 44, 1110–1117, 1979.

83. T. Taneda, T. Watanabe, and N. Seki. Purification and some properties of a calpain from carp muscle. *Bull. Jap. Soc. Sci. Fish,* 49, 219–228, 1983.

84. I. Papa, C. Alvarez, V. Verrez-Bagnis, J. Fleurence, and Y. Benjamin. Evidence for time dependent alpha-actinin delocalisation and proteolysis from post mortem fish white muscle (*Dicentrarchus labrax* and *Salmo trutta*). In J.B. Luten, T. Borresen, and J. Oehlenschlager, Eds. *Seafood from Producer to Consumer, Integrated Approach to Quality*. Amsterdam: Elsevier Science, 1997, 247–252.

85. S.T. Jiang, Y.T. Wang, and C.S. Chen. Purification and some properties of calpain II from tilapia muscle (*Tilapia nilotica* × *Tilapia aurea*). *J. Agric. Food Chem.*, 39, 237–241, 1991.

86. H. Toyohara and Y. Makinodan. Comparison of calpain I and II from carp muscle. *Comp. Biochem. Physiol.*, 92B, 577–581, 1989.

87. G.H. Geesink, J.D. Morton, M.P. Kent, and R. Bickerstaffe. Partial purification and characterization of Chinook salmon (*Oncorhyuchus tshawytscha*) calpains and an evaluation of their role in postmortem proteolysis. *J. Food Sci.*, 65, 1318–1324, 2000.

88. H. Tsuchiya and N. Seki. Action of calpain on α-actinin within and isolated from carp myofibrils. *Nippon Suisan Gakkaishi*, 57, 1133–1139, 1991.

89. H. Tsuchiya, S. Kita, and N. Seki. Postmortem changes in α-actinin and connectin in carp and rainbow trout muscles. *Nippon Suisan Gakkaishi*, 58, 793–798, 1992.

90. I. Papa, C. Mejean, M.C. Lebart, C. Astier, C. Roustan, Y. Benjamin, C. Alvarez, V. Verrez-Bagnis, and J. Fleurence. Isolation and properties of white skeletal-muscle α-actinin from sea-trout (*Salm trutta*) and bass (*Dicentrarchus labrax*). *Comp. Biochem. Physiol.*, 74, 225–237, 1995.

91. D.E. Goll, V.F. Thompson, R.G. Taylor, and J.A. Christiansen. Role of calpain system in muscle growth. *Biochimie*, 74, 225–237, 1992.

92. M. Koohmaraie. The role of endogenous proteases in meat tenderness. *Proc. Recip. Meat Conf.*, 41, 89–100, 1989.

93. J.H. Wang, W.C. Ma, J.C. Su, C.S. Chen, and S.T. Jiang. Comparison of the properties of m-calpain from tilapia and grass shrimp muscles. *J. Agric. Food Chem.*, 41, 1379–1384, 1993.

94. J.H. Wang, J.C. Su, and S.T. Jiang. Stability of calcium-autolyzed calpain II from tilapia muscle (*Tilapia nilotica* × *tilapia aurea*). *J. Agric. Food Chem.*, 40, 535–539, 1992.

95. H. Toyohara, Y. Makinodan, K. Tanaka, and S. Ikeda. Detection of calpastatin and a trypsin inhibitor in carp muscle. *Agric. Biol. Chem.*, 47, 1151–1154, 1983.

96. S. Ishiura, S. Tsuji, H. Hurofushi, and K. Suzuki. Purification of an endogenous 68,000-Dalton inhibitor of Ca^{++}-activated neutral protease from chicken skeletal muscle. *Biochim. Biophys. Acta*, 701, 216–223, 1982.

97. G. Guroff. A neutral, calcium-activated proteinase from the soluble fraction of rat brain. *J. Biol. Chem.*, 239, 149, 1964.

98. M. Koohmaraie. Inhibition of postmortem tenderization in ovine carcasses through infusion of zinc. *J. Anim. Sci.*, 68, 1476–1483, 1990.

99. M. Koohmaraie, A.S. Babiker, A.L. Schroeder, R.A. Merkel, and T.R. Dutson. Acceleration of postmortem tenderization in ovine carcass through activation of Ca^{++}-dependent proteases. *J. Food Sci.*, 53, 1638–1641, 1988.

100. T.H. Kuo, F. Giacomelli, D. Kithier, and A. Malhotra. Biochemical characterization and cellular localization of serine protease in myopathic hamster. *J. Mol. Cell. Cardiol.*, 13, 1035, 1981.

101. R.G. Woodbury, G.M. Gruzenski, and D. Lagunoff. Immunofluorescent localization of a serine protease in rat small intestine. *Proc. Natl. Acad. Sci. USA*, 75, 2785, 1978.

102. Y. Makinodan, M. Yamamoto, and W. Shimidu. Studies on muscle of aquatic animals. XXXIX. Protease in fish muscle. *Bull. Jap. Soc. Sci. Fish,* 29, 776–780, 1963.

103. C.S. Cheng, D.D. Hamann, and N.B. Webb. Effect of thermal processing on minced fish gel texture. *J. Food Sci.*, 44, 1080–1086, 1979.

104. Y. Makinodan, Y. Yokoyama, M. Kinoshita, and H. Toyohara. Characterization of an alkaline proteinase of fish muscle. *Comp. Biochem. Physiol.*, 87B, 1041–1046, 1987.

105. K. Iwata, K. Kobashi, and J. Hase. Studies on muscle alkaline protease. III. Distribution of alkaline protease in muscle of freshwater fish, marine fish and in internal organs of carp. *Bull. Jap. Soc. Sci. Fish*, 40, 201–209, 1974.

106. E.J. Folco, L. Busconi, C.B. Martone, R.E. Trucco, and J.J. Sanchez. Activation of an alkaline proteinase from fish skeletal muscle by fatty acids and sodium dodecyl sulfate. *Comp. Biochem. Physiol.*, 91B, 473–476, 1988.

107. L. Busconi, E.J. Folco, C.B. Martone, R.E. Trucco, and J.J. Sanchez. Identification of two alkaline proteases and a trypsin inhibitor from muscle of white croaker (*Micropogon opercularis*). *FEBS Lett.*, 176, 211–214, 1984.

108. E.J. Folco, L. Busconi, C.B. Martone, R.E. Trucco, and J.J. Sanchez. Action of two alkaline proteases and a trypsin inhibitor from white croaker skeletal muscle (*Micropogon opercularis*) in the degradation of myofibrillar proteins. *FEBS Lett.*, 176, 215–219, 1984.

109. L. Busconi, E.J. Folco, C.B. Martone, R.E. Trucco, and J.J. Sanchez. Action of a serine proteinase from fish skeletal muscle on myofibrils. *Arch. Biochem. Biophys.*, 252, 329–333, 1987.

110. I. Stoknes and T. Rustad. Proteolytic activity in muscle from Atlantic salmon (*Salmo salar*). *J. Food Sci.*, 60, 711–714, 1995.

111. I. Stoknes, R. Rustad, and V. Mohr. Comparative studies of the proteolytic activity of tissue extract from cod (*Gadus morhua*) and herring (*Clupea harengus*). *Comp. Biochem. Physiol.*, 106B, 613–619, 1993.

112. T. Lin and T.C. Lanier. Properties of an alkaline protease from the skeletal muscle of Atlantic croaker. *J. Food Biochem.*, 4, 17–28.

113. K. Iwata, K. Kobashi, and J. Hase. Studies on muscle alkaline protease. I. Isolation, purification and some physio-chemical properties of an alkaline protease from carp muscle. *Bull. Jap. Soc. Sci. Fish*, 39, 1325–1337, 1973.

114. S. Benjakul, W. Visessanguan, and K. Leelapongwattana. Purification and characterization of heat-stable alkaline proteinase from bigeye snapper (*Priacanthus macracanthus*) muscle. *Comp. Biochem. Physiol.*, 134B, 579–591, 2003.

115. Y. Makinodan, N.N. Kyaw, and S. Ikeda. Classification of carp muscle alkaline proteinase and its action against intracellular proteins. *Bull. Jap. Soc. Sci. Fish,* 48, 479, 1982.

116. J. Hase, K. Kobashi, N. Nakai, E. Mitsui, K. Iwata, and T. Takedera. The quaternary structure of carp muscle alkaline protease. *Biochim. Biophys. Acta,* 611, 205–213, 1980.

117. Y.J. Choi, T.C. Lanier, H.G. Lee, and Y.J. Cho. Purification and characterization of alkaline proteinase from Atlantic menhaden muscle. *J. Food Sci.,* 64, 768–771, 1999.

118. Y.J. Choi, Y.J. Cho, and T.C. Lanier. Purification and characterization of proteinase from Atlantic menhaden muscle. *J. Food Sci.,* 64, 772–775, 1999.

119. H. Su, T.S. Lin, and T.C. Lanier. Contribution of retained organ tissues to the alkaline protease content of mechanically separated Atlantic croaker (*Micropogon undulatus*). *J. Food Sci.,* 46, 1650–1653, 1658, 1981.

120. N.B. Webb, E.R. Hardy, G.G. Giddings, and J.J. Howello. Influence of mechanical separation upon proximate composition, functional properties and textural characteristics of frozen Atlantic croaker muscle tissue. *J. Food Sci.,* 41, 1277, 1976.

121. JR. Dingle and J.A. Hines. Protein instability in minced flesh from fillets and frames of several commercial Atlantic fishes during storage at −5°C. *J. Fish Res. Bd., Canada,* 32, 775, 1975.

122. T.C. Lanier. Suitability of red hake, *Urophycis chuss*, and silver hake, *Merluccius bilinearis*, for processing into surimi. *Mar. Fish Rev.,* 46(2), 43, 1984.

123. H. Toyohara, M. Kinoshita, and Y. Shimizu. Proteolytic degradation of threadfin-bream meat gel. *J. Food Sci.,* 55, 259–260, 1990.

124. C.S. Cheng, D.D. Hamann, N.B. Webb, and V. Sidwell. Effect of species and storage time on minced fish gel texture. *J. Food Sci.,* 44, 1087–1092, 1979.

125. K. Iwata, K. Kobashi, and H. Hase. Study on muscle alkaline protease. VII. Effect of the muscular alkaline protease and protein fractions purified from white croaker and horse mackerel on the "Himodori" phenomenon during kamaboko production. *Nippon Suisan Gakkaishi,* 45, 157–161, 1979.

126. H. Toyohara, T. Sakata, K. Yamashita, M. Kinoshita, and Y. Shimizu. Degradation of oval-filefish meat gel caused by myofibrillar proteinase(s). *J. Food Sci.,* 55, 354–368, 1990.

127. Y. Shimizu, A. Nomura, and F. Nishioka. Modori (fish gel degradation occurring at around 60°C)-inducing property of croaker myosin preparation. *Bull. Jap. Soc. Sci. Fish,* 52, 2027–2032, 1986.

128. M. Kinoshita, H. Toyohara, and Y. Shimizy. Diverse distribution of four distinct types of modori (gel degradation)-inducing proteinases among fish species. *Nippon Suisan Gakkaishi,* 56, 1485–1492, 1990.

129. M.-J. Cao, K. Hara, K. Osatomi, K. Tachlbara, T. Izumi, and T. Ishihara. Myofilbril-bound serine proteinase (MBP) and its degradation of myofibrillar proteins. *J. Food Sci.,* 64, 644–647, 1999.

130. M. Ohkubo, K. Miyagawa, K. Osatomi, K. Hara, Y. Nozaki, and T. Ishihara. Purification and characterization of myofibril-bound serine protease from lizard fish (*Saurida undosquamis*) muscle. *Comp. Biochem. Physiol.,* 137B, 139–150, 2004.

131. J. Yongsawatdigul, J.W. Park, P. Virulhakul, and S. Viratchakul. Proteolytic degradation of tropical tilapia surimi. *J. Food Sci.,* 65, 129–133, 2000.

132. G. Salvesen and H. Nagase. Inhibition of proteolytic enzymes. In R.J. Beynon, and J.S. Bond, Eds. *Proteolytic Enzymes: A Practical Approach.* New York: IRL Press, 1989, 83–104.

133. F.L. Garcia-Carreno and P. Hernandez-Cortes. Use of protease inhibitors in seafood products. In N.F. Haard and B.K. Simpson, Eds. *Seafood Enzymes.* New York: Marcel Dekker, 2000, 531–547.

134. F.L. Garcia-Carreno. Proteinase inhibitors. *Trends Food Technol.,* 7, 197–204, 1996.

135. L. Stevens. Egg white proteins. *Comp. Biochem. Physiol.,* 110B, 1–9, 1991.

136. W.-H. Liu, G.E. Means, and R.E. Feeney. The inhibitory properties of avian ovoinhibitors against proteolytic enzyme. *Biochim. Biophys. Acta,* 229, 176–185, 1971.

137. Y. Suzuki, K. Yoshida, T. Ichimiya, T. Yamamoto, and H. Sinohara. Trypsin inhibitors in guinea pig plasma: isolation and characterization of contrapsin and two isoforms of α-1-antiproteinase and acute phase response of four major trypsin inhibitors. *J. Biochem.*, 107, 173179, 1990.

138. L. Lorand. *Methods in Enzymology, Vol. XLV: Proteolytic Enzymes Part B.* New York: Academic Press, 1976, 695–739.

139. K. Fossum and J.R. Whitaker. Ficin and papain inhibitor from chicken egg white. *Arch. Biochem. Biophys.*, 125, 367–375, 1968.

140. A. Anastasi, M.A. Brown, A.A. Kembhavi, M.J.H. Nicklin, C.A. Sayers, D.C. Sunter, and A.J. Barrett. Cystatin, a protein inhibitor of cysteine proteinase. *Biochem. J.*, 211, 129–138, 1983.

141. V. Turk and W. Bode. The cystains: protein inhibitor of cysteine proteinases. *FEBS Lett.*, 285, 213–219, 1991.

142. J.-J. Lee, S.-S. Tzeng, and S.-T. Jiang. Purification and characterization of low molecular weight kininogen from pig plasma. *J. Food Sci.*, 65, 81–86, 2000.

143. T. Morean, N. Gutman, D. Faucher, and F. Gauthier. Limited proteolysis of T-kinninogen (thiostatin): release of comparable fragments by different endopeptidases. *J. Biol. Chem.*, 264, 4298–4330, 1989.

144. S. Higashiama, I. Ohkubo, H. Ishiguro, M. Kunimatsu, K. Sawaki, and M. Sasaki. Human high molecular weight kinninogen as a thiol proteinase inhibitor: presence of the entire inhibitions capacity in the native form of heavy chain. *Biochemistry*, 25, 1669–1675, 1986.

145. R.L. Heinrikson and F.J. Kezdy. Acidic cysteine protease inhibitors from pineapple stem. In L. Lorand, Ed. *Methods in Enzymology Vol. XLV.* New York: Academic Press, 1976, 740–751.

146. H. Nagase, E.D. Harris, J.F. Woessner, and K. Brew. Ovostatin: a novel proteinase inhibitor form chicken egg white. 1. Purification, physicochemical properties, and tissue distribution of ovostatin. *J. Biol. Chem.*, 258, 7481–7489, 1983.

147. S.-T. Jiang, J. Wu, M.-J. Su, and S.-S. Tzeng. Purification and characterization of calpastatin from grass prawn muscle (*Penaeus monodon*). *J. Agric. Food Chem.*, 48, 3851–3856, 2000.

148. R. Ueno, T. Kanayama, K. Tomiyasu, A. Fujikami, and T. Nakashima. Ueno Fine Chemicals Industry, Ltd., Osaka, Japan. Processes for Production of Refrigerated Minced Fish Flesh and Fish Paste Product Having Improved Quality. U.S. Patent No. 4,464, 1984.

149. V.C. Weerasinghe, M.T. Morrissey, and H. An. Characterization of active components in food-grade protease inhibitors for surimi manufacture. *J. Agric. Food Chem.*, 44, 2584-2590, 1996.

150. D.H. Wasson, K.D. Repond, J.K. Babbitt, and J.S. French. Effects of additives on proteolytic and functional properties of arrowtooth flounder surimi. *J. Aquat. Food Prod. Technol.*, 1, 147–164, 1992.

151. T.C. Lanier, T.S. Lin, D.D. Hamann, and F.B. Thomas. Effects of alkaline protease in minced fish on texture of heat processed gels. *J. Food Sci.*, 46, 1643–1645, 1981.

152. M.T. Morrissey, J.W. Wu, D.D. Lin, and H. An. Effect of food grade protease inhibitor on autolysis and gel strength of surimi. *J. Food Sci.*, 58, 1050–1054, 1993.

153. K.D. Reppond and J.K. Babbitt. Protease inhibitors affect physical properties of arrowtooth flounder and walleye pollock surimi. *J. Food Sci.*, 58, 96–98, 1993.

154. K.P. Kaiser and H.D. Belitz. Specificity of potato isoinhibitors towards various proteolytic enzymes. *Z. Lebensm. Unters. Forsch.*, 19, 18–22, 1973.

155. H. Akazawa, Y. Miyauchi, K. Sakurada, D.H. Wasson, and K.D. Reppond. Evaluation of protease inhibitors in Pacific whiting surimi. *J. Aquat. Food Prod. Tech.*, 2, 79–95, 1993.

156. V. Weerasinghe, M.T. Morrissey, Y. Chung, and H. An. Whey protein concentrate as a proteinase inhibitor in Pacific whiting surimi. *J. Food Sci.*, 61, 367–371, 1996.

157. H. An, T.A. Seymour, J.W. Wu, and M.T. Morrissey. Assay systems and characterization of Pacific whiting (*Merluccius productus*) protease. *J. Food Sci.*, 59, 277–281, 1994.

158. D.H. Wasson, J.K. Babbitt, and J.S. French. Characterization of a heat stable protease from arrowtooth flounder; *Atheresthes stomias. J. Aquat. Food Prod. Tech.*, 1, 167–182, 1992.

159. E.J. Folco, L. Busconi, C.B. Martone, and J.J. Sanchez. Fish skeletal muscle contains a novel serine proteinase with an unusual subunit composition. *Biochem. J.*, 263, 471–475, 1987.

160. M.L. Izquierdo-Polido, T.A. Haard, J. Hung, and N.F. Haard. Oryzacystatin and other proteinase inhibitors in rice grain: potential use as a fish processing aid. *J. Agric. Food Chem.*, 42, 616–622, 1994.

161. J.W. Wu and N.F. Haard. Use of cysteine protease inhibitors from injured tomato leaves in whiting surimi. *J. Food Biochem.*, 22(5), 383–398, 1998.

162. B.Y. Lamb-Sutton. Possible Role of alpha-2-Macroglobulin in the Inhibition of Heat-Stable Fish Muscle Proteases. MS thesis, North Carolina State University, 1995.

163. T.C. Lanier and I.S. Kang. Plasma as a Source of Food Additives. Presented at the *Kinsella Symposium,* Las Vegas, NV, 1997.

164. T.A. Seymour, M.Y. Peters, M.T. Morrissey, and H. An. Surimi gel enhancement by bovine plasma proteins. *J. Agric. Food Chem.*, 45, 2919–2923, 1997.

165. I.S. Kang and T.C. Lanier. Bovine plasma protein functions in surimi gelation in comparison to cysteine protease inhibitors. *J. Food Sci.*, 64, 842–846, 1999.

166. M. Abrahamson, H. Dalboege, I. Olafsson, S. Carlsen, and A. Grubb. Efficient production of native, biologically active human cystatin C by *Escherichia coli*. *FEBS Lett.*, 236, 14–18, 1988.

167. K. Abe, Y. Emori, H. Kondo, S. Arai, and K. Suzuki. The NH_2-terminal 21 amino acid residues are not essential for the papain-inhibitory activity of oryzacystatin, a member of the cystatin superfamily. *J. Biol. Chem.*, 263, 2755–2759, 1988.

168. K. Abe, Y. Emori, H. Kondo, K. Suzuki, and S. Arai. Molecular cloning of a cysteine proteinase inhibitor of rice (Oryzacystatin). *J. Biol. Chem.*, 262, 16793–16797, 1987.

169. A.D. Rowan, J. Brizin, D.J. Buttle, and A.J. Barret. Inhibition of cysteine proteinases by a protein inhibitor from potato. *FEBS Lett.*, 269, 328–330, 1990.

170. M.E. Hines, C.I. Osuala, and S.S. Nielsen. Isolation and partial characterization of a soybean cystatin cysteine proteinase inhibitor of coleopteran digestive proteolytic activity. *J. Agric. Food Chem.*, 39, 1515–1520, 1991.

171. M. Abe and J.R. Whitaker. Purification and characterization of a cysteine proteinase inhibitor from the endosperm of corn. *Agric. Biol. Chem.*, 52, 1583–1584, 1988.

172. K. Abe, H. Kondo, and S. Arai. Purification and characterization of a rice cysteine proteinase inhibitor. *Agric. Biol. Chem.*, 51, 2763–2768, 1987.

173. Y. Zhao, M.A. Botella, L. Subramanian, X. Niu, S.S. Nielsen, R.A. Bressan, and P.M. Hasegawa. Two wound-inducible soybean cysteine proteinase inhibitors have greater insect digestive proteinase inhibitory activities than a constitutive homolog. *Plant Physiol.*, 111, 1299–1306, 1996.

174. S. Lalitha, R.E. Shade, J. Huesing, P.M. Hasegawa, R.A. Bressan, and S.S. Nielsen. Recombinant Soybean Cystatine Proteinase Inhibitors Reduce Growth and Development of Select Crop Pests. Paper #23A-7. Presented at the *Annual Meeting of IFT*, Orlando, FL, June 14–18, 1997.

175. I.S. Kang and T.C. Lanier. Inhibition of Protease in Intact Fish Fillets by Recombinant Cystatin. Paper # 50D-14. Presented at the *Annual Meeting of IFT*, Chicago, IL, 1999.

176. I.S. Kang, J.-J. Wang, J.C.H. Shih, and T.C. Lanier. Extracellular production of a functional soy cystatin by *Bacillus subtilis*. *J. Agric. Food Chem.*, 52, 5052–5056, 2004.

177. G.H. Chen, S.J. Tang, C.S. Chen, and S.T. Jiang. Overexpression of the soluble form of chicken cystatin in *Escherichia coli* and its purification. *J. Agric. Food Chem.*, 48, 2602–2607, 2000.

178. G.H. Chen, S.J. Tang, C.S. Chen, and S.T. Jiang. High-level production of recombinant chicken cystatin by *Pichia pastoris* and its application in mackerel surimi. *J. Agric. Food Chem.*, 49, 641–646, 2001.

179. J.W. Park, J. Yongsawatdigul, and E. Kolbe. Proteolysis and gelation of fish proteins under ohmic heating. In *Process-Induced Chemical Changes in Food*. New York and London: Plenum Press. 25-34, 1998.

180. D.H. Green and J.K. Babbitt. Control of muscle softening and protease-parasite interactions in Arrowtooth flounder, *Atheresthes stomias*. *J. Food Sci.*, 55, 579, 1990.

181. R. McLay. Activities of cathepsins A and D in cod muscle. *J. Sci. Food Agric.*, 31, 1050–1054, 1980.

182. A.A. Iodice, V. Leong, and I.M. Weinstock. Separation of cathepsins A and D of skeletal muscle. *Arch. Biochem. Biophys.*, 117, 477–486, 1966.

183. J.K. McDonald, B.B. Zeitman, T.J. Reilly, and S. Ellis. New observations on the substrate specificity of cathepsin C (dipeptidyl aminopeptidase I). *J. Biol. Chem.*, 244, 2693–2709, 1969.

184. H. An, M.Y. Peters, and T.A. Seymour. Roles of endogenous enzymes in surimi gelation (Review). *Trends Food Sci. Technol.*, 7, 321–327, 1996.

185. W. Visessanguan, S. Benjakul, and H. An. Porcine plasma proteins as a surimi protease inhibitor: Effects on actomyosin gelation. *J. Food Sci.*, 65, 607–611, 2000.

186. S. Rawdkuen, S. Benjakul, W. Visessanguan, T.C. Lanier. Chicken plasma protein affects gelation of surimi from bigeye snapper (*Priacanthus tayenus*). *Food Hydrocolloids*, 18, 259–270, 2004.

187. J.A. Ramirez, F.L. Garcia-Carreno, O.G. Morales, and A. Sanchez. Inhibition of modori-associated proteinases by legume seed extracts in surimi production. *J. Food Sci.*, 67, 578–581, 2000.

188. M. Saito, K. Saito, N. Kunisaki, and S. Kimura. Green tea polyphenols inhibit metalloproteinase activities in the skin, muscle, and blood of rainbow trout. *J. Agric. Food Chem.*, 50, 7169–7174, 2002.

189. H. Toyohara, M. Kinoshita, K. Sasaki, S. Yamaguchi, Y. and Shimizu, M. Sakaguchi. Occurrence of a modori inhibitor in spinach leaves. *Nippon Suisan Gakkaishi*, 58, 1705–1710, 1992.

7

Waste Management and By-Product Utilization

MICHAEL T. MORRISSEY
Oregon State University Seafood Lab, Astoria, Oregon

JOHN LIN
Pacific Surimi Co., Warrenton, Oregon

ALAN ISMOND
Aqua-Terra Consultants, Bellevue, Washington

CONTENTS

7.1 Introduction 281
7.2 Surimi Waste Management and Compliance 283
    7.2.1 Measurements Needed for Compliance 284

	7.2.1.1	Accurate Wastewater Flow Meters.... 284
	7.2.1.2	Correct Measurement of Solid Concentration..................................... 285
	7.2.1.3	Correct Reporting of Tonnage 286
7.2.2		How to Implement a Waste Management Program... 286
	7.2.2.1	Plant Audit and Mass Balance.......... 286
	7.2.2.2	Water Reduction/Reuse 287
	7.2.2.3	Waste Solids Recovery....................... 287

7.3 Solid Waste 288
    7.3.1 Fish Meal and Fish Protein Hydrolysates 288
    7.3.2 Fish Oil Recovery 290
    7.3.3 Specialty Products.. 292
7.4 Surimi Wastewater.. 293
    7.4.1 Chemical Methods 295
    7.4.2 Biological Methods 297
        7.4.2.1 Aerobic Process 297
        7.4.2.2 Anaerobic Process 298
    7.4.3 Physical Methods.. 299
        7.4.3.1 Dissolved Air Flotation....................... 299
        7.4.3.2 Heat Coagulation................................ 300
        7.4.3.3 Electrocoagulation 301
        7.4.3.4 Centrifugation..................................... 301
        7.4.3.5 Membrane Filtration 303
7.5 Recovery of Bioactive Components and Neutraceuticals.. 306
    7.5.1 Bioactive Compounds 306
        7.5.1.1 Enzymes 307
        7.5.1.2 Other Waste Compounds.................... 309
    7.5.2 Recovery of Bioactive Compounds 310
7.6 Opportunities and Challenges................................... 311
    7.6.1 The Limitations of Fish Solids Recovery........ 311
        7.6.1.1 Quality Impediments.......................... 311
        7.6.1.2 Environmental Limitations................ 312
        7.6.1.3 Marketing Impediments..................... 312
        7.6.1.4 Proximity to Market........................... 312
        7.6.1.5 Labor and Maintenance Considerations..................................... 313

7.6.2 Current and Future Potential 313
7.7 Summary 315
References 316

7.1 INTRODUCTION

Surimi processing, which includes several unit operations, can be divided into two major stages (Figure 7.1). The first stage includes heading, gutting, deboning, and mincing, which prepare the fish mince for the washing and refining operations of the second stage. Streams of water are injected into the heading, gutting, and deboning machines during the first stage to remove fish fluid and the muscle meat that adheres to the machines. The injected water also provides lubrication to promote smooth operation. A large amount of water is often used for transporting skins, viscera, and backbones from filleting and deboning machines to a scrap delivering system. For this purpose, using fresh water is not necessary. The overall fresh-water consumption can be largely reduced using recycled water for transporting scraps. In addition, fish are comminuted during the first stage into mince, 3 to 4 mm in diameter, which is then pumped to unit operations for washing and refining.

In the second stage, fish mince is repeatedly washed with chilled water and dewatered to produce high-quality surimi. The water/mince ratio and the number of washing cycles can differ, depending on the desired surimi quality. For at-sea surimi operations, where fresh fish are readily available and water is limited, only one washing cycle may be used, while washing may be repeated two or three times for shore-based production.[1,2] The major objective of washing and dewatering is to remove the soluble sarcoplasmic proteins, lipids, fish blood, and other water-soluble matter in the flesh, as well as to concentrate the myofibrillar proteins. These soluble components are in the wastewater in varying concentrations, depending on the wash ratios and the processing step. Wu et al.[3] and Lin and Park[4] reported that myofibrillar proteins are

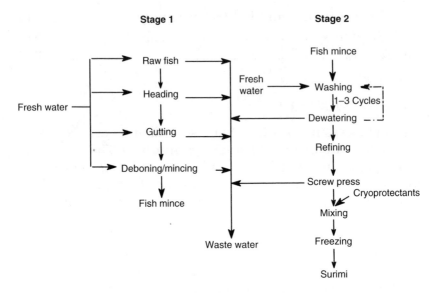

Figure 7.1 Flow chart of surimi processing waste water.

also readily solubilized under certain conditions and lost to wastewater during processing.

The largest percentage of solid waste that can be readily separated from wastewater is produced during the first stages of processing. Solid waste consists of the head, viscera, skin, meat, and frames that remain after the muscle meat is extracted from the fish. Washwater, on the other hand, is utilized during the second stage and contains insoluble components such as myosin fractions, scales, fat, and other tissue fibers; and soluble compounds such as enzymes, polypeptides, blood components, and inorganic minerals. Wastewater is generated in the second stage of the operation from surimi process water after washing the minced fish meat and from the final press.

This chapter is divided into different sections. The next section looks at surimi waste management, types of waste measurement that are important, and general methods that companies can use to implement a waste management program in their operations. The remaining sections discuss the uses of solid wastes, wastewater treatments, and the potential for using by-products as neutraceuticals in this growing field.

There is a movement in all food industries not to categorize what remains after primary processing as waste, but rather as by-products. Certainly, the increased use of fish body parts for different foods and functional ingredients, as well as the processing of by-products into potentially value-added components, have created an increased interest in the surimi industry for improving the bottom line. Hopefully, this chapter will spur some additional thinking and new research ideas along these lines.

7.2 SURIMI WASTE MANAGEMENT AND COMPLIANCE

There are three principle reasons why the surimi industry needs to address solid and liquid processing waste:

1. Environmental regulations will only become more stringent over time, and disposal surcharges will be ever-increasing, as will the cost for fresh water. Consequently, it will become cost prohibitive to maintain the current levels of pollutant discharge.
2. In the near future, there will be niche markets for companies that can produce surimi in a controlled environment, resulting in improved quality and using "greener" production methods.[5]
3. Waste in the surimi industry is synonymous with unrecovered resource. Competitive pressures will favor companies that can maximize the recovery of food and non-food-grade products.

The most pressing reasons to improve waste management in a surimi plant are to comply with government regulations that govern discharge and waste disposal. To assess compliance with an existing permit or pending new regulations, it is important to have an accurate assessment of the current plant discharge. Permits can be structured in many ways, including limits on:

1. Gallons of waste water discharged
2. Concentration of pollutants in the wastewater discharged
3. Tonnage of pollutants in the wastewater discharged
4. Tonnage of pollutants discharged per unit of seafood processed

Generally, the limits are specified for a given period, such as daily, monthly, per operating season, or annually. There are three critical elements involved in correctly assessing whether a plant is in compliance with a permit limit: (1) measuring the flow rate of wastewater discharged, (2) correctly determining the pollutant concentration in the wastewater, and (3) accurately reporting the tonnage processed, when applicable.

7.2.1 Measurements Needed for Compliance

7.2.1.1 Accurate Wastewater Flow Meters

The flow rate of wastewater must be accurately measured for the time period of concern. The best option is to install a flow meter on the pipe that discharges the final effluent. Not all flow meters are the same; and for surimi processing wastewater, meters with non-moving flow sensors and no obstructions in the pipe are preferable. Flow meters must be installed per the manufacturers' recommendations in order to operate correctly.

Table 7.1 shows the breakdown of water use at a typical shore plant surimi operation. At least once per year, the meter should be checked against a known volume of water. This can be done during downtime by filling a pump with a known volume of water and pumping it through the meter. Inaccuracies in flow measurements due to malfunctioning meters can result in unnecessary noncompliances if the meter is overreporting the actual flow. Meter readings should be recorded at the same time every day, and the time period should correspond with the production cycle. If the tonnage processed must be reported for compliance and the production

TABLE 7.1 Water Usage in Different Processing Steps of a Shore-Side Surimi Operation

Processing Step	Type of Water	% Water Usage
Holding tanks	Fresh water	10.2
Fillet machines	Fresh water	7.6
Backbone discharge	Recycled water	2.0
Skin discharge	Recycled water	13.6
1st Washing	Fresh water	9.8
2nd Washing	Fresh water	9.8
1st Washing for recovered meat	Fresh water	2.7
2nd Washing for recovered meat	Fresh water	1.7
Sanitation	Fresh water	21.2
Floor wash	Recycle water	21.2
Total		100.0

day runs from midnight to midnight, meter readings should also be recorded at midnight.

7.2.1.2 Correct Measurement of Solid Concentration

Accurately determining the solid concentration in the final wastewater requires proper sampling and accurate lab analyses. A single sample, or grab sample, can be collected at a recorded time. For surimi operations, the variability in the wastewater is such that collecting a single sample in 24 hr is not likely to yield an accurate assessment of the discharge for the entire day. A composite sample is preferable. This sampling strategy requires taking equal volumes of wastewater spaced evenly over the entire day and combining them to make one sample.

Despite taking care to accurately measure the flow of wastewater and collecting a representative sample of wastewater, compliance can be affected by the accuracy of the analyses performed on the sample. If the samples are sent to an external laboratory, split samples should be sent on occasion to two separate laboratories. If the results from the two laboratories differ by more than 10%, additional split samples should be sent out, and the laboratories should be requested to supply their bench notes for comparison. Again, an erroneous result from the lab can make the difference between being in or out of compliance with a discharge permit. When discharging to a municipal wastewater treatment plant, erroneously high results can equate to significant sewer surcharges. When designing a new wastewater treatment facility, it is imperative to have an accurate assessment of the maximum and average pollutant load and flow; otherwise, the facility may be over- or under-sized.

7.2.1.3 Correct Reporting of Tonnage

In the case where compliance is dictated by pounds of solids per unit of fish processed, it is important to accurately assess production tonnage. Some facilities record pounds round based on boat offloads for a given day. This may or may not correspond to the pounds round processed on that day. It is important to match the pollutant load measured for a given time period with the tonnage processed for that same time period. Otherwise, it may be possible to over- or understate the pollutant load per unit of fish processed

7.2.2 How to Implement a Waste Management Program

7.2.2.1 Plant Audit and Mass Balance

The most useful tool in waste management is to perform a waste and wastewater audit on a regular basis.[6] This involves identifying key areas of the plant where wastewater flows can be measured and samples collected for analyses such as BOD (biological oxygen demand), TS (total solids), salinity (if seawater is used), TSS (total suspended solids), and oil and grease. Waste streams can be classified by flow, solids concentration, and pollutant mass loading. The integrity of the flow measurements and pollutant concentrations can be verified by setting up a mass balance for each pollutant. This involves calculating the pounds per day of pollutant at each sampling point and adding them together to check whether they equal the pollutant load in the final effluent.

7.2.2.2 Water Reduction/Reuse

The audit will reveal areas of high water use that should be scrutinized in a water conservation program. Reducing water use may involve turning off water to areas of the plant during breaks and downtime, adjusting water flows to a practical minimum, and ensuring that hoses have convenient shut-off valves. Water reduction must be balanced with water use to maintain phytosanitary requirements and machinery function. Water reuse is sometimes possible, such as recycling decanter effluent for fluming waste solids; however, bacterial issues may limit this option.

7.2.2.3 Waste Solids Recovery

Recycling waste solids depends on several factors. Waste streams that contact nonsanitary surfaces such as the floor cannot be considered for food-grade product recovery, but may have potential for non-food-grade by-products. By comparing the TS and TSS loading for each waste stream, it is possible to assess the difficulty in recovering additional products. Suspended particles can generally be recovered more economically through mechanical means such as screening or centrifuging. Dissolved solids (the difference between TS and TSS, accounting for salinity) are more costly and difficult to recover. Options might include precipitation through pH adjustment and chemical addition, heat coagulation, or membrane filtration. Decanter centrifuges are a good example of how to increase food-grade product yield. However, once installed, the decanters should be monitored on a regular basis to ensure that they are being operated for maximum recovery. This may involve sampling the decanter influent and effluent, and using a benchtop centrifuge to assess the amount of separable solids recovery.

In some cases, increasing the recovery of food-grade products may actually increase the pollutant load in the final effluent. A case in point would be to divert waste fish solids that would have been sent to rendering or disposal and converting them into additional surimi product. In the process of washing, screening, and pressing, what were previously waste solids are now lost to the wastewater, thereby increasing the pollutant load. For plants that have rendering facilities, it has been estimated that a pound of waste solids diverted to the meal plant produces less pollution than a pound of solids processed into surimi.

7.3 SOLID WASTE

7.3.1 Fish Meal and Fish Protein Hydrolysates

Solid wastes constitute 50 to 70% of the original raw material, depending on the method of meat extraction from the carcass. These wastes, as shown in Figure 7.2, are a mixture of heads, viscera, skin, frames, and left-over muscle meat. Over the past 50 years, the seafood industry has developed several options for waste utilization. In the larger plants, much of the solid waste is changed into fish meal, which has a ready market for aquaculture and agricultural feeds.[7,8] Fish meal operations involve high heat drying and a reduction of 1:6 in the final product form. The final product is a dried, stable product with an extended shelf life. Fish meal operations, however, usually require a large capital investment and therefore need substantial volume for a return on the investment. Additionally, the liquid waste stream from rendering known as stickwater may present disposal problems of its own. For processing facilities that use saltwater for fish storage and solids fluming, the resultant stickwater may have a salt content that makes it undesirable to recycle the stickwater solids into the fish meal. In addition, the stickwater contains a high level of soluble proteins that may or may not be desirable in the finished meal product.

Research has also been done regarding the development of fish protein hydrolysates (FPH).[9,10] FPH are defined products from a process that breaks down the fish flesh by proteolytic digestion. FPH are in liquid or dried form and can be used for aquaculture diet formulations, as well as agricultural feeds. Most FPH operations use commercial (rather than endogenous) enzymes to produce a hydrolyzed product with good protein functionality. Depending on whether the final

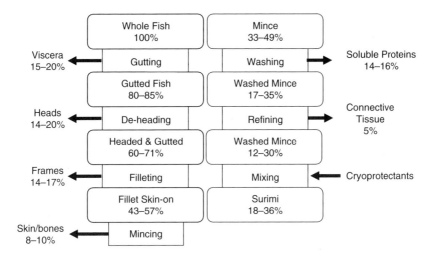

Figure 7.2 Flow chart of surimi processing solid waste.

product is dried or left in a semi-solid state, FPH operations need a lower capital investment. FPH operations also have the advantage of being able to scale down or up, depending on the processing volume.

The pet food industry in the United States utilizes large volumes of solid wastes. The wastes are ground and thoroughly mixed, reduced to 77% moisture, and then frozen in blocks. Another form of waste utilization is fish silage. Fish silage is produced by an acid hydrolysis of fish flesh and can be accelerated with heat. The product is stable due to the low pH and there are few odors associated with the process.

There are increasing markets for fish by-products as organic agricultural fertilizers. They are made into a liquid form through acid hydrolysis, with post-process pH adjustment to neutrality or by enzymatic hydrolysis. Several plants are making a dried pellet fertilizer from an enzymatic hydrolysis process. Fertilizers can also be made by composting surimi solid waste with other organic waste.[11] Several farms in the United States, for example, are adding fish waste to timber waste on a large scale and composting over several months. The compost piles are periodically turned and the final product is bagged for either home gardens or commercial agricultural needs. There is little capital investment needed for compost operations and the raw material costs are minimal.

Research is also underway to explore ways to use fish wastes for human food. FPH have been investigated as a possible product for a low-cost human food. Studies have shown the feasibility of utilizing solid wastes from surimi operations for fish sauce fermentation. Upgrading from fish waste to human food, and/or complete utilization of whole fish, using traditional Asian methods has been successfully evaluated. Good-quality fish sauce was developed from Pacific whiting and its by-products.[12,13] The nature of the abundant proteolytic enzymes in Pacific whiting contributed to accelerated fermentation of the fish sauce.

Consequently, by-product utilization can be a profit center for seafood plants, including surimi operations. There is a growing need for animal protein for aquaculture and agricultural feeds from solid waste. Fish fertilizers are also becoming more popular for both large farms and small house gardens. In addition, new research is discovering different uses for enzymes and other compounds from fish solid waste and surimi washwater. Extraction of these compounds, for example, might be commercially feasible in the future for use in food or as a neutraceutical. The major concern in by-product utilization, however, is choosing the right products to produce, developing solid markets for these products, and product distribution.

7.3.2 Fish Oil Recovery

Fish oils are becoming increasingly important in the food industry as both a nutraceutical and a food ingredient. The omega-3 fatty acids ($\omega 3$s) are of the most interest, and eicosapentanoic acid (20:5 $\omega 3$) and docosahexaenoic acid (22:6 $\omega 3$) are currently receiving much attention because they play a key role in the prevention and treatment of a wide range of human diseases and disorders. These include the prevention of heart disease by lowering cholesterol and low-density lipoproteins (LDL), lowering blood pressure, reducing clot formation, and general reduction in coronary heart disease.[14,15] They also have been found to inhibit cancer and tumor growth, improve the immune response, and have a predominant effect on brain development. For these and other reasons, they have been termed "essential" fatty acids.[16] The majority of fish oils are produced as a by-product from fish meal production. Omega Protein is a Texas-based company and the largest producer of fish oil in the United States. It was formerly a by-product of the company's menhaden fishmeal production but with the increased demand for high-quality omega-3 oils, the company has focused on improving production of its fish oils as a primary product. Most of the world's fish oil production is from fatty fish.

Surimi production has traditionally been from low-fat pollock and other gadoids. However, due to the shear volume of production, even low-fat pollock (2 to 4% fat) will produce commercial quantities of fish oil on a daily basis.[17] A 1-million-lb surimi operation in Alaska will theoretically produce 10,000 lb per day of fish oil. In the traditional surimi process, however, most of the oil in the muscle meat is removed in the washing process and not recoverable. Fish oil is extracted from the liver and viscera, which contain higher amounts of fat, and is used primarily as a partial substitute for diesel oil in plant operations as the cost of fuel is high. The economics of fish oil utilization for human consumption from Alaska resources is problematic due to transportation costs, market competition, and infrastructure needs for refining and stabilizing the oil. Currently, most of the fish oil produced at surimi plants in Alaska is used for heat/energy generation and is mixed with other fuel oils.

As noted in Chapter 1, more pelagic species (often high-fat fish) are being used for surimi production. The use of mackerel, which can run as high as 15% fat, can be potentially used for fish oil production as well. This may, in part, offset the lower prices for nontraditional species such as mackerel. The advantage of fish oil production from surimi is the potential of having a higher-quality product. Surimi production is a nonthermal process, unlike fish meal operations that require high heat and the cooking of the meat to facilitate separation of the oil from the protein. This can lead to oxidation of the oils and substantial refinement of the oils in downstream processing. Although there are currently no commercial operations producing fish oil as a surimi processing by-product, it still has potential, especially for fatty fish species, and one that can help the bottom line of some companies that until now focused solely on surimi.

7.3.3 Specialty Products

During the early, heady times of U.S. surimi production, the focus was to produce high-quality surimi as fast as possible. It was a race for the fish and processing was forced to keep up. There was high demand for SA grade product (highest quality) from Japan, and surimi yields were often less than 20%. What was not used for surimi was sent to the fish meal operation and little thought was given to specialty products made from different organs or cuts of the animal. This has changed dramatically over the past decade as researchers, processors, and sales people have developed new products and new markets for numerous by-products that were considered waste just a few years ago.

There are several reasons why this has occurred. The American Fisheries Act, which has allowed better programming of harvest and processing rather than a race for the fish, allows processors to strategize to obtain both maximum yield and expanded markets. In addition to recovering more product in the surimi process, the industry can also look more closely at the by-product market and develop processing options for organs and other parts of the fish.[18] For example, many pollock processors have developed markets for pollock stomachs and intestines in the Asian market. Pollock milt is also being used for specialty markets that were not explored a decade ago. Although these markets may be marginally profitable at this stage, they do allow the processor to offer a portfolio of products and, more importantly, reduce waste. Other potential markets exist for tongues and pollock skin. There has been considerable research on the use of fish skin for extraction of collagen. Several of these products are dis-

TABLE 7.2 Composition of Surimi Wastewater Discharged from Various Processing Steps

Waste Steam	Moisture (%)	Protein (%)	Non-Protein Nitrogen (%)	Fat (%)	Ash (%)
1st Dehydrator	97.1	2.34	0.13	0.19	0.41
2nd Dehydrator	98.8	0.97	0.08	0.11	0.38
1st Rotary Screen	98.9	0.99	0.04	0.06	0.17
3rd Dehydrator	98.9	0.86	0.05	0.04	0.21
2nd Rotary Screen	98.9	0.99	0.04	0.05	0.09
Screw Press	99.2	0.89	0.04	0.08	0.14

Source: Adapted from Reference 2.

cussed in more detail in the section ahead on "Recovery of Bioactive Ingredients and Nutraceuticals."

7.4 SURIMI WASTEWATER

Surimi manufacturing consumes large volumes of fresh water during washing procedures and generates large amounts of wastewater with low concentrations of organic materials. Huang[19] measured water usage in a shore-side plant and found that 5.7 L wastewater were generated for every kilogram of fish processed. The composition of wastewater from various processing steps is illustrated in Table 7.2. Surimi wastewater usually contains a relatively high protein concentration as compared to wastewater from other types of seafood operations. In the early washing stages, fat and insoluble compounds, such as skin particles, are also contained in the washwater. The wastewater discharged from the first dehydrator (D1) contained the highest level of protein, nonprotein nitrogen, fat, and ash. The wastewater in successive discharge points revealed decreasing protein, nonprotein nitrogen, fat, and ash. D1 is a prewash step and constitutes only a small portion of overall wastewater (less than 2%). Most of the proteins in the wastewater from the prewash step mainly contain sarcoplasmic protein with no noticeable myofibrillar proteins. As the washing process progresses, the myofibrillar proteins are more likely to be lost in the waste streams.

Myofibrillar proteins constitute approximately 66 to 77% of the total protein in fish flesh. Therefore, recovery of myofibrillar proteins is expected to be in this range. However, in a typical surimi operation, only 50 to 60% of the myofibrillar proteins are retained through the washing and dewatering process. Approximately 40 to 50% myofibrillar proteins are lost in soluble or insoluble forms due to factors such as changes in pH and ionic strength, proteolysis, and mechanical forces in mincing, washing, screening, and screw pressing. Proper processing control for minimizing the loss of functional proteins in the waste stream (Chapter 2) is the first and most important step in managing surimi wastewater.

Surimi wastewater streams, in contrast to solid wastes, present a different set of problems for waste disposal because of their high volumes and low concentrations of solid material. If plants are located close to municipalities, wastewater must go through publicly owned treatment plants to reduce the BOD (biological oxygen demand) and TS (total solids) before being released into local rivers. Charges for each wastewater treatment facility are based on the levels of released solids and the total volume of wastewater treated. Limited direct discharge of wastewater into local rivers and bays may be allowed in certain areas with the permission of environmental regulatory authorities. Consequently, disposing of surimi wastewater poses more challenges for shore-side surimi plants than for at-sea operations, which are allowed to release wastewater into the open ocean. Water shortages, stringent environmental regulations, and the rising costs of water disposal have caused concern among surimi manufacturers. Because wastewater is high in organic loads, direct discharge can cause potential negative environmental impacts, thus threatening aquatic organisms in the water. The wastewater effluent must rapidly mix with the currents and the solid waste must be diluted to a level that will not harm aquatic life. If direct discharge is not permitted, wastewater must be treated and these costs will impact the profitability of any processing operation.

In general, the objective of wastewater treatment is to remove the organic and inorganic solids.[20] However, if the effluent also contains valuable solids that are worth recovering, wastewater treatment can be a product recovery process as well. Chemical, biological, and physical methods can be used to treat wastewater. Usually, a combination of these methods can be used to achieve effective removal of materials in the wastewater. Factors to consider when selecting a wastewater treatment facility include the effectiveness of solids removal, the cost of chemicals, and the cost of equipment and maintenance.

Chemical methods involve the application of chemical agents to coagulate the soluble and insoluble materials from the wastewater. Biological methods, on the other hand, utilize naturally occurring microorganisms to digest the organic materials. Upon separation of the microbiological cells from the treated wastewater, the organic load in the effluent is significantly reduced. Physical methods, however, use screens, sand bed filters, membrane filters, and other mechanical actions to remove the solids.

7.4.1 Chemical Methods

Adjusting the pH of wastewater to the isoelectric point of the proteins is the simplest approach for pretreating wastewater. Fish proteins have different isoelectric points but the majority range from 4.2 to 5.0. As the pH approaches the isoelectric point, due to the lack of the electrostatic repulsion between the protein particles, they tend to aggregate, grow in size, and finally, precipitate. The precipitated proteins can then be separated from the wastewater.[21]

High charge density synthetic polymers (polyelectrolytes), such as polyacrylamide, polyDADMAC (dimethyl diallyl diammomiumchloride), and EPIXDMA (epichorohydrin/dimethyl amine), are widely used in place of aluminum- and iron-based coagulants.[22,23] In the surimi industry, polyacrylamide is used in conjunction with other technologies for wastewater treatment, such as dissolved air flotation. Application of polyacrylamide to the wastewater causes the proteins to flocculate

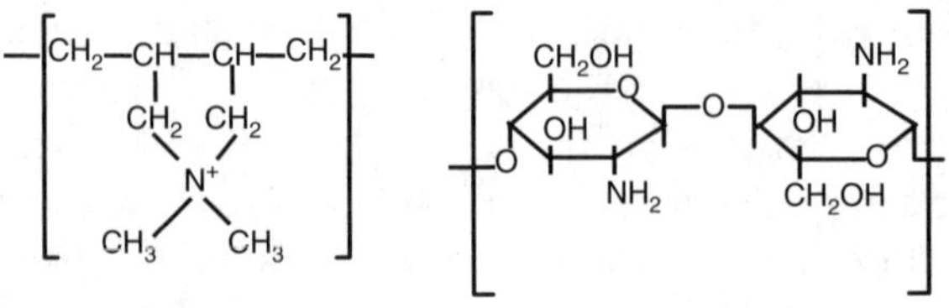

Figure 7.3 Polymeric coagulants used in waste treatment.

and float to the surface, improving the efficiency of the dissolved air flotation system (Figure 7.3).

Polymeric coagulants are basically polymeric chains containing large amounts of amine groups (Figure 7.3). Although polymeric coagulants are highly efficient (~10 ppm), less expensive, and suitable for a wider range of pollutants, some of the polymeric coagulants are considered potentially toxic to humans and animals and are not biodegradable. In addition, no synthetic chemical coagulants have been approved for human consumption.

Similar to synthetic polyelectrolytes, chitosan is a naturally occurring polysaccharide and a by-product of seafood processing. Chitosan is partially deacetylated chitin, which is a natural polymer found in the shells of shellfish and contains positively charged amine groups (Figure 7.3). Because it carries positive charges at acidic conditions, it can attract and destabilize negatively charged pollutants in the water. Chitosan is biodegradable and nontoxic in nature and is therefore an ideal candidate for recovering protein from wastewater.[24] Depending on the nature of the wastewater, however, the effectiveness of chitosan in protein recovery may differ. In comparison to other synthetic coagulant, however, a large-scale application of chitosan for wastewater treatment is cost prohibitive (M. Ludlow, personal communication, 2003).

Johnson and Gallanger[25] compared the efficiency of ferric sulfate and chitosan in removing organic substances from the wastewater of seafood processing plants. Effluents of crab, shrimp, and salmon processing were tested with different levels of ferric salts and chitosan. Up to 99% removal of TSS was observed, with chitosan being most effective. They concluded, however, that although effective, treating seafood wastewater with either coagulant was not economical for seafood processors. Savant and Torres[26] report that a lower-cost chitosan–alginate complex was effective for protein removal from surimi wash water.

7.4.2 Biological Methods

Biological treatment has been a successful method for treating large volumes of domestic and industrial wastewater. Wastewater solids from the seafood industry are mainly organic solids, such as protein, peptides, amino acids, fish oils, and some carbohydrates, all of which are readily degraded by microorganisms. The basic principle of a biological treatment is to grow cells of microorganisms, usually bacteria, in aerobic or anaerobic conditions. In both systems, soluble organic materials are converted to clusters of biological cells that can settle out and easily separate from the wastewater by sedimentation or other conventional separation method.[27] Although very effective in reducing organic loads in the wastewater, all the potentially useful nutritive components, such as proteins, are completely lost. The use of biological methods may be constrained by low temperatures, potential high volumes, and salinity of the wastewater.

7.4.2.1 Aerobic Process

The activated sludge method is the most commonly used aerobic method in wastewater treatment (Figure 7.4). In Japan, surimi wastewater is treated with this method after a primary treatment.[28] With a sufficient supply of oxygen to the nutrient-rich wastewater, bacteria and other microorganisms rapidly consume organic carbohydrates and proteins. These organic compounds are then converted to CO_2 and water-soluble inorganics (NH_4^+). The biomass, called "activated sludge," accumulates in the form of flakes containing a rich flora of aerobic bacteria, fungi, and bacteria-eating protozoa. After treatment, the majority of the sludge is dewatered and can be used as an agricultural fertilizer. Some of the sludge, which contains

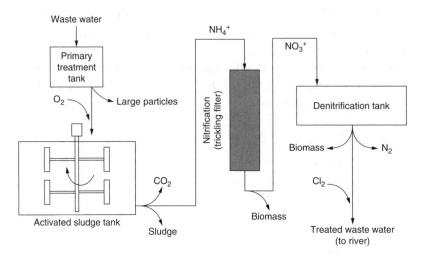

Figure 7.4 Activated sludge method.

active bacteria, is recycled back to the treatment tank and used as a bacteria source.

In a wastewater treatment plant, usually another aerobic process follows to grow lower aerobic priority bacteria. This process takes place in a trickling filter and is called a nitrification process in which NH_4^+ formed during the activated sludge treatment tank is converted to NO_3^+. The latter can be further converted into N_2 in an anaerobic process. After complete treatment, organic matter and harmful inorganic substances in the wastewater are removed. Following disinfection, the treated wastewater can be directly discharged into rivers.[27]

In a case study by Mikkelson and Lowery,[29] a six-phase sequencing batch reactor (SBR) system was designed to continuously treat high-strength wastewater from a meat processing plant. The system was capable of handling 150,000 gal wastewater per day and reduced BOD to 1% of the original value.

7.4.2.2 Anaerobic Process

Seafood industry wastewater can be treated with anaerobic treatment methods because it contains high concentrations of organic matter suitable for the growth of anaerobic bacteria. In addition, because the process does not require aeration, it is a more economical process for the treatment of wastewater with high solids content (COD > 4000 mg/L) and also a more suitable treatment for small quantities of wastewater.[27] Furthermore, the volume of sludge formed in an anaerobic process is smaller than that generated in an aerobic process and the generated methane can be used as an energy source. Compared to aerobic treatment processes, however, anaerobic treatment is a slower process due to the slow growth rate of methanogenic bacteria.

7.4.3 Physical Methods

Physical methods are usually used in the primary treatment of wastewater and employ one or more physical principles to separate waste materials from the water phase. For seafood processing wastewater, fish heads, bones, and undersized fish in the wastewater stream are examples of large matter that is separated by physical methods. The simplest and most widely used physical method is the use of screens to trap and prevent large materials from entering the waste stream. Screens are useful in surimi processing operations, especially for materials in the waste streams from stage I, which are involved in the fluming of fish, as well as the heading and gutting operations.

7.4.3.1 Dissolved Air Flotation

For smaller materials in surimi washwater, such as protein particles and particularly fats/oils, dissolved air flotation (DAF) can be used. In a DAF process, air is added to the recirculated treated effluent or injected into the wastewater with or without chemical addition prior to the DAF tank. This causes some of the undissolved solids to float to the surface. As a result, a foam-like layer of solids floats upward, accumulates on the surface of the wastewater stream, and is easily skimmed off.[27] Depending on the types of solids in the wastewater, the DAF process can be very effective, particularly for treating fats/oils in the wastewater. DAF has been adopted in several surimi processing plants and a typical DAF schematic

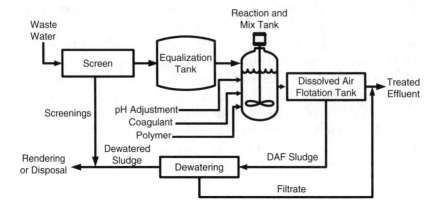

Figure 7.5 Dissolved air flotation (DAF) scheme.

is shown in Figure 7.5. The wastewater is screened to remove larger particles and then pumped to an equalization tank. The tank buffers fluctuations in flow and pollutant load. In the United States, polyacrylamide is added to the reaction and mix tank, which causes proteins in the wastewater to flocculate. The protein flocci, having a lower density than water and with added air, float to the surface of the flotation tank. The resultant sludge can be further dewatered for rendering or disposal with the screenings.

7.4.3.2 Heat Coagulation

Heat coagulation represents another category of physical treatments that can recover proteins from wastewater for animal consumption. In a heat coagulation process, heat is supplied to elevate the temperature of the wastewater, causing heat-sensitive proteins to coagulate and precipitate. The coagulated proteins can be easily removed from the wastewater by conventional solid–liquid separation techniques, such as sedimentation, filtration, and centrifugation. In the dairy industry, thermal coagulation of whey protein is used as a pretreatment step to coagulate and recover whey protein.[30] For surimi wastewater, heat treatment can be used to readily remove insoluble proteins.

In general, heat can be directly introduced, via steam injection, to the wastewater (direct heating) or transferred to the wastewater through a heat exchanger (indirect heating). When an indirect heating method is used, steam is supplied inside the heat exchange tubes or plates of a heat exchanger. Heat is then transferred from the steam through the wall of the heat exchanger to the wastewater. Given the low temperature and the high volume of the wastewater, the energy costs can be substantial.

As protein coagulation occurs, considerable amounts of proteins are deposited on the surface of the heating walls, causing the heat transfer resistance to increase.[31] Although scraped-surface heat exchangers can significantly reduce the coated foulants by a rotating shaft, these exchangers are generally very expensive and require higher operating and maintenance costs. Huang et al.[32] heated surimi wastewater by ohmic heating and found that 92.1% protein of the total suspended solids (TSS) were removed from the washwater when the temperature reached 70°C.

7.4.3.3 Electrocoagulation

Electrocoagulation is a combined physical and chemical process involving an alternating or direct current that passes through the wastewater to initiate coagulation. Ions released from the electrodes can neutralize the charge-carrying particles in the wastewater, causing them to destabilize and coagulate. Electrocoagulation is effective in removing metal ions and has demonstrated the ability to remove BOD, TSS, oil, and grease from wastewater as well. Application of electrocoagulation to treat three streams of wastewater from a fish processing plant showed that more than 93% TSS and 35.5 to 76.9% COD could be removed.[33] However, there has been only limited success in using electrocoagulation in surimi plants.

7.4.3.4 Centrifugation

Centrifugation is another conventional method widely used for solid–liquid separation.[34] In large-volume Alaska pollock surimi plants, several centrifuges may be installed to recover suspended solids from the washwater. The centrifuge, however, requires a high capital investment due to the complexity of the equipment and the high maintenance costs resulting from the high-speed moving parts. It also consumes a relatively large amount of energy and generates a slurry concentrate containing 5 to 20% dry solids. In addition, the separation capacity depends significantly on the particle size and density, which results in an incomplete separation for the industrial centrifuges. Nonetheless, several plants use centrifuges to recover lost myofibrillar protein in the wastewater and to increase overall surimi yield, as well as to reduce solids in the wastewater streams.

Decanter centrifuges have also been adopted by the seafood industry to recover large particles from the waste streams. Typically, a range of 1700 to 3000x g centrifuge force is used for recovering protein particles from the surimi wastewater streams. Swafford et al.[35] reported the use of a decanter centrifuge (Figure 7.6) to recover insoluble solids from rinsing and dewatering wastewater. Up to 80% insoluble solids were recovered. The surimi produced through decanter is assigned as recovery-grade surimi. Recovery-grade surimi possesses fairly good color and low impurities, but is usually of lower gel strength.

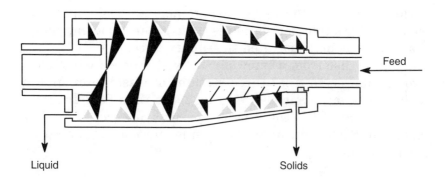

Figure 7.6 Flow of mass in decanter centrifuge.

7.4.3.5 Membrane Filtration

Membrane filtration employs a positive pressure as the driving force to push the liquid phase through the membrane pores without involving a phase change. It is particularly suitable for recovering and concentrating thermally sensitive components, such as protein, in the food and biotechnology industry.[36] It has been widely used in the dairy industry to concentrate milk, partially remove water from the milk prior to cheese making, and recover whey protein and lactose from the waste steams of cheese making.[37] Although there are few examples of membrane use in the surimi industry, trial runs with membranes for protein removal from stickwater plants at surimi operations have been hopeful.[38]

A membrane is a thin sheet of artificial or natural polymeric material with pores distributed throughout the material. Under a certain pressure, the membrane rejects large particles while species smaller than the pores pass through it. The type and performance of a membrane are determined by pore size and pore size distribution. There are two types of membrane filtration: dead-end filtration and cross-flow filtration. In a dead-end filtration process, fluid flows perpendicular to the membrane surface. Due to a rapid accumulation of solutes on the membrane surface, the permeate flux declines dramatically (Figure 7.7A). This type of filtration is widely used in laboratories for filtering small-volume liquids. It is also used in the pharmaceutical industry for sterile filtration.[39]

Cross-flow filtration, however, is a major, revolutionary concept in the development of membrane filtration technology. In cross-flow filtration, fluid flows in a direction tangential to the membrane surface (Figure 7.7B). In addition to the pressure applied normally to the membrane surface, a shear force is exerted tangentially to the membrane surface, removing the solutes that accumulate on that surface. A steady-state filtration can develop after the shearing force is in equilibrium with the forces that lead to the accumulation of the solutes on the membrane surface. The permeate flux in cross-flow filtration is much higher than that of dead-end filtration.

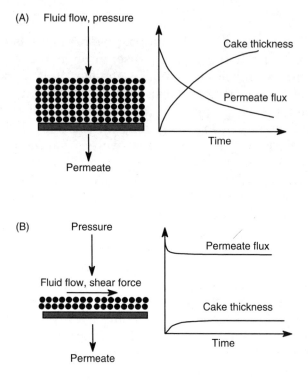

Figure 7.7 Dead-end filtration (A) and cross-flow filtration (B).

Cross-flow filtration is also predominately used in reverse osmosis, ultrafiltration, and microfiltration.

The application of membrane filtration to recover solids from the wastewater of seafood processing has been studied over the past 30 years. Miyata[40] used a bench-scale ultrafiltration system to concentrate proteins from the washwater of red muscle fish. After centrifugation to remove coarse particles, the wastewater was pumped to a tubular filter until a 10/1 volume reduction was achieved. Results showed that about 90% of the protein in the feed could be concentrated and the recovery was independent of the feed concentration.

A more comprehensive study was conducted by Ninomiya et al.[41] using ultrafiltration to recover waste-soluble proteins in the wastewater from surimi processing plants. Ceramic membranes with pore sizes ranging from 0.05 to 1.5 mm were used, in addition to microporous membranes, to recover solids from the washwater of fish paste processing.[42-44] In addition, wastewater from different fish species was used in their investigation and an ultrafiltration unit with a molecular weight cut-off (MWCO) of 20,000 Da was used to concentrate the wastewater samples, which contained 0.1 to 2% protein to 0.4 to 18% protein.

French scientists also conducted research on the application of membrane filtration to recover soluble proteins from surimi washwater.[45-47] Ultrafiltration membranes with MWCO ranging from 10 to 100 kDa were used in their laboratory and pilot experiments. Surimi washwater reconstituted from freeze-dried sarcoplasmic proteins was used to select suitable ultrafiltration membranes and fresh surimi washwater was used in the pilot studies. Results showed that more than 75% of the effluent polluting loads (expressed in BOD_5 and COD) could be reduced with a 100% retention of protein. A clear filtrate, which contained dissolved amino acids and other lower molecular weight compounds, was also obtained. Permeate fluxes started from 50 L/m²hr at the beginning of filtration and dropped to 10 L/m²hr in the end. Regenerated cellulose membranes performed better than polysulfone membranes due to their superior hydrophilicity. However, regenerated cellulose membranes are not a good choice for wastewater treatment because they cannot withstand the harsh chemical conditions during cleaning and sanitation operations.

Ultrafiltration to recover proteins from Alaska pollock surimi washwater was investigated by Peterson.[48] A plate-and-frame configuration was found superior to other configurations for the filtration of highly viscous product in an economically feasible manner. The gel strength of recovered protein concentrate was comparable to second-grade surimi.

Lin et al.[2] used a microfiltration unit and a spiral-wound polysulfone (30 kDa MWCO) membrane filtration unit to recover solids from surimi washwater. They found that proteins recovered by microfiltration showed very high gelling functionality and had a composition comparable with proteins in regular surimi. A 10% substitution with recovered proteins did not affect the quality of surimi. Solids recovered by ultrafiltration, however, had considerable dark color as well as strong odor characteristics, and therefore were not suitable as a surimi ingredient. An 89 to 94% reduction in COD was achieved after ultrafiltration treatment of surimi wastewater. The ultrafiltration permeates had a high degree of clarity and could be potentially recycled.[2]

One of the major constraints with membrane use in surimi operations is membrane fouling, which dramatically reduces the efficiency of the process. Huang and Morrissey[49] investigated mechanisms of membrane fouling during microfiltration of surimi washwater to recover myofibrillar proteins. Experimental results showed that the initial membrane fouling process could be modeled by the standard pore blocking law and the development of fouling continued with a continuous process of cake formation. The filtration resistance of the cake layer increased with the feed concentration and was found one order of magnitude higher than the initial dominant pore blocking resistance.

7.5 RECOVERY OF BIOACTIVE COMPONENTS AND NEUTRACEUTICALS

7.5.1 Bioactive Compounds

Surimi washwater and processing discards can be thought of as a biochemical mixture of potentially valuable compounds. Several compounds have value for chemical and pharmaceutical applications.[50] There is also an increasing interest in the use of marine fish and shellfish for neutraceuticals and bioactive components.[51] Much of this interest centers on enzymes from marine organisms and their unique properties.[52,53] Although there are several uses for solid waste materials from surimi operations, there have been very few applications for the dissolved and solute components of surimi washwater.

The premise of surimi operations is to wash out components that interfere with the manufacture of good-quality surimi. These include compounds that hinder gelation of the myofibrillar proteins, cause discoloration, or might add to the impurity count during surimi testing. Although there are some nonsoluble myofibrillar proteins that make it through the sieves or centrifuges, the majority of the material is soluble protein. Estimates of the amount of protein lost during the washing process range from 10 to 20% of the protein in the original raw material. The concentration of solids in the washwater depends on the washing ratios as well as the stage of washing, and can range from 0.5 to 2% of the waste stream.[54]

The soluble compounds are the most difficult to recover and can be classified into several subcomponents. These include sarcoplasmic proteins with a molecular weight range from 10 to 100 kDa. They make up 3.8 to 8.8 g protein per 100 g muscle tissue, depending on the species.[55] These are compounds that are soluble in the muscle sarcoplasm and are represented by enzymes, heme proteins, albumins, and Ca^{2+}-binding proteins called parvalbumins (e.g., calmodulin).[56] There are also highly soluble non-protein nitrogenous compounds, which include free amino acids, dipeptides, nucleotides, guanidino compounds, urea, quaternary ammonium compounds, and their derivatives.

7.5.1.1 Enzymes

There have been several reviews about the types of enzymes present in fish, especially fish processing discards.[56-58] These can be loosely divided into two groups, as either derived from eviscera or muscle protein. The majority of the research has been undertaken on proteinases, although other enzymes (e.g., lipoxygenases and transglutaminase) are of increasing interest. Proteinases are responsible for many of the deteriorative reactions in seafood, such as texture softening in fillets, gaping between myotomes, and the modori phenomenon in surimi gels. However, under different circumstances, they can also be used for a number of seafood processes, such as production of fish protein hydrolysates, fish sauce, or to facilitate roe processing and flavor development.

The viscera proteinases, including the gastric protease, pepsin, and the intestinal proteinase trypsin, are found in the solid processing waste fraction. The enzymes have several unique qualities, including high molecular activity at low temperature for cold-water species. These proteinases can be used as feed or processing feeds. In addition, several tests have shown potential use in the pharmaceutical industry for specific food processes. Much of this work, however, has been done on cod and very little on surimi resources, such as pollock, whiting, or tropical species. It is estimated that in the Alaska pollock surimi industry alone, 100 million kg viscera is removed each year, the majority of which is used for fish meal or discarded at sea.

Muscle proteinases have been largely studied for their modori-producing effect. These proteases are classified into several groups, including alkaline proteinases and the cathepsins, primarily acid proteases. The cathepsins are lysosomal proteinases that can exist in muscle cells or macrophages. Research on Pacific whiting and arrowtooth flounder has shown that cathepsin L is responsible for gel softening in these species. Morrissey and co-workers[59] have shown that although 90% of the enzyme can be removed through normal washing procedures, the remaining proteinase is sufficient to cause significant gel weakening; therefore, protease inhibitors must be used. Because of the high levels of proteinases in Pacific whiting, efforts have been made to recover the enzyme in the surimi washwater. A combination of heat treatment and membrane filtration is being investigated to aid in enzyme recovery.

Collagenase from marine sources has been isolated from several species and may be more specific and better suited for use in meat tenderization than the plant proteinases currently used.[56] Transglutaminase (TGase) has also been identified in a number of fish species.[60,61] TGase has been shown to form strong cross-linking bonds in myofibrillar proteins.[62] Potential recovery of TGase from surimi washwater would allow technicians to add it back to final product form to improve gel strength. Currently, TGase, from a microbial source is being tested for use in improving low-grade surimi with good results.

Haard[53] has reviewed potential uses for different proteinases. These include their use in stickwater treatment to lower viscosity, development of fish protein concentrates or hydrolysates, development of specific flavor compounds, pigment isolation from shrimp shell waste, tenderization of squid, use in skin and scale removal in fish, and membrane removal in roe sacks. The majority of fish used for surimi come from polikilotherms adapted to cold-water habitats. These conditions affect the isoenzyme expression, substrate binding, thermal stability, and thermodynamic properties. The main advantage of cold-adapted enzymes is that they can run at a reduced reaction temperature, thus minimizing the effects of heat and bacterial growth. They are often inactivated at lower temperatures as well.

7.5.1.2 Other Waste Compounds

7.5.1.2.1 Collagen

Most surimi operations have skinning machines to remove the skins while extracting the flesh. Fish skins are of interest in the food processing industry for their use in the production of fish gelatin from the extracted collagen.[8,63,64] Fish gelatin differs from animal gelatin in that it has a lower content of the amino acids proline and hydroxyproline. Consequently, fish gelatin forms a gel at 8°C in comparison to mammal-derived gelatin, which gels at 30°C. The gels, however, tend to be weaker than those made from mammalian gelatin. Fish gelatin has been used in capsules for the pharmaceutical industry and light-sensitive coatings on photographic films. In addition, collagen from cod swim bladders has been used as a clarifier in the Japanese brewing and wine industry.[52] Currently, work is underway to modify fish gelatin for use as an acceptable food ingredient in kosher foods.[65]

7.5.1.2.2 DNA and Nucleotides

Ockerman and Hansen[66] have shown that nucleotides can be produced from testes, and nucleic acids can be produced from salmon milt. Protamine, a basic peptide containing 80% arginine, is also present in fish milt.[51] These are used in the cosmetic industry. No attempts, however, have yet been made to produce these products from fish used in surimi, but the pollock A season, which is used primarily for roe processing, could be a potential source of production.

7.5.1.2.3 Omega-3 Fatty Acids

Most of the muscle meat of white fish used for surimi production is low in fat (less than 2%). However, fish discards, especially the liver, are considerably higher in fat content as well as fat-soluble vitamins. There have been numerous studies concerning the health benefits of long-chain polyunsaturated fatty acids.[14,67] Extraction of these compounds can occur during solid waste processing, that is, fish meal or fish protein hydrolysate production. In addition, production of surimi from high-fat fish, such as mackerel, sardines, or menhaden, will allow direct recovery of fish oil from the primary process.

7.5.1.2.4 Antioxidants

Carnosine and anserine were recently recovered from surimi washwater.[68] These dipeptides are present in the muscle tissue and are readily removed during the washing process. Chan and Dekker[69] have shown that these compounds are active antioxidants in different food systems. Work is currently underway to recover these compounds in whiting washwater and determine their potential use in food systems.

7.5.2 Recovery of Bioactive Compounds

Little research has been done on recovering bioactive compounds from surimi washwater. Some initial studies undertaken by Jaouen and Quemeneur[47] described how compounds might be recovered using different membranes. Huang and Morrissey[70] studied the use of membranes in-depth and discovered that recovery was hindered by membrane fouling and pore blocking. Pretreatment with heat or acid precipitation of the insoluble protein was necessary to remove the myofibrillar-like proteins, as well as facilitate separation and removal of the soluble proteins from the washwater. This type of pretreatment, however, decreases the functionality of the precipitated protein and lessens the chance of using this fraction for functional foods.

DeWitt and Morrissey[70] have shown that cathepsin L, a protease in relatively high concentration in Pacific whiting, can be concentrated through a combination of acid precipitation and membrane concentration. Smaller compounds, such as amino acids, are more readily separated from larger macromolecules in surimi washwater by either membrane concentration or column chromatography. Research showed that the dipeptides anserine and carnosine, active antioxidants with commercial value, could also be recovered from surimi washwater. However, problems arise with increased recovery of unwanted small molecular weight compounds, such as trimethylamines (TMA). Consequently, to use the dipeptides, further concentration and purification steps are required.[68]

The major challenge for the recovery of bioactive compounds is the separation of different protein fractions. Because it is possible that a percentage of the insoluble proteins can be added back into the surimi or used as a food ingredient in other products, the ideal situation would be to separate the insoluble protein so that it retains some functionality. Once this separation occurs, the soluble components can then be removed, usually through membrane separation by molecular weight. Often, however, the insoluble protein must be removed by a method that denatures the proteins and leaves fewer options for protein use (i.e., fish meal, fish protein hydrolysates, or fertilizers). However, if specific soluble proteins are in high enough concentrations and can be targeted for selective removal, the economic benefits can be worthwhile.

7.6 OPPORTUNITIES AND CHALLENGES

7.6.1 The Limitations of Fish Solids Recovery

7.6.1.1 Quality Impediments

For smaller surimi operations that do not have an integrated fish meal operation, recovering more non-food-grade waste fish solids can add more cost, not value, to the operation. Larger operations that do render their own solid waste may benefit financially from increasing solids recovery. However, the quality and composition of the solids must be carefully examined so that the existing rendered product does not deviate from the established product quality specifications. For example, recovering solids from the final effluent may increase the amount of protein and fat added to the rendering plant. In some cases, this can compromise the proximate analysis of the existing rendered product.

7.6.1.2 Environmental Limitations

Another limiting factor for operations that produce fish meal, fertilizers, or other products is the requirement for additional energy costs and a corresponding increase in air emissions. Large operations that generate their own electricity and operate under an air quality permit may find that the requirements for recovering more waste solids are in conflict with permitted limitations for air pollution. The controls to prevent excess air emissions can be costly and can therefore negate the financial benefit of recovering more fish solids.

7.6.1.3 Marketing Impediments

If a surimi operation recovers food and non-food products that are compatible with the existing product specifications, market acceptance should not be a problem. However, if the recovered solids result in new products, market research may be necessary to determine if there is a demand for these products.[71] Many surimi operations might not have the product development and marketing expertise to complement the effort for increased recovery. Another consideration is the potential size of the market relative to the projected tonnage for the new food or non-food product that could be recovered. In some cases, the projected tonnage might exceed the existing market demand or might be large enough to suppress market prices.

7.6.1.4 Proximity to Market

Although there are a variety of technologies that can aid in the recovery of fish solids, it is preferable to recover solids that can be marketed profitably. The proximity of the surimi plant to potential customers for the recovered solids may dictate the selection of the technology and product form. The greater the moisture content of the recovered product and the greater the distance to market, the greater the transportation cost. This is especially true for Alaska surimi plants that are far from potential markets and can have significant transportation costs.

7.6.1.5 Labor and Maintenance Considerations

When considering adding new technologies to recover more fish solids, the skill level of the labor force must be assessed. Some technologies require a higher level of skill to operate and maintain. For plants located in remote areas, this may be an issue, as well as access to spare parts and servicing.

7.6.2 Current and Future Potential

Much of the solid waste is already utilized and converted into fish meal, hydrolysates, and fertilizers. The challenge is to break away from these traditional waste product streams into new products that can be directed toward human consumption and have greater profit potential (Table 7.3). Some of these

TABLE 7.3 Surimi Processing By-Products

Type	Specific Compounds	Applications, Products
Organs	Roe	Food
	Milt	Food, Nucleotides
	Skin	Collagen, Gelatin
	Livers	Food, Omega-3s, Fuel oil
	Stomachs	Food, Niche markets
Nutraceuticals	Omega-3 fatty acids	Pharmaceuticals, Food additive
	Collagen	Gelatin, Food additive
	Polypeptides	Hormone-like additives
	Dipeptides	Antioxidants, Bioactive ingredients
	Hydrolysates	High-end foods, Niche markets
Enzymes	Proteases	Fish sauce, Protein digestion
	Collagenase	Tissue softening

include upgrading fish protein hydrolysates to highly functional foods that suit certain needs in the food industry. Work has already begun in the production of fish sauce from surimi by-products that can be utilized as a food condiment. The pH-shift method and recovery of protein might also lend itself to the production of human food from different by-product streams such as fish oils.

Soluble proteins in wastewater streams are more difficult to recover and may need specialized equipment and methods. Membrane technology has improved tremendously over the past decade but there is still the problem of membrane fouling, especially with myofibrillar protein in the wastewater. However, with different pretreatments, this can be minimized and, in the future, there might be selective recovery of different soluble proteins that have economic value. Recovery of enzymes and bioactive components from surimi waste presents both a challenge and an opportunity. Work must be done to determine the economic feasibility of component recovery and whether markets justify the required capital investment.

Wastewater treatment in the seafood industry is a very complicated issue, raising concerns from environmental regulatory agencies, lawmakers, the general public, and seafood processors. The seafood industry must be willing to treat and reduce the volume of wastewater. The challenge is to determine whether the priority of companies is to reduce the cost of wastewater treatment and/or recover solids (i.e., protein) for use as a by-product. Cooperation with federal and local environmental regulatory agencies is also necessary to develop short-term and long-term wastewater reduction and treatment strategies.

Technically, wastewater from the seafood or other sectors of the food industry is not difficult to treat but costs for the treatment can be high. It is important for seafood processors to develop environmental awareness among their employees and to educate them not to waste fresh water.[6] It would be helpful for companies to establish a water saving plan. This plan must include guidelines for freshwater usage and spent water recycling in places where fresh water is not needed. In addition, activities leading to water usage savings should be encouraged and activities leading to increased wastewater generation should be discouraged.

From a technical standpoint, to achieve cost-effective and successful wastewater treatment, it is also necessary to develop a prioritized plan. There are many sources in a company that generate wastewater, but not all the points produce high levels of waste. There are usually only a few point sources that generate waste water high in BOD, COD, TS, and TSS. Wastewater from these points must be separated from the other wastewater streams, and to do so requires a prioritized treatment plan.

7.7 SUMMARY

Utilization of by-products from surimi processing is an important issue for the seafood industry. There is increasing concern worldwide about natural resource utilization. There is also pressure from environmental groups to have less waste in the seafood industry as a whole. Combined political and environmental pressures make it imperative that the seafood industry utilize as fully as possible what is harvested from the seas. An equally important concern for the industry is whether waste utilization makes economic sense. In the past, Olympic fisheries required that harvesting and processing be as fast as possible to assure high production rates. It was not uncommon to have yields below 15% prior to 1992 as more profit could be made in the higher-grade surimi. This practice, however, has changed because fisheries management has divided the seasons, allowing certain segments of the industry to have quota management within their sector. This management allows fishing companies to determine a best-use scenario for both their processing and waste utilization. This rationalization of the fisheries has improved yields and allowed the industry to better use by-products and explore new markets.

Wastewater remains the biggest problem for the surimi industry. It is important for companies to develop long-range strategies for wastewater treatment and protein recovery. The driving force in the past was primarily environmental; for the purpose of reducing protein loads in waste streams and minimizing costs for its disposal. The recovery of myofibrillar protein in surimi washwater that is then put back into primary products has been an economic gain for the industry. There has gradually been a shift in strategies to try to incorporate both removal and utilization of other proteins and bioactive compounds although the economics of cost recovery of these compounds from wastewater are still not profitable. It is necessary to clearly define goals for by-product recovery, as they directly determine the technology that will be employed and the costs for operation and maintenance. A technology adopted to recover solids for human consumption will be more expensive than one developed for producing animal feeds or landfills. One strategy might include a combination of different technologies for achieving different goals and developing more cost-effective and profitable surimi by-product utilization operations.

REFERENCES

1. C.M. Lee. Surimi process technology. *Food Technol.*, (11), 69–80, 1984.
2. T.M. Lin, J.W. Park, and M.T. Morrissey. Recovered protein and reconditioned water from surimi processing waste. *J. Food Sci.*, 50(1), 4–9, 1995.
3. Y.J. Wu, M.T. Ataliah, and H.O. Hultin. The proteins of washed, minced fish muscle have significant solubility in water. *J. Food Biochem.*, 15, 209–218, 1991.
4. J. Lin and J.W. Park. Extraction of proteins from Pacific whiting mince at various washing conditions. *J. Food Sci.*, 61, 432–438, 1996.
5. A. Ismond. Best Management Practices and Low-Tech Solutions for Increasing the Efficiency of Seafood Processing Plants. Report prepared by Aqua-Terra Consultants, for the Department of Agriculture, Fisheries and Food of British Columbia. 24 pages. ISBN 0-7726-2365-1, 1994a.
6. A. Ismond. How to Do a Seafood Processing Plant Water, Waste, and Wastewater Audit. Report prepared by Aqua-Terra Consultants, for the Department of Agriculture, Fisheries and Food of British Columbia. ISBN 0-7726-2366-X, 1994b, 1–30.

7. S. Barlow. World market overview of fish meal and oil. In P.J. Bechtel, Ed. *Advances in Seafood ByProducts: 2002 Conference Proceedings.* Fairbanks, AK: University of Alaska, Alaska Sea Grant College Program, 2003.

8. R. Hardy. Marine byproducts for aquaculture use. In: P.J. Bechtel, Ed. *Advances in Seafood ByProducts: 2002 Conference Proceedings.* Fairbanks, AK: Alaska Sea Grant College Program, University of Alaska, 2003.

9. S. Benjakul and M.T. Morrissey. Protein hydrolysates from Pacific whiting solid wastes. *J. Agric. Food Chem.,* 45, 3423–3430, 1997.

10. J.M. Regenstein, S. Goldhur, and D. Graves. Increasing the value of Alaskan pollock byproducts. In P.J. Bechtel, Ed. *Advances in Seafood ByProducts: 2002 Conference Proceedings.* Fairbanks, AK: Alaska Sea Grant College Program, University of Alaska, 2003.

11. K. Hilderbrand. Unpublished data. Oregon State University Hatfield Marine Science Center, Newport, OR, 1998.

12. K. Lopetcharat and J.W. Park. Biochemical and microbial characteristics of Pacific whiting and surimi by-product during fish sauce fermentation. Presented at the *Annual Meeting of the Institute of Food Technologists,* Chicago, IL, 1999.

13. S. Tungkawachara and J.W. Park. Development of Pacific whiting fish sauce; market potential and manufacturing in the United States. In P.J. Bechtel, Ed. *Advances in Seafood ByProducts: 2002 Conference Proceedings.* Fairbanks, AK: Alaska Sea Grant College Program, University of Alaska, 2003.

14. J. Nettleton. *Omega-3 Fatty Acids and Health.* New York: Chapman & Hall, 1995.

15. R. Katz and J. Nettleton. Omega-3 fatty acids in health, nutrition, and disease: future U.S. market considerations. In P.J. Bechtel, Ed. *Advances in Seafood ByProducts: 2002 Conference Proceedings.* Fairbanks, AK: Alaska Sea Grant College Program, University of Alaska, 2003.

16. J.R. Hibbeln. Seafood consumption, the DHA content of mothers' milk and prevalence rates of postpartum depression: a cross-national, ecological analysis. *J. Affect. Disorder,* 69(1–3), 15–29, 2002.

17. A.C.M. Oliveira and P.J. Bechtel. Alaska pollock byproducts: lipid content and composition. In *Proceedings of the First Joint Trans Atlantic Fisheries Technology Conference (TAFT)*. Reykjavik: Icelandic Fisheries Laboratories, 2003.

18. C. Crapo and P. Bechtel. Utilization of Alaska's seafood processing byproducts. In P.J. Bechtel, Ed. *Advances in Seafood ByProducts: 2002 Conference Proceedings*. Fairbanks, AK: Alaska Sea Grant College Program, University of Alaska, 2003.

19. L. Huang. Application of Membrane Filtration to Recover Solids from Protein Solutions. Ph.D. thesis, Oregon State University, Corvallis, OR, 1997.

20. S.H. Goldhor and J.D. Koppernaes. *A Seafood Processor's Guide to Water Management*. New England Fisheries Development Association, Inc., Charlestown, MA, 2001.

21. J.C. Cheftel, J.L. Cuq, and D. Lorient. Amino acids, peptides, and proteins. In O.R. Fennema, Ed. *Food Chemistry*. New York: Marcel Dekker, 1985, 245–369.

22. D.R. Kasper and J.C. Reichenberger. Use of polymers in the coagulation process. *Proceedings of AWMA Seminar on Use of Organic Polyelectrolytes in Waste Treatment*, Las Vegas, NV, 1983.

23. C. Lind. A coagulant road map. *Public Work*, (3), 36–38, 1995.

24. D. Knorr. Recovery and utilization of chitin and chitosan in food processing waste management. *Food Tech.*, 1, 114–122, 1991.

25. R.A. Johnson and S.M. Gallanger. Use of coagulants to treat seafood processing waste waters. *J. Water Poll. Control Fed.*, 56, 970–976, 1984.

26. V. Savant and J.A. Torres. Protein adsorption on chitosan-polyanion complexes: application to aqueous food processing wastes. *Proceedings from the 7th Conference of Food Engineering (CoFE 2001)*, A Topical Conference of 2001 AIChE Annual Meeting. Nov. 5–9, 2001.

27. W. Fresenius, W. Schneider, B. Bohnke, and K. Poppingaus. *Waste Water Technology — Origin, Collection, Treatment and Analysis Of Waste Water.* Berlin, Germany: Springer-Verlag, 1989.

28. M. Okada. History of surimi technology. In T.C. Lanier and C.M. Lee, Eds. *Surimi Technology.* New York: Marcel Dekker, 1992, 3–39.

29. K.A. Mikkelson and K.W. Lowery. Designing a sequential batch reactor system for the treatment of high strength meat processing wastewater. *Proceedings of 47th Industrial Waste Conference,* West Lafayette, IN, 1992.

30. R.J. Pearce. Whey protein recovery and whey protein fractionation. In J.G. Zalow, Ed. *Whey and Lactose Processing.* New York: Elsevier Science, 1992.

31. C. Sandu and R.K. Singh. Energy increase in operation and cleaning due to heat exchanger fouling in milk pasteurization. *Food Tech.,* 45(12), 84–91, 1991.

32. L. Huang, Y. Chen, and M.T. Morrissey. Coagulation of fish proteins from frozen fish mince wash water by ohmic heating. *J. Food Process. Eng.,* 20, 285–300, 1997.

33. C.W. Dalrymple. Use of electrocoagulation for wastewater treatment. *Proceedings of Wastewater Technology Conference and Exhibition,* Vancouver, BC, 1994.

34. R.H. Perry, D.W. Green, and J.O. Maloney. *Perry's Chemical Engineers' Handbook.* New York: McGraw-Hill, 1984.

35. T.C. Swafford, J. Babbitt, K. Reppond, A. Hardy, C.C. Riley, and T.K.A. Zetterling. Surimi process yield improvement and quality contribution by centrifuging. In R.E. Martin and R.L. Collette, Eds. *Proceedings of the International Symposium on Engineered Seafood including Surimi.* National Fisheries Institute, Seattle, WA, 1990, 483–496.

36. J.D. Dziezak. Membrane separation technology offers processors ultimate potential. *Food Tech.,* (9), 108–113, 1990.

37. E. Renner, and M.H. Abd El-Salam. *Application of Ultrafiltration in the Dairy Industry.* New York: Elsevier Science, 1991.

38. L.D. Petersen, C. Crapo, J. Babbitt, and S. Smiley. Membrane filtration of stickwater. In P.J. Bechtel, Ed. *Advances in Seafood Byproducts.* Fairbanks, AK: Alaska Sea Grant College Program, University of Alaska 2003.

39. V. Goel, M.A. Accomazzo, A.J. Dileo, P. Meier, A. Pitt, and M. Pluskal. Dead-end microfiltration: application, design, and cost. In W.S.W. Ho and K.K. Sirkar, Eds. *Membrane Handbook*. New York: Van Nostrand Reinhold, 1992.

40. Y. Miyata. Concentration of protein from the wash water of red meat fish by ultrafiltration membrane. *Bull. Japan Soc. Sci. Fish.*, 50(4), 659–663, 1984.

41. K. Ninomiya, T. Ookawa, T. Tsuchiya, and J. Matsumoto. Recovery of water soluble protein in waste wash water of fish processing plants of ultrafiltration. *Bull. Japan Soc. Sci. Fish.*, 51(7), 1133–1138, 1985.

42. A. Watanabe, T. Ohtani, H. Horikita, H. Ohya, and S. Kimura. Recovery of soluble protein from fish jelly processing with self-rejection dynamic membrane. In L. Le Maguer and P. Jelen, Eds. *Food Engineering Process Applications, Vol 2*, 1986, 4th International Conference on Engineering and Food, University of Alberta, Edmonton, 225–236.

43. A. Watanabe, T. Shoji, M. Nakajima, H. Nabetani, and T. Ohtani. Electronmicroscopic observation of self-rejection type dynamic membrane formed with water soluble proteins in waste water from fish paste process. *Nippon Nogeikagaku Kaishi*, 62(7), 1061–1066, 1988.

44. T. Shoji, M. Nakajima, H. Nabetani, T. Ohtani, and A. Watanabe. Effect of pore size of ceramic support on the self-rejection characteristics of the dynamic membrane formed with water soluble proteins in waste water. *Nippon Nogeikagaku Kaishi*, 62(7), 1055–1060, 1988.

45. P. Jaouen, M. Bothorel, and F. Quemeneur. Recovery of soluble proteins by membrane separation process: performances and membrane cleaning. Preprints of *6th International Symposium on Synthetic Membranes in Science and Technology*, Thubingen, Germany, 1989.

46. P. Jaouen, M. Bothorel, and F. Quemeneur. Treatment of seafood processing effluents through microporous membranes — applications and prospects. *Proceedings of International Congress on Membranes and Membrane Process*, Chicago, IL, 1990.

47. P. Jaouen and F. Quemeneur. Membrane filtration for wastewater protein recovery. In G.M. Hall, Ed. *Fish Processing Technology*. New York: VCH Publishers, 1992.

48. L.D. Peterson. Product recovery from surimi wash water. In S. Keller, Ed. *Making Profits out of Seafood Wastes. Proceedings of the International Conference on Fish By-Product,* Anchorage, AK, 1990.

49. L. Huang and M.T. Morrissey. Fouling of membranes during microfiltration of surimi wash water: roles of pore blocking and surface cake formation. *J. Membr. Sci.,* 144, 113–123, 1998.

50. M.T. Morrissey. Marine biotechnology. In *Proceedings of the First Joint Trans Atlantic Fisheries Technology Conference (TAFT).* Reykjavik: Icelandic Fisheries Laboratories, 2003.

51. T. Ohshima. Recovery and use of nutraceutical products from marine resources. *Food Tech.,* 52(6), 50–54, 1998.

52. T. Ohshima. By-products and seafood production in Japan. *J. Aquat. Food Prod. Tech.,* 54, 27–42, 1996.

53. N.F. Haard. Specialty enzymes from marine organisms. *Food. Tech.,* 52(7), 65–67, 1998.

54. L. Huang, M. Santos, P.R. Singh, and M.T. Morrissey. Characterization of Waste Water from the Surimi Processing Industry. Report, Oregon State University Seafood Laboratory, Astoria, OR, 1996.

55. T. Suzuki. *Fish and Krill Protein Processing Technology.* London, U.K.: Applied Science Publishers, 1981.

56. N.F. Haard, B.K. Simpson, and B.S. Pan. Sacroplasmic proteins and other nitrogenous compounds. In Z.E. Sikorski, B.S. Pan, and F. Shahidi, Eds. *Seafood Proteins.* New York: Chapman & Hall, 1994.

57. N.F. Haard. A review of proteolytic enzymes from marine organisms and their application in the food industry. *J. Aquat. Food Prod. Tech.,* 1(1), 17–36, 1992.

58. A. Gildberg. Enzymes and bioactive peptides from fish waste related to fish silage, fish feed and fish sauce production. In *Proceedings of the First Joint Trans Atlantic Fisheries Technology Conference (TAFT).* Reykjavik: Icelandic Fisheries Laboratories, 2003.

59. H. An, V. Weerasinghe, T.A.S. Seymour, and M.T. Morrissey. Cathepsin degradation of Pacific whiting surimi proteins. *J. Food Sci.,* 59, 1013–1017 and 1033, 1994.

60. N. Seki, H. Uno, N.H. Lee, I. Kimura, K. Toyoda, T. Fujita, and K. Arai. Transglutaminase activity in Alaska pollock muscle and surimi, and its reaction with myosin B. *Nippon Suisan Gakkaishi*, 56(1), 125–132, 1990.

61. H. Araki and N. Seki. Comparison of reactivity of transglutaminase to various fish actomyosins. *Nippon Suisan Gakkaishi*, 59, 711–716, 1993.

62. H. An, M.Y. Peters, and T.A. Seymour. Roles of endogenous enzymes in surimi gelation. *Trends Food Sci. Tech.*, 7, 321–327, 1996.

63. M. Gudmindsson, H. Hafsteinson. Gelatin from cod skins as affected by chemical treatments. *J. Food Sci.*, 62, 37–39, 47, 1997.

64. J.S. Kim and J.W. Park. Characterization of acid-soluble collagen from Pacific whiting surimi processing by-products. *J. Food Sci.*, 89, C637–642, 2004.

65. J.M. Regenstein. Characterization of several fish gelatins. In F. Shahidi, Y. Jones, and D.D. Kitts, Eds. *Seafood Safety, Processing, and Biotechnology.* Lancaster, PA: Technomic Publications 1997, 187–197.

66. H.W. Ockerman and C.L. Hansen. Seafood by-products. In *Animal By-Product Processing.* New York: VCH Publications, 1998, 279–308.

67. D.J. Garcia. Omega-3 long-chain PUFA nutraceuticals. *Food Tech.*, 52(6), 45–49, 1998.

68. J. Kaur, M.T. Morrissey, L. Yang, and E.A. Decker. Concentration of Anserine and Carnosine in Surimi Wash Water. Presented at the *Annual Meeting of the Institute of Food Technologists,* Atlanta, GA, 1998.

69. K.M. Chan and E.A. Dekker. Endogenous skeletal muscle antioxidants. *Crit. Rev. Food Sci. Nutr.*, 34, 403–426, 1994.

70. C. DeWitt and M.T. Morrissey. Optimization of parameters for the recovery of cathepsin proteases from surimi waste water. Presented at the *50th Pacific Fisheries Technologist Meeting,* Parksville, BC, 1999.

71. A. Ismond. The impact of food safety and competitive markets on byproduct recovery strategies. In P.J. Bechtel, Ed. *Advances in Seafood ByProducts: 2002 Conference Proceedings.* Fairbanks, AK: Alaska Sea Grant College Program, University of Alaska, 2003.

8

Freezing Technology

EDWARD KOLBE
Oregon and Alaska Sea Grant, Portland, Oregon

CONTENTS

8.1 Introduction 326
8.2 Horizontal Plate Freezers 327
8.3 Airflow Freezers 331
    8.3.1 Spiral Freezer 331
    8.3.2 Tunnel Freezer 332
    8.3.3 Blast Freezer... 334
8.4 Brine Freezers 336
    8.4.1 Sodium Chloride (NaCl).................................... 337
    8.4.2 Calcium Chloride ($CaCl_2$)................................ 337
    8.4.3 Glycols 338
    8.4.4 Other 338

8.5 Cryogenic Freezers...... 338
    8.5.1 Liquid Nitrogen (LN) 339
    8.5.2 Carbon Dioxide (CO_2)...... 340
8.6 Freezing the Product...... 342
8.7 Freezing Capacity...... 344
8.8 Freezing Time...... 347
8.9 Some "What-If" Effects on Freezing Time...... 356
    8.9.1 Block Thickness 357
    8.9.2 Cold Temperature Sink, T_a 357
    8.9.3 Heat Transfer Coefficient, U 359
8.10 Energy Conservation...... 362
    8.10.1 Freezer Design and Operation 363
    8.10.2 Refrigeration Machinery Options...... 366
    8.10.3 A Blast Freezer Case 367
8.11 Conclusions 368
Acknowledgments...... 368
References...... 369

8.1 INTRODUCTION

The role of a commercial freezer is to extract heat from a stream of product, converting most of the free moisture to a solid. This extraction must be sufficiently rapid so the product will experience minimum quality degradation, the rate keeps pace with the production schedule, and the average product temperature upon exit roughly matches the subsequent storage temperature. This chapter addresses the general approaches used to freeze surimi and surimi seafoods.

    Surimi, which is stabilized, uncooked, washed mince, is frozen almost exclusively as rectangular blocks in plate freezers. Surimi seafoods, on the other hand, are cooked and formed products made from surimi and other ingredients. Freezing of these products most commonly occurs in contact with cold air or cryogenic gases in blast or spiral freezers. Before addressing the issues of freezing rates and predictions, the freezing systems currently in use are reviewed.

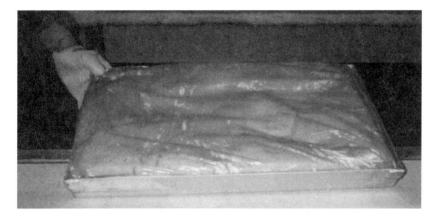

Figure 8.1 A block of surimi, weighed, bagged and placed in a stainless steel pan on its way to the plate freezer.

8.2 HORIZONTAL PLATE FREEZERS

Horizontal plate freezers used by the surimi industry are well suited for surimi blocks having predictable dimensions, rectangular shapes, and large flat surfaces. Typically, a 10-kg block of surimi is metered into 2- or 3-mil polyethylene bags, with final block dimensions of approximately 56 × 310 × 590 mm. The end of the bag is then folded under the block, which is placed within an aluminum or stainless steel tray (Figure 8.1). After loading onto the plate freezer shelves, hydraulic rams press the upper shelf hard onto the package and appropriate spacers. This ensures uniform thickness of the frozen package (Figure 8.2). Consequently, both underfilling and overfilling can create difficulties in the process. Underfilling allows for air gaps and poor heat transfer at the upper surface of an individual pan, whereas overfilling may lead to package distortion or tearing and possible air gaps at the top of adjacent packages on the plate.

The movable shelves in a typical plate freezer are most commonly made of extruded aluminum, with internal channels to pass refrigerant from one side of the plate to the other (Figure 8.2). Plates move up and down as the operator loads and unloads blocks; thus, flexible hoses supply refrigerant to

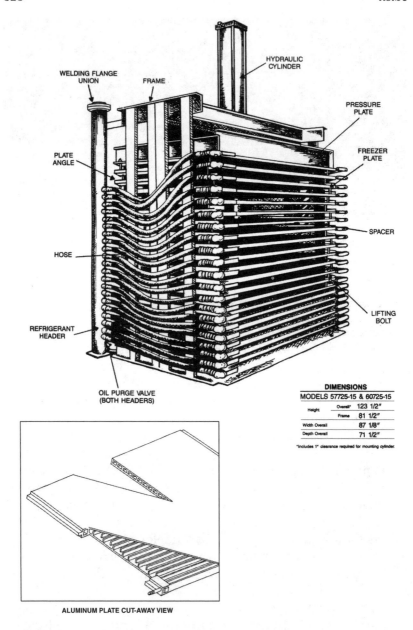

Figure 8.2 A horizontal plate freezer shown without the enclosure housing. Inset shows cut-away of an extruded aluminum plate. (Courtesy Dole Refrigerating Co.)

the plates. An older plate design consists of a serpentine grid of square steel tubing housed within two parallel steel sheets making up the surfaces of the plate. An outer border seals this thin box, while a vacuum and filler material ensure good heat transfer between tubing and the outer plate. The steel plates tend to be on the order of 15% less expensive than those made of extruded aluminum. However, the one-piece aluminum structure ensures better heat transfer from the outer plate surface to the refrigerant flowing within the passages.

The refrigerant used in most shore-based surimi plate freezers is ammonia, which floods the plates from reservoirs installed outside the freezer enclosure. Heat flows from the freezing product to the relatively colder refrigerant, which continually vaporizes at a low pressure. The liquid/vapor mixture in this flooded system then flows, either by gravity (called a flooded system) or by means of a pump (in a liquid overfeed system), to one of the external reservoirs where the liquid/vapor mixture is separated. The liquid then goes back through the freezer plates. Meanwhile, the vapor is pumped first to the compressor where it is pressurized (to the "high side"), and then to the condenser where it is liquefied and subsequently returned to the low-pressure reservoir.

The cold refrigerant temperature inside the plates is uniquely related to the vapor pressure because the mixture is saturated (i.e., the liquid and vapor exist in equilibrium). Thus, from reading the suction pressure on a gauge and then finding the unique saturation conditions for that particular refrigerant, the boiling refrigerant temperature (called the saturated suction temperature) can be determined. In virtually all cases, the suction pressure gauge will have the corresponding saturation temperatures printed on its face for the particular refrigerant used in the freezer. During product freezing, a few degrees difference exists between the plate surface and the flowing refrigerant. However, by reading the suction pressure gauge, the approximate plate temperature can be determined.

Refrigerants other than ammonia are also used in various circumstances. Common halocarbon examples are R-502, R-402, R-404a, and R-22, which are sometimes referred to by the generic name (and former trade name of Dow Chemical Co.) "Freon." Refrigerant R-502 was long used for low-temperature freezing; but due to its atmospheric ozone-depletion characteristics, it is now replaced by R-402 and R-404a. R-22, however, is a cheaper and more commonly available refrigerant. R-22 is often used in liquid overfeed plate-freezing systems on surimi factory ships to avoid some safety and regulatory difficulties with ammonia, although the per-kilogram cost of R-22 is greater (by a factor of 7) than that of ammonia.

Halocarbons are also common in small systems. These small systems use these refrigerants in a self-contained compressor/condenser unit and require a minimum volume of refrigerant to operate. The refrigerants are injected into the plates through a restriction valve in what is called a dry expansion (vs. flooded or liquid overfeed) supply system. This means that high-pressure liquid is metered into the low-pressure plate passages by a thermostatically controlled expansion valve. The expansion valve enables only enough liquid needed to absorb the freezing load at that moment. Thus, the liquid becomes completely vaporized shortly before reaching the end of the passage, where it is then sucked up by the compressor. A major disadvantage of a dry expansion system is that temperature and heat transfer conditions inside the plates tend to vary throughout the freezing cycle. Initially, the thermostatic expansion valve cannot supply an adequate flow rate and the temperature tends to increase. In a liquid overfeed system however, pumps continually supply 3 to 4 times the refrigerant needed to absorb the heat load. The temperatures and conditions, therefore, remain relatively uniform.[1]

A third type of system, relatively less common than liquid overfeed, uses a secondary refrigerant, such as propylene glycol, that is refrigerated in a heat exchanger outside the freezer and then pumped through the plates. A somewhat uniform passage temperature can be maintained if the flow rate is sufficiently high. The advantage of this system is that potential leaks (e.g., at the flexible supply hoses) present less risk. The disadvantage is the increased energy consumption caused by additional pumping and by the lower (and, thus, less-efficient) evaporating refrigerant temperature in the external heat exchanger required by the secondary refrigerants.

8.3 AIRFLOW FREEZERS

Three different air-blast systems are typically used for freezing surimi seafoods: spiral, tunnel, and blast freezers. The type of freezer is defined by the nature of the cabinet where air and product come in contact. In each system, however, the refrigeration machinery pumps liquid, whether ammonia or halocarbons, to an evaporator. In the evaporator, powerful fans blow air through a bank of finned tubing. Inside the tubing, refrigerant boils off, absorbing heat from the recirculating air. The cold air then heads off to collect more heat from the freezing product.

Unless the products are packaged, all of these systems must be frequently defrosted because any moisture lost from freezing product deposits as frost on the colder coil surfaces. Consequently, frost quickly restricts the amount of airflow and diminishes the transfer of heat. Defrost cycles are anywhere from once a week to once every freeze cycle, depending on the product (wet or packaged) and freezer design.

8.3.1 Spiral Freezer

In a spiral freezer, product enters the freezer on a belt, which travels in a spiral motion through a near-cube-shaped room (Figure 8.3). The airflow direction, depending on the manufacturer, is horizontal, vertical, or some combination thereof as it flows over the product riding along on the belt. Both the belt speed and loading density can be controlled to ensure complete freezing. Most manufacturers also include an option for some kind of continuous belt-cleaner outside the freezer.

Spirals enable a continuous-feed, well-controlled freezing process at very low air temperatures (typically approaching –40°C). Spirals can also support a variety of product shapes and sizes. In addition, spiral freezers occupy a relatively small amount of floor space in the plant.

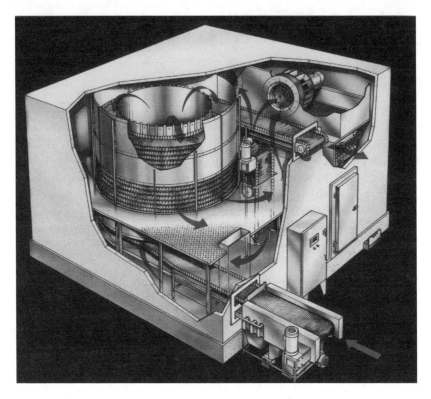

Figure 8.3 Spiral blast freezer. (Courtesy Northfield Freezing Systems.)

8.3.2 Tunnel Freezer

The tunnel freezer, in its simplest form, is a straight, continuous-link belt carrying product through a tunnel. In the tunnel, the product is pummeled with high-velocity cold air supplied by evaporator/fan units housed alongside (Figure 8.4). Different manufacturers present a number of variations to this theme. For example, airflow can be horizontal or vertical. The route of the product through the tunnel can also be made up of several passes to prolong the time in the freezer. In addition, different belts can operate at different speeds. For example, wet product is loaded in a single layer on the first belt to effect a quick-frozen outer crust, thus forming a

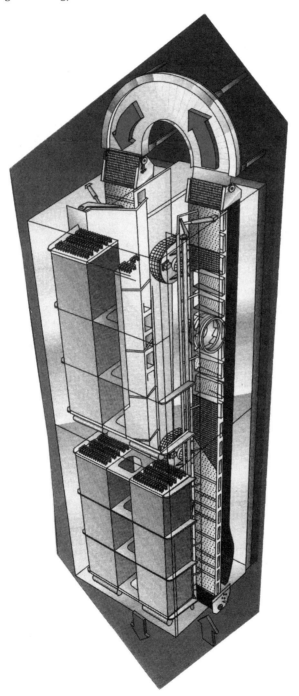

Figure 8.4 Tunnel blast freezer. (Courtesy Frigoscandia, Inc.)

good moisture barrier. It then dumps onto a second, slower belt where the product completes its requisite freeze time.

In a fluidized-bed variation, small "individually quick-frozen" products, such as shrimp, surimi seafood chunks or flakes, peas, and berries, riding on this second belt are literally suspended in a blast of cold air directed upward through the belt and product mass. In another variation, some tunnels accommodate wet or runny products, which are deposited on a solid belt having cold contact from below.

Tunnels perform the same function as spirals but take up more floor space. Tunnel freezers also tend to be mechanically simpler and require perhaps less attention to adjustment and maintenance.

8.3.3 Blast Freezer

The term "blast freezer" commonly refers to a batch freezing operation during which product is wheeled into a room or large cabinet where it remains until frozen. Blast freezers are sometimes called "batch-continuous" systems if the trolleys are periodically removed row-by-row on a first-in-first-out basis. Powerful fans blow circulating air through evaporators located above the ceiling (Figure 8.5). The cold air then flows uniformly over the product loaded on the trolleys below. Large products that take a long time to freeze are more suited to a batch process than to a continuous process in a spiral or tunnel freezer.

Assuming proper design, a major impediment to performance involves product loading and uniform airflow. Without a balanced resistance and/or the proper vanes to direct air as it first hits the trolleys, flow velocities over products in different locations will vary significantly. As shown later (see Figure 8.15), such a variation can seriously affect freezing time. Without uniform resistance to the air over the cross-section of the room, the air will seek the shortcut, or path of least resistance, resulting in imbalanced velocities (Figure 8.6). A second flow problem results when air turning the first corner and hitting the trolley is not properly directed (Figure 8.7). Figure 8.7 shows how airflow in one blast freezer is

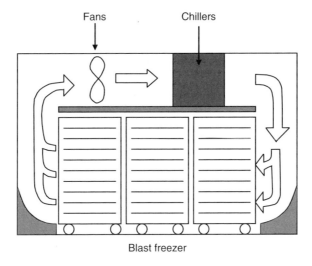

Figure 8.5 Batch blast freezer with product loaded on trolleys.

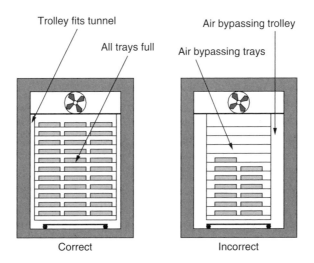

Figure 8.6 Loading an air blast freezer. (Adapted from Reference 22.)

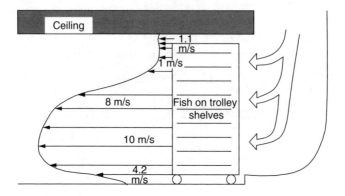

Figure 8.7 Unbalanced air velocities in an empty blast freezer. Profile shows air velocity, in meters per second. (Adapted from Reference 9.)

distributed over a "load" represented by just one lightly loaded trolley.

8.4 BRINE FREEZERS

Brines typically refer to salt solutions, but glycols and other fluids are mentioned as well. These liquids, after flowing over the freezing product, circulate through a heat exchanger where they give up heat to the vaporizing refrigerant. The cold brine (sometimes referred to as a secondary refrigerant or secondary coolant) is then pumped back to sprayers or immersion tanks where it removes heat from the freezing product. In general, the product is wrapped in a sealed package to prevent direct contact of the brine with the food. A common exception is on-board tuna boats and some shrimp boats in southern waters, where whole fish or shrimp freeze in direct contact with refrigerated sodium chloride brine.

Brines, glycols, or other heat transfer fluids can be sprayed directly on a package (or on a fish in the case of albacore tuna boats). For some brines, in spray systems, antifoaming agents are necessary. The fluid might alternatively circulate through a tank in which the products are immersed. In some older systems, brines are sprayed against the bottom surface of a continuous stainless steel, sheet metal belt, rapidly freezing fillets and similar products. For packages frozen in direct contact with fluids, some washing of the surface may be necessary.

One major advantage of brine freezers over air blast freezers is that they can operate with higher energy efficiency. This relates to better heat transfer and higher operating temperatures. The "heat transfer coefficient" is a term that describes the relative ease with which heat flows from the freezing surface of the product to the surrounding cold medium. In brines and other liquids, this coefficient tends to be higher than in air blast systems. As a result, it can be shown that for some product sizes, freezing can be about as rapid in relatively warm (say, −20°C) brines, as in the much colder (perhaps −40°C) air blast systems.[2] This means that the refrigerant evaporating temperatures, and thus the energy efficiency, will be higher. This is further explained in a subsequent section (see Figure 8.11).

8.4.1 Sodium Chloride (NaCl)

Mixing sodium chloride with water at a weight ratio of 23.3% results in the eutectic point of −20.6°C, the lowest freezing temperature of the brine.[3] It is not practical, however, to operate at a temperature this low, caused in part by the temperature drop that must exist between the brine and the vaporizing refrigerant in the heat exchanger. Attempts to freeze at such a low temperature would cause the brine to begin freezing on the relatively lower-temperature tube walls. Consequently, a practical minimum operating temperature for sodium chloride brine is around −17°C.

8.4.2 Calcium Chloride (CaCl$_2$)

The theoretical eutectic point of calcium chloride brine is −55°C in a solution of 29.9% by weight.[4] Thus, calcium chloride brine can operate at a lower temperature than sodium chloride without the same risk of freeze-up. However, the brine becomes viscous at such low temperatures. Handbooks indicate a minimum application temperature closer to −20°C. In addition, unlike sodium chloride, calcium chloride in contact with food is considered problematic.

8.4.3 Glycols

Both ethylene glycol and propylene glycol are used as secondary refrigerants, although the former is shunned for food applications because of its greater toxicity. The advantage of ethylene glycol is that it is less viscous at low temperatures. Both glycols, however, have added corrosion inhibitors, as do many of the salt brines, and this creates some environmental or treatment plant problems regarding disposal. When mixed with water, the freezing point of propylene glycol can be as low as −50°C. However, as with calcium chloride brine, the fluid is extremely viscous at such low temperatures, demanding high pumping energy and diminished heat transfer. Consequently, a −18°C operating temperature at a mixture of roughly 50:50 is a minimum temperature target.[1,4]

8.4.4 Other

A few manufacturers advertise proprietary formulations of "heat transfer fluids," possibly petroleum or silicone oil-based, that can operate easily at −40°C. Although quite expensive, such a cold temperature, combined with any agitation at all, would ensure a rapid freezing rate of small, individual packages.

8.5 CRYOGENIC FREEZERS

The term "cryogenic" simply means very low temperatures; and in food freezing, it refers almost exclusively to liquid nitrogen and liquid carbon dioxide systems. Cryogenic freezers differ dramatically from mechanical refrigeration. The cryogenic refrigerant, sprayed or pumped directly into the freezing cabinet or tank, collects heat as it vaporizes and eventually discharges into the atmosphere. Temperatures can be very low. Liquid nitrogen at atmospheric pressure vaporizes at −196°C; whereas, liquid carbon dioxide first turns to "snow" when vented into the atmosphere and then vaporizes at −78°C. As the result of these low temperatures, freezing is very rapid and under some conditions can even be too fast. For example, a whole fish fillet dropped into a liquid nitrogen bath is likely to be damaged because of fracture.

Fast freezing and the resulting low dehydration loss are major advantages of cryogenics. In addition, cryogenic freezers have a high reserve freezing capacity. They can be easily turned up or down if the production rate is uncertain. However, in current costs per kilogram frozen product, freezing with liquid nitrogen and liquid carbon dioxide is more expensive than mechanically refrigerated freezing due to the high cost of the fluids. Analyses that estimate these costs tend to be sensitive to product value, perceived value enhancement due to rapid freezing, and the expected (or assumed) reduction of dehydration loss with cryogenic freezing. To optimize these cost factors, a few firms now market a combination "cryomechanical" freezer, particularly for unpackaged foods. Cryomechanical freezers use cryogenics in the first stage (for rapid crust freeze), followed by a mechanical air blast system.

8.5.1 Liquid Nitrogen (LN)

A plant using a liquid nitrogen (LN) system would need to erect a holding tank while setting up a contract with a supplier. The supply must be reliable and close enough so that delivery costs are manageable. One supplier noted that about 150 to 200 km by road was a reasonable operating distance. Most companies can also set up automatic monitoring and refill options.

Storage tank size depends on the production rate and proximity of the supplier. Common capacities range from 3400 to 45,000 L (900 to 12,000 gal). A conservative estimate is that 1.5 kg LN is used as each kilogram of product is frozen. [1 L LN has a mass of 0.8 kg (or 50.4 lb.m/ft^3)] The tank is also designed for 200 to 1700 kPa (20 to 250 psi) of pressure. The tanks should be well insulated to minimize heat leakage from outside. However, the temperature difference between LN and outside air is extremely large, and a "perfect" insulator is not available. Therefore, some small but continuous rate of evaporation and loss will occur. Users report losses that can exceed 1% per day. For small tanks, the loss would be higher. Additional losses may occur by heat leakage into the pipes supplying the freezer, thus minimizing that distance and using good insulation is significant. Vacuum-jacketing of supply lines longer than 30 to 50 m may also be cost effective.

One typical LN freezer is designed as a tunnel, with LN spray heads located near the exit end (Figure 8.8). As the evaporating liquid spray absorbs heat from the freezing product, fans blow the cold vapor in a direction counter to that of the conveyed product. This enables the vapors to rapidly pre-cool and crust-freeze the entering product. Consequently, the tunnel system allows for efficient use of the refrigerant and avoids radical temperature changes and stress that might cause the product to crack.

Liquid nitrogen is also used in spiral freezers; and for small lots, liquid nitrogen is used in cabinet freezers (Figure 8.9). In both systems, fans are used to maintain high gas velocity over the product surface.

8.5.2 Carbon Dioxide (CO_2)

As liquid CO_2 vents into the atmosphere, it converts first to a solid/vapor mixture that might appear as "snow" in the freezer cabinet. (If compressed into blocks, this snow would become "dry ice.") The solid snow then converts into a vapor as it absorbs the heat of sublimation from the freezing product. The vapor can collect additional heat and then exhaust into the atmosphere.

Typical CO_2 insulated storage tanks range in capacity from 13,000 to 48,000 L (3500 to 12,000 gal) and store $-18°C$ (0°F) liquid at 2100 kPa (300 psi). [1 L liquid CO_2 has a mass of 1 kg (63.4 lb.m/ft^3).] Because the storage temperature is much higher than LN, losses caused by heat leakage can be controlled to zero. This is aided in hot-weather/low-use periods by a small mechanical refrigeration unit that will switch on to refrigerate coils lining the tank walls. Furthermore, as

Figure 8.8 Liquid nitrogen tunnel freezer. (Courtesy Air Products and Chemicals, Inc.)

Figure 8.9 Liquid nitrogen cabinet freezer. (Courtesy Martin/Baron, Inc.)

with LN, the supply company can provide full service, remote monitoring and troubleshooting, and timely refill as needed.

8.6 FREEZING THE PRODUCT

Some concerns regarding freezing operations include (1) controlling the freezing rate, (2) the time required to freeze the product, (3) the temperature, and (4) freezer capacity. In dealing with seafood, the rate of freezing is important to ensure quality, although this statement must be somewhat tempered for surimi and surimi seafood. For raw surimi blocks in common commercial plate freezers, the rate of freezing (or inversely, the freezing time) will have a minimum effect on product quality for short-term storage. This is because freezing will be sufficiently fast ("hours" vs. "days") to minimize protein denaturation or ice crystal damage in the surimi, which is a cryoprotectant-rich medium.

Surimi seafood, on the other hand, has a lower concentration of cryoprotectants because of the addition of water, salt, and other ingredients (e.g., modified starch). These cooked and frozen-stored products appear to have some sensitivity to freezing rate. Slowly frozen surimi seafood exhibits excessive free moisture upon thawing.

A rapid rate of freezing at the beginning of the process is important for unpackaged products. A quickly formed frozen crust will minimize moisture loss in air blast and cryogenic freezers. A rapid finish of the freezing process will then limit any further desiccation (i.e., sublimation of the ice layer) and loss of moisture (weight) to the high-velocity air swirling around the product.

Minimizing the time in the freezer is also important for production reasons. Freezers are often the "bottleneck" in the process, and hustling the product through helps keep pace with the line. In addition, the freezer is an expensive piece of machinery. Therefore, accelerating more products through eliminates the need to invest in more machines.

It is extremely important, however, that the product be kept in the freezer long enough to ensure complete freezing. Consider the consequences of taking out product before it is completely frozen:

- The refrigeration machinery in the cold store will become overloaded and the room will begin to warm. The machinery is sized not to freeze things, but to maintain low temperature.
- Fluctuating cold store room temperature will cause fluctuating temperatures on the surface of products that are already there. This leads to sublimation or ice transforming directly to water vapor, which causes dried-out surface patches, or "freezer burn."[5]
- Incompletely frozen product will finish freezing in the cold store. Long freezing times seriously affect quality.

One particularly bad example was measured to be on the order of "weeks."

Consequently, knowing how long the product must be held in the freezer is very important. This knowledge depends on an understanding of two terms: "freezing capacity" and "freezing time."

8.7 FREEZING CAPACITY

Freezing capacity is the rate at which energy is removed. It relates to the amount of heat that must be taken out of each kilogram, multiplied by the production rate in kilograms per hour. "Rate of transferring energy" is the engineering definition of "power;" thus, freezing capacity also dictates the size of the machinery in terms of power usage (expressed as horsepower or kilowatts).

To calculate capacity and power, observe the graphic display of heat removal at each temperature from a freezing block of surimi (Figure 8.10). The area under the curve represents the total amount of heat energy that must be removed to freeze a 10-kg block of surimi. As indicated, approximately 475 kiloJoules (kJ) of heat energy must be removed to prechill this block to its initial freezing point of $-1.5°C$. Then, over a short temperature span, from -1.5 to about $-7°C$, most of the water (about 72%) freezes to ice as 2220 kJ, the latent heat of fusion, is removed.

Depending on the specifications of the individual companies, surimi storage temperatures can fall in the range of -18 to $-30°C$. Only another 735 kJ are then required to take the now-mostly-frozen block down to its final average temperature, which is assumed here to be $-25°C$. Therefore, 3425 kJ must be removed from this 10-kg block to freeze it. The least required capacity of the freezer (i.e., of the refrigeration machinery) will be this unit of heat multiplied by the desired rate of production. If the plate freezer holds 60 blocks and it is desired to completely freeze a batch every 2 hr, the machinery must remove heat from the surimi at a rate of:

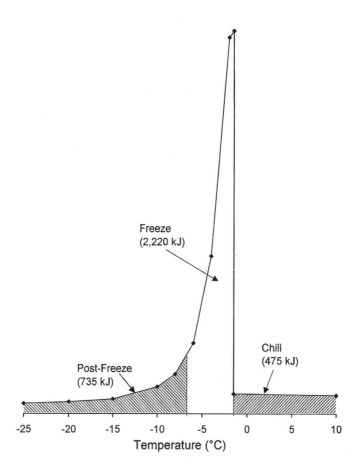

Figure 8.10 Heat capacity of a 10 kg block of surimi with 8% cryoprotectants; 80% moisture content. (Adapted from Reference 8.)

$$\frac{3,425 \text{ (kiloJoule s)}}{\text{Block}} \times \frac{(60 \text{ Blocks})}{2 \text{ hours}} = 102,750 \text{ (kiloJoule s / hr)}$$

This is equivalent to 28.5 kJ/sec (28.5 kW) of power.

Note: Terminology becomes confusing when English and Système Internationale (SI) units are interchanged. Many U.S. manufacturers still use the English term "BTU" (British thermal unit) for a unit of heat or energy, and "degrees

Fahrenheit" for a unit of temperature. These relate to SI units by the following conversions:

A temperature change of 1°C is the same as a 1.8°F change

1 Btu = 1055 Joules = 1.055 kJ

Horsepower is a rate of transferring energy

1 Horsepower (HP) = 0.7068 Btu/sec
= 0.7457 kJ/sec = 0.7457 kilowatt (kW)

1 Refrigeration ton = 12,000 Btu/hr = 3.52 kW

The term "one refrigeration ton" is the rate of heat removed when a ton (2000 lb) of ice melts in 24 hr

A refrigeration plant capable of removing 28.5 kW of heat energy will not adequately freeze the 60 blocks in this example. There are other heat loads on the system, such as heat that is in the structure (relatively warm plates, frames, housing), heat leakage from the outside room, and heat added by refrigerant pumping, which the machinery must remove as well. For plate freezers, anywhere from 5 to 20% of the total load might be contributed by these sources.[6,7] The fans used in blast freezers also represent additional loads that can be 30 to 40% of the total refrigeration requirement, because all of the energy that goes into the fans ends up as heat in the freezer. Consequently, with heat leakage, defrost, and structural cooling, only about 50% of the refrigeration machinery capacity is available for freezing the product in an air blast system.

The temperature at which the refrigeration unit is operating also influences the ability of the machinery to freeze at the target production rate. One of these temperatures is the condensing (high side) temperature, although this results in only a 1% loss of capacity for each approximately 1°C increase in the condensing temperature.[6] The saturated suction temperature (inside the plates, or inside the air chilling unit), however, has the most significant influence on capacity (Figure 8.11). The temptation might be to adjust (i.e, lower) the plate or refrigerant temperature to speed up freezing. However, decreasing a −34°C operating temperature by 5°C will diminish the refrigeration capacity in this example by 17% (Figure 8.11).

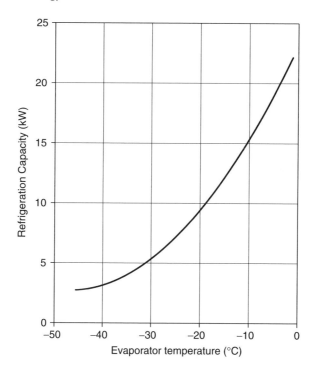

Figure 8.11 Refrigeration capacity of the machinery varies with saturated refrigerant suction temperature in this representative example.

8.8 FREEZING TIME

Assuming the system has the needed refrigeration capacity, the second concept to consider is freezing time. Freezing time is the elapsed time from the placement of the block or package in the freezer to the time its average temperature reaches some desired final value. This final temperature should be close to the expected storage conditions.

The core temperature of a freezing product will change with time (Figure 8.12). Note how much time it takes to prechill the core to the flat, near-horizontal plateau called the thermal arrest zone. Prechilling could remove much of this heat, decrease the overall freezing time, and thus increase the capacity of the freezer. If attempting to measure freeze time,

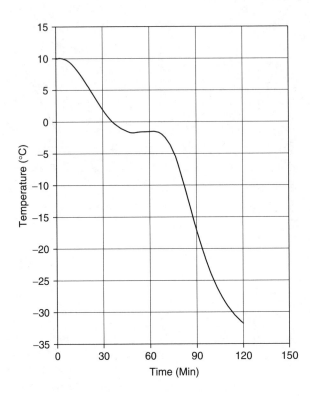

Figure 8.12 Typical core temperature curve in a freezing surimi block, 56 mm thick.

the temperature sensor should be positioned at the point that will freeze last and all measured freezing curves should show the flat thermal arrest zone. If there is no flat thermal arrest zone, the temperature sensor is likely in the wrong spot.

Although capacity determines how big the compressor should be, freezing time dictates how big the box or freezer compartment should be. Consider two different-sized freezers (Figure 8.13); both have the same capacity (kg/hr and kW of power). In the top freezer, thin packages freeze in 1 hr, whereas in the bottom freezer, thicker packages can take 3 hr. Although both have the same compressor size, the thick package freezer has to hold 3 times as much as the one freezing thin packages.

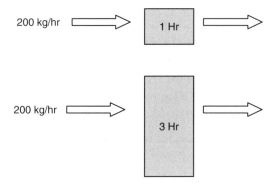

Figure 8.13 Freeze time affects freezer cabinet size.

As explained previously, the time a product is left in a freezer is critical to production, cost, and, potentially, quality. To control the freezing time, it must be determined for each situation, either by measuring directly or by calculating. Before discussing measurement, it is valuable to first look at Plank's equation. Plank's equation is one of the early equations used to estimate freezing time and, although not precise, is instructive because it includes most of the influencing parameters.

$$T = \frac{\rho L}{(T_F - T_A)} \left[\frac{Pd}{h} + \frac{Rd^2}{k} \right]$$

where T is the freezing time in minutes or hours.

Several parameters cannot be controlled, including:

ρ = product density (kg/m^3)
L = heat of fusion (kJ/kg). This is the heat that must be removed to freeze the water in the product (Figure 8.10).
k = thermal conductivity of the product (kW/mC). This is a measure of how quickly heat flows through the frozen product. It is mostly a function of moisture content but can be influenced by porosity and the direction of muscle fibers.

T_F = temperature at which product begins to freeze (°C). This is generally around −1.5°C for surimi and surimi seafood.[8]

P, R = geometry terms, which are fixed in value depending on the shape of the product, whether a flat slab, a long cylinder, or a sphere.

The remaining parameters can be controlled:

d = product thickness or diameter (m). As this increases, the heat path from the interior to the surface of the product lengthens and the freezing time increases. Not much can be done in plate freezers for blocks of fixed thickness. However, there may well be more options with the various kinds of freezers for surimi seafoods whose packages are varied.

T_A = ambient (or surrounding) temperature (°C). T_A can represent the temperature of the blast air swirling around a surimi seafood package. It can also be the temperature of refrigerant as it flows and vaporizes inside the horizontal plates of a plate freezer. The lower (colder) this value, the higher the "driving force" ($T_F - T_A$) pushing heat from the product, and the faster freezing takes place. However, as described previously, adjusting T_A downward will also reduce the refrigeration capacity.

Although T_A is often assumed to be some set value, it will in fact typically increase at the beginning of a freezing cycle. If the product is warm, the rate of heat flowing into the refrigeration system can briefly exceed its capacity and the saturated suction temperature, which relates to T_A initially rises. T_A then reaches a value at which the increased refrigeration capacity (Figure 8.11) matches the heat flow from the warm product. Later, as freezing proceeds and the heat flow rate decreases, the saturated suction temperature falls back to its set point. In plate freezers, this happens more dramatically in smaller, dry expansion systems than in the large liquid overfeed systems where it may not be noticeable.

Other things can locally affect T_A and freezing time. In a blast freezer, individual packages shielded from the blast may experience a local ambient temperature that is warmer than that of neighboring packages. Oil clogging a flow passage inside a freezer plate would result in a local "hot spot" and a freezing time increase in the adjacent block.

U = the heat transfer coefficient (W/m²C). This is a measure of how quickly heat is removed from the surface of the product through direct contact of the product with a freezer plate or by the motion of a surrounding air, gas, or brine over the outside surface. U can have an influence on freezing time that is greater than that of the other parameters, but it is the most difficult-to-predict parameter in the freeze-time model. Its value generally results from a combination of surface resistances. For plate freezing, it depends on the contact between block and plate, which in turn depends on pressure. Its value can be diminished by:

- A layer of frost on the plates
- Layers of packaging material, such as polyurethane and cardboard
- Air voids at the surface caused by bunched-up packaging, an incompletely filled block, excessive thickness of adjacent blocks, or warped plates
- Decreasing flow rate of boiling refrigerant inside the plate as freezing progresses and refrigerant demand falls; this is minimal for large flooded systems, but can be quite significant for small dry-expansion plate freezers (Figure 8.14)

For air blast or brine freezers, the value of U depends primarily on the local velocity and is generally of higher value in brine than in air. The U value can also be diminished by resistances caused by packaging layers, air gaps caused by packaging or product geometry, and shading of fluid flow

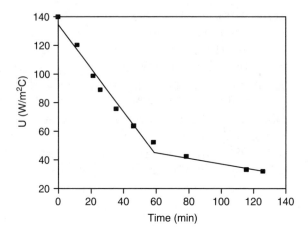

Figure 8.14 Heat transfer coefficient variation measured in a small horizontal plate freezer. (Adapted from Reference 17.)

velocities, which creates a local dead spot or poor circulation patterns in the freezer. An example of the last effect appears in Figure 8.15. Headed-and-gutted 3-kg salmon were placed on racks within a blast freezer similar to that previously diagrammed (see Figure 8.5). Because of the poor balance of airflow (low on top, high at the bottom), freezing times on the top shelves exceeded those on the bottom shelves by 4 hr.[9]

There are large ranges of U values in commercial freezers, and engineers must judiciously select, derive, or measure realistic values for prediction and design. Table 8.1 represents one fairly conservative set of values. The literature also reports wide ranges of values measured for particular circumstances.[6] For example, U for naturally circulating air (free convection) was reported as 5 to 10 W/m^{2}C, U for air blast freezers as 10 to 67 W/m^{2}C, U for brine freezing as 300 to 500 W/m^{2}C; and U for plate freezing as 50 to 500 W/m^{2}C. The use of packaging materials can also reduce these coefficients in various ways.

The influence of this uncertainty on calculated freezing time can be dramatic. Consider, for example, freezing a 25-mm-thick vacuum-packaged surimi seafood product from

Freezing Technology

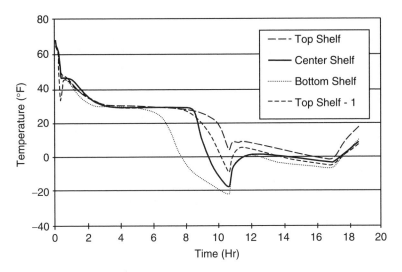

Figure 8.15 Measured variation of salmon freezing time with shelf distance from the floor in a batch blast freezer. (Adapted from Reference 9.)

TABLE 8.1 Heat Transfer Coefficients, U

Condition	Coefficient, U (W/m²C)
Naturally circulating air	5
Air blast	22
Plate contact freezer	56
Slowly circulating brine	56
Rapidly circulating brine	85
Liquid nitrogen:	
Low side of horizontal plate where gas blanket forms	170
Upper side of horizontal plate	425
Boiling water	568

Source: Adapted from Ref. 21.

16°C to a core temperature of −23°C in an air blast freezer. Using a mathematical model derived from the Plank equation format,[10,11] Figure 8.16 shows how freezing time varies with air velocity. For this model, it is assumed that a representative commercial freezer is used in which the blast air temperature is −23°C at the beginning of the freezing cycle and falls rapidly to −32°C. As velocity increases toward 5 m/sec, the freezing time falls from about 2.5 hr and begins to level off at 1.25 hr. It is not a good idea, however, to increase air velocity much past 5 m/sec because more powerful fans begin to contribute a substantial heat load to the system.[7]

Varying other conditions can further lower freezing times. Adjusting the settings and loading rate so the air temperature is a steady −34°C and prechilling the product so it enters at about 4°C, instead of 16°C, are both effective ways of shortening the freezing time. Furthermore, if the product is removed from the freezer when the core reaches −12°C, the freezing time could be reduced to approximately 15 min. Thus, by manipulating the external conditions, it is possible to reduce the freezing time for this product from 2.5 hr to 15 min.

There are two ways to determine freezing time: (1) calculation or (2) measurement. Many mathematical models have been devised to simulate freezing[6] and some of these are available in the form of relatively user-friendly computer programs. One of these, developed at Oregon State University, uses a complex modification of the closed-form Plank's equation, developed by Cleland and Earle,[10] to accurately describe freezing times.[11] The program allows a selection of conditions covering a range of freezers and seafood product types, although it is not generally applicable to cryogenic freezing. A second computer tool, developed by Mannapperuma and Singh[12,13] at the University of California at Davis, uses an enthalpy-based finite difference numerical model adapted to a user-friendly input format. It covers both freezing and thawing for a range of food products and allows one to visualize the internal temperature profiles. The program is distributed by the World Food Logistics Organization.[14]

These models enable one to predetermine the effect of changing certain variables, such as making the package thin-

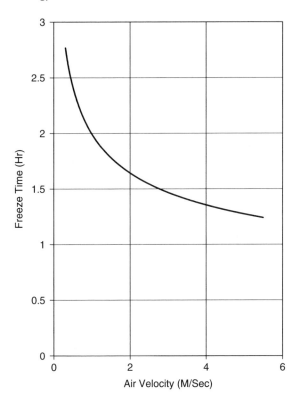

Figure 8.16 Predicted influence of air velocity on freezing time. Product is a fish muscle slab of thickness 25 mm, vacuum packaged; Initial temperature = 16°C; Final core temperature = –23°C; Air temperature range = –23 to –32°C.

ner, adding a cardboard cover, increasing air velocity, or decreasing initial temperature. It is difficult, however, for the operator to know, for example, the exact air velocity flowing over each package in a rack. (This relates to the difficult-to-predict value of U.) Consequently, many uncertainties exist and these will cause freezing time predictions to be off by typically 25% or more.[11] Sometimes, ±25% is close enough, especially if the goal is to make comparisons. For fine-tuning, the best way to know the freezing time is to measure it.

Many self-contained dataloggers currently on the market can measure freezing time. It is necessary to use one with separate probes to allow the sensor to be inserted into the core of the product. Thermister probes, however, are nonexpendable and must be recovered from the product after freezing. In addition, continuous freezers (e.g., tunnels or spirals) require self-contained loggers. For stationary freezers, on the other hand, such as cabinets, blast rooms, and plate freezers, it is often simpler to use a hand-held thermocouple meter that an operator might read every 10 or 15 min. Once freezing is complete, the wire can be snipped off if the product is to be discarded, and a new thermocouple can be quickly made by retwisting and soldering the ends.

Another common industrial technique used for larger products or blocks is to drill a small-diameter hole in the frozen block after it has been removed from the freezer. Then insert a probe, such as a properly calibrated dial thermometer, to contact the bottom of the hole and hold it there until the measured temperature reaches a minimum value. If the hole diameter is sufficiently small, the hole is sufficiently deep (10 cm, or 4 in.), and the time period is no longer than 2 or 3 min, the measured temperature can be accurate to within 1°C. Errors of 20°C, however, are possible if the procedure is done incorrectly.[15, 16]

8.9 SOME "WHAT-IF" EFFECTS ON FREEZING TIME

It is impossible, or at least not cost effective, to measure the results of each combination of packaging, geometry, product, and temperature schedule. This highlights the value of simulation models, which can indicate the results of such combinations as long as the user can systematically check them against selected experiments.

In this section, a numerical model is used to investigate some "what-if" questions that might influence the production rate using a plate freezer. The model is typical of several commercial software versions that solve partial differential equations and successfully simulate the freezing of seafood.[17,18] The following cases resulted from simulations using software "PDEase" (Macsyma Inc., Arlington, MA). Thermal property algorithms were previously developed for surimi having an 80% moisture content and a 4% sucrose/4% sorbitol cryoprotectant mix.[8,19]

8.9.1 Block Thickness

Normal freezing profiles can be determined for a standard block thickness (assumed here to be 56 mm), a single-layer poly bag, and reasonably good plate contact (assume U = 100 W/m²C) (Figure 8.17). For this example, the initial uniform temperature is given as 10°C at 0 min. The vertical dotted lines represent the outer surfaces of the block, and the horizontal dotted line represents the average block temperature after 90 min of freezing (Figure 8.17). Such an average would result if freezing were to suddenly stop and internal temperatures were allowed to equilibrate.

The temperature at the center of the block falls quickly to the initial freeze point of about −1.5°C and remains there as heat is extracted from the outer areas, converting water to ice (Figure 8.18). The temperature–time curves for the two off-center positions, however, do not show the "thermal arrest" plateau (Figure 8.18). If the temperature probe were located at one of these sites, an erroneously fast freezing time would be measured.

The freezing time for the center of the block is approximately 90 min, after which the block could be placed in a −20°C cold store with little further change in temperature. Further running of this program for different product thicknesses, assuming that the final average temperature in all cases is about −20°C, demonstrates how freezing time varies with product thickness (Figure 8.19).

8.9.2 Cold Temperature Sink, T_a

Block freezing time is expected to vary with the evaporating refrigerant temperature inside the plates (Figure 8.20). For a particular combination of equipment, these curves could be used, rather than the refrigeration capacity curves (Figure 8.11), to arrive at a maximum throughput for the plate freezing operation.

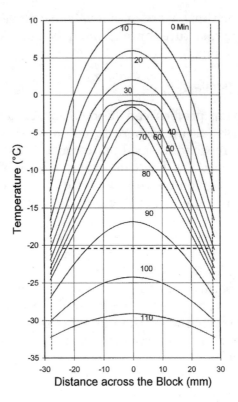

Figure 8.17 Predicted temperature profiles at 10-minute intervals within a block of surimi placed in a horizontal freezer. Initial temperature = 10°C; Block thickness = 56 mm; Refrigerant temperature inside the plates = –35°C; Overall heat transfer coefficient, U, on the block surface = 100 W/m²C. The vertical dashed lines on the left and right represent the bottom and top of the block, respectively.

Internal plate temperatures may actually increase somewhat at the beginning of a freeze cycle and this could extend the expected freezing time. Normally, for flooded or liquid-overfeed systems, this increase would be quite small.[20] However, when the product capacity greatly exceeds the refrigeration capacity (as in an overloaded blast freezer, for example), the suction temperature will rise appreciably as the rate of heat flow from the freezing product adjusts to match the rate of heat absorbed by the refrigeration machinery.

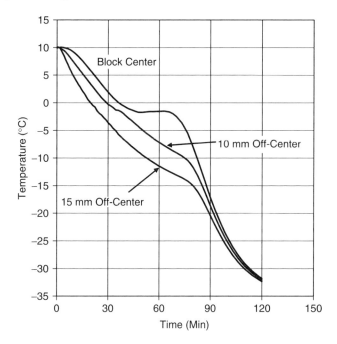

Figure 8.18 Simulated temperature vs. time for three positions within the surimi block of Figure 8.17.

Once blocks have been frozen and removed from the freezer, the external room temperature can begin warming the blocks. Suppose some blocks from the previous example (Figure 8.17) are removed from the freezer after 90 min and placed on a conveyor in a 15°C room where they sit for the next 30 min. After 30 min, the average temperature will be −13°C, increasing at the rate of about 1°C every 5 min (Figure 8.21).

8.9.3 Heat Transfer Coefficient, U

Simulating freeze time for a range of external heat transfer coefficients U is also instructive (Figure 8.22). The same surimi block is used to show how various combinations of external velocities, media, and packaging could affect freezing time. As indicated, freezing times are less influenced by U as U increases; the time approaches a near-minimum of

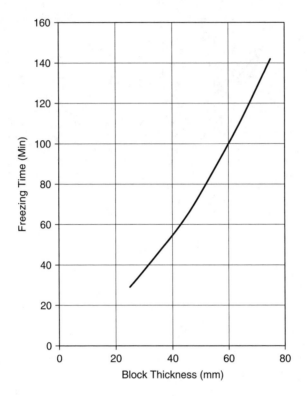

Figure 8.19 Predicted freezing time vs. block thickness for surimi. 80% moisture content; 8% cryoprotectants; Plate temperature = −35°C; Initial product temperature = 10°C; U = 100 W/m²C. Freezing time in this example results in an average block temperature of −20°C.

55 min when U = 500 (not shown). In fact, all products will have some value of U where the internal resistance to heat flow begins to "control" the process. That is, even if heat leaves the outer surface so readily that the surface becomes equal to the surroundings, there will still remain an internal resistance to heat flow. A complicating factor in any freezer is that conditions causing low U values and excessive freezing times may not be uniform throughout the batch, leading to some incompletely frozen product, which is then shipped to cold storage.

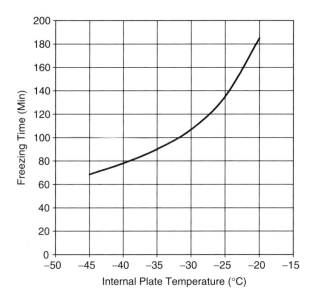

Figure 8.20 Predicted freezing time vs. saturaged suction temperature within plates in a horizontal plate freezer. (Surimi properties and conditions given in Figure 8.19.)

A more common malady for plate freezers may be an unevenness of contact, which can decrease the value of U on one surface and prolong freezing time. As an illustration, consider the previous freezing simulation (Figure 8.17), assuming a liquid-overfeed system with fixed plate temperature. In comparison, a block with good contact on the bottom side, but having a 2-mm air gap on the top side, will be assumed. This condition could be caused by a warped pan, an incompletely filled block, a heavy frost layer, a bunched poly bag, or over-filled adjacent blocks.

The results highlight several things. First, they show that it took essentially twice as long (180 min) to achieve the same average temperature of –20°C as did the balanced contact case (Figure 8.17). Second, the block freezes unevenly, as expected (Figure 8.23). Third, the geometric center is not the last point to freeze. Therefore, measuring the temperature at the center of these blocks (Figure 8.24) would fail to display

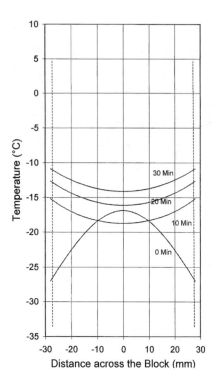

Figure 8.21 Simulated temperatures in blocks of Figure 8.17 removed from the freezer at 90 min, then exposed to 15°C room air. Packaging is a single layer polyethylene bag.

the plateau or "thermal arrest zone" that was seen for the geometric center of the block frozen with balanced contact on both sides (Figure 8.18). Consequently, it would indicate to the operator that freezing was complete quite a while before it actually was.

8.10 ENERGY CONSERVATION

In surimi plants, refrigeration is typically the largest energy user in the process.[23] And there are a number of things that can be done, both with freezing technology and with the refrigeration equipment, to lower the consumption and cost of energy.[24,25] Some measures require a major investment,

Freezing Technology

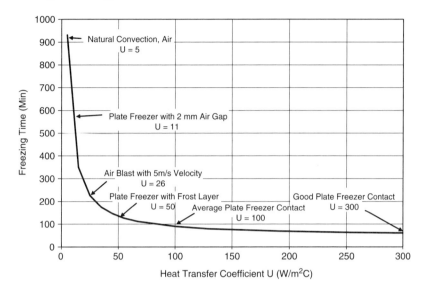

Figure 8.22 Simulated freezing time vs. U for a surimi block in a range of environmental surroundings at −35°C. Initial block temperature = 10°C; block thickness = 56 mm.

with the payoff period depending strongly on the local unit cost of electrical energy. Other measures involve simpler alterations of the process that create energy payoffs at relatively minor cost.

The following measures, applying primarily to vapor-compression refrigeration systems, will get results.

8.10.1 Freezer Design and Operation

Earlier sections noted that a good share of the freezer's heat load results from sources other than freezing products. These could be fans, defrosting, and leakage of air or heat into the freezer compartment. This suggests several measures that will reduce energy consumption:

- *Reduce the freezing time.* Anything that reduces time in the freezer will also reduce the contribution of extra heat loads. Modifications that can lower freeze times were discussed in previous sections.

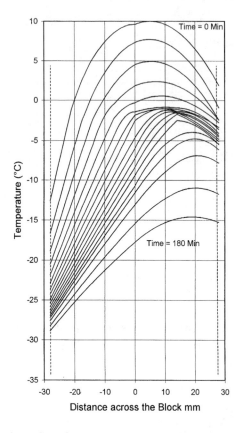

Figure 8.23 Simulated temperature profiles at 10-minute intervals within a surimi block in a horizontal plate freezer. On the lower surface (left dashed line): good contact (U = 100 W/m²C); on the upper surface (right dashed line): a 2 mm air gap (U = 11 W/m²C). Initial product temperature = 10°C; Plate temperature = –35°C.

- *Reduce the rate of heat leakage.* Erecting enclosures (e.g., around a plate freezer), increasing insulation thickness, redesigning doors to allow quicker loading and less air infiltration will all improve energy efficiency.
- *Select an oversized evaporator.* This applies primarily to blast freezers. The larger the heat transfer area in the evaporator, the smaller the temperature difference between the freezing medium (such as air) and refrig-

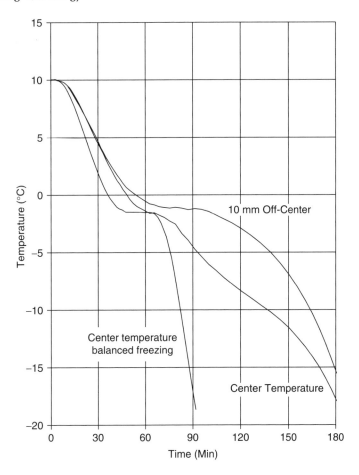

Figure 8.24 Simulated block temperature vs. time for two points within the block of Figure 8.23. Superimposed is the core freezing curve of Figure 8.18.

erant. This enables several possibilities: faster freeze with lower temperature, lower air velocity (and thus fan energy), and operation at a higher refrigerant temperature. The last measure increases the refrigeration capacity (Figure 8.11) while decreasing the power (kW) needed to remove a ton of refrigeration. [*Note:* the same kind of benefit would result if a packaged product

could be frozen in a brine freezer (vs. plate or blast) that operates at a significantly higher refrigerant temperature.]
- *Modify the defrost schedule.* Defrosting adds heat to the system. Often, defrosting occurs on a regular basis — for example, after each freeze cycle. And in some cases, this could be modified for greater energy efficiency. Measure the freeze-time effects of delaying defrost, and then adjust the manual or automatic control schedule accordingly.
- *Control the speed of evaporator fans.* Fan heat in blast freezers can be decreased with the use of fan speed control; more details appear below.

8.10.2 Refrigeration Machinery Options

Today's refrigeration engineers are particularly conscious of maximizing energy efficiency. For existing systems that might have developed piecemeal over the years, the list of changes/retrofits can cover a vast range. For new systems, there are some basic energy conservation design measures that frequently make the list. The involvement of refrigeration and energy efficiency engineers will fill in the details.

- *Maximize the refrigerant evaporating temperature.* Each 1°C increase in refrigerant temperature in the evaporator can decrease energy consumption by roughly 2 to 3%.[26] Look for such opportunities in systems that support multiple-temperature freezers and in freezers that could operate at slightly warmer temperatures.
- *Minimize the refrigerant condensing temperature.* Anything that can reduce the condensing temperature can reduce the energy consumed — 2 to 3% for each 1°C reduction.[25,26]
- *Control condenser size and fan speed.* Ammonia condensers are typically evaporative air-cooled units; an oversized condenser will remove the necessary heat load using less horsepower from fan and pump motors. Capacity control using variable-speed drives on the fan

Freezing Technology

motors is also important; fan power falls as the cube of the speed. And choose axial vs. centrifugal fans for greater efficiency.[24]

- *Use automatic gas purgers.* Noncondensable gases in ammonia refrigeration systems will hurt performance in several ways.
- *Diversify compressor selection.* Capacity control typically used on screw compressors makes them very inefficient in the low-capacity range. However, screw compressors are very efficient at full speed and capacity. Energy efficiency will result from a system design that uses a larger number of smaller compressors, one or two of which have capacity controlled by variable-speed motors.
- *Employ automatic controls.* The experience of designers has shown significant savings with centralized computer control of the system. Programs can sense changing loads and ambient conditions to continually optimize the operation.
- *Use adequate piping.* Both size and insulation can be important. Long runs of piping can allow heat leakage and pressure drops that will depress efficiency. In one example, allowing pressure drop in the suction line to increase from 7 to 14 kPa (1 to 2 psi) increased energy use by 5%.[25] Allowing an excessive increase in refrigerant temperature leaving the evaporator (i.e., superheat) will cost energy as well; about 0.4% for each 1°C.
- *Recover waste heat.* Condenser heat can be used to complement energy needs elsewhere in the plant — for room heating, foundations under cold storage rooms, boiler supply water, cleanup water, and others.

8.10.3 A Blast Freezer Case

A recent project with blast freezing demonstrated significant savings that could result from an improved control of airflow and from the use of speed-control devices on fans.[27] A stationary blast freezer processing 22-lb cartons of sardines in 19,000-lb lots was modified to improve efficiency and to conserve energy. Baffles were first added to produce uniform airflow. The maximum measured freeze time of 12.6 hr fell to 10.5 hr, and the total electrical energy savings were estimated at 12%. A variable frequency drive (VFD) was then installed to slow the evaporator fans during the freeze cycle. In one example run, fans were slowed to 75% speed after an initial freezing period; freezing times then increased by about an hour, while overall energy consumed fell by an additional 10%. Based on the effect of the VFD alone and $0.05/kWh power in this case, the payback period would be in the range of 850 freezing cycles.

Mathematical models calibrated by measurements analyzed further design and fan control options. One option assumed periodic reversing of fans, which would create a more uniform and faster freezing rate. Such an arrangement predicts further energy savings of 8%. A second option assumed the fans to slow continuously as the heat load fell. This predicted a 2% increase in freezing time with predicted energy savings of 11%.

8.11 CONCLUSIONS

Although critical to plate freezing of surimi blocks or blast freezing of surimi seafood, the practices and parameters presented here are just as important to all food freezing operations. The simulations presented cannot describe the exact results in any one plant. However, they represent an expected variation of results with critical factors. An understanding of such influences can help formulate equipment decisions with refrigeration contractors, load and operate freezers for maximum efficiency, conduct monitoring of freezer performance to avoid problems, and ensure maximum product quality.

ACKNOWLEDGMENTS

The author acknowledges funding support for this project from the Sea Grant programs of Oregon, Alaska, and Washington, and from the National Fisheries Institute.

REFERENCES

1. ASHRAE (American Society of Heating, Refrigerating, and Air Conditioning Engineers). *Refrigeration Handbook*. Atlanta, GA: ASHRAE, 1994.

2. E. Kolbe, C. Craven, G. Sylvia, and M. Morrissey. 2004. Chilling and Freezing Guidelines to Maintain Onboard Quality and Safety of Albacore Tuna. Oregon State University Agric. Exp. Station Special Report 1006. Web address: http://eesc.oregonstate.edu/agcomwebfile/EdMat/SR1006.pdf.

3. K.S. Hilderbrand. Preparation of Salt Brines for the Fishing Industry. SG22. Oregon State University Extension Sea Grant, Newport, 1979.

4. ASHRAE (American Society of Heating, Refrigerating, and Air Conditioning Engineers). *Fundamentals Handbook*. Atlanta, GA: ASHRAE, 1997.

5. E. Kolbe and D. Kramer. Planning Seafood Cold Storage (MAB-46). Alaska Sea Grant College Program, University of Alaska, Fairbanks, 1993, 54.

6. D.J. Cleland and K.J. Valentas. Prediction of Freezing Time and Design of Food Freezers. In K.J. Valentas, E. Rotstein, and R.P. Singh, Eds. *Handbook of Food Engineering Practice*. Boca Raton, FL: CRC Press, 1997.

7. J. Graham. Planning and Engineering Data. 3. Fish Freezing. FAO Fisheries Circular 771. Rome, Italy: Food and Agriculture Organization of the United Nations, 1984.

8. D.Q. Wang and E. Kolbe. Thermal properties of surimi analyzed using DSC. *J. Food Sci.*, 56(2), 302–308, 1991.

9. E. Kolbe and D. Cooper. Monitoring and Controlling Performance of Commercial Freezers and Cold Stores. Final report submitted to Alaska Fisheries Development Foundation, 1989.

10. D.C. Cleland, R.L. Earle. Freezing time prediction for different final product temperatures. *J. Food Sci.* 49, 1230–1232, 1984.

11. E. Kolbe. An interactive fish freezing model compared with commercial experience. *Proc. XVIII Int. Congress of Refrig.*, Aug. 1991, Vol. IV, pp. 1902–1905. http://seagrant.oregonstate.edu/sgpubs/ onlinepubs.html#kolbe, 1991

12. J.D. Mannapperuma and R.P. Singh. Prediction of freezing and thawing times of foods using a numerical method based on enthalpy formulation. *J. Food Sci.*, 53(2), 626–630, 1988.

13. J.D. Mannapperuma and R.P. Singh. A computer-aided method for the prediction of properties and freezing/thawing of foods. *J. Food Eng.*, 9, 275–304, 1989.

14. 14. WFLO (World Food Logistics Organization). Industrial-Scale Food Freezing — Simulation and Process. P. O. Box 79753, Baltimore, MD 21279-0753. Web address: http://www.wflo.org/hq/forms/freeze.pdf.

15. J. Graham. Temperature and Temperature Measurement. Torry Advisory Note No. 20, Ministry of Agriculture, Fisheries and Food, Torry Research Station, Aberdeen, U.K., 1977.

16. W.A. Johnston, F.J. Nicholson, A. Roger, and G.D. Stroud. Freezing and refrigerated storage in fisheries. FAO Fisheries Technical Paper 340. Rome: Food and Agriculture Organization of the United Nations, 1994.

17. D.Q. Wang and E. Kolbe. Analysis of food block freezing using a PC-based finite element package. *J. Food Eng.*, 21, 521–530, 1994.

18. Y. Zhao, E. Kolbe, and C. Craven. Simulation of onboard chilling and freezing of albacore tuna. *J. Food Sci.*, 65(5):751–755, 1998.

19. D.Q. Wang and E. Kolbe. Thermal conductivity of surimi — measurement and modeling. *J. Food Sci.*, 55(5), 1217–1221 and 1254, 1990.

20. H. Takeko. Processing and refrigeration facilities for a fish factory ship. *Proceedings of the Int. Inst. of Refrig.*, Vol, 1, 1974, 203–213.

21. D.R. Heldman, R.P. Singh. *Food Process Engineering.* Second Ed. Westport, CT: AVI Publishing Co., 1981.

22. J. Graham. Installing an Air-Blast Freezer? Torry Advisory Note No. 35, Torry Research Station, Aberdeen, U.K., 1974.

23. E. Kolbe. Estimating energy consumption in surimi processing. *J. Appl. Eng. Agric.*, 6(3), 322–328, 1990.

24. M.H. Wilcox. State of the Art Energy Efficiency in Refrigerated Warehouses. Technical Paper #15. *Proc. of the IIAR Ammonia Refrigeration Conf.,* Dallas, 1999, 327–340.

25. W. Gameiro. Energy Costs — Are Changing Refrigeration Design. Technical Paper #6, *Proc. of the 2002 IIAR Ammonia Refrigeration Conf.,* Kansas City, MO, 2002, 207–233.

26. W.F. Stoecker. *Industrial Refrigeration Handbook.* New York: McGraw-Hill, 1998.

27. E. Kolbe, Q. Ling, and G. Wheeler. Conserving energy in blast freezers using variable frequency drives. *Proc. 26th National Industrial Energy Technology Conference,* Houston, pp. 47–55, 2004. http://seagrant.oregonstate.edu/sgpubs/onlinepubs.html#kolbe.

Part II
Surimi Seafood

9

Surimi Seafood: Products, Market, and Manufacturing

JAE W. PARK

Oregon State University, Astoria, Oregon

CONTENTS

9.1 Introduction 376
    9.1.1 Surimi-Based Products in Japan and the
        United States 377
        9.1.1.1 Japanese Market 377
        9.1.1.2 The U.S. Market 380
    9.1.2 Market Developments in France 383
    9.1.3 Surimi Seafood Products in Other
        Countries 385
9.2 Manufacture of Surimi-Based Products 388
    9.2.1 Kamaboko 388

    9.2.2 Chikuwa 390
    9.2.3 Satsuma-age/Tenpura... 390
    9.2.4 Hanpen 392
    9.2.5 Fish Ball 393
    9.2.6 Surimi Seafood 395
            9.2.6.1 Filament Meat Style 396
            9.2.6.2 Solid Meat Style 417
9.3 Other Processing Technology 419
    9.3.1 Ohmic Heating 419
    9.3.2 High Hydrostatic Pressure 424
    9.3.3 Least-Cost Linear Programming 425
Acknowledgments 428
References 430

9.1 INTRODUCTION

During the 8th century, surimi-based products in Japan were described as *chikuwa*. For a long time afterward, surimi-based products became more familiar to the Japanese people as *kamaboko*, *chikuwa*, and *satsuma-age*. In the 1970s, a new generation product called crabstick (crab leg) was developed (N. Kato, personal communication, 1997).

In Japan, based on the regional availability of fish, it is common to see different shapes, colors, and textures of surimi-based products from region to region. The various kinds of fish harvested in the sea around Japan that are utilized as surimi (raw material for surimi-based products) include croaker, threadfin bream, lizardfish, conger eel, flatfish, flounder, sardine, mackerel, and Alaska pollock. In addition, the production of surimi-based products in Japan is seasonal. The industry produces more kamaboko for the winter season as a result of the increased consumption during the New Year holidays.

The market in the United States, however, is exactly the opposite. Here, surimi seafood is consumed largely during the summer as a salad ingredient. Since the late 1970s, surimi-based crabstick (hereafter, surimi seafood) has become increasingly popular in North America. The consumption of surimi seafood has proliferated mainly in cold salads with an

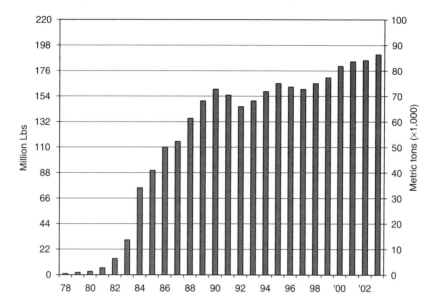

Figure 9.1 Changes of surimi seafood consumption in the United States (1978–2003).

annual growth rate of 10 to 100% in the 1980s. In 1991, however, consumption suffered a negative growth rate due to the price of surimi and, thereafter, U.S. consumption has remained steady but with a slight downward trend until the late 1990s. In addition, in 2000, it reached 82,000 metric tons (t) (Figure 9.1).

This section discusses surimi-based products in Japan, the United States, and other major countries, along with their processing technologies. Additionally, other processing technologies such as ohmic cooking, high hydrostatic pressure, and the least cost linear program for surimi blending are described.

9.1.1 Surimi-Based Products in Japan and the United States

9.1.1.1 Japanese Market

The traditional Japanese surimi-based products are divided into six categories. They are satsuma-age (fried), chikuwa

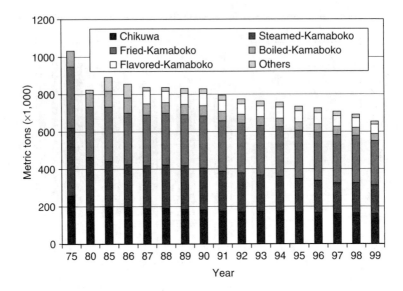

Figure 9.2 Changes of surimi-based product consumption in Japan (1975–1999).

(baked), kamaboko (steamed), flavored kamaboko (fish sausage/ham), hanpen/naruto (boiled), and other surimi-based products, such as surimi seafood (crab leg). Over the past 30 years, production of surimi-based products in Japan has changed significantly (Figure 9.2). In 1975, 1,034,262 t of surimi-based products were manufactured in Japan in descending order: kamaboko (steamed) 35.0%, satsuma-age (fried) 31.6%, chikuwa (baked/broiled) 25.0%, hanpen/naruto (boiled) 8.2%, fish sausage/ham <1.0%, and others, including surimi seafood (crabmeat) (<1.0%). In 1999, the consumption of surimi-based products reduced significantly to about 654,000 t (Figure 9.2).[1] These changes indicate an alteration in the diet pattern of the young generation in Japan as more Western fast-foods, such as hamburgers and pizza, are being consumed.

The seasonal consumption trend of surimi-based products in Japan has been fairly consistent year after year, peaking in the winter season and dropping in the summer. The

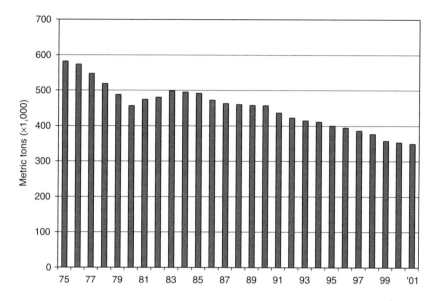

Figure 9.3 Trend of surimi usage in Japan (1975–2001).

main dish in winter is *oden,* which is a hot soup served in a bowl. Kamaboko is an essential ingredient for New Year's Day foods. In the summertime, however, most surimi-based products are used in a cold dish, such as salad.

The usage of surimi in Japan was about 580,000 t in 1975 and decreased to 350,000 t in 1999 (Figure 9.3). Reduced surimi production by Japan and a change in species used for surimi have affected the processing technology of traditional surimi-based products. However, Japan still uses about 60% of the surimi produced in the world.[2]

Another reason for the continuous reduction of consumption in Japan is that no new, exciting products have been introduced since the invention of crabstick. In recent years, however, many studies have been initiated to search for the great health benefit of kamaboko products to bring new attention to it as being part of a healthy diet.[1] The industry, consequently, now attempts to advertise the positive effect of kamaboko on human health to consumers and tries to revitalize their business under the leadership of Zenkama (All Japan Kamaboko Association).

9.1.1.2 The U.S. Market

Surimi seafood introduced in North America during the late 1970s was in the form of crabmeat leg (crabstick). A small modification from crabmeat stick to a ready-to-use flake created a huge market with more than 100% growth between 1983 and 1984 (Figure 9.1). Consequently, flake-type products cover more than 90% of the U.S. market. There are four typical types of surimi seafood: crabmeat sticks, flakes, chunks, and combo as shown in Figure 9.4. The terminology used to describe chunk and flake is often confused by Asian manufacturers and traders in the United States and the European Union (EU). Flake described in Figure 9.4 is "mistakenly" called chunk. When the crabstick is cut straight, it is a chunk; when cut diagonally, it is a flake. The correct terminology should be used for better communication and less confusion.

Unfortunately, in the United States, surimi-based seafood is legally called imitation crabmeat. In the 1980s, the FDA offered some guidance for labeling surimi seafood. If the surimi-based product purports to be or is represented as any specific type of natural seafood, including shape or form representations, but is nutritionally inferior to that of seafood, it must be labeled as imitation in accordance with 21 CFR 101.3.[3] The size of the word "imitation" on the package should not be smaller than half the size of the brand name on the package. Surprisingly, the United States is the only country in the world that requires "imitation" in the labeling of surimi-based seafood.

In the past 15 years, the U.S. surimi seafood industry has tried to get approval from the FDA to use the words "surimi," "surimi seafood," or "crab-flavored fish" as a common name. However, it is still debated between the U.S. industry and the FDA as to whether or not this label misleads consumers. According to a survey regarding consumers' attitudes toward "imitation" crabmeat,[4] almost 25% were seriously con-

Surimi Seafood: Products, Market, and Manufacturing

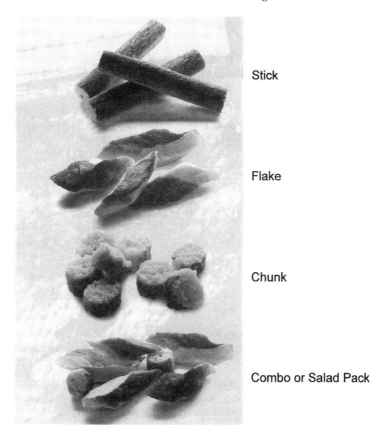

Figure 9.4 Four types of cut in crab-flavored surimi seafood.

cerned with the word "imitation." More recently, many consumers are turning away from surimi seafood because "imitation" is used in the labeling of these products.

In 1998, the Canadian government approved the removal of the word "imitation" from the labeling of surimi seafood. Cananda's labeling is very similar to the European method. The products are labeled as crab-flavored fish, shrimp-flavored fish, lobster-flavored fish, etc. This new labeling has helped rebuild the surimi seafood market in Canada (R. Chambers, personal communication, 1999) and could be used to boost U.S. markets as well.

However, every country in the world uses its own labeling system to describe surimi seafood (crabstick) without using the term "imitation." According to a survey conducted in 2003, a wide range of descriptions are used without the term "imitation," including *kanikama* (crab-flavored stick) in Japan; *gehmahtsahl* (crab-flavored meat) in Korea; *delices a la chair de crabe* (delicacy with crabmeat), *bâtonnets de poisson goût crabe, saveur crabe, au goût de crabe* (fish stick with crab flavor, surimi stick with crab taste, and stick with fish and crab flavor, respectively) in France; *preparazione alimentare a base di pesce/surimi al gusto di granchio* (crab-flavored, fish/surimi-based food) in Italy; *palitos de cangrejo* (crabstick), *palitos de mar* or *palitos del océano* (seastick), or *delicias de surimi* (surimi delicacy) in Spain; and simply *crabstick* in Russia.

As for the use of multiple species of surimi, most countries do not require the manufacturer to identify the name of species — except in the United States. In September 1999, the U.S. FDA approved the use of disjunctive labeling, which allowed the use of "fish protein (Alaska pollock, Pacific whiting, and/or any other species)" in case multiple species are used. In other countries, fish, fish meat, or surimi is simply listed in the ingredient statement.

When crabstick was invented, it might have been developed to resemble shellfish. But it is impossible to make a product identical to the natural counterpart. Several marketing attempts have been made to introduce other value-added surimi-based products, such as breaded scallop, pizza topping, clam, crab cakes, and salmon. These were not successful, however, primarily because of the simple eating patterns of surimi seafood by American consumers, who primarily regard surimi seafood as crabmeat salad. Consequently, there is a great need for product development of surimi seafood as something other than a salad ingredient, perhaps as a stand-alone food such as a snack food or hot meal.

According to a U.S. industry source, it is the best estimate that in 2003, the U.S. annual consumption of surimi seafood reached about 90,000 t (Figure 9.1). This industry has been consolidated to only eight manufacturers and faces a significant challenge from domestic competition with a matured market and low-quality/cheap Chinese imports. However, there is a little positive effort to market high-quality surimi seafood by the leading manufacturers.

In conclusion, surimi seafood is unique and, therefore, a distinction must be made with regard to its great values in taste, price, safety, and convenience, instead of being labeled as imitation shellfish.

9.1.2 Market Developments in France

Surimi seafood consumption in Europe has increased by 81% since 1997, reaching approximately 120,000 t in 2002. This development mostly resulted from French and Spanish markets in the early years. In France, the sales of surimi seafood products increase each year by over 10%, resulting in gains of over 40,000 t in 2003 (Figure 9.5). The French market consists of its own "chilled" products, which is estimated at more than 35,500 t, and "frozen" imports (6100 t) from Lithuania, Korea, and Thailand. As for the market segment, 63% is sold in supermarkets, 10 to 15% is sold in food services/catering, and 15 to 20% is used by industrial producers that manufacture salad and other ready-to-eat foods. Sticks and ministicks (for snacks) represent 85% of the volume sold in supermarkets, whereas shredded products represent less than 10% (J. Beliveau, personal communication, 2004).

French manufacturers have positively contributed to the development of a variety of surimi seafood products, more than manufacturers in any other countries outside Japan. This success is based on two important factors: (1) the French consumer's willingness to try something interesting, and (2) industrial efforts in research and development by two leading manufacturers (Fleury Michon and Cuisimer). These two companies have continuously innovated their product lines to increase surimi seafood consumption and launched new products in 2002 such as spreadable surimi seafood with a creamy texture and surimi flakes or "Pinces Océanes," which is a new product molded in the shape of crab claws and designed to serve as a hot entrée (J. Beliveau, personal communication, 2004).

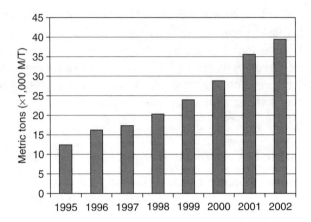

Figure 9.5 Change of surimi seafood consumption in France (1995–2002).

In France, both innovation and communication combine together for market development and French leaders, consequently, launch two or three new products every year. They also benefit from advertising campaigns on TV and strong consumer education by emphasizing the origin of the products, with special references on their packaging such as "naturally rich in proteins, prepared with wild fish." In addition, ADISUR (Association for the Development of Surimi Industry) strengthened its communication to reassure consumers about the origins of the fish and the use of fresh fillets for the production of the surimi base. ADISUR's primary mission is to make consumers consider surimi seafood as real food, thus utilizing it in a variety of applications (J. Beliveau, personal communication, 2004).

Another reason for the success of surimi seafood in France is innovative packaging. Following the launch of products individually wrapped in flow-packs, which became very successful owing to their modernity and finger-food aspects, the market for family-size products (500 g rigid and resealable pack) was also expanded by 30%. The convenient aspect of this kind of packaging enabled consumers to differentiate private labels from foreign products.

In conclusion, multidisciplinary efforts made by the major French manufacturers (high quality levels, market innovation with products beyond imported sticks, consumer-friendly packaging, and media campaigns) have kept the market for surimi seafood at an extremely successful pace. However, there are also signs indicating that the market in France is getting close to its maturity level. In 2001, supermarkets launched their own brands due to price competitiveness. In addition, a recent survey carried out by a French consumer study group (SECODIP) showed that surimi seafood stick prices dropped by 5.1% between 2001 and 2002 (J. Beliveau, personal communication, 2004).

9.1.3 Surimi Seafood Products in Other Countries

South Korea is the second largest manufacturer of surimi-based products in the world. In South Korea, the largest surimi-based product is *ah-mook* (fried), which is traditionally mixed with vegetables. Like *oden* in Japan, *ah-mook* is also used in hot soup. Crabstick production for the domestic market and export was around 50,000 t in 1998. Due to continuous drawbacks in export markets, however, the total production of crabstick, including crab claw in 2003, was 47,000 t, with one third export and two thirds domestic consumption. However, the traditional products reached 172,000 t, with the majority being *ah-mook*, followed by sausage and ham (J.H. Kim, personal communication, 2004). One of the most negative signs of the domestic market in Korea was that all crabstick products are uniform (in terms of diameter, length, and applied color), regardless of the manufacturer. However, a newly introduced up-scale crabstick, *crami* has made a big hit in the domestic market. Its market steadily grows. This *crami* product is popular as a snack food due to its excellent texture.

In 1996, the European Union (EU) imported more than 32,000 t finished products.[5] The main supplier was South Korea, followed by Thailand, Malaysia, and China. The largest consumer in the EU is France, with a total consumption of 40,000 t in 2002. Spain is the second-largest consumer in

Europe, with a consumption of 17,000 t in 1997, and typically imports low-priced (low-quality) frozen crabstick from Asia. Italy is the third-largest consumer, with an estimated consumption of 6000 t.[5] In 2003, the size of the market covering the EU and Eastern Europe reached more than 150,000 t.[2]

During the past few years, however, a sharp increase in crabstick production by European companies has been observed. France has been the most successful, due to the high quality of its products and the continuous introduction of new refrigerated items to the market. One reason for the success of the surimi seafood business in France is the establishment of industry-driven quality assurance. French manufacturers of surimi seafood and ingredients, under the ADISUR (Association for the Development of Surimi Industry), established a set of quality standards in 1997.[5] The standards determined the minimal composition requirement of raw-material surimi as 85% fish flesh and a minimum 35% surimi, or 30% fish flesh for surimi seafood products.

The consumption of surimi seafood in Eastern Europe, including Russia, has grown significantly from 2000 t (in 1995) to 85,000 t (in 2003) (Figure 9.6). Over the past 3 years, the consumption increased by 100%. The production of crabstick is now spread over all of Eastern Europe, including Estonia, Lithuania, Belarus, and Russia. This industry started with Chinese and Korean imports. Now, domestic production has begun and covers this demand and supplies in some parts of the EU. The world's largest crabstick manufacturer, Viciunai that produces approximately 34,000 t a year, is also located multinationally in Estonia, Lithuania, and Russia.

The export of surimi seafood (crabstick) from China reached 15,745 t in 1997, and the major markets of China are Russia and Europe.[6] In 2002, there were more than 25 crabstick manufacturing companies in China, including several joint ventures with Korean manufacturers. Most of them are located in either Dalian or Qingdao. Annual production reached more than 85,000 t in 2002 (J.H. Lee, personal communication, 2004). However, the Chinese industry has struggled since 2002. The EU has banned all seafood imports from

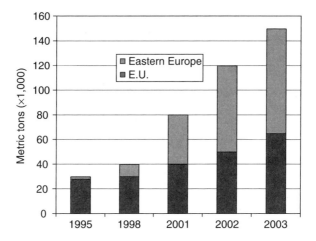

Figure 9.6 Consumption of surimi seafood crabstick in Europe and Eastern Europe.

China due to residual contents of antibiotics for over a year. The EU also actively participated in the crabstick price war in Russia, reaching below U.S.$1.00/kg delivered in Russia in 2003. As a result, the quality of crabstick sold in Russia is extremely low, and Chinese crabstick manufacturing industry is undergoing a serious business struggle. The production of crabstick in China was consequently reduced to 60,000 t in 2003, with one third for domestic consumption and two thirds for export (J.H. Kim, personal communication, 2004).

The establishment of the crabstick industry in Australia has been quite long and involves two manufacturers (6000 t a year). Its market, however, has not significantly grown and somewhat struggles to this day. In contrast, a new area for success of the surimi seafood (crabstick) market is South America. Five manufacturers in Peru, Brazil, Argentina, and Uruguay are all new establishments and are looking for growth in the domestic market.

Thailand is the leading manufacturer of crabstick in Southeast Asia. The production of crabstick in Thailand is 40,000 t a year. Domestic consumption is about 3000 t and the balance is exported (A. Law, personal communication, 2004).

Fish ball is popular in Southeast Asia but the quality characteristics are varied among countries. For example, fish balls found in Singapore and Malaysia are typically whiter in color and more elastic, while those sold in the Philippines and Hong Kong have a firmer texture (i.e., less elastic) but darker color and stronger fishy note.

In Singapore, the most popular surimi-based product is the fish ball, which is used in a local application, *Yong Tau Foo*. Some 4 million people consume approximately 70 t fish ball/fish cake a day, resulting in about 6 kg per-capita consumption. About 14,000 t surimi are imported to Singapore a year to produce fish ball/fish cake and crabstick (S.M. Tan, personal communication, 2004). Fish ball consumption in Thailand is about 12,000 t a year, while the consumption of low-grade chicken balls made from by-products is about 25,000 t. These chicken balls are primarily used in noodle soup and often grilled on charcoal (A. Law, personal communication, 2004).

9.2 MANUFACTURE OF SURIMI-BASED PRODUCTS

As a raw material, surimi from various species has been used in Asia and Europe, while only Alaska pollock and Pacific whiting surimi are primarily used in the United States. Fresh surimi is used in very small amounts in Japan and other Asian countries, while only frozen surimi is used in the United States and Europe. Sequential processes for surimi-based product manufacturing are described in detail later in this chapter including Japanese traditional products, Southeast Asian products, and new-generation surimi seafood.

9.2.1 Kamaboko

Kamaboko is thought to have existed as early as the Nara period (710 to 784) in Japan. It is the most typical surimi-based product in Japan. The term often refers to all surimi-based products in Japan. Surimi paste is formed into a quonset hut shape on a wood board before any thermal treatment

Surimi Seafood: Products, Market, and Manufacturing

(Figure 9.7). Sometimes, its surface is coated with colored paste for appearance. These processes can be done manually (traditionally by trained artisans) or using machinery (for mass production). In addition, the shape and texture of kamaboko vary, depending on the geographical region in Japan.

After its unique shape is formed, the surimi paste is subjected to a low-temperature setting process (20 to 40°C for

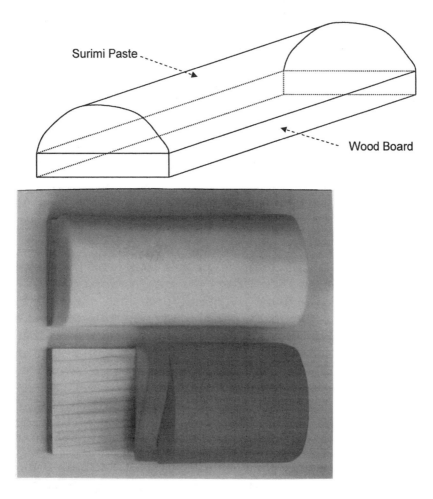

Figure 9.7 Kamaboko in Quonset hut shape and products. (Courtesy of Kibun, Japan.)

30 to 60 min), depending on species. During this process, the gel-forming ability of solubilized myofibrillar proteins is highly enhanced, which yields a very strong and elastic gel. Cooking by either steaming or baking is carried out to complete the gelation of fish proteins. The finished steamed product is called *mushi* (steaming) kamaboko, which is widely manufactured in eastern Japan. Odawara City is especially famous for this steamed kamaboko. On the other hand, the baked kamaboko is called *yaki-ida* (baked on the board), which is mainly produced in western Japan.

Another type of kamaboko is called molded kamaboko, which is also processed in a quonset hut-shaped mold. The molding technique is applied for the utilization of low-grade surimi that commonly has a low gelling ability. The process is almost the same as regular kamaboko, but surimi paste is poured into a plastic mold and "cooked" at 90°C (baking or steaming) after setting at 10 to 15°C for 10 hr. The finished products are packed, pasteurized, and chilled before entering their marketing channel (N. Kato, personal communication, 1997).

9.2.2 Chikuwa

Chikuwa, an original model of surimi-based products, is listed in literature of the 8th century (N. Kato, personal communication, 1997). Its shape is typically that of a pipe or tube (Figure 9.8). Surimi paste is placed onto a grooved hole in a rectangle shape on the surface of a drum. The paste is then rolled onto a metal stick on the conveyor. The rolled paste on the stick is baked rotationally in the oven on the screw conveyer for gelation. The finished products are packed, pasteurized, and chilled before entering their marketing channel.

9.2.3 Satsuma-age/Tenpura

Satsuma-age is fried kamaboko (Figure 9.9A) and has various shapes and characteristics. It has several different names, depending on the region. In Kagoshima prefecture, where satsuma-age was first developed, it is called *tsuke-age,* while in Tokyo it is *satsuma-age* and in Osaka it is *tenpura* (N.

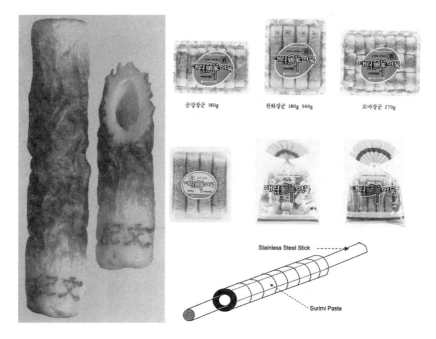

Figure 9.8 Chikuwa: its shape and products. (Courtesy of Daerim Fishery, Korea and Kibun, Japan.)

Kato, personal communication, 1997). Additional ingredients such as vegetables, shrimp, squid, and minced fish are sometimes mixed into the surimi paste for satsuma-age. The paste is then molded into various shapes (stick, patty, ball, or nugget) before frying. These fried products are also very common in Korea where they are traditionally called *twighin ahmook*.

Although frying is the primary cooking method for satsuma-age, three types of precooking are often used to differentiate the finished products. They are *yude-age* (boiled-fried), *mushi* (steamed-fried), and *ki* (fried-fried) (N. Kato, personal communication, 1997). In recent years, however, the majority of *satsuma-age* is manufactured using a two-step frying process because it yields high gel strength and productivity. The first frying is done at 130°C and the second frying is subsequently performed at 170°C.

Figure 9.9 Tenpura (satsuma-age) (A) and hanpen (B). (*Source:* Courtesy of of Kibun, Japan.)

9.2.4 Hanpen

Hanpen (Figure 9.9B) was first made in Tokyo and Chiba prefecture areas. Its texture is soft like a marshmallow (N. Kato, personal communication, 1997) or soft tofu. For the development of soft texture, whipping is required at the final step of mixing using *yamaimo* (a species of yam). Recently, gums or polysaccharides have been used as whipping and stabilizing agents. Vegetable oil is commonly mixed as well for the development of soft texture. The surimi paste is traditionally whipped using the pestles of the stone mixer at high speed. In recent years, however, the surimi paste is aerated compulsorily by a continuous mixer. The whipped paste is then boiled in hot water (80 to 85°C) to fix the soft gel texture.

9.2.5 Fish Ball

Fish ball (Figure 9.10) is the most popular surimi-based products in Southeast Asia. Ingredient preparation for fish ball manufacturing is similar to that for *kamaboko* and crabstick. Therefore, for detailed manufacturing technology up to the comminution step, refer to the next section ("Surimi Seafood"). The remaining manufacturing steps are discussed as shown in the flow diagram (Figure 9.11)

Typical ingredients used for fish ball, in addition to surimi, are salt, sugar, monosodium glutamate (MSG), starch, and water. The proportion of starch is relatively small compared to crabstick. No flavors or protein additives are added to the formulations.

Once the paste is prepared, it is extruded (formed) into a ball shape using a special device and dropped into warm water (20 to 40°C, depending on species) for setting (Figure 9.12) for 30 to 60 min. Unlike other surimi seafood manufacturing, in the manufacturing of fish ball, salt is added at the last minute of comminution. Otherwise, fish balls sink to the bottom of the setting container or float on the surface of the water. Either case can lead to deformation of fish balls, resulting in flat or oval-shaped products, or sometimes tangling with each other. Therefore, keeping a uniform shape, in a mass production, is critical. Traditionally, the last-minute addition of salt during comminution has kept extruded fish balls floating freely in the middle of the setting water to prevent deformation. Due to limited studies in fish ball manufacturing, there is no known relationship between the fish

Figure 9.10 Most common display of fish ball. (*Source:* Courtesy of Thong Siek Foods, Singapore.)

balls' floating in the middle of the water and addition of salt during comminution. It most likely has something to do with the density of comminuted paste and/or the thermal stability of species.

Once fish balls are set, they are placed in hot water (95 to 98°C) for 10 to 30 min (until the core temperature reaches 80°C), followed by chilling under running tap water. The fish ball, after draining water, is then packed in a poly bag before checking with metal detection.

Unlike cooked fish balls, raw fish balls, which are partially set, are sold in a poly bag containing cold water. They are perceived as fresher and tastier by consumers. The most

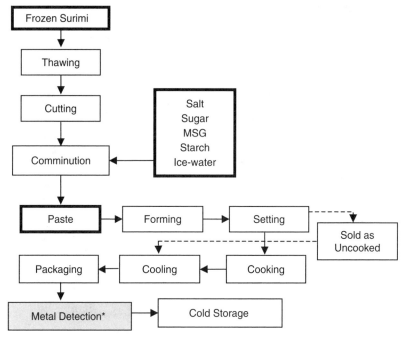

Figure 9.11 Processing flow of fish ball.

challenging endeavor is to deal with a shorter shelf life and to prevent oversetting while in the market.

9.2.6 Surimi Seafood

Surimi-based seafood analog products (hereafter called surimi seafood) are developed in several styles, but particularly as crabmeat. There are two categories of crabmeat style, depending on the manufacturing methods: filament meat and solid meat. In addition, stick (leg), flake, and chunk are made using two different methods (Figure 9.4). For shrimp and lobster styles, products are made using a mold.

Figure 9.12 Setting/cooking of fish ball. (*Source:* Courtesy of Yeap Soon Eong, MFRD, Singapore.)

9.2.6.1 Filament Meat Style

During the forming step of surimi seafood, the blended surimi paste is extruded either directly or using a meat transfer unit, which is commonly equipped with a stuffing machine, onto a cooking belt. Crab- and scallop-style products, which exhibit aligned fibers in the finished products, begin as a very thin sheet of paste extruded onto a belt (or drum). The thickness and width of the paste ribbon determines the diameter of the filament-style products and also affects productivity. A flow diagram of filament-style surimi seafood is shown in Figure 9.13.

9.2.6.1.1 Inspection and Storage of Raw Materials

The functional and compositional properties of surimi largely influence the quality of the finished product. Upon receiving,

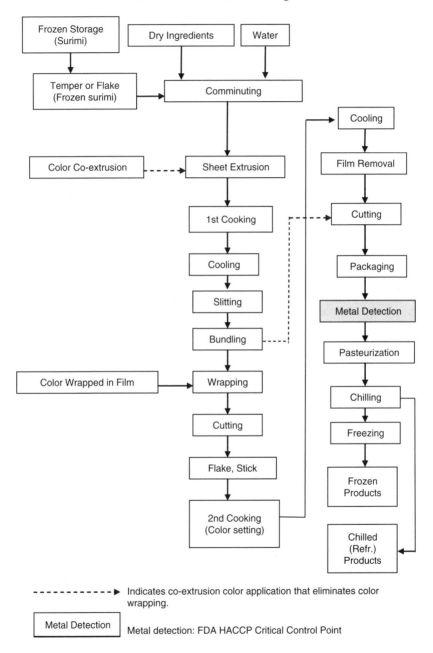

Figure 9.13 Flow diagram of the manufacturing of crabstick.

the temperature of randomly selected surimi blocks must be measured to monitor temperature abuse during transportation. Prior to use, the functional and compositional properties must be determined for possible optimized blending of surimi lots. This approach allows manufacturers to control the quality and consistency of the finished product. In addition to surimi, other raw materials such as starches, protein additives, food-grade chemicals, flavorings, and colorings should also be carefully inspected upon receipt. Most of these can be stored at room temperature, with the exception of frozen or refrigerated egg white and volatile flavorings. The shelf life of flavorings can be significantly expanded if they are stored at less than 5°C.

Is it necessary to have metal detection before thawing or breaking frozen blocks? Do you have to rely on the metal detection by surimi manufacturers? These questions can be evaluated based on industry experience. However, it would be wiser to have metal detection at the beginning as well, although the HACCP requires metal detection only for finished products.

9.2.6.1.2 Comminution and Ingredient Blending

Before frozen blocks of surimi are comminuted, they must be either partially thawed or broken into smaller pieces using a frozen meat breaker or hydroflaker. This step reduces the load on the equipment during the subsequent fine comminution step. Placing frozen blocks between warm plates typically accelerates thawing. Special attention, however, must be given to avoid over-thawing, which may induce protein denaturation. If a frozen meat breaker is used, care must be taken to remove any imbedded plastic film from the bag that encased the block.[7] Radio frequency or microwave could also be used for uniform thawing in a short period. However, further studies are needed for industrial-scale application.

The sequence of ingredient incorporation during comminution affects the textural quality of the final product and varies with the type of equipment used. In Japan and other

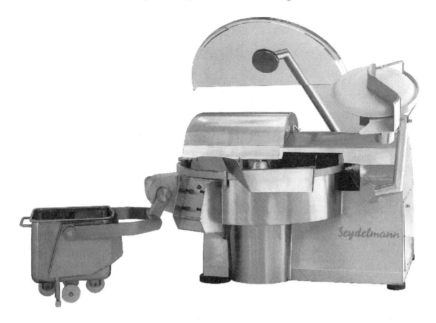

Figure 9.14 Vacuum silent cutter. (*Source:* Courtesy of Seydelmann & Robert Reiser & Co.)

Asian countries, a traditional stone mixer is still used by small operators. A 300-L bowl can hold up to 650 lb finished paste. A 500-L bowl unit with vacuum is commonly used, which can hold more than 900 lb of paste. A chopping time of approximately 25 min is normally required with open bowl cutters, compared to less than 15 min using a vacuum silent cutter (Figure 9.14). Thus, there has been a recent trend within the U.S. industry to adopt the more efficient vacuum cutting equipment.

The two major functions of comminution are to chop (extract) the (salt-soluble) myofibrillar proteins as much as possible to yield a smooth-textured paste and to uniformly incorporate (mix) other ingredients into the surimi paste. The maximum solubilization of available myofibrillar proteins depends on time, temperature, and other mechanical functions of the silent cutters, such as blade configuration, sharpness and/or vacuum, etc.

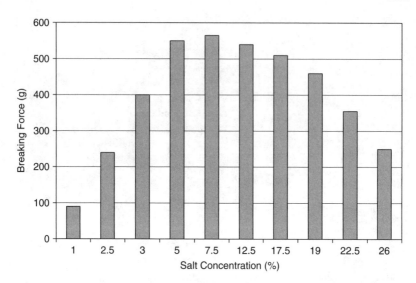

Figure 9.15 Effects of salt concentration on breaking force of surimi. (*Source:* Adapted from Ref. 45.)

Comminution first reduces the particle size of the fish meat almost to a powder. In this first stage of comminution, salt is added and chopping proceeds to facilitate solubilization of the myofibrillar proteins. By chopping surimi with salt only, the concentration of salt becomes relatively high, resulting in more salt-soluble proteins being utilized (Figure 9.15). In the second stage of comminution, the remaining ingredients are added and blended. With non-vacuum equipment, it is very important that the dry ingredients are carefully sprinkled onto the surimi paste, and starch should be premixed with water before addition. This is not required for vacuum equipment, into which all ingredients can be simultaneously added and blended in a short period of time.[7]

In addition, when the paste is prepared, it goes through a fine filtering screen (perforations less than 1.0 mm) to remove any plastic pieces or broken pin bones (particularly from surimi made from small fish).

The final paste temperature at the completion of chopping, in industrial practice, is generally near 10°C. Lee,[8] however, has suggested that the temperature of the paste must be kept at or below the temperature above which fish actomyosin becomes unstable. For Alaska pollock and Spanish mackerel, the paste should be kept below 10 and 16°C, respectively. In addition, Park[9] compared the effects of temperature during chopping on the shear stress and shear strain values of Alaska pollock surimi gels. The gelling property of surimi paste chopped at 20°C was extremely low. Maintaining chopping temperatures between 0 and 5°C, however, provided maximum gelling functionality.[9] Therefore, it was a general rule of thumb for chopping temperature that the lower the chopping temperature, the better the gel texture.

Park and co-workers[10] extended their efforts to study the effect of thermal sensitivity of fish proteins on gel texture using a variety of species harvested from various habitats with different water temperatures (Figure 9.16). This study indicated that there is a strong relationship between habitat temperature and protein stability. While cold-water fish species (Alaska pollock: AP) showed maximum gel strength and cohesiveness at lower chopping temperatures (0°C) and temperate-water species (Pacific whiting: PW) at 5 to 10°C, warm-water fish species (bigeye snapper: BE, lizardfish: LF, and threadfin bream: TB) had their maximum gel strength and cohesiveness at higher chopping temperatures (i.e., 20 to 25°C). This study also indicated that myofibrillar proteins from warm-water fish species (BE, LF, and TB) were more heat tolerant than that of cold-water fish species, and surimi gels should therefore be prepared at higher chopping temperatures for obtaining better protein functionality.

The old chopping principle — "The colder the chopping temperature, the better the gel texture" — is consequently no longer acceptable. The optimum chopping temperature must instead be determined based on the species used. Therefore, the use of a single species is likely to maximize the gelation properties of the surimi. However, in the case of mixing multiple species of surimi (i.e., cold-water species and warm-water species), it would be difficult to find an optimum chopping temperature for fish proteins due to their different thermal sensitivities. Therefore, the mixing of either species with temperate-water species would be the alternative solution.

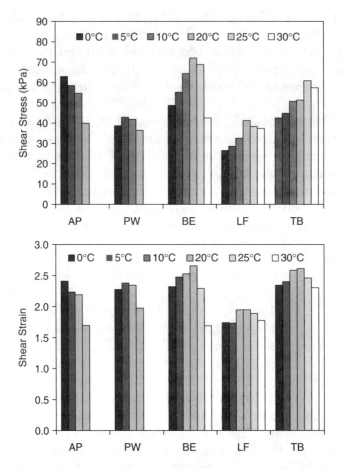

Figure 9.16 Effects of chopping temperature on gel texture of surimi gels. AP: Alaska pollock; PW: Pacific whiting; BE: bigeye snapper; LF: lizardfish; TB: threadfin bream. (*Source:* Adapted from Ref. 10.)

9.2.6.1.3 Cooking and Chilling

Most surimi seafood products are made on machinery manufactured by one of four major equipment manufacturers: Yanagiya-Bibun, Young Nam, IKM, and Tono. Unlike the conventional crabstick line with a continuous cooking belt, a cooking drum has been introduced recently and wisely used

(Figure 9.17). This line is very energy efficient and fits well in a compact place. Yanagiya and IKM also introduced an ohmic cooker in the mid-1990s. Ohmic heating, however, is discussed in detail later in this chapter.

There are slight differences between the machines, but the basic processing steps are similar. The type and sequencing of the heating elements can vary. Most commonly, the sequence for conventional cooking on the continuous flat belt is radiant heat, followed by steam heat, and then radiant heat again. Another common option consists of steam heat followed by radiant heat. Some machines employ steam heat only.

A drum cooker, which is designed to save space, has a slightly different heating system. Surimi seafood paste is extruded on a large drum that contains high-temperature steam inside. Regardless of the heating method, however, the ambient temperature in the tunnel where the paste is exposed to radiant or steam heat is 90 to 95°C.

Whether it is a drum or belt, the total cooking time depends on the product specification. For the thin (1.2 to 2.2 mm) sheets required to make filament-style surimi seafood, the cooking time generally varies from 30 to 100 sec, depending on production speed and other mechanical adjustments. This short cooking process time induces gelation of the surimi proteins, but is not sufficient to gelatinize (swell) the starch and/or to gel other protein additives.

Immediately upon completion of this first cooking step, the product is cooled by air (sometimes forced air) at room temperature or below. The cooling system employed is very different between manufacturers. Texture development is finalized during a later pasteurization step that adds rigidity and water-holding properties to the gel.

9.2.6.1.4 Fiberization

Fiberization for filament-style surimi seafood, which requires aligned fibers, is accomplished by elongated cuts running lengthwise on the gelled sheet. Slitting is obtained by passing the sheet through two rollers with teeth (slitters). The space between the teeth controls the number and width of the

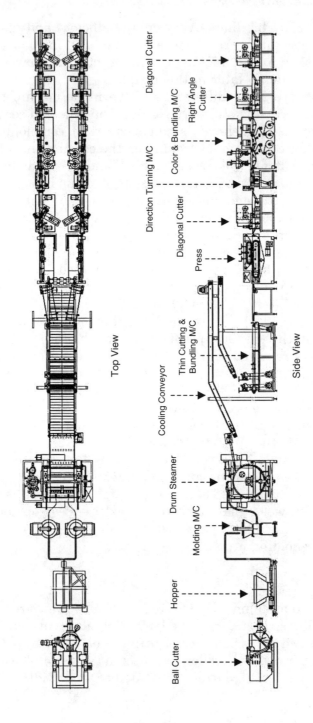

Figure 9.17 LIBETTI crabstick line with drum cooker. (*Source:* Courtesy of Young Nam Machinery, Korea.)

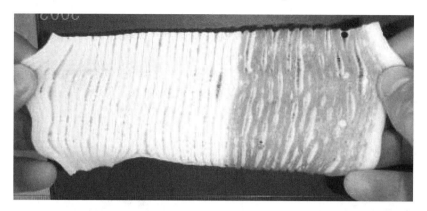

Figure 9.18 Degree of slitting controls the texture. (*Source:* Courtesy of Fleury Michon, France.)

individual fibers. A water mist or vegetable oil drip is often used to facilitate smooth passage of the sheet through the cutting rollers.

As shown in Figure 9.18, depending on the degree of slitting, it can be classified as light, medium, and heavy.

9.2.6.1.5 Bundling

Bundling is a process that rolls the cooked product sheet tightly into a rope shape (Figure 9.19). However, it is not uncommon to see a hole in the center of crabstick. There are a few factors that affect the occurrence of the hole, including the (1) bundling angle (as the angle α gets larger, the tighter the roll can be obtained); (2) thickness of the sheet (the thinner the sheet, the tighter the roll); (3) wetness of the bundling belt; (4) formula; and (5) speed of bundling belt (v_1) and conveyor belt (v_2). To ensure tight rolling and to eliminate a hole in the center, it is quite common to use liquid or meat binders as a paste on the sheet while forming the rope. The rope is then passed through a color application machine.

A polyethylene plastic film, onto which colored fresh paste is applied, wraps around the product rope, which is then cut to a specified length. The wrapping film is a single-layer

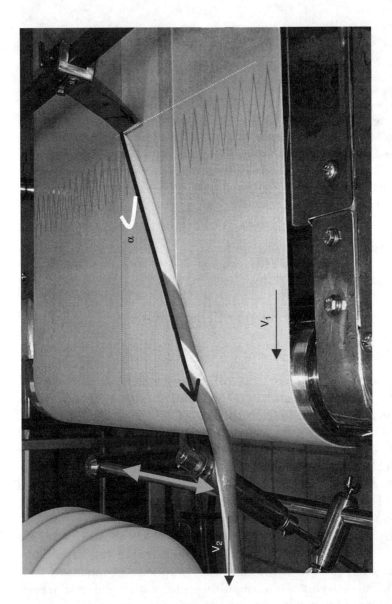

Figure 9.19 Bundling process affects the tightness of the rope. (*Source:* Courtesy of Fleury Michon, France.)

plastic made of high-density polyethylene (HDPE) (95%) and low-density polyethylene (LDPE) (5%). The colored paste is then set to a gel by cooking under steam for 15 to 30 min, followed by rapid chilling.

9.2.6.1.6 Color Application, and Wrapping

Color application using a plastic wrap film, however, often causes color flaking when the finished product is cut into a different shape. Therefore, color co-extrusion was initiated in the United States during the late 1980s (Figure 9.20). Colored paste is co-extruded on the edge of a sheet of uncolored surimi paste prior to the initial cooking step. This obviates the need for an intermediate wrapping step (and subsequent cooking for color setting and unwrapping). Co-extruded color, however, can sometimes bleed or transfer into the white portion of the product when vacuum-packed products are pasteurized at high temperatures. This problem, however, has been solved by leading manufacturers of color blending. A tiny amount of PGPR (polyglycerol polyricinoleate) derived from castor oil apparently works well as an excellent emulsifier to prevent color bleeding. This ingredient is not listed as illegal but has not been approved in certain countries (i.e., United States) simply because no application for GRAS (generally recognized as safe) has been filed as yet. It is also used successfully in chocolate manufacturing, simply because the chocolate industry has submitted its GRAS application to the U.S. FDA. Details can be found in the color chapter in this book (see Chapter 15).

9.2.6.1.7 Cutting

Filament products can be cut into different dimensions while in the rope shape. The most popular form of surimi seafood in the United States is termed "flake" cut. The rope is cut diagonally at a 25 to 30° angle in 5- to 8-cm-long pieces, tip to tip. Another shape is the stick shape, which is cut straight at a 90° angle, and usually cut 3, 5, or 8 inches in length. This is the most popular cutting in countries outside the United States. For filament chunk products, the pieces are cut 1 to 2

Figure 9.20 Co-extruded color.

cm long. In France, a crabmeat mosaic product (Figure 9.21) is made by packing 10 to 15 filament ropes in a large casing after dipping in a gelatin solution. When the gelatin solution is set in a refrigerator, the multiple ropes are cut straight.

In the manufacturing of crabstick wrapped in plastic, a wrapper longer than the crabstick is often observed. This results from the tension of the rope given on the cutting machine. However, this problem can be solved by running the continuous rope slowly rather than tightly or too fast.

9.2.6.1.8 Packaging

Most surimi seafood is packed in plastic films under full or partial vacuum to enhance the appearance of the product and extend its shelf life. Depending on product specification, the product is often individually quick-frozen (IQF) and packaged in bulk. As far as packaging materials are concerned, polymers commonly used for surimi seafood include polyester (nylon), polypropyrene, polyvinylidene chloride (PVDC), eth-

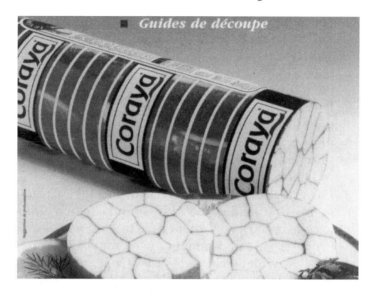

Figure 9.21 Surimi seafood in a mosaic shape. *(Source:* Courtesy of Cuisimer, France.)

ylenevinyl alcohol (EVOH), polyethylene (PE), polyester (PET), and adhesive.

The film, depending on the application, is often composed of four or five layers of these seven components with varying thickness. Nylon provides strength when used as the outermost layer. EVOH or PVDC gives gas and moisture vapor impermeability. PE is required for sealability. Details on individual components are as follows:

1. Polyethylene (PE) provides soft, flexible but tough, odorless, and tasteless films. It is an excellent barrier for moisture, but not for gas. It is required for surimi seafood packaging film because it provides good heat sealability. Because the surface is nonpolar, printing on PE is difficult. There are two types of PE: low density polyethylene (LDPE) and high-density polyethylene (HDPE). LDPE provides good tensile strength and burst strength, retains physical properties to as low as –60°C, and has the highest gas

permeability. HDPE provides higher tensile and burst strength than LDPE. Its linear nature gives directional tearing. It also gives an opaque appearance, superior chemical resistance, and lower gas permeability than LDPE.
2. Polypropylene (PP) is glossy, clear, odorless, tasteless, heat-stable, and noisy. It stretches less and has higher tensile strength than LDPE and is used for juice bottles. It gives excellent clarity and is difficult to heat seal.
3. Polystyrene (PS) is one of the most versatile polymers. It provides metallic sounds (when dropped), excellent clarity, low heat resistance, and poor barrier properties. It is commonly used for foamed trays.
4. Poly vinylidene chloride (PVDC) is very soft, transparent, and heat-sealable. It provides the lowest moisture permeability of all films. It is a good material for wrapping because it tends to cling to itself easily.
5. Ethylenevinyl alcohol (EVOH) has an extremely high barrier property to gas transport, including flavors. It is often laminated with a moisture barrier (i.e., PE) to protect barrier properties. It is very expensive.
6. Polyester (PET) is low cost, glossy, transparent, odorless, tasteless, tough, and heat-sealable. It provides very good moisture and gas barrier properties, very high tensile strength, and can withstand hot filling and boil-in-bag application. It can withstand a wide temperature range (−60 to 220°C).
7. Nylon, which is also known as polyamide (PA), provides high strength and toughness. It is flexible at low temperature and shows high thermal stability. In addition, it is expensive.

The thickness of the film is currently measured in micrometers (μm), but in many cases as mils, where 1 mil equals 0.001 inch. Basis weight is a combined measure of density and thickness, and is measured in grams per square meter (gsm). For example, newspaper is 50 gsm, while cardboard box is 195 to 586 gsm. Tensile strength is the measure of the force required to produce failure in a strip of sheet.

Because surimi seafood is pasteurized at high temperature for a long period of time and/or continues to be frozen, the selection of the correct film is not simple. Table 9.1 shows typical films used in surimi seafood packaging.

For the most commonly used roll stock package (Figure 9.22), two components (i.e., forming film and non-forming film) are used. Both forming and non-forming films are commonly made of nylon and linear low-density PE with a boilable adhesive as glue. However, the forming film is twice as thick as the non-forming film. Film B and C are non-forming film, while Film A and D are forming film (Table 9.1).

9.2.6.1.9 Metal Detection

According to the FDA HACCP guideline,[11] metal detection must be administered when the following operation/equipment is associated with food during processing: metal-to-metal contact, especially in mechanical cutting or blending operations; and other equipment with metal parts that can break loose, such as moving wire mesh belts, injection needles, screens, portion control equipment, metal ties and can openers. Surimi seafood operation is heavily associated with continuous mechanical actions and food is therefore exposed to metal hazard. However, FDA HACCP does not specify what metal detection controls must be used. It is left to the discretion of the manufacturer, although there is regulatory action against metal fragments exceeding 7 to 25 mm in the HACCP guideline. This limit (i.e., 7 to 25 mm) is very wide. Therefore, most U.S. surimi seafood manufacturers use their own "more strict" calibration metals criteria; they are 2.0 to 2.5 mm for ferrous, 2.5 to 3.0 mm for non-ferrous, and 3.0 to 4.0 mm for stainless.

9.2.6.1.10 Pasteurization and Rapid Chilling

The pasteurization step assures the microbiological quality of products, assuming sanitary ingredients and packaging

TABLE 9.1 Typical Films Used in Surimi Seafood Packaging

Structure	Density	Thickness		Basis Weight		Tensile Strength				Number of Layers	Materials
						MD		TD			
		μm	mil	g/m²	In.²/lb	N/cm²	lb/in.²	N/cm²	lb/in.²		
A	0.997	150	6	149.5	4707	4200	6100	4300	6200	7+	Nylon/PE/nylon/EVOH/Nylon/PE/LLDPE
B	0.976	90	3.6	87.8	8015	5800	8400	4800	6950	3	BOPA(nylon)/adhesive/LLDPE
C	1.005	87	3.5	87.4	8051	4000	5800	4200	6100	3	PET(polyester)/adhesive/LLDPE
D	0.989	125	5	123.6	5693	5000	7250	4500	6500	3	Nylon/adhesive/LLDPE

Note: MD = machine direction; TD = transverse direction.

Surimi Seafood: Products, Market, and Manufacturing 413

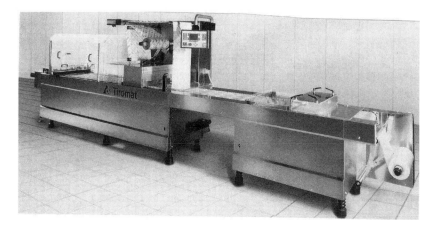

Figure 9.22 Packaging machine. (*Source:* Courtesy of Tiromat.)

materials have been used. Surimi seafood is cooked in a high-moisture environment to a temperature that is sufficient to cook and pasteurize, but not sufficient to sterilize. Pasteurization with the proper heat treatment provides a greater advantage over sterilization with regard to sensory attributes. The longer the cooking at a higher temperature, the more negative the sensory attributes.

The U.S. surimi seafood industry experienced textural softness, brownish discoloration, and/or off-odor when surimi content was largely replaced by higher starch content and pasteurized for a longer time at a higher temperature.[9] The pasteurization methods used by all U.S. surimi seafood manufacturers, however, are different. Therefore, there is great need to develop standard pasteurization methods that are scientifically valid. The U.S. surimi seafood industry and the National Food Processors Association (Seattle, WA) jointly conduct a wide range of pasteurization studies using crabsticks with inoculated *Clostridium botulinum*.

When surimi seafood paste was held at 90°C, deterioration of shear strain values and whiteness upon extended heat treatment were significant (Figure 9.23). According to Shie and Park,[12] the time required to obtain a zero aerobic plate count at 93, 85, and 75°C was 5, 15, and 15 min, respectively.

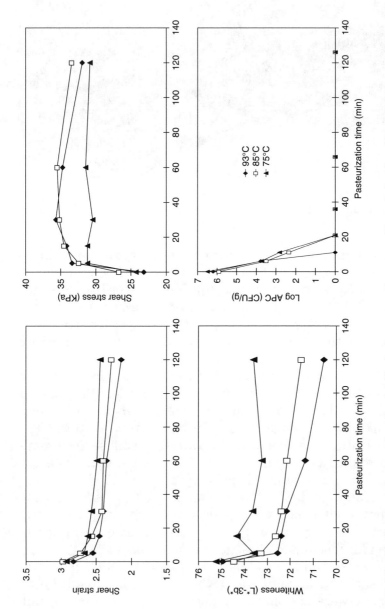

Figure 9.23 Texture, color, and microbiological properties as affected by heating time and temperature. (*Source:* Adapted from Ref. 12.)

Therefore, an efficient and/or optimum pasteurization procedure is needed for the industry to use as a model. Microbiological safety related to pasteurization and chilling is discussed in more detail in Chapter 12.

Recently, Jaczynski and Park[13] constructed and validated an interactive temperature prediction model for surimi seafood during pasteurization based on the Gurney-Lurie chart with a heat transfer coefficient and thermal diffusivity of surimi seafood. The model allows the input of four variables of a product: (1) processing temperature, (2) processing time, (3) initial temperature (product), and (4) product thickness. These variables can be changed at any time, resulting in respective changes. In this manner, thermal processing of surimi seafood can be optimized.

This model can be used to design a new thermal processing method for surimi seafood. It can also be used to verify existing thermal processing methods. However, it must be noted that a surface response model (Figure 9.24) was based on the heat transfer coefficient (h) and thermal diffusivity (α) determined based on our experimental setup. To apply this model in a different system, the specific h and α values for the specific product must be determined and incorporated into the spreadsheet.

9.2.1.6.11 Rapid Freezing and Storage

Refrigerated products are packed into cartons after chilling individually packaged units to 4°C. Most refrigerated, partial or full vacuum-packaged products distributed in U.S. supermarkets have a shelf life of 60 to 90 days. For longer distribution times, the frozen form is recommended. Chilled products are typically frozen in a blast or contact freezer, or in a liquid nitrogen tunnel. To reach an internal temperature of –18°C for a 2.5-lb package (10°C) usually takes 60 to 70 min in a spiral freezer and 15 to 20 min in liquid nitrogen. The freezing rate affects product quality and shelf life. At customary freezer temperatures (–10 to –20°C), about 90% of the moisture freezes out, resulting in approximately a tenfold

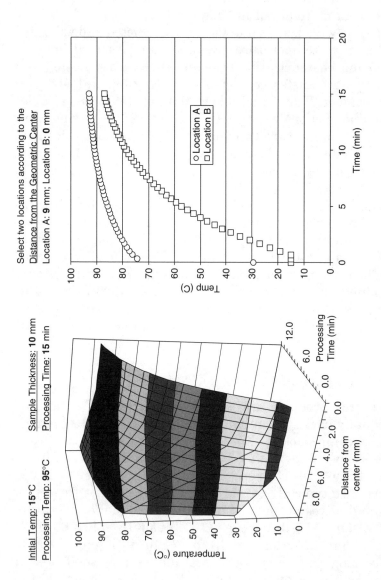

Figure 9.24 Surface response model for temperature prediction while heating for pasteurization. (*Source*: Adapted from Ref. 13.)

increase in the concentration of solutes in the remaining free liquid.[14]

Fast freezing, which generates numerous small ice crystals rather than fewer large ice crystals, is very important in maintaining good quality during long-term storage; otherwise, wet and brittle products are commonly obtained upon thawing. Minimal fluctuation in the storage temperature is also helpful in this regard. Formulation adjustments must be made, which can vary depending upon the freezing system, to prevent the damaging effects of freezing on product quality. For example, the use of modified starches, especially those prepared by hydroxypropylation and cross-linking, instead of native starches, can reduce the weeping (moisture release) that often accompanies the thawing of frozen products.[7,15] Where fast freezing is applied using liquid nitrogen, the use of this type of specialty starch may not be necessary (if the product contains good-quality surimi at greater than 40%). Detailed information regarding freezing technology is discussed in Chapter 8.

9.2.6.2 Solid Meat Style

9.2.6.2.1 Block (Solid Meat) Formation

Preincubation before cooking, as a final heat treatment, greatly increases the strength (shear stress) and, to a somewhat lesser degree, the cohesive nature (shear strain) of gels. The effects of preincubation on gel functionality depend on the species and proliferate using the habitat temperature of the fish.[16,17] The relationship between the water temperature of the fishing grounds or habitat and the best gel-setting temperature of surimi are discussed by Kim[18] and Park.[19] Presetting at 5°C overnight or 40°C for 40 min are alternative preincubation treatments that can also be used to obtain a marked improvement in gel rigidity and cohesiveness.

Kim et al.[20] found that the gel texture of surimi prepared from Atlantic croaker (*Micropoga undulatus*) and Alaska pollock responded similarly to a 40°C preincubation, but responded differently to a 4°C overnight preincubation. The strongest pollock surimi gels were obtained with a 4°C presetting for 24 hr. The strongest croaker surimi gels, however, were formed at a 40°C presetting for 30 min, followed by a final heating at 90°°C for 15 min.

Klesk et al.[21] reported that tropical tilapia performed best when it was set at 40°C for 1 hr. In addition, Kamath et al.[22] reported that maximum production of cross-linked polymers occurred at the optimum setting temperatures, that is, at 25°C for Alaska pollock surimi and 40°C for Atlantic croaker. Furthermore, the optimum gelling ability of Alaska pollock, Southern blue whiting, and hoki were obtained by presetting gels at 25°C for approximately 4 hr.[23] Finally, Park et al.[19] revealed that the optimum temperature of presetting for Pacific whiting surimi is 25°C. Park and colleagues[10] evaluated various species from different regions and found the optimum setting temperatures were approximately 5, 25, 40, 40, and 25°C for Alaska pollock, Pacific whiting, bigeye snapper, lizardfish, and threadfin bream, respectively. Setting at 60°C induced gel weakness for all fish species, indicating that all tested surimi contained proteolytic enzymes; in particular, Pacific whiting and lizardfish, which resulted in no measurable gels, appeared to have a significant concentration of proteolytic enzymes.

9.2.6.2.2 Fiber (Shred) and Binder

The most important processing step in this category is to develop excellent texture of random fibers. For random fiber products, such as shrimp and lobster, a cooked (gelled) block is shredded into short fibers before blending with fresh paste, which acts as a binder (7:3 or 6:4), and then extruded.

9.2.6.2.3 Extrusion

The mixture is extruded onto a stainless steel belt in a continuous rope shape (Figure 9.25). After cooking for 10 to 15 min under steam (>95°C) and cooled (typically by fans), the product can be cut into various lengths: solid meat sticks or solid meat chunks. Sticks and chunks are either individually quick frozen (IQF) or pasteurized in plastic bags. Solid meat chunks are often packaged with filament flakes for a salad

Figure 9.25 Extruded (solid) meat.

pack combo (Figure 9.4). Color application for extruded products is accomplished using co-extrusion.

9.2.6.2.4 Molding

Surimi paste, which is commonly mixed with pre-prepared fibers, is formed in a molding machine or cold-extruded in a three-dimensional shape, such as in the manufacture of shrimp- or lobster-shaped surimi seafood (Figure 9.26). For color application, a color solution is sprayed onto the inside of the mold before stuffing or sprayed on the surface of cooked molded products. In the latter case, additional heating is needed to set the color.

9.3 OTHER PROCESSING TECHNOLOGY

9.3.1 Ohmic Heating

Ohmic heating is a method in which alternating electrical current is passed through an electrically conducting food

Figure 9.26 Molded products (lobster and shrimp).

product. Heat is internally generated due to the electrical resistance of the food sample, resulting in a rapid heating rate. This differs from microwave heating, during which energy transferred to polar molecules in food (mainly water) is converted to heat.[24] This may cause uneven heating in a number of foods with solid chunks. In particulate food products, ohmic heating provides uniform temperature distribution because both the liquid and solid phases are simultaneously heated.[25] These developments have led to the evolution of commercial ohmic sterilization for particulate foods such as chunky soups and stews.[26]

The application of ohmic heating in seafood has also been investigated. When ohmic heating was compared with traditional 90°C waterbath heating, the gel strength of Alaska pollock, threadfin bream, and sardine surimi improved.[27,28] Yongsawatdigul et al.[29,30] investigated the feasibility of ohmic heating to maximize the gel functionality of Pacific whiting surimi without enzyme inhibitors. The ohmically heated gels showed more than a twofold increase in shear stress and shear strain over gels heated in a waterbath.[29] Ohmic heating can also generate very uniform heating. When enzyme-laden Pacific whiting surimi is subjected to slow heating, no gels are formed. However, when fast heating is provided, strong gels are formed from Pacific whiting surimi (Figure 9.27). Degradation of myosin heavy chain and actin was also minimized by ohmic heating, resulting in a continuous network structure of the gels.[29,30] Shear stress values of ohmically heated pollock surimi gels were also better than gels heated in a waterbath (90°C for 15 min). Shear strain values of pollock surimi gels, however, were not affected by ohmic heating. When beef plasma protein was added, however, the effects of this heating method became neutralized.

The use of ohmic heating for surimi with heat-stable endogenous proteases can overcome some of the negative effects of commercial enzyme inhibitors, including high cost, off-odor, off-color, and labeling concerns. However, corroded electrodes were commonly observed after heating at 60 Hz. Wu et al.[31] investigated the effect of various frequencies up

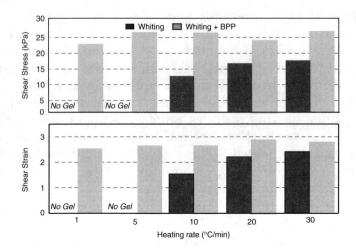

Figure 9.27 Shear stress and strain of ohmically heated surimi gels. BPP: with 1% beef protein. (*Source:* Adapted from Ref. 30.)

to 200 kHz on the formation of corrosion. When a frequency greater than 5 kHz is applied, corrosion disappeared.

Two Japanese manufacturers have introduced a commercial-scale ohmic cooker (Figure 9.28). For the manufacture of filament-style surimi products, a continuous thin sheet of surimi paste is extruded on a wet fabric conveyor belt located above the ohmic heater. Many stainless steel electrodes, which consist of anodes and cathodes placed one after another at a 10- to 20-mm distance, revolve to move the belt as a roller. The electric current is sent from one cathode to the next anode through the wet fabric belt and the heat is induced between the electrodes. The thin surimi paste sheet, which conducts heat very effectively with increased moisture and salt content,[32] is cooked to form an elastic gel sheet. Because the heat is conducted within the surimi paste, the removal of air pockets in the sheet is very critical for uniform heat generation. Therefore, the use of a vacuum silent cutter is highly recommended for ohmic cooking process.

It is known that formulations containing the least surimi content can be successfully run on the ohmic cooker. Subsequently, products cooked under the ohmic cooker give better

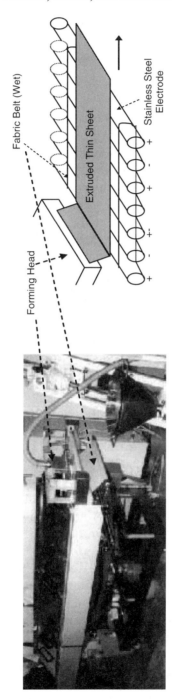

Figure 9.28 Outline of commercial-scale ohmic cooker.

texture when the product is aged during frozen storage (A. Law, personal communication, 1996). However, further studies are needed to verify this observation.

9.3.2 High Hydrostatic Pressure

High hydrostatic pressure (HHP) is a simple concept. The food product is sealed in a plastic bag, inserted into a chamber, and subjected to pressures up to 7000 atmospheres (atm). For comparison, a typical autoclave cooker is at 2.5 atm (120°C) and an extruder cooker is at 50 atm (180°C), while a French press goes as high as 1500 atm and the deepest part of the ocean has approximately 1200 atm of pressure.[33] An HHP unit applies extremely high pressures, causing physical and chemical changes without the presence of heat. This pressure affects cell membranes, microorganisms, and enzymes, all of which are important constituents of food.[34] A number of food materials, such as egg yolk, meat, and soy protein, have been shown to gel under HHP. Surimi, which is stabilized myofibrillar proteins, readily gels at pressures as low as 2 kbar (2000 atm). Gelation of pollock surimi by HHP was attributed to increased cross-linkage of the myosin heavy chain.[35] Okazaki and Nakamura[36] have shown that the gelation of sarcoplasmic proteins from different fish depended on species, pH, protein concentration, and pressure treatment. The freezing point of water at 2 kbar is –20°C. This would allow a product to be kept at temperatures below 0°C without the formation of damaging ice crystals.

Chung et al.[37] investigated the effects of HHP on the gel strength of Pacific whiting and Alaska pollock. A threefold increase in strain and stress was found for HHP-treated whiting gels made without an enzyme inhibitor (i.e., beef plasma protein), as compared with gels heated in a 90°C waterbath. When pressure treatment was undertaken at 50°C, Pacific whiting surimi gels without beef plasma protein were too weak to measure. The protease enzyme in Pacific whiting surimi has been shown to have an optimum temperature of 55°C.[38] These results would indicate that the enzyme remains active during pressure treatments at 55°C. Pollock surimi also showed significant increases in both strain and stress values for all pressure treatments except for those run at 50°C. A direct relationship between pressure and stress values was inversely related to the amount of pressure used.[37] HHP represents a potential processing technology for surimi-based seafood. At present, the only HHP treatments conducted on a commercial scale are for the preservation of high-valued jams and jellies in Japan.

9.3.3 Least-Cost Linear Programming

Although there are other functions in addition to water/fat/particle binding and texturization, such as emulsification and foaming, these properties have not been traditionally important in the production of muscle foods.[39] Surimi offers significantly greater functionality than competing proteins in terms of its gelling properties. It is recognized that each functional property — such as water, fat, and particle binding, as well as texturization — is linked to the formation of a stable gel network structure in food gels.[39,40] High-quality surimi will form very deformable gels covering a wide range of shear stress values.

Because surimi forms the base protein matrix in which all other ingredients are imbedded as a filler or interacts with as a binder, the combination of surimi controls the color, flavor, and texture of surimi-based products. There have been efforts to use a least cost formulation as an effective means of controlling the quality, while minimizing ingredient cost in order to utilize various grades of surimi.[41,42] Figure 9.29 provides a practical application of blending surimi for the production of consistent-quality surimi seafood products. In formula optimization, the main goal is to find the best combination for each key ingredient or component. In addition, the levels for critical processing variables must be discussed. Ingredients are the independent variables. The dependent variables are always the entity to be optimized (i.e., maximized or minimized); they are functional properties such as shear stress, shear strain, and whiteness, and compositional properties such as moisture, protein content, pH, impurity, and cost.

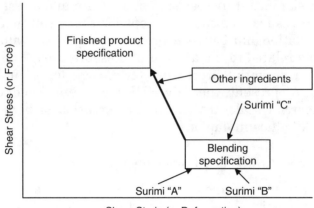

Figure 9.29 Control of finished product specification using a blending technique.

Blending different grades of pollock and/or whiting surimi in different proportions shows the linearity of shear stress, shear strain, and whiteness (Figure 9.30). There was no interaction between blending different grades of surimi lots as shown in high r^2 (>0.99) and low p-value (<0.001) according to Yoon et al.[43] The different ratios of high to low grade used were 1:0, 0.5:0.5, and 0:1, respectively. Because the linear effect of blending surimi on each functionality was determined in the range of 0 to 100%, the first-order canonical model was developed and employed for a linear programming model.[43] The functionality of each surimi is individually measured before blending, and the coefficient of each variable is calculated in canonical form from the linear regression model. The final equation for the optimum formulation was developed as follows.

Functional properties of the blend include:

= [(% surimi 1) × (Functional property of surimi 1) + (% surimi 2) × (Functional property of surimi 2) + … + (% surimi n) × (Functional property of surimi n)]

= Σ [(% surimi i) × (Functional property of surimi i)]

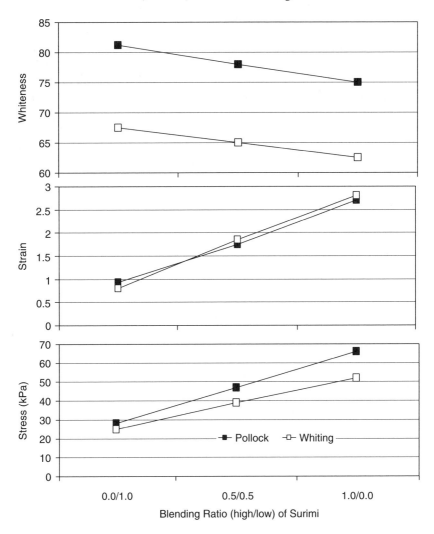

Figure 9.30 Effects of blending different surimi lots on shear stress (kPa), shear strain, and whiteness (L* – 3b*). 0.0/1.0 indicates 100% low-grade surimi; 1.0/0.0 denoted 100% high-grade surimi.

where i = 1, 2, ..., n.

The stress values of blended pollock (high and low grade) at any proportion were estimated, according to Yoon et al.[43], by the following canonical model form:

Stress (kPa) of the blend = 64.5 × (% of high grade) + 28.24 × (% of low grade)

where 64.5 and 28.24 represents the stress values of 100% pollock high and low grade, respectively (Figure 9.30).

Because the functionality of blended surimi showed a linear relationship and proportionality to the quantity of each surimi (Figure 9.30), surimi from ten different lots (A through J) can be blended to make formulations with specific target constraints, as established in Table 9.2. Responses are quite different, depending on the quality of the surimi. Using linear programming, the optimum blending for the least cost consists of 0.7% surimi B, 72.6% surimi F, 6.3% surimi G, and 20.4% surimi H. This blending formulation, which utilizes different levels of surimi, is not only equal to or exceeds the target constraints in functionality, but also provides the least cost ($0.97/lb).

The use of all high-grade surimi certainly provides high-quality products but also results in higher cost. When all low-grade surimi is used to reduce cost, quality is sacrificed. In addition, neither method provides the consistency of quality because they are not optimized for the selection of surimi. The use of linear programming can play an important role in blending different surimi lots while providing the least cost and consistent quality of surimi.[41,43,44] Once surimi is controlled for consistent quality and least cost, the development of the least-cost surimi seafood formulation is rather easy (Figure 9.29).

ACKNOWLEDGMENTS

The author would like to acknowledge Dr. Noboru Kato for his early contributions to Japanese traditional surimi-based products and Jean-Luc Beliveau for his invaluable information regarding the French market.

TABLE 9.2 Least-Cost Linear Programming for Blending Various Surimi

Available Surimi Price ($/lb)	A 0.95	B 0.8	C 1.05	D 0.75	E 1.15	F 1.1	G 0.75	H 0.58	I 1.5	J 10.05	Target Constraints
Responses											
Shear Stress	64.5	28.24	52.5	23.5	39	43	24	25	50	50	≥ 38
Shear Strain	2.65	0.925	2.75	0.85	2.85	2.75	1.8	0.9	2.6	2.55	≥ 2.3
Whiteness	75.5	81	78	67	75	82	5	53	57	62	≥ 75
Purity	8	9	7	8	8	8	0	8	8	9	≥ 7.5
Weight	1	1	1	1	1	1	1	1	1	1	1
Solution ($/lb)	0.970										
Fraction	0.000	0.007	0.000	0.000	0.000	0.726	0.063	0.204	0.000	0.000	
Lower	0	0	0	0	0	0	0	0	0	0	0
Upper	0	1	0	1	0	1	1	1	1	1	1

Constraint Values of Blended Surimi

Shear Stress	38.02
Shear Strain	2.30
Whiteness	75.00
Purity	7.50
Weight	1

REFERENCES

1. Zenkama. Trends of production of surimi-based products. In Zenkama homepage http://www.zenkama.com/zenkama/seisan.html. All Japan Kamaboko Association (Zenkama). Tokyo, Japan, 2004.

2. P. Guenneuges. Update on the World Surimi Market: Supply and Demand. Presented at the *4th Surimi Industry Forum*, Astoria, OR, 2004.

3. FDA. Code of Federal Regulations 21. Food and Drug Agency. Washington, D.C., 1992.

4. J.W. Park. Consumer Perception towards Nutritionally Fortified Surimi Crabmeat: Survey. Unpublished data, OSU Seafood Lab, Astoria, OR, 1994.

5. Anonymous. Surimi takes over Europe by storm. *Produits de la Mer en Europe,* April-May, 141–142, 1998.

6. W. Court. Surimi market report (19/11/98). Fish Info Service. Website at www.sea-world.com/fis/reports/surimi/surimi.html. 1998.

7. J.W. Park and T.C. Lanier. Surimi and surimi seafoods. In R.E. Martin, E.P. Carter, G.J. Flicks, Jr., and L.M. Davis, Eds. *Marine and Fresh Water Products Handbook.* Lancaster, PA: Techno Publishing Co., 2000, 417–443.

8. C.M. Lee. Surimi processing technology. *Food Technol.,* 38(11), 69–80, 1984.

9. J.W. Park. Surimi seafood: products, market, manufacturing. In J.W. Park, Ed. *Surimi and Surimi Seafood.* New York: Marcel Dekker, 2000, 201–235.

10. O. Esturk, J.W. Park, and S. Thawornchinsombut. Thermal sensitivity of fish proteins from various species on rheological properties. *J. Food Sci.,* 69(7), E412–416, 2004.

11. FDA. *Fish & Fisheries Products Hazards & Control Guidance, 3rd ed.* Office of Seafood, FDA. Washington, D.C., 2001.

12. J.S. Shie and J.W. Park. Physical characteristics of surimi seafood as affected by thermal processing conditions. *J. Food Sci.,* 64, 287–290, 1999.

13. J. Jaczynski and J.W. Park. Temperature prediction during thermal processing of surimi seafood. *J. Food Sci.*, 67(8), 3053–3057, 2002.

14. W.D. Powrie. Characterization of food myosystems and their behavior during freeze-preservation. In O.R. Fennema, W.D. Powrie, and E.H. Marth, Eds. *Low Temperature Preservation of Foods and Living Matter.* New York: Marcel Dekker, 1973, 282–336.

15. H. Yang and J.W. Park. Effects of starch properties and thermal-processing conditions on surimi-starch gels. *Lebensmittel Wissenschaft & -Technologie,* 31(4), 344–353, 1998.

16. K. Arai, K. Kawamura, and C. Hayashi. The relative thermostabilities of the actomyosin-ATPase from the dorsal muscles of various fish species. *Bull. Jap. Soc. Sci. Fish.,* 39, 1077–1082, 1973.

17. T. Misima, H. Mukai, Z. Wu, K. Tachibana, and M. Tsuchimoto. Resting metabolism and myofibrillar Mg^{++}-ATPase activity of carp acclimated to different temperatures. *Nippon Suisan Gakkaishi,* 59, 1213–1218, 1993.

18. J.M. Kim and C.M. Lee. Effect of starch on textural properties of surimi gel. *J. Food Sci.,* 52, 722–725, 1987.

19. J.W. Park, J. Yongsawatdigul, and T.M. Lin. Rheological behavior and potential cross-linking of Pacific whiting (*Merluccius productus*) surimi gel. *J. Food Sci.,* 59, 773-776, 1994.

20. B.Y. Kim, D.D. Hamann, T.C. Lanier, and M.C. Wu. Effects of freeze-thaw abuse on the viscosity and gel-forming properties of surimi from two species. *J. Food Sci.,* 51, 951–956 and 1004, 1986.

21. K. Klesk, J. Yongsawatdigul, J.W. Park, S. Viratchakul, and P. Virulhakul. Functional properties of tropical tilapia surimi as compared to Alaska pollock and Pacific whiting surimi. Presented at the *Annual Meeting of IFT,* Chicago, IL, 1999.

22. G.G. Kamath, T.C. Lanier, E.A. Foegeding, and D.D. Hamann. Nondisulfide covalent cross-linking of myosin heavy chain in "setting" of Alaska pollock and Atlantic croaker surimi. *J. Food Biochem.,* 16,151–172, 1992.

23. G.A. MacDonald, J. Stevens, and T.C. Lanier. Characterization of New Zealand hoki and Southern blue whiting surimi compared to Alaska pollock surimi. *J. Aquat. Food Prod. Technol.*, 3(1), 19–38, 1994.

24. C.R. Buffler. *Microwave Processing and Cooking*. New York: Van Nostrand Reinhold, 1993.

25. D.L. Parrot. Use of ohmic heating for aseptic processing of food particulates. *Food Technol.*, 46(12), 68–72, 1992.

26. C.H. Biss, S.A. Coombes, and P.J. Skudder. The development and application of ohmic heating for the continuous processing of particulate food stuffs. In R.W. Field and J.A. Howell, Eds. *Processing Engineering in the Food Industry*. Essex, England: Elsevier Applied Science, 1989, 17–27.

27. M. Shiba. Properties of kamaboko gels prepared by using a new heating apparatus. *Nippon Suisan Gakkaishi*, 58, 895–901, 1992.

28. M. Shiba and T. Numakura. Quality of heated gel from walleye pollock surimi by applying joule heat. *Nippon Suisan Gakkaishi*, 58, 903–907, 1992.

29. J. Yongsawatdigul, J.W. Park, E. Kolbe, Y. AbuDagga, and M.T. Morrissey. Ohmic heating maximizes gel functionality of Pacific whiting surimi. *J. Food Sci.*, 60, 10–14, 1995.

30. J. Yongsawatdigul and J.W. Park. Linear heating rate affects gelation of Alaska pollock and Pacific whiting surimi. *J. Food Sci.*, 61, 149–153.

31. H. Wu, B. Flugstad, E. Kolbe, J.W. Park, and J. Yongsawatdigul. Electric properties of fish mince during multi-frequency ohmic heating. *J. Food Sci.*, 63, 1028–1032, 1998.

32. J. Yongasawatdigul, J.W. Park, and E. Kolbe. Electrical conductivity of Pacific whiting surimi during ohmic heating. *J. Food Sci.*, 60, 922–925, 935, 1995.

33. M.T. Morrissey, J.W. Park, and J. Yongsawatdigul. Innovative processing in the seafood industry: The potential for ohmic heating and high hydrostatic pressure. Presented at the *First International Symposium of Biochemical Engineering and Food Technology at ITESM-Campus Quertaro*, Quertaro, Mexico, 1994.

34. D. Farr. High pressure technology in food industry. *Trends Food Sci. Technol.,* 1(1), 14–16, 1990.

35. S. Shoji, H. Saeki, A. Wakemeda, M. Nakamura, and M. Nonaka. Gelation of salted paste of Alaska pollock by high pressure and change in myofibrillar protein in it. *Nippon Suisan Gakaishi,* 56, 2069–2076, 1990.

36. E. Okazaki and K. Nakamura. Factors influencing texturization of sarcoplasmic protein of fish by high pressure treatment. *Nippon Suisan Gakkaishi,* 58, 2197–2206, 1992.

37. Y.C. Chung, A. Gebrehiwot, D.F. Farkas, and M.T. Morrissey. Gelation of surimi by high hydrostatic pressure. *J. Food Sci.,* 59, 523–524 and 543, 1994.

38. M.T. Morrissey, P.S. Hartley, and H. An. Proteolytic activity in Pacific whiting and effects of surimi processing. *J. Aquat. Food Prod. Technol.,* (4), 5–18, 1995.

39. T.C. Lanier. Functional properties of surimi. *Food Technol.,* 40(3), 107–114, 124, 1986.

40. J.C. Acton, G.R. Ziegler, and D.L. Burge. Functionality of muscle constituents in the processing of comminuted meat products. *Crit. Rev. Food Sci. Nutri.,* 18(2), 99, 1983.

41. T.C. Lanier and J.W. Park. *Application of Surimi Quality Measurements to Least Cost Linear Programming of Surimi Product Formulations.* Anchorage, AK: Alaska Fisheries Development Foundation, 1990.

42. J.W. Park. Use of various grades of surimi with an application of least cost formulation. In G. Sylvia and M.T. Morrissey, Eds. *Pacific Whiting: Harvesting, Processing, Marketing, and Quality Assurance.* Corvallis, OR: Oregon Sea Grant, 1993, 17–19.

43. Y.B. Yoon, B.Y. Kim, and J.W. Park. Linear programming in blending various components of surimi seafood. *J. Food Sci.,* 62, 564–567, 1997.

44. W.B. Yoon, B.Y. Kim, and J.W. Park. Surimi-starch interactions based on mixture design and regression models. *J. Food Sci.,* 62, 555–560, 1997.

45. M. Okada. *Science of Kamaboko.* Tokyo: Seizando Publisher, 2000, 60–62.

10

Surimi Gelation Chemistry

TYRE C. LANIER and PATRICIO CARVAJAL
North Carolina State University, Raleigh, North Carolina

JIRAWAT YONGSAWATDIGUL
Suranaree University of Technology, Thailand

CONTENTS

10.1 Introduction 436
10.2 Protein Components of Surimi 437
    10.2.1 Myofibrillar Proteins 437
        10.2.1.1 Myosin 439
        10.2.1.2 Actin 441
        10.2.1.3 Other Myofibrillar Proteins 441
        10.2.1.4 Thick Filament Assembly 442
    10.2.2 Stroma Proteins 444

  10.2.3 Sarcoplasmic Proteins 444
    10.2.3.1 Heme Proteins 446
    10.2.3.2 Enzymes 447
10.3 Lipid Components of Fish Muscle 450
10.4 Bonding Mechanisms during Heat-Induced
  Gelation of Fish Myofibrillar Proteins 451
  10.4.1 Hydrogen Bonds 451
  10.4.2 Ionic Linkages (Salt Bridges) 452
  10.4.3 Hydrophobic Interactions 455
  10.4.4 Covalent Bonds 456
    10.4.4.1 Disulfide Bonds 456
    10.4.4.2 Rheological Behavior of
         Cross-Linked Protein Gels 459
    10.4.4.3 Role of Disulfide Bonding in
         Myosin/Actomyosin Gelation 459
    10.4.4.4 Covalent Cross-linking during
         Setting 461
    10.4.4.5 Protein Stability Effects on
         Setting 467
    10.4.4.6 Endogenous Transglutaminase
         (TGase) 468
    10.4.4.7 Exogenous TGase Addition 469
10.5 Factors Affecting Fish Protein Denaturation and
  Aggregation 471
  10.5.1 The Importance of Muscle pH (Acidity) 475
  10.5.2 The Frozen Storage Stability of Surimi 476
10.6 Summary: Factors Affecting Heat-Induced
  Gelling Properties of Surimi 476
References 477

10.1 INTRODUCTION

Natural foods, despite their high water content, are made solid by confining the water within cells. In contrast, fabricated foods with solid-like (viscoelastic) properties are almost always hydrogels (water confined in a polymer matrix). Most gelling carbohydrates and gelatin (protein) form hydrogels when their concentrated solutions are cooled; subsequently,

Surimi Gelation Chemistry

however, these melt upon heating (i.e., are thermo-reversible). Surimi, on the other hand, like the muscle proteins of other animal species, as well as egg white, wheat gluten, and milk β-lactoglobulin, forms a thermo-irreversible gel upon heating that does not melt with further temperature change. Surimi is also known to produce gels of very high gel strength and deformability. The excellent heat-induced gelation properties of surimi make it useful as a food ingredient. This chapter reviews the chemistry of muscle proteins with regard to surimi gelation, as affected by various factors associated with the manufacturing of surimi and surimi seafoods.

10.2 PROTEIN COMPONENTS OF SURIMI

10.2.1 Myofibrillar Proteins

Striated fish muscle is composed of muscle fibers (the muscle "cells"), which in turn contain myriad myofibrils. The myofibrils are constructed of end-on-end contractile units called sarcomeres, which contain three types of filaments — thick, thin, and connecting — arranged in such a fashion as to impart the striated appearance of muscle under the microscope (Figure 10.1). The predominant protein in the sarcomere, found in the thick filament system, is myosin (approximately 55 to 60% of the total myofibrillar proteins). A number of studies have found it largely responsible for the functional properties of muscle tissues, including gelation and water-binding. However, postmortem myosin binds tightly to the predominant protein of the thin filaments, actin, resulting in the formation of a complexed protein termed actomyosin. Thus, in surimi, actomyosin, rather than myosin, is the predominant protein component, and its concentration and properties largely dictate the heat-induced gelation properties of surimi. It is useful, however, to consider the properties of myosin and actin separately because they exist in different filaments of the sarcomere under physiological conditions. Also, addition of polyphosphate compounds to surimi or any post-rigor meat has been shown to induce some dissolution of actomyosin into its two primary components, when residual ATPase activity exists in the myosin.

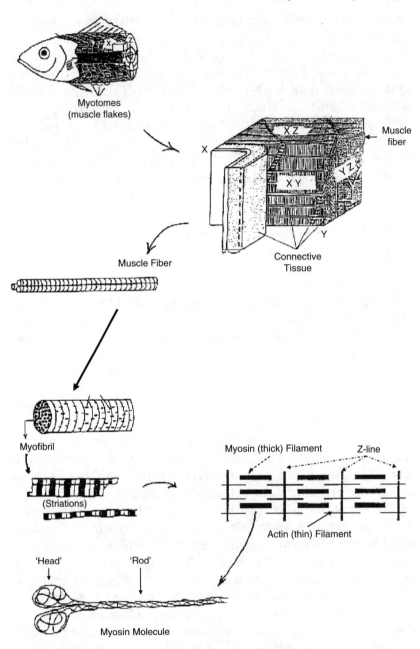

Figure 10.1 Diagram depicting successively greater detail in fish muscle microstructure. (*Source:* From Ref. 136. With permission.)

10.2.1.1 Myosin

Myosin is a relatively large protein, with a molecular weight of 470 kDa[1] and is unusual in that it has both fibrous (long, extended shape) and globular (spherical shaped) properties (Figure 10.1). By contrast, most protein ingredients used in foods, such as those from egg, whey, and soy, are globular and have a molecular mass of about 520 kDa. Each myosin molecule is composed of two 220-kDa heavy amino acid chains and two pairs of different light chains (LCs), ranging from 17 to 22 kDa.[2] The entire molecule is approximately 160 nm in length. The heavy chains interact to form two distinct domains: a pair of globular "heads" (S1) and a fibrous or elongated domain, the so-called "rod."

Globular Head Domain. The N-terminal pair ends of the myosin heavy chain (about 800 amino acid residues) — or heavy meromyosin subfragment-1 (HMM-S1)[3,4] — folds into elongated pear-shaped heads typical of a globular protein, being 60% α-helix and 15% β-sheet structure. It has ATPase activity (which releases the energy for muscle contraction) and binds to actin (thin filament) in the absence of ATP (postmortem). ATPase activity has often been monitored as a sensitive indicator of the extent of myosin unfolding (denaturation) and aggregation. On each of the globular heads, two small light chains are noncovalently attached. These light chains are thought not to play an important role in the gelation of muscle proteins.

Fibrous Rod Domain. The myosin rod domain is approximately 150 nm long and 2 nm in diameter, and is divided in two main regions: (1) the C-terminal end, comprising two thirds of the rod, is termed light meromyosin (LMM) and (2) the N-terminal third, termed heavy meromyosin subfragment 2 (HMM-S2), connects the globular head and LMM. It is believed that the LMM portion of the myosin molecules confers the solubility and aggregation properties to form the backbone of the thick filament, whereas the HMM-S2 portion acts as a fairly flexible link to the myosin head (which allows it to "ratchet" the sliding filaments together during contraction).

Structurally, 90% of the rod (LMM and S2 domains) consists of two right-handed α-helical coils that intertwine around one another, forming a left-handed coiled-coil. Residues from each helix form a "knobs-into-holes" type of packing arrangement in which a residue from one helix (knob) packs into a space surrounded by four side chains of the facing helix (hole), directly to the side of the equivalent residue from the facing helix (Figure 10.2).[5,6] This geometry contrasts with the more irregular packing of globular helix proteins, in which a residue packs above or beneath the equivalent residue from the facing helix.

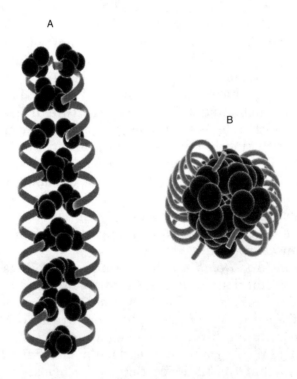

Figure 10.2 Side (A) and top (B) views depicting the "knobs-into-holes" arrangement of hydrophobic side chains in the core of the coiled-coil myosin rod. The ribbons represent the helical backbone while the balls represent side chains at the hydrophobic core. (Source: From Ref. 6. With permission.)

The geometry of the α-helices requires that the sequence of the polypeptide chain be made of seven-residue repeats (heptad repeat). The seven positions in a heptad are conventionally labeled as $(a\text{-}b\text{-}c\text{-}d\text{-}f\text{-}g)_n$. The knobs and holes are the sides chains at the *a* and *d* positions, which are predominantly occupied by hydrophobic residues. Highly charged residues occupy the remaining positions with negative and positive patches spaced 14 residues apart. The seven-residue repeat of hydrophobic (H) and charged (C) residues, HCCHCCC, is the accepted hallmark of coiled-coil molecules. Thus, a hydrophobic stripe is formed on the surface of the α-helix that winds around the axis. This becomes internally located and buried along the axis of the coiled-coil molecules formed when the α-helices wind around one another in a water environment.

Numerous studies have shown that hydrophobic core packing is the dominant determinant for coiled-coil stability. The charged residues *e* and *g*, located close to the hydrophobic ones, further stabilize the coiled-coil structure by a salt bridge between the two α-helices.[7-9] On the other hand, residues *b*, *c*, and *f* lie on the outermost surface of the coiled-coil molecule exposed to the solvent. They have a special role in the packing of coiled-coil molecules into thick filament.[4,10] The helical-wheel diagram, as shown in Figure 10.3A, illustrates the arrangements of residues in coiled-coils. Figure 10.3B illustrates side chains packing in the interhelical hydrophobic core.

10.2.1.2 Actin

Actin comprises 15 to 30% of the myofibrillar protein. The monomer form of actin is globular in shape, being termed "globular actin" (G-actin), and has a molecular weight of 43 kDa. However, globular actin molecules polymerize to form the actin filament, a form that is referred to as "fibrous actin" (F-actin), which resembles a "string of pearls" in shape.

10.2.1.3 Other Myofibrillar Proteins

Other small fractions of proteins associated with either actin or myosin are tropomyosin, troponin complexes, actinins, M-

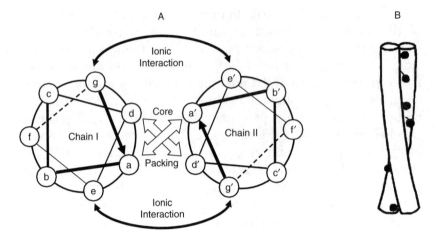

Figure 10.3 (A): Helical wheel diagram of the myosin rod coiled coil, showing the helical cross-section. The seven positions in each heptad are labeled a through g. Positions a and d are primarily occupied by hydrophobic side chains while positions e and g are mostly charged side chains. Side chains at the a and d positions from each chain form the inter-helical hydrophobic core while side chains at the e and g positions from each chain form inter-helical ionic interactions; these collectively stabilize the coiled-coil conformation of the rod. (*Source:* From Ref. 137.) (B): Side view of rod coiled-coil showing the periodicity of the core hydrophobic residues. (*Source:* From Ref. 11. With permission.)

proteins, and C-proteins.[3] These fractions play important roles in the structural integrity of the sarcomere. During processing of surimi into surimi seafood, the disassembly of the sarcomeres, which is an important prerequisite to obtain even distribution of proteins in the heat-induced gel structure, may require their selective solubilization or degradation.

10.2.1.4 Thick Filament Assembly

Natural thick filaments each consist of around 300 myosin molecules with their rod-like tails (LMM) packaged end-to-end in a regular staggered array (Figure 10.4). The globular myosin heads project from either end, leaving a bare central

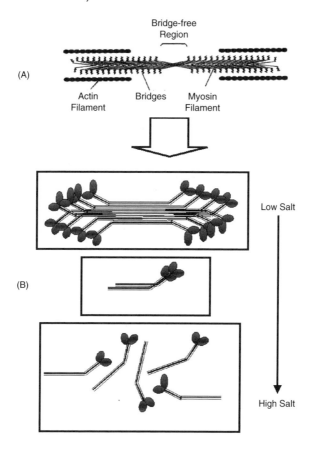

Figure 10.4 (A): Further detail of arrangements between thin filaments (actin) and thick filaments (myosin) in muscle. (*Source:* Ref. 138. With permission.) (B): Successive disassembly of thick filaments in solution as the ionic strength (salts concentration) is increased. (*Source:* From Ref. 139. With permission.)

region. It is these myosin heads that form the cross-bridges that interact with the thin filaments in intact myofibrils during contraction. Myosin and actomyosin can be extracted from fresh muscle by treatment with solutions of high ionic strength (>0.5 M KCl) because this causes the thick filaments to depolymerize (and other minor protein constituents to solubilize as well), leading to disassembly of the sarcomeres of the myofibrils.[4,11] The recently developed process for preparing a new type of surimi by alkaline/acid pH-shifting (see Chapter 3) likely results in more effective disassembly of the sarcomeres, perhaps partially explaining the enhanced gelation properties of fish proteins prepared by this method.

Myosin and actin are soluble in mild salt (NaCl) solutions (1 to 8%) but are largely insoluble in water of lower ionic strength (~0.05 to ~0.5%). The solubility of myofibrillar proteins, however, is enhanced when the ionic strength approaches zero.[12,13] In preparing the meat for gel formation by cooking, salt is added to enhance solubility and aid dispersion of the proteins. The increased solubility of myofibrillar proteins near zero ionic strength can possibly contribute to their loss when excessive washing is applied during surimi manufacturing.

10.2.2 Stroma Proteins

Connective tissue (*stroma*) proteins, primarily collagen, are almost totally insoluble in water or saline. Because the basis of the conventional surimi manufacturing process is the leaching out of water-soluble components, collagen is retained along with the myofibrillar proteins when surimi is made. Collagen can convert to gelatin when heated, depending on the structure of the collagen present. This soluble gelatin is thought to interfere with the gelation of the myofibrillar proteins. If present in high concentrations, such as in meats from land animals, it can accumulate in unsightly pockets in a heat-induced gelled meat product. However, fish have only a small percentage of stroma proteins relative to their myofibrillar protein content, and therefore the presence of collagen likely has a negligible effect on the gelling ability of surimi. Unfortunately, the opposite is true for mammalian and avian species. Unless a new processing step is introduced to reduce connective tissue content, surimi prepared from these species by conventional washing methods can be negatively affected (Table 10.1).

10.2.3 Sarcoplasmic Proteins

In addition to the lipid fraction (fats, oils, and membrane phospholipids; see next section), which is largely separated

TABLE 10.1 Comparison of Protein Composition of Fish and Animal Meats

Animal Species	(% of Total Proteins)		
	Sarcoplasmic	Myofibrillar	Stroma
Cod	21	76	3
Carp	23–25	70–72	5
Flatfish	18–24	73–79	3
Beef	16–28	39–68	16–28

by flotation during the washing step of surimi manufacture, the remainder of the muscle consists of water-soluble (soluble at low ionic strength) components, primarily the sarcoplasmic (nonmyofibrillar) proteins. Most sarcoplasmic proteins, in contrast to the largely fibrillar or rod-like conformations of myosin and actomyosin, are globular in tertiary structure. Being water soluble, sarcoplasmic proteins are largely removed by the conventional leaching process of surimi manufacturing.

It was originally thought that sarcoplasmic proteins, if not removed, would dilute the concentration of the better-gelling myofibrillar proteins and adversely affect surimi gelation. It was also thought that sarcoplasmic proteins, if present in high concentration when the myofibrillar proteins were heated during food manufacture, would unfold (denature) and attach to the myofibrillar proteins, thus blocking potential sites of protein–protein interaction between myofibrillar proteins and also leading to a decrease in the gelling ability of surimi.[14] However, some researchers have reported that sarcoplasmic proteins do not interfere with the gelation of myofibrillar proteins.[15]

Sarcoplasmic proteins comprise several types of proteins, including enzymes and heme proteins (myoglobin and hemoglobin). Some enzymes, such as certain heat-stable proteinases, certainly have a negative effect on the gelation of myofibrillar proteins due to their ability to cleave and thereby weaken the protein structures (see Chapter 6). However, other sarcoplasmic enzymes, such as transglutaminase (TGase), promote protein cross-linking, resulting in stronger textural properties. Proteinase and TGase activities in the sarcoplasmic fraction vary with fish species, and proteinase activity can be mitigated by proper heat processing or the addition of food-grade proteinase inhibitors (see Chapter 6). Surimi prepared by the alkaline/acid pH shifting method contains a much higher content of sarcoplasmic proteins than surimi manufactured by the conventional washing process, yet exhibits gelling properties superior to that of conventionally washed surimi (see Chapter 3).

Denaturation is an unfolding of the long amino acid (polypeptide) chain that comprises every protein. This chain is normally folded upon itself to constitute the secondary (internal structural features), tertiary (gross shape of the unit or subunits), and quaternary (presence of subunits) structures of the native protein. Denaturation exposes more reactive surfaces of a protein. This can lead to protein–protein interactions, termed "aggregation," which under the proper conditions will result in gel formation. Denaturation and attachment of sarcoplasmic proteins to myofibrillar proteins can also occur during the leaching process of surimi manufacturing if the temperature rises too high or if oxidation of the proteins occurs via excessive aeration. This leads to retention of sarcoplasmic proteins that would otherwise be removed through the leaching process.

10.2.3.1 Heme Proteins

Sarcoplasmic proteins include the heme proteins, which are responsible for the pigmentation of unleached mince. The heme (iron-containing) proteins of blood and red muscle cells are hemoglobin and myoglobin, respectively. Removal of the colored heme-containing moiety from the muscle during refining depends on maintaining the heme proteins in a relatively undenatured state. Denaturation of the heme proteins, before or during processing, can result in their binding to myofibrillar proteins and cause discoloration of the surimi.

Retention of heme proteins in the leached surimi also introduces ferric iron, which is a known catalyst of lipid oxidation. Oxidation of residual lipid (primarily cell membrane phospholipid) can, in turn, contribute to premature denaturation and aggregation of the myofibrillar proteins. If this occurs during the manufacture or subsequent storage of surimi, prior to its use, it will reduce the ultimate gelling ability of the surimi.

Animals contain two main types of muscle fibers, dark (red) and light, their coloration depending on the respective amount of myoglobin present in the muscle fiber.[16] It is extremely difficult to leach all the myoglobin from dark, red muscle because it is deposited within the muscle cells. In contrast, the water leaching process more easily removes hemoglobin, which is contained in the free-floating red blood cells of the bloodstream. The newer alkaline/acid pH shifting process for surimi manufacture, which removes relatively little of the sarcoplasmic protein fraction, still effects considerable removal of heme proteins due to their greater solubility under conditions where other proteins are precipitated (see Chapter 3).

In fish, dark and light fiber types are largely segregated into dark and light muscles, respectively. Conversely, in many land animals, the different fiber types can be dispersed within a particular muscle. Consequently, in fish, much of the heme pigment can be easily removed by physical separation of dark and light muscles prior to mechanical deboning and leaching, whereas separating the fiber types of many land animal species would prove more difficult. Because dark muscle has both a higher myoglobin content and higher ratio of sarcoplasmic to myofibrillar proteins, avoidance of dark muscle (when possible) is desirable to improve the gelling ability and appearance of surimi. Dark muscle is more predominant in pelagic fish such as sardine and mackerel and their use in surimi manufacture has thus been limited. The most practical approach to obtaining a light-colored surimi therefore is to use species that are mainly composed of light muscle.

10.2.3.2 Enzymes

Some of the sarcoplasmic proteins of surimi are enzymes, the biological catalysts of the chemical reactions that muscle cells carry out in life. Nakagawa et al.[17] observed that the gelling properties of surimi were negatively related to the residual aldolase activity, a glycolytic enzyme. Some other studies, however, have suggested that sarcoplasmic proteins may actually contribute positively to gel strength.

For example, as the content of sarcoplasmic protein was increased in surimi made from Pacific mackerel, the puncture strength of the gel also progressively increased.[18,19] The 94-, 64-, 40-, and/or 26-kDa molecular weight components of sarcoplasmic proteins seemed most important to the gelling ability of the sarcoplasmic fraction.[18,20] Sarcoplasmic components of sardine obtained from an 80 to 100% ammonium sulfate fractionation were found to increase the rigidity of a gel made from its myofibrillar proteins.[21] Ko and Hwang[22] also reported that the addition of the sarcoplasmic protein fraction to surimi improved the heat-induced gelation properties, by exhibiting a restrictive effect on the proteolytic softening of the gels. In addition, Siang and Miwa[23] suggested that there is a fraction in the kidney of sardine and coral fish that actually enhances gel strength. Addition of sarcoplasmic proteins also increased the breaking force (strength) of gels made from threadfin bream and seabass myofibrillar proteins.[15,24] Nowsad et al.[25] further examined the sarcoplasmic fraction of fish and discovered that the presence of a cross-linking enzyme (transglutaminase; see below) had a gel-enhancing effect. Addition of a concentrated sarcoplasmic fraction of tilapia (*Oreochromic niloticus*), which contained high TGase activity, increased the breaking force of lizardfish surimi gels.

10.2.3.2.1 TMAO Demethylase

All marine fish species contain trimethylamine oxide (TMAO), a water-soluble nitrogenous compound used by fish for osmoregulation. TMAO demethylase is an enzyme that is especially prevalent in gadoid (cod-like) species, such as whiting, hoki, and pollock, that degrades TMAO to formaldehyde (FA) and dimethylamine during frozen storage. FA is a strong protein denaturant and thus the gelling properties of surimi or minced fish can deteriorate rapidly if this enzyme system is active and present at sufficient concentration.

Much of the TMAO demethylase activity resides in organ tissues such as the kidney, liver, and pyloric caeca. If these are thoroughly removed from the fish prior to deboning the meat, the problem is somewhat alleviated. Fortunately, if properly conducted, the leaching process acts to remove the majority of the TMAO from the meat and appears to inactivate or remove the demethylase enzyme as well.[26]

10.2.3.2.2 Proteolytic Enzymes

The majority of fish also possess heat-stable proteolytic (protein-degrading) enzymes (proteases). The source, type, and content of proteases can vary greatly for each species (see Chapter 6). Such enzymes disintegrate the protein network formed by the gelation of the myofibrillar proteins, resulting in a mushy, rather than firm gel texture. These proteases attack the muscle proteins most actively during the cooking of surimi seafood, when the temperature is between 50 and 70°C. Some of these heat-stable proteases (e.g., in croakers, tilapia, and some pelagic fish) are most active at higher pH (alkaline or neutral proteases, most active near pH 8.0), whereas others, such as those in Pacific whiting, are most active at pH 5.5 (cathepsin L). However, all are still quite active over the pH range of most surimi and minced fish, which is between 6.5 and 7.5.

In many species the origin of heat-stable proteases appears to be gut tissues (particularly kidney and stomach). Rapid evisceration and thorough cleaning of the fish before processing, especially when fish are feeding prior to harvest, will reduce proteolytic activity in the meat and reduce the chance of enzyme migration from the gut tissue into the muscle. Although most heat-stable proteases are water soluble, the leaching step of surimi manufacturing rarely eliminates the activity. This suggests that the enzyme binds to the myofibrillar proteins and in some cases the protease may be closely associated with the myofibrils.

In some fish species, notably sardine, Atlantic menhaden, Pacific whiting, and arrowtooth flounder, the levels of heat-stable proteolytic enzymes can be high even when the fish are gutted and cleaned from the live state. Levels of protease in

Pacific whiting have been linked to the presence of *Myxosporidia* parasites in the flesh.

Practical approaches to controlling the problem include rapid and thorough evisceration of fish following harvest, avoidance of parasitized fish (Pacific whiting), and minimization of time held in the temperature zone of activation while processing the surimi into surimi seafood. The enzyme is inactivated by heating at 80°C or greater; therefore rapid cooking, such as ohmic or microwave heating, eliminates the problem.[27] In addition, blood plasma protein and some other naturally derived proteins (such as those from egg, potato, and whey), when added to surimi, inhibit the degradative activity of the enzyme.[28] This is the primary means of enzyme control in surimi made from parasitized Pacific whiting and perhaps a safe precaution when using surimi or mince of any species.

10.3 LIPID COMPONENTS OF FISH MUSCLE

Triacylglycerides (fats and oils) are largely removed during surimi manufacture by flotation, aided by mechanical action and possible melting/softening. Most of the depot fat is removed when fish are headed, gutted, and skinned because fish generally deposit most of their fat in these regions. There is, however, a small percentage of membrane phospholipids in fish muscle that is difficult to remove by washing. These phospholipids are highly unsaturated and often in contact with muscle heme iron and are therefore very sensitive to spoilage by oxidation. Such oxidation causes off-flavors and may hasten denaturation of the myofibrillar proteins. More phospholipids are contained in dark muscle fibers than in light muscle fibers, such that removal of dark muscle prior to leaching helps eliminate this unstable lipid fraction. The new alkaline acid-aided processes for refining fish muscle proteins promise to largely remove this unstable fraction (see Chapter 3).

Lipids in surimi are even more unstable if pro-oxidants such as iron (e.g., from water pipes, machinery, or residual heme proteins) are present. The mincing and washing procedures generally incorporate a large amount of oxygen into the surimi, making lipid oxidation even more likely. Lipid oxidation does not seem to be a problem in surimi processed from lean white-fleshed species, but is reported to be a very distinct problem in surimi made from some dark-fleshed fish and particularly surimi from mammalian and avian muscle. In the latter, muscle lipid oxidation may be the primary factor for limiting storage life, causing the formation of disagreeable flavors and leading to the denaturation of proteins and decreased gelling ability through peroxide formation. In such cases, addition of antioxidants early in the surimi process is advisable.

10.4 BONDING MECHANISMS DURING HEAT-INDUCED GELATION OF FISH MYOFIBRILLAR PROTEINS

Myofibrillar proteins have highly reactive surfaces once the protein is unfolded (denatured). During heating of salted surimi pastes, the proteins unfold, exposing the reactive surfaces of neighboring protein molecules, which then interact to form intermolecular bonds. When sufficient bonding occurs, a three-dimensional network is formed, resulting in a gel. Four main types of chemical bonds can link proteins: (1) hydrogen bonds, (2) ionic linkages, (3) hydrophobic interactions, and (4) covalent bonds.

10.4.1 Hydrogen Bonds

Hydrogen bonds are weaker dipole bonds that, mainly by virtue of their great numbers rather than individual bond strength, can be important in the stabilization of bound water within the hydrogel and add gel strength during cooling and aging of surimi seafoods. During heating, a large number of hydrogen bonds that maintain the folded protein structure are broken between the carbonyl and amide groups in the peptide backbone. This in turn allows the peptide backbone to become extensively hydrated and "structure" (reduce the mobility of) the water with which it is in contact. This hydration of exposed peptide backbones is a key factor in the water-holding capacity of the gel that is subsequently formed by protein–protein aggregation.

Hydrogen bonds between proteins are more numerous when the gel is colder; this is why surimi gels become firmer at colder temperatures (Figure 10.5).[29] This also explains why sols of dessert gelatin gel when they are cooled and melt when heated; a gelatin gel is formed almost entirely by thermo-reversible hydrogen bonds. Similarly, starches gel or "retrograde" due to hydrogen bonding, as in the "staling" of bread that leads to loss of softness in the texture. Therefore, cooling adds to the rigidity of surimi seafoods, which contain starches, and further contributes to the retrogradation of the starch over time and a strengthening of the surimi seafood texture.

Hydrogen bonds between amino acids also stabilize the internal (secondary) structure of individual protein molecules in water. The α-helix of native and partially denatured proteins and the β structure that forms on heating and cooling are both stabilized by hydrogen bonds.[30]

10.4.2 Ionic Linkages (Salt Bridges)

Ionic linkages are the attraction of positively charged sites to negatively charged sites on the protein surface. At the normal pH of surimi (near neutral) the carboxyl groups (COO-) of glutamic acid and aspartic acid, two amino acids on the protein chain, are negatively charged, while the amino groups (NH_2+) of lysine and arginine are positively charged. An ionic attraction will be formed between these groups, called salt bridges, and the myofibrillar proteins associate with each other to form an aggregate that is insoluble in water. Salt bridges are especially abundant all along the myosin rod at position e and g of the seven-residue repeat of the coiled-coil rod (Figure 10.3). Ionic (electrostatic) interactions are thought to be the most important forces involved in the assembly of myosin thick filaments[31] and the addition of salt interferes with this electrostatic attraction, leading to the disassembly of the thick filaments and better dispersion of the myosin or actomyosin.

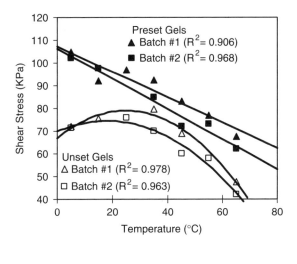

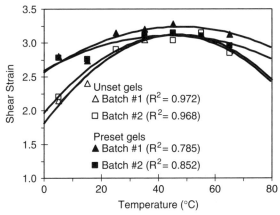

Figure 10.5 Influence of test temperature on fracture shear stress and shear strain of Alaska pollock gels. (*Source:* Adapted from Ref. 93.)

For surimi to gel well, salt (usually sodium chloride) must be added to break ionic linkages and assist in dispersion of the proteins, as an even dispersion of the proteins is necessary for the development of an elastic structure in the heat-set gel.[32] The salt ions (Na^+, Cl^-) selectively bind to the oppositely charged groups exposed on the protein surface. Intermolecu-

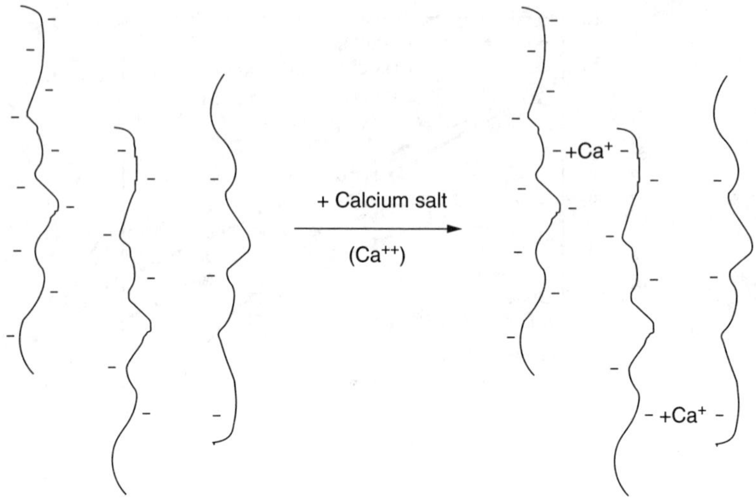

Figure 10.6 Formation of calcium cross-links between proteins.

lar ionic linkages among the myofibrillar proteins are ruptured and the proteins dissolve because of their increased affinity for water.[32] Salt addition, however, must be accompanied by a sufficient degree of grinding or comminution to enhance protein solubilization and dispersion.

$$\text{Protein - COO}^- \;{}^+\text{H}_3\text{N - Protein} \xrightarrow{\text{NaCl}}$$
(aggregate)

$$\text{Protein - COO}^-\text{Na}^+ + \text{Cl}^-\;{}^+\text{H}_3\text{N}^-\text{Protein}$$
(solubilized)

Myofibrillar proteins carry an overall net negative charge at the normal pH of surimi. Calcium ions, having a divalent positive charge (Ca^{2+}), can thus form ionic linkages between negatively charged sites on two adjacent proteins (Figure 10.6). The addition of Ca^{2+} ions may contribute to the strengthening of surimi gels but the intermolecular ionic bonds alone will not induce gelation of surimi. The commercial practice of adding calcium salts to improve the gelling prop-

Surimi Gelation Chemistry

erties of surimi is actually based more on the effect of calcium as a co-factor for an endogenous cross-linking enzyme (transglutaminase) in the muscle (see later discussion).

The total net negative charge on the proteins increases as the meat pH and ionic strength increases. This causes repulsion between proteins that helps stabilize them to aggregation (and also leads to better protein dispersion) until such time as sufficient heating imparts sufficient energy to the vibrating molecules to overcome this repulsion, and their high concentration and unfolded condition lead to aggregation and gelation.

10.4.3 Hydrophobic Interactions

In contrast to hydrogen bonds, which dissipate upon heating, hydrophobic interactions (which are effectively "bonds" between proteins in the aqueous environment of surimi gels) are strengthened by rising temperature, at least to near 60°C. The formation of intermolecular hydrophobic interactions among proteins, in addition to the disulfide bonding that occurs during heating, is presently thought to be a primary mechanism for surimi gel formation that results from heating or high pressure (≥300 MPa) treatment.[33] Hydrophobic interactions may also play an important role in the observed strengthening of surimi gels exposed to ultraviolet light.[34]

The formation of intra- or intermolecular hydrophobic interactions results from the thermodynamic response of protein surfaces exposed to the water in which they are dispersed or solubilized. The interior of the folded protein chain has a greater density of hydrophobic amino acids that are "fat-like," in that they shed water like lipids. Conversely, the amino acids on the surface of the folded protein are largely hydrophilic. By this arrangement the folded protein achieves thermodynamic equilibrium in water. This is because exposure of those internally buried hydrophobic residues minimizes the exposure to, and ordering of, water (a thermodynamically unfavorable decrease in entropy) that would occur were they exposed at the surface.

When a protein unfolds (typically as a response to heating), the hydrophobic core is exposed to water. Water molecules near these exposed hydrophobic groups become oriented (ordered) into hydrogen-bonded clathrates, which are ice-like structures. Such orientation or ordering decreases the mobility of the water molecules; thus, the system is less random and entropy is decreased. To minimize their exposure to water and create a more thermodynamically stable system, the hydrophobic portions of the protein associate closely with the other hydrophobic portions, similar to the manner in which fat or oil droplets placed in water tend to clump (Figure 10.7). Association of the hydrophobic portions of two neighboring proteins to reduce system entropy results in an effective "binding" (via protein–protein hydrophobic interactions) of the proteins, thus leading to protein aggregation and (under the proper conditions) the formation of a gel network.

10.4.4 Covalent Bonds

Covalent bonds are rigid chemical bonds formed by the sharing of electrons between proteins, which are not easily broken once formed. In contrast to the other types of protein bonds previously discussed, covalent bonds are largely temperature insensitive once formed.

10.4.4.1 Disulfide Bonds

During heating at high temperatures (cooking at >40°C), disulfide bonding (S–S) is the predominant covalent bond thought to contribute to gel formation of proteins. An intermolecular disulfide bond is formed by the oxidation of two cysteine (an amino acid) molecules on neighboring protein chains, which have reactive sulfhydryl (-SH) groups:

$$\text{Protein-SH} + \text{HS-Protein} \xrightarrow{O_2} \text{Protein-S-S-Protein} + H_2O \text{ (water)}$$

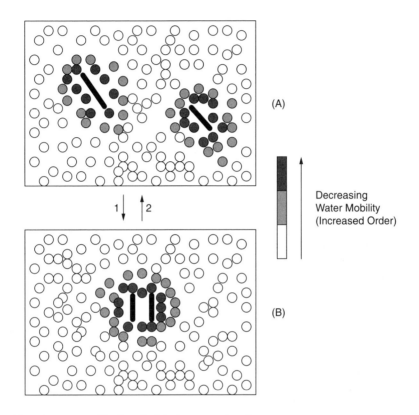

Figure 10.7 Hydrophobic interaction in an aqueous environment. Hydrophobic groups (dark bodies) dispersed in water (A) destabilize the system by decreasing the entropy (randomness; S) because the mobility of neighboring water molecules is decreased by their presence. Although the enthalpy (heat content; H) of the system is slightly increased when the hydrophobic groups associate closely (i.e., $\Delta H_1 > \Delta H_2$) as in (B), entropy is greatly increased ($\Delta S_1 \gg \Delta S_2$). Because ΔG (free energy of the system) = $\Delta H - T\Delta S$, where T = temperature, the free energy of the system decreases when hydrophobic groups associate in (B); that is, $\Delta G_1 < \Delta G_2$. Systems of the lowest free energy are the most stable.

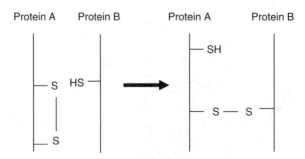

Figure 10.8 Formation of intramolecular and intermolecular disulfide bonds.

Consequently, surimi gels are strengthened by the addition of oxidants, such as potassium bromate (now an illegal additive)[35] and ascorbic acid/dehydroascorbate (at the time of gel formation), which accelerate oxidation of sulfhydryl groups.[36–38] Kishi et al.[39] showed that copper ions also promote the formation of disulfide bonds in fish myosin, as does hypochlorite in treated tap water.[40]

The formation or natural occurrence of disulfide bonds between amino acids within a protein (*intra*molecular disulfide bonds) can be converted to protein–protein (*inter*molecular) disulfide bonds (cross-links) through disulfide interchange (Figure 10.8). If the intramolecular groups are predominantly cystine or free sulfhydryl groups, addition of cysteine or cystine, respectively (-SH and S–S- sulfur-containing amino acids, respectively), can promote movement of the disulfide bond from an intramolecular to an intermolecular (i.e., protein cross-link) locus. Kim[41] found that cystine addition to pollock surimi pastes prior to heating induced the formation of stronger gels.

Although the formation of covalent bonds is generally considered an irreversible event, the oxidation of surimi proteins that has occurred prior to heating of surimi during food manufacture has been shown to be partially reversible by the addition of reducing agents. The gelling quality of both freeze-damaged[42] and ozonated (for decoloring of heme pigments) surimi[43,44] was improved by addition of either chemical (e.g.,

NaHSO$_3$, cysteine, ascorbic acid) or enzymatic (e.g., NADPH-sulfite reductase) reducing agents.

10.4.4.2 Rheological Behavior of Cross-Linked Protein Gels

Gels that consist of polymers cross-linked only by covalent bonds display a type of rheological behavior known as rubber elasticity. The elastic modulus (stress/strain) of the gel is uniform at any deformation up to the point of failure; the gels show no relaxation of stress upon being held indefinitely at a fixed deformation; and as the temperature of the gel is increased, the elastic modulus increases, primarily due to a decreasing strain with increasing temperature. Such gels are totally elastic and therefore have no viscous element (see Chapter 11).

Surimi gels, like most food gels, display viscoelasticity in that the elastic modulus might change with deformation up to the point of failure, the gels display marked stress relaxation upon being deformed (primarily because of weakening hydrogen bonds), and the elastic modulus decreases (stress decreases) as the temperature increases. However, Niwa et al.[45] prestrained heat-set surimi gels to remove the viscous element caused by movable bonds such as hydrogen bonds and hydrophobic interactions, and observed rubber elastic behavior. Lee et al.[46] further demonstrated that the degree of this rubber elastic behavior increased in proportion to the number of covalent bonds introduced into a surimi gel by addition of a variety of agents, both chemical and enzymatic.

10.4.4.3 Role of Disulfide Bonding in Myosin/Actomyosin Gelation

During heating, discrete regions of the myosin molecule, which exhibit less thermal stability, initially unravel and then aggregate before the entire molecule completely denatures.[47] Samejima et al.[48] and Yasui and Samejima[49] showed that the total number of free sulfhydryl groups in the myosin heads decreases as temperature increases. They also reported that the rigidity of gelled myosin S-1 (globular head) fragments decreased upon adding the sulfhydryl-blocking reagent dithiothreitol (DTT). From this evidence they postulated that the cross-links that initiate the formation of a myosin–actomyosin gel network are disulfide bonds. Disulfide bonding is also involved in the further development of the gel network during heating, resulting in the S-1 (globular head) and rod (fibrillar tail) fractions of myosin being primarily involved in the dimerization and polymerization of myosin, respectively.[50]

However, Sano et al.[51] postulated that formation of carp myosin first begins to develop by interactions of rod light meromyosin (LMM) at 30 to 45°C. The second step of gel formation involves aggregation among heavy meromyosin (HMM; head region of myosin) by hydrophobic interactions at 50°C. Higuchi et al.[52] similarly showed that myosin gelation is primarily mediated by interactions between the myosin rod regions. Others support the view that aggregation/gelation of the proteins during heating would occur even without an initial head-to-head aggregation caused by SH group oxidation, due to the formation of intermolecular hydrophobic interactions.[53,54] Sharp and Offer[55] postulated that head-to-head aggregation of myosin was actually due to the hydrophobic interactions resulting from the removal of certain light chains from the head region during heating, which revealed hydrophobic patches on the surface. The formation of hydrophobic "cross-links" (interactions) between proteins is a logical consequence of the general increase in surface hydrophobicity, which accompanies heat-induced unfolding and seems to precede the gelation of myofibrillar proteins during heating.[56] Smyth et al.[57] also showed that disulfide bonding is not a requirement for the gelation of actomyosin, but that such intermolecular covalent bonding does contribute to gel network formation.

These studies indicate the important role of discrete regions of the myosin molecule in forming gel networks. Discrepancies in the relative importance and order of reaction during heating for these various myosin subfragments is probably due to the structural differences in these discrete portions among different fish species.[58]

Although actin does not form gels when heated to the 60–70°C range, it is reported to have a synergistic role in the gelation of myofibrillar proteins.[59–61] The optimal weight ratio of myosin to actin that yields the highest shear modulus (rigidity) is 15:1. The complex formed between F-actin and myosin seemed to behave as a cross-link between the rod portion of the myosin molecules, and consequently increased the rigidity of the gel.

Actomyosin from carp was reported to unfold around 30°C, with extensive unfolding at 30 to 50°C.[62] Aggregation of actomyosin occurred simultaneously with unfolding at 30°C via hydrophobic interactions and disulfide linkages. When the heating temperature exceeded 40°C, myosin molecules dissociated from the actin filaments. The dissociated myosin molecules aggregated with one another via disulfide linkages and hydrophobic interactions, resulting in large aggregates at temperatures above 60°C.

10.4.4.4 Covalent Cross-Linking during Setting

Surimi can form a less hard, but very deformable gel when comminuted with sodium chloride and held at a low temperature (0 to 40°C, depending on the species) without cooking. Subsequent cooking of these gels at higher temperatures results in the formation of stronger gels than gels cooked without a low-temperature preincubation.[63–65] This gelation and textural strengthening of salted surimi paste at low temperatures is termed "setting." Setting is often utilized in the manufacture of surimi seafood.[66]

The unique setting ability of surimi is thought to mainly result from the enzymatically catalyzed formation of non-disulfide covalent bonds between protein molecules. These bonds form between the amino acids glutamine and lysine [ε-(γ-glutamyl) lysine dipeptide cross-links], each on a neighboring protein chain, from the action of the enzyme *transglutaminase* (protein-glutamine γ-glutamyltransferase; EC 2.3.2.13) present in (endogenous to) the fish muscle. This cross-linking results in the formation of myosin polymers with a concurrent decrease in myosin heavy chain monomers[67] (Figure 10.9).

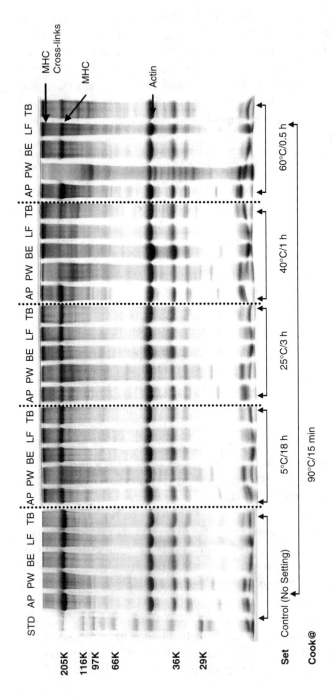

Figure 10.9 Formation of myosin heavy chain (MHC) cross-links of various surimi during pre-incubation at various temperatures and times. Sample was solubilized in an SDS-urea-mecaptoethanol buffer to ensure breaking of all disulfide bonds and electrophoresed on 10% polyacrylamide gels. Note that no polymer formation was observed (bands forming above MHC) for control (no setting). AP = Alaska pollock; PW = Pacific whiting; BE = bigeye snapper; LF = lizardfish; and TB = threadfin bream. (*Source:* Adapted from Ref. 67.)

Cross-linking seems to occur between actomyosin and the other myofibrillar protein constituents of the muscle to yield huge aggregates.[68] The content of ε-(γ-glutamyl) lysine dipeptide cross-links generally correlates with the level of increase in gel strength.[69] However, Lee et al.[46] found that the rate at which the cross-linking reaction proceeds seems also to be an important factor in the ultimate strength attained.

A correspondence between cross-link type(s) and numbers and the ultimate gel mechanical properties (strength, deformability) would not be expected in all cases. This is because other factors — such as the distribution of the cross-links (within and between types), the distribution of the proteins being cross-linked, the geometry of the gel matrix formed, and the relative lengths of non-cross-linked proteins between points of cross-linking — all likely contribute to the ultimate mechanical/fracture properties of protein gels. An analogy would be that many different structures could be built using a certain number of nails and board feet of lumber, but factors such as the geometric arrangement of the boards in the structure, their thickness and length at any location, and the distribution of the nails (as well as their respective size and strength) would collectively influence the stability of the structure.

Role of Hydrophobic Interactions in Setting. Covalent dipeptide linkages are not the only protein–protein interactions that stabilize the low-temperature induced "set" gel. Several experimental findings suggest that intermolecular hydrophobic interactions contribute to the low-temperature setting reaction as well. An increase in protein surface hydrophobicity can be measured during setting.[32,56] Also, substitution of hydrophobic groups onto the surface of proteins from poorly setting species greatly enhanced their gelling ability at low temperature.[70,71] Involvement of hydrophobic interactions in setting was also observed in the Raman spectroscopy of set surimi gels, resulting in the decreased intensity of a band near 2930 cm^{-1}, assigned to C–H stretching vibrations.[30] The requirement of salt addition to surimi paste to induce setting also supports a role of hydrophobic interactions in setting, because certain salts, such as sodium chloride, act with water molecules to strengthen the hydrophobic interactions between proteins.[32,72]

An argument against the predominant role of TGase in setting of surimi[73] was made via demonstration that highly washed surimi, for which the actomyosin was shown to have negligible TGase activity, or surimi treated with a sulfhydryl agent to inhibit TGase, still exhibited setting with accompanying formation of non-disulfide cross-linked myosin.[74,75] Some (greatly reduced) gelation and myosin cross-linking was observed when TGase was ostensibly removed or inactivated, indicating that mechanisms other than (but certainly in addition to) endogenous TGase may participate in gelation of surimi during setting.

Shoji et al.[76] conducted another study that seemed to contradict the role of TGase in low-temperature-induced setting of surimi. Although they found that pressure-induced (300 MPa) gelation of surimi also inactivated endogenous TGase (no measurable TGase activity), subsequent setting of these pressure-induced gels at low temperature enhanced their strength. Contrary to these findings, other workers measured TGase activity following pressure-induced gelation under similar conditions, which induced gel strength enhancement and non-disulfide cross-linking of myosin[33] (Figure 10.10). The increase in gel strength noted during 25°C incubation after pressure treatment was prevented by the addition of EDTA, which chelates calcium and thus, inactivates TGase.

However, when another research group repeated this experiment at this and higher pressures,[77] subsequent holding at 5°C showed an enhancement of gel strength. The greatest setting effect was observed following the 300-MPa pressure treatment. Measurement of myosin cross-linking was only reported for the 500-MPa treatment, in which a 60% increase in breaking strength during 5°C setting exhibited no corresponding cross-linking of myosin. The pressure-induced gel, which is largely formed due to hydrophobic interactions facilitated by the pressure-induced unfolding of actomyosin, can evidently develop further hydrophobic interactions during subsequent low-temperature holding. In the presence of residual or added TGase activity, however, such a gel is further

Surimi Gelation Chemistry

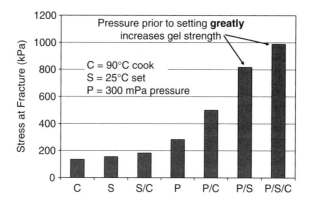

Figure 10.10 Effect of setting and/or cooking subsequent to high-pressure treatment on gelation of pollock surimi. (*Source:* Adapted from Ref. 33.)

strengthened by the formation of ε-(γ-glutamyl) lysine covalent cross-links.[33] Evidently, unfolding the proteins by high-pressure treatment (resulting in cold-induced gelation) renders more cross-linking sites available to TGase such that a stronger texture results.[78] The combination of high pressure treatment plus low temperature setting could, therefore, be used to either strengthen gels, or produce gels of normal strength more quickly or with less protein in the formulation.

The effect of added sugars and polyols can help illuminate the respective roles of covalent dipeptide bonds and hydrophobic interactions in setting. Niwa[32] showed that sucrose addition suppressed the exposure of hydrophobic sites on the surface of surimi proteins (Figure 10.11). Both sucrose and sorbitol have this effect because they stabilize the protein structure (see Chapter 5). Their addition also leads to a measurable decrease in myosin ATPase activity.[79]

If greater exposure of hydrophobic sites were correlated to an increased setting effect (i.e., stronger gel under setting conditions), a corresponding decrease in the strength of set gels containing sucrose or sorbitol would also be expected. However, Arai et al.[73] reported that, in the presence of protein stabilizers like sorbitol and monosodium glutamate, the

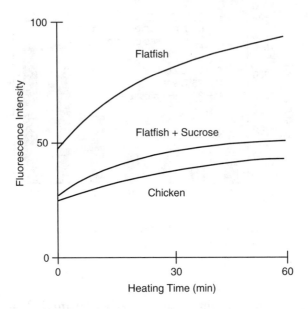

Figure 10.11 Effect of heating actomyosin solutions at 40°C on the fluorometric intensity measured at 470 nm in the presence of sodium anilino-naphthalene 8-sulfonate (λ_{exc} = 365 nm). Protein concentration: 1 mg/ml; solvent: 0.45 M KCl/phosphate buffer (pH 7.5). (*Source:* Adapted from Ref. 32.)

strength of set-only gels slightly increased, while the strength of set and cooked gels containing these stabilizers greatly increased. Myosin cross-linking was slightly depressed, especially the formation of highly polymerized species, but the content of mildly cross-linked myosin greatly increased in the presence of sorbitol.[80]

The addition of calcium chelating agents such as EDTA or citrate[81,82] or the addition of specific TGase inhibitors such as certain ammonium salts[83] almost totally eliminates the setting phenomenon in surimi pastes. Also, the addition of beef plasma has a strengthening effect on gels subjected to low-temperature setting prior to cooking.[84] Plasma, in addition to being a gelling adjunct and protease inhibitor, is high in fibrinogen, which is a natural substrate of TGase. These data suggest that covalent cross-linking by endogenous transglutaminase is the main contributor to gelation induced by low-temperature setting of salted surimi pastes. However, intermolecular hydrophobic interactions must also form and reinforce the gel matrix. Hydrogen bonds will additionally affect the strength of these gels, particularly at colder temperatures.[29]

10.4.4.5 Protein Stability Effects on Setting

Setting proceeds best at higher pH[85,86] and requires the addition of salt, with NaCl favoring subsequent cross-linking compared to KCl.[87] The salt destabilizes (partially unfolds) the native proteins, exposing more sites for dipeptide cross-linking by TGase, as well as for the formation of hydrophobic interactions. The extent and optimal temperature of setting can vary, depending on the fish species, sex, harvest season, fishing grounds, and environmental water temperature.[88–90]

The optimum setting temperature varies according to species, mainly because the myosin or actomyosin, which is the substrate for the reaction, has a different *thermal stability* depending on the fish species.[67,91–93] Generally, fish that inhabit colder waters have the least stable proteins compared to fish living in warmer waters. Yet fish from cold, but very deep waters may have more stable proteins than fish in the same waters nearer the surface, as a protective mechanism to the higher pressure of their habitat. Kamath et al.[94] have shown that Alaska pollock surimi exhibits maximum gel strength when preincubated at 25°C for 2 to 3 hr. Similar gel strength enhancement could be achieved at lower temperature (5°C) with a much longer setting time; Lee and Park[95] found that pollock gels achieved highest gel strengths when set at 5°C, while setting of Pacific whiting gels was most effective at 25°C. Optimum setting temperature for warm-water species was quite different: 25 to 40°C for threadfin bream, bigeye snapper, and lizardfish.[67]

Kim[41] found that the strongest gels were obtained with 40°C setting for Atlantic croaker, a more temperate water species; gels did not set at all for this species below 25°C. Setting of tropical fish species is also typically optimal at higher temperatures.[96–98] Yongsawatdigul et al.[99] found that setting of threadfin bream, a major resource for surimi production in Southeast Asia, could be induced at both 25 and 40°C, but setting at 25°C took longer (4 hr) than at 40°C (2 hr). Because threadfin bream actomyosin exhibited its major conformational changes at greater than 35°C,[100] thermally induced unfolding of protein might not be the sole prerequisite for setting. The partial unfolding of actomyosin, which is induced only by NaCl addition, appeared to be sufficient to induce TGase cross-linking of reactive amino acids at 25°C.

In the manufacture of Southeast Asian fish balls, setting is typically carried out commercially at temperatures ranging between 30 and 35°C, near the ideal setting temperature for the tropical fish species historically used in fish ball production. However, by substituting a cold-water species such as Alaska pollock surimi into the formulation, the setting ability under these conditions would be quite poor.

10.4.4.6 Endogenous Transglutaminase (TGase)

TGase, which is found endogenous to fish meat and surimi, is water soluble and can be removed if washing is too extensive.[74] Sufficient calcium ions must be present for the endogenous TGase to be active and induce setting.[95,101,102] In cases where endogenous TGase activity is not desired, such as in the holding of paste prior to later extrusion or forming, a *chelating* agent can be added, such as sodium citrate or EDTA, which will bind calcium and prevent the TGase from cross-linking proteins. In contrast, addition of calcium salts to surimi enhances the TGase-mediated setting reaction, resulting in stronger gels.[95,102–104] Calcium addition, however, potentiates greater denaturation during frozen storage of surimi; therefore, it is best to add calcium only during surimi seafood manufacturing.[102]

High variability in the TGase activity of surimi is common and is due to a combination of factors. First, TGase is a water-soluble enzyme and its content can vary greatly with the type and extent of purification process employed during surimi manufacture. Nowsad et al.[25] showed that the sarcoplasmic fraction of fish can actually enhance the gelling ability when added back to surimi due to its higher TGase activity. Yongsawatdigul et al.[99] indicated that the first washing cycle removed a large portion of the endogenous TGase from threadfin bream muscle. Only about 44% of the original TGase activity was retained in the final surimi.

Second, it is likely that different fish species, and perhaps different individuals within species, could vary in natural content of the enzyme, possibly affected by habitat, feed, and physiological condition. Tilapia, rohu, and threadfin bream are among tropical species exhibiting high TGase activity.[105] In addition to the varied TGase activity among fish species, the extent of setting will also depend on the numbers of reactive glutamine residues on the myosin molecules. Maruyama et al.[106] indicated that myosin heavy chain (MHC) of carp possessed less reactive glutamine residues than rainbow trout and mackerel. As a result, only protein dimers were formed in carp MHC, whereas MHC of rainbow trout and mackerel were able to form large polymers at a faster rate.

A third possible contributing factor is that the α_2-macroglobulin component of fish blood plasma (or added beef plasma) has the ability to form ε-(γ-glutamyl) lysine crosslinks in fish protein.[84] Also, it has been shown that in certain species, such as salmon, the water-soluble fraction of muscle also contains factors that inhibit TGase activity.[103]

10.4.4.7 Exogenous TGase Addition

The commercial development of a low-cost source of TGase from microbial culture offers a means of upgrading the gelling quality of surimi.[107] This microbial TGase (MTGase) is not calcium sensitive; therefore, neither chelating agents nor calcium salts have any marked effect on its activity. While some reports have indicated that the benefits from MTGase addition are mainly noticeable in products from lower-quality surimi, there is also a substantial increase in the strength of gels made with high-quality pollock surimi (Figure 10.12). Only five units per gram of surimi protein is effective in increasing the gel strength during setting (most commercial preparations of MTGase contain at least 50 units of enzyme

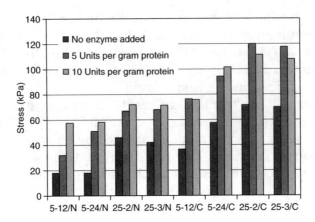

Figure 10.12 Effect of MTGase on the strength (stress at fracture) of pollock surimi gels. 5 and 25 are the setting temperatures, followed by the number of hours setting was carried out. Cooking was at 90°C for 20 min. N = no cook, set only; and C = set and cook.

per gram). There are reports, however, that added MTGase leads to a harder, less "natural" texture in surimi gels than the endogenous enzyme.[108,109] Despite these published reports, MTGase is widely used in Japan to strengthen surimi gels and many other protein foods.[110]

In Pacific whiting gels, which normally exhibit lower gel strength than gels from Alaska pollock surimi, the addition of MTGase has even more pronounced effects, especially when combined with beef plasma (1%), which inhibits the parasite-related heat-stable protease content (Figure 10.13). Additionally, the fibrinogen in beef plasma is an excellent substrate for the enzyme.[111]

An experimental fungal TGase was also commercially developed that, like the endogenous tissue enzyme, is calcium sensitive. Initially, concern was expressed that such an enzyme might be less effective than MTGase in gel strength enhancement of surimi, because 0.3% of polyphosphates, which is normally added as a cryoprotectant, also chelates calcium. However, it was later demonstrated that the fungal enzyme was just as effective in surimi as MTGase, provided

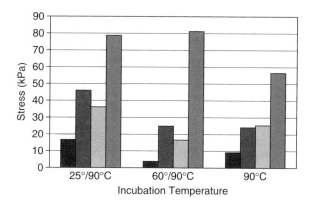

Figure 10.13 Synergistic effect of beef plasma and MTGase in gelation of Pacific whiting surimi gels. From left to right for each incubation temperature: control, plasma, MTGase, and plasma+MTGase.

a small amount of calcium ion was also added.[82] TGase from tissue sources other than surimi, such as plasma or by cloning the genetic material of animal or plant tissue into microorganisms, could also be developed in the future.[84,112–114]

Although TGase-induced cross-linking can greatly enhance the strength of surimi gels, excessive cross-linking can produce a gel that is much less deformable, and therefore more brittle. This is a similar effect to that of the natural cross-linking of collagen molecules that occurs in tendon, bones, and skin as animals age. The accumulating cross-links make the collagen steadily less deformable and more brittle. As a result, bones and tendons become easily snapped and the skin loses much of its elasticity.

10.5 FACTORS AFFECTING FISH PROTEIN DENATURATION AND AGGREGATION

Fish proteins will denature (unfold) at any temperatures, although the rate of denaturation is slowed as the temperature is lowered. For this reason, rapid chilling of fish after harvest is important, particularly because fish temperature tends to rise by 5 to 10°C following death due to continuing muscle metabolism. The time/temperature history and the thermal (heat) stability of the muscle proteins determine the rate of protein denaturation in the fish. Lower muscle pH accelerates denaturation at any time/temperature of holding prior to surimi processing. Any such denaturation should be avoided, as it will inevitably lead to premature aggregation of the proteins and decreased gelling ability.

ATP (adenosine triphosphate) is the "fuel" used by muscles for contraction. When the animal dies, ATP begins to degrade. Because the rate of ATP breakdown is governed by time and temperature, the time/temperature history, or "freshness," of fish muscle can be estimated by the extent of ATP breakdown. The breakdown of ATP produces the following order of compounds:

$$ATP \rightarrow ADP \rightarrow AMP \rightarrow IMP \rightarrow HxR \rightarrow Hx$$

where
 ATP = adenosine triphosphate
 ADP = adenosine diphosphate
 AMP = adenosine monophosphate
 IMP = inosine monophosphate
 HxR = inosine
 Hx = hypoxanthine

The "K value" is a measurement of the extent of enzymatic breakdown of ATP and is defined as:

$$K \text{ value } (\%) = \frac{HxR + Hx}{ATP + ADP + AMP + IMP + HxR + Hx}$$

Thus, the K value is an enzymatic measure of the freshness of the muscle. The use of fresh muscle for surimi manufacture ensures that prior denaturation and aggregation of muscle proteins will be minimized.

The rate at which the K value increases with storage time on ice is different for each fish species, and can even vary depending on the harvest season.[115–117] Likewise, the rate of denaturation and aggregation of the proteins at any time and temperature also varies by species, according to its protein stability. Therefore, the relationship between "freshness," as measured by the K value, and the state of the proteins (gelling ability of the surimi) is different for each species. Consequently, some fish can be stored several days on ice and still produce a surimi with excellent gelling properties, while others must be processed quickly to avoid loss of gelling properties.[118] Fish from warmer waters or greater depths would be expected to have more stable proteins than those from shallower and colder waters.

Ablett et al.[119] found that white hake (*Urophycis tenuis*) kept on ice for 2 days resulted in relatively high surimi gel quality; thereafter, gel strength decreased after fish was stored in ice for up to 8 days. The gelling quality of hoki (*Macruronus novaezelandiae*) surimi was still acceptable after holding on ice up to 10 days.[120] A 50% decline in gel strength was observed in surimi produced from lizardfish after 3 days' storage in ice.[121] Gel strength of surimi produced from threadfin bream (*Nemipterus tolu*) kept on ice for 3 days was still acceptable.[122,123]

Oxidation of SH groups to disulfide linkages and the exposure of hydrophobic amino acids, both of which lead to protein aggregation, typically can occur during iced storage of fish.[124,125] Sompongse et al.[126] found that, during iced storage of fish, oxidation of SH groups occurred primarily on the myosin rod, accompanied by unfolding, as evidenced by a decrease in rod α-helical content.

Myosin is also an enzyme, and has the ability to split ATP. Consequently, the ATPase (ATP-splitting) activity of myosin is another good measure of the quality of surimi proteins because as myosin denatures and aggregates, it also loses ATPase activity[79,127–130] (Figure 10.14).

The active site of actomyosin Ca^{2+}-ATPase involves sufhydryl groups.[129] Oxidation of these SH groups during iced and frozen storage of fish was reported to inactivate the Ca^{2+}-ATPase of actomyosin.[126,131] Katoh et al.[132] reported that surimi with high gelling quality exhibited high myofibrillar Ca^{2+}- and Mg^{2+}-ATPase activities. A decrease of Ca^{2+}-ATPase of actomyosin during iced storage of fish was accompanied by reduced gel strength of surimi prepared from it.[123]

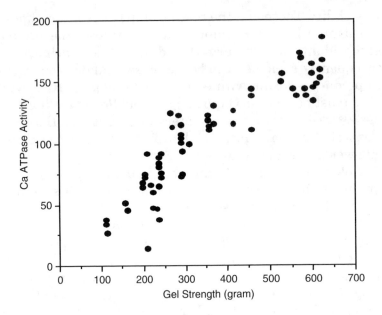

Figure 10.14 Correspondence of $Ca^{2+}ATPase$ activity of myosin and gel-forming ability of surimi, measured as the product of punch force and deformation. (Adapted from Ref. 119.)

The rate of denaturation of the proteins at any time and temperature also varies by species, according to its protein stability. Protein denaturation and aggregation, induced by heating under the correct conditions, drives the gelation of surimi. However, denaturation of the proteins before the surimi is made into surimi seafood will initiate interactions and the formation of bonds between proteins (premature protein aggregation) that will prevent good gelation later. Denaturation of fish proteins during handling or storage of fish can be caused by the following factors:

- Excessively high temperatures or excessive time held at refrigerated temperatures
- Generation or introduction of denaturants, such as metal ions (from heme proteins or outside sources), formaldehyde (from TMAO breakdown), peroxides (from lipid oxidation), or salts

Surimi Gelation Chemistry

- Oxidation by excessive mechanical agitation in air, pro-oxidant compounds, or when proteins dehydrate (freezer burn)
- Low pH/high acidity
- Freezing and storage in the absence of cryoprotectants

10.5.1 The Importance of Muscle pH (Acidity)

Pelagic fish, specifically species that are migratory and surface feeding, generally contain a higher percentage of dark muscle. Their muscles contain higher quantities of glycogen needed for sustained swimming, which upon death is converted to lactic acid. This results in a greater post-mortem drop in muscle pH (increase in acidity).[133] The drop in pH (rise in acidity) can dramatically affect the heat gelation properties of surimi by leading to accelerated denaturation of the protein. For example, the denaturation rate constant of Pacific mackerel myofibrils at pH 5.8 is twice that measured at pH 6.5.[134] While light-fleshed species normally do not decrease in pH below 6.2 to 6.5, darker fleshed fish attain pH values well below 6.0, similar to red meats. It is well known that mammalian and avian meats gel optimally while pre-rigor (near pH 7.0); similarly, the highest-quality surimi is processed from pre-rigor fish, regardless of species.[135]

The acid version of the newly developed pH shifting process for surimi manufacture employs a step that significantly lowers the pH (to 3.0 or below) before later neutralizing with a base to recover, and reportedly restores full gelation ability to the muscle proteins (see Chapter 3). The alkaline version of this process resembles the method used to make soy protein isolates or calcium caseinate. In either configuration, the myofibrillar (and also sarcoplasmic, but not stroma) proteins become soluble at high or low pH. Then the unwanted components are removed by centrifugation prior to neutralizing the mixture to precipitate and recover the desired protein fraction. The surimi-like material produced by this process is light in color and high in gelling ability, even when made from dark-meat species. Chapter 3 presents a more complete explanation of this process and the quality of surimi produced by it.

10.5.2 The Frozen Storage Stability of Surimi

Myofibrillar proteins of surimi will denature and aggregate (and thus lose their gelling ability) during extended frozen storage. Fish proteins are particularly susceptible to denaturation and aggregation, unless mixed with cryoprotectants before freezing (see Chapter 5). Cryoprotectants are ingredients (mainly sugars and polyols) that, when intimately mixed with wet proteins, protect them from many of the deleterious influences during freezing and storage. Cryoprotectants, however, may not be effective against the presence of some denaturing compounds, such as formaldehyde (from TMAO breakdown) or peroxide (from lipid oxidation), although sugar addition has been shown to slow TMAO demethylase activity.[136] Consequently, these denaturing compounds must be controlled by proper cleaning of the fish and thorough water leaching of the flesh.

10.6 SUMMARY: FACTORS AFFECTING HEAT-INDUCED GELLING PROPERTIES OF SURIMI

Surimi manufactured by the conventional water leaching process is primarily composed of myofibrillar proteins, water, and cryoprotectants. Its gelling properties are primarily affected by the:

1. Extent of denaturation and premature aggregation of the myofibrillar proteins before manufacturing into surimi seafood
2. Species and habitat of the fish, which determines the thermal stability of the myofibrillar proteins
3. Activity of proteolytic enzymes, which will cleave proteins and disrupt the gel
4. Activity of endogenous or added protein oxidants, as well as cross-linking enzymes, which contribute to protein cross-linking
5. Relative concentration of myofibrillar vs. sarcoplasmic and/or stroma proteins

This last factor is not only a matter of the presence and concentration of water-soluble proteins and stroma proteins, but also the total concentration of myofibrillar proteins. If the water content is too high, gels will be weak. Thus, proper dewatering of surimi is important to the gelling properties of surimi (see Chapters 2 and 3). Additionally, the particular types and concentrations of sarcoplasmic or stroma proteins likely affects their roles in heat-induced gelation of surimi. Effects of other food ingredients of surimi seafood will also be superimposed on the gelling properties of the myofibrillar proteins. For example, the type and concentration of salts, sugars or sugar alcohols, hydrocolloids, etc. can alter the characteristics of the gel formed (see Chapter 13).

REFERENCES

1. P.J. Bechtel. Muscle development and contractile proteins. In P.J. Bechtel, Ed. *Muscle as Food.* Orlando, FL: Academic Press, 1986, 2–31.

2. S. Lowey and D. Risby. Light chains from fast and slow muscle myosins. *Nature,* 234, 81–85, 1971.

3. A. Asghar and A.M. Pearson. Influence of ante- and post-mortem treatments upon muscle composition and meat quality. *Adv. Food Res.,* 26, 53–213, 1980.

4. G. Offer. Myosin filaments. In J.M. Squire and P.J. Vibert, Eds. *Fibrous Protein Structure.* New York: Academic Press, 1987, 307–356.

5. F.H.C. Crick. The packing of alpha-helices — simple coiled-coils. *Acta Crystallogr.,* 6(8-9), 689–697, 1953.

6. J. Walshaw and D.N. Woolfson. Extended knobs-into-holes packing in classical and complex coiled-coil assemblies. *J. Struct. Biol.,* 144(3), 349–361, 2003.

7. W.D. Kohn, C.M. Kay, and R.S. Hodges. Salt effects on protein stability: two-stranded alpha-helical coiled-coils containing inter- or intrahelical ion pairs. *J. Mol. Biol.,* 267(4), 1039–1052, 1997.

8. P.E. Hoppe and R.H. Waterston. Hydrophobicity variations along the surface of the coiled-coil rod may mediate striated muscle myosin assembly in *Caenorhabditis elegans*. *J. Cell. Biol.*, 135(2), 371–382, 1996.

9. P. Burkhard, S. Ivaninskii, and A. Lustig. Improved coiled-coil stability by optimizing ionic interactions. *J. Mol. Biol.*, 318(3), 901–910, 2002.

10. M.J. Arrizubieta and E. Bandman. The role of interhelical ionic interactions in myosin rod assembly. *Biochem. Biophys. Res. Commun.*, 244(2), 588–593, 1998.

11. A. Lupas. Coiled coils: New structures and new functions. *Trends Biochem. Sci.*, 21(10), 375–382, 1996.

12. G. Steffansson and H.O. Hultin. On the solubility of cod muscle proteins in water. *J. Agric. Food Chem.*, 42, 2656–2664, 1994.

13. M. Lin and J.W. Park. Extraction of proteins from Pacific whiting mince at various washing conditions. *J. Food Sci.*, 61, 432–438, 1996.

14. M. Okada. Effect of washing on the jelly forming ability of fish meat. *Nippon Suisan Gakkaishi*, 30, 255–261, 1962.

15. K. Morioka, Y Shimizu. Relationship between the heat-gelling property and composition of fish sarcoplasmic proteins. *Nippon Suisan Gakkaishi*, 59(9), 1631, 1993.

16. Y.L. Xiong. Myofibrillar protein from different muscle fiber types: implications of biochemical and functional properties in meat processing. *Crit. Rev. Food Sci. Nutr.*, 34(3), 292–320, 1994.

17. T. Nakagawa, F. Nagayama, H. Ozaki, S. Watabe, and K. Hashimoto. Effect of glycolytic enzymes on the gel-forming ability of fish muscle. *Nippon Suisan Gakkaishi*, 55, 1045–1050, 1989.

18. K. Morioka and Y. Shimizu. Contribution of sarcoplasmic proteins to gel formation of fish meat. *Nippon Suisan Gakkaishi*, 56, 929–933, 1990.

19. K. Morioka and Y. Shimizu. Heat-coagulation property of fish sarcoplasmic proteins. *Nippon Suisan Gakkaishi*, 58, 1529–1533, 1992.

20. K. Morioka, T. Nishimura, A. Obatake, and Y. Shimizu. Relationship between the myofibrillar protein gel strengthening effect and the composition of sarcoplasmic proteins from Pacific mackerel. *Fisheries Sci.*, 63(1), 111–114, 1997.

21. M. Karthikeyan, S. Mathew, B.A. Shamasundar, and V. Prakash. Fractionation and properties of sarcoplasmic proteins from oil sardine (*Sardinella longiceps*): influence on the thermal gelation behavior of ashed meat. *J. Food Sci.*, 69(3), 79–84, 2004.

22. W.C. Ko, M.S. Hwang. Contribution of milkfish sarcoplasmic protein to the thermal gelation of myofibrillar protein. *Fisheries Sci.*, 61, 75–78, 1995.

23. N.C. Siang and K. Miwa. Gel strength enhancing effect of aqueous extract of fish kidney tissue. *Nippon Suisan Gakkaishi*, 58, 805, 1992.

24. E. Okazaki, K. Kanna, and T. Suzuki. Effect of sarcoplasmic protein on rheological properties of fish meat gel formed by retort-heating. *Bull. Japan Soc. Sci. Fish*, 52, 1821–1827, 1986.

25. A. Nowsad, E. Katoh, S. Kanoh, and E. Niwa. Effect of sarcoplasmic proteins on the setting of tranglutaminase-free paste. *Fisheries Sci.*, 61(6), 1039–1040, 1995.

26. J.F. Holmquist. Interrelations between Salt-Extractable Protein, Actomyosin, Ca-Atpase Activity, and Kamaboko Quality Prepared from Frozen Red Hake Fillets, Mince and Surimi. M.S. thesis, University of Massachusetts, Amherst, 1982, 91.

27. J. Yongsawatdigul, J.W. Park, D. Kolbe, Y. AbuDagga, and M. Morrissey. Ohmic heating maximizes gel functionality of Pacific whiting surimi. *J. Food Sci.*, 60(1), 10–14, 1995.

28. D.D. Hamann, P.M. Amato, M.C. Wu, and E.A. Foegeding. Inhibition of modori (gel weakening) in surimi by plasma hydrolysate and egg white. *J. Food Sci.*, 55, 665–669, 1990.

29. J.R. Howe, D.D. Hamann, T.C. Lanier, and J.W Park. Fracture of Alaska pollock gels in water: effects of minced muscle processing and test temperature. *J. Food Sci.*, 59, 770–780, 1994.

30. M. Bouraoui, S. Nakai, and E. Li-Chan. *In situ* investigation of protein structure in Pacific whiting surimi and gels using Raman spectroscopy. *Food Res. Int.*, 30, 65–72, 1997.

31. N.S. Miroshinichenko, I.V. Balanuk, and D.N. Nozdrenko. Packing of myosin molecules in muscle thick filaments. *Cell Biol. Int.*, 24(6), 327–333, 2000.

32. E. Niwa. Chemistry of surimi gelation. In T.C. Lanier and C. Lee, Eds. *Surimi Technology*. New York: Marcel Dekker, 1992, 389—427.

33. G.M. Gilleland, T.C. Lanier, and D.D. Hamann. Covalent bonding in pressure-induced fish protein gels. *J. Food Sci.*, 62(4), 713–716, 1997.

34. S. Ishizaki, M. Ogasawara, M. Tanaka, and T. Taguchi. Ultraviolet denaturation of flying fish myosin and its fragments. *Fisheries Sci.*, 60(5), 603–606, 1994.

35. M. Okada and M. Nakayama. The effect of oxidants on jelly strength of kamaboko. *Bull. Jap. Soc. Sci. Fish*, 27, 203, 1961.

36. R. Yoshinaka, M. Shiraishi, and S. Ikeda. Effect of ascorbic acid on the gel formation of fish meat. *Bull. Jap. Soc. Sci. Fish*, 38, 511, 1972.

37. H.G. Lee, C.M. Lee, K.H. Chung, and S.A. Lavery. Sodium ascorbate affects surimi gel-forming properties. *J. Food Sci.*, 57, 1343, 1992.

38. K. Nishimura, M. Goto, and Y. Itoh. Influence of oxygen radicals produced by ascorbic acid on amino acid composition of muscle proteins during the formation of a heat-induced gel of fish meat. *Fisheries Sci.*, 60(6), 799–800, 1994.

39. A. Kishi, Y. Itoh, and A. Obatake. The polymerization of protein through disulfide bonding during the heating of carp myosin. *Nippon Suisan Gakkaishi*, 61(1), 75–80, 1995.

40. A. Kishi, Y. Itoh, and A. Obatake. Effects of tap water on the polymerization through SS bonding during the heating of carp myosin. *Nippon Suisan Gakkaishi*, 63(2), 242–243, 1997b.

41. B.Y. Kim. Rheological Investigation of Gel Structure Formation by Fish Proteins during Setting and Heat Processing. Ph.D. dissertation, North Carolina State University, Raleigh, 1987.

42. S. Jiang, C. Lan, and C. Tsao. New approach to improve the quality of mince fish products from freeze-thawed cod and mackerel. *J. Food Sci.*, 51, 310–312, 1986.

43. S. Jiang, M. Ho, S. Jiang, L. Lo, and H. Chen. Color and quality of mackerel surimi as affected by alkaline washing and ozonation. *J. Food Sci.*, 63(4), 652–655, 1998a.

44. S. Jiang, M. Ho, S. Jiang, and H. Chen. Purified NADPH-sulfite reductase from *Saccharomyces cerevisiae* effects on quality of ozonated mackerel surimi. *J. Food Sci.*, 63(5), 777–781, 1998b.

45. E. Niwa, E. Chen, T. Wang, S. Kanoh, and T. Nakayama. Extraordinarity in the temperature-dependence of physical parameters of kamaboko. *Nippon Suisan Gakkaishi*, 54, 1789–1793, 1988.

46. H.G. Lee, T.C. Lanier, and D.D. Hamann. Covalent cross–linking effects on thermo-rheological profiles of fish protein gels. *J. Food Sci.*, 62(1), 25–28, 32, 1997.

47. J.K. Chan, T.A. Gill, and A.T. Paulson. The dynamics of thermal denaturation of fish myosins. *Food Res. Int.*, 25, 117–123, 1992.

48. K. Samejima, M. Ishioroshi, and T. Yasui. Relative roles of the head and tail portions of the molecule in heat-induced gelation of myosin. *J. Food Sci.*, 46, 1412–1418, 1981.

49. T. Yasui and K. Samajima. Recent advances in meat science in Japan: functionality of muscle proteins in gelation mechanism of structured meat products. *Japan Agric. Res. Q.*, 24, 131–137, 1990.

50. A. Kishi, Y. Itoh, and A. Obatake. The subfragment responsible for the polymerization of myosin heavy chain through SS bonding during the heating of carp myosin. *Nippon Suisan Gakkaishi*, 63(2), 237–241, 1997.

51. T. Sano, S.F. Noguchi, J.J. Matsumoto, and T. Tsuchiya. Thermal gelation characteristics of myosin subfragments. *J. Food Sci.*, 55, 55–58, 1990.

52. T. Higuchi, T. Ojima, and K. Nishita. Heat-induced structural changes and aggregation of walleye Pollack myosin in the light meromyosin region. *Fisheries Sci.*, 68, 1145–1150, 2002.

53. J.C. Acton and R.L. Dick. Functional roles of heat protein gelation in processed meat. In J.E. Kinsella, W.G. Soucie, Eds. *Food Proteins*. Champaign, IL: Am. Oil Chem. Soc., 1989.

54. H.G. Lee and T.C. Lanier. The role of covalent crosslinking in the texturizing of muscle protein sols. *J. Muscle Foods*, 6, 125–138, 1996.

55. A. Sharp and G. Offer. The mechanism of formation of gels from myosin molecules. *J. Sci. Food Agric.*, 58, 63–73, 1992.

56. L. Wicker, T.C. Lanier, DD. Hamann, and T. Akahane. Thermal transitions in myosin-ANS fluorescence and gel rigidity. *J. Food Sci.*, 51, 1540–1543 and 1562, 1986.

57. A.B. Smyth, D.M. Smith, and E. O'Neill. Disulfide bonds influence the heat-induced gel properties of chicken breast muscle myosin. *J. Food Sci.*, 63(4), 584–588, 1998.

58. J.K. Chan, T.A. Gill, and A.T. Paulson. Thermal aggregation of myosin subfragments from cod and herring. *J. Food Sci.*, 58, 1057–1061 and 1069, 1993.

59. M. Ishioroshi, K. Samejima, Y. Arie, and T. Yasui. Effect of blocking the myosin-actin interaction in heat induced gelation of myosin in the presence of actin. *Agric. Biol. Chem.*, 44, 2185, 1980.

60. K. Samejima, M. Ishioroshi, and T. Yasui. Heat induced gelling properties of actomyosin: effect of tropomyosin and troponin. *Agric. Biol. Chem.*, 46, 535–540, 1982.

61. T. Yasui, M. Ishioroshi, and K. Samejima. 1982. Effect of actomyosin on heat-induced gelation of myosin. *Agric. Biol. Chem.*, 46, 1049–1059.

62. T. Sano, T. Ohno, H. Otsuka-Fuchino, J.J. Matsumoto, and T. Tsuchiya. Carp natural actomyosin: thermal denaturation mechanism. *J. Food Sci.*, 59, 1002–1008, 1994.

63. T.C. Lanier, T.S. Lin, Y.M. Liu, and D.D. Hamann. Heat gelation properties of actomyosin and surimi prepared from Atlantic croaker. *J. Food Sci.*, 47, 1921–1925, 1982.

64. E. Niwa and G. Nakajima. Difference in protein structure between elastic kamaboko and brittle one. *Nippon Suisan Gakkaishi*, 41, 579, 1975

65. M. Okada. Application of setting phenomenon for improving the quality of kamaboko. *Bull. Tokai Reg. Fish Res.*, 24:67–70, 1959.

66. T.C. Lanier. Functional properties of surimi. *Food Technol.*, 40, 107–111, 1986.

67. O. Esturk and J.W. Park, and S. Thawornchinsombut. Thermal sensitivity of fish proteins from various species on rheological properties. *J. Food Sci.*, in press, 2005.

68. Y. Funatsu, N. Katoh, and K. Arai. Aggregate formation of salt-soluble proteins in salt-ground meat from walleye pollock surimi during setting. *Nippon Suisan Gakkaishi,* 62(1), 112–122, 1996.

69. C. Imai, Y. Tsukamasa, M. Sugiyama, Y. Minegishi, and Y. Shimizu. The effect of setting temperature on the relationship between ε-(γ-glutamyl)lysine crosslink content and breaking strength in salt-ground meat of sardine and Alaska pollock. *Nippon Suisan Gakkaishi,* 62(1), 104–111, 1996.

70. E. Niwa, T. Nakayama, and I. Hamada. Arylsulfonylatio chloride induced setting of dolphinfish sol. *Nippon Suisan Gakkaishi,* 47, 179–182, 1981.

71. E. Niwa, R. Suzuki, K. Sato, T. Nakayama, and I. Hamada. Setting of flesh sol induced by ethylsulfonylation. *Nippon Suisan Gakkaishi,* 47, 915–919, 1981.

72. L. Wicker, T.C. Lanier, J.A. Knopp, and D.D. Hamann. Influence of various salts on heat-induced ANS fluorescence and gel rigidity development of Tilapia myosin. *J. Agric. Food Chem.,* 37, 18–22, 1989.

73. K. Arai, Y. Funatsu, S. Nanbu, N. Yamada, and N. Kato. Some food additives as affecting materials of heat-induced gel formation with crosslinking of myosin heavy chain of frozen surimi. *Proc. 206th Am. Chem. Soc. Natl. Meet.,* Chicago, 1993.

74. A.A. Nowsad, S. Kanoh, and E. Niwa. Setting of transglutaminase-free actomyosin paste prepared from Alaska pollock surimi. *Fisheries Sci.,* 60(3), 295–297, 1994.

75. A.A. Nowsad, S. Kanoh, and E. Niwa. Setting of surimi paste in which transglutaminase is inactivated by *p*-chloromeruribenzoate. *Fisheries Sci.,* 60(2), 185–188, 1994.

76. T. Shoji, H. Saeki, A. Wakameda, M. Nakamura, and M. Nonaka. Gelation of salted paste of Alaska. *Nippon Suisan Gakkaishi,* 56(12), 2069–2076, 1990.

77. T. Ueda, R. Kusaba, S. Kamimura, E. Okazaki, Y. Fukuda, and K. Arai. Effect of heating on quality of pressure-texturized gels from chum salmon mince. *Nippon Suisan Gakkaishi,* 63, 600–607, 1997.

78. M. Nonaka, R. Ito, A. Sawa, M. Motoki, and N. Nio. Modification of several proteins by using Ca-independent microbial transglutaminase with high-pressure treatment. *Food Hydrocolloids*, 11(3), 351–353, 1997.

79. G.A. MacDonald, T.C. Lanier, and F.G. Giesbrecht. Interaction of sucrose and zinc for cryoprotection of surimi. *J. Agric. Food Chem.*, 44(1), 113–118, 1996.

80. Y. Funatsu, H. Hosokawa, S. Nanbu, and K. Arai. Effects of sorbitol on gelation and crosslinking of myosin heavy chain of salt-ground meat from walleye pollock. *Nippon Suisan Gakkaishi*, 59(9), 1599–1607, 1993.

81. H. Takeda and N. Seki. Enzyme-catalyzed crosslinking and degradation of myosin heavy chain in walleye pollock surimi paste during setting. *Fisheries Sci.*, 62(3), 462–467, 1996.

82. S. Wang and T.C. Lanier. Effects of endogenous, fungal and microbial transglutaminase. *J. Food Sci.*, submitted, 2005.

83. T. Shoji, H. Saeki, A. Wakameda, and M. Nonaka. Influence of ammonium salt on the formation of pressure-induced gel from walleye pollock surimi. *Nippon Suisan Gakkaishi*, 60(1), 101–109, 1994.

84. I.S. Kang, J.J. Wang, J.C.H. Shih, and T. Lanier. Extracellular production of a functional soy cystatin by *Bacillus subtilis*. *J. Agric. Food Chem.*, 52(16), 5052–5056, 2004.

85. P.J. Torley and T.C. Lanier. Setting ability of salted beef/pollock surimi mixtures. In E. Graham Gligh, Ed. *Seafood Science and Technology*. Oxford U.K.: Fishing News Books Ltd,, 1992, 305–316.

86. M. Matukawa, F. Hirata, S. Kimura, and K. Arai. Effect of sodium pyrophosphate on gelling property and crosslinking of myosin heavy chain in setting-heating gel from walleye pollock surimi. *Nippon Suisan Gakkaishi*, 62(1), 94–103, 1996.

87. J. Wan and N. Seki. Effects of salts on transglutaminase-mediated crosslinking of myosin in suwari gel from walleye *Nippon Suisan Gakkaishi*, 58(11), 2181–2187, 1992.

88. Y. Shimizu, R. Machida, and S. Takenami. Species variations in the gel-forming characteristics of fish meat paste. *Nippon Suisan Gakkaishi*, 47:95–104, 1981.

89. S. Ishikawa. Manufacture of kamaboko and surimi from sardines. II. Effect of temperature during manufacture on the jelly strength of kamaboko. *Bull. Tokai Reg. Fish. Res.,* 94, 37, 1978.

90. T. Ikeuchi and W. Simidu. Study on cold storage of brayed fish meat for the material of kamaboko. I. Effects of setting phenomenon on the jelly forming ability of frozen brayed fish meat. *Bull. Jap. Soc. Sci. Fish,* 29, 51, 1963.

91. H. Araki and N. Seki. Comparison of reactivity of transglutminase to various fish actomyosins. *Bull. Jap. Soc. Sci. Fish,* 59(4), 711–716, 1993.

92. E. Niwa, T. Suzumura, A. Nowsad, and S. Kaho. Setting of actomyosin paste containing few amount of tranglutaminase. *Nippon Suisan Gakkaishi,* 59(12), 2043–2046, 1993.

93. D. Joseph, T.C. Lanier, and D.D. Hamann. Temperature and pH effects of tranglutaminase-catalyzed "setting" of crude fish actomyosin. *J. Food Sci.,* 59(5), 1018–1023, 1036, 1994.

94. G.G. Kamath, T.C. Lanier, E.A. Foegeding, and D.D. Hamann. Non-disulfide covalent cross-linking of myosin heavy chain in "setting" of Alaska pollock and Atlantic croaker surimi. *J. Food Biochem.,* 16, 151–172, 1992.

95. N.G. Lee and J.W. Park. Calcium compounds to improve gel functionality of Pacific whiting and Alaska. *J. Food Sci.,* 63(6), 969–974, 1998.

96. S. Benjakul, W. Visessanguan, S. Riebroy, S. Ishizaki, and M. Tanaka. Gel-forming properties of surimi produced from bigeye snapper, *Priacanthus tayenus* and *P. Macracanthus*, stored in ice. *J. Sci. Food Agric.,* 82, 1442–1451, 2002

97. K. Klesk, J. Yongsawatdigul, J.W. Park, S. Viratchakul, and P. Virulhakul. Gel forming ability of tropical tilapia surimi gels at various thermal treatments as compared to Alaska Pollock and Pacific whiting surimi. *J. Aqua. Food Prod.,* 9(3), 91–104, 2000.

98. O.G. Morales, J.A. Ramírez, D.I. Vivanco, and M. Vázquez. Surimi of fish species from the Gulf of Mexico: evaluation of the setting phenomenon. *Food Chem.,* 75, 43–48, 2001.

99. J. Yongsawatdigul, A. Worratao, and J.W. Park. Effect of endogenous transglutaminase on threadfin bream surimi gelation. *J. Food Sci.,* 67, 3258–3263, 2002.

100. J. Yongsawatdigul and J.W. Park. Thermal denaturation and aggregation of threadfin bream actomyosin. *Food Chem.*, 83, 409–416, 2003.

101. R. Ofstad, E. Grahl-Madsen, B. Gundersen, K. Lauritzen, T. Solberg, and C. Solberg. Stability of cod (*Gadus morhua* l.) surimi processed with $CaCl_2$ and $MgCl_2$ added to the wash water. *Int. J. Food Sci. Technol.*, 28(5), 419–427, 1993.

102. H. Saeki. Gel-forming ability and cryostability of frozen surimi processed with $CaCl_2$ washing. *Fisheries Sci.*, 62(2), 252–256, 1996.

103. J. Wan, I. Kimura, M. Satake, and N. Seki. Effect of calcium ion concentration on the gelling properties and transglutaminase activity of walleye pollock surimi paste. *Fisheries Sci.*, 60(1), 107–114, 1994.

104. H. Saeki and F. Hirata. Behavior of fish meat components during manufacture of frozen surimi through processing with $CaCl_2$ washing. *Fisheries Sci.*, 60(3), 35–340, 1994.

105. A. Worratao and J. Yongsawatdigul. Cross-linking of actomyosin by crude tilapia (*Oreochromis niloticus*) transglutaminase. *J. Food Biochem.*, 27, 35–51, 2003.

106. N. Maruyama, H. Nozawa, I. Kimura, M. Satake, and N. Seki. Transglutaminase-induced polymerization of a mixture of different fish myosins. *Fisheries Sci.*, 61(3), 495–500, 1995.

107. H. Ando, M. Adachi, K. Umeda, A. Matsuura, R.M. Nonaka, R. Uchino, H. Tanaka, and M. Motoki. Purification and characteristics of novel transglutaminase derived from microorganisms. *Agric. Bio. Chem.*, 53, 2613–2617, 1989.

108. Y. Abe, K. Yasunaga, S. Kitakami, Y. Murakami, T. Ota, and K. Arai. Quality of kamaboko gels from walleye pollock frozen surimi of different grades on applying additive containing TGase. *Nippon Suisan Gakkaishi*, 62(3), 439–445, 1996.

109. K. Yasunaga, Y. Abe, M. Yamazawa, and K. Arai. Heat-induced change in myosin heavy chains in salt-ground meat with a food additive containing transglutaminase. *Nippon Suisan Gakkaishi*, 62(4), 659–668, 1996.

110. T. Asagami, M. Ogiwara, A. Wakameda, and S. Noguchi. Effect of microbial transglutaminase on the quality of frozen surimi made from various kinds of fish species. *Fisheries Sci.*, 61(2), 267–272, 1995.

111. T.C. Lanier and I.S. Kang. Plasma as a source of functional food additives. *International Conference of Food System Functionality* (http://www.msstate.edu/org/fsfa/web-compilations.htm), Las Vegas, NV, 1997

112. S. Nakamura, M. Ogawa, M. Saito, and S. Nakai. Application of polymannosylated cystatin to surimi from roe-herring to prevent gel weakening. *FEBS Lett.*, 427, 252–254, 1998.

113. A.J. Whitmore and P. Torley. The Transglutaminase Mediated Setting Reaction in Surimi and Purification of a Transglutaminase from Plasma. Publication of Meat Industry Research Institute of New Zealand (MIRINZ), 1992.

114. K. Washizu, K. Ando, S. Koikeda, S. Hirose, A. Matsuura, H. Takagi, M. Motoki, and K. Takeuchi. Molecular cloning of the gene for microbial transglutaminase from *Streptoverticillium* and its expression in *Streptomyces lividans*. *Biosci. Biotech. Biochem.*, 58(1), 82–87, 1994.

115. M.E. Surette, T.A. Gill, and P.J. LeBlanc. Biochemical basis of postmortem nucleotide catabolism in cod (*Gadus morhua*) and its relationship to spoilage. *J. Agric. Food Chem.*, 36, 19–22, 1988.

116. P.T. Lakshmanan, P.D. Antony, and K. Gopakumar. Nucleotide degradation and quality changes in mullet *(Liza corsula)* and pearlspot *(Etroplus suratensis)* in ice and at ambient temperatures. *Food Control*, 7(6), 227–283, 1996.

117. K. Gopakumar. Enzymes and enzyme products as quality indices. In N.F. Haard and B.K. Simpson, Eds. *Seafood Enzymes Utilization and Influence on Postharvest Seafood Quality*. New York: Marcel Dekker, 2000, 337–364.

118. G.A. MacDonald, J. Lelievre, and N.D.C. Wilson. Strength of gels prepared from washed and unwashed minces of hoki (*Macruronus novaezelandiae*) stored in ice. *J. Food Sci.*, 55(4), 976–978, 982, 1990.

119. R.F. Ablett, E.G.Bligh, and K. Spencer. Influence of freshness on quality of white hake (*Urophycis tenuis*) surimi. *Can. Inst. Food Sci. Technol. J.*, 24(1/2), 36–41, 1991.

120. G.A. MacDonald, J. Lelievre, and N.D. Wilson. The strength of gels prepared from washed and unwashed minces of hoki (*Macruronus novaezelandiae*) stored in ice. *J. Food Sci.*, 54(4), 976, 1990.

121. T. Kurokawa. Kamaboko-forming ability of frozen and ice stored lizard fish. *Bull. Japan Soc. Sci. Fish.*, 45(12), 1551–1555, 1979

122. Y.S. Yean. The quality of surimi made from Threadfin bream stored in ice for different periods. *Int. J. Food Sci. Technol.*, 28, 343–346, 1993.

123. J. Yongsawatdigul and J.W. Park. Biochemical and conformation changes of actomyosin from threadfin bream stored in ice. *J. Food Sci.*, 67, 985–990, 2002.

124. S.I. Roura, J.P. Saavedra, R.E. Truco, and M. Crupkin. Conformational change in actomyosin from post-spawned hake stored on ice. *J. Food Sci.*, 57(5), 1109–1111, 1992.

125. S. Benjakul, T.A. Seymour, M.T. Morrissey, and H. An. Physicochemical changes in Pacific whiting muscle proteins during iced storage. *J. Food Sci.*, 62(4), 729–733, 1997

126. W. Sompongse, Y. Itoh, S. Nagamachi, and A. Obatake. Effect of the oxidation of SH groups on the stability of carp myosin during ice storage. *Fisheries Sci.*, 62, 468–472, 1996.

127. T. Kawashima, K. Arai, and T. Saito. Studies on muscular proteins of fish. IX. An attempt on quantitative determination of actomyosin in frozen surimi from Alaska pollock. *Bull. Jap. Soc. Sci. Fish*, 39, 207–214, 1973.

128. T. Ooizumi, K. Hashimoto, J. Ogura, and K. Arai. Quantitative aspect for protective effect of sugar and sugar alcohol against denaturation of fish myofibrils. *Bull. Jap. Soc. Sci. Fish*, 47, 901, 1981.

129. T. Ooizumi, Y. Nara, and K. Arai. Protective effect of carboxylic acids, sorbitol and Na-glutamate on heat denaturation of chub mackerel myofibrils. *Bull. Jap. Soc. Sci. Fish.*, 50(5), 875, 1984.

130. M. Yamaguchi and T. Sekine. Sulfhydryl groups involved in the active site of myosin A adenosine triphosphatase. *J. Biochem.*, 59, 24–33, 1966.

131. S.T. Jiang, D.C. Hwang, and C.S. Chen. Denaturation and changes in SH group of actomyosin from milkfish (Chanos chanos) during storage at −20°C. *J. Agric. Food Chem.*, 36:433–437, 1988.

132. N. Katoh, H. Nozaki, K. Komatsu, and K.I. Arai. A new method for evaluation of quality of frozen surimi from Alaska pollock. Relationship between myofibrillar ATPase activity and kamaboko forming ability of frozen surimi. *Nippon Suisan Gakkaishi*, 48(8), 1027–1032, 1979.

133. Y. Fujii, N. Nakamura, and Y. Ishikawa. Suitability of fresh and frozen fish to fish jelly products. In General Report of Research of Effective Utilization of Abundantly Caught Dark Fishes. Japan: Fisheries Agency, 1978, 119–127.

134. A. Hashimoto and K. Arai. Freshness of fish, manufacturing condition and quality of surimi. In General Report of Research of Effective Utilization of Abundantly Caught Dark Fishes. Japan: Fisheries Agency, 1978, 119–127.

135. J.W. Park, R.W. Korhonen, and T.C. Lanier. Effects of rigor mortis on gel-forming properties of surimi and unwashed mince prepared from tilapia. *J. Food Sci.*, 55, 353–355, 360, 1990.

136. G.A. MacDonald, N.D. Wilson, and T.C. Lanier. Stabilized mince: an alternative to the traditional surimi process. In *Chilling and Freezing of New Fish Products*. Paris: International Institute of Refrigeration, 1990, 69–76.

137. T. Suzuki. *Fish and Krill Protein Processing Technology*. London: Applied Science Publishers, 1981, 1–56.

138. B.Y. Yu. Coiled-coils: stability, specificity, and drug delivery potential. *Adv. Drug Deliver. Rev.*, 54(8), 1113–1129, 2002.

139. C. Cohen. Why fibrous proteins are romantic. *J. Struct. Biol.*, 122(1-2):3–16, 1998.

140. M. Wick. Filament assembly properties of the sarcomeric myosin heavy chain. *Poultry Sci.*, 78(5), 735–742, 1999.

11

Rheology and Texture Properties of Surimi Gels

BYUNG Y. KIM
Kyung Hee University, Seoul, Korea

JAE W. PARK
Oregon State University Seafood Lab, Astoria, Oregon

WON B. YOON
CJ Foods Research Center, Seoul, Korea

CONTENTS

11.1 Introduction 493
11.2 Fundamental Test 496
    11.2.1 Force and Stress 496

- 11.2.2 Deformation and Strain ... 498
- 11.2.3 Flow and Rate of Strain ... 499
- 11.2.4 Rheological Tests Using Small Strain (Deformation) ... 501
  - 11.2.4.1 Compressive Test for Surimi Gel ... 501
  - 11.2.4.2 Shear Type Test for Surimi Paste and Gels ... 501
  - 11.2.4.3 Stress Relaxation Test ... 503
  - 11.2.4.4 Oscillatory Dynamic Test ... 508
- 11.2.5 Rheological Testing Using Large Strain (Failure Test) ... 518
  - 11.2.5.1 Axial Compression for Cylinder Type Gels ... 519
  - 11.2.5.2 Compressive Test for Convex Shape Samples ... 520
  - 11.2.5.3 Compressive Test for Rod-Type Samples ... 521
  - 11.2.5.4 Torsion Test ... 523
- 11.3 Empirical Tests ... 528
  - 11.3.1 Punch (Penetration) Test ... 529
  - 11.3.2 Texture Profile Analysis (TPA) ... 534
  - 11.3.3 Relationship between Torsion and Punch Test Data ... 537
- 11.4 Effects of Processing Parameters on Rheological Properties of Surimi Gels ... 540
  - 11.4.1 Effects of Fish Freshness/Rigor Condition ... 540
  - 11.4.2 Effect of Refrigerated Storage of Gels ... 540
  - 11.4.3 Effect of Sample Temperature at Measurement ... 541
  - 11.4.4 Effect of Moisture Content ... 543
  - 11.4.5 Effect of Low Temperature Setting ... 544
  - 11.4.6 Effect of Freeze-Thaw Abuse ... 546
  - 11.4.7 Effect of Functional Additives ... 546
  - 11.4.8 Texture Map ... 548
- 11.5 Viscosity Measurements ... 549
  - 11.5.1 Measurement of Dilute Extract ... 550
  - 11.5.2 Measurement of Surimi Seafood Pastes ... 550

        11.5.3   Rheological Behavior of Surimi Paste..........556
11.6 Practical Application of Dynamic Rheological
        Measurements 558
        11.6.1   Gelation Kinetics of Surimi Gels 559
                11.6.1.1   Thermorheological Properties and
                             Kinetic Model..................................559
                11.6.1.2   Non-Isothermal Kinetic Model......561
                11.6.1.3   Rubber Elastic Theory...................562
                11.6.1.4   Dependence of Gelation
                             Temperature on Moisture
                             Content 563
                11.6.1.5   Activation Energy during
                             Gelation 565
        11.6.2   Estimation of Steady Shear Viscosity of
                   Fish Muscle Protein Paste 567
                11.6.2.1   Steady Shear Viscosity of Surimi
                             Paste and the Cox-Merz Rule 567
                11.6.2.2   Superimposed Viscosity Using the
                             Cox-Merz Rule................................568
                11.6.2.3   Concentration Dependence of the
                             Viscosity of Surimi Paste 569
11.7 Summary..574
Acknowledgements...575
References..576

11.1 INTRODUCTION

Texture is a major component in measuring the functional characteristics of raw surimi materials, as well as the effects of manufacturing conditions and properties of the finished surimi seafood products. Fish myofibrillar proteins, a major surimi constituent, possess the unique ability to "set" into gels of high cohesiveness at low temperatures (0 to 40°C) and produce particularly hard gels when processed at higher temperatures following setting. These gel-forming properties are what make surimi a valuable texture-building agent in formulated muscle foods. Surimi can be used to bind natural muscle fibers together or may be the base material for the

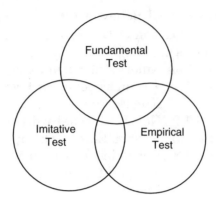

Figure 11.1 Three classes of instrumental analysis.

forming of processed meat fibers into analog products. In either case, the properties of the surimi gel will have a marked effect on the perceived texture of the product. Efforts to measure food texture by instrumental methods have resulted in three classes of rheological tests: fundamental, empirical, and imitative (Figure 11.1).

Fundamental tests measure properties that are familiar to engineers, e.g., ultimate strength, Poisson's ratio, and various moduli such as Young's modulus, shear modulus, and bulk modulus. Results of fundamental rheological tests are measured in terms of kilograms (kg), meters (m), and seconds (sec). The results are objective; therefore, regardless of the method used, the same results are obtained within the experimental error. Converting all results to these three basic units is advantageous because they are easily accessible as standards. Unfortunately, food, in rheological terms, is extremely complicated. Consequently, fundamental testing is often laborious and time-consuming and may not give simple answers.

Empirical testing, on the other hand, is quicker and simpler than fundamental testing and, although loosely defined, has been determined from practical experience to correlate well with textural quality. A number of empirical tests have been developed to give relatively good correlation with sensory evaluation of texture on numerous foods. These tests include puncture, shear, and extrusion. The disadvantages of empirical tests are that the results are specific to a particular instrument and cannot be compared between different testing methods.

Imitative tests use instruments to imitate the conditions to which food is subjected to in the mouth or on the plate. Texture profile analysis and shear cell (i.e., Kramer, Warner-Bratzler) fall into this category. The texture profile quantifies specific characteristics that can be directly related to the overall acceptance or hedonic ratings.[1]

The behavior of material under applied forces defines its mechanical properties. Accordingly, the stress-strain behavior of a material under static and dynamic loading, as well as the flow characteristics of the material, can be classified as mechanical properties. The physical approach to define mechanical behavior is called rheology. Rheology is the science of material deformation and flow. Because action of force results in deformation and flow in the material, the mechanical properties will be referred to as rheological properties. In addition, rheology considers the time effect during the loading of a body. Thus, mechanical behaviors of a material, such as time-dependent stress and strain behavior, creep, stress relaxation, dynamic mechanical modulus, and viscosity, are expressed in terms of three parameters: force, deformation, and time.

Rheological measurements can aid food developers in controlling the chemical interactions of food components to produce particular food structures with desired textural attributes.[2] The mechanical characteristics of the texture of homogeneous, isotropic (properties of independent upon orientation) gels made from surimi by heating are mostly accounted for by measuring the stress or gel strength (stress at fracture), and strain, cohesiveness, or deformability (strain at fracture). Concerning the fluid properties of raw material, such as surimi paste, viscosity, which may be time or shear dependent, is another rheological property of interest. In relation to the formation of thermo-irreversible gels, such as crabmeat analogs (hereafter referred to as surimi seafood), "small strain testing" (deforming a sample only by a small percentage of deformation required to break) and "large strain testing" (deforming to the point of permanent structural change: fracture) may yield important information regarding the structure of the surimi gel. Large strain instrumental testing is required to consistently correlate with sensory texture,[3-5] which in turn can determine the acceptability of a product and influence quality control.

11.2 FUNDAMENTAL TEST

For basic studies or when definitive results are desirable, a fundamental test is recommended. Such a test will yield results in a form that is theoretically independent of instrument and specimen geometry. Mitchell[6] stated that it is desirable to employ fundamental test methods when the aim is to use rheological experiments to elucidate gel structure.

11.2.1 Force and Stress

A force, although defined in terms of its power to produce acceleration, is also an agency capable of deforming a material. When an external force acts on a body, several different cases can be distinguished, including tension, compression, and shear. Normal force is perpendicular to the plane on which it operates and can be either tensile or compressive according to the direction in which it acts. Shear force, however, is parallel to the plane on which it operates (Figure 11.2). What is the difference between force and stress? If someone sits on either a chair or a pin, the force (the load) that is pressed down is the same, but the area of application is extremely decreased with the pin. Therefore, it is more meaningful to use force divided by the unit area.

Stress is the internal force acting upon an area of the body. The normal stress (σ), which can be either compressive or tensile, is calculated by dividing the magnitude of the applied force by the cross-sectional area of the specimen and has the units of force per unit area (dyne/cm^2, Newton/m^2 or Pascal):

$$\sigma = F(t)/A_o \qquad (11.1)$$

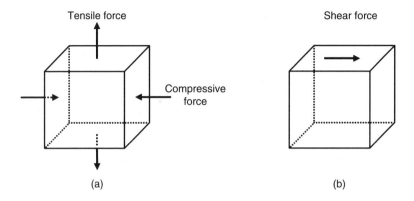

Figure 11.2 Normal force (a) and shear force (b).

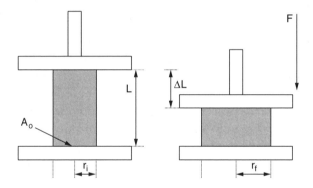

Figure 11.3 Normal force, normal stress, and normal deformation.

where F(t) is the normal force and A_o is the area of contact (Figure 11.3).

A normal stress can be presented either as apparent (nominal) stress, which is computed using the initial cross-sectional area of the specimen, or as true (natural) stress, which is computed using the actual cross-sectional area of the deformed specimen. Stress refers only to an idea of averaging the intensity of force. For many practical purposes, this simple way to calculate stress is satisfactory. However, the actual internal stress distribution can be far from uniform in the

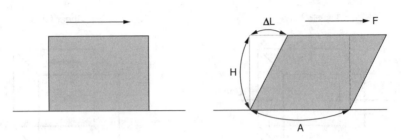

Figure 11.4 Shear force, shear stress, and shear strain.

loading of a real body, especially when structural irregularities are present. In certain instances, the non-uniform stress distribution can lead to a stress concentration where the intensity far exceeds the average stress.

Figure 11.4 represents the side view of a deformed element of the sample specimen with its original rectangular parallel shape (ΔL for deformed length and H for height). The top has been pushed to the right by a tangential force (F). This type of deformation changes the shape but does not change the volume of the material element. It is called shear and is the preferable type of deformation for fundamental measurements. If the motion of the top surface is continuous, a shear type equation can be developed. Shear stress (τ) is the applied force divided by the area of the surface to which it is applied tangentially and has the same units as normal stress:

$$\tau = F(t)/A_0 \tag{11.2}$$

where $F(t)$ is the shear force, and A_0 is the area, which remains constant.

11.2.2 Deformation and Strain

Normal deformation (ΔL) is the absolute elongation or contraction of the specimen in the direction of the applied force. The normal strain (ε: apparent strain or engineering strain) is the ratio between the normal deformation and the initial length (L) of the specimen (in the same direction as the applied force) (Figure 11.3):

$$\varepsilon = \Delta L/L \tag{11.3}$$

where ΔL is normal deformation and L is the initial length.

One approach to develop an appropriate axial strain equation for the compression of cylinders at a large axial strain is to consider the strain as the sum of an infinite number of small axial strain changes. Integrating these small strains (each based on the previous specimen height) results in the true strain (ε_T) and is given by:

$$\varepsilon_T = \int (dL/L) \tag{11.4}$$
$$= \ln (L/L_o)$$
$$= \ln (L_o + \Delta L)/L_o$$
$$= \ln (1 + \varepsilon)$$

A positive sign (+) is used for the axial tensile test, whereas a negative sign (−) is used for the compressive test. Shear deformation (ΔL) is demonstrated in Figure 11.4. For small deformations, the shear strain (γ) is the ratio of the deformation (ΔL) to the height or thickness (H) of the specimen:

$$\gamma = \Delta L/H \tag{11.5}$$

11.2.3 Flow and Rate of Strain

Because most solid food materials, including surimi seafood products, are viscoelastic in nature, the rate at which they are loaded or deformed may significantly affect their mechanical response. Many testing machines that are used in food testing, such as the Instron Universal Testing Machine, TA Texture Analyzer, Bohlin rheometer, and the Brookfield or Haake viscometer, operate at one or more constant speeds. This produces constant deformation rates; that is, the displacement of the moving part is the same for any given lapse of time, irrespective of specimen dimension or its inherent characteristics. For deformation tests with a Universal Testing Machine, if the deformation is small and changes in the specimen length are small, a constant deformation rate can be used for an approximation of a constant strain rate, such as normal strain rate (ε') or shear strain rate (γ'). For larger deformation under a compressive test, the strain rate increases as the deformation increases. For small deformation, the normal strain rate is defined as:

$$\varepsilon' = \varepsilon/t \quad (11.6)$$
$$= (\Delta L/L)/t$$
$$= (\Delta L/t)/L$$
$$= v/L$$

where v is the cross-head speed of the machine (mm/sec), L is the initial length (mm) of the specimen, and t is the lapse of time that the cross head travels to deform the specimen. A positive sign (+) is used for axial tensile experiments and a negative sign (−) is used for compressive experiments.

The shear strain rate ($\dot{\gamma}$) for small deformation is defined as:

$$\dot{\gamma} = \gamma/t$$
$$= (\Delta L/H)/t$$
$$= (\Delta L/t)/H$$
$$= v/H \quad (11.7)$$

where H is the height (mm) of the specimen.

A sinusoidal type of deformation is imposed on a specimen during dynamic testing. Dynamic testing experiments commonly subject a material to small-amplitude sinusoidal strains (γ) as a function of time (t):

$$\gamma = \gamma_0 \sin(\omega t) \quad (11.8)$$

where γ_0 is the maximum strain amplitude and ω is the angular frequency expressed in radians/second. The shear strain rate ($\dot{\gamma}$), the first derivative of strain with respect to time, is given by:

$$\dot{\gamma} = \gamma_0 \omega \cos(\omega t) \quad (11.9)$$

11.2.4 Rheological Tests Using Small Strain (Deformation)

Small strain rheological measurements (strains less than required for structural breakdown) do not directly produce food texture data, but instead monitor physical property changes in the gel that relate to molecular changes.

11.2.4.1 Compressive Test for Surimi Gel

A few basic requirements are fundamental for compressive tests of surimi gels. They are:

1. Apply a truly concentric or axial load so that bending stresses are not set up because of irregularities in alignment.
2. Avoid friction between the end surface of the specimen and the bearing plate of the testing machine due to expansion of the specimen.
3. Select a specimen with such a length-to-diameter ratio that a proper degree of stability is obtained and buckling is avoided. The modulus of elasticity was found by taking the ratio of conventional stress to conventional strain as follows:

$$E = (F/A)/(\Delta L/L_o) \qquad (11.10)$$

Considering the fact that in biological materials a part of the deformation is always nonrecoverable, the slope of the stress–strain curve is taken assuming that the slope is "approximately" linear (Figure 11.5).

11.2.4.2 Shear Type Test for Surimi Paste and Gels

Heat-set surimi gels exhibit both viscous and elastic properties. Measurement of these properties can yield data that, when properly interpreted, increases our understanding of the chemistry, that is, changes during processing events involving mechanical, thermal, or chemical reactions. As the solubilized myofibrillar proteins of surimi bond with one

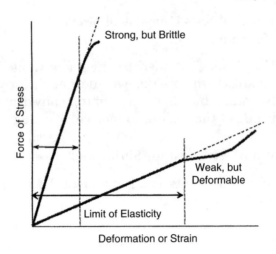

Figure 11.5 Typical stress–strain behavior of surimi gels under compression with a small strain.

another under the influence of heat, liquid movement within the material is altered. Because of the changing solute concentration, the viscosity of liquid can also change. Such changes affect the mechanical properties of the hydrogel so that more or less force may be required to deform it, and a different proportion of the deformation energy will be inelastic and dissipated as a heat.[2]

In the case of fluids behaving in a viscous manner, the viscosity (η) of surimi paste is:

$$\eta = \text{shear stress } (\tau)/\text{shear rate } (\dot{\gamma}) \text{ (Pa·sec)} \quad (11.11)$$

For solids (surimi gels), shear strain is assumed small and shear modulus (G) is defined as:

$$G = \text{shear stress } (\tau)/\text{shear strain } (\gamma) \text{ (Pa)} \quad (11.12)$$

G is also referred to as the modulus of rigidity.[7]

The flow properties of surimi paste have been used as a measure of quality and machinability as affected by certain processing conditions. Flow behavior can be expressed in terms of shear stress at a specific shear rate, as shown in

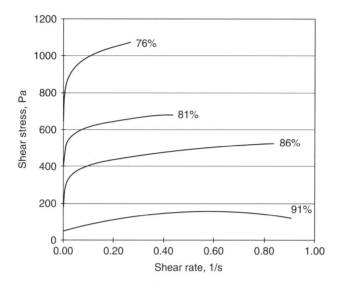

Figure 11.6 Changes in the viscosity of surimi paste as a function of moisture content. All samples were prepared at 2% salt.

Equation 11.11. It is of interest to determine shear stress vs. shear rate changes in surimi paste caused by various moisture and salt concentrations (Figure 11.6 and Figure 11.7). Measurements were carried out at 5°C isothermal. The flow behavior of sols made from Alaska pollock changed significantly with increasing moisture and salt content. The apparent viscosity, obtained from shear stress divided by shear rate, decreased due to moisture and salt being added. Because the plot of shear stress vs. shear rate was not a straight line and did not pass through the origin, the raw surimi paste demonstrated a non-Newtonian flow behavior with a significant yield stress.

11.2.4.3 Stress Relaxation Test

In the stress relaxation test, the specimen is suddenly brought to a given deformation (strain) and the stress required to keep the deformation constant is measured as a function of time. The recorded relaxation curve is in the form of a force vs.

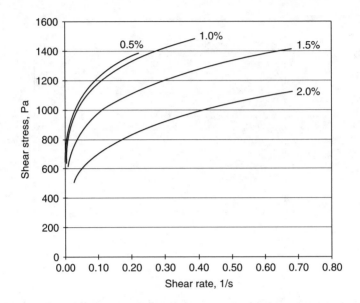

Figure 11.7 Changes in the viscosity of surimi paste as a function of salt content. All samples were prepared at 76% moisture.

time, force vs. stress, or modulus decay vs. time relationship. An ideal elastic body, however, does not relax at all. In contrast, an ideal viscous body cannot maintain any stress in the absence of motion, so it will relax instantaneously (Figure 11.8). In viscoelastic materials, like surimi gels, the effect is intermediate. The stress relaxes at a definite, although not necessarily, constant rate.

The observed mechanical relaxation phenomenon is a result of structural and molecular reorientation. The relaxation modulus $\sigma(t)$ is known to be a monotonously decreasing function of time and can be written as:

$$\sigma(t) = \sigma_e + (\sigma_o - \sigma_e)e^{-t/\tau} \qquad (11.13)$$

where
$\sigma(t)$ = stress at time t,
σ_e = equilibrium stress at time $(t = \infty)$
σ_o = initial stress
e = Napierian logarithm base (2.72)

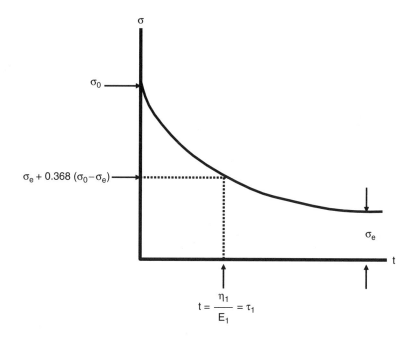

Figure 11.8 Stress relaxation curve for a typical viscoelastic surimi gel.

τ = relaxation time, the time for the decay to reach (1/2.72) of ($\sigma_o - \sigma_e$) or accomplish 63.2% of its decay

Equilibrium stress (σ_e) represents the "degree of solidity" of the material; therefore, it assumes a zero value for a liquid-like material. This can also serve as a demonstration that elasticity, viscoelasticity, and viscosity form a continuous rheological domain. A further distinction can be made between a viscoelastic solid, which contains some permanent cross-links, and a viscoelastic liquid, in which no permanent cross-links are present. Viscoelastic solids, like surimi gels, show an initial compressed stress decay (σ_o) and reach equilibrium stress (σ_e) under a rapid axial compression.

Stress decay plots for a surimi gel heated in two ways are shown against relaxation time (Figure 11.9): either (1) cooked at 90°C for 15 min, or (2) set at 25°C for 2 hr and then

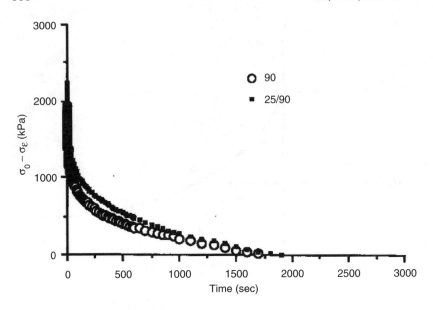

Figure 11.9 Stress relaxation curves of surimi gels made by different thermal processing. One cooked at 90°C for 15 min and the other heated at 25°C for 2 hr, then cooked at 90°C for 15 min.

cooked at 90°C for 15 min. The 25/90°C heat-treated surimi gel had a higher initial stress and was firmer than the sample cooked directly at 90°C (Table 11.1).

11.2.4.3.1 Rheological Model for Stress Relaxation

A rheological model is a convenient way to represent data and shed some light on the time scale of the molecular mechanisms contributing to the viscoelastic response. A three-element Maxwell model is used to represent the viscoelastic response to a step strain because the use of more complex models makes interpreting the data more difficult (Figure 11.10).

Linear viscoelastic material, where the ratio of stress to strain is a function of time and not the stress magnitude, can be described by combining the Hookean (elastic) and Newtonian (viscous) bodies. Table 11.1 provides a comparison of the

TABLE 11.1 Comparison of Three-Element Maxwell Model Constants Derived from Stress-Relaxation Data of Surimi Gels

Constant	90°C	25°C/90°C
Initial stress (σ_o), kPa	3.9	5.5
Equilibrium stress (σ_e), kPa	1.91	3.23
Elastic element (E_1), kPa	19.5	22.3
Elastic element (E_e), kPa	19.1	32.3
Viscous element (η), kPa.sec	2731	4453
Relaxation time (τ_{rel}), sec	140	200

Note: 90°C: Gels were prepared at 90°C for 15 min cooking. 25°C/90°C: Gels were prepared at 25°C pre-incubation for 2 hr, followed by cooking at 90°C for 15 min.

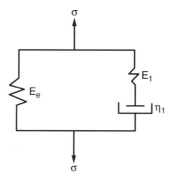

Figure 11.10 Schematic diagram of the three-element Maxwell model.

three-element model values of surimi gels cooked in two different ways. The gels set at 25°C before heating at 90°C produced a higher elastic element, viscous element, and a longer relaxation time. The viscous element (η) relieves stress in the Maxwell portion of the model. This energy is lost as heat until the equilibrium elastic axial compression modulus is reached. A higher ratio of viscous constant to elastic constant in the Maxwell element of the 25/90°C gel is reflected in a 40% longer relaxation time, as compared to surimi gel heated at 90°C. The high viscous element in the 25/90°C heated gel may be due to the better alignment of protein chains and from the dissociation and re-association of noncovalent bonds as the protein chains pass each other.[2]

11.2.4.3.2 Relationship between Rupture Properties and Stress Relaxation

Small deformation tests, such as stress relaxation, relate to texture in that the axial compression modulus (E) and shear modulus (G), as well as the relaxation time, tend to increase as the number of cross-links in the food increases. Because it is reasonable to assume that on rupture the network will break at these cross-links, it is expected that the rupture strength will increase with the apparent modulus. This is not always the case, however, as rupture and non-rupture tests depend on the primary molecular weight of the polymer involved in different ways. Thus, large (rupture) and small (non-rupture) deformation tests may not rank a series of gels in the same order.[8] In fact, Wood[9] showed that the consumer assessment of the hardness of gels correlated with the rupture strength rather than the elastic modulus. In general, modulus values from small strain tests are not good predictors of rupture stress because protein degradation may not influence modulus values, while strongly influencing rupture values.[10]

11.2.4.4 Oscillatory Dynamic Test

Despite the simplicity of creep and stress relaxation, oscillatory dynamic measurements need scales to obtain complete information about the viscoelastic behavior of a material. In addition, chemical and physiological changes may affect the physical behavior of biological materials during the experiment. This disadvantage can be overcome by dynamic mechanical tests where the specimen is deformed by a stress that varies sinusoidally with time. Because a periodic experiment at an angular frequency (ω) is equivalent to tests at time ($1/\omega$), it is possible to provide rheological information corresponding to very short times.

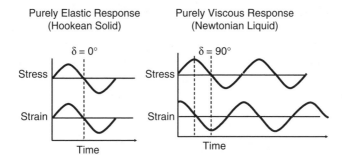

Figure 11.11 Comparison of the shear stress response of an elastic solid (left) and a viscous liquid (right) under an oscillatory shear strain.

11.2.4.4.1 Viscoelastic Properties of Materials

The dynamic tests allow the calculation of the viscoelastic properties of material such as the elastic modulus and mechanical damping over a wide range of frequencies. Mechanical damping is associated with the energy loss, heating, and toughness of the material. It also provides the molecular structure and chemical composition of high polymers.

When a linear viscoelastic material is subjected to periodically varying stress, the strain will also vary periodically, but out of phase with the stress (Figure 11.11). This behavior results in a complex frequency-dependent modulus (G^*) that can be separated into an in-phase or real component (G'), which is associated with the storage of energy, and an out-of-phase or imaginary component (G''), which is associated with the loss of energy. A complex number coordinate system is often used for mathematical clarity and easy manipulation. The y-axis is the imaginary axis and the real part of the number is represented on the x-axis.

$$G^* = \sigma_{max}/\varepsilon_{max} (\cos \delta + i \sin \delta) = G' + iG'' \quad (11.14)$$

where i is the imaginary number, G' is the storage modulus, G'' is the loss modulus, and G^* is the complex modulus.

The complex modulus G^* is experimentally given by the ratio of peak stress to peak strain in the sinusoidal stress–strain curve. The ratio ($\sigma_{max}/\varepsilon_{max}$) is the absolute modulus ($|G^*|$) and this is a measure of the total unit shear resistance to deformation. The phase angle δ between stress and strain can be obtained from the sinusoidally varying curves:

$$\delta = \omega \Delta t \qquad (11.15)$$

where ω is the angular frequency (radians/sec) and Δt is the shift in time of the peak values of the stress and strain curves.

For simplicity, experimental results from oscillatory shear are often presented in terms of G' and G'' or $\tan \delta$ as functions of temperature at a constant frequency or as functions of frequencies at a constant temperature. Most food gelation studies have been performed at a constant frequency (often less than 1 Hz) with temperature and/or time varying.[11] Other known relationships include:

$$G' = |G^*| \cos \delta$$
$$G'' = |G^*| \sin \delta$$
$$G''/G' = \tan \delta \qquad (11.16)$$

This is proportional to the energy dissipated/stored per cycle. Specifically, the energy dissipated due to viscous behavior per cycle, divided by the maximum elastic energy stored, is $2\pi \tan \delta$.[12] A perfectly elastic material would exhibit $\delta = 0°$, whereas for a perfectly viscous material, $\delta = 90°$. In the latter case, $G' = 0$ and all the work of deformation is converted to heat. Surimi and other protein gels are normally quite elastic so values of δ are less than $10°$.

The price of commercial instruments for this dynamic test is generally high, and any purchased instrument should be designed to cover the desired shear stress and strain rate ranges. Modern rheometers distinguish between viscous and elastic resistance to forces, as well as quantify the magnitudes of resistance, deformation, and rate of deformation. Bohlin et al.[13] used a specially designed instrument (model VOR; Bohlin Reologi AB, Lund, Sweden) to continuously measure the time-dependent viscoelastic properties of coagulating milk. The

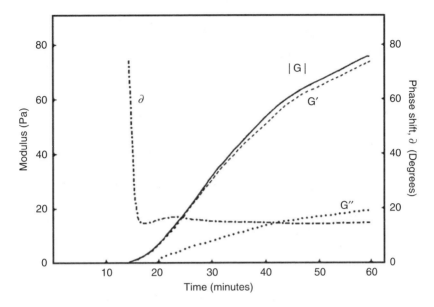

Figure 11.12 The secondary phase of milk coagulation at standard conditions: 31°C, 0.1 g $CaCl_2$/kg milk, and 0.3 ml rennet/kg mild; frequency, 0.5 Hz; shear strain amplitude, 0.075. (*Source:* From Ref. 21. With permission.)

instrument employed a concentric cylinder geometry with the angle of rotation kept small and changing sinusoidally. Absolute values of the shear modulus and the energy loss phase angle for various calcium chloride and rennet concentrations were determined as functions of time during the sol-to-gel transformation. The transition from sol to gel was evident by the sudden drop in the phase angle from near 80° (purely viscous) to near 20° (elasticity dominant) (Figure 11.12). The structures that changed the material from viscous to elastic formed almost instantaneously. At this point, the structure was barely rigid enough to yield a discernible value of G′, but G″ increased slowly with time.

This is similar to what other researchers have observed for Pacific whiting myosin (Figure 11.13). In the case of Pacific whiting myosin, the change from viscous to elastic behavior was fairly complete at 30°C, whereas most of the increase in

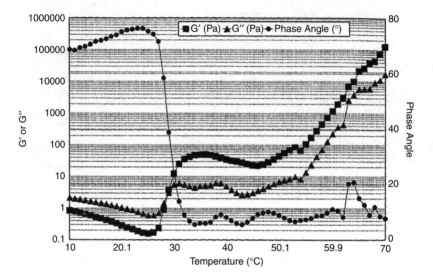

Figure 11.13 Dynamic test curves of Pacific whiting myosin.

G' occurred after 30°C, and continued to increase up to 70°C. It was also noteworthy that the phase angle dropped to below 10°, indicating a highly elastic behavior for high-quality whiting myosin. In the case of muscle sol, it exhibited some elasticity before cooking; but it is a viscous material, so a typical phase angle would be 30 to 45°, depending on the moisture content. When fish myosin solution was subjected, the phase angle was initially about 70° and then dropped significantly to 10° (Figure 11.13). The transition from a sol to a gel is evident from the changes in δ.

11.2.4.4.2 Stress and Frequency Sweep

A rheological study of fish protein gelation can be done using an oscillatory test mode (Figure 11.14) while the temperature varies during the gel forming history. It is important that the test be conducted under small stress so that data is obtained in the linear viscoelastic range. Less than 100-Pa stress should be applied to Alaska pollock surimi sol (12.08% protein content) at 10°C, while any stress up to 1000 Pa can be used in the same sample stored at 80°C within the linear viscoelas-

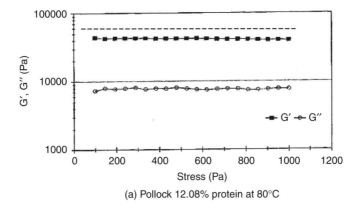

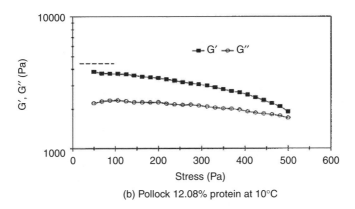

Figure 11.14 Dynamic properties of Alaska pollock surimi sol (12.08% protein concentration) at constant frequency amplitude as affected by temperatures of (1) 80°C and (b) 10°C. Dotted line indicates the linear viscoelastic region.

tic range (Figure 11.14). Cooked gels require higher stress for structural deformation.

The linear viscoelastic region was extended significantly as the moisture concentration decreased from 81 to 75% (Figure 11.15). This indicated that increasing protein concentration increased the required stress for structural deformation. The linear region ranged from 20 to 150 Pa, 20 to 75 Pa, and 20 to 46 Pa for 75, 78, and 81% moisture, respectively.

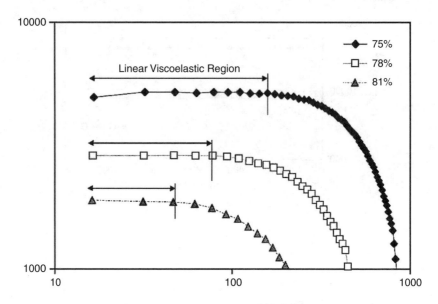

Figure 11.15 Effect of stress on G′ of Pacific whiting surimi as affected by moisture content at 20°C and 0.1 Hz.

Most food gelation studies were performed at constant frequency (often less than 1 Hz) with temperature and/or time varying.[13–16] Egelandsdal et al.[16] found significant effects on G′ and δ of myosin gels when the strain amplitude varied from 0.003 to 0.1. It is necessary, however, to know and record instrument-controlled variables, including oscillation frequency and strain amplitude. G′ and G″ data, as a function of frequency, can be used to define the important parameters affecting the formation of gel, such as gelling temperature and time. Surimi pastes (sol) with low protein concentration show a crossover point at which G′ equals G″. Surimi gels cooked at 90°C show the same proportional change for G′ and G″ in relation to frequency, over a wide range of frequencies (Figure 11.16).

11.2.4.4.3 Temperature Sweep

Storage modulus (G′) values of Pacific whiting surimi paste during the temperature sweep are shown in Figure 11.17. There

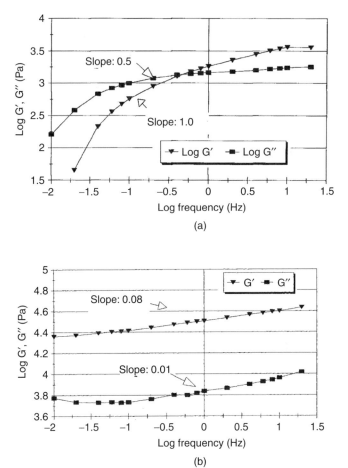

Figure 11.16 Dynamic properties (G', G") of Alaska pollock surimi sol (A) and gel (B) as a function of frequency.

was a small drop of the G' value starting at 30°C. G' reached its lowest value at around 45°C, and then increased rapidly until reaching 60 to 70°C. Thermal gelation was completed at the temperature where G' reached a maximum plateau. When fish myosin solution was subjected to the temperature sweep (Figure 11.13), the gelation profile was the same as those of surimi paste. Myosin gels were typically formed at two major stages

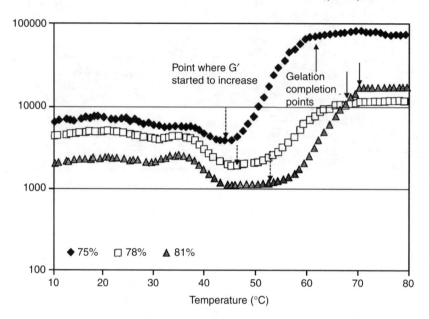

Figure 11.17 Changes in storage modulus (G′) of Pacific whiting as function of moisture content (75, 78, 81%).

during heating. At 30 to 40°C, α-helices in the tail portion of myosin molecules unfold; and then, the unfolded hydrophobic regions interact with each other above 50°C.

As G′ started to increase, the moisture content also affected the rate of formation of G′. The higher the moisture content, the slower the G′ formation rate. The G′ of surimi paste reached its lowest value, regardless of the moisture concentration at approximately 45°C. While G′ of surimi paste containing 75% moisture increased immediately after reaching its lowest value, G′ started to increase at 52°C for surimi containing 81% moisture.

Manufacturing meat products is traditionally accomplished by comminution with salt and other functional ingredients, as well as thermal heating. This process determines the texture of meat products through protein–protein or protein–water interactions. A new binding system in restructuring portioned meats with nonthermal treatment has been

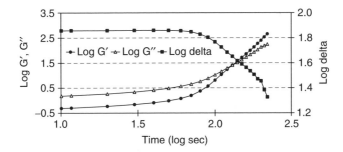

Figure 11.18 Dynamic properties of fibrinogen–thrombin mixture (20:1) at 5°C holding temperature.

recently developed using fibrinogen and thrombin extracted from bovine plasma proteins. Because this mixture creates an adhesion system and is very reactive, the gelation time of a fibrinogen–thrombin solution at low temperatures is determined by monitoring changes in the elastic and viscous element. From the beginning of the reaction time up to 630 sec during holding at 5°C, the magnitude of G″ was much greater than G′. This order, however, was reversed as the reaction time was extended. The crossing point of G′ and G″ is called the "gel point" of the cross-linking polymer and is an indication of the transition from a viscoelastic liquid to a solid state.[17–20] After the gel point, all protein aggregates instantaneously bind together into one continuous molecular structure. As the holding temperature increased from 5 to 35°C, the sol–gel transition point of the fibrinogen–thrombin mixture changed from 630 to 120 sec. This trend indicates that gelation of fibrinogen–thrombin is temperature dependent.[21] Overall changes in gelation times and modulus values at G′ = G″ are compared as a function of temperature (Figure 11.18). Best results are obtained at low frequencies because at low strain rates, the molecular properties are elucidated.[11]

Changes in the dynamic properties of Alaska pollock surimi paste are shown in Figure 11.19. A cone and plate Bohlin dynamic tester was used at a frequency of 0.1 Hz and shear stress amplitude of 350 Pa. The temperature history was as follows: 1°C/min and 0.5°C/min rise from 10 to 80°C.

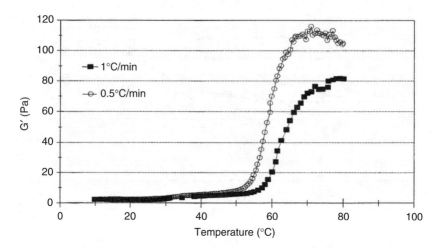

Figure 11.19 Dynamic properties of Alaska pollock surimi sol-to-gel transition at two different heating rates. Frequency fixed at 0.1 Hz.

Intermolecular disulfide bonds increase at the higher temperatures,[18] and it can be speculated that the decrease in slope near 70°C is due to the completion of most covalent bonding.[22]

11.2.5 Rheological Testing Using Large Strain (Failure Test)

Texture is a critical preference factor for most surimi-based foods. Large deformation testing of surimi gels has been found useful in the objective measurement of the texture of such materials. In this application, an important goal is to relate the instrumental measurement to the sensory experience of mastication. Texture is a multifaceted concept, and it is unlikely that measurements made from a single instrument can adequately describe all the textural characteristics of a specific food form. However, if considerations are limited to those characteristics that have been termed "mechanical," instrumental results can be correlated with sensory textural attributes quite well. Mechanical characteristics are those "manifested by the reaction of the food to stress" and include, among others, hardness (force required to bite through), cohesiveness (deformation of the sample by the teeth before breaking), and springiness (rate at which a deformed food piece returns to its original shape).

11.2.5.1 Axial Compression for Cylinder Type Gels

Axial compression of a cylindrical specimen between parallel plates causes the specimen height to decrease and its diameter to increase. Using the undeformed height and diameter in equations, based on the assumption of small deformation, is inappropriate if the ratio of the decrease in height divided by the initial height is larger than 0.1. Most surimi gels are very deformable and this ratio can approach unity. Therefore, if fundamental units, independent of a specific test, are used to specify fracture conditions, equations based on larger strain conditions should be used. Uniaxial compression or tension of specimens of known size and shape are the most common tests used that yield stress and strain at failure. If the specimen is in the form of a cylinder, the computations are simple, with the maximum axial compressive stresses and strains being:

$$\sigma_{max} = F/\pi r^2 \qquad (11.17)$$

$$\varepsilon_{max} = \Delta L/L_o \qquad (11.18)$$

At levels for failure to occur, the change in length and cross-sectional area can be large. Continuously modifying the L_o in the denominator for each change in length (ΔL) is used to calculate a strain called "true strain."

$$\sigma_{true\ max} = F/\pi r^2\ (1 + \upsilon\varepsilon)^2 \qquad (11.19)$$

$$\varepsilon_{true\ max} = -\ln\ (1 - \varepsilon_{max}) \qquad (11.20)$$

where υ is Poisson's ratio.

Although a uniaxial compression test may not be the best test for obtaining data on fracture stress and strain conditions in terms of fundamental units, it is nonetheless an easy test to perform. If care is taken, valid information can be obtained for soft gels with a value of shear strain up to around 1.0. Montejano et al.[23] evaluated the fracture of fish gels at room

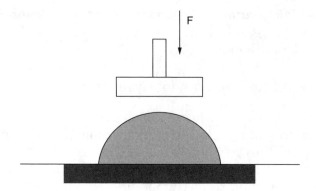

Figure 11.20 Application of convex shape in compression.

temperature using axial compression and compared results with data from the torsion method. Results from the two methods were found to be in general agreement. One critical point using axial compression test is that elastic surimi gels are not fractured even when 99% deformation is applied. The gels are simply spread out flat without fracture.

11.2.5.2 Compressive Test for Convex Shape Samples

The purpose of this section is to determine the mechanical attributes of gel texture, resistance to mechanical injury, and force-deformation behavior of convex-shaped food materials, such as fruits, vegetables, seeds, grains, and fish balls (Figure 11.20). Compression tests of intact, biological materials provide an objective method for determining mechanical properties that are significant in quality evaluation and control, maximum allowable load needed for minimizing mechanical damage, and minimum energy requirements needed for size reduction. Determination of compressive properties requires the production of a complete force-deformation curve. From the force-deformation curve, various rheological information can be obtained. These include: stiffness, modulus of elasticity, modulus of deformation, force and deformation to the point of inflection (or to bioyield and to rupture), work to the point of inflection (or to bioyield and to rupture), and maximum normal contact stress or a stress index at low levels of deformation. Based on the work of Hertz,[24] an equation derives for contact stresses of biomaterials having spherical or other ellipsoidal bodies:

$$E = \frac{0.531 \, F(1-v^2)}{D^{3/2}} [(1/R_1 + 1/R_1')^{1/3} + (1/R_2 + 1/R_2')^{1/3}]^{3/2} \quad (11.21)$$

where D = deformation, v = Poisson's ratio, E = Young's modulus, R_1 or R_1' = curvature of top, and R_2 or R_2' = curvature of bottom.

If a single plate contact is used, the last R_2 terms should be eliminated; and if an indent is used, 1/d (indent diameter) should be added to the above equation. The Hertz theory has been shown to give good results when combined with a Universal Testing Machine, such as Instron, Sintech/MTS, or a penetrometer.

11.2.5.3 Compressive Test for Rod-Type Samples

The mathematical theory of beam curves can be used for the elastic bending of a rod-type food (Figure 11.21). The rod-type biomaterial, such as forage, some vegetable stalks, rectangular bars, and surimi seafood (crabstick), are limited in their ability to produce a force-deformation curve. Knowing the load and deflection, the following formula for a simple beam can be used to calculate an apparent stiffness or modulus of elasticity. Before demonstrating the equation, it is necessary to examine some of the fundamental assumptions that were originally made in beam theory. The body of the material must be homogeneous and isotropic. Hooke's law can be approximated at very low levels of load, and stress is recovered when it is unloaded. Finally, compressive and tensile phenomena occur at the same time when the sample is bent.

The mathematical expression for the beam curve is:

$$E \, I \, d^2y/dx^2 = -M \quad (11.22)$$

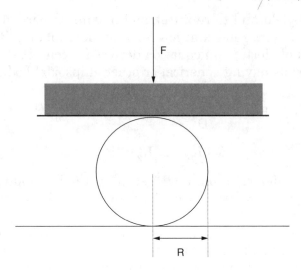

Figure 11.21 Application of rod shape in bending.

where y = downward vertical displacement, d^2y/dx^2 = 2nd derivative of y with respect to x, E = modulus of elasticity, I = moment of inertia (given by I= $\pi r^4/4$ for a beam with circular section) and M = torque momentum. A negative (−) sign means that deflection occurred downward. The value of y, using the boundary condition (y = 0 at x = 0, dy/dx = 0 at x = a half the length of rod) and maximum deflection at the center, the modulus of elasticity (E) and failure stress (σ) can be calculated as follows:

$$E = P\,l^3/(48\,I\,\delta) \tag{11.23}$$

$$\sigma = P\,l/\pi R^3 \tag{11.24}$$

where P = force, δ = maximum deflection, R = radius of rod, and l = total length of rod.

When food materials in the form of rod bodies are considered, the bending theory can be used to determine the modulus of elasticity and failure stress. Moreover, methods for calculating the above equation can vary, depending on cross-sectional area and the shape of the rod, such as circular, rectangular, etc.

11.2.5.4 Torsion Test

For moderately deformable gels, shear stresses at failure calculated from fracture forces during axial compression[25] agree with shear stresses calculated from torsion testing.[4] Shear strains at fracture calculated from axial compression deformation are also in agreement with those calculated from the angle of twist in torsion. However, at shear strains higher than about 1.2, axial compression shear stresses and strains are in error because of excessive change in the shape of the specimen that cannot be easily compensated for. Even when the sample fails at shear strains below 1.2, numerical values from the axial compression tests may be somewhat different from values obtained in torsion due to various reasons[26]:

1. Change in shape being imperfectly accounted for in the axial compression specimen
2. Lack of perfect homogeneity and isotropy in the specimens with axial compression and torsion evaluating different locations and fracture planes
3. Fracture occurring in tension in the torsion test, but in shear in the axial compression test
4. Friction on the axial compression specimen ends, causing departure from a cylinder producing a barrel or hourglass shape

Stress and strain are often separated into those called "dilatational" and those called "deviatory." Those called dilatational cause volume changes and those called deviatory change shape, but not volume. This becomes important in studying material failure because some foods, particularly those that are incompressible, may not be sensitive to volume-changing stresses, but can be very sensitive to shape-changing stresses. If nearly incompressible materials, like a surimi gel product, are subjected to hydrostatic pressure, there will be no change in shape. If, however, they are deformed in shape, they eventually break. Shear stresses are deviatory; and if there is a condition called "pure shear," the overall effect is a change in shape with negligible change in volume. Torsion of

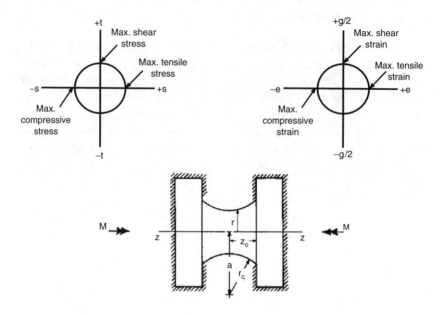

Figure 11.22 Stress and strain Mohr's circle torsion and critical dimensions of the torsion specimen.

a specimen having circular cross-sections produces pure shear (Figure 11.22).

The Hamann torsion gelometer (Figure 11.23) has given meaningful evaluations of a variety of food gels, including those prepared from surimi, egg, cheese, whey, gums, bologna, and frankfurters. Using the torsion (twisting) test, many of the uncertainties in calculating fundamental fracture parameters can be eliminated.[2] Consider the following advantages:

1. This test produces pure shear — a stress condition that does not change the specimen volume even if the material is compressible.
2. The specimen shape is maintained during the test, minimizing geometric considerations.

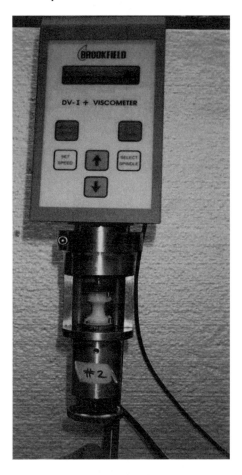

Figure 11.23 Hamann torsion gelometer.

3. Because of items 1 and 2, the calculated shear stress and strain are true values up to large twist angles (45°, equivalent to shear strain near 1.0). Applying a large twist angle correction method can be extended to shear strains of about 3.0.
4. There is no restriction on the criterion for fracture. The material can fail in shear, tension, compression, or a combination mode.

5. Principle (maximum) shear, tension, and compression stresses all have the same magnitude (Figure 11.22) but act in different directions.[7] Therefore, it is easy to determine if the material failed due to shear, tension, or compression.
6. Friction between the specimen and test fixture does not have to be considered.

From experimental results, the shear strain limit, for the torsion test, occurs when the specimen shape at the critical cross-section noticeably changes. For surimi gels, depending on thermal treatments, the shear strain can be as high as 3.0.[11] Gels are known for their elastic behavior and this is generally true for protein gels. Twist graphs of torque vs. angle are normally linear. Assuming a linear torque vs. angle of twist, a rotational rate of 2.5 rpm, a grove width of 1.2 cm, and a 1.0-cm cross-section diameter at the center of the groove, Hamann[27] developed a typical equation for fracture shear stress (τ) and uncorrected shear strain (γ), respectively:

$$\tau_f = 1580 \times \text{Torque in the Brookfield viscometer, (Pa)} \quad (11.25)$$

$$\gamma_f = 0.150, \text{sec}^{-1} \times \text{Time to fracture} - 0.00848 \times \text{Torque in the Brookfield viscometer (dimensionless)} \quad (11.26)$$

The last term in the strain equation is due to spring wind-up in the 5xHBTD Brookfield viscometer. For an instrument with a very stiff sensing element, this term would not be present. If shear strain is above about 1.0, γ should be corrected using an equation given by Nadai.[28] Returning to the notation that γ_t is the true shear strain of interest at fracture:

$$\gamma_t = \ln [1 + \gamma^2/2 + \gamma (1 + \gamma^2/4)^{1/2}] \quad (11.27)$$

Three planes of fracture in torsion are described in Figure 11.24. If a crack in the torsion gels is diagonal, it indicates a failure in tension. If the break is perpendicular to the cylindrical axis of the specimen, it is shear fracture.[27] The torsion test has been considerably accurate and well adapted in the United States as a method to measure the fracture properties of surimi gels. However, it requires extra time to

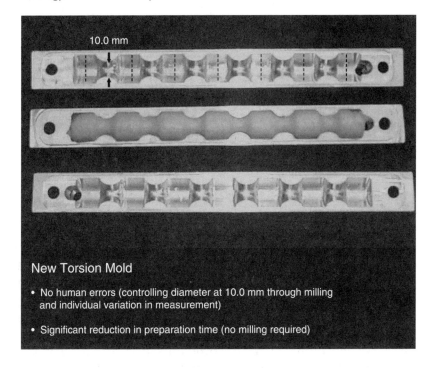

Figure 11.24 Torsion mold eliminating milling and measurement errors.

mill the sample specimens into a dumbbell shape with a precise 1.0-cm cross-sectional diameter at the center of the groove. Although not an easy task, maintaining an exact 1.0-cm cross-section diameter is required for the equations developed by Hamann.[34]

Shear stress values are extremely affected by the accuracy of the diameter. In the case of Alaska pollock surimi gels, stress values can fluctuate between 40 and 70 kPa for a range of cross-section diameters from 0.90 to 1.10 cm.[29] The effect of the cross-section diameter on shear strain is not as large as in shear stress. There is a trend, however, of a reduction in strain as the diameter increases. Therefore, it is extremely important to keep the cross-section diameter at 1.0 cm.

Another example of problems associated with the torsion method is the milling process. In addition to time consump-

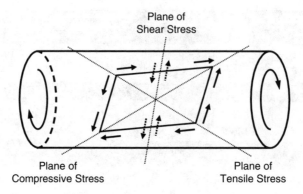

Figure 11.25 Three planes of gel fracture by torsion.

tion, there is the consideration of what effect the milling process has on the gels. In all previous studies, it has been assumed that there is no structural damage to the gels during milling. Hoffman and Park[29] also revealed a significant variation among individuals in measuring the diameter using a caliper. They eliminated the problems associated with the torsion test by using molded gels instead of milled gels (Figure 11.25). It was speculated, however, that molded gels might form a skin, resulting in inaccurate measurements. A previous study[30] investigated and found that skin formation of low-fat mozzarella cheese in pizza can be prevented by lightly coating a cheese surface with a hydrophobic material (i.e., Pam™ spray oil) prior to baking. Hoffman and Park[29] also assured that no skin was formed due to the use of a lecithin-based spray by coating the mold.

11.3 EMPIRICAL TESTS

Various types of sample geometries have been used for instrumental texture evaluations of muscle foods. Most somewhat imitate the cutting or chewing of the mouth; however, the data cannot be matched quantitatively with that from another testing geometry. These empirical tests can yield single or multiple measurements.

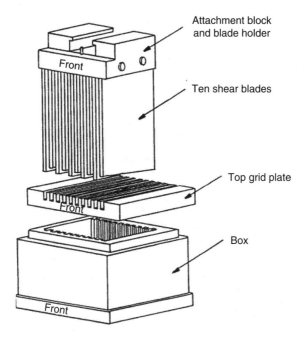

Figure 11.26 Kramer shear cell.

Choosing between the punch-and-die and the slot-and-blade test fixture depends on what information is desired. The punch-and-die method evaluates a small area, making it useful for showing differences due to location within a sample. This fixture, because it is symmetrical about its central axis, is not suitable for determining anisotropy. The advantages of a multiple slot-and-blade fixture, such as the Kramer shear cell (Figure 11.26), are that it can sense anisotropy and the test sample mass and area are large such that the influence of local irregularities is reduced.

11.3.1 Punch (Penetration) Test

The punch test, although considered an empirical test, is the single most popular gel measurement technique used in the surimi industry for evaluating "gel strength" or stiffness. The punch test imitates the large deformations to failure involved in mastication. Many studies have been reported that correlate puncture methods with the sensory properties of surimi gels. This attribute of the test, coupled with its convenience, has made it popular for quality control within the surimi industry.

The test was initially developed by Matsumoto and Arai[31] and later modified by Okada and Yamazaki.[32] The "Okada gelometer" became the standard instrumental method used in the Japanese surimi industry. In this test, a punch probe of a specific diameter (3.0 mm) and length (25 mm) is used to compress the surface of a gel specimen at a constant deformation rate (10 to 60 mm/min) until puncture occurs. Many of the modern penetrometers used in industry operate at a fixed 60 mm/min and a 5.0-mm probe is commonly used.

The recorded peak force (F) at break and the depth of penetration are used to describe the gel properties. Often, these two values are multiplied together to give the "jelly strength." The jelly strength is the value that is used in the Japanese grading standards. This type of measurement has been frequently used for surimi gel samples and offers good correlation with attributes such as first-bite hardness.

Although it is very simple and the most widely used for objective measurement within the surimi industry, several restrictions are given, as follows:

1. For this test to be adequate, either the range of cohesiveness must be small enough that it is not considered an important sensory note, or the hardness and cohesiveness of the samples are related in a consistent way so they are not independent. This is often true if protein concentration, species, filler ingredients, and process variables are invariant. The cohesiveness of most surimi seafoods varies significantly, depending on surimi quality, and cohesiveness and hardness vary independently.
2. The punch is usually conducted using a sphere (5 to 30 mm diameter) driven by a small-diameter shaft at a speed of 10 to 60 mm/min. A change in test variables can lead to different results for the same sample. For example, by increasing the downward speed of the

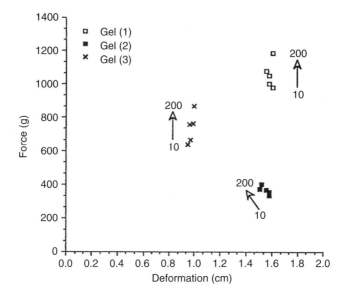

Figure 11.27 Effects of probe penetration speeds (10 to 200 mm/min) on force and deformation. (*Source:* Adapted from Ref. 2.)

probe (Figure 11.27), the puncture force increases. When the size of the probe increases (Figure 11.28), both the puncture force and deformation increase.

3. Fracture property measurements of foods are much less reproducible than small deformation parameters, and a coefficient of variation (standard deviation of the mean) on the order of 10% is common. For the punch test, which measures gel properties at a point of penetration into the gel, coefficients of variation can be even higher because failure takes place at defects in the sample where the number and extent of such defects may vary from sample to sample. Usually, firmer gels show a higher trend for increased error in the force value. Vacuum chopping may be one option to improve the precision of the test.

4. The use of the term "gel strength," which is often referred to as jelly strength in Japan, has misrepresented the quality of surimi. Gel strength, calculated

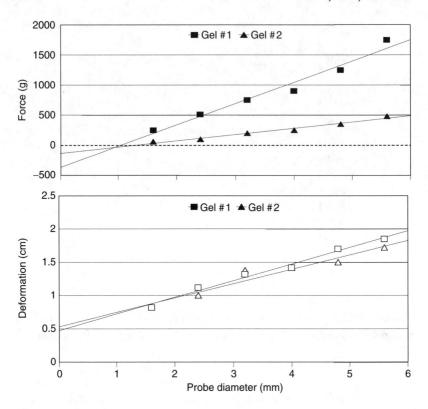

Figure 11.28 Effects of probe diameter on deformation and force values of gels. *(Source:* Adapted from Ref. 2.)

based on force multiplied by deformation using units of gram-centimeters (g.cm), does not provide any significant meaning to the rheological properties of gels. However, it has been arbitrarily (perhaps wrongly) used in the surimi industry as a symbol of surimi quality. Hardness (strength of gels) denotes breaking force values better than the scientifically wrong term, gel strength.

As illustrated in Figure 11.29, it is obvious that five different gels could have the same gel strength (i.e., 960 g.cm), but the protein quality of the gels is significantly different. When located in a texture map

Rheology and Texture Properties of Surimi Gels

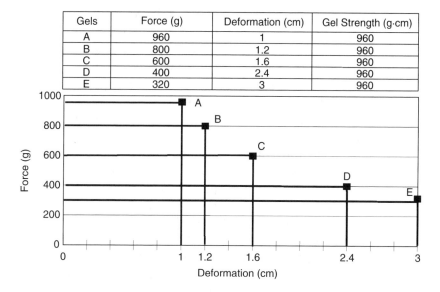

Figure 11.29 Five different surimi gels with the same gel strength (jelly strength).

based on force and deformation, they are extremely different. Therefore, if all surimi is equally priced, the purchase of surimi "E" would be ideally the best value because deformation indicates the quality of surimi proteins. Because force values depend on the quantity of proteins (inversely moisture), the purchase of surimi "A" would give the least value. Therefore, force and deformation values must be expressed individually, and not in the form of gel strength (or jelly strength, g.cm), to indicate the gel functionality of surimi.

5. Most punch test units used worldwide in the surimi industry are made by two leading companies in Japan. These conventional instruments are easy to operate and give fairly decent results. However, when gels are too soft or too elastic, wrong values can be detected. When soft gels are subjected, the machine is not sensitive enough to detect the point of pene-

tration. When very elastic gels are subjected, the shaft of the probe is touched by the gel matrix before breaking, resulting in a larger area of contact (like using a larger-diameter probe) and giving higher force values (see Figure 11.28). These conventional instruments cannot be calibrated on site and must be shipped to the manufacturer in Japan for calibration services.

Park and colleagues, along with the U.S. surimi industry leaders, evaluated 12 samples of surimi gels with a wide range of quality using five units of conventional rheometer or rheotex and two units of TA XT plus Texture Analyzer (Figure 11.30). As shown in Table 11.2, soft gel (B5LF) could not be measured using any of the five conventional units, while two units of TA XT Plus were able to detect the gel values. Indeed, the accuracy of the measurement, based on the standard deviation, was much higher with the TA Plus instrument. This accuracy might have been due to the ability to calibrate the instrument as needed.

11.3.2 Texture Profile Analysis (TPA)

A group at the General Foods Corporation Technical Center pioneered the development of the texture profile analysis. Their test involved compressing a bite-size piece of food, a cube approximately 1 cm, to 25% of its original height (75% compression) two times in a reciprocating motion, which imitates the action of the human jaw. From the resulting force–time curve, a number of textural parameters that correlate well with the sensory evaluations of those parameters were extracted.[1]

The TPA has been widely used for the empirical determination of a number of textural attributes of muscle foods and surimi gels. TPA involves the repeated compression of a sample to its original height between two parallel surfaces recording force vs. displacement (Figure 11.31). The maximum force of the first compression determines hardness, and the ratio of the area under the second cycle compression curve

Figure 11.30 TA XT Plus texture analyser. (*Source:* Courtesy of Stable Micro System, U.K.)

to the area under the first cycle compression curve determines cohesiveness. The hardness and cohesiveness of TPA can be expected to correlate with the sensory texture profile evaluation. Montejano et al.[4] found a correlation of r = 0.74 and r = 0.81, respectively, for a set of eight food gels, including two surimi gels. The procedure works for cohesion because gels that are more cohesive are springier and fracture into larger fragments, thus requiring a relatively high energy for the second compression. A problem with using TPA on surimi products is that they may be so cohesive that they do not

TABLE 11.2 Comparison of Accuracy in Empirical Measurement

Sample Code	Conventional				TA Plus			
	Mean	Std. Dev.	% CV	Range	Mean	Std. Dev.	% CV	Range
B1PW	13.66	1.3	9.55	2.97	11.56	0.2	1.70	0.28
B5LF		No Detection			4.87	0.19	3.93	0.27
B3AP	15.82	1.14	7.18	2.69	13.29	0.79	5.98	1.12
B8PW	15.83	1.56	9.82	3.69	13.42	0.04	0.29	0.05
B6AP	12.35	1.27	10.31	3.19	9.24	1.36	14.69	1.92
B9PWNAP	13.59	2.31	16.98	5.0	11.17	0.09	0.82	0.13
B2BE	12.63	1.99	15.80	4.85	10.05	0.21	2.11	0.3
B10LFPP	11.36	2.83	24.90	6.31	7.85	0.21	2.70	0.3
B11AP	15.93	1.76	11.04	4.28	13.44	0.06	0.42	0.08
B4TB	12.63	1.85	14.67	4.36	9.57	0.06	0.65	0.09
B12PW	15.52	2.27	14.60	5.8	12.15	0.49	4.07	0.7
B7AP	14.12	1.69	11.94	3.7	10.31	0.56	5.39	0.79
			Mean 13.34	Mean 4.26			Mean 3.56	Mean 0.5

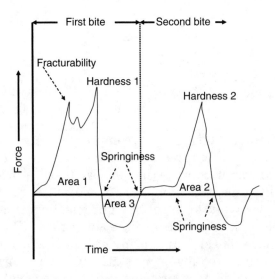

Figure 11.31 Texture profile analysis (TPA) curves.

fracture and fragment, even when compressed to less than 10% of their original height.

11.3.3 Relationship between Torsion and Punch Test Data

Knowing stress and strain conditions that cause breaking are important because they relate to sensory texture. Gels made from 14 different surimi lots using three different processing conditions[33] were subsequently evaluated by sensory measurement and each of three instrumental methods: TPA, punch, and torsion (Table 11.3). Torsional values (shear stress and shear strain) were more highly correlated with the sensory terms firmness and cohesiveness than values from the punch test and TPA.

TABLE 11.3 Correlation of Instrumental Parameters with Selected Sensory Terms for Standard Gels Made from 14 Lots of Surimi

		Cooking Method Parameters		
		40/90°C	60°C	90°C
Torsion	Stress, kPa	70.90[a]	44.70[b]	60.70[c]
	Strain, m/m	2.54[a]	2.09[b]	2.40[c]
	Rigidity, kPa	27.90[a]	21.40[b]	25.30[c]
Punch	Force, gms	815.00[a]	500.00[b]	572.00[c]
	Deformation, cm	1.53[a]	1.28[b]	1.37[c]
	Stiffness, gm/cm	532.00[a]	417.00[b]	418.00[b]
ITPA*	Hardness, N/gm	29.90[a]	24.70[b]	26.90[c]
	Cohesiveness, m²/m²	0.50[a]	0.46[b]	0.49[a]
Sensory**	Firmness	9.30[a]	6.80[b]	8.60[c]
(14-pt. scale)	Cohesiveness	8.80[a]	6.10[b]	8.00[a]
	Gel persistence (at 5 chews)	9.90[a]	7.50[b]	9.30[a]
	Rigidity***	10.10[a]	8.20[b]	9.50[c]

Note: Different superscripts denote a significant difference in means ($p < 0.01$)

* Instrumental (Instron) texture profile analysis.
** Texture profile panel.
*** Evaluated with the fingers.

Source: Adapted from Ref. 40.

The three instrumental methods deform the gels beyond the point of fracture, as occurs in sensory evaluation. The tests usually showed significant differences in the same direction between the processing treatments. The torsion specimens, however, do not undergo major changes in shape or volume. Therefore, unit shear stress and unit shear strain can be calculated directly from the twisting moment and angle of twist, respectively. In contrast, punch penetration by a solid cylinder can be accompanied by gross shape changes if the penetration force is large and/or the sample is too elastic. The influence of these changes on test results cannot be easily determined quantitatively;[34] thus, unit stress and strain cannot be accurately computed from punch test measurements.

Alaska pollock gels (>1000 samples) were prepared with 2% salt and 78% moisture adjustment and analyzed using the torsion and punch tests. As illustrated in Figure 11.32 and Figure 11.33, an interesting linear relationship was found.[35] Based on the torsion data, force and deformation values were obtained using the following equations:

1. Shear stress (X) → Force (Y):

$$Y = 5.433X + 90.239 \qquad (11.28)$$

2. Shear strain (X) → Deformation (Y):

$$Y = 0.266X + 0.345 \qquad (11.29)$$

The punch and torsion tests are the methods most widely used to determine the quality of surimi gels, both in the surimi industry and in academia. The Codex Alimentarius Commission of the Food and Agriculture Organization of the United Nations and the World Health Organization[36] have officially accepted these two methods (see Appendix). The commission allows surimi buyers and sellers the option to choose between the two methods. It is important to know that fundamental errors can be made with the punch test, while operational uneasiness is always a concern with the torsion test.

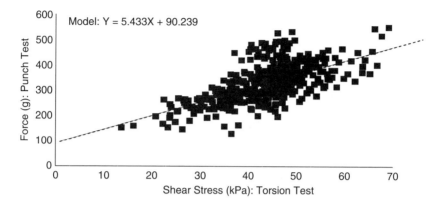

Figure 11.32 Relationship between shear stress (torsion) and force (punch). Gels were prepared with 78% moisture and 2% salt (312 data points were used). Each data point was obtained on the basis of a ten-sample specimen using a Rheo Tex for the punch test and a Hamann gelometer for the torsion test.

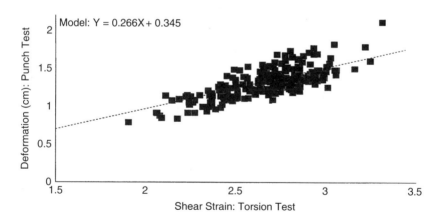

Figure 11.33 Relationship between shear stress (torsion) and shear strain (torsion). Gels were prepared with 78% moisture and 2% salt (312 data points were used). Each data point was obtained on the basis of a ten-sample specimen using a Rheo Tex for the punch test and a Hamann gelometer for the torsion test.

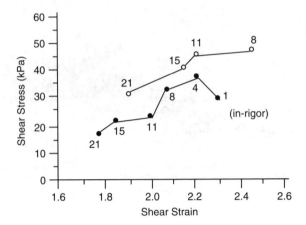

Figure 11.34 Loss of gel-forming ability as hoki are stored in ice. Gels (80–81%) were prepared at 90°C for 40 min from unwashed (•) and washed mince (○). The number of days the fish was stored in ice is shown beside each data point. (*Source:* Adapted from Ref. 2)

11.4 EFFECTS OF PROCESSING PARAMETERS ON RHEOLOGICAL PROPERTIES OF SURIMI GELS

11.4.1 Effects of Fish Freshness/Rigor Condition

There is a general decline in both the cohesiveness and hardness of gels made from fish stored for increasing time periods in ice (Figure 11.34). The rate of decline depends on both the denaturation of the myofibrillar proteins and the extent of proteolytic degradation of myofibrillar proteins for hoki[37] and tilapia.[38] Both are controlled post-mortem to some extent by low-temperature storage. The denaturation rate is often a function of muscle pH. The post-mortem pH of tilapia changed from 6.89 to 6.21 after 24-hr storage and 6.23 after 72-hr storage in ice.[38]

11.4.2 Effect of Refrigerated Storage of Gels

Another common concern regarding gel analysis is how soon after preparation the gel must be tested. During 6 days of refrigerated storage, both shear stress and shear strain gradually increased for Alaska pollock and Pacific whiting surimi (Figure 11.35). This increase is likely due to the function of hydrogen bonds, which are reinforced during refrigeration and become weaker at increased temperatures. Consequently, it is recommended to keep gels consistently in an equal time of storage before gel testing.

11.4.3 Effect of Sample Temperature at Measurement

The temperature dependency of pollock gels is clearly demonstrated in Figure 11.36. Regardless of salt concentration, shear stress values of all pollock surimi gels increased as the temperature reached 20 to 25°C and gradually decreased as the temperature increased even further. Shear strain values, however, continued to increase and reached a maximum at 50 to 60°C before gradually decreasing. Howe et al.,[39] Niwa et al.,[40] and Park and Lindwall[41] also found this trend.

Several major forces stabilize fish protein gels, including covalent bonds, hydrophobic interactions, and hydrogen bonds. Foegeding et al.[42] described the force of covalent bonds as being relatively independent of temperature. Hydrogen bonds become stronger as temperatures decrease, while hydrophobic interactions tend to maximize near 60°C.[43,44] The trend shown in Figure 11.37 suggests that stress is strongly influenced by hydrogen bonds. Shear strain, however, may not be strongly affected by hydrogen bonds. Instead, shear strain is affected by hydrophobic interaction because the increase of strain values up to 60°C corresponds with the optimum temperature for hydrophobic interaction.[45,46]

Surimi seafood mixed in a salad is served cold; but if the surimi seafood is applied as part of a casserole or soup, it may be served warmer at around 50 to 60°C. Interestingly, Howe et al.[39] found that stress values of set gels were inversely related to test temperature in a linear manner, while unset gels were less strong and responded differently (Figure 11.37). The rheological values of the unset gels measured at various temperatures were also similar to those found by Park.[45]

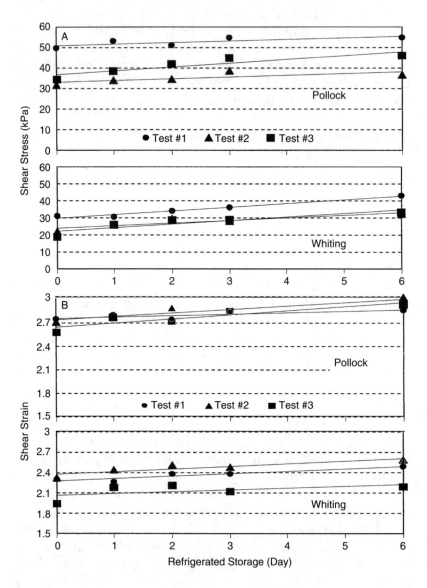

Figure 11.35 Changes in shear stress (A) and shear strain (B) during refrigerated storage of pollock and whiting surimi gels.

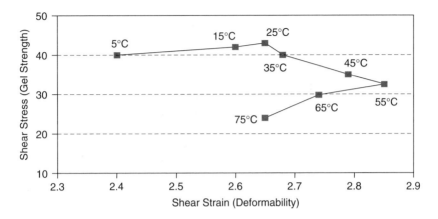

Figure 11.36 Effect of sample temperature at measurement on the rheological properties of pollock gels. Gels were prepared with 78% moisture and 2% salt content.

11.4.4 Effect of Moisture Content

The addition of water to surimi seafood is indispensable in maintaining acceptable texture and minimizing the cost of raw materials. The shear stress of both high- and low-grade pollock linearly decreased from ~65 to ~3 kPa and from ~48 to ~3 kPa, respectively, as the moisture content increased from 75 to 85.5% (Figure 11.38). Pacific whiting showed the same trends as pollock. Shear stress of both high and low grades of Pacific whiting decreased linearly from ~53 to ~3 kPa and from ~45 to ~3 kPa, respectively (Figure 11.38). Shear strain, commonly referred to as an indicator of protein quality,[2] was not affected within a certain range of moisture content. Moisture contents between 75 and 81% did not affect shear strain values for both high- and low-grade pollock surimi, However, their values decreased linearly from ~2.6 to ~1.9 and from ~2.1 to ~1.5, respectively, as the moisture content changed from 81 to 85.5%. Unlike pollock, the shear strain of Pacific whiting gels behaved differently with varying moisture contents. For whiting, the values decreased linearly from ~2.75 to ~2.1 for high grade and from ~2.35 to ~1.65 for low grade as moisture content increased from 75 to 85.5% (Figure 11.38).

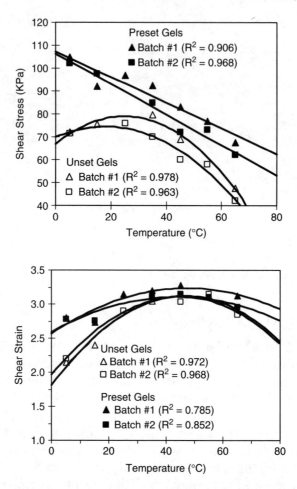

Figure 11.37 Influence of test temperature on fracture shear stress and strain of pollock surimi gels as affected by setting. (*Source:* Adapted from Ref. 39.)

11.4.5 Effect of Low Temperature Setting

Kim[46] found that the effects of low-temperature setting before cooking primarily affect the stress measurement. Two thermal treatments in forming gels from Alaska pollock and Atlantic croaker surimi were: (1) 90°C for 15 min (without setting) and (2) setting at 40°C for 30 min prior to heating at

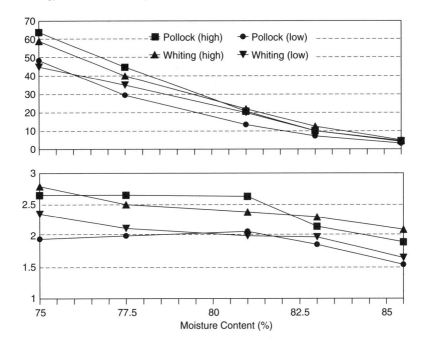

Figure 11.38 Effect of moisture content on the rheological properties of gels prepared from Alaska pollock and Pacific whiting surimi. Top: shear stress (kPa). Bottom: shear strain (dimensionless). (*Source:* From Ref. 51. With permission.)

90°C for 15 min. Stress values doubled with the 40 to 90°C treatment, but strain values hardly changed. Figure 11.39 shows the incremental effect of increasing the setting time at 4°C, before cooking at 90°C, on the stress of Alaska pollock surimi gels. Strain of the samples throughout the setting period was hardly affected.

The optimum setting temperature often corresponds well with the water temperature of the fishing grounds.[47–49] The strongest pollock gel was obtained with 4°C setting for 48 hr, followed by cooking at 90°C for 15 min; while Atlantic croaker surimi performed best at a treatment of 40°C setting for 30 min and 90°C cooking for 15 min. Lee and Park[47] confirmed that pollock gels were more effective at a 5°C setting, while whiting gels were more effective at a 25°C setting.

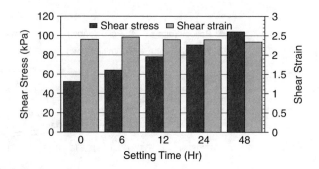

Figure 11.39 Effect of setting time at 4°C before cooking at 90°C on gel fracture of Alaska pollock.

11.4.6 Effect of Freeze-Thaw Abuse

Surimi gels made from low-quality Alaska pollock surimi showed a significant and continuous reduction in stress and strain when subjected to 0, 3, 9, and 15 cycles of freeze-thaw abuse.[12] However, when the superior-quality pollock surimi gels were subjected to freeze-thaw cycles (0, 3, 6, 9), it was noted that the first three cycles contributed more damage to the gels[50] (Figure 11.40).

11.4.7 Effect of Functional Additives

The effect of starch on the shear stress of both pollock and whiting surimi was highlighted when the starch content increased from 0 to 6%.[51] When 10% starch was added, failure shear stress values were reduced greatly for all starches, indicating that starch inhibited gelation of fish proteins by competing for the available water.[50] Shear strain of surimi gels with added starch responded differently than shear stress. Shear strain values of pollock-starch gels, except modified waxy cornstarch, slightly increased with starch content from 0 to 3%. However, strain values remained constant with starch levels up to 10%, except for wheat starch and modified waxy cornstarch. In general, at a higher level of starch (6 to 10%), shear stress decreased while shear strain remained

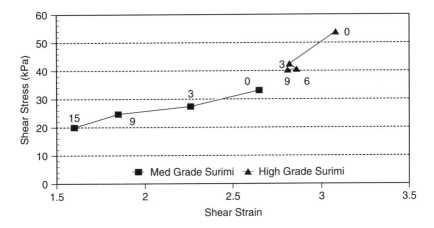

Figure 11.40 Effect of freeze–thaw cycles on stress and strain on gel fracture. Filled squares represent medium-grade pollock surimi gels subjected to 0, 3, 9, and 15 cycles. Fill triangles represent high-grade pollock surimi gels subjected to 0, 3, 6, and 9 cycles.

constant, indicating the presence of a nonlinear interaction between surimi and starch (Figure 11.41).[51]

Numerous protein additives can be categorized into two segments: (1) functional filler and (2) functional binder. Whether it is a functional binder or filler, it must be hydrolyzed to be functional. Therefore, understanding the protein–water, protein–protein, and protein–lipid–water interactions is important for the formation of a stable gel. Park[52] demonstrated that egg whites and beef plasma proteins acted as functional binders in surimi gels; and others, such as soy protein isolate, wheat gluten, and whey proteins, were functional fillers. Because the functionality of whey proteins depends on the nature of cheese processing and further modification, the rheological properties are extremely different among them. Park[53] investigated 14 different whey protein concentrates from a commercial company and found that their functions as a binder and/or enzyme inhibitor varied from extremely good to poor. Detailed information on ingredient technology can be found in Chapter 13.

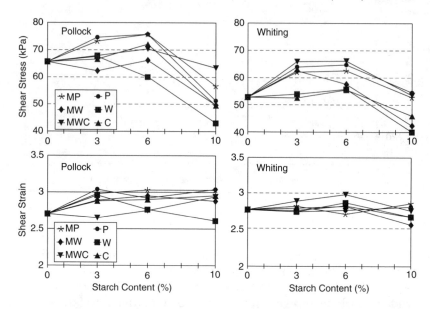

Figure 11.41 Effect of added starches on the fracture properties of pollock and whiting gels. All treatments were prepared at 78% moisture and 2% salt content. MP: modified potato starch; P: potato starch; MW: modified wheat starch; W: wheat starch; MWC: modified waxy corn starch; and C: corn starch. (*Source:* From Ref. 51. With permission.)

11.4.8 Texture Map

The texture map has been used extensively in research to relate to both the underlying chemical mechanism of surimi gelation and the interactions of surimi with other food constituents. Four typical characteristics of surimi gels have been identified: (1) brittle, (2) rubbery, (3) tough, and (4) mushy.[54] When gels break without resisting deformation, but need sufficient force, they are described as brittle. When gels strongly resist against deformation, but do not require a large force, they are categorized as rubbery. When gels are both brittle and rubbery, they are tough. When gels are neither brittle nor rubbery, they are soft or mushy. This map can be used as an excellent R&D tool for food manufacturers, especially when

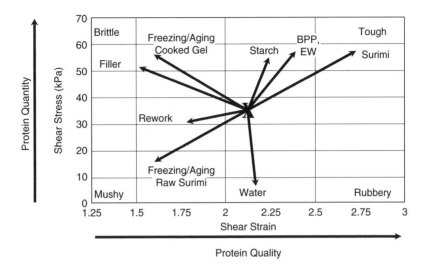

Figure 11.42 Texture map demonstrates a relationship between sensory textural descriptors and rheological properties as affected by ingredients and other factors.

the product requires optimization. The use of appropriate food ingredients and/or physical treatment can change the textural properties of the finished products (Figure 11.42).

11.5 VISCOSITY MEASUREMENTS

Viscosity measurements can have two important uses in surimi seafood manufacturing:

1. The apparent viscosity of dilute extracts of fish proteins has been shown to correlate with their degree of denaturation, which in turn will affect their gelling properties.[55,56] Such a measurement can serve as a rapid means of quality assessment for raw surimi. It can also be used to assess the quality of certain species of fish prior to surimi manufacture.[57]
2. The flow property of surimi pastes determines the ability to pump the material within the manufacturing plant and affects the extrusion properties of the

material during forming operations. Thus, monitoring the viscosity of the surimi seafood paste obtained directly from the chopping bowl can be useful in process control.

11.5.1 Measurement of Dilute Extract

Following extensive investigation regarding the factors affecting dilute extract measurement, Borderias et al.[58] reported the following conditions as being optimum when the Brookfield viscometer was used:

- Ratio of surimi to 5% NaCl solution at 1:4
- Blending time (Omnimixer, setting 7) of 1 min
- Blending and measuring temperature range of 2 to 5°C
- pH between 6.5 and 7.0
- Holding time between blending and measurement of 60 min

A Japanese standard test[59] for determining the viscosity of a dilute extract has also been elaborated to evaluate surimi quality. The test involves mixing 857 ml of 3.5% salt water (10°C) with 143 g thawed surimi, yielding a final salt concentration of 3%. This solution is then placed in a mixer (Mitsubishi M310) equipped to maintain a cold temperature (10°C), which minimizes foaming and blended at scale 1 for 8 min. Finally, the mixture is left to stand for 40 min and then the viscosity is measured in a Brookfield viscometer at a temperature of 10 ± 0.05°C.

11.5.2 Measurement of Surimi Seafood Pastes

Viscosity measurements can be made directly on the paste of surimi, salt, water, and other ingredients produced from blending in the silent cutter or on a plain surimi–salt–water paste as a quality control check prior to production. The recommended procedure is to use a small tube piston-type (capillary) viscometer (Figure 11.43). The extrusion tubes can be interchanged to adjust the internal diameter to match the viscosity range of the material being tested. In practice, tubes

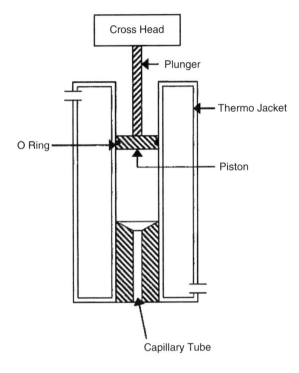

Figure 11.43 Capillary extrusion viscometer.

of the same diameter that have the same entrance and exit design, but vary in length, are run consecutively for samples from the same paste in order to account for the entrance and exit effects. The material is then extruded at several different rates by incrementing the speed of the cross head. At each speed, the maximum sustained force reading is obtained.

The apparent viscosity at any cross head speed is obtained by:

$$\eta = \frac{\pi R^2 \Delta P}{8q\Delta L} \tag{11.30}$$

where

η = apparent Newtonian viscosity
R = radius of the extrusion tube

ΔL = difference in extrusion tube lengths
ΔP = pressure difference between using the long and short tubes force difference/piston area
q = flow rate in cubic meters per second cross-head speed × piston area

The parameter measured is termed "apparent viscosity" because surimi pastes are non-Newtonian fluids, which exhibit a viscosity that depends on the shear rate (rate of flow). The paste typically exhibits a high yield stress because a certain force must be applied to initiate the flow of the product through a pipe or orifice. This shear stress can be calculated from the following equation:

$$\tau_y = \frac{R\Delta P_y}{8\Delta L} \qquad (11.31)$$

where ΔP_y is the difference in the initial peak pressure for the two tube lengths. As the rate of shear (movement) stabilizes, the pressure stabilizes.

Surimi seafood pastes are also thixotropic. This means the viscosity corresponds not only to the rate of flow, but also to the time of shearing. The viscosity of Alaska pollock surimi paste changes at different shear rates as a function of moisture (Figure 11.44) and salt content (Figure 11.45). The viscosity proportionally decreased as the moisture content increased from 76 to 86%.

In a rotational viscometer, the shear rate is derived from the rotational speed of a cylinder or a cone. In general, the concentric cylinder and cone-plate geometry is widely used to measure viscosity. In the concentric cylinder viscometer, a cylinder (bob) is located concentrically (coaxially) inside a cup containing a selected volume of the test material (Figure 11.46). The material in the cup could possibly be ruptured when the bob is inserted into the cup, especially when the viscosity of the material is very high. Obviously, choosing the geometry to measure the viscosity of surimi paste depends on the moisture content and the composition of the paste. By

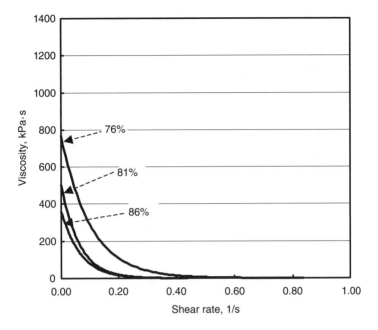

Figure 11.44 Viscosity changes in Alaska pollock surimi paste as a function of moisture content. All samples were prepared at 2% salt.

changing the rotational speed (shear rate) and measuring the resulting shear stress, it is possible to obtain viscosity data over a wide range of shearing conditions. In the concentric cylinder geometry, the shear stress can be determined from the total torque (M):

$$\tau = \frac{M}{2\pi r_i^2 h} \qquad (11.32)$$

where r_i is the radius of the rotating bob. In the case of a Newtonian fluid, the shear rate (γ'_N) is expressed as:

$$\gamma'_N = \frac{2\Omega}{\left[1 - \left(\frac{r_i}{r_o}\right)^2\right]} \qquad (11.33)$$

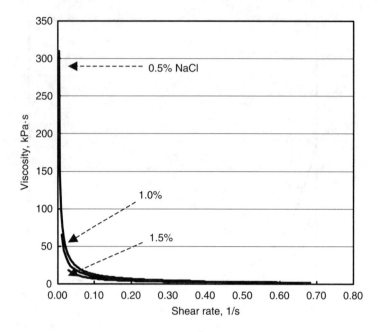

Figure 11.45 Viscosity changes in Alaska pollock surimi paste as a function of salt content. Surimi paste was prepared at 76% moisture content.

where Ω is the angular velocity of the rotating bob, r_i is the radius of the bob, and the r_o is the radius of the cup. Then, the viscosity from this geometry is given by:

$$\eta = \left(\frac{M}{4\pi h \Omega}\right)\left(\frac{1}{r_i^2} - \frac{1}{r_o^2}\right) \quad (11.34)$$

where h is the length of the bob in contact with the fluid.

In general, for surimi paste, the relationship between shear stress and shear strain is described by the power law model with a yield stress. Then, the angular velocity is expressed by the Reiner-Riwlin equation:

$$\Omega = \frac{M}{4\pi h \eta'}\left(\frac{1}{r_i^2} - \frac{1}{r_o^2}\right) + \frac{\sigma_0}{\eta'}\ln\frac{r_o}{r_i} \quad (11.35)$$

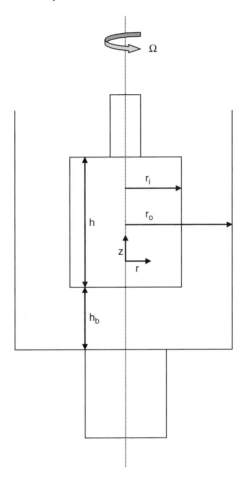

Figure 11.46 Schematic diagram of the concentric cylinder geometry.

where σ_o = yield stress and η' is the plastic viscosity.

Because equipment for small-amplitude oscillatory shear tests is affordable, the cone-plate geometry can be easily applied for measuring the viscosity of semi-solid materials (e.g., high-viscosity paste). This cone-plate device minimizes the damage to the sample structure that is likely to occur by inserting the bob in the cup in the concentric cylinder geometry. The shear stress and shear rate are given by the following equations, respectively:

$$\text{Shear stress, } \sigma_{\theta\phi} = \frac{3T_{CN}}{D}$$

$$\text{Shear rate, } \dot{\gamma} = \frac{\Omega}{\theta_0}$$

(11.36a, b)

where T_{CN} is the torque per unit area, D is the diameter of the rotating cone or plate, and θ_0 is the cone angle in radians.

11.5.3 Rheological Behavior of Surimi Paste

Investigating the rheological properties of the paste state is interesting. Many studies on the rheological behavior of surimi seafood have focused on the gel state. However, understanding the rheological properties of surimi paste is necessary to design the processing equipment or to find optimum operating conditions. It is a challenge, however, to estimate the gel fracture properties using rheological properties of the paste in a way that will significantly shorten the testing procedure.

Often, measuring the viscosity at two extremes, extremely low (e.g., <0.01 sec^{-1}) and high shear rate (e.g., >10^2 sec^{-1}), requires considerable effort.[60] Especially at high shear rate, the measurements potentially include experimental errors due to the slip between the surface of the experimental apparatus and surimi paste. Such slip can be observed while measuring viscosity from the phase separation of semi-solid foods, such as yogurt. The SAOS test produces the complex viscosity data defined as the complex modulus (Equation 11.14) divided by frequency. Such small amplitude rotational movement might minimize the slip effect. Thus, it provides the viscosity data measured in the high shear rate region. The Cox-Merz relation is useful for correlating the steady shear viscosity with the complex viscosity.[61] The useful application from the Cox-Merz rule to estimate the steady shear viscosity at high shear rate is introduced in the next section of this chapter.

The steady shear viscosity of surimi paste proportionally decreased with increased moisture content.[62] The apparent steady shear viscosity (400 sec^{-1}) was used to measure the interaction between preactivated iota-carrageenan and surimi.[63] The viscosity dramatically increased from 110 to 170 kPa for 5% (g solid/100g) preactivated iota-carrageenan. This increase might result from the carrageenan interaction (i.e., hydrogen bonding and ion interaction).

The applied stress dependence of G′ and G″ at the paste (or sol) state is shown in Figure 11.47, and the moisture content dependence of σ_0 for G′ and G″ are plotted in Figure 11.48. The σ_0 from both G′ and G″ at paste showed a strong linear relationship with the protein concentration in the paste (R^2 = 0.94 and 0.91, respectively). Based on this linear dependence, another relationship between nonlinear failure shear (FS) stress and SAOS results can be drawn:

$$\sigma_{0,s} \approx C$$
$$\text{FS Stress} \approx \sigma_{0,s} \approx C \quad (11.37\text{a,b})$$

where C is the polymer concentration and $\sigma_{0,s}$ indicates the applied stress at the limit of linear viscoelastic region (LVER) of the paste state.

In the LVER, G′ of surimi paste is higher than G″ (Figure 11.49), although the linear viscoelastic behavior of the liquid-like material typically shows G″ >> G′.[62] The G′ >> G″ behavior may be due to the entanglement of protein molecules in the surimi paste. The entangled parts of molecules may act as temporary networks under infinitesimal displacement.[64] Those temporary networks can store the mechanical energy during deformation, but the networks will disentangle after a certain timeframe. The linear dependence of the $\sigma_{0,s}$ of surimi paste on protein concentration is likely due to the disentanglement of the temporary network in protein molecules. The G′ of surimi paste at 0.01 Hz was plotted as moisture content changed (Figure 11.50). The linear dependence of G′ on moisture content (R^2 = 0.90) was slightly weaker at the paste condition than at the gel condition (R^2 = 0.95).

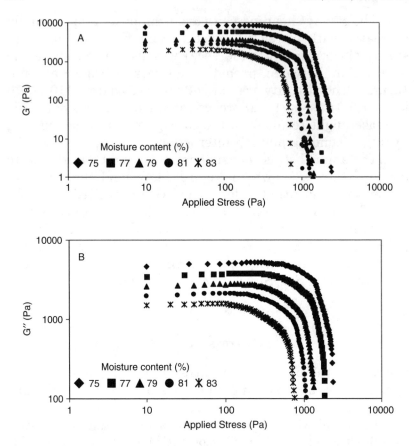

Figure 11.47 Stress dependence of storage (A) and loss (B) modulus of surimi paste as a function of moisture content. All samples were prepared using 2% salt.

11.6 PRACTICAL APPLICATION OF DYNAMIC RHEOLOGICAL MEASUREMENTS

The use of dynamic rheological data, such as G′ and G″, is now popular to characterize food gel properties, including surimi gels. In this section, the dynamic rheological data for both surimi paste and the gel state are introduced and applied to characterize the surimi gel's interesting physical behaviors, such as gelation kinetics and steady-state viscosity.

Rheology and Texture Properties of Surimi Gels

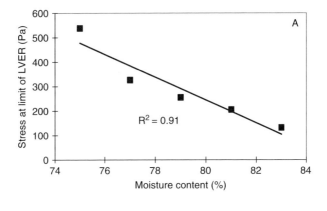

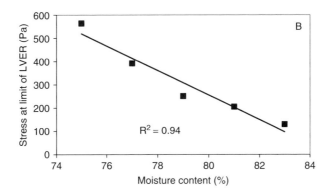

Figure 11.48 Applied stress at the limit of LVER for storage (A) and loss (B) modulus of surimi paste as a function of moisture content. All samples were prepared using 2% salt.

11.6.1 Gelation Kinetics of Surimi Gels

11.6.1.1 Thermorheological Properties and Kinetic Model

Due to its gel-forming ability, the textural properties of surimi seafood are the most important quality factors. In addition to texture, the thermal properties of surimi are also important because surimi gelation is heat induced. Most surimi seafood products are manufactured by various thermal treatments

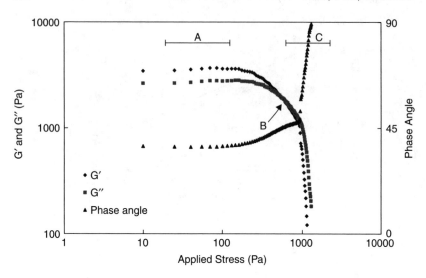

Figure 11.49 Shear stress dependence of storage and loss modulus of surimi paste at 79% moisture content. A: linear viscoelastic region; B: critical; and C: rupture state. All samples were prepared using 2% salt.

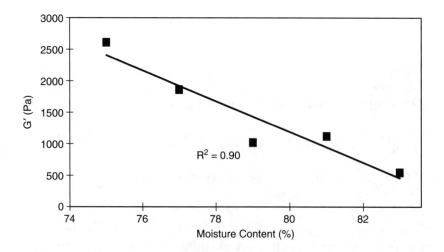

Figure 11.50 Storage modulus of surimi paste at 0.1 Hz as a function of moisture content. All samples were prepared using 2% salt.

(see Chapter 9). Thus, both rheological and thermal properties are the primary control variables of product quality during surimi seafood processing.

Small amplitude oscillatory shear (SAOS) measurements are widely used to characterize the rheological behavior of food gels during gelation.[65] Many studies have reported changes in G′ and G″ during surimi gelation, but the results have only been compared qualitatively. Based on the rubber elastic theory, the elastic modulus is proportional to the concentration (the number density) of the elastically active network chain (EANC), which is defined as the number of crosslinks per unit volume.[66] If the elastic modulus is solely explained by entropic contribution, the relationship between concentration and modulus can directly represent the EANC of different species, which is further explained in the subsequent section ("Rubber Elastic Theory").

In general, kinetic models describe the extent of reaction due to changes in the initial concentration under either isothermal or non-isothermal conditions. In the case of gelation, because the modulus is directly proportional to the concentration involved in gelation, the changes in concentration can be replaced with rheological properties, such as G′. Gelation experiments are readily performed under non-isothermal heat treatment by heating (or cooling) at a constant rate. Under non-isothermal conditions, kinetic models are useful in characterizing changes in rheological behavior during gelation. The non-isothermal kinetic model could represent mechanisms involved in chemical reactions, and help estimate the reaction rate and energy barrier for the reactions.

11.6.1.2 Non-Isothermal Kinetic Model

The general expression for non-isothermal kinetics is as follows:

$$\int_{C_0}^{C} \frac{dC}{C^n} = k_0 \int_{0}^{t} \exp\left(-\frac{Ea}{RT(t)}\right) dt \qquad (11.38)$$

where C = concentration, C_o = initial concentration, t = time, k_o = frequency factor, Ea = the activation energy (J/mole), T = absolute temperature (K), and R = universal gas constant (8.314 J/mol.K).

The evaluation of this general expression depends on the heating rate. If the heating rate is linear, the non-isothermal kinetic relationship, based on the experimental data and regression analysis, is obtained by following the steps described by Rhim and others.[67]

For the decomposition reaction, the rate of reaction is given by:

$$-\frac{dC}{dt} = kC^n \qquad (11.39)$$

The Arrhenius relation gives the temperature dependence of the reaction rate constant k:

$$k = k_0 \exp\left(-\frac{Ea}{RT}\right) \qquad (11.40)$$

After solving the differential equation, the kinetic model under a non-isothermal state with linear heating rate becomes:

$$\ln\left(-\frac{1}{C^n}\frac{dC}{dt}\right) = \ln k_0 - \frac{Ea}{RT} \qquad (11.41)$$

The kinetic parameters k_o and Ea are determined from an Arrhenius-type plot of Equation 11.33). Yoon[62] and Yoon et al.[68] described the detailed derivation procedure for Equation 11.34.

11.6.1.3 Rubber Elastic Theory

According to rubber elastic theory, the plateau modulus (Ge) is expressed as:

$$Ge = \nu KT \qquad (11.42)$$

where ν = number density of EANC and K = Boltzmann constant (1.38×10^{-23} J/K). Because shear modulus represents energy stored during deformation, G can be qualitatively replaced by G'. The number density (or concentration) of EANC is directly proportional to the shear modulus (G). This relationship implies that a change in concentration (dC) in the kinetic model is equivalently replaced by a change in modulus (dG) during gelation.

11.6.1.4 Dependence of Gelation Temperature on Moisture Content

Changes in storage modulus (G') of Alaska pollock (AP) surimi during heating from 20 to 80°C at 1°C/min are presented in Figure 11.51. The data followed the prototypical behavior of thermosetting gelation.[2] The kinetic process during gelation was specifically investigated at various moisture contents using the non-isothermal kinetic model.

Theoretically, the gel point in a polymerizing system is defined as the point at which an infinite three-dimensional (3D) network first appears.[69] The gelation temperature (T_{gel}) was defined as the temperature where a substantial increase in G' was observed during the temperature sweep.[65] Thus, T_{gel} corresponds to the temperature at which a G' minimum was observed.[70,71] The myofibrillar protein in the surimi should be unfolded to be a flexible actomyosin and form a 3D network with rubber-like elasticity. The T_{gel} of AP depends on moisture content that is is inversely proportional to polymer concentration. The dependence of T_{gel} on moisture content is clearly shown in Figure 11.51. For moisture contents less than 92.5%, the G' minimum of each curve was estimated by setting the first derivative of the polynomial equation (Equation 11.43), representing data from 34 to 51°C, with respect to the temperature at zero (i.e., $dG'/dT = 0$).

$$G' = \beta + \alpha_1 T^1 + \alpha_2 T^2 + \alpha_3 T^3 \tag{11.43}$$

For moisture contents greater than or equal to 92.5%, the graphical extrapolation method was used, because G' monotonically increased in this temperature range. Coefficients of the polynomial equation (Equation 11.43) and T_{gel}

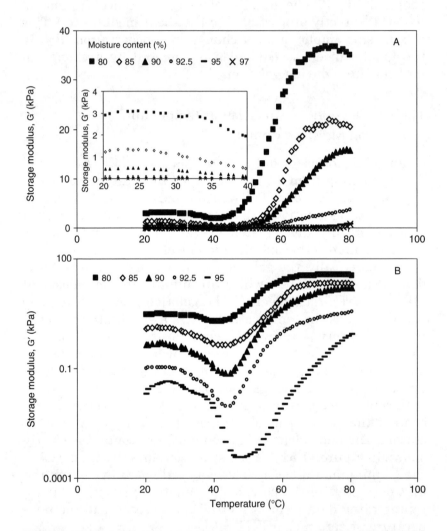

Figure 11.51 Changes in G′ during gelation of Alaska pollock surimi on a linear scale (A). The semi-logarithmic scale (B) shows the moisture dependence of the gelation temperature. The insert figure in A describes the local maxima observed at approximately 34°C. All samples were prepared using 2% salt.

TABLE 11.4 Coefficients of Polynomial Equation 11.42 and Gelation Temperature (T_{gel})

Moisture Content (%)	Alaska Pollock Surimi					
	β	α_1	α_2	α_3	R^2	T_{gel} (°C)
80	−82491	7357	−207	1.92	0.99	41.2[a]
85	−14095	1355	−39	0.37	0.99	41.8[a]
90	−21571	1736	−45	0.39	0.99	43.8[a]
92.5	N/A	N/A	N/A	N/A	N/A	44.1[b]
95	N/A	N/A	N/A	N/A	N/A	46.2[b]

[a] Determined by setting $dG'/dT = 0$.
[b] Determined by graphical extrapolation between two linear segments.

values are summarized in Table 11.4. The T_{gel} for AP surimi increased with increasing moisture content (Table 11.4).

11.6.1.5 Activation Energy during Gelation

The kinetic relationship in Equation 11.41 can be used for gelation kinetics, and the concentration (C) and change in concentration (dC), as discussed in the previous section, can be replaced with G' and dG'

$$\ln\left[\left(\frac{1}{G'^n}\right)\left(\frac{dG'}{dt}\right)\right] = \ln k_0 - \left(\frac{Ea}{RT}\right) \quad (11.44)$$

The negative sign on the left-hand side of Equation 11.41 was changed to positive because the rate of changes of G' is positive in this case.[67,72]

The theoretical value of the reaction of disulfide bonds during gelation of myofibrillar protein was determined to be a second-order reaction (i.e., n = 2). Wu and et al.[73] and Yongsawatdigul and Park[74] also reported a second-order reaction rate for surimi during ohmic heating.

To evaluate n from the experimental data in this study, multiple linear regression analysis was performed after rewriting Equation 11.44 as:

$$\ln\left[\frac{dG'}{dt}\right] = \ln k_0 + n \ln G' - \frac{Ea}{RT} \quad (11.45)$$

The reaction order was statistically determined and averaged to be about 1.34 (±0.09). It is lower than the expected value of 2. However, because the n value we determined was greater than 1 (i.e., higher than a first-order reaction), the second-order reaction (n = 2) was assumed for the kinetic analysis. This assumption (i.e., n = 2) was further verified by determining the correlation coefficient (R^2) for actual fit of the experimental data per Equation 11.43; for n values of 1, 1.34, and 2 the corresponding R^2 values were 0.04, 0.37, and 0.75, respectively. Among many gelation kinetic models, this nonisothermal model is the most useful because most gelation experiments are performed *in situ* at a constant rate of heating or cooling. In addition, changes in G' easily provide an estimate of the structure development rate (SDR = dG'/dt). The Ea of gelation is calculated from the slope of the Arrhenius-type plot according to Equation 11.44 (see Figure 11.52).

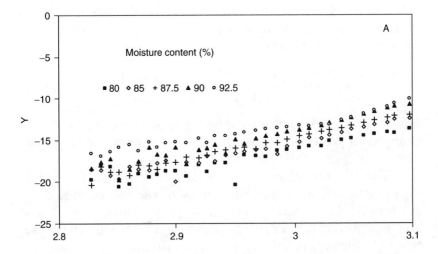

Figure 11.52 Arrhenius-type plot for the changes in G' of Alaska pollock surimi during gelation; n = 2 and Y = ln[(1/G'^{2} × (dG'/dt)]. All samples were prepared using 2% salt.

TABLE 11.5 Comparison of the Activation Energy of Gelation of AP Surimi

Moisture Content (%)	R^2	Activation Energy (kJ/mol)
80	0.81	176.5
85	0.81	204.3
90	0.97	213.6
92.5	0.93	210.8
95	0.95	232.9

The "almost" linear relationship in the high-temperature region (lower region of 1/T) in Figure 11.52 indicates a unique gelation mechanism during heating. The Ea of AP increased from 176.5 to 232.9 kJ/mol as moisture content increased from 80 to 95% (Table 11.5). The lower Ea at lower moisture contents (i.e., higher protein concentration) implies that they were more favorable for gelation. The Ea of gelation of surimi at different moisture contents are relatively higher than those for thermo-reversible gel systems.

11.6.2 Estimation of Steady Shear Viscosity of Fish Muscle Protein Paste

11.6.2.1 Steady Shear Viscosity of Surimi Paste and the Cox-Merz Rule

Steady shear viscosity (η) is the most useful parameter to characterize the rheological behavior during processing of polymeric materials.[75,76] Even for cooking, especially in the case of continuous radiation heating, the heating rate is strongly related to the shear rate dependence of steady shear viscosity. In general, the surimi paste is highly concentrated (i.e., 75 to 90% moisture concentration) so that it shows a "weak" solid-like behavior (i.e., storage modulus (G′) > loss modulus (G″)) at low temperature (<50°C), although there are no cross-links.[62,77]

Due to such solid-like properties, it is difficult to measure the steady shear viscosity at high shear. Dynamic tests are therefore necessary to measure the rheological properties of such semi-solid materials without breaking the material's internal structure. The correlation between steady shear viscosity (η) and dynamic shear viscosity (η^*) is empirically described by an empirical relationship known as the Cox-Merz rule.[9] The Cox-Merz relation and/or its modified forms have been applied to many fluid and semi-solid foods.[78–80]

11.6.2.2 Superimposed Viscosity Using the Cox-Merz Rule

The η and η^* values of 95% moisture content AP surimi paste are plotted vs. shear rate and angular velocity in Figure 11.53. Both the η of η^* values of surimi paste showed strong shear thinning behavior according to the power law relations:

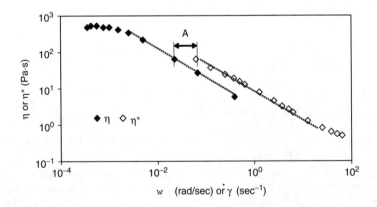

Figure 11.53 Viscosity of Alaska pollock surimi at 95% moisture content obtained from continuous shear measurement and dynamic shear measurement. The dotted lines on the linear segments indicate shear thinning regions. (ω = angular velocity (rad/sec); $\dot{\gamma}$ = shear rate (sec^{-1}); η = steady shear viscosity (Pa.sec); and η^* = dynamic shear viscosity (Pa.sec)). All samples were prepared using 2% salt.

$$\eta = m(\dot{\gamma})^{n-1} \qquad (11.46)$$

$$\eta^* = m^*(\varpi)^{n^*-1} \qquad (11.47)$$

where, $\dot{\gamma}$ = shear rate (sec^{-1}), ω = frequency (sec^{-1}), n and m = power-law parameters for steady shear viscosity measurement, and n* and m* = power-law parameters for dynamic viscosity measurement. The magnitude of η^* was higher than η at a given shear rate (between 10^{-2} and 10^{-1} sec^{-1}). As shown in Figure 11.53, it is impossible to estimate the η value at a high shear rate according to the Cox-Merz rule, which states that:

$$\eta^*(\varpi) = \eta(\dot{\gamma})\big|_{\varpi=\dot{\gamma}} \qquad (11.48)$$

However, the power law exponents (n*–1 and m*–1) in the shear thinning region, shown as dotted lines in Figure 11.53, were nearly identical (n* = –0.798 and m* = –0799). This implies a possibility of superimposing η^* on to η along the x-axis by introducing a horizontal shift factor (A). This will give the following modified Cox-Merz relation:

$$|\eta^*|(\varpi) = \eta(\dot{\gamma}), \quad \text{with } \varpi = (10^A)\dot{\gamma} \qquad (11.49)$$

The steady shear viscosity is superimposed well by the dynamic shear viscosity by choosing a suitable A (Figure 11.54). The A values were determined graphically by matching the superimposed line such that the highest R^2 value was obtained for the superimposed line. The dotted line in Figure 11.54 represents the η^* shifted along the frequency axis with a horizontal shift factor (A = –0.5135), and the R^2 value of the linear segment (i.e., shear rate between 10^{-2} and 10^1) on the superimposed line was 0.999.

11.6.2.3 Concentration Dependence of the Viscosity of Surimi Paste

The η and η^* values of surimi paste at different moisture contents (95, 90, 85, 80, and 75%) were superimposed by apply-

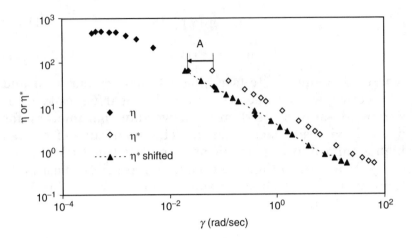

Figure 11.54 Illustration of the modified Cox-Merz rule. The steady and dynamic shear viscosity of surimi paste at 95% moisture content was superimposed by intorducing a horizontal shift factor (A). (ω = angular velocity (rad/sec); γ = shear rate (sec^{-1}); η = steady shear viscosity (Pa.sec); and η^* = dynamic shear viscosity (Pa.sec).) All samples were prepared using 2% salt.

ing the frequency shift factor (A) to each concentration (Figure 11.55 and Figure 11.56). Table 11.6 shows the A values for each concentration and the R^2 values of the linear segment of each superimposed η curve. The A values increased with a decrease in moisture content up to 85%, then remained nearly constant. A change in A for surimi paste at high moisture content (>85%) may imply some structural changes due to the chemical denaturation of myofibrillar protein by added salt.

In this section, the moisture content used in the measurement was converted to total polymer concentration to study the effect of concentration on the macroscopic behavior of the paste. In the low shear rate region (approximately <10^{-2} sec^{-1}), a limiting value of steady shear viscosity (i.e., zero shear viscosity (η_0) was observed (Figure 11.57). The onset shear rate (1/λ) is defined as the shear rate at which the limiting steady shear viscosity begins. The 1/λ at each moisture content (95, 90, 85, 80, and 75%) was graphically determined as

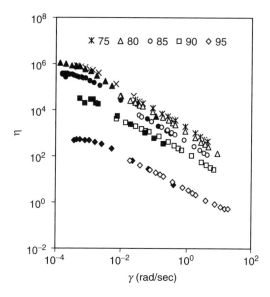

Figure 11.55 Moisture content dependence of the viscosity of surimi pastes. The viscosity of high shear rate was superimposed from the dynamic shear viscosity by the modified Cox-Merz rule (γ = shear rate (sec^{-1}); η = steady shear viscosity (Pa.sec); **X** indicates the steady shear viscosity at 75% moisture content; and open symbols indicate steady shear viscosity superimposed from dynamic viscosity). All samples were prepared using 2% salt.

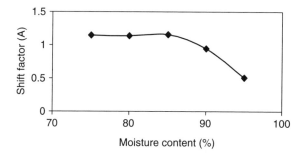

Figure 11.56 Moisture content dependence of the horizontal shift factors. All samples were prepared using 2% salt.

TABLE 11.6 Frequency Shift Factor (A) for Different Moisture Contents of Surimi Paste, and the R^2 Values of the Linear Segment of Each Superimposed Steady Shear Viscosity (η) Curve

Moisture Content (%)	A	R^2
75	1.1482	0.99
80	1.1385	0.97
85	1.1585	0.98
90	0.9432	0.99
95	0.5135	0.99

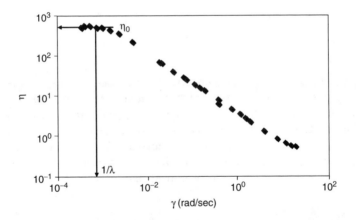

Figure 11.57 Illustration of zero viscosity and the onset shear rate of surimi paste at 95% moisture content (γ = shear rate (sec^{-1}); η = steady shear viscosity (Pa.sec)).

2.46, 1.65, 0.695, 0.596, and 0.302 (x 10^{-3} sec^{-1}), respectively, and the η_0 was plotted vs. the total polymer concentration (g/ml) in Figure 11.58.

The zero shear viscosity (η_0) increased with protein concentration (c). The solution state of the macromolecule is generally distinguished to be in three states, depending on the onset of coil overlap concentration (c*): such as dilute (c < c*), semi-dilute (c > c*), and concentrated (c >> c*) regime, depend-

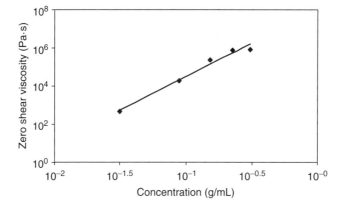

Figure 11.58 The concentration dependence of the zero shear viscosity of surimi paste.

ing on the interactions between molecules.[68] In general, the polymer chains are considered a random coil configuration and the concentration dependence of η_0 is expressed as[81]:

$$\eta_0(c) \propto c^{1.3} (c \leq c^*)$$

$$\propto c^{3.3} (c \geq c^*)$$

The power law exponent of surimi paste was determined as 3.51 (Figure 11.58). This implies that the concentration used in our experiment (5 to 25%) is in the semi-dilute regime. It must also be in a highly entangled state and the molecular chains are highly interactive. Thus, the macroscopic behavior (i.e., viscosity) corresponds not only to the motion of the individual molecules (i.e., internal degree of freedom), but also to the short- and long-range intermolecular interactions (i.e., relaxation process). These interactions strongly correspond to the shear thinning behavior of surimi pastes, as previously observed (see Figure 11.55). In addition, the onset shear rate $(1/\lambda)$, at which the zero shear viscosity is reached, decreased with polymer concentration (Table 11.7). This is a reflection of the fact that the time-scale for macromolecular disentanglements increases with concentration.[81]

TABLE 11.7 Concentration Dependence of the Carreau Model Parameters (λ and η)

Moisture Content (%)	λ	η
75	3311.258	0.155
80	1677.852	0.133
85	1438.849	0.117
90	606.0606	0.301
95	406.5041	0.333

11.7 SUMMARY

Gelation of surimi proteins and their textural properties are the primary factors that determine the intrinsic and extrinsic parameters involved in the manufacture of surimi and surimi seafood. Unlike other animal proteins, the gelation properties of surimi proteins are extremely variable, depending on species, fish habitat, harvesting conditions, processing procedures, and storage conditions. Therefore, accurate measurements of the rheological properties of surimi under the intended conditions of use and in combination with desired additives are critical in controlling product quality and new product development. Least-cost linear programming can use these measurements to obtain a balance between ingredient cost and product quality. The goal is to minimize the cost of ingredients without sacrificing product quality. In addition, it could lead to better prediction and control during processing operations.[51,52,54]

Shear strain at fracture is a stable measure of functional protein quality, whereas stress is strongly influenced by dilution, ingredients, and processing variables (Table 11.8). This is another reason why the torsion test is superior to the empirical test (punch), because in the empirical test the strain-related measurements are greatly influenced by the stress measurement.

In the day-to-day processing of surimi and surimi seafood, the rapid assessment of textural properties is very much

TABLE 11.8 Degree of Influence of Several Processing and Experimental Factors on Texture Values of Surimi Gels

Factors	Torsion Data Punch Data	Stress (Force)	Strain (Deformation)
Protein denaturation		Moderate	Strong
Proteolysis		Strong	Strong
Protein (moisture) concentration		Strong	Moderate
Filler ingredients		Strong	Weak
Binder ingredients		Strong	Strong
Low temperature setting		Strong	Moderate
Chemical oxidant		Strong	Weak
Refrigerated age of gels		Strong	Moderate
Chopping temperature		Strong	Strong
Gel temperature at measurement		Strong	Strong
Freeze-thaw abuse		Strong	Strong
Diameter of punch probe		Strong	Strong
Speed of punch probe		Strong	Weak

needed to take the necessary steps to control product quality. Currently, when gel quality data is obtained, it is too late to make any adjustments. Either the product has been shipped or the counterpart products have already been processed. Therefore, it would be of great advantage to develop a method to determine texture quality as the product is manufactured. This goal, however, remains a challenge to those researching this field.

ACKNOWLEDGEMENTS

The authors would like to dedicate this chapter to the late Dr. Donald Dale Hamann for his development of the rheological measurement of surimi gels during his tenure at North Carolina State University (Raleigh, NC). He was a great teacher and engineer, with great thoughts on food science. His contribution to the rheology of surimi gels will be remembered always.

The authors also extend their gratitude to Joodong Park for his assistance in developing experimental data and graphic figures.

REFERENCES

1. A.S. Szczesniak. Development of standard rating scales for mechanical parameters of texture and correlation between objective and the sensory methods of texture evaluation. *J. Food Sci.*, 28, 397–403, 1963.

2. D.D. Hamann and G.A. MacDonald. Rheology and Texture Properties of Surimi and Surimi-Based Foods. In T.C. Lanier and C.M. Lee, Eds. *Surimi Technology*. New York: Marcel Dekker, 1992, 429–500.

3. D.D. Hamann and N.B. Webb. Sensory and instrumental evaluation of material properties of fish gels. *J. Texture Studies*, 10, 117–121, 1979.

4. J.F. Montejano, D.D. Hamann, and T.C. Lanier. Comparison of two instrumental methods with sensory texture of protein gels. *J. Texture Studies*, 16, 403–424, 1985.

5. A.S. Szczesniak. Rheological basis for selecting hydrocolloids for specific applications. In G.G. Phillips, D.J. Wedlock, and P.A. Williarms, Eds. *Gums and Stabilisers for the Food Industry 3*. London: Elsevier Applied Science Publishers, 1985, 311.

6. J.R. Mitchell. Rheology of gels. *J. Texture Studies*, 7, 313–339, 1976.

7. N.H. Polakowski and E.J. Ripling. *Strength and Structure of Engineering Materials*. Englewood Cliffs, NJ: Prentice Hall, 1966.

8. J.R. Mitchell. The rheology of gels. *J. Texture Studies*, 11, 315–337, 1980.

9. F.W. Wood. Psychophysical studies on liquid foods and gels. In P. Sherman, Ed. *Food Texture and Rheology*. London: Academic Press, 1979, 21.

10. D.D. Hamann. Methods for measurement of rheological changes during thermally induced gelation of proteins. *Food Technol.*, 41(3), 100–108, 1987.

11. D.D. Hamann, S. Purkayastha, and T.C. Lanier. Applications of thermal scanning rheology to the study of food gels. In V.R. Harwalker and C.Y Ma, Eds. *Thermal Analysis of Foods*. Essex, England: Elsevier, Barking, 1990, 306–332.

12. R.W. Whorlow. *Rheological Techniques.* Chichester, UK: Ellis Horwood, 1980.
13. L. Bohlin, P. Hegg, and H. Ljusberg-Wahren. Viscoelastic properties of coagulating milk. *J. Dairy Sci.,* 67, 729–734, 1984.
14. T. Beveridge, L. Jones, and M.A. Tung. Progel and gel formation and reversibility of gelation of whey, soybean, and albumen gels. *J. Agric. Food Chem.,* 32, 307–313, 1984.
15. K. Samejima, B. Egelandsdal, and K. Fretheim. Heat gelation properties and protein extractability of beef myofibrils. *J. Food Sci.,* 50, 1540–1543, 1985.
16. B. Egelandsdal, K. Fretheim, and O.J. Harbitz. Dynamic rheological measurements on heat-induced myosin gels. I. An evaluation of the method's suitability for the filamentous gels. *J. Sci. Food Agric.,* 37, 944–954, 1986.
17. H.H. Winter and F. Chambon. Analysis of linear viscoelasticity of a crosslinking polymer at the gel point. *J. Rheology,* 30, 367–382, 1986.
18. H.H. Winter. Can the gel point of a cross-linking polymer be detected by the G'-G" crossover? *Polymer Engineering and Science,* Dec. 27, 1698–1702, 1987.
19. R. Muller, E. Gerard, P. Dugand, D. Remp, and Y. Gnanou. Rheological characterization of the gel point: A new interpretation. *Macromolecules,* 24, 1321–1326, 1991.
20. Y.L. Hsieh, J.M. Regenstein, and M.A. Rao. Gel point of whey and egg proteins using dynamic rheological data. *J. Food Sci.,* 58, 116–119, 1993.
21. W.B. Yoon, B.Y. Kim, and J.W. Park. Rheological characteristics of fibrinogen-thrombin solution and its effects on surimi gels. *J. Food Sci.,* 64 (2), 291–294, 1999.
22. E. Niwa, E. Chen, T. Wang, S. Kanoh, and T. Nakayama. Extraordinarity in the temperature-dependence of physical parameters of kamaboko. *Nippon Suisan Gakkaisi,* 54, 241–244, 1988.
23. J.G. Montejano, D.D. Haman, and T.C. Lanier. Final strengths and rheological changes during processing of thermally induced fish muscle gels. *J. Rheol.,* 57, 557–579, 1983.

24. H. Hertz. Miscellaneous paper. New York: Macmillan and Company, 1896.

25. P.W. Voisey, C.J. Randall, and E. Larmond. Selection of an objective test of weiner texture by sensory analysis. *Can. Inst. Food Sci. Technol. J.,* 8(1), 24, 1975.

26. D.D. Christianson, E.M. Casiraghi, and E.B. Bagley. Uniaxial compression of bonded and lubricated gels. *J. Rheol.,* 29, 671–677, 1985.

27. D.D. Hamann, Structural failure in solid foods. In M. Peleg and E. Bagley, Eds. *Physical Properties of Foods.* Westport, CT: AVI, 1983, 351–383.

28. A. Nadai. Plastic behavior of metals in the strain-hardening range. Part I. *Appl. Physics,* 8, 205–123, 1937.

29. J.J. Hoffman and J.W. Park. Improved torsion testing using molded surimi gels. *J. Aquat. Food Prod. Technol.,* 10(2), 75–84, 2001.

30. M.A. Rudan and D.M. Barbano. A model of mozzarella cheese melting and browning during pizza baking. *J. Dairy Sci.,* 81, 2312–2319, 1998.

31. J.J. Matsumoto, T. Arai. *Bull. Japan Soc. Sci. Fish.,* 17, 377, 1952.

32. M. Okada and A. Yamazaki. *Bull. Jap. Soc. Sci. Fish.,* 52, 1261, 1958.

33. T.C. Lanier, D.D. Hamann, and M.C. Wu. Development of methods for quality and functionality assessment of surimi and minced fish to be used in gel type food products. Report to Alaska Fisheries Development Foundation, Inc, Anchorage, AK, 1985.

34. D. Kilcast, M.M. Boyar, and J.B. Hudson. Gelation photoelasticity: a new technique for measuring stress distributions in gels during penetration testing. *J. Food Sci.,* 49, 654–657, 666, 1984.

35. J.W. Park. Unpublished data, SeaFest/JAC Creative Foods, Motley, MN, 1991.

36. Anonymous. Codex Code for Frozen Surimi. Rome, Italy: FAO/WHO, 1999.

37. G.A. MacDonald, J. Lelievre, and N.C. Wilson. Strength of gels prepared from washed and unwashed minces of hoki stored in ice. *J. Food Sci.*, 54, 976–980, 1990.

38. J.W. Park, R.W. Korhonen, and T.C. Lanier. Effects of rigor mortis on gel forming properties of surimi and unwashed mince prepared from tilapia. *J. Food Sci.*, 55, 353–355 and 360, 1990.

39. J.R. Howe, D.D. Hamann, T.C. Lanier, and J.W. Park. Fracture of Alaska pollock gels in water: effects of minced muscle processing and test temperature. *J. Food Sci.*, 59, 777–780, 1994.

40. E. Niwa, T.T. Wang, S. Kanoh, and T. Nakayama. Temperature dependence of elasticity of kamaboko. *Nippon Suisan Gakkaoshi*, 53, 2255–2257, 1987.

41. J.W. Park and W. Lindwall. Unpublished data, SeaFest Products, Motley, MN, 1989.

42. E.A. Foegeding, C. Gonzales, D.D. Hamann, and S.E. Case. Polyacrylamide gel as an elastic model for food gels. *Food Hydrocolloids*, 8, 125–134, 1994.

43. E. Niwa. The chemistry of surimi gelation. In T.C. Lanier and C.M. Lee, Eds. *Surimi Technology*. New York: Marcel Dekker, 1991.

44. H.A. Scheraga, G. Nemethy, and I.Z Steinberg. The contribution of hydrophobic bonds to the thermal conformation. *J. Biol. Chem.*, 237, 2506–2508, 1962.

45. J.W. Park. Effects of salt, surimi and/starch content on fracture properties of gels at various test temperatures. *J. Aquat. Food Product Technol.*, 4(2), 75–84, 1995.

46. B.Y. Kim. Rheological Investigation of Gel Structure Formation by Fish Proteins during Setting and Heat Processing. Ph.D. thesis, North Carolina State University, Raleigh, NC, 1987.

47. N.G. Lee and J.W. Park. Calcium compounds to improve gel functionality of Pacific whiting and Alaska pollock surimi. *J. Food Sci.*, 63, 969–974, 1998.

48. T. Mishima, H. Mukai, Z. Wu, K. Tachibana, and M. Tsuchimoto. Resting metabolism and myofibrillar Mg^{++}-ATPase activity of carp. *Nippon Suisan Gakkaishi*, 59, 1213–1218, 1993.

49. J.W. Park, J. Yongsawatdigul, and T.M. Lin. Rheological behavior and potential cross-linking of Pacific whiting surimi gel. *J. Food Sci.*, 59, 773–776, 1994.

50. J.W. Park, H. Yang, and S. Patil. Preparation of temperature-tolerant fish protein gels using special starches. In A.M. Spanier, Tamura, Okai, and Mills, Eds. *Chemistry of Novel Foods*. New York: Allured Publishing, 1997.

51. W.B. Yoon, J.W. Park, and B.Y. Kim. Linear programming in blending various components of surimi seafood. *J. Food Sci.*, 62(2), 561–564 and 567, 1997.

52. J.W. Park. Functional protein additives in surimi gels. *J. Food Sci.*, 59, 525–527, 1994.

53. J.W Park. A Report: The Functionality of Whey Protein Concentrates. Union, NJ: MD Foods USA, 1996.

54. T.C. Lanier. Functional properties of surimi. *Food Technol.*, 39(3), 107–114, 1986.

55. J.J. Matsumoto. Chemical deterioration of muscle proteins during frozen storage. In J.R. Whitaker and M. Fujimaki, Eds. *Chemical Deterioration of Proteins*. Advances in Chemistry Series, No. 123. Washington, D.C.: American Chemical Society, 1980, 95–124.

56. F. Jimenez-Colmenero and A.J. Borderias. A study of the effect of frozen storage on certain functional properties of meat and fish protein. *J. Food Technol.*, 18, 731, 1983.

57. T. Suzuki. *Fish and Krill Protein: Protein Technology*. London: Applied Science Publishers 1980.

58. A.J. Borderias, F. Jimenez-Colmenero, and M. Tajada. Parameters affecting viscosity as a quality control for frozen fish. *Mar. Fisheries Rev.*, 47(4), 43–45, 1985.

59. Anonymous. Surimi Workshop III. Primary Processing of Surimi. Seattle, WA: Japan Deep Sea Trawlers Association, 1984.

60. M.A. Rao. *Rheology of Fluid and Semisolid Foods, Principles and Applications*. Gaithersburg, MD: An Aspen Publication, 1999, 59–151.

61. W.P. Cox and E.H. Merz. Correlation of dynamic and steady flow viscosities. *J. Polym. Sci.*, 23, 619–622, 1958.

62. W.B. Yoon. Rheological Characterization of Biopolymer Gel Systems. Ph.D. thesis University of Wisconsin–Madison, WI, 2001.

63. I. Filipi and C.M. Lee. Preactivated iota-carrageenan and its rheological effects in composite surimi gel. *Lebensm-Wiss U-Technol*, 31, 129–137, 1998.

64. J.D. Ferry. *Viscoelastic Properties of Polymer, 3rd ed.* New York: John Wiley & Sons, 1980, 641.

65. S. Gunasekaran and M.M. Ak. Dynamic oscillatory shear testing of foods — selected applications. *Trends Food Sci., Technol.*, 11(3), 115–127, 2000.

66. L.R.G. Treloar. *The Physics of Rubber Elasticity, 3rd ed.* Oxford, U.K.: Clarendon Press, 1975.

67. J.W. Rhim, R.V. Nunes, V.A. Jones, and K.R. Swartsel. Determinant of kinetic parameters using linearly increasing temperature. *J. Food Sci.*, 54, 446–450, 1989.

68. W.B. Yoon, S. Gunasekaran, and J.W. Park. Characterization of thermorheological behavior of Alaska pollock and Pacific whiting surimi. *J. Food Sci.*, 69, E388–343, 2004.

69. J.D. Ferry. Protein gels. *Adv Protein Chem.*, 4, 1–78, 1948.

70. A.H. Clark. Structural and mechanical properties of biopolymer gel modulus. In E. Dickinson, Ed. *Food Polymers, Gels, and Colloids. Royal Society of Chemistry,* Special Pub. No 82, 1991, 322–350.

71. A.J. Steventon, L.F. Gladden, and PJ. Fryer. Gelation of whey protein concentration. *J. Texture Stud.*, 22, 201–218, 1991.

72. L.J.A. da Siva, M.A. Rao, and J.-T. Fu. Rheology of structure development and loss during gelation and melting. In M.A. Rao and RW. Hartel, Eds. *Phase/State Transition in Foods.* New York: Marcel Dekker, 1998, 111–157.

73. J.Q. Wu, D.D. Hamann, and E.A. Foegeding. Myosin gelation kinetic study based on rheological measurements. *J. Agric. Food Chem.*, 39, 299–236, 1991.

74. J. Yongsawatigul and J.W. Park. Linear heating rate affects gelation of Alaska pollock and Pacific whiting surimi. *J. Food Sci.*, 61, 149–153, 1996.

75. J.F. Agassant, P. Avenas, J.-Ph. Sergent, and P.J. Carreau. *Polymer Processing — Principles and Modeling.* New York: Hanser Publishers, 1991.

76. D.G. Baird and D.I. Collias. *Polymer Processing — Principles and Design.* New York: John Wiley & Sons, 1998.

77. B.Y. Kim and J.W. Park. Rheology and texture properties of surimi gels. In J.W. Park, Ed. *Surimi and Surimi Seafood.* New York: Marcel Dekker, 267–324, 2000.

78. S.J. Dus and J.L. Kokini. Prediction of the nonlinear viscoelastic properties of a hard wheat flour dough using Bird-Carreau constitutive model. *J. Rheol.*, 34, 1069–1084, 1990.

79. S. Berland and B. Launay. Rheological properties of wheat flour doughs in steady and dynamic shear: effect of water content and some additives. *Cereal Chem.*, 72, 48–52, 1995.

80. C. Yu and S. Gunasekaran. Correlating dynamic and steady flow viscosity of food materials [abstract]. *IFT Annual Meeting Book of Abstracts,* Dallas, TX, 2000, 85, Abstract #49-4.

81. E. Dickison. *An Introduction to Food Colloids.* New York: Oxford Science Publication, 1992.

12

Microbiology and Pasteurization of Surimi Seafood

YI-CHENG SU, PH.D.
Oregon State University, Astoria, Oregon

MARK A. DAESCHEL, PH.D.
Oregon State University, Corvallis, Oregon

JOE FRAZIER
National Food Processors Association, Seattle, Washington

JACEK JACZYNSKI, PH.D.
West Virginia University, Morgantown, West Virginia

CONTENTS

12.1 Introduction 585
12.2 Growth of Microorganisms in Foods 585
12.3 Surimi Microbiology 587
12.4 Microbial Safety of Surimi Seafood 590
    12.4.1 *Listeria Monocytogenes* 591

12.4.2 *Clostridium Botulinum* 593
12.5 Pasteurization of Surimi Seafood 596
12.6 Process Considerations and Pasteurization
 Verification for Surimi Seafood 600
  12.6.1 Principles of Thermal Processing to
    Surimi Seafood Pasteurization 601
  12.6.2 D-Value 602
  12.6.3 z-Value 602
  12.6.4 F-Value (Lethality Value) 605
  12.6.5 General Considerations for Heat Process
    Establishment or Verification 606
  12.6.6 Study Design and Factors Affecting
    Pasteurization Process 607
  12.6.7 Temperature Distribution Test Design 608
  12.6.8 Heat Penetration Test Design 610
  12.6.9 Initial Temperature (IT) and
    Product Size 611
  12.6.10 Product Preparation/Formulation 611
  12.6.11 Heat Resistance of Selected "Target"
    Microorganism 614
  12.6.12 Analyzing the Pasteurization Penetration
    Data 617
12.7 Temperature Prediction Model for Thermal
 Processing of Surimi Seafood 624
12.8 Predictive Model for Microbial Inactivation
 during Thermal Processing of Surimi Seafood 626
12.9 New Technologies for Pasteurization: High-Pressure
 Processing and Electron Beam 628
  12.9.1 High-Pressure Processing 628
  12.9.2 Food Irradiation 629
  12.9.3 Electron Beam 630
  12.9.4 Electron Penetration in Surimi Seafood 631
  12.9.5 Microbial Inactivation in Surimi Seafood.... 633
  12.9.6 Effect of E-Beam on Other Functional
    Properties of Surimi Seafood 634
12.10 Packaging Considerations 637
References 638

12.1 INTRODUCTION

The quality of surimi seafood largely depends on the types and levels of microbial contaminants in fish and the ingredients used for production. The natural microbial flora associated with marine species are usually related to the environment and vary greatly due to geographic location and water temperature. Gram-positive mesophiles are known to be frequently associated with warm-water fish while Gram-negative psychrotrophs have been reported to be the predominant species of cold-water fish.[1] This chapter discusses the microbiology of surimi seafood and the processes for reducing microbial contamination.

12.2 GROWTH OF MICROORGANISMS IN FOODS

Foods are excellent growth media for microorganisms and can serve as vehicles for food-borne diseases. Fresh food products usually contain a variety of microorganisms and the degree of contamination depends on the growth environment and the methods of harvesting and handling. Although the microorganisms commonly found in foods are mainly spoilage bacteria, human pathogens — including *Salmonella enteritidis*, *Listeria monocytogenes*, and *Vibrio parahaemolyticus*, as well as microbial toxins such as staphylococcal enterotoxins and mycotoxins — can sometimes be present in foods. Controlling the growth of microorganisms and the production of microbial toxins in foods is currently the main approach to preserve quality and ensure product safety.

The growth of microorganisms in foods can be affected by many factors, particularly nutrients, water activity (a_w), oxygen content, temperature, and pH value.[2] Most fresh foods contain essential nutrients (sugar, amino acid, vitamins, and minerals) with water activities (a_w) higher than 0.99 and will therefore support the growth of all kinds of microorganisms. Reducing the water activity in food generally results in retardation or inhibition of microbial growth. Most bacteria, including pathogens, require a minimum a_w of 0.90 to grow.

However, some bacteria, including *Staphylococcus aureus*, are salt tolerant and can grow at $a_w = 0.86$. In addition, yeasts can grow at $a_w = 0.88$ or higher, and molds are usually more dry resistant and can grow at a_w as low as 0.80.

Microorganisms can be classified into two main categories, aerobic and anaerobic, based on the requirement of oxygen for growth. Many microorganisms, including mold and most bacteria, are aerobic and require oxygen to grow. Therefore, their growth in products is inhibited by vacuum packaging. However, anaerobic bacteria such as *Clostridium* spp. that do not require oxygen to grow can grow in vacuum-packed products and cause spoilage as well as toxin formation in certain situations. In addition, facultative anaerobic bacteria such as *Staphylococcus aureus* can also grow without oxygen. However, their growth is usually much better when oxygen is available.

Temperature is another important parameter affecting the growth of microorganisms. Most bacteria can grow over a wide range of temperatures. However, they all have their optimal temperatures for growth. Psychrotrophs are a group of bacteria that grow best between 20 and 30°C, but can grow at refrigeration temperatures (0 to 7°C). This group of bacteria is the main causative agent for product spoilage during refrigerated storage. Mesophiles are bacteria that grow best at 30 to 40°C. Most human pathogens belong to this group because normal human body temperature is at or around 37°C. Thermophiles are those that prefer higher temperatures (50 to 60°C) to grow and are usually not a concern for product spoilage or safety.

The pH value of a food product can also affect the growth of microorganisms. Most bacteria can grow well at pH between 5.0 and 9.0. However, they all grow best at or near neutral pH (between 6.5 and 7.5). Bacterial growth is usually inhibited when the pH value drops below 4.0. However, microorganisms that can grow below pH 4.0 include lactic acid bacteria, yeasts, and molds. In addition, there are a few microorganisms of public health significance that can also grow below pH 4.0.

Other factors, such as the presence of competing microorganisms or naturally occurring antimicrobial substances, can also influence the growth of microorganisms in foods. Therefore, the time required for bacteria to multiply varies, depending on several factors. In general, most bacteria can multiply every 20 to 40 min under optimal growth conditions, which means a single cell can become more than 2 million cells within 7 hr under favorable conditions.

Many methods can be used to preserve foods for human consumption. Among them, thermal processing, such as cooking and canning, is the most common method used for reducing bacterial contamination in foods. Other processes, including dehydration, refrigeration, freezing, and addition of preservatives, can also be used to inhibit the growth of microorganisms. These processes can either be used alone or in combination to increase product shelf life. Many foods can be formulated to contain low water activity ($a_w < 0.85$), high salt content (>10%), or low pH (<4.6) for extended storage periods.

12.3 SURIMI MICROBIOLOGY

Marine and freshwater fish harbor indigenous microbial populations that for the most part participate in noninvasive commensal relationships with the species. The skin, gills, and intestinal tract contain microorganisms that exist on surfaces as biofilm–slimes. These biofilms, in fact, can be beneficial to fish as they can form a barrier against parasites and other fish pathogens. The microbial flora is diverse, with freshwater fish generally having a greater number of mesophilic Gram-positive bacteria as compared to marine species where Gram-negative bacteria are predominant.[3]

These indigenous microbial communities may contain species that are nonpathogenic to the host but potentially pathogenic to humans. Examples include *Aeromonas hydrophila, Edwrardsiella tarda, Klebsiella* spp, *Plesiomonas shigelloides, Streptococcus iniae,* and *Vibrio* spp. These pathogens can pose threats to health and safety through several mechanisms. Improperly cooked or processed seafood may contain these microorganisms and, when ingested, can elicit gastroenteritis, sepsis, and chronic infections. A route of exposure just as important as ingestion is through skin punctures and lesions.[4] Both commercial and recreational fisherman can contract infections from punctures caused by fish spines and teeth.

Certain fish species (primarily the scombroids, such as tuna, sardine, and mackerel) can also harbor bacteria that produce large amounts of histamine if fish are improperly cooled. The indigenous *Enterobacteriaceae* associated with these fish are responsible for decarboxylating fish histidine to histamine. *Morganella* and *Klebsiella* species are bacteria most commonly identified with scombroid poisoning.[5] This food-borne intoxication manifests itself as hypotension, headache, palpitation, and facial flushing. Although symptoms are usually of short duration and most cases are self-resolved, people who have chronic respiratory problems are at higher risk for more serious complications.

Gloria et al.[6] identified the intestinal tract of albacore tuna as containing the greatest amount of histamine as compared to light or dark muscle, suggesting that the decarboxylating bacteria are primarily located in the intestine. Thermal processing (canning or pasteurization) of histamine-contaminated fish, however, is not effective in destroying or reducing histamine levels.[7,8] The presence of histamine has not been reported to occur with surimi or surimi seafood. This is likely because the species of fish most often used for surimi products are non-scombroid. However, this does not preclude the possibility of histamine being formed if, in the future, other fish species are considered for making surimi. Nevertheless, proper cooling and refrigeration of freshly caught fish is paramount in any food processing situation to maximize product quality and safety.

On-board fish handling practices must be adequate to ensure that the catch arrives at the processing plant without any significant increase in microbial numbers. Debate exists as to whether it is preferable to head and gut fish or to leave fish whole during transport to shore. Work by Scott et al.[9] indicated little difference in microbial quality between the two methods with orange roughy stored on ice. However, with some species, which have pronounced enzymatic digestive activity, it may be preferable to gut to prevent flesh deterioration. A thorough discussion of on-board handling practices as it relates to microbial quality was presented by Mayer and Ward.[10]

There exists a chain of product custody relating to who is in charge of maintaining the quality and safety of seafood. Starting with the fisherman and extending to the consumer, there are several points at which product safety can be compromised. Unfortunately, pollution of marine and fresh waters is a fact of life and hence fish or shellfish product, when harvested, may already contain pathogens originating from fecal contamination. Regulation and monitoring of harvest areas is one approach to prevent fecal pollution from becoming a food safety concern. The use of food sanitizing agents such as chlorine, ozone, and potassium sorbate has also been utilized; however, there has been only partial success in actually reducing the microbial loads on fish.[11,12] Moreover, there is growing reluctance by consumers to accept foods that have been chemically preserved. Clearly, the best approach to minimize food safety concerns with either fresh fish or that destined for further processing, such as surimi, is to quickly lower the holding temperature to 0°C and to move it to retail markets or processing facilities as quickly as possible. Fish, when brought to the processing facility, again may be quickly contaminated if proper evisceration and cleaning protocols are not in place. The intestines of fish contain a diverse population of microorganisms that can contaminate the finished product if not adequately contained. Spore-forming bacterial pathogens are of particular concern because they require elevated processing temperatures to be inactivated. Species such as *Clostridium perfringens* (enterotoxigenic) have been isolated from fish guts.[13]

Surimi production involves a series of processing steps, each of which can serve as an opportunity for microbial contamination. Himelbloom et al.[14,15] reviewed the literature regarding the incidence of microbial contamination with surimi and surimi seafood. The first studies provided by Yoon and Matches[16] and Yoon et al.[17] demonstrated that common food-borne pathogens (*Staphylococcus aureus, Salmonella* spp., *Yersina enterocolitica,* and *Aeromonas hydrophila*) could all grow to populations greater than 1×10^8 CFU/g if intentionally introduced (10^3 to 10^4 cells/g) to surimi seafood crab legs and held at temperatures of either 10 or 15°C. This study established that surimi can support the growth of pathogens if the product is temperature abused. The companion study documented the survival of the indigenous surimi microflora during storage and reported that these populations increased significantly over time, regardless of storage temperature (0, 5, 10, and 15°C).

Similar to the first study, the important observation was that surimi must be handled and stored under conditions destined to preclude contamination and to prevent growth of existing microorganisms. Subsequently, in a series of studies by Himelbloom et al.,[18–20] it was shown that microbial populations associated with Alaska pollock surimi increase as fish is processed into surimi at each step. Therefore, among all processing steps, the use of clean fish at the time of deboning is the most important point in controlling the microbial load. The major groups of microorganisms were identified as *Bacillus* and *Pseudomonas* spp. The sanitary quality of processing water and surimi ingredient additives is also believed to be important in ensuring the microbial safety and quality of finished surimi seafood.

12.4 MICROBIAL SAFETY OF SURIMI SEAFOOD

The microorganisms commonly found during seafood processing are mainly those associated with water, fish (skin, gill, intestine), the fishing vessel, processing equipment, and the people handling materials and products. Although the degree of contamination depends largely on the methods of harvesting and on-board handling, cross-contamination can easily occur during post-harvest processing. The types of bacteria commonly associated with seafood are mainly psychrotrophs and can cause product spoilage during refrigerated storage. However, a few pathogens, such as *Listeria monocytogenes*, *Vibrio parahaemolyticus*, and *C. botulinum* type E spores, may also be present in seafood products.

Surimi seafoods are generally free of pathogens because of the thermal processes involved in production. However, microbial hazards could be introduced into finished products through post-processing contamination. The risks are influenced by the original microbial loads, handling and processing procedures, storage conditions, and preparation for consumption. Two of the most serious safety concerns for surimi seafood are the potential contamination by *Listeria monocytogenes* and *Clostridium botulinum* spores.

12.4.1 Listeria Monocytogenes

Listeria monocytogenes is widely distributed in the environment and has been isolated from both fresh and marine water.[21,22] The organism can cause severe diseases in humans, especially pregnant women, newborns, the elderly, and people with weakened immune systems. Numerous outbreaks of listeriosis have occurred in North America since the early 1980s involving the consumption of contaminated coleslaw,[23] pasteurized milk,[24] Mexican-style cheese,[25] hot dogs and undercooked chicken,[26] turkey franks,[27] and processed deli meat.[28] The U.S. Department of Agriculture estimates that 2500 cases of listeriosis occur in the United States each year, with 2300 hospitalizations and 500 deaths.[29]

Raw fish and the processing environment are known to be the major sources for *L. monocytogenes* contamination in seafood. Many categories of seafoods, including a variety of frozen and minimally processed products, have been shown to be frequently contaminated with *L. monocytogenes*.[30–35] Incidences of *L. monocytogenes* contamination in frozen seafood, including raw shrimp and lobster tail, scallops, cooked shrimp and crabmeat, and surimi-based seafood, have been reported to be as high as 26%.[36,37]

Listeria monocytogenes can be destroyed by pasteurization and adequate thermal processing. However, it often enters cooked, ready-to-eat products as a post-processing contaminant. Contamination of *L. monocytogenes* in surimi seafood is a safety concern because surimi seafood analogs are often used for preparing cold sandwiches that are usually consumed without further heating. Although the levels of *L. monocytogenes* in contaminated foods are usually low, the organism can survive and multiply at refrigeration temperature.[38,39] Populations of *L. monocytogenes* in contaminated products can

TABLE 12.1 Seafood and Surimi Seafood Linked to Outbreaks of Listeriosis

Seafood Products	Location	Year	Cases	Ref.
Raw fish or shellfish	New Zealand	1980	22	40
Shrimp	Connecticut (U.S.A.)	1989	9	41
Smoked mussels	Tasmania (Australia)	1991	2	42
Smoked mussels	New Zealand	1992	2	43
Gravad and cold-smoked rainbow trout	Sweden	1994–1995	9	44
Imitation crabmeat	Canada	1996	2	38

increase to significant levels during refrigerated storage and cause listeriosis if products are subsequently consumed.

While there have been no reported outbreaks of listeriosis involving surimi seafood consumption in the United States, seafood products such as shrimp, cold-smoked fish, and smoked mussels have been linked to outbreaks of listeriosis (Table 12.1). Recently, two cases of listeriosis associated with the consumption of surimi-based crabmeat occurred in Canada.[38] A husband and wife were hospitalized with symptoms of nausea, fever, vomiting, and diarrhea shortly after the consumption of the product. *Listeria monocytogenes* serotype 1/2b was isolated from the stool of both patients and the surimi-based crabmeat that was eaten.

The zero-tolerance regulatory policy for *L. monocytogenes* in ready-to-eat products imposed by the U.S. Food and Drug Administration (FDA) presents a significant challenge to the seafood industry. A survey conducted by the FDA on both domestic and imported crabmeat and smoked fish showed that minimally processed seafoods were frequently contaminated with *L. monocytogenes* (Table 12.2). The first reported class I recall of surimi seafood due to *L. monocytogenes* contamination occurred in 1988 for crabstick manufactured in Japan and distributed in three states.[45] Since then, several surimi seafoods have been recalled due to the increased surveillance for *L. monocytogenes* (Table 12.3).

TABLE 12.2 FDA Survey of *L. monocytogenes* in Crab and Smoked Fish (1991–1996)

Sample	Year	No. of Samples Analyzed	No. of Samples Positive	Percentage (%) Positive
Crab	1991	260	12	4.6
	1992	358	34	9.5
	1993	400	37	9.3
	1994	348	28	8.0
	1995	297	25	8.4
	1996	223	6	2.7
Total		1,886	142	7.5
Smoked fish	1991	133	16	12.0
	1992	207	32	15.5
	1993	233	38	16.3
	1994	264	33	12.5
	1995	177	23	13.0
	1996	198	22	11.1
Total		1,212	164	13.5

Source: Data adapted from Ref. 34.

Contamination of *L. monocytogenes* in ready-to-eat food products generally results in expensive costs due to product recalls, investigation of contamination, extensive clean-up, and loss of consumer confidence. Between October 1993 and September 1998, a total of 1328 food products recalls were reported to the FDA because of microbial contamination.[47] Among them, *L. monocytogenes* accounted for the greatest number of total recalls (Table 12.4). Because *L. monocytogenes* is expected to be eliminated by the cooking step involved in production, the presence of *L. monocytogenes* in surimi seafood would either indicate an inadequate heat process or post-processing contamination. These problems can be eliminated if products are heat-pasteurized after packaging. However, more sophisticated packaging is needed to accomplish this.

12.4.2 *Clostridium Botulinum*

Most surimi seafoods are packed in reduced oxygen packaging and sold as ready-to-eat (RTE) products under refrigerated

TABLE 12.3 Class I Recall of Seafood and Surimi Seafood Contaminated with *Listeria monocytogenes*

Product	Manufacturer	Distribution	Date Recalled
Imitation breaded scallops	Korea	New York, New Jersey	09/19/1990
Imitation crabmeat salad	Nebraska	Iowa, Nebraska	08/06/1992
Cold-smoked sablefish	New York	New York	11/17/1992
Smoked mussels	New Zealand	California, Colorado, Florida, Georgia, Illinois, Kansas, Maryland, Maine, Michigan, North Carolina, Nevada, Oklahoma, Oregon, Pennsylvania, Texas, Utah	12/28/1992
Alaska snow crab legs and claws	Washington	Northwest U.S. and Canada	01/27/1993
Fresh crabmeat	Florida	Florida	03/15/1993
Imitation crabmeat (chunks)	Washington	Arizona, California, Nevada, Oregon, Washington	03/29/1996
Imitation King crab meat	Washington	Washington	06/02/1997
Surimi (imitation crab) spread	Idaho	Idaho, Nevada, Oregon, Utah, Wyoming	08/12/1997
Smoked salmon dip	New Jersey	Indiana, Kentucky, New Jersey, North Carolina, Pennsylvania, Virginia	04/02/1999
Cold-smoked Atlantic salmon	Massachusetts	Massachusetts	06/14/1999
Imitation crabmeat	Washington	Canada	07/28/1999
Shrimp (cooked and peeled)	Washington	California, Maryland, Montana, Oregon, Washington	05/14/2001
Fresh crab claw meat	Florida	Georgia, Maryland, New York, Pennsylvania, Texas	04/10/2003
Smoked salmon	Oregon	Oregon, Washington	06/05/2003

Source: Adapted from Ref. 46.

TABLE 12.4 Microbial Agents Responsible for Food Product Recalls Reported to U.S. FDA between October 1993 and September 1998

Microbial Agents	No. of Products Recalled	Percentage
Listeria monocytogenes	813	61.2
Salmonella	143	10.8
Yeast and/or mold	96	7.2
Staphylococcus aureus	69	5.2
Norwalk-like virus	61	4.6
Clostridium botulinum	38	2.9
Coliforms (other than *E. coli* O157:H7)	36	2.7
Hepatitis A virus	33	2.5
Escherichia coli O157:H7	16	1.2
Lactobacillus	12	0.9
Bacillus cereus	3	0.2
Other	8	0.6
Total	1328	100

Source: Adapted from Ref (47).

or frozen storage conditions. Products that are distributed under refrigerated temperatures may encounter a potential hazard from the growth of *Clostridium botulinum* if the product is contaminated with bacterial spores. Although keeping vacuum-packed surimi seafood at temperatures of 3.3°C (38°F) and below provides an adequate barrier to the growth of *C. botulinum*, temperature abuse during storage may not always be avoided. *Clostridium botulinum* is a strict anaerobic, spore-forming bacterium that can produce a series of neurotoxins causing human paralyses and, in many cases, death. This organism is ubiquitously distributed in the environment and can be found in birds, fish, and ocean sediments. Seven serologically different toxins, designated as type A through G, are produced by *C. botulinum*. Among them, type E is of great concern to the seafood industry because it is mainly associated with the marine environment and spores have been isolated from a number of marine species, particularly from the gills and intestinal tract.[48–50]

Clostridium botulinum is classified into two categories based on enzymatic activities. Proteolytic strains of *C. botulinum* (all type A, some type B and F) produce heat-resistant spores and generate putrid odors during growth. They require a minimal temperature of 10°C (50°F) to grow, and their growth can be inhibited by a water-phase salt (WPS) equal to or greater than 10% or a water activity (a_w) equal to or less than 0.93.[51,52] On the other hand, nonproteolytic strains (all type E, some type B and F) do not produce putrid odors from growth and their spores are less heat resistant. Growth of nonproteolytic *C. botulinum* can be inhibited by a WPS equal to or greater than 5% or an a_w less than or equal to 0.96 if no thermal processing is combined.[52] However, they can grow and produce toxins at temperatures as low as 3°C (37.4°F).[53,54] Despite the physiological characteristics of *C. botulinum*, all the toxins they produced can be inactivated by heating at 60°C (140°F) for 5 min.[55]

The ability of *C. botulinum* spores to survive cooking processes and germinate in vacuum-packed products under elevated temperatures is a significant safety concern for temperature-abused, vacuum-packed surimi seafood. Although outbreaks of human botulism caused by the consumption of surimi seafood have never been reported, many outbreaks have been linked to fishery products such as smoked, salted, or fermented products that were eaten without prior heating.[56] Ready-to-eat, vacuum-packed surimi seafood should therefore be stored at temperatures below 3°C (37.4°F) to prevent the growth of nonproteolytic types of *C. botulinum*, particularly type E strains.

12.5 PASTEURIZATION OF SURIMI SEAFOOD

Pasteurization is a heat process designed to destroy pathogens and reduce spoilage bacteria in products. Surimi seafoods that have been properly pasteurized should contain no pathogens and only low levels of vegetative spoilage bacteria, except for heat-resistant bacterial spores. Rapid cooling of pasteurized surimi seafood from 60°C (140°F) to 21.1°C (70°F) or below within 2 hr, and to 3°C (37.4°F) or below within an additional 4 hr, helps to reduce the growth of spoilage bacteria and prevent spore germination of *Bacillus* and *Clostridium* species.[15]

Various time and temperature combinations, ranging from 20 min at 70°C to 30 min at 95°C, have been used by surimi seafood manufacturers for pasteurizing crab analogs.[52] A pasteurization process of 90°C for 10 min or a process that would achieve a 6-log reduction of C. *botulinum* spores is currently recommended for hermetically sealed surimi products. However, this process may cause adverse effects in the nutritional value and overall product quality. Other intervention strategies such as water-phase salt content can be used in combination with a heating process to inhibit the growth of nonproteolytic *C. botulinum* and toxin formation.

The resistance of *C. botulinum* spores to heat processing is strain dependent and can be affected by product components. Studies have shown that *C. botulinum* type E spores inoculated into hot-smoked fish were inactivated by a heat process at 85°C (185°F) for 85 min, at 88.9°C (192°F) for 65 min, or at 92.2°C (198°F) for 55 min.[57] No type E toxins were produced in the heat-processed products during 21 days of storage at 25°C. Spores of nonproteolytic type B, however, were found to be more heat resistant than type E spores. Longer heating times of 175, 85, and 65 min at 85, 88.9, and 92.2°C, respectively, were required to prevent nonproteolytic type B strains from producing toxins.

In a similar study to develop a pasteurization process for hand-picked Dungeness crabmeat packed in oxygen-impermeable flexible pouches, mixture of three strains of *C. botulinum* nonproteolytic type B (total 10^7 spores) were completely inactivated by a heat process at 88.9°C for 90.3 min or at 94.4°C for 20.3 min. The D values for the nonproteolytic type B mixtures were calculated as 12.9 at 88.9°C and 2.9 at 94.4°C, respectively.[58] The investigators concluded that this pasteurization process inactivated spores of nonproteolytic *C. botulinum* types B, E, and F, and extended the refrigerated shelf life of the products. However, these processes did not inactivate the heat-resistant proteolytic strains of *C. botulinum* or other, more heat-resistant spore-forming bacteria.

Recently, Peterson et al.[52] investigated the effects of combining water-phase salt (WPS) and heat pasteurization on controlling nonproteolytic *C. botulinum* types B and E in crab analogs. Crab analogs containing different WPS contents (2.1, 2.4, and 2.7%) were inoculated with type B or E spores at levels ranging from 10^2 to 10^4 spores/g and vacuum-packed in oxygen-impermeable bags. The vacuum-sealed bags were placed in a hot water bath for heat treatments at 80, 85, and 90°C, respectively, for 15 min (after the product reached the desired internal temperature). The heat-processed samples were stored at 10°C for 120 days and at 25°C for 15 days to determine the effectiveness of each process on spore inactivation.

Results indicated that samples inoculated with type B or E spores at 10^2 spores/g and processed at 80°C remained nontoxic after 120 days of storage at 10°C (Table 12.5). When samples were inoculated with a higher level of spores (10^4 spores/g) and processed at 80°C, type E toxin was in samples containing 2.1% WPS after 80 days of storage at 10°C while type B toxin was in all three groups of samples containing different amounts of WPS beginning on day 80. The minimal inhibitory level of WPS decreased when the processing temperature increased to 85°C. None of the samples inoculated with 10^4 type E spores became toxic after 120 days of storage at 10°C.

For samples stored at 25°C, no type E toxins were produced in samples inoculated with 10^2 spores/g and processed at 80°C (Table 12.6). However, the toxins were in samples containing WPS of 2.1 and 2.4% when a higher level of spore (10^4 spores/g) was used as the inoculum. When the processing temperature was increased to 85°C, no type B toxins were found in any samples and type E toxins were produced only in samples containing the lowest WPS (2.1%). All the crab analogs inoculated with 10^3 type E spores/g and processed at 90°C for 15 min remained free of toxin at 25°C for 15 days.

The combination of WPS and heat pasteurization could be applied to surimi seafood for controlling the growth and toxin production of nonproteolytic type *C. botulinum*. Heating surimi seafood to an internal temp of 85°C for 15 min and a WPS equal to or greater than 2.4% could therefore be used

TABLE 12.5 Production of Toxins by Nonproteolytic *Clostridium botulinum* Type B and E in Vacuum-Packed Heat-Processed Surimi Seafood (Crabstick) Stored at 10°C

C. botulinum	Inoculation (spores/g)	Processing Temp. (°C)	Initial Toxin Production in Heat-Processed Samples during Storage (days)		
			2.1% (WPS)	2.4%	2.7%
Type E	10^2	80	0/6[a] (120 d)	0/6 (120 d)	0/6 (120 d)
		85	0/6 (120 d)	0/6 (120 d)	0/6 (120 d)
	10^4	80	1/6 (80 d)	0/6 (120 d)	0/6 (120 d)
		85	0/6 (120 d)	0/6 (120 d)	0/6 (120 d)
Type B	10^2	80	0/6[a] (120 d)	0/6 (120 d)	0/6 (120 d)
		85	0/6 (120 d)	0/6 (120 d)	0/6 (120 d)
	10^4	80	1/6 (100 d)	2/6 (80 d)	1/6 (90 d)
		85 (not tested)			

[a] Number of samples became toxic/number of samples tested.

Source: Adapted from Ref. 52.

TABLE 12.6 Production of Toxins by Nonproteolytic *Clostridium botulinum* Type B and E in Vacuum-Packed Heat-Processed Surimi Seafood (Crabstick) Stored at 25°C

C. botulinum	Processing Temp. (°C)	Inoculation (spores/g)	Initial Toxin Production in Heat-Processed Samples during Storage (days)		
			2.1% (WPS)	2.4%	2.7%
Type E	80	10^2	0/3[a] (15 d)	0/3 (15 d)	0/3 (15 d)
		10^4	1/3 (7 d)	1/3 (7 d)	0/3 (15 d)
	85	10^3	1/3 (11 d)	0/3 (15 d)	0/3 (15 d)
Type B	85	10^3	0/6 (15 d)	0/6 (15 d)	0/6 (15 d)

[a] Number of samples became toxic/number of samples tested.

Source: Adapted from Ref. 52.

to process products that will be stored at refrigeration temperatures. However, the heat resistance of *C. botulinum* varies among strains and can be affected by different ingredients. Therefore, the heat process should be evaluated for individual products. The best control is to keep products at 3.3°C (38°F) and avoid temperature abuse.

12.6 PROCESS CONSIDERATIONS AND PASTEURIZATION VERIFICATION FOR SURIMI SEAFOOD

Surimi seafoods are pasteurized with a mild heating process to produce a product free of pathogens. This means that the pasteurization process must be designed to ensure an appropriate reduction in the numbers of microorganisms of public health concern to a "safe" level. The two primary pathogenic microorganisms of concern in surimi seafood products — *L. monocytogenes* and nonproteolytic *C. botulinum* — have been discussed. Selecting which of these two "target" organisms to control by pasteurization will be dictated by the packaging style and the method of storage/distribution of the finished product as follows:

1. *L. monocytogenes* is the target organism of concern for product that is either (a) packaged in air and stored/distributed in either refrigerated or frozen form, or (b) vacuum-packaged and stored/distributed in frozen form only.
2. Nonproteolytic *C. botulinum* type B is the target organism of concern for product that is vacuum-packaged and stored/distributed in refrigerated form.

But what exactly is considered a "safe reduction in numbers" for these two pathogens? How do surimi seafood manufacturers know or determine that their pasteurizer is "doing its job" in actually delivering the right amount of heat to the product or to achieve the necessary level of reduction? Furthermore, what HACCP critical limits and monitoring procedures surrounding the pasteurization step should be in place to ensure consistent and continued control of the process?

This section attempts to provide answers to these questions and provide additional guidance intended to assist the surimi seafood manufacturer to ensure that the adequacy of the pasteurization process is consistent with both FDA HACCP guidance and recognized scientific principles and practices of thermal processing.

12.6.1 Principles of Thermal Processing to Surimi Seafood Pasteurization

Although the history of thermal processing or canning of foods can trace its roots back to 1810, it was not until around 1870 that scientists began to truly understand the science of thermal processing.[59] It was then that Louis Pasteur demonstrated the role that certain microorganisms (bacteria, yeasts, and molds) played in food spoilage and fermentation. Pasteur showed that the microorganisms responsible for spoiling perishable foods could be destroyed by mild heat treatment and the food would remain preserved and unspoiled if it was packaged in such a manner as to avoid recontamination. Hence, the term "pasteurization" was borne out of his work.

Not long after, it was demonstrated that certain microorganisms were also capable of causing food-borne disease (the bacterial spore was discovered in 1876). The science of thermal processing advanced rapidly thereafter — studies that defined the heat resistance of numerous food spoilage and food pathogenic microorganisms, as well as studies describing the time–temperature parameters necessary for their destruction. It was also observed that microorganisms exposed to lethal temperatures died at a constant rate (logarithmic) and this led to the 1920s development of mathematical expressions for predicting the time and temperature necessary to destroy microorganisms.

The principles of thermal processing first described nearly a century ago still serve as the basis for today's thermal processes for food products. The rate of product heating (heat penetration, or HP) is combined with thermal death time (TDT, or heat resistance data) for a particular target microorganism in the product to mathematically calculate a "safe" time–temperature heat process. When establishing or verifying a pasteurization process for surimi seafood products, knowledge and information in this area are needed. For this reason, it is important to be familiar with the following terms and definitions in the science of thermal processing.[59,60]

12.6.2 D-Value

The D-value is known as the "decimal reduction time" and is defined as the time (usually in minutes) required to kill 90% of the spores or vegetative cells of a given microorganism at a specific temperature in a specific medium. A 90% reduction in bacteria is equivalent to a reduction from 10,000 to 1000 bacteria/g (1-log cycle). D-values can be determined from survivor curves when the log of the population is plotted against time (Figure 12.1), or by the formula:

$$D_T = \frac{T}{(\log A - \log B)}$$

where:
    T = time of heating
    A = the initial number of microbial cells
    B = the final number of surviving microbial cells after heating time

12.6.3 z-Value

The z-value is the number of degrees (°F or °C) that results in a tenfold change (1-log cycle) in an organism's heat resistance; that is, an organism's D-value will be either increased or decreased by 1-log unit when it is heated at "z degrees F" higher or lower, respectively. The z-value, therefore, gives an indication of the relative impact of different temperatures on a microorganism, with smaller z-values indicating greater sensitivity to increasing heat.

On a thermal death time curve where the logarithms of at least 2 D-values are plotted against temperature, the z-value is the number of degrees required to traverse 1-log cycle (Figure 12.2). The z-value can also be calculated by the formula:

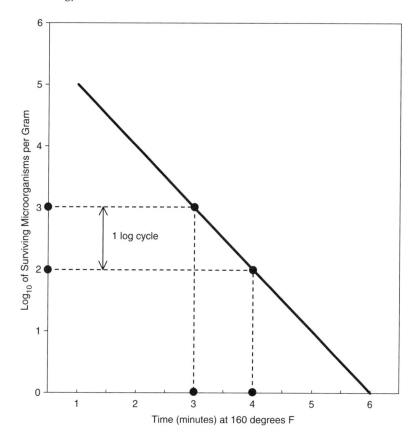

Figure 12.1 Decimal reduction time graph (hypothetical data) illustrates a decimal reduction value of 160°F of 1 min ($D_{160} = 1.0$ min).

$$z = \frac{(T_1 - T_2)}{(\log D_2 - \log D_1)}$$

where:

T_1 and T_2 are temperatures
D_1 and D_2 are D-values at temperatures T_1 and T_2

Using the z-value of a particular organism and by expressing the above equation in a different way, D-values at different temperatures can also be determined as follows:

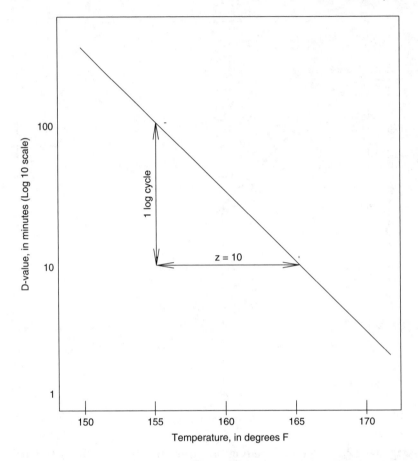

Figure 12.2 Example thermal death time curve depicting an organism with z = 10°F.

$$\log D_2 - \log D_1 = \frac{(T_1 - T_2)}{z}$$

where:
D_1 = Known D-value at temperature T_1
D_2 = Unknown D-value at temperature T_2

For example, using the known D-value and z-value for *L. monocytogenes* in surimi seafood from Mazzotta,[61] where

$D_{151} = 0.40$ min and z = 10.26°F, a different D-value at 140°F can be calculated as follows:

$$\log D_{140} - \log D_{151} = [(151 - 140)/10.26]$$

$$\log D_{140} - \log(0.4) = 1.072$$

$$\log D_{140} = 1.072 + \log(0.4) = 1.072 + (-0.398) = 0.674$$

By raising both sides of the equation by "power 10," to eliminate the log expression, the result of the equation becomes:

$$D_{140} = 10^{(0.674)} = 4.72 \text{ min}$$

Thus, the equivalent D-value at 140°F is 4.72 min for *L. monocytogenes* in surimi seafood. Equivalent D-values should not be calculated for temperatures far higher or lower than those used in the original laboratory studies or errors may result due to the nonlinearity of some survivor curves.

12.6.4 F-Value (Lethality Value)

The F-value, or lethality value, is defined as the time, usually in minutes (min), necessary to destroy a given number of microorganisms at a specific constant temperature or the number of equivalent minutes spent at a given reference temperature. The F-values and D-values are related in that the F-value of a heat process generally represents multiple D-values; and can be described by the formula:

$$F_T = D_T \times \log(A)$$

where *A* is the target number of microorganism to destroy

Using the same example values from above, if the $D_{151} = 0.40$ min for *L. monocytogenes* in surimi seafood, then a pasteurization process with an F-value of 2.4 min ($F_{151} = 2.4$) would achieve 6-log reduction for the target microorganism, or a 99.9999% kill (*A = 1,000,000* in the above equation). The lethality, F, of the process would be described as being a "6D" process — a reduction of the target pathogen by six orders of magnitude.

F-values allow the direct comparison of two or more heat processes (having the same z-value). The equivalent lethality of a process at a different temperature can be calculated in basically the same manner that D-values are calculated for different temperatures:

$$\log F_2 - \log F_1 = \frac{(T_1 - T_2)}{z}$$

For example, a "6D" process for *L. monocytogenes* in surimi seafood, as described above, has a lethality value F_{151} = 2.4 min (z = 10.26°F). Stated another way, if the product could be instantly heated to an internal temperature of 151°F, held for 2.4 min, and then instantly cooled back down again, a 6-log reduction for *L. monocytogenes* would be achieved. A heat process that delivers the same equivalent lethality, however, can be achieved in 0.033 min (about 2 sec) at a temperature of 170°F, or F_{170} = 0.033 min (2 sec), respectively.

Of course, in commercial practice, surimi seafood pasteurization systems do not "instantly heat" or "instantly cool" product. As product heats up during the pasteurization process, it builds up or accumulates lethality during the "come-up" time, with more and more lethality being delivered as the internal product temperature continues to rise toward the reference temperature and beyond. In this regard, the terms "F-value," "equivalent lethality," "total lethality," "total cumulative lethality," or "total accumulated lethality" are, for the most part, used interchangeably.[59,61,62] These principles will be used later in this section to calculate the total or cumulative lethality of a surimi seafood pasteurization process.

12.6.5 General Considerations for Heat Process Establishment or Verification

According to guidance presented in Chapters 13 and 17 of the FDA *Fish & Fishery Products Hazards and Controls Guidance*,[63,64] an acceptable "safe" pasteurization process is one resulting in a 6-log reduction of the target pathogen (i.e., a 6D process). As discussed earlier, the target organism will be either *L. monocytogenes* or nonproteolytic *C. botulinum*, depending on how the final packaged product is stored/distributed. When nonproteolytic *C. botulinum* is the target organism, the FDA recognizes an alternative to the 6D heat process, specifically for surimi seafood product that has been formulated to contain a certain minimum level of water-phase salt. Further discussion of both these control options for non-proteolytic *C. botulinum* follows later in this section.

FDA guidance also states that the adequacy of the pasteurization process should be established by scientific study. Pasteurization process systems come in a variety of shapes and sizes (i.e., batch, single- or multiple-belt continuous conveyance, steam or hot water immersion, etc.). Therefore, designing and conducting the scientific studies necessary to establish or verify the adequacy of the process will be system-specific in most cases, depending upon the pasteurization system employed.

The primary steps associated with establishing or verifying the adequacy of a pasteurization process are:

1. Design a study that accurately and adequately considers all the factors/conditions that affect the rate of product heating.
2. Execute the study by gathering data on the rate of product heating — referred to as heat penetration (HP) data — under a specific set of processing conditions.
3. Analyze the HP data and determining the appropriate pasteurization process for the product — the minimum pasteurization time/temperature to achieve the minimum 6D process for the selected target pathogen. This involves identifying other critical factors or conditions that may affect the adequacy of the pasteurization process and establishing appropriate critical limits (CLs), controls, and monitoring procedures.

12.6.6 Study Design and Factors Affecting Pasteurization Process

Scientific studies to establish or verify a pasteurization process must demonstrate that the proper heat process is delivered to each and every unit of product. The studies must therefore be designed appropriately with knowledge and information in the following three areas:

1. ***Type of pasteurization system.*** This would include information on the adequacy of temperature distribution in the system, the heating medium used, and how the equipment is designed, operated, and maintained. Verifying the uniformity of the temperature distribution of the pasteurization system is essential to study design because the existence of "cold spots" will influence the rate of heat penetration into the product, which, in turn, can affect the adequacy of the process.
2. ***Heating characteristics of the product and container.*** This information is acquired by gathering product heat penetration data. Previous studies conducted by the National Food Processors Association indicated that heat penetration in surimi seafood products followed simple conduction, with the slowest heating portion being at the geometric center (center core) of the container or package.
3. ***Selecting the appropriate target microorganism and information of the heat resistance of that particular organism in the product.*** This information is generally available for a variety of different seafood products from existing scientific literature. Where this information is not available, thermal death time (TDT) studies may be needed to determine heat resistance.

12.6.7 Temperature Distribution Test Design

In general, before product heat penetration testing is done, the uniformity of temperature distribution within the pasteurization system should be characterized. This is accomplished using temperature-measuring devices called thermocouples (TCs) positioned in various locations within the pasteurizer system. Where continuous pasteurization process systems are employed, wireless temperature-sensing devices or dataloggers are particularly useful (also useful for product heat penetration testing).

Depending on the size of the system, several test runs should be made to profile the temperature distribution of the system. Temperature distribution testing of the pasteurization system should always be done under full load conditions, as this is the worst case with respect to temperature distribution uniformity. The data must then be analyzed to determine if any "cold spots" exist in the pasteurization system, that is, areas or regions within the unit that show slower heating (lower temperature profile) than other areas. If a cold spot is identified, it is important that subsequent product heat penetration testing be done in this area or region of the pasteurization system.

For example, for a continuous conveyance pasteurization system having a 12-ft-wide stainless steel mesh belt, several wireless temperature dataloggers could be spread out along the width of the belt in between packages of product and allowed to run through the system. For example, one datalogger is located in the middle of the belt among the packages of product and two others are located in the vicinity of packages that are at the extreme edges (left and right side) of the conveyor belt. The temperature data is downloaded to a computer and then analyzed to determine temperature uniformity. If the data shows that the left side of the 12-ft conveyor belt is consistently 4–5°F lower in temperature compared to the other areas tested along the belt, subsequent product heat penetration testing would be conducted in that "low temperature area" of the pasteurizer belt.

Surimi seafood processors unfamiliar with designing and conducting temperature distribution tests may wish to consult with a process authority to ensure that their test protocol takes into account the worst-case heating conditions of the pasteurizer system.[65] While temperature distribution testing is generally done prior to product heat penetration testing, it can be done concurrently provided there are enough thermocouples or dataloggers to undertake both tests. An adequate number of test runs must be conducted to simultaneously determine both the uniformity of temperature distribution within the system and the slowest heating product. Doing both tests simultaneously may be more efficient and/or cost effective for the processor.

The processor should also bear in mind that major changes to the pasteurization process system (e.g., modifying the steam spreader configuration or hole size/spacing of a continuous steam pasteurization system) can negatively impact temperature distribution and therefore the rate of product heating, which could ultimately affect the adequacy of the pasteurization process. Changes to related operations can also impact temperature distribution, such as product/package size, product loading patterns, changes to custom-built product holding trays/baskets, and conveyor belts with substantially different perforation or mesh size. The processor should evaluate any such changes to determine the potential (negative) effect on adequate temperature distribution. In addition, it is important to have a prerequisite, preventive maintenance program in place, which includes periodic inspections of the system components involved in the distribution/circulation of the heating medium (i.e., hot water pumps, steam spreaders and perforations, etc.) to ensure that adequate temperature distribution is being maintained.

12.6.8 Heat Penetration Test Design

Gathering product heat penetration data is also done using thermocouples, which are positioned in the slowest heating portion of product (i.e., the geometric center or center core of the container or package for surimi seafood products). In general, a somewhat conservative approach should be followed in designing the heat penetration study to show that the slowest heating unit/portion of product under the worst set of heating conditions receives the appropriate heat necessary to control the target microorganism. As with temperature distribution testing, processors that are not familiar with designing and conducting heat penetration tests may wish to consult with a process authority when developing their test protocol.[65]

When conducting the heat penetration study, attention must be paid to the specific set of pasteurization/operating conditions. Such conditions will likely be the focus of subsequent monitoring of the pasteurization process step under HACCP and may include additional critical factors that can affect the product heat penetration. In addition to the length of time and temperature at which the product is heated, the rate of heat penetration into the product may be influenced by a number of other processing or packaging conditions/factors that should be considered in designing the study. Critical factors that can affect the rate of heat penetration for surimi seafoods include but are not limited to:

- Temperature distribution uniformity within the pasteurization system
- Initial temperature (IT) of product
- Size/shape, thickness, density or "mass" of product
- Product preparation/formulation (particularly as it affects the percent solids)

12.6.9 Initial Temperature (IT) and Product Size

The rate at which a product heats and the time it takes to achieve the desired 6D reduction of the target microorganism are influenced by the starting initial temperature (IT) of the product. A longer time would be needed for a product to reach the desired 6D reduction endpoint with a lower IT (e.g., 45°F) than a warmer IT (e.g., 60°F). Similarly, thicker and/or denser products would heat more slowly. A 2-in.-thick vacuum package of surimi seafood product will heat more slowly than a 1.5-in.-thick package (Figure 12.3). When conducting product heat penetration testing, thermocouples would be positioned at the geometric center (the slowest heating portion of product) such as at 1 in. and 0.75 in. in these two examples, respectively.

12.6.10 Product Preparation/Formulation

Surimi seafood products typically contain ingredients such as starches and gums that can affect the density or viscosity of the product. These ingredients, as well as a variety of others, can also affect the solids content of the product. Both variables — viscosity and solids content — are generally recognized as factors that affect the rate of product heating. Because surimi seafood products are already in a fully cooked state of fish

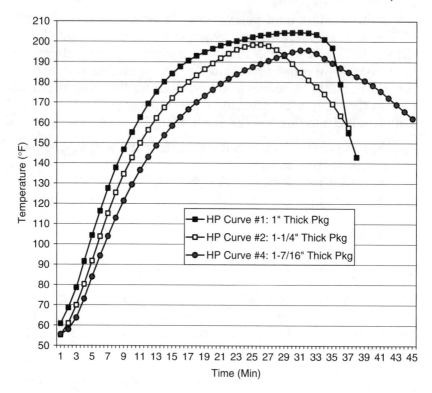

Figure 12.3 Heat penetration curves for surimi seafood with various package thicknesses.

proteins prior to final packaging and pasteurizing, viscosity is not a significant factor affecting the rate of product heating at the final pasteurization step, despite the presence of ingredients that might indicate otherwise.

However, changes to product formulation or preparation that result in significant changes to the solids content of the product, particularly after the initial gel setting and cooking step(s) and prior to final packaging and pasteurizing, can potentially impact the product heat penetration rate. In general, new product heat penetration data should be generated for new or different formulations resulting in a 5% increase or more in solids contents (i.e., a less moist product) compared to the original formulation.

In summary, it is recommended that a conservative study be designed to demonstrate the adequacy of the pasteurization process — one that accounts for the worst-case conditions or factors encountered. The study should examine factors that commonly influence the rate of product heating during pasteurization and should include:

- Coldest portion/region within the pasteurization system (under full load conditions)
- Lowest pasteurization process temperature
- Shortest pasteurization process time
- Coldest product (lowest IT)
- Thickest (or largest mass) product
- Product preparation/formulation

In designing the study, it may be possible (even desirable at times) to set up extremely conservative conditions for certain factors that affect product heat penetration although such condition(s) would never exist in the actual process. The intent of creating such conditions would be to preclude the factor from being considered a critical factor of the pasteurization process and eliminate the need to establish a critical limit for the factor and subsequent monitoring of the factor.

Similarly, the existence of inherent equipment design, self-limiting, and self-controlling conditions of the process may preclude an otherwise critical factor from consideration. For example, if inherent in the process is the fact that the initial temperature of the product never falls below 50°F, then conducting the pasteurization process study using refrigerated product (38 to 40°F) would be sufficient justification to preclude initial temperature as a critical factor of the pasteurization process because the artificially created condition would never occur during "real" production. Another example would be the inherent design of certain makes/models of vacuum packaging equipment using top and bottom roll-stock poly film. For certain models, the depth of the filling pocket, which defines the package thickness, is physically self-limited to the depth of the metal forming heads on the equipment itself. This maximum package thickness cannot be exceeded and still maintain the proper package seal/seam integrity.

Thus, if the thickest packages of product that the metal forming heads are capable of producing were used in the surimi seafood heat penetration study, the package thickness may be precluded as a critical factor of the pasteurization process because the equipment itself precludes thicker packages from being produced. In such a case, a prerequisite program should be in place that ensures the vacuum packaging equipment is set to the appropriate package size/specification.

When performing temperature distribution and/or product heat penetration tests, proper records are essential. It is important to record all relevant information about the test and potential critical factors, including the test conditions that involve the setup and operation of the pasteurization process system equipment. While it is not possible to list all possible types of information that might be needed, the processor may wish to consider recording the following test parameters:

- Test date and test ID
- Name(s) of individual(s) conducting the test
- Description of the pasteurization system and heating medium
- Product or product style
- Product preparation/formulation, including viscosity, percent solids, fill procedure, and weight
- Size, type, and thickness of package
- Test product or package ID and corresponding thermocouple ID
- Thermocouple position in the product
- Initial temperature of product
- Pasteurization system operating parameters or settings
- Temperature distribution "maps," including the location of various thermocouples in the system

12.6.11 Heat Resistance of Selected "Target" Microorganism

When product-specific heat resistance data is not available for a particular target organism, the FDA has provided a wide range of recommended process times/temperatures adequate to achieve a 6D process for *L. monocytogenes* and non-proteolytic *C. botulinum* type B. These recommendations[66] are found in Tables 12.7 and 12.8. These tables are best described as "F-value" or "equivalent lethality" process tables because the process time/temperature recommendations contained therein represent the equivalent number of minutes at a given reference temperature and z-value to provide for a 6D process.

TABLE 12.7 FDA Fish and Fishery Products Hazards and Controls Guidance for Inactivation of *Listeria monocytogenes*

Internal Product Temp. (°F)	Internal Product Temp. (°C)	Lethal Rate	Time for 6D Process (min)
145	63	0.117	17.0
147	64	0.158	12.7
149	65	0.215	9.3
151	66	0.293	6.8
153	67	0.398	5.0
154	68	0.541	3.7
156	69	0.736	2.7
158	70	1.000	2.0
160	71	1.359	1.5
162	72	1.878	1.0
163	73	2.512	0.8
165	74	3.415	0.6
167	75	4.642	0.4
169	76	6.310	0.3
171	77	8.577	0.2
172	78	11.659	0.2
174	79	15.849	0.1
176	80	21.544	0.09
178	81	29.286	0.07
180	82	39.810	0.05
182	83	54.116	0.03
183	84	73.564	0.03
185	85	10.000	0.02

Note: z = 13.5°F (7.5°C).

Source: Adapted from Ref. 66.

TABLE 12.8 FDA Fish and Fishery Products Hazards and Controls Guidance for Inactivation of Nonproteolytic *Clostridium botulinum* type B

Internal Product Temp. (°F)	Internal Product Temp. (°C)	Lethal Rate	Time for 6D Process (min)
185	85	0.193	51.8
187	86	0.270	37.0
189	87	0.370	27.0
190	88	0.520	19.2
192	89	0.720	13.9
194	90	1.000	10.0
196	91	1.260	7.9
198	92	1.600	6.3
199	93	2.000	5.0
201	94	2.510	4.0
203	95	3.160	3.2
205	96	3.980	2.5
207	97	5.010	2.0
208	98	6.310	1.6
210	99	7.940	1.3
212	100	10.000	1.0

Note: For temperatures less than 194°F ([90°C)], $z = 12.6$°F ([7.0°C)]; for temperatures above 194°F ([90°C)], $z = 18$°F ([10°C)].]

* Note: these lethal rates and process times may not be sufficient for the destruction of nonproteolytic *C. botulinum* type B in Dungeness crabmeat, because of the potential that substances that may be naturally present, such as lysozyme, may enable the pathogen to more easily recover from heat damage.

Source: Adapted from Ref. 66.

For *L. monocytogenes*, Mazzotta[61] has already described the specific heat resistance of this organism in surimi seafood products. Therefore, the D- and z-values established by Mazzotta's work can be used in calculating an appropriate 6D pasteurization process in lieu of the conservative time/temperature processes recommended by the FDA for controlling *L. monocytogenes* in surimi seafood.

Conversely, TDT studies are currently underway by the U.S. surimi seafood manufacturers under coordination of the

Microbiology and Pasteurization of Surimi Seafood 617

National Food Processors Association. They will describe the heat resistance of nonproteolytic *C. botulinum* type B in surimi seafood products. The 6D process times/temperatures recommended by the FDA[66] should be used until such studies are completed. When nonproteolytic *C. botulinum* type B is the target organism, the FDA specifically recommends that surimi seafood be pasteurized to a cumulative lethality equivalent to F_{194} of 10.0 min. The FDA also recognizes that past studies demonstrated that certain combinations of product salt content and heat can prevent growth and toxin formation by nonproteolytic *C. botulinum* type B in refrigerated, vacuum-packaged surimi seafood. Surimi seafood products containing a minimum of 2.4% water-phase salt can be pasteurized to an internal temperature of 185°F for at least 15 min to produce a safe product.[52,63,67]

12.6.12 Analyzing the Pasteurization Penetration Data

The lethality of a pasteurization process in destroying a particular target pathogen can be calculated using what is known as the General Method.[68,69] The General Method calculation is the "reference" for all other thermal process calculations (e.g., Ball formula method). Using the General Method, time/temperature data obtained directly from the product heat penetration test is coupled with thermal death time information (D- and z-values) for the target pathogen to directly calculate the lethality of the pasteurization process. The General Method takes into account all the lethality accumulated under the heat penetration curve, including both the product heating and cooling phases (Figure 12.4).

Because the area under the curve cannot be easily integrated directly over time, the General Method uses a summation technique to approximate the solution, that is, the total F-value. This involves breaking down the area under the heat penetration curve into smaller "time segments of heating" and calculating the lethality for each segment or section using the following formula:

$$\text{Lethality or F-value} = (\text{Lethal rate}) \times (\text{Time})$$

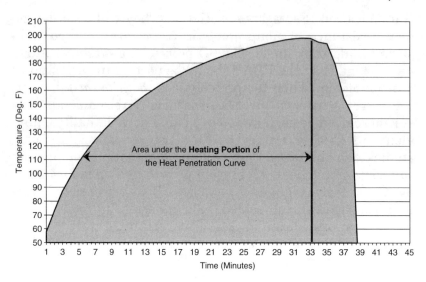

Figure 12.4 Illustration of "Area under the Heating Curve" for Calculating Cumulative Lethality by the General Method.

where:
Lethal rate = $\log^{-1}[(T - T_r)/z] = 10^{[(T - T_r)/z]}$
T = product temp. (°F) T_r = reference temp. (°F)
z = z-value (°F)

Starting at a given time, the lethal rate is calculated for each time interval based on the lowest temperature for that time interval. After the lethality for each and every time interval has been calculated, all the values are summed, resulting in the total cumulative lethality (F-value) for the process.

The "trapezoidal" General Method calculation[69] is a relatively simple and easy method to use when dealing with even time intervals. This method optimizes the cumulative lethality under the heating curve (although there is a small amount of lethality unaccounted for). Lethalities are calculated using the following formula:

$$F = t \{ [(L_1+L_2)/2] + [(L_2+L_3)/2] + \ldots + [(L_{n-1}+L_n)/2] \}$$

where:
    t = time interval

L_1 = lethal rate at temperature at time 1
L_2 = lethal rate at temperature at time 2

Figure 12.5 is a graphical representation of calculating the total cumulative lethality under the heating curve using the trapezoidal General Method formula above.

These calculations are easily performed today by a number of readily available "computation" or "spreadsheet" PC programs (Excel, Lotus, etc.). Table 12.9 provides an example of a total cumulative F-value calculation for a surimi seafood pasteurization process (for heating phase only) where nonproteolytic *C. botulinum* is the target organism. Note in the

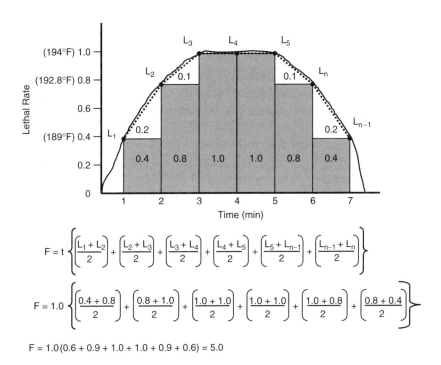

$$F = t\left\{\left(\frac{L_1+L_2}{2}\right) + \left(\frac{L_2+L_3}{2}\right) + \left(\frac{L_3+L_4}{2}\right) + \left(\frac{L_4+L_5}{2}\right) + \left(\frac{L_5+L_{n-1}}{2}\right) + \left(\frac{L_{n-1}+L_n}{2}\right)\right\}$$

$$F = 1.0\left\{\left(\frac{0.4+0.8}{2}\right) + \left(\frac{0.8+1.0}{2}\right) + \left(\frac{1.0+1.0}{2}\right) + \left(\frac{1.0+1.0}{2}\right) + \left(\frac{1.0+0.8}{2}\right) + \left(\frac{0.8+0.4}{2}\right)\right\}$$

$F = 1.0(0.6 + 0.9 + 1.0 + 1.0 + 0.9 + 0.6) = 5.0$

Total F_{194} Value = 5.0

Figure 12.5 Graphical representation of the General Method calculation (trapezoidal method) for even time intervals. *(Source:* Adapted from Ref. 69.)

calculations that a z-value of 12.6°F is used for temperatures less than 194°F, and z = 18°F is used for temperatures above 194°F. Potential additional lethality attributable to the cooling phase is not considered in this example (the pros and cons of which are discussed later in this chapter).

Pasteurization time/temperature can usually be optimized using the General Method for determining the total F-value of the process and therefore minimize undesirable quality changes from overcooking. The results of the process total F-value calculations are typically used along with HACCP principles to establish the critical limits for minimum product pasteurization time and temperature necessary to achieve a 6D process for the target organism. The processor must keep in mind that the method depends on the specific set of conditions used to generate the product heat penetration data, which will likely mean including additional critical factors in monitoring the pasteurization process. As previously mentioned, it is advisable to design and conduct the heat penetration tests under conservative conditions because those conditions will need to be controlled in a manner that assures the process delivers the minimum necessary lethality to the product. Subsequent HACCP monitoring procedures, in most instances, are designed to continuously monitor the time/temperature of the pasteurization process step and other critical factors that may affect the pasteurization process with sufficient frequency to achieve and assure control.

When calculating the total F-value of the process, it is generally not recommended to include any lethality that may be attributed to the "cooling" phase of the process, that is, lethality accumulated after exiting the pasteurizer due to the significant amount of heat still present in the product. There are several reasons for this. First, it is important to cool the product as rapidly as possible to minimize the opportunity for heat-tolerant microorganisms that survive the pasteurization process (e.g., injured cells of thermo-tolerant yeasts and molds, heat-resistant spores of *Bacillus* spp. or proteolytic *Clostridium* spp., etc.) to grow and cause a food safety hazard or reduce the shelf life of the product. Therefore, the focus should be on the product attaining the necessary total lethal-

Microbiology and Pasteurization of Surimi Seafood 621

TABLE 12.9 Example Lethality, F_{194}, Calculation for Surimi Seafood, for Target Organism Nonproteolytic *C. botulinum* (heating phase only)

	HP Test Information:		
Product ID : Crabmeat flakes, Formula 001	Net Wt: 2.5 lb vac pack	Test date/Time: 12/5/03, 9:45AM	
Past. ID/Heat Medium: #1 - Continuous/Steam	Pkg ID/TC loc.: #1, TC at GC	Pkg thickness : 1.5"	
Pkg loc. on past. belt : Middle	Belt speed set point (ft/min): 2.00	Belt speed set point (ft/min): 1.99	
Past. Temp. Set Points: #1 Front: 210°F #2 Mid: 210°F #3 Back: 208°F	Pasteurizer temp, actual: 210.5°F	Past. dwell time, actual: 34 min.	
	Ref. Temp, (°F) (Tref)	Z value (°F) (when Tp < ref)	Z value (°F) (when Tp > Tref)	Time Interval (min.)
Target Organism Data for: Nonproteolytic C. bot	194	12.6	18	1
Elapsed Time (min)	Internal Product Temp (Tp)	Lethal Rate: og^{-1} [(Tp – Tr)/z]	Cumulative Lethality- (F @ 194°F)
0	57.88	0.000	
1	73.07	0.000	0.00
2	87.10	0.000	0.00
3	98.19	0.000	0.00
4	108.19	0.000	0.00
5	116.83	0.000	0.00
6	124.32	0.000	0.00
7	130.99	0.000	0.00
8	136.97	0.000	0.00
9	142.37	0.000	0.00
10	147.38	0.000	0.00
11	152.08	0.000	0.00
12	156.50	0.001	0.00
13	160.60	0.002	0.00

TABLE 12.9 Example Lethality, F_{194}, Calculation for Surimi Seafood, for Target Organism Nonproteolytic *C. botulinum* (heating phase only) (Continued)

Elapsed Time (min)	Internal Product Temp (Tp)	Lethal Rate: og^{-1} [(Tp − Tr)/z]	Cumulative Lethality- (F @ 194°F)
14	164.39	0.004	0.01
15	167.92	0.009	0.01
16	171.15	0.015	0.02
17	174.15	0.027	0.05
18	176.89	0.044	0.08
19	179.45	0.070	0.14
20	181.80	0.108	0.23
21	183.99	0.161	0.36
22	186.00	0.232	0.56
23	187.90	0.328	0.84
24	189.65	0.452	1.23
25	191.29	0.609	1.76
26	192.84	0.809	2.47
27	194.22	1.029	3.39
28	195.54	1.218	4.51
29	196.74	1.420	5.83
30	197.63	1.591	7.33
31	198.00	1.668	8.96
32	197.83	1.632	10.61
33	195.23	1.170	12.01
34	194.02	1.003	13.10
		0.000	13.60

ity from just the heating phase. Second, the processor has these advantages by not including any cooling lethality in the total F-value calculation:

1. There is an inherent "safety factor" built into the process that provides not only additional assurance of process adequacy, but also an important consideration if pasteurization time/temperature deviations were to occur.
2. Process time/temperature controls should be established only for the heating phase of the process, that

is, for only the pasteurization step (and not for the subsequent cooling step).

There may be certain situations, however, where the processor deems it necessary or desirable to consider the lethality contributed by the cooling process in the total F-value of the process. In such instances, the processor will also need to establish another set of time/temperature control parameters surrounding the cooling portion(s) of the process. Thus, both pasteurizing (heating) and cooling parameters must be controlled and maintained in the same manner as when the product heat penetration test runs were performed to assure delivery of the minimum necessary lethality to the product.

In most surimi seafood plants, the overall cooling process can usually be broken down into two distinct portions: (1) a brief "ambient-air" or room-temperature cooling portion immediately after the product exits the pasteurizer, and (2) a cooling phase where the product is cooled by a mechanically chilled medium (e.g., mechanically chilled/ice-water bath or spray, refrigerated air blast tunnel, etc.). Of these portions, the most significant contribution to the total F-value will come from the time the product is in the ambient-air cooling portion of the overall cooling process. This time can be up to 4 or 5 min in length in some surimi seafood plants. In contrast, no more than the first 1 to 1.5 min will contribute significantly to the total F-value of the process if the product is cooled with a mechanically chilled cooling system.

When a processor does choose to consider lethality contributed by the cooling process, it is recommended that it be limited to lethality contributed from only the ambient-air cooling portion of the overall cooling process. However, if the processor also chooses to include potential lethality during the mechanical chilling/cooling portion of the overall cooling process, then testing should also be done to profile/characterize the temperature distribution of this stage of the process to ensure the total F-value calculation accounts for the worst-case (i.e., the fastest) cooling scenario and to provide information needed to establish appropriate time/temperature controls and monitoring procedures for the mechanical chilling/cooling step. Subsequent HACCP monitoring procedures, in most instances, are designed to continuously monitor the time/temperature of the pasteurization process step and to monitor other critical factors that may affect the pasteurization process with sufficient frequency to achieve and assure control.

As previously mentioned, there are two pasteurization options available for controlling growth when nonproteolytic *C. botulinum* type B is the target organism of concern in surimi seafood:

1. Pasteurizing the product to achieve a minimum total equivalent lethality (F_{194} = 10.0 min)
2. Pasteurizing the product to a minimum internal temperature of 185°F and holding it at or above this temperature for at least 15 min when product contains a minimum 2.4% water-phase salt

When the latter control is used, the processor will be analyzing the time/temperature data obtained directly from the product heat penetration test to determine the number of minutes that the product is at or above 185°F, instead of calculating the total F_{194} value for the pasteurization process. Similar to using the total F-value, this alternative method still dictates that appropriate pasteurizing parameters be identified and controlled/maintained in the same manner as when the product heat penetration test runs were performed. Of course, inherent with this method of control is also identifying appropriate critical factor(s) and HACCP controls with respect to achieving and maintaining a minimum 2.4% water-phase salt content in the product.

12.7 TEMPERATURE PREDICTION MODEL FOR THERMAL PROCESSING OF SURIMI SEAFOOD

Figure 12.6 shows an example output from the temperature prediction model for surimi seafood heated in water. This predictive model was constructed using the Gurney-Lurie charts with heat transfer coefficient (h) and thermal diffusiv-

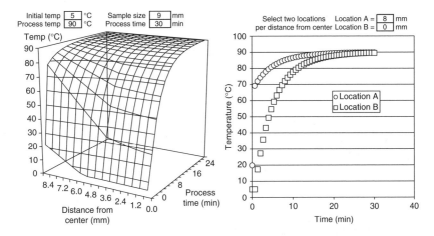

Figure 12.6 Temperature prediction model as affected by various processing parameters. This figure represents a model prediction under given conditions: process temperature 90°C, process time 30 min, package size 9 mm, and product initial temperature 5°C. (*Source:* From Ref. 70. With permission.)

ity (α) determined experimentally.[70] This model allows the input of four variables of a product: processing temperature, processing time, initial product temperature, and product thickness. These variables can be changed at any time, resulting in respective display changes. In this manner, thermal processing of surimi seafood can be optimized. It can be used to design a new thermal processing for surimi seafood or verify an existing one. However, it should be noted that this surface response model was developed based on the heat transfer coefficient (h) and thermal diffusivity (α) determined by the experimental setup. To apply this model in a different system, the specific h and α values must be determined and incorporated into the model.

Product size is typically determined as one half of the smallest dimension of the product.[71] Therefore, Figure 12.6 represents the temperature prediction for a 1.8-cm thick package. The surface response shows the temperature profile for the entire thermal processing time and also the temperature changes that occur at different locations within the package.

In this manner, thermal changes that occur in the package can be fully demonstrated.

The surface response model also allows the user to select two locations within the package to determine the temperature profile (Figure 12.6). Using the two locations — (1) 1.0 mm under the package surface (8 mm from the center) and (2) at the center of the package (0 mm from the center) — the user can determine how the center reaches the desired temperature (90°C) within the thermal processing time (30 min) relative to the other location.

12.8 PREDICTIVE MODEL FOR MICROBIAL INACTIVATION DURING THERMAL PROCESSING OF SURIMI SEAFOOD

The microbial inactivation model is a useful tool for predicting thermal destruction of microorganisms in surimi seafood.[72] With known D-values, destruction of a particular bacterium during the thermal process can be predicted with five variables: product size, initial product temperature, processing time, processing temperature, and initial microbial concentration (Figure 12.7). Based on the first four input variables (product size, initial product temperature, processing time, and processing temperature), the temperature prediction model (Figure 12.6) estimates temperatures, which are subsequently used to determine D-values at intermediate temperatures between 55 and 95°C (Table 12.10). The estimated D-values, in combination with process time and initial microbial concentration (input variables), are used to calculate the microbial concentration at any location of the surimi seafood product and at any time of the thermal processing. The model also calculates the average microbial concentration in a surimi seafood package after it has been thermally processed.

Figure 12.7 shows an example of changes in microbial load in a surimi seafood package (2 cm thick, 5°C initial temperature, initial micro population of $S.\ aureus$ of 2.8×10^5 CFU/g) heated at 70°C for 15 min. It must be noted that the model was based on the laboratory data for inactivation of $S.$

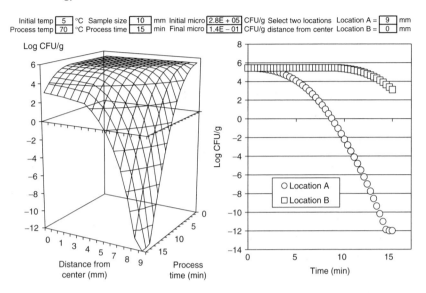

Figure 12.7 Microbial inactivation model as affected by various processing parameters. This figure represents a model prediction under given conditions: process temperature 70°C, process time 15 min, package size 10 mm, product initial temperature 5°C, and initial population of *S. aureus* 2×10^5 CFU/g. *(Source:* From Ref. 72. With permission.)

TABLE 12.10 Thermal D-Values for *S. aureus* in Surimi Seafood

Temperature (°C)	D-value (s)
55	971.54
65	49.46
75	6.52
85	1.53
95	0.65

Source: From Ref. 72. With permission.

aureus in surimi seafood heated in a water bath (55 to 95°C). Commercial settings may give different simulation results.

12.9 NEW TECHNOLOGIES FOR PASTEURIZATION: HIGH-PRESSURE PROCESSING AND ELECTRON BEAM

12.9.1 High-Pressure Processing

The emerging alternative food processing technology of high-pressure processing (HPP) is one of the more promising approaches to ensure product safety without the traditional drawbacks of conventional thermal processing. A fresher flavor and aroma, enhanced texture, and higher nutrient retention are all benefits of HPP. Another advantage is the opportunity to use a diverse assortment of packaging materials and formats.

The lethal effect of HPP on microbial cells and spores is well described and is essentially the result of structural changes in the cellular membrane with essential enzyme denaturation an important contributing effect.[73] Typical pressure values range from 500 to 700 MPa with a variety of times and pulse cycles. Bacterial spores are much more resistant to pressure effects; thus, most applications have involved acidified foods or products that require only pasteurization-equivalent treatments. However, progress is being made in the development of technologies that can achieve high-pressure sterilization of foods. This approach utilizes combinations of pressure, varying pulse time, and elevated product temperature (105°C) to achieve commercial sterility.[74]

HPP has been evaluated as an alternative technology with a variety of seafood products, including cod,[75] salmon,[76] and catfish.[77] Most studies thus far have focused on how pressure affects product tenderness, juiciness, color, flavor, and oxidative stability. Both positive and negative influences from pressurization have been observed, such as increases in lipid oxidation[78] and improved tissue firmness of fresh fish.

Lanier[79] also observed that pressure (300 MPa) could be used to form surimi gels without heat and high pressure produced, strong gels of acceptable quality. However, concern was voiced that pressure could degrade fish quality by destabilizing proteins. Moreover, it is uncertain whether pressure is an effective means to inactivate proteolytic enzymes that compromise product texture.[79,80]

There is also a lack of specific studies that address the microbial inactivation kinetics in HHP-treated seafoods. Morrissey et al.[80] reported that bacterial vegetative cells, as assessed by total plate count, were eliminated from Pacific whiting surimi gels formed at a pressure of 4 kbar. Miyao et al.[81] evaluated the survival of a diverse group of microorganisms isolated from surimi and reported that greater than 300 MPa was sufficient for inactivating most bacterial strains. However, the Gram-positive bacteria were significantly more resistant. There is a need, therefore, for more systematic studies that integrate microbial challenge studies in concert with product quality determinations.

12.9.2 Food Irradiation

Food irradiation is the use of ionizing radiation or ionizing energy to eliminate microbial contaminants in foods. The energy applied to foods exists in the form of waves and is defined by its wavelength. The electromagnetic spectrum identifies types of energy and their respective wavelengths. As the wavelength of energy gets shorter, the energy increases. Electric power, microwave, and light emit low-energy radiation (i.e., long wavelengths) that is sufficient to cause movement of molecules. Therefore, heat is generated through the friction between the molecules exposed to these types of low-energy radiation. However, low-energy radiation cannot structurally change the atoms in molecules.

Ionizing radiation has higher energy, sufficient to change atoms by ejecting an electron from them to form an ion, and hence the name ionizing radiation. However, the energy of ionizing radiation is not sufficient to split atoms and cause exposed foods to become radioactive. Therefore, the sources of ionizing radiation allowed for food processing (cobalt-60, cesium-137, accelerated electrons, and x-rays) cannot and do not make food radioactive. The three sources of ionizing radiation — radioisotopes (cobalt-60 and cesium-137), electron accelerators, and x-rays — have similar effects because they fall in the shortwave, high-energy region of the electromagnetic spectrum.

12.9.3 Electron Beam

The electron beam (e-beam) is one of the major sources of ionizing radiation. An e-beam, in contrast to thermal and high-pressure processing, utilizes high-energy electrons accelerated to the speed of light by a linear accelerator for a sterilization effect. The dose used in measuring radiation activity is the quantity of radiation energy absorbed by the food as it passes through the radiation field during processing. The dose is generally measured in Grays (Gy) or kiloGrays (kGy), where 1 Gray = 0.001 kGy = 1 Joule (J) of energy absorbed per kilogram (kg) of irradiated food. Dose can also be measured in Rads (100 Rads = 1 Gray).

Unlike gamma radiation (cobalt-60 and cesium-137), an e-beam enables application. Therefore, food processing time with an e-beam is typically much shorter than with gamma radiation (0.01 to 1 Gy/sec).[82] Because radiation processing does not increase the temperature of processed food, the e-beam is likely to minimize the degradation of food quality.[83] Another advantage of the e-beam over the gamma rays is that the electron source for the e-beam is electricity and therefore it does not use radioisotopes (cobalt-60 and cesium-137).[84] A major disadvantage of the e-beam is that it has limited penetration depth.[85] However, the overall antimicrobial effects of gamma radiation and e-beam are comparable.[85,86] To increase electron penetration, the e-beam is commonly applied to two sides (top and bottom) of the product. This e-beam configuration is referred to as two-sided e-beam.

Ionizing radiation has been applied to various foods to improve food safety and reduce microbial spoilage. Fresh and frozen red meats are irradiated up to 4.5 and 7.0 kGy, respectively, to reduce *Escherichia coli* O157:H7. Irradiation of poultry and poultry products up to 3 kGy and up to 7 kGy for fresh and frozen products, respectively, is used to reduce *Salmonella*, *Campylobacter*, and other food poisoning bacteria. In addition, dried herbs and spices are processed up to 10 kGy to reduce food poisoning bacteria. Doses up to 2 kGy are also used to reduce spoilage microorganisms in certain fruits and vegetables. For example, low doses up to 1 kGy are used in strawberries to inactivate enzymes responsible for fruit ripening. Tubers, such as potatoes and onions, are irradiated up to 1 kGy to prevent sprouting. Irradiation at 1 kGy is also used as a quarantine measure to kill insects in cereals, grains, and certain fruits. The Joint Expert Committee on Food Irradiation representing the Food and Agriculture Organization/International Atomic Energy Agency/World Health Organization (FAO/IAEA/WHO) concluded that irradiation of any food up to 10 kGy caused no toxicological hazards and introduced no nutritional or microbiological problems.[87]

12.9.4 Electron Penetration in Surimi Seafood

Alaska pollock surimi was used to prepare gels at 78% moisture for electron penetration experiments using one-sided e-beam at a fixed energy of 10 MeV (million electron volts). The absorbed dose increased up to 2 cm deep from the gel surface, followed by a decrease, reaching a minimum value at approximately 5 cm from the gel surface.[88] It is typical that the dose absorbed increases under the surface of the processed product and then decreases.[89] This phenomenon has been attributed to the formation of secondary electrons (because of their lower energy) that are more effectively absorbed than the primary electrons.[90]

The dose absorbed (kGy) by products followed a polynomial function regardless of the dose applied and product thickness, according to Jaczynski and Park.[88] In their electron penetration experiments, e-beam at 3 and 20 kGy was, respectively, applied to 7- and 9-cm-thick surimi gels. In all cases, the dose absorbed followed the same polynomial function ($P > 0.05$). Absorbed doses were plotted against distance from the surimi gel surface to create a dose map. Based on the dose

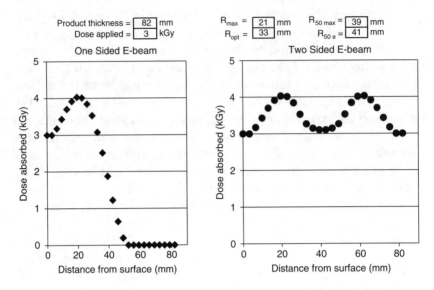

Figure 12.8 Predictive model for dose absorbed of one- and two-sided e-beam as affected by various processing parameters. This figure represents a model prediction under given conditions: product thickness 82 mm, e-beam dose applied 3 kGy. (*Source:* From Ref. 91. With permission.)

map, a predictive model for electron penetration in surimi seafood using one- or two-sided e-beam was developed (Figure 12.8). This model allows the estimation of dose absorbed during e-beam processing of surimi seafood by entering product thickness and dose applied.[88]

The dose map allowed determination of parameters such as R_{opt} (depth of surimi gel at which the absorbed dose equals the dose at the surface of surimi gel), R_{50e} (depth of surimi gel at which the absorbed dose decreases by 50% of the absorbed dose at the surface of surimi gel), R_{max} (depth of surimi gel at which the absorbed dose reaches its maximum value), and R_{50max} (depth of surimi gel at which the absorbed dose decreases by 50% of its maximum value). These parameters are useful when designing e-beam processing of food products. They allow quick and easy determination of "critical depths" in a food product processed with e-beam.

The R_{opt}, R_{max}, R_{50e}, and R_{50max} were calculated as 33, 21, 41, and 39 mm, respectively. According to the R_{opt} and R_{50e} of 33 and 41 mm, respectively, one- and two-sided e-beam can efficiently penetrate 33 and 82 mm of surimi seafood, respectively.[88] Efficient penetration is defined as penetration that results in the dose absorbed in the entire surimi seafood is equal to or greater than the dose applied. Therefore, two-sided e-beam represents better utilization of the dose applied. This finding is in accordance with the literature, which suggests that two-sided e-beam at a fixed energy of 10 MeV could be applied to foods up to 8 to 10 cm thick that have a specific density of 1 g/cm^3.[90]

12.9.5 Microbial Inactivation in Surimi Seafood

Commercial surimi seafood (crabstick style) was ground and inoculated with six strains of *Staphylococcus aureus*. Inoculated samples were incubated to reach a concentration of approximately 10^9 CFU/g and subjected to one-sided e-beam with energy fixed at 10 MeV. E-beam at 1, 2, and 4 kGy resulted in 2.9-log reduction, 6.1-log reduction, and no detectable colonies of *S. aureus*, respectively, in processed samples (Figure 12.9). The D_{10}-value was determined as 0.34 kGy.[91] Effects of radiation are reported to be linear with doses up to 15 kGy.[92] Therefore, the application of 4 kGy in our tests may have resulted in a 12-log reduction, as verified by no colonies recovered at 4 kGy (Figure 12.9). The absence of oxygen under vacuum packaging did not affect microbial inactivation ($P > 0.05$). However, the sample temperature during e-beam processing had a significant effect ($P < 0.05$) on microbial inactivation.[91] When frozen samples were subjected to 1 kGy, microbial reduction was about 1-log lower than in unfrozen surimi seafood. E-beam at 2 kGy resulted in the best inactivation for samples at room temperature, followed by chilled and frozen, respectively.

The determination of the D_{10}-value and model for electron penetration in surimi seafood allowed the development of a predictive model for microbial inactivation (Figure 12.10). Simulation of microbial inactivation by two-sided e-beam for

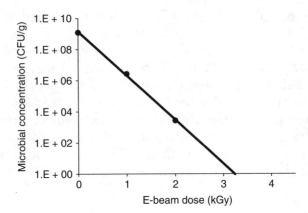

Figure 12.9 Survival of *S. aureus* in surimi seafood subjected to e-beam. (*Source:* From Ref. 91. With permission.)

surimi seafood thicker than 82 mm demonstrates under-processing starting at a depth of 33 mm (R_{opt}) from both the top and bottom surfaces. If the thickness of the surimi seafood processed with two-sided e-beam exceeds 82 mm, then the maximum under-processing occurs at the geometrical center of the package.[91]

12.9.6 Effect of E-Beam on Other Functional Properties of Surimi Seafood

Commercial surimi seafood (crabstick style) was subjected to e-beam processing with energy fixed at 10 MeV. The L* and a* values were not affected (P > 0.05) by e-beam. The b* value of the crabsticks decreased (P < 0.05) from 1.4 for nonirradiated samples to 0.9, 0.8, and 0.8 for samples subjected to e-beam at 1, 2, and 4 kGy, respectively. The lower b* value resulted in whiter color.[93] Ozone, which can be generated during e-beam processing,[90] might have bleached the yellow hue, resulting in a reduced b* value.

Shear stress of surimi gels increased in proportion to e-beam dose up to 6 to 8 kGy and then decreased. Shear stress increased from 47 kPa for nonirradiated samples to 87 and 100 kPa for samples subjected to e-beam at 6 and 8 kGy,

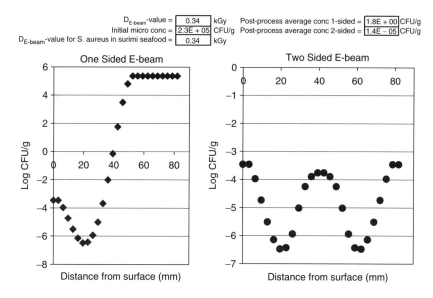

Figure 12.10 Predictive model for microbial inactivation by one- and two-sided e-beam as affected by various processing parameters. The figure represents a model prediction under given conditions: product thickness 82 mm, dose applied 3 kGy, microbial target *S. aureus*, initial population 2.3×10^5 CFU/g. (*Source:* From Ref. 91. With permission.)

respectively. Shear strain of surimi gels was not affected by e-beam.[94] Shear stress and shear strain indicate the strength and cohesive nature of the surimi gels, respectively.[95] Therefore, the results suggest that e-beam treatment up to 6 to 8 kGy improved the strength of surimi gels.[94]

Electrophoresis of Alaska pollock surimi and surimi gels prepared from Alaska pollock surimi was conducted under denaturing conditions of sodium dodecyl sulfate (SDS) and β-mercaptoethanol (β-ME).[96] SDS-PAGE in 12% polyacrylamide gel (Figure 12.11) showed a gradual degradation of myosin heavy chain (MHC) with increasing e-beam dose.[94] Gradual disappearance of MHC resulted in an increase of smaller molecular weight proteins (200 to 50 kDa) in each lane below MHC. The complete disappearance of the MHC band was observed at 25 kGy for Alaska pollock surimi (23°C) and

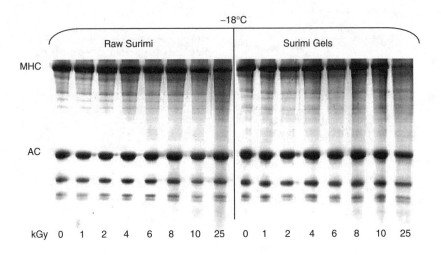

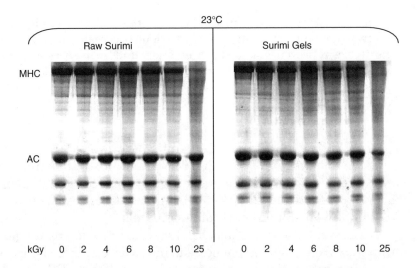

Figure 12.11 SDS-PAGE of proteins from Alaska pollock surimi (left lanes) and surimi gels (right lanes) applied to 12% polyacrylamide gel at 25 µ/well. Sample temperature was –18°C (top) and 23°C (bottom) during e-beam treatment. MHC = myosin heavy chain; AC = actin. (*Source:* From Ref. 94. With permission.)

surimi gels (23°C). However, samples subjected to 25 kGy while at −18°C showed a thin MHC band, suggesting slower degradation at the lower temperature. Actin (AC) was only slightly affected in surimi gels subjected to e-beam at 25 kGy (Figure 12.11, bottom right).

12.10 PACKAGING CONSIDERATIONS

The packaging of surimi seafood has undergone many changes as new methods and materials have been developed. Various types of packaging, including aseptic packaging, vacuum packaging, modified atmosphere packaging, air-included packaging, and retort packaging, can be used for surimi seafood, depending on the types of products and processing procedures.[97] Aseptic packaging reduces the concern of postprocessing contamination and extends the shelf life of products. However, this type of packaging requires a special system, including an aseptic environment. Products must also be cooled down in an aseptic room and aseptically packed in the room before being transported to the storage room.

Retort packaging can produce a commercially sterile product. However, the process can result in overcooked products and loss of quality, such as changes of color, loss of elasticity, and development of off flavors. Ueno[98] investigated the color changes of fish meat sausage that occurred during retort sterilization and found that the color changes caused by amino-carbonyl reaction became more pronounced as the temperature and time of the retort process increased. The study also showed that a retort process at 120°C for 30 min greatly reduced the rupture strength and elasticity of the fish sausage.

The off-odors that developed in retort-sterilized products were due to the generation of hydrogen sulfide and other chemicals from fish proteins. More off-odors were generated in retort-sterilized kamaboko as the sterilization temperature and time increased.[99] Therefore, high temperature and short time should be used to minimize the production of off-odors. Kamaboko retort-sterilized at 127°C for 6 min and kept at 10°C or below during distribution did not produce off-odors or lose elasticity.

High-capacity ultraviolet (UV) sterilization can be applied to surimi seafood to reduce surface contamination and extend the shelf life of products. Studies have shown that *Escherichia coli* contamination on fish sausage surfaces could be completely destroyed with a 5-sec exposure to UV light at an intensity of 40 mW/cm^2.[100] However, the bactericidal effects of UV light differ among bacterial species, and many packaging materials do not allow the transmittance of UV light. Higher doses of UV beams are required to destroy molds and spore-forming bacteria such as *Bacillus subtilis*.[101]

Modified atmosphere packaging (MAP) is a system in which the air inside a package is replaced either totally or in part by other gases (mainly nitrogen and carbon dioxide).[102] This type of packaging has been widely used for surimi seafood products due to a number of preservative effects. The mixture of nitrogen and carbon dioxide promotes an anaerobic environment, which prevents oxidation of fats and pigments and inhibits the growth of aerobic microorganisms. Although MAP is frequently used to preserve the quality and extend storage period of surimi seafood, the potential growth of nonproteolytic types *C. botulinum* in the reduced oxygen package is a safety concern. A storage temperature of less than 3°C (<37.4°F) is recommended for MAP products to ensure the safety of surimi seafood.[103,104]

REFERENCES

1. J.M. Shewan. The bacteriology of fresh and spoiling fish and the biochemical changes induced by biochemical action. *Proceedings of the Conference of Handling, Processing, and Marketing of Tropical Fish*. London, 1977, 51.

2. J. Jay. Intrinsic and extrinsic parameters of foods that affect microbial growth. In *Modern Food Microbiology, 6th ed.* Gaithersburg, MD: Aspen Publishers, 2000, 35–56.

3. J. Jay. Seafoods. In *Modern Food Microbiology, 6th ed.* Gaithersburg, MD: Aspen Publishers, 2000, 101–105.

4. G.J. Tsai and T.H. Chen. Incidence and toxigenicity of *Aermonas hydrophila* in seafood. *Int. J. Food Microbiol.*, 31, 121–131, 1996.

5. J.E. Stratton and S.L. Taylor. Scombroid poisoning. In D.R. Ward and C. Hackney, Eds. *Microbiology of Marine Food Products*. New York: AVI Publishers, 1991, 331–351.

6. M.B.A. Gloria, M.A. Daeschel, C. Craven, and K.S. Hildebrand. Histamine and other biogenic amines in albacore tuna. *J. Aquat. Food Product. Technol.,* 8, 55–69, 1999.

7. C. Ienistea. Significance and detection of histamine in food. In B.C. Hobbs and J.H.B. Christian, Eds. *The Microbiological Safety of Foods. Proceeding of the 8th International Symposium of Food Microbiology,* 1973, 427–439.

8. J. Luten, W. Bouquet, L.A. Sauren, M.M. Burggraff, G. Rieekwel-Booy, P. Durand, M. Etienne, J.P. Gouyou, A. Landrein, A. Ritchie, M. Leclerq, and R. Guinet. Biogenic amines in fishery products: standardization methods within EC. In H.H. Huss, Ed. *Quality Assurance in the Fish Industry*. London: Elsevier, 1992, 427–439.

9. D.N. Scott, G.C. Fletcher, M.G. Hogg, and J.M. Ryder. Comparison of whole and headed and gutted orange roughy stored in ice: sensory, microbiology and chemical assessment. *J. Food Sci.,* 49, 79–86, 1986.

10. B.K. Mayer and D.R. Ward. Microbiology of finfish and finfish processing. Poisoning. In D.R. Ward and C. Hackney, Eds. *Microbiology of Marine Food Products*. New York: AVI publishers, 1991, 3–17.

11. E.M. Ravesi, J.J. Licciardello, and L.D. Raicot. Ozone treatments of fresh Atlantic cod, *Gadus morhua*. *Marine Fish Rev.,* 49, 37–42, 1987.

12. M.S. Fey and J.M. Regenstein. Extending the shelf-life of fresh wet red hake and salmon using CO_2-O_2 carbon dioxide-oxygen modified atmosphere and potassium sorbate ice at 1 degree Celcius. *J. Food Sci.,* 47, 1048–1054, 1982.

13. J.R. Matches, J. Liston, and D. Curran. *Clostridium perfringens* in the environment. *Appl. Microbiol.,* 28, 655–660, 1974.

14. B.H. Himelbloom, J.S. Lee, and R.J. Price. Microbiology and HACCP in surimi manufacturing. In J.W. Park, Ed. *Surimi and Surimi Seafood*. New York: Marcel Dekker, 2000, 79–90.

15. B.H. Himelbloom, J.S. Lee, and R.J. Price. Microbiology and HACCP in surimi seafood. In J.W. Park, Ed. *Surimi and Surimi Seafood*. New York: Marcel Dekker, 2000, 325–341.

16. I.H. Yoon and J.R. Matches. Growth of pathogenic bacteria on imitation crab. *J. Food Sci.*, 53, 688–690,1582, 1988.

17. I.H. Yoon, J.R. Matches, B. Rasco. Microbiological and chemical changes of surimi based imitation crab during storage. *J. Food Sci.*, 53, 1343–1346, 1988.

18. B.H. Himelbloom, J.K. Brown, and J.S. Lee. Microbiological investigation of Alaska shore based surimi production. *J. Food Sci.*, 56, 291–293, 1991.

19. B.H. Himelbloom, J.K. Brown, and J.S. Lee. Microorganisms isolated from surimi processing operations. *J. Food Sci.*, 56, 299–301, 1991.

20. B.H. Himelbloom, J.K. Brown, and J.S. Lee. Microorganisms on commercially processed Alaskan finfish. *J. Food Sci.*, 56, 1279–1281, 1991.

21. K.G. Colburn, C.A. Kaysner, C. Abeyta, and M.M. Wekell. *Listeria* species in a California coast estuarine environment. *Appl. Environ. Microbiol.*, 56, 2007–2011, 1990.

22. M.L. Motes. Incidence of *Listeria* spp. in shrimp, oysters, and estuarine waters. *J. Food Prot.*, 54, 170–173, 1991.

23. S. Sabanadesan, A.M. Lammerding, and M.W. Griffiths. Survival of *Listeria innocua* in salmon following cold-smoke application. *J. Food Prot.*, 63, 715–720, 2000.

24. C.B. Dalton, C.C. Austin, J. Sobel, P.S. Hayes, W.F. Bibb, L.M. Graves, B. Swaminathan, M.E. Proctor, and P.M. Griffin. An outbreak of gastroenteritis and fever due to *Listeria monocytogenes* in milk. *N. Engl. J. Med.*, 336, 100–105, 1997.

25. M.J. Linnan, L. Mascola, X.D. Lou, V. Goulet, S. May, C. Salminen, D.W. Hird, M.L. Yonekura, P. Hayes, R. Weaver, A. Audurier, B.D. Plikaytis, S.L. Fannin, A. Kleks, and C.V. Broome. Epidemic listeriosis associated with Mexican-style cheese. *N. Engl. J. Med.*, 319, 823–828, 1988.

26. W.F. Schlech, P.M. Lavigne, R.A. Bortolussi, A.C. Allen, E.V. Haldane, A.J. Wort, A.W. Hightower, S.E. Johnson, S.H. King, E.S. Nicholls, and C.V. Broome. Epidemic listeriosis — evidence for transmission by food. *N. Engl. J. Med.,* 308, 203–206, 1983.

27. Anonymous. Epidemiologic notes and reports listeriosis associated with consumption of turkey franks. *Morb. Mortal. Wkly. Rep.,* 38(15), 267–268, 1989.

28. Centers for Disease Control and Prevention. Multistate outbreak of listeriosis — United States. *JAMA,* 285, 285–286, 2001.

29. Food Safety and Inspection Service. Revised Action Plan for Control of *Listeria monocytogenes* for the Prevention of Foodborne Listeriosis. U.S. Department of Agriculture http://www.fsis.usda.gov/oa/topics/lm_action.htm, 2000.

30. T. Autio, S. Hielm, M. Miettinen, A.-M. Sjöberg, K. Aarnisalo, J. Björkroth, T. Mattila-Sandholm, and H. Korkeala. Sources of *Listeria monocytogenes* contamination in a cold-smoked rainbow trout processing plant detected by pulsed-field gel electrophoresis typing. *Appl. Envir. Microbiol.,* 65, 150–155, 1999.

31. M. Baker and M. Brett. Listeriosis and mussels. *Commun. Dis. New Zealand,* 93, 13, 1993.

32. R.M. Dillon and T.R. Patel. *Listeria* in seafoods: a review. *J. Food Prot.,* 55, 1009–1015, 1992.

33. J.M. Farber. *Listeria monocytogenes* in fish products. *J. Food Prot.,* 54, 922–924,934, 1991.

34. K.C. Jinneman, M.M. Wekell, and M.W. Eklund. Incidence and behavior of *Listeria monocytogenes* in fish and seafood. In E.T. Ryser and E.H. Marth, Eds. *Listeria, Listeriosis and Food Safety,* 2nd ed. New York: Marcel Dekker, 1999, 601–630.

35. L.M. Rørvik, D.A. Caugant, and M. Yndestad. Contamination pattern of *Listeria monocytogenes* and other *Listeria* spp. in a salmon slaughter-house and smoked salmon processing plant. *Int. J. Food Microbiol.,* 25, 19–27, 1995.

36. S.D. Weagant, P.N. Sado, K.G. Colburn, J.D. Torkelson, F.A. Stanley, M.H..Krane, S.C. Shields, and C.F. Thayer. The incidence of *Listeria* species in frozen seafood products. *J. Food Prot.,* 51, 655–657, 1988.

37. H.-C. Wong, W.-L. Chao, and S.-J. Lee. Incidence and characterization of *Listeria monocytogenes* in foods available in Taiwan. *Appl. Environ. Microbiol.*, 56, 3101–3104, 1990.

38. J.M. Farber, E.M. Daley, M.T. Mackie, and B. Limerick. A small outbreak of listeriosis potentially linked to the consumption of imitation crab meat. *Lett. Appl. Microbiol.*, 31, 100–104, 2000.

39. S. Guyer and T. Jemmi. Behavior of *Listeria monocytogenes* during fabrication and storage of experimentally contaminated smoked salmon. *Appl. Environ. Microbiol.*, 57, 1523–1527, 1991.

40. D. Lennon, B. Lewis, C. Mantell, D. Becroft, K. Dove, S. Farmer, S. Tonkin, N. Yeates, R. Stamp, and K. Mickleson. Epidemic perinatal listeriosis. *Pediatr. Infect. Dis.*, 3, 30–34, 1984.

41. F.X. Riedo, R.W. Pinner, M. De Lourdes Tosca, M.L. Cartter, L.M. Graves, M.W. Reeves, R.E. Weaver, B.D. Plikaytis, and C.V. Broome. A point-source foodborne outbreak: documented incubation period and possible mild illness. *J. Infect. Dis.*, 170, 693–696, 1994.

42. D.L. Mitchell. A case cluster of listeriosis in Tasmania. *Commun. Dis. Intell.*, 15, 427, 1991.

43. M.S.Y. Brett, P. Short, and J. McLauchlin. A small outbreak of listeriosis associated with smoked mussels. *Int. J. Food Microbiol.*, 43, 223–229, 1998.

44. H. Ericsson, A. Eklöw, M.-L. Danielsson-Tham, S. Loncarevic, L.-O. Mentzing, I. Persson, H. Unnerstad, and W. Tham. An outbreak of listeriosis suspected to have been caused by rainbow trout. *J. Clin. Microbiol.*, 35, 2904–2907, 1997.

45. E.T. Ryser and E.H. Marth. Incidence and behavior of *Listeria monocytogenes* in fish and seafood. In E.T. Ryser and E.H. Marth, Eds. *Listeria, Listeriosis and Food Safety*. New York: Marcel Dekker, 1991, 496–512.

46. FDA. The FDA Enforcement Report (http://www.fda.gov/opacom/Enforce.html). Center for Food Safety and Applied Nutrition, U.S. Food and Drug Administration, Washington, D.C., 2004.

47. S. Wong, D. Street, S.I. Delgado, and K.C. Klontz. Recalls of foods and cosmetics due to microbial contamination reported to the U.S. Food and Drug Administration. *J. Food Prot.*, 63, 1113–1116, 2000.

48. M.W. Eklund and F.T. Poysky. *Clostridium botulinum* type E from marine sediments. *Science,* 149, 306, 1965.

49. M.W. Eklund and F.T. Poysky. Incidence of *C. botulinum* type E from the Pacific coast of the United States. In M. Ingram and T.A. Roberts, Eds, *Botulism 1966.* London: Chapman & Hall, 1967, 49–55.

50. J.T.R. Nickerson, S.A. Goldblith, G. Digioia, and W.W. Bishop. The presence of *Clostridium botulinum* type E in fish and mud taken from the Gulf of Maine. In M. Ingram and T.A. Roberts, Eds. *Botulism 1966.* London: Chapman & Hall, 1967, 25–33.

51. J.P.P.M. Smelt and H. Haas. Behavior of proteolytic *Clostridium botulinum* types A and B near the temperature limit of growth. *Eur. J. Appl. Microbiol. Biotechnol.,* 5, 143–154, 1978.

52. M.E. Peterson, R.N. Paranjpye, F.T. Poysky, G.A. Pelroy, and M.W. Eklund. Control of non-proteolytic *Clostridium botulinum* types B and E in crab analogs by combinations of heat pasteurization and water phase salt. *J. Food Prot.,* 65, 130–139, 2002.

53. K. Abrahamsson, B. Gullmar, and N. Molin. The effect of temperature on toxin formation and toxin stability of *Clostridium botulinum* type E in different environments. *Can. J. Microbiol.,* 12, 385–394, 1966.

54. H.M. Solomon, D.A. Kautter, and R.K. Lynt. Effect of low temperature on growth on nonproteolytic *Clostridium botulinum* type B and F and proteolytic type G in crab and broth. *J. Food Prot.,* 45, 516–518, 1982.

55. G. Sakaguchi. Botulism. In H. Reimann, Ed. *Food-Borne Infections and Intoxications.* New York: Academic Press, 1979, 390–433.

56. M.W. Eklund. Significance of *Clostridium botulinum* in fishery products preserved short of sterilization. *Food Technol.,* 36, 107–112 and115, 1982.

57. M.W. Eklund, M.E. Peterson, R. Paranjpye, and G.A. Pelroy. Feasibility of a heat-pasteurization process for the inactivation of nonproteolytic *Clostridium botulinum* types B and E in vacuum-packaged, hot-process (smoked) fish. *J. Food Prot.,* 51, 720–726, 1988.

58. M.E. Peterson, G.A. Pelroy, F.T. Poysky, R.N. Paranjpye, F.M. Dong, G.M. Pigott, and M.W. Eklund. Heat-pasteurization process for inactivation of nonproteolytic types of *Clostridium botulinum* in picked Dungeness crabmeat. *J. Food Prot.*, 60, 928–934, 1997.

59. R. Bibek. Control by heat. In *Fundamental Food Microbiology*. Boca Raton, FL: CRC Press, 1996, 381–391.

60. Food Processors Institute and National Food Processors Association. *Thermal Process Development Workshop*. The Food Processors Institute, Washington, D.C., 2000.

61. A.S. Mazzotta. Thermal inactivation of stationary-phase and salt adapted *Listeria monocytogenes* during postprocess pasteurization of Surimi-based imitation crabmeat. *J. Food Prot.*, 64, 483–485, 2001.

62. T. Rippen, C. Hackney, G. Flick, G. Knobl, and D. Ward. Thermal Processing: Principles and Definitions. In *Seafood Pasteurization and Minimal Processing Manual,* Virginia Cooperative Extension Publication 600-0061, Virginia Sea Grant Publication VSG 93-09, Virginia Polytechnic Institute and State University, Blacksburg, VA, 1993, 3–12.

63. FDA. Pathogen survival through pasteurization. In *Fish and Fishery Products Hazards and Controls Guidance, 3rd ed.* U.S. Food and Drug Administration, Center for Food Safety and Applied Nutrition, Office of Seafood, Washington, D.C., 2001, 219–226.

64. FDA. *Clostridium botulinum* toxin formation. In *Fish and Fishery Products Hazards and Controls Guidance, 3rd ed.* U.S. Food and Drug Administration, Center for Food Safety and Applied Nutrition, Office of Seafood, Washington, D.C., 2001, 167–190.

65. National Food Processors Association. NFPA Bulletin 26-L: Thermal Processes for Low-Acid Foods in Metal Containers. Washington, D.C.: National Food Processors Association, 1996, 72–84.

66. FDA. Appendix 4: Table A-3 Inactivation of *Listeria monocytogenes* and Table A-4 Inactivation of nonproteolytic *Clostridium botulinum* type B. In *Fish and Fishery Products Hazards and Controls Guidance, 3rd ed.* U.S. Food and Drug Administration, Center for Food Safety and Applied Nutrition, Office of Seafood, Washington, D.C., 2001, 283.

67. National Food Processors Association. Industry Advisory Notice: Important Notice to All Surimi Seafood Producers. National Food Processors Association, Seattle, WA, August 3, 2001.

68. W.D. Bigelow, G.S. Bohart, A.C. Richardson, and C.O. Ball. Heat penetration in processed canned foods. *The National Canners Association Bulletin 16-L* (1920, out of print). Reprinted in *Introduction to Thermal Processing of Foods*. Westport, CT: AVI Publishing, 1961.

69. M. Patashnik. A simplified procedure for thermal process evaluation. *Food Technol.*, 7,1–6, 1953.

70. J. Jaczynski and J.W. Park. Temperature prediction during thermal processing of surimi seafood. *J. Food Sci.*, 67, 3053–3057, 2002.

71. C.J. Geankoplis. *Transport Processes and Unit Operation, 3rd ed.* Engelwood Cliffs, NJ: PTR Prentice Hall, 1993.

72. J. Jaczynski and J.W. Park. Predictive models for microbial inactivation and texture degradation in surimi seafood during thermal processing. *J. Food Sci.*, 68, 1025–1030, 2003.

73. D.G. Hoover, C. Metrick, A.M. Papineau, D.F. Farkas, and D. Knoor. Biological effects of high pressure processing on food microorganisms. *Food Technol.*, 43, 99, 1989.

74. R.S. Meyer, K.L. Cooper, D. Knoor, and H.L.M. Lelieveld. High pressure sterilization of foods. *Food Technol.*, 54, 67–72, 2000.

75. K. Angsupanich and D.A. Ledward. High pressure effects on (*Gadus morhua*) muscle. *Food Chem.*, 63, 39–50, 1998.

76. J.R. Lakshmanan, J.R. Piggott, and A Paterson. Potential applications of high pressure for improvement in salmon quality. *Trends in Food Sci. Technol.*, 14, 354–363, 2003.

77. D.R. McKenna, K.E. Nanke, and D.G. Olson. The effects of irradiation, high hydrostatic pressure and temperature during pressurization on the characteristics of cooked-reheated salmon and catfish fillets. *J. Food Sci.*, 68, 368–377, 2003.

78. T. Ohshima, H. Ushio, and C. Koizumi. High-pressure processing of fish and fish products. *Trends in Food Sci. Technol.*, 4, 370–375, 1993.

79. T.C. Lanier. High Pressure processing effects on fish proteins. *Adv. Exp. Med. Biol.*, 434, 45–55, 1998.

80. M.T. Morrissey, Y. Karaibrahimoglu, and J. Sandhu. Effect of high hydrostatic pressure on pacific whiting surimi. *Adv. Exp. Med. Biol.*, 434, 57–65, 1998.

81. S. Miyao, T. Shindoh, K. Miyamori, and T. Arita. Effects of high pressurization on the growth of bacteria derived from surimi. *J. Japan Soc. Food Sci. Technol.*, 40, 478–484, 1993.

82. S.J. Lewis, A. Velasquez, S.L. Cuppett, and S.R. McKee. Effect of e-beam irradiation on poultry meat safety and quality. *Poult. Sci.*, 81, 896–903, 2002.

83. G.G. Giddings. Radiation processing of fishery products. *Food Technol.*, 38, 61–65, 1984.

84. S.E. Luchsinger, D.H. Kropf, C.M. Garcia Zepeda, M.C. Hunt, J.L. Marsden, E.J. Rubio Canas, C.L. Kastner, W.G. Kuecker, and T. Mata. Color and oxidative rancidity of gamma and electron beam-irradiated boneless pork chops. *J. Food Sci.*, 61, 1000–1005, 1996.

85. T. Hayashi. Comparative effectiveness of gamma rays and electron beams in food irradiation. In S. Throne, Ed. *Food Irradiation*. London: Elsevier Applied Sciences, 1991, 169–206.

86. W.M. Urbain. *Food Irradiation*. Orlando, FL: Academic Press, 1986.

87. World Health Organization. Wholesomeness of Irradiated Food. Technical Report Series 651. Geneva, 1981.

88. J. Jaczynski and J.W. Park. Application of electron beam to surimi seafood. In V. Komolprasert and K. Morehouse, Eds. *Irradiation of Food and Packaging: Recent Developments*. Washington, D.C.: American Chemical Society, 2004, 165–179.

89. J.F. Diehl. *Safety of Irradiated Foods, 2nd ed.* New York: Marcel Dekker, 1995.

90. V. Venugopal, S.N. Doke, and P. Thomas. Radiation processing to improve the quality of fishery products. *Crit. Rev. Food Sci. Nutr.*, 39, 391–440, 1999.

91. J. Jaczynski and J.W. Park. Microbial inactivation and electron penetration in surimi seafood during electron beam processing. *J. Food Sci.*, 68, 1788–1792, 2003.

92. D.A.E. Ehlerman. Food irradiation: a challenge to authorities. *Trends Food Sci. Technol.*, 4, 184–189, 1993.

93. J. Jaczynski and J.W. Park. Physicochemical properties of surimi seafood as affected by electron beam and heat. *J. Food Sci.*, 68, 1626–1630, 2003.

94. J. Jaczynski and J.W. Park. Physicochemical changes in Alaska pollock surimi and surimi gel as affected by electron beam. *J. Food Sci.*, 69(1), FCT53–57, 2004.

95. J.W. Park, J. Yongsawatdigul, and T.M. Lin. Rheological behavior and potential cross-linking of Pacific whiting (*Merluccius productus*) surimi gel. *J. Food Sci.*, 59:773–776, 1994.

96. D.M. Bollag, M.D. Rozycki, and S.J. Edelstein. *Protein Methods, 2nd ed.* New York: Wiley-Liss, 1996.

97. M. Yokoyama. Packaging of surimi-based products. In T.C. Lanier and C.M. Lee, Eds. *Surimi Technology*. New York: Marcel Dekker, 1992, 317–334.

98. S. Ueno. Quality control and safety of fish meat sausage during retort sterilization. *New Food Ind.*, 18, 21–22, 1976.

99. M. Yamazawa, M. Murase, and K. Shiga. Study on quality of retort sterilized kamaboko. *Bull. Japan Soc. Sci. Fish.*, 45, 187–192, 1979.

100. K. Hirose, J. Hoya, K. Satomi, and M. Yokoyama. Sterilization of sausage surface by high intensity UV lamp system. *Nippon Shokuhin Kogyo Gakkaishi*, 29, 518–521, 1982.

101. M. Yokoyama. Packaging of fishery processed foods. *Shokuhin-Kogyo*, 30, 30–38, 1987.

102. L.E. Lampila. Modified atmosphere packaging. In D.R. Ward and C. Hackney, Eds. *Microbiology of Marine Food Products*. New York: Van Nostrand Reinhold, 1991, 373–393.

103. M.W. Eklund. Effect of CO_2-modified atmospheres and vacuum packaging on *Clostridium botulinum* and spoilage organisms of fishery products. *Proceedings of the First National Conference on Seafood Packaging and Shipping,* Washington, D.C., 1982, 298–331.

104. L.S. Post, D.A. Lee, M. Solberg, D. Furgang, J. Specchio, and C. Graham. Development of botulinal toxin and sensory deterioration during storage of vacuum and modified atmosphere packaged fish fillets. *J. Food Sci.,* 50, 990–996, 1985.

13

Ingredient Technology for Surimi and Surimi Seafood

JAE W. PARK, PROFESSOR

Oregon State University Seafood Laboratory, Astoria, Oregon

CONTENTS

13.1 Introduction 650
13.2 Ingredient Technology 653
    13.2.1 Water 653
    13.2.2 Starch 657
        13.2.2.1 What Is Starch? 657
        13.2.2.2 Modification of Starch 657
        13.2.2.3 Starch as a Functional Ingredient for Surimi Seafood 659
    13.2.3 Protein Additives 668
        13.2.3.1 Whey Proteins 669

            13.2.3.2   Egg White Proteins......................... 673
            13.2.3.3   Plasma Proteins 676
            13.2.3.4   Soy Proteins 677
            13.2.3.5   Wheat Gluten and Wheat
                       Flour... 681
      13.2.4 Hydrocolloids.. 681
            13.2.4.1   Carrageenan.................................. 682
            13.2.4.2   Konjac 683
            13.2.4.3   Curdlan.. 684
            13.2.4.4   Alginate 685
      13.2.5 Cellulose.. 686
      13.2.6 Vegetable Oil and Fat Replacer..................... 686
      13.2.7 Food-Grade Chemical Compounds 689
            13.2.7.1   Oxidizing Agents........................... 689
            13.2.7.2   Calcium Compounds...................... 689
            13.2.7.3. Transglutaminase (TGase)............ 692
            13.2.7.4   Phosphate 695
            13.2.7.5   Coloring Agents............................. 696
13.3 Evaluation of Functional Ingredients....................... 699
      13.3.1 Texture 699
      13.3.2 Color 700
      13.3.3 Formulation Development and
             Optimization 700
Acknowledgements.. 701
References.. 702

13.1 INTRODUCTION

The three major functional properties of surimi seafood are color, flavor, and texture. Controlling color and flavor is relatively easy because of linear responses. However, the addition of ingredients influences the textural properties of the product mostly in a nonlinear fashion, thus making texture more difficult to control. Consequently, there exists a wide range of textural properties (Figure 13.1). The main ingredients used for the development and modification of textural characteristics of surimi seafood are surimi, water, starch, protein additives, and hydrocolloids.

Ingredient Technology for Surimi and Surimi Seafood 651

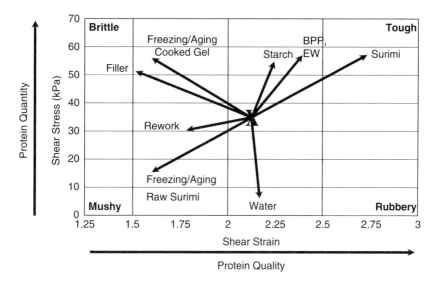

Figure 13.1 Texture map demonstrating effects of ingredients on gel texture.

Lanier[1] defined the relationship between the physical parameters and sensory characteristics of texture. When the human mouth is able to perceive the relatively high stiffness compared to the cohesiveness of a product, it produces a "brittle" sensation in the mouth. For brittle foods, the food structure strongly resists deformation, but upon subjection to sufficient force will collapse before appreciable deformation of the food has occurred. A low value of the stiffness/cohesiveness ratio indicates a "rubbery" material. The overall magnitude of the two textural parameters places the textural description on a continuum moving from a perception of "mushy" (neither stiff nor rubbery) upward to one of "toughness" (both stiff and rubbery).[2] For surimi seafood, a relative balance exists between the textural stiffness and cohesiveness.

Ziegler and Foegedding[3] proposed five possible models for the spatial partitioning of a gelling protein and a gelling or nongelling co-ingredient (Figure 13.2). The first two models (A, B) are termed "filled gels." The two types of filled gels can be distinguished, depending on the phase state of the system:

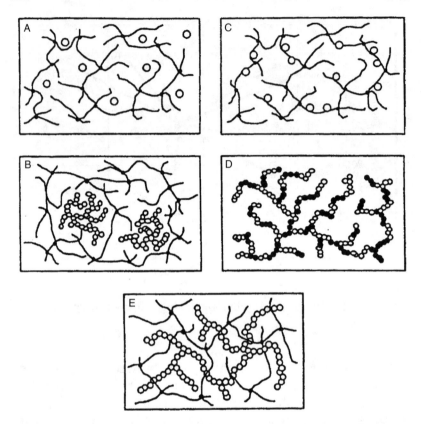

Figure 13.2 Five models of protein–protein or protein–carbohydrate interactions. (*Source:* From Ref. 3. With permission.)

single-phase gels and two-phase gels. Single-phase gels contain a filler that remains soluble in the interstitial fluid of the gel matrix (Figure 13.2A). In a two-phase gel, however, thermodynamic incompatibility causes a phase separation to occur, with the gel filler existing as dispersed particles of liquid or as a secondary gel network, presumably unassociated with the myofibrillar gel matrix (Figure 13.2B). Surimi gel with added starch results in the formation of a two-phase gel.

The third model represents "complex" gels, which form when interactions among the components lead to their physical association. In this model a "nongelling" component may associate with the primary network in a random fashion via nonspecific interactions (Figure 13.2C). Such interactions may reduce the flexibility of the primary network chains and add to the rigidity of the gel. In the fourth model, however, two or more proteins may co-polymerize to form a single, heterogeneous network (Figure 13.2D). Bovine serum albumin and ovalbumin form gels of this type.[4] The fifth model represents an interpenetrating gel network (Figure 13.2E). The addition of beef plasma protein results in a significant increase in shear stress and no change or slight increase in shear strain,[5] which could result from the formation of an interpenetrating gel network.

A minimum concentration of myofibrillar proteins is necessary for gelation. Below this critical concentration, heating will effect only coagulation and precipitation. As the concentration of myofibrillar proteins is increased above this concentration, the gel becomes stronger and more rigid until water, which is required for protein hydration and solubilization, becomes limited. In addition, gel matrix development by myofibrillar proteins can be directly influenced by chemical interactions between the non-muscle proteins and the myofibrillar proteins.[6] The myofibrillar gel matrix is also affected indirectly by changes in the molecular environment (contribution to total protein concentration, water state and availability, ionic strength and types, pH) brought on by the presence of non-muscle proteins.

This chapter discusses the technology of various functional binders and fillers used in surimi seafood. In addition, cryoprotectants used in surimi are discussed.

13.2 INGREDIENT TECHNOLOGY

13.2.1 Water

The addition of water to surimi seafood is required to maintain acceptable texture as well as to minimize the cost of raw materials. The polar nature of water favors clustering of the hydrophobic residues within the folded polypeptide chain to minimize the entropy that would result from the exposure of the hydrophobic residues to water at the surface.[6] This also contributes to the conformational stability of protein molecules before heating and may become the basis for intermolecular bonding when hydrophobic sites on adjacent protein molecules are exposed to the surface during heating. Thus, water serves to initially disperse the myofibrillar protein molecules, allowing a more expanded network to develop as protein–protein bonds form during heating.

The charged myofibrillar proteins are also soluble in water in the presence of salt, being stabilized in a particular three-dimensional conformation by the balance between the intramolecular forces and surface interactions with water. When ingredients other than myofibrillar protein and salt are included, the water content may not reflect the quantity of water available to hydrate and suspend the myofibrillar protein molecules due to a portion of water being tied up by these added ingredients through chemical and/or physical means. It is, therefore, critical to develop a formula that can hold the maximum amount of water.

Addition of water to surimi, however, causes a great reduction in shear stress (gel hardness), while it has a slight impact on shear strain (gel cohesiveness).[2] Shear stress of both high- and low-grade pollock linearly decreased from ~65 to ~3 kPa and from ~48 to ~3 kPa, respectively, as the moisture content increased from 75 to 85.5% (Figure 13.3). Pacific whiting showed the same trends as pollock. The shear stress of both high- and low-grade Pacific whiting decreased linearly from ~53 to ~3 kPa and from ~45 to ~3 kPa, respectively (Figure 13.3). This trend is also demonstrated in the texture map (Figure 13.1).

Shear strain, commonly referred to as an indicator of protein quality,[7] however, was not affected within a certain range of moisture content. Shear strain values of both high- and low-grade pollock surimi were unaffected by moisture contents between 75 and 81%. However, their values decreased linearly from ~2.6 to ~1.9 and from ~2.1 to ~1.5, respectively, as the moisture content changed from 81 to 85.5%. Unlike pollock, the shear strain of Pacific whiting gels behaved differently with varied moisture contents. Their val-

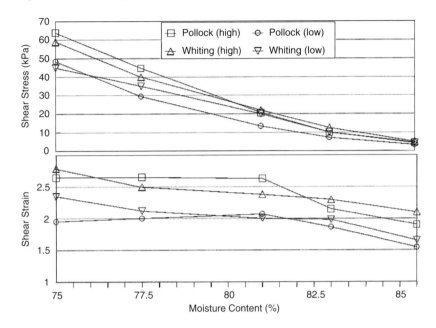

Figure 13.3 Effects of moisture content on shear stress and shear strain of gels from low- and high-grade surimi (Alaska pollock and Pacific whiting). Gels were prepared with 2% salt. (*Source:* From Ref. 17. With permission.)

ues decreased linearly from ~2.75 to ~2.1 for high grade and from ~2.35 to ~1.65 for low grade as the moisture content increased from 75 to 85.5% (Figure 13.3).

Addition of water to surimi also changes the color hues of the gels (Figure 13.4). L* values (lightness) increased and b* values (yellow hue) decreased as the amount of water increased.[8] A difference in moisture content of 7.5% increased the L* values by ~3 and decreased the b* values by ~2 in both pollock and whiting surimi gels. The water effects on the a* value, however, were not pronounced.

The quality of water affects the color, texture, and flavor characteristics of finished products. Lee[9] emphasized the importance of mineral contents in water. Ca^{2+} and Mg^{2+} are responsible for texture changes, while Fe^{2+} and Mn^{2+} are responsible for color changes. Okada[10] hypothesized

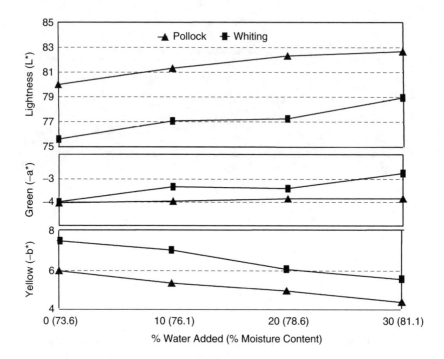

Figure 13.4 Effects of moisture content on color values (L*, a*, and b*) of surimi gels (Alaska pollock and Pacific whiting). Gels were prepared with 2% salt. *(Source:* From Ref. 8. With permission.)

that the superior quality of kamaboko in Odawara, Japan, was likely due to a calcium content of 20 to 40 mg/100g in the water. The positive effect of calcium ion is pronounced when calcium is added during chopping and surimi paste is going through setting (*suwari*). However, the inclusion of calcium in frozen surimi, as added in surimi washwater or mixed along with cryoprotectants, affects gel texture negatively.

Water, however, is often not considered a food ingredient in certain countries. It is common in Asia to find products without water listed as an ingredient on the label. In most Western countries, however, water is the second largest ingredient in most surimi seafood.

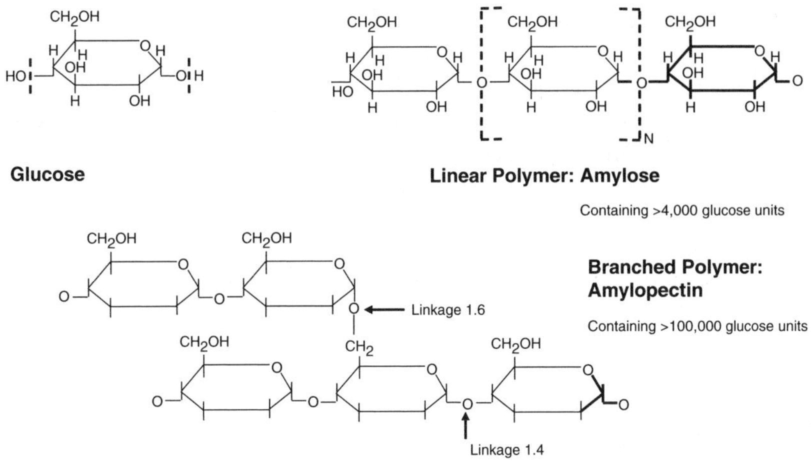

Figure 13.5 Two basic components of starch: amylase and amylopectin.

13.2.2 Starch

13.2.2.1 What Is Starch?

Starch is composed of tiny microscopic granules in stratified layers. The granules consist of a radially oriented crystalline aggregate of two major polymers: amylose and amylopectin. One granule may contain millions of molecules of these D-glucose polymers. Amylose is a straight-chain polymer and is a relatively small polymer consisting of several hundred glucose units linked by α-1-4-glucosidic linkages. Amylopectin, on the other hand, is a branched polymer of glucose units with α-1-6-glucosidic linkages at the branch points and α-1-4 linkages in the linear regions (Figure 13.5).

13.2.2.2 Modification of Starch

The nature of the modification, such as cross-linking, hydroxypropylation, acetylation, and pregelatinization, also determines the physical properties of starch. Cross-linking, for example, provides stable viscosity and tolerance against excessive processing conditions (heat, acid, and shear), which raises the gelatinization temperature of starch. Cross-linking is a chemical method that builds a covalent bridge between two starch molecules using a di- or polyfunctional reagent, such as phosphorus oxychloride:

$$\text{Starch-OH} + \text{POCl}_3 \xrightarrow{\text{NaOH}} \text{Starch-O}\underset{\underset{\text{O-Na}^+}{|}}{\overset{\overset{\text{O}}{\|}}{\text{P}}}\text{O-Starch}$$
(Phosphorus oxychloride)

Another common chemical modification is substitution. The primary purpose of substitution is to impart resistance to retrogradation and gelling of amylose, as well as eliminate the association of the linear segments of amylopectin at low temperatures. Other effects of substitution include lowering the gelatinization temperature (unlike cross-linking), increased viscosity, and improved colloidal properties. The introduction of substitute groups on the starch by treatment

$$\text{Starch-OH} + \underset{\text{(Propyleneoxide)}}{\text{CH}_3\text{-CH-CH}_2 \atop \diagdown\!\!\!\diagup \atop \text{O}} \xrightarrow{\text{NaOH}} \text{Starch-O-CH}_2\text{-CHOH-CH}_3$$

$$\text{Starch-OH} + \underset{\text{(Aceticanhydride)}}{[\text{CH}_3\text{-}\overset{\overset{\text{O}}{\|}}{\text{C}}]_2\text{-O}} \xrightarrow{\text{NaOH}} \text{Starch-O-}\overset{\overset{\text{O}}{\|}}{\text{C}}\text{-CH}_3 + \text{CH}_3\text{-C-ONa} + \text{H}_2\text{O}$$

with monofunctional reagents, which react with the hydroxyl groups on the starch, produces starch esters and ethers. Substitution is performed by treatment with reagents, such as acetate, succinate, phosphate, hydroxypropyl, and octenylsuccinate:

If cross-linking and substitution are chemical methods for modification, pregelatinized (instant) starch, in contrast, is categorized as physically modified starch. The process of pregelatinization is to make the starch granules fully swollen before drying so that grown granules easily absorb water, regardless of heating. Both native starch and chemically modified starch can be processed into pregelatinized starch.

13.2.2.3 Starch as a Functional Ingredient for Surimi Seafood

Starch plays an important role in the formation of the network structure of surimi-starch gels and therefore is an important functional ingredient in surimi seafood. The most commonly used native starches are wheat, corn, potato, waxy maize, and tapioca. The functions of starches, however, are quite different (Table 13.1). High amylose starches, such as corn and wheat, form somewhat brittle gels, while high amylopectin starches, such as tapioca and waxy maize, form adhesive and cohesive gels. High amylose starches with a hydroxypropylated, cross-linking modification also make gels adhesive and cohesive. In addition, the color of high amylose starch gels is opaque, while high amylopectin starches make gels translucent. A cross-linking modification can make gels more translucent or transparent.

Different botanical sources of starches behave differently with regard to the texture of surimi-starch gels. Potato starch, for example, increases the gel strength more than other native starches because of its ability to bind a larger amount of water or swell to a bigger granule size.[11] In addition, the functional properties of a modified starch are different compared to the native starch. Other factors such as heating temperature, degree of swelling, and water uptake of the starch granule influence gelatinization of starch in the protein gel.[12] Inadequate gelling time and temperatures result in some granules being un-swollen or un-gelatinized and thus affect the flavor, color, and texture of the final products. The optimum level of starch must be determined with all possible considerations in product development as follows: characteristics of the final product (texture, color, or flavor), surimi content, available free water, salt or sugar content, refrigerated or frozen storage, and expected cost.

Starch is an important ingredient to provide good sensory characteristics as long as it is not abused. It is commonly used

TABLE 13.1 Compositional and Functional Properties of Native Starch

	Wheat	Corn	Potato	Waxy Maize	Tapioca
Protein, %	0.4	0.6	0.06	0.15	0.1
Fat, %	0.8	0.44	0.05	0.15	0.1
Ash, %	0.2	0.1	0.4	1.1	0.2
Phosphorus, %	0.06	0.02	0.095	0.01	0.02
Amylose, %	25	26	20	1	17
Amylopectin, %	75	74	80	99	83
Average granule size, μm	30	15	40	15	20
Gelatinization temp, °C	77	67	61	72	65
Peak viscosity (BU) at 6%	80	250	800	600	600
Solubility (95°C),%	41	25	82	23	48
Swelling, g water/g	21	24	>1000	64	71
Paste clarity	Opaque	Opaque	Translucent	Translucent	Translucent
Paste texture	Short	Short	Long	Short	Long
Paste flavor	Cereal	Cereal	Mild	Cereal	Mild

Note: BU: Brabender amylogram unit.

either to maintain gel strength with a reduction of surimi content or to secure the storage stability of refrigerated and/or frozen products. The type and content of starch profoundly affect the texture and color of the surimi gels. In the surimi seafood industry, a level of 4 to 12% combined starches is commonly used. However, the usage of starch has been abused. Some cheap crabsticks currently marketed in Russia contain more than 20% starch. Without formula optimization, sticky and soft-textured products are produced when a large portion of surimi is replaced with a large quantity of starch. Understanding the chemistry and physical properties of starch is therefore the most important issue in formulating surimi-based products.

Based on granule size, potato starch has the highest swelling capability, followed by tapioca, waxy maize, corn, and wheat (Table 13.1). As starch granules are heated, irreversible swelling of the granules occurs. If heating continues to reach the gelatinization temperature, the granules absorb more water and the viscosity dramatically increases.[13] Starch granules continue to absorb water and expand until the gel matrix limits them. The expansion of starch granules results in a reinforcing or pressuring effect on the gel matrix, as well as higher gel strength. Gel strength, however, can decrease if too much starch ($\geq 8\%$) is added.[14]

Thermal changes of the starch granule in the surimi–starch system are different from those in the starch/water system. Gelatinization of starch occurs concomitantly with the thermal gelation of the fish proteins. However, starch gelatinization is delayed by the presence of myofibrillar proteins, salt, sugar, and/or sorbitol in the surimi–starch system. The myofibrillar proteins are thermally denatured before the starch is completely gelatinized.[15] Additionally, water entrapped in the protein gel network limits the availability of water for starch gelatinization and results in competition for water between starch and protein.[16] Although the starch granules expand in the surimi gel matrix, they cannot expand as much as in the starch–water system because the fish proteins limit the availability of water.

Starch increases surimi gel strength more effectively at low concentration (up to 3%) than at high concentration (6 to 9%). The effect of added starch on the shear stress of both pollock and whiting surimi were highlighted when the starch content increased from 0 to 6%.[17] When 10% starch was added, failure shear stress values were significantly reduced for all starches, indicating that starch inhibited the gelation of fish proteins by competing for the available water.[14] Shear strain of surimi gels with added starch, on the other hand, responded differently. Shear strain values of pollock–starch gels, except modified waxy corn starch, slightly increased at 3% starch content. However, strain values remained constant with starch up to 10%, except for wheat starch and modified waxy corn starch. In general, at a higher level of starch (6 to 10%), the shear stress decreased while shear strain remained constant, indicating the presence of a nonlinear interaction between surimi and starch (Figure 13.6).[17]

The ratio of amylose and amylopectin also varies among different starches. In addition, amylose and amylopectin behave differently during gelatinization.[13] Konoo et al.[18] evaluated the effects of the amylose/amylopectin ratio and the degree of pre-gelatinization of the starch on Alaska pollock surimi gels. For native starch, the amylose/amylopectin ratio did not significantly affect the breaking strength (force). However, breaking strength increased with increasing amylose content for pre-gelatinized starch. At low temperature, the amylopectin in fish muscle did not retrograde for at least 60 min, indicating the delay of gelation. However, when left to stand for more than 24 hr, the amylopectin shrank into smaller particles as it retrograded.

During refrigerated storage, the expressible moisture and compressive force of surimi gels also increased with a higher amylose fraction. This was attributed to retrogradation of the gelatinized starch.[16] During frozen storage, high amylose starches undergo severe retrogradation, resulting in gels with higher expressible moisture as well as increased brittleness.[19] Hydroxypropylated and cross-linked waxy maize starch (1.0 to 2.0%), however, are popularly used to overcome severe expressible moisture (dripping) problems upon thaw-

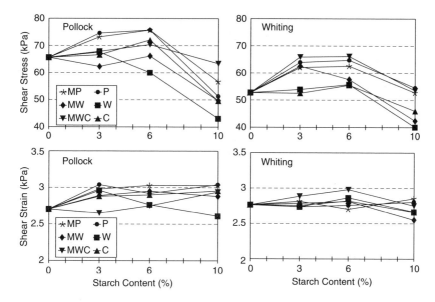

Figure 13.6 Effects of various starches on texture of two surimi gels (Alaska pollock and Pacific whiting). MP: modified (acetylated) potato starch; P: native potato starch; MW: modified (cross-linked) wheat starch; W: native wheat starch; MWC: modified (cross-linked/hydroxypropylated) waxy maize corn starch; C: native corn starch. (*Source:* From Ref. 17. With permission.)

ing after frozen storage. Acetylated potato starch is also used for longer shelf life of frozen products, although it does not effectively control the dripping problem.

The stickiness of refrigerated surimi seafood manufactured with a high concentration of starch is also a problem during the first 7 to 10 days of storage. Therefore, it is common to age products in the refrigerator for a week before shipping. The use of pre-gelatinized, cross-linked, or hydroxypropylated starch often solves the stickiness problem immediately. Park and co-workers[20] presented the effective control of stickiness of crabstick containing starch more than 6% (Table 13.2). A significantly higher stickiness was measured for crabstick containing acetylated potato starch. A reduction in stickiness (as starch granules retrograde) was noticed for native and

TABLE 13.2 Control of Stickiness Using Various Potato Starches in Crabstick Formula during 10-day Refrigerated Storage

Type of Starch	%	Day 1	Day 2	Day 3	Day 7	Day 10
Control	0	8.97	10.97	8.74	9.60	7.54
Native	3	10.30	10.00	8.36	8.26	6.40
	6	10.84	12.12	8.16	5.89	6.74
	12	48.66	30.53	20.42	11.23	11.34
Acetylated	3	32.37	19.53	18.53	21.23	21.90
	6	33.50	37.52	29.72	25.13	21.78
	12	50.57	32.94	44.52	39.95	34.16
Pre-gelled/cross-linked	3	12.53	5.89	6.06	6.43	6.42
	6	3.68	4.51	4.53	4.55	4.35
	12	3.82	2.53	3.15	3.21	3.46

Note: [Adhesiveness measured as an area under the tensile mode using TA.XT Plus texture analyzer (Micro Stable System, Surrey, U.K.)].

acetylated potato starch during 10-day refrigerated storage. However, the stickiness of crabstick containing acetylated potato starch was still high. At 3%, pre-gelled/cross-linked potato starch controlled stickiness well. A minimum level of pre-gelled starch, however, must be determined in regard to its effect on gel texture.

Starch also plays an important role in the textural behavior of surimi gels when they are reheated.[21] The "rubberiness" of a heated surimi-based product can be somewhat reduced by increasing the level of starches or, more effectively, by adding hydroxypropyrated and cross-linked waxy maize starch. Wu et al.[15] demonstrated that higher salt and, to a lesser extent, higher sucrose concentrations delay the starch gelatinization temperature and affect the rigidity modulus (firmness).

Pre-gelatinized starches, on the other hand, are cold-water swelling starch products. The granular starch is converted into a cooked/gelatinized starch and then dried. Thus, the swollen granules absorb water instantly, regardless of temperature. Native starch, chemically modified starch, or cross-linked starch can also be made as pre-gelatinized starch. When pre-gelatinized starch is used in large quantity (>2 to 3%) in surimi seafood, it generally inhibits the gel formation of surimi seafood because it absorbs water before the fish proteins can bind with water.[15] Accordingly, the breaking force and deformation of pollock surimi gels were depressed as the amount of pre-gelatinized corn, potato, and tapioca starch was increased.[22] Such depression in the elasticity was accelerated by pre-soaking starches in water before addition and prevented by delaying starch addition to surimi during grinding.

Temperature and time are two of the most important factors that influence the gelatinizing properties of starch as well as the gelation properties of fish proteins in the surimi-starch system. Most studies on surimi–starch gels have concentrated on the effect of an individual starch at a fixed heating temperature and time. Yang and Park,[23] however, investigated the effects of both heating temperature and time on the physical properties of surimi–starch gels. The results showed that a high heating temperature (90°C) caused granules to become larger when native corn starch was used, but did not affect modified starch or potato starch, which have lower gelatinization temperatures (Figure 13.7).[23] Corn starch, which gelatinizes at 67 to 70°C in water, was not gelatinized at 70°C (Figure 13.7A), indicating that the gelatinization temperature increased in the presence of surimi and other ingredients. However, potato starch, which has a gelatinization temperature around 60°C, was gelatinized at 70°C and showed no difference in granule size at the two temperatures (Figure 13.7B). Waxy maize starch, on the other hand, gelatinized at 72°C (Table 13.1). However, when modified (i.e., hydroxypropylated), its gelatinization temperature was reduced to 70°C (Figure 13.7C).

The color of surimi–starch gels depends not only on the starch concentration, but also on the starch properties. If granules were not fully swollen in the gel, the gel became more opaque (higher L* value) and slightly more yellow (higher b* value) as the starch concentration increased (Figure 13.8). If the granules were fully swollen in the gel, the gel became more translucent (lower L* value) and slightly blue (negative b* value).[22] Higher L* values of gels with corn starch, cooked at 70°C (Figure 13.8), clearly indicate that corn starch was not gelatinized.

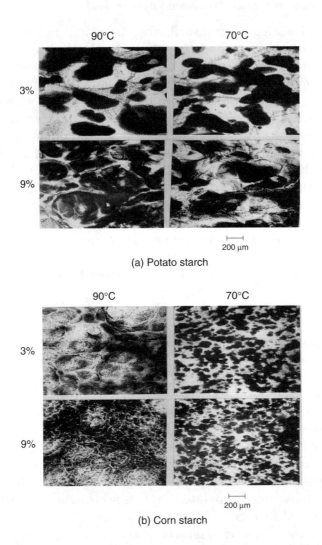

Figure 13.7 Microstructure of (A) native potato starch (3 and 9%) in surimi gels cooked at 70 and 90°C for 30 min; (B): native corn starch (3 and 9%) in surimi gels cooked at 70 and 90°C for 30 min; and (C) hydroxypropylated waxy maize starch (3 and 9%) in surimi gels cooked at 70 and 90°C for 30 min. *(Source:* From Ref. 23. With permission.) (Continued)

Ingredient Technology for Surimi and Surimi Seafood   667

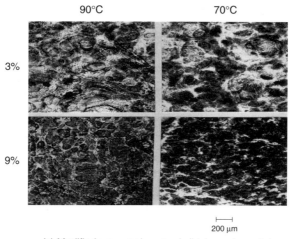

(c) Modified waxy maize starch (high amylopectin)

Figure 13.7 (Continued.)

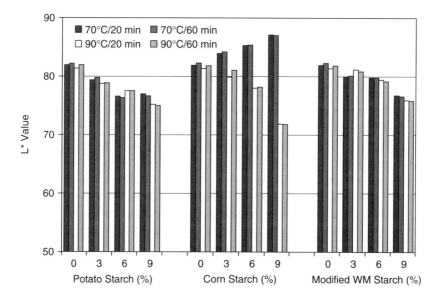

Figure 13.8 Color values (L*) of surimi gels mixed with various starches and cooked at various thermal conditions. WM: waxy maize.

13.2.3 Protein Additives

Protein additives are globular proteins while surimi proteins are fibrillar. They must have at least one special property, such as functionality, nutrition, or economic benefits, to be beneficial. The nutritional quality of protein can be measured using a protein efficiency ratio (PER). The protein efficiency ratios of animal proteins are relatively high, at 2.5 to 3.0, while plant proteins are as low as 1.0 to 2.0. Dried protein additives have approximately 60 to 95% protein content and therefore have potential as a nutritional fortifier in surimi seafood.

Regardless of the benefits imparted by a protein additive, the overriding factor is economics. Therefore, if a protein additive can offer an advantage, without adverse effects or additional cost, the addition is considered favorable. There are three major functional advantages of addition of protein additives in surimi seafood. They are (1) improvement of gel firmness and elasticity, (2) enzyme inhibition for surimi containing heat-stable protease, (3) anti-retrogradation of starch during refrigerated or frozen storage.

With respect to the functional properties of protein additives, protein–water, protein–protein, and protein–lipid–water interactions are very important for the formulation of the stable gel network structure.[24] Park ([25]) evaluated seven commercially available protein additives (1% dried weight) to investigate their interaction with surimi gels in the presence of 2% salt. Frozen egg white, dried egg white, and beef plasma protein acted as functional binders, while wheat gluten, soy protein isolate, whey protein concentrate, and whey protein isolate acted as functional fillers. The functional binders increased both the shear stress and shear strain values of surimi gels, whereas the functional fillers increased the shear stress while the shear strain values of the gels decreased. All protein additives had an impact on the color of the surimi gels with a slight reduction in the L* value (lightness) and a large increase in the b* value (yellow hue).

Lanier[6] summarized the interactions of muscle and protein additives (non-muscle proteins) affecting heat-set gel rheology. Gel matrix development by myofibrillar proteins can also be directly influenced by chemical interactions between the protein additives and myofibrillar proteins, as well as indirectly by changes in the molecular environment brought on by the presence of protein additives. These changes include the contribution to total protein concentration, water availability, ionic strength/types, and pH.

Most recently, attention has been given to protein additives with regard to their allergenic properties. Eight major allergens are milk, egg, soy, wheat, peanut, shellfish, fruits, and tree nuts.[26] The Japanese government also established a mandatory labeling law on five designated foods (wheat, buckwheat, egg, milk, and peanut).[27] Among them, protein additives from milk, egg, and wheat are commonly used as a functional ingredient for surimi seafood manufacturing. Therefore, if used, a warning label must be present.

13.2.3.1 Whey Proteins

There are numerous whey proteins on the market that have different functional and nutritional characteristics because whey protein is a by-product of various dairy products (Figure 13.9). Two major proteins compose whey proteins: β-lactoglobulin (75%) and α-lactoalbumin (25%). In addition, the gelation properties of whey proteins depend on temperature, pH, and ionic strength. α-Lactoalbumin sets a gel at 62 to 68°C, while β-lactoglobulin sets at 78 to 83°C. Elastic gels are formed at pH 6 to 7.5, whereas brittle gels form at pH 3 to 4 and opaque aggregates at pH 4 to 6. Additionally, weak and translucent gels form at low concentration and/or low temperature (55 to 70°C), while strong and opaque gels form at high concentration and/or high temperature.

Technical advancements by leading manufacturers of whey proteins have been significant. Two new products are a high-pH WPC and a processed WPC. The processed WPC (PWP) was developed through concentration, refining, heating under special conditions, and spray drying or retort heating.[28] PWP shows extremely better functionality than normal WPC. It is a modified whey protein arranged in a linear chain of numerous globular proteins (Figure 13.10).

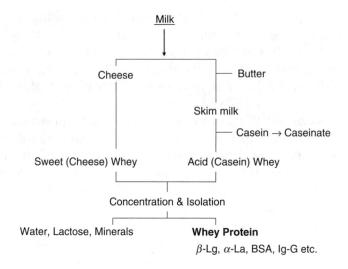

Figure 13.9 Processing diagram of whey proteins.

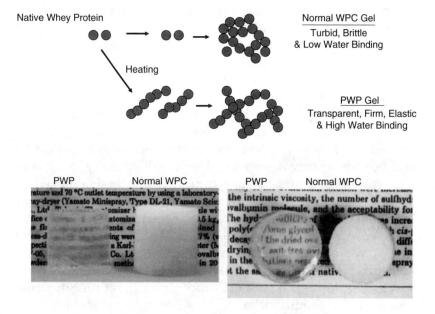

Figure 13.10 Gelation mechanism model of normal WPC and PWP. (*Source:* Adapted from Ref. 28.)

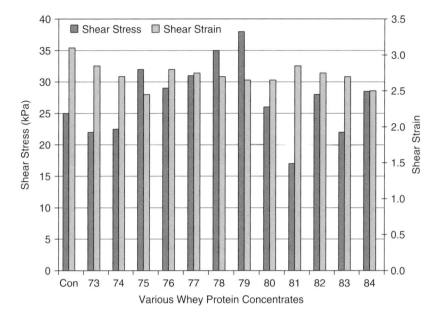

Figure 13.11 Effects of various whey protein concentrates (2%) on the texture of Pacific whiting surimi gels. Con: control; 73–86: various whey protein samples obtained from a single source. All gels were prepared at 2% salt and 78% moisture.

Various studies on the gelation of whey proteins have been extensively conducted. Park[8] evaluated 14 different whey proteins from a commercial company to compare their effectiveness as gelling agents (Figure 13.11) and enzyme inhibitors (Figure 13.12). Surprisingly, the functionality was extremely different for all whey proteins studied. Because whey proteins are by-products of cheese processing (Figure 13.9), the functional properties depend on the nature of the cheese and further chemical modification. The resulting physicochemical properties of whey proteins are therefore extremely diverse. The selection of whey protein must be carefully conducted based on the desired result. Whey proteins, in general, do not form fine gels in the presence of salt; therefore, whey proteins perform poorly in surimi seafoods, which contains 1.5 to 2.0% salt. In contrast, PWP

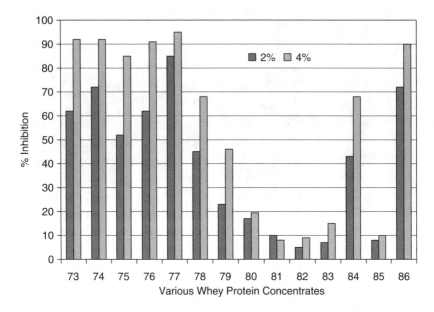

Figure 13.12 Effect of various processed whey proteins (PWP, 2 and 4%) on the inhibition of autolytic enzyme activity of Pacific whiting surimi. 73–86: various whey protein samples obtained from a single source.

(Nutrilac SU-7723 from Arla Foods) can form a fine and translucent gel at cold temperature and/or in the presence of salt.

According to a survey in 1995, every surimi seafood manufacturer in the United States used a different chopping procedure (Figure 13.13). However, a 12-min chopping time was common when a vacuum silent cutter was used, while 18-min chopping was common when using a conventional open bowl cutter. The chopping procedure, especially with regard to the time of processed whey proteins (PWP) (before or after salt addition), made a significant difference in gel strength (shear stress). Certain chopping procedures, where PWP was added at the very beginning of chopping before addition of salt (C, D, I, and J in Figure 13.12), resulted in gels with exceptionally high shear stress values (Figure 13.14).

Ingredient Technology for Surimi and Surimi Seafood

	0	2	4	6	8	10	12	14	16	18
A	Su	Salt+N		Others (Vacuum) ▶						
B	Su	Salt			Others+N (Vacuum) ▶					
C	Su+N	Salt		Starch		Others				▶
D	N	Su	Salt		Starch		Others			▶
E	Su	Salt+N		Starch		Others				▶
F	Su	Salt		Starch+N		Others				▶
G	Su	Salt		Starch		Others+N				▶
H	Su	Others+N								▶
I	Su	Others+N		Salt (Vacuum)						▶
J	Su+N	Others (Vacuum)								▶

Su: Frozen surimi (partially broken)
Nu: Nutrilac Su-7723

Figure 13.13 Various chopping procedures used in the U.S. surimi seafood industry to incorporate ingredients.

13.2.3.2 Egg White Proteins

Two forms of egg white proteins are commonly used in surimi seafood: liquid and dried. Liquid egg white (11% protein, 88% moisture, and 1% other components) is commercially available as a frozen or refrigerated product. The major protein components in egg white are ovalbumin (54%), conalbumin (12%), and ovomucoid (11%). The gelation temperatures of these components vary significantly. Conalbumin gels at 62°C and ovalbumin gels at 75°C. The addition of small amounts of salt (less than 0.5%) and higher pH are also known to improve egg white gels. When high-temperature pasteurization is applied, a sulfur odor is often detected as a negative characteristic. However, the odor dissipates within seconds after the vacuum pouch is open.

Liquid egg white is the most commonly used protein additive in surimi seafood. The level of usage is 2 to 5%. It is not uncommon, however, to add 6 to 7% for a low-cost formula containing less than 30% surimi. When two types of egg white were used to replace surimi content, while keeping moisture

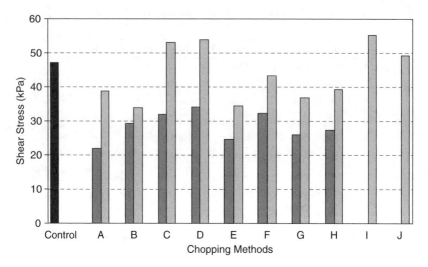

Figure 13.14 Effect of various chopping methods on shear stress. Processed whey protein (PWP) was used at 3.5% to reduce surimi content by 12.2 and 17.2%.

values at 78%, no significant change in shear stress or shear strain value was observed. Shear strain values actually increased gradually as LEW increased to 10% and surimi content decreased from 80 to 75%. However, when DEW was used, the shear stress continuously increased as surimi reduced from 80 to 42% and DEW increased to 10% (Figure 13.15). It was clear that LEW affects shear strain more, whereas DEW affects shear stress more while moisture content was equally adjusted.

There are several kinds of dried egg white products for various food applications. Most dried egg white products show greater gelling properties compared to liquid egg white with the same protein concentration. The functions of dried egg white in surimi or surimi seafood are to modify the texture, inhibiting "modori" (gel-softening), add whiter color, and prevent freeze-thaw damage. Handa et al.[29] evaluated seven

Ingredient Technology for Surimi and Surimi Seafood

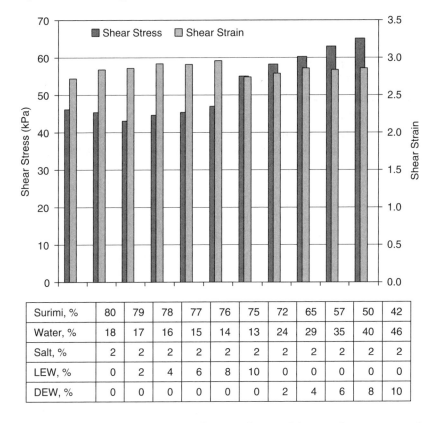

Figure 13.15 Effects of two forms of egg white on the texture of surimi gels. LEW: liquid egg shite; DEW: dried egg white. All samples contained 2% salt and 78% moisture.

dried egg whites (DEW) and found that the average molecular weight affected hydrophobicity and surface sulfhydryl groups.

Dry heating (storage in a hot room under controlled temperature and relative humidity) is one of the most promising approaches for improving the gelling properties of DEW.[30] The partially unfolded conformation formed by dry heating may contribute to the improved gelling properties of DEW.[31] Commercial DEW products, typically with 4 to 8% moisture content, are generally dry-heated at 60 to 80°C for 3 to 30 days for pasteurization or improving the gelling properties.[29]

13.2.3.3 Plasma Proteins

Beef plasma protein is the most functional protein additive and has been successfully used in surimi and surimi seafoods as an enzyme inhibitor and/or gel enhancer until recent outbreaks of BSE (bovine spongiform encephalopathy) in the EU, Japan, Canada, and the United States. Starting during the 2004 fishing season, no manufacturers are expected to use beef plasma protein. However, new efforts should be made to manufacture food-grade plasma proteins from swine, poultry, or farm salmon.

Plasma protein, which is mostly in a dried form, is used as a gelling agent and/or protease inhibitor. The major proteins are water-soluble albumins, salt-soluble globulins, and fibrinogen. The gelation property of these proteins also depends on temperature. Albumin gels at 85°C and fibrinogen gels at 50°C. Food-grade plasma proteins currently available are beef plasma and pork plasma. However, the availability of food-grade pork plasma protein (PPP) is limited. A recent study conducted at the OSU Seafood Lab demonstrated that PPP performed slightly better than beef plasma protein (BPP) for enzymatic inhibition (Figure 13.16) and gel enhancement (Figure 13.17). This study indicates that PPP can replace BPP as an effective enzyme inhibitor. A minimum of 0.05% PPP effectively inhibited more than 70% of the autolytic enzyme.

Fibrinogen and thrombin are purified from beef plasma protein. Fibrinogen is converted to fibrin in the presence of thrombin at lower temperatures. This reaction can be enhanced as the temperature increases. Yoon et al.[32] investigated the gel-enhancing effects of a fibrinogen and thrombin mixture on surimi gels. The suggested use of fibrinogen to thrombin was at a ratio of 1:20. The results showed that low-quality surimi produced higher textural values when a fibrinogen and thrombin mixture (3 to 5%) was added. The shear strain value, which indicates the quality of surimi proteins, also increased from 1.63 to 2.33 when surimi paste with 5% fibrinogen–thrombin mixture was set and cooked. The gelation time of surimi proteins was also significantly reduced when a mixture of fibrinogen and thrombin was used, regardless of holding temperature.

Ingredient Technology for Surimi and Surimi Seafood

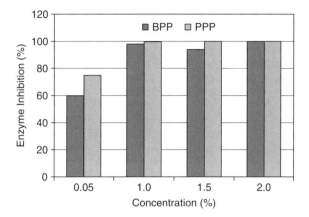

Figure 13.16 Enzyme inhibition of pork plasma protein (PPP) as compared to beef plasma protein (BPP).

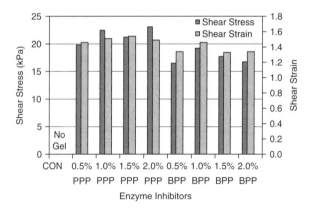

Figure 13.17 Comparison of pork plasma protein (PPP) and beef plasma protein (BPP) on gel texture of Pacific whiting surimi. CON: contained no enzyme inhibitor and show no gel.

13.2.3.4 Soy Proteins

Soy proteins are popular in Asia, but not in Western countries. Since the U.S. FDA authorized a health claim for soy protein in controlling heart disease in October 1999, soy proteins have received increased attention as a functional ingredient for surimi seafood manufacturing. Commercial soy protein products used in the surimi seafood industry include functional soy protein concentrate and isolated soy protein; both derived from defatted soy flakes having a protein content of roughly 50%. Soy protein concentrate is the product of soluble sugar removal from the defatted soy flake; functional soy protein concentrates (FSPC) are prepared by modifying the traditional soy protein concentrate material. Generally, modification of traditional soy protein concentrate to form FSPC results in the FSPC being more hydrophobic compared to the traditional soy protein concentrate material. Isolated soy protein is derived from defatted flakes through isoelectric solubilization of the soy protein, followed by decanting or centrifugation of the solid material with subsequent acid precipitation of the desired protein fraction and removal of the soy whey (D. Cantrell, personal communication, 2004).

Major proteins in soy protein are globulins, which consist of several subunits, such as 2S, 7S, 11S, and 15S. Among these fractions, 7S (conglycinin) and 11S (glycinin) are considered of major importance in relation to the gelation of soy proteins. Soy protein fraction gelation is dramatically influenced by both heating temperature and ionic strength. Both glycinin and conglycinin form a gel at 95°C, but in the presence of salt (0.2 M) glycinin forms aggregates while conglycinin gels when heated to at 85°C.

Most ISPs (isolated soy proteins) utilized throughout the surimi seafood manufacturers are in the form of emulsion curd. This curd is prepared by mixing ISP, water, and vegetable oil at an approximate ratio of 2:9:1 (protein:water:oil). Choi et al.[33] evaluated the functionality of ISP and functional soy protein concentrate (FSPC) in surimi seafood formulations. When ISP and FSPC were evaluated in both dried powder and emulsion curd forms, they produced different gel characteristics. Shear stress values of surimi seafood gels were slightly better when soy protein emulsions were used in the surimi seafood gel formulations, while shear stress values demonstrated that the use of emulsion curd produced surimi seafood gels that tended to possess greater cohesiveness (higher shear strain values) compared to product made with

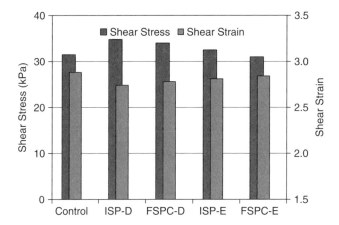

Figure 13.18 Effects of two soy proteins (concentrate and isolate) on texture of surimi gels. They were applied as a dried (D) or emulsion (E) form. ISP: isolated protein; FSPC: functional soy protein concentrate.

direct dry addition of either ISP or FSPC powders (Figure 13.18). Additionally, data from this study demonstrated that ISP and FSPC products contribute to the surimi seafood gel matrix in a similar fashion; regardless of the means of addition to the surimi seafood formulation (either emulsion curd or direct dry addition) during gel preparation, shear strain and shear stress values were similar (Figure 13.18). Additionally, it is important to note that the use of emulsion curd tends to improve the lightness of surimi seafood gels, as determined by CIE L-value (Figure 13.19).

Park[34] measured the storage modulus (G′) of soy emulsion made using both FSPC and ISP in the presence and absence of 2% salt, respectively. Both FSPC emulsions showed higher G′ than ISP emulsion without salt until the temperature reached 80°C. However, G′ values of ISP emulsion with 2% salt collapsed when the temperature reached 74°C (Figure 13.20). This study clearly indicates that ISP is very sensitive to salt, resulting in poor performance in the presence of salt. But it has been noted that advanced technology has developed a new ISP that is significantly less sensitive to salt (D.

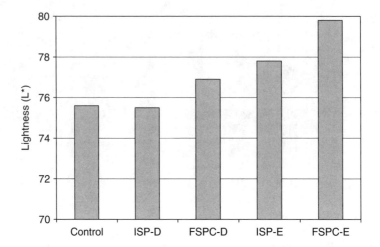

Figure 13.19 Effects of two soy forms (concentrate and isolate) on the color (lightness) of surimi gels. They were applied to surimi as a dried (D) or emulsion (E) form. ISP: isolated soy protein; FSPC: functional soy protein concentrate.

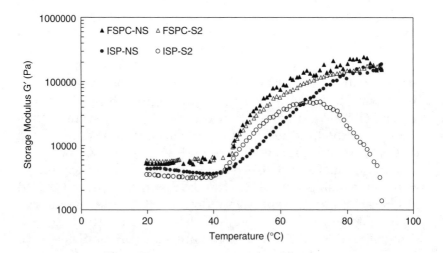

Figure 13.20 Storage modulus (G′) of emulsion made of FSPC and ISP with salt (2%) and with no salt (NS). FSPC: functional soy protein concentrate; ISP: siolated soy protein.

Cantrell, personal communication, 2004). It is necessary, however, to confirm salt stability before using ISP in a surimi seafood formulation.

13.2.3.5 Wheat Gluten and Wheat Flour

Wheat gluten is the least expensive protein additive available to surimi seafood. It is a by-product from wheat starch manufacturing. Major flour proteins are categorized as gluten (85%) and non-gluten (15%). Non-gluten proteins include albumins and globulins. The basic mechanism for the gel formation of gluten is disulfide bonds. The uniqueness of gluten is due to the dough-forming ability contributed by gliadin and glutenin. Gliadin is a low molecular weight protein and is soluble in ethanol, while glutenin is a high molecular weight protein and is soluble in water or dilute acid or alkali. Because gliadin is not water soluble, mixing wheat gluten in surimi seafood paste is not easy unless a high-speed vacuum silent cutter is used or solubilized in sorbitol solution. In addition, gliadin forms films upon hydration, whereas glutenin forms strands.

Chung and Lee[35] reported that 2% gluten addition significantly improved the freeze-thaw stability by reducing expressible moisture, although it also reduced the strength of surimi gels. According to Yamashita and Seki,[36] soy or wheat protein (5 or 10%) added to pollock surimi considerably increased gel strength but decreased the setting response because of the dilution of myosin and decreased heavy chain cross-linking.

When wheat flour was used instead of wheat starch and corn starch, the shear stress values of surimi seafood gels were not different (Figure 13.21). This result indicates that the use of wheat flour, instead of wheat starch and corn starch, does not sacrifice the gel values while obviously achieving a great saving in formulation cost. When various modified starches were used along with wheat flour, stress values increased.

13.2.4 Hydrocolloids

Gums are a group of polymeric compounds that can be dissolved or dispersed in water to give a thickening or gelling

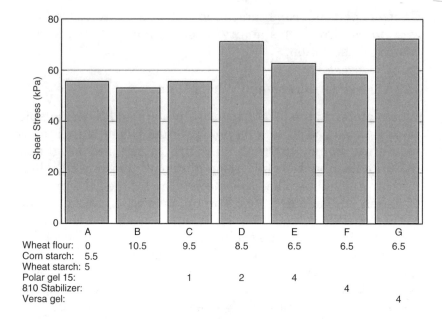

Figure 13.21 Effects of wheat flour, in combination with various starches, on the gel strength of surimi seafood gel during frozen storage. Test formula: 45% pollock surimi, 42.5% water, 10.5% starch and/or flour, and 2% salt. Polar gel 15, 810 Stabilizer, and Versa gel are modified waxy maize starches.

effect. Because they are colloidal in nature, they are also known as hydrocolloids. The thickening or gelling properties vary with the type of gum and the preparation conditions (i.e., pH, temperature, hydration technique, ionic strength, and the presence of synergistic compounds).[11]

13.2.4.1 Carrageenan

Carrageenan and konjac are typical hydrocolloids used in surimi seafood. Carrageenan is a gelling agent extracted from certain species of red sea plants. It is widely used in the food industry for its unique stabilizing and texturizing effects, as well as a fat replacement. To compensate for retarded carrageenan solubility, it is generally recommended to disperse the carrageenan with agitation into water. Heating (82 to 85°C) also ensures full solubility of the polysaccharide.

Three common types of carrageenan are manufactured commercially: (1) kappa-carrageenan, extracted from *Euchema cottonii*; (2) iota-carrageenan, from *Euchema spinosum*; and (3) lambda-carrageenan, from *Gigartina acicularis*. The extracts are primarily sulfated polysaccharides of varying ester content in descending order of lambda, iota, and kappa. Their molecular weights generally range between 100,000 and 500,000.

Kappa- and iota-carrageenans form thermally reversible gels upon heating and cooling, while lambda-carrageenan is basically used for thickening and suspending applications. The gels formed by kappa-carrageenan are also brittle, whereas iota-carrageenan gels exhibit elasticity. In addition, the strongest gels of kappa-carrageenan can be formed in the presence of K^+ ions, while the most elastic gels of iota-carrageenan form in the presence of Ca^{2+} ions. Kappa-carrageenan gels, however, do not have freeze/thaw stability, whereas iota-carrageenan gels do. Iota-carrageenan provides the best functionality, considerably improving the gelling potential and freeze/thaw stability of Atlantic pollock (*Pollachius virens*) surimi,[37] as well as Alaska pollock surimi and red hake (*Urophycis chuss*) surimi.[38] Da Ponte et al.[39] also reported that iota- and lambda-carrageenans improved the water retention capacity of raw minced cod and decreased toughening during frozen storage. In addition, iota-carrageenan exhibits good synergy with starch.[40]

13.2.4.2 Konjac

Konjac flour, which primarily contains glucomannan, is water soluble. It cannot form a gel alone, even if heated and cooled to room temperature, due to the polysaccharide of konjac flour, which contains acetyl groups that inhibit the formation of a gel network. Introducing a mild alkali into a konjac solution, however, results in a strong, elastic gel that retains its structure under various heat conditions, such as boiling water, retort, or microwave.[40,41] The traditional Oriental method of preparing thermally stable konjac noodles requires that the konjac paste be soaked in mild lime water.

Using rapid freezing to promote gelation, Wu and Suzuki[42] patented the use of konjac to simulate the strong texture of shellfish meat. Their method was to mix calcium hydroxide with surimi paste containing solubilized konjac slurry and then form it into a thin block before rapidly freezing it. The frozen block was thawed in cold water before shredding into fibers. Due to a strong off-odor generated from the chemical reaction of calcium hydroxide, extremely long washing is required. Therefore, the production yield of the recovery is significantly low. However, fibers generated through this process are very strong and chewy, therefore making them retortable.

Park[43] investigated changes in shear stress and strain values upon cooling, reheating, and freeze-thawing of surimi-konjac gels at various test temperatures. Konjac at 5% showed the ability to reinforce gel hardness by 8 to 10 times in both whiting and pollock surimi. When konjac flour was added at 4%, stress values were minimally affected by temperature, indicating that gels were most heat tolerant in both whiting and pollock surimi. These gels also exhibited an ability to maintain consistent shear strain values against freeze-thaw. Konjac up to 2% increased the lightness of surimi gels as well, while the yellow hue gradually increased.

Konjac gel prepared with a mixture of fillers is often frozen and sold as a frozen block (10 kg) in Asia as a yield extender. It is also noted to give less stickiness to low-cost crabstick containing a higher level of starch. However, further studies must be conducted to scientifically verify its function.

13.2.4.3 Curdlan

Curdlan is a polysaccharide consisting of 400 to 500 D-glucose residues and is produced by the microorganism *Alcaligenes faecalis var. myogenes*. Curdlan has the unique characteristic of forming a gel when heated in an aqueous suspension. It was discovered in 1966 by Harada and named for its ability to curdle when heated.[44] Curdlan is insoluble in water, alcohol,

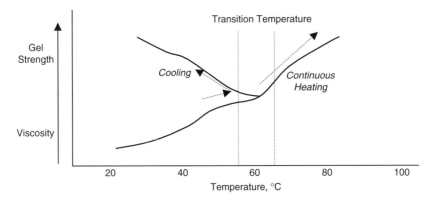

Figure 13.22 Rheological transition of curdlan gels on heating and cooling.

and most organic solvents; however, it dissolves in alkali solutions such as sodium hydroxide.

As a functional ingredient in surimi seafood, curdlan is a new addition. Heating an aqueous suspension of curdlan above 80°C forms a thermally irreversible gel. Curdlan gels can also be formed by heating to 60°C followed by cooling, which is similar to agar-agar and carrageenan gels (Figure 13.22).

13.2.4.4 Alginate

Alginate is extracted from brown sea plants. It has the unique ability to form a thermally irreversible gel by reaction with polyvalent cations, especially calcium. This property can be useful in preparing a variety of fabricated food products.[45] Despite its ability to form a firm gel, it typically weakens surimi gels when incorporated into the formulation.[11] If chelating agents such as sodium phosphate are used in alginate solution, the alginate is prevented from reacting with the polyvalent ions (i.e., calcium). Because all commercial surimi contains sodium phosphate as a cryoprotectant, a mixture of surimi, alginate, and calcium compounds will not produce positive results.

Alginate can be used as a fibrous filler or other texturizer. Extruding an alginate solution into a calcium ion bath produces fibers. These calcium–alginate fibers, after adjusting colors similar to surimi gels, can be used for molded products such as shrimp- or lobster-style surimi seafood.

13.2.5 Cellulose

Cellulose, the most abundant biopolymer, consists of β-1,4-glucosidic linkages and is available to the surimi seafood industry in powder form or as frozen block containing water and wheat starch. Unlike starch and protein, cellulose does not undergo any thermal transitions and thus volume expansion does not alter its viscoelastic properties. In addition, due to its large surface area and polymeric nature, it has the ability to entrap a large amount of water (like a sponge).

When determined by a centrifugation method, the water entrapped by cellulose fiber was not as tightly bound as in biopolymers that strengthen surimi gels.[46] The lack of tightly bound water and mass expansion during heating may explain why cellulose does not strengthen surimi gels.[46] The addition of 1 to 2% cellulose powder with a particle size of 17 to 20 μm significantly improved the frozen storage quality of surimi gels by keeping them from becoming brittle. Cellulose addition to minced fish also increased whiteness, while the gel strength and water holding capacity decreased. Addition of agar, however, increased gel strength but reduced whiteness.[47]

Methyl cellulose is manufactured by chemical modification of cellulose. Methyl cellulose or hydroxypropyl cellulose, unlike other gums, has the unique ability to form a heat-induced gel. Upon cooling, however, the gel effect begins to reverse and viscosity rapidly drops.

Like a frozen konjac gel block, a mixture of cellulose, wheat starch, and water is frozen and sold in block (10 kg) form as a yield extender in Asia. However, this ingredient should be further studied for its functionality.

13.2.6 Vegetable Oil and Fat Replacer

Soybean and canola oils are commonly used in surimi-based crabmeat as a texture modifier, color enhancer, or processing

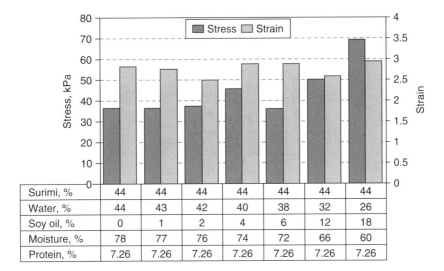

Figure 13.23 Effect of soybean oil on the texture of surimi gels. Oil replaced water by 1:1.

aid. Lee and Abdollahi[48] suggested that the addition of oil/fat modified the texture of fish protein gels in the following manner: it prevents sponge-like texture development during extended frozen storage; reduces brittleness; and minimizes textural variations resulting from cooking. Vegetable oil can replace water by 1:1 up to 6% without changing the shear stress and shear strain values (Figure 13.23). A formula that contains 36% water, 30% surimi, and 34% remaining ingredients can therefore perform similarly to a formula that contains 30% surimi, 30% water, 6% vegetable oil, and 34% remaining ingredients. Substituting oil for water gives surimi seafood manufacturers the option to keep surimi as the number-one ingredient for a product containing less than 30% surimi.

Vegetable oil also makes products whiter through a light scattering effect that results from the emulsion that is created when oil is comminuted with surimi proteins and water. As the percentage of oil increased, the lightness and yellow hue linearly increased as well (Figure 13.24). In addition, vegeta-

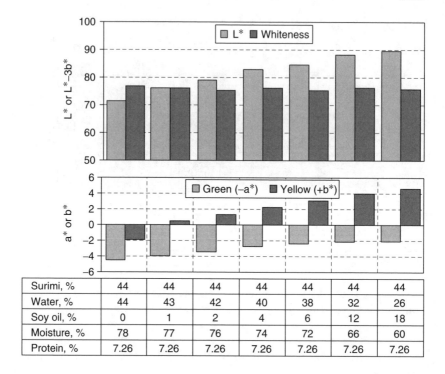

Figure 13.24 Effect of soybean oil on the color of surimi gels. Oil replaced water by 1:1.

ble oil is often used as a processing aid on the cooking machine, especially when the paste becomes sticky due to increased starch content in the formula.

In an effort to reduce the fat content in a product while maintaining the same taste and creamy "mouth-feel," a few fat replacers have been introduced into surimi seafood. They are typically either carbohydrate based or protein based. Carbohydrate-based fat replacers include maltodextrin, modified food starch, polydextrose, zanthan/guar gum, and inulin. Protein-based fat replacers are generally made from egg white, whey protein, or soy protein. They typically help coat the tongue like oil to give more residence time on the tongue for the flavor to have an impact. The oil-included formula exhibited a richer, fuller flavor.

13.2.7 Food-Grade Chemical Compounds

13.2.7.1 Oxidizing Agents

Oxidizing agents are commonly used in bread dough to improve textural properties by the formation of S–S bonds through the oxidation of sulfhydryl (-SH) groups. A similar effect was proposed for the gelation of surimi.[11,49,50] The use of L-ascorbic acid, sodium ascorbic acid, and erythorbic acid has therefore been suggested for gel-strengthening in thermally induced surimi gels. Lee et al.[51] found that sodium-L-ascorbate (SA) significantly improved the compressive force of the gel, as well as the sensory firmness of molded and fiberized products, with the maximum effect at a level of 0.2%.

The gel-enhancing effect of ascorbate in surimi produced from fish during the spawning season was not as high as in regular surimi. The effectiveness of SA, therefore, was directly related to the original surimi quality regardless of vacuum treatment, which indicates the unimportance of airborne oxygen. SA also promoted freeze syneresis, resulting in increased expressible moisture and brittle texture development. However, the freeze-induced syneresis can be minimized by the use of hydroxypropylated starch.

Pacheco-Aguilar and Crawford[52] determined that the reinforced oxidation of sulfhydryl groups on the myofibrillar proteins in gel formation was due to disulfide bonds by potassium bromate. Potassium bromate, however, has not been approved as an additive in surimi seafood products.

13.2.7.2 Calcium Compounds

Calcium compounds were commonly added as gel enhancers, especially to Pacific whiting surimi, until the late 1990s. Since the patent of Yamamoto et al.[53] on the use of a mixture of sodium bicarbonate, calcium citrate, and calcium lactate as gel quality improving agents, calcium compounds are often used in surimi gels without a complete understanding of how the calcium ions function. When commercial surimi processing of Pacific whiting started in 1991, all manufacturers of whiting surimi used either one or a combination of the following compounds as gel enhancers: calcium lactate, calcium citrate, calcium sulfate, and calcium caseinate. The usage level, however, varies from company to company (i.e., 0.1 to 0.3%). Manufacturers, however, do not have a clear understanding of why these chemicals are used for processing Pacific whiting surimi.

In addition to the endogenous enzyme problem in Pacific whiting, which is negated by the use of enzyme inhibitors, the low gel strength compared to Alaska pollock was thought to be due to a lower concentration of calcium ions in the flesh. Gordon and Roberts[54] reported that the calcium content of Pacific whiting was 8.7 mg per 100 g meat, while pollock contained 63 mg calcium ions per 100 g meat.[55]

Lee and Park[56] studied the effects of calcium compounds on the gel functionality of Pacific whiting and Alaska pollock surimi at three different thermal treatments. Calcium acetate, chloride, and caseinate were very soluble, while the others were less soluble. In addition, calcium lactate showed the highest solubility at a concentration of 0.2%, whereas the solubility of the other compounds, in descending order, was phosphate, citrate, sulfate, and carbonate. When a 25°C setting was applied to surimi, the addition of calcium was the most effective, regardless of fish species.

Addition of specific calcium compounds is most favorable when surimi seafood processing needs pre-incubation (setting) for fiber production. The effect of calcium on shear stress depended on species, setting temperature, and the specific compound. Shear strain values, however, were not affected by adding calcium compounds, regardless of species and thermal treatments. Lee and Park[56] concluded that the textural properties of surimi can be improved maximally with a 25°C pre-incubation and the addition of 0.1% calcium lactate or 0.05% calcium acetate for Alaska pollock (Figure 13.25), and 0.2% calcium lactate for Pacific whiting (Figure 13.26).

Since the introduction of Pacific whiting surimi in 1991, there has been a rumor that Pacific whiting surimi has a much shorter frozen shelf life compared to Alaska pollock surimi. This was simply because calcium salt was added during surimi manufacturing. Similar to calcium addition with

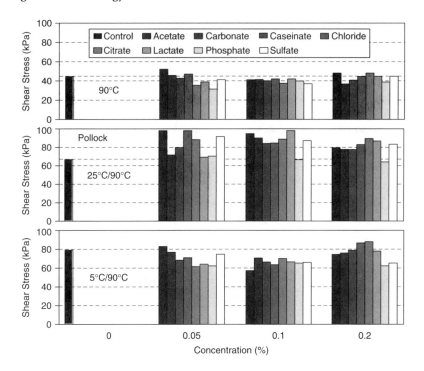

Figure 13.25 Effect of various calcium compounds on shear stress of Alaska pollock surimi gels at various thermal treatments. 90°C: cooked at 90°C for 15 min; 25°C/90°C: set at 25°C for 2 hr, followed by cooking at 90°C for 15 min; 5°C/90°C: set at 5°C for 18 hr, followed by cooking at 90°C for 15 min. (*Source:* Adapted from Ref. 56.)

cryoprotectants, if mince is washed with $CaCl_2$ for easy dewatering, the added calcium accelerates the chemical denaturation during frozen storage, resulting in reduced gel strength. Calcium can only be added to obtain a positive result when gel or surimi seafoods are manufactured with slow cooking (not in crabstick production where rapid heating is applied). Otherwise, it gives a negative effect. Negative effects of calcium on frozen surimi were also reported by Saeki.[57] They studied the gel-forming ability and cryostability of frozen surimi processed with $CaCl_2$ washing and found a significant reduction in frozen shelf life when the added calcium concentration exceeded 5 mM/kg.

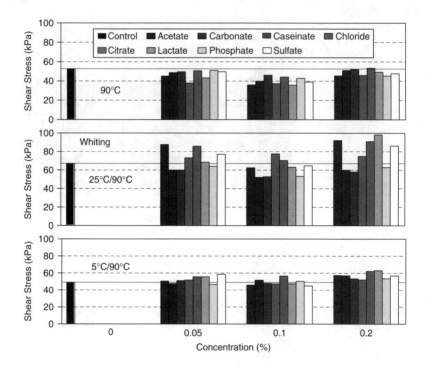

Figure 13.26 Effect of various calcium compounds on shear stress of Pacific whiting surimi gels at various thermal treatments. 90°C: cooked at 90°C for 15 min; 25°C/90°C: set at 25°C for 2 hr, followed by cooking at 90°C for 15 min; 5°C/90°C: set at 5°C for 18 hr, followed by cooking at 90°C for 15 min. (*Source:* Adapted from Ref. 56.)

Starting in the early 2000s, calcium compounds are not directly added to washed Pacific whiting mince along with cryoprotectants. Therefore, the shelf life of Pacific whiting surimi should not be shorter than that of Alaska pollock surimi.

13.2.7.3. Transglutaminase (TGase)

Transglutaminase (TGase) is an enzyme that catalyzes the polymerization and cross-linking of proteins through the formation of covalent bonds between protein molecules. Non-disulfide covalent bonds are formed between glutamic acid and lysine residues in proteins (Figure 13.27). This link enhances

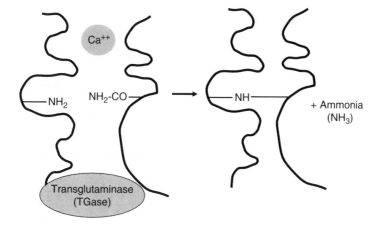

Figure 13.27 Formation of ε(γ-glutamyl)lysine bonds by transglutaminase.

the physical strength (hardness and cohesiveness) of surimi gels. It has been proven that endogenous TGase in fish also initiates setting (*suwari* in Japanese).[58] This endogenous TGase is Ca^{2+} dependent.[57] Ajinomoto Japan, however, succeeded in creating a fermentation method for the mass production of this functional enzyme from microorganisms, which is Ca^{2+} independent and therefore easy to use in food processing.[59]

Wang and Lanier[60] evaluated three TGases, from various sources, on pollock gelation: endogenous TGase, microbial TGase, and fungal TGase. Fungal TGase induced stronger gels (higher stress) in surimi when 5 mM Ca^{2+} was added, whereas calcium had little effect on the performance of microbial TGase. For both enzymes, 25°C incubation for 2 to 3 hr produced stronger gels than 5°C incubation for 12 to 24 hr or 40°C incubation for 60 to 90 min. According to Soeda et al.,[61] the effect of microbial TGase on the formation of cross-links was observed at >1.5 to 3.0 unit/g protein for high-grade pollock surimi and 3.0 to 5.0 unit/g protein for low-grade pollock surimi.

Lanier[62] evaluated the effects of the concentration of microbial transglutaminase (MTGase) from Ajinomoto. A

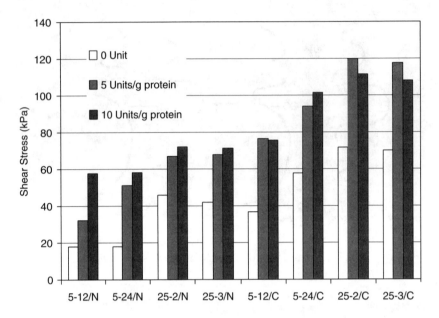

Figure 13.28 Effects of MTGase on the strength of pollock surimi gels: 5 and 25 are the setting temperatures, followed by the number of hours that setting was carried out. Cooking was at 90°C for 20 min. N: no cook (set only); C: set and cook. *(Source:* Adapted from Ref. 62.)

significant effect was found at 5 units of MTGase per gram of surimi protein. However, when the concentration was doubled to 10 units, the effectiveness was hardly noticed, particularly with cooked samples (Figure 13.28). This study clearly indicates that a search for optimum concentration is required.

Thermal stability of TGase is demonstrated in Figure 13.29. The enzyme activity is 100% when the temperature reaches 30 to 40°C and 90% when 50°C. However, the activity significantly reduced to 10% when the temperature reached 60°C, indicating the use of TGase may not work functionally in a fast heating process (i.e., crabstick or tempura [frying]). Further information related to the chemistry of TGase is available in Chapter 10.

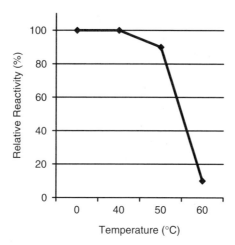

Figure 13.29 Thermal stability of transglutaminase (Tgase).

13.2.7.4 Phosphate

Phosphates are added to surimi as a cryoprotectant at 0.25 to 0.3%, traditionally as a mixture (50:50) of sodium tripolyphosphate or tetrasodium pyrophosphate. However, it is still not clear how phosphates work as a cryoprotectant. The most likely explanation is the specific function on actomyosin as a metal chelator, resulting in antioxidant properties. Although minced fish flesh goes through extensive washing and dewatering, there are still small quantities of metal ions remaining in the dewatered meat. Oxidation reactions, these metal ions, if not removed completely or inactivated (chelated), can also accelerate the denaturation of fish myofibrillar proteins at freezing and during frozen storage. In addition, because of the strength of the phosphates to raise pH, the water binding of the gel improves and more salt-soluble myofibrillar proteins are extracted. The pH of fish decreases from neutral 7 to 6.5 while the fish goes through *rigor mortis*. By adding 0.25 to 0.3% polyphosphate, the pH of washed fish meat can be neutralized to between 7.0 and 7.2. This pH neutralization is a critical step for keeping the texture-forming ability of fish proteins. The overall function of cryoprotectants in surimi during frozen storage is described in detail in Chapter 5.

For surimi seafood, the use of phosphates reduces the viscosity of the paste, allowing for better machinability. Texture and water retention is also increased when phosphates are properly added. Due to the limited solubility of traditional phosphates in cold water, it is recommended that these phosphates are added as a pre-prepared solution. Hunt et al.[63] demonstrated that the use of potassium phosphates, which also contains high solubility, could enhance the shelf life of frozen surimi, as indicated by the gel-forming ability of surimi after nine freeze–thaw cycles. Chemical blends (50:50) of tetrasodium pyrophosphate and sodium tripolyphosphate also performed better than the conventional mechanical blend as a cryoprotectant (Figure 13.30). This study indicates phosphate combinations containing better soluble potassium phosphate can be added directly to the batch. Either addition of pre-prepared sodium phosphate solution or direct addition of potassium phosphate would guarantee the full function of phosphate in the comminuted meat system. There is no strict limit, however, on the level of phosphate for seafood in the United States, whereas meat and poultry products are allowed to contain a residual concentration of added phosphate at 0.5% maximum.

13.2.7.5 Coloring Agents

There are two, obviously different color issues in surimi seafood. One is the color of the body meat, while the other is the color applied on the surface of surimi seafood. Applied color (pigments) and measurement are discussed in detail in Chapter 16. This section discusses the colors of body meat.

Three ingredients are commonly used to modify the physical appearance of body meat. They are calcium carbonate ($CaCO_3$), titanium dioxide (TiO_2), and vegetable oil, which make products chalky and opaque white. The usage levels vary, depending on product specification, but common levels are 0.5 to 1.5% for $CaCO_3$, 0.02 to 0.06% for TiO_2, and 2 to 6% for vegetable oil. When the finished product is used as salad meat

Ingredient Technology for Surimi and Surimi Seafood

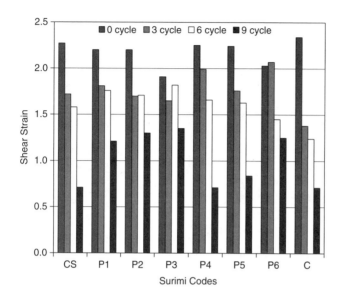

Codes	Sample outline
CS	Conventional surimi (conventional phosphate; mechanical blends of 50% TSPP, tetrasodium pyrophosphate, and 50% STPP, sodium tripolyphosphate)
P1	Surimi added P1 (chemical blends of 50% TSPP and 50% STPP for faster/higher solubility)
P2	Surimi added P2 (chemical blends of 50% TSPP and 50% STPP for higher viscosity)
P3	Surimi added P3 (mechanical blends of 50% tetrapotassium pyrophosphate and 50% STPP to check the influence of potassium)
P4	Surimi added P4 (mechanical blends of 40% TSPP, 40% STPP and 20% sodium hexametapolyphosphate; SHMP to check the influence of SHMP)
P5	Surimi added P5 (only long chained SHMP treated with trisodium phosphate; TSP to adjust pH to 10)
P6	Surimi added P6 (only middle chained SHMP treated with TSP to adjust pH to 10).
C	Control: surimi without phosphate

Figure 13.30 Effects of various phosphate blends on the quality of fish proteins during nine freeze–thaw cycles.

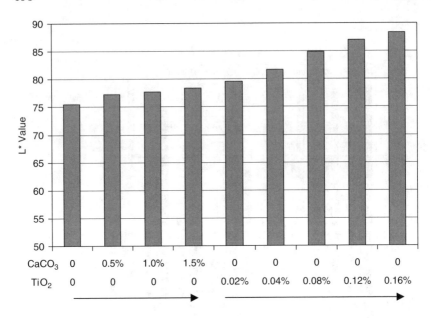

Figure 13.31 Effect of calcium carbonate and titanium dioxide on the color (L*) of surimi gels. The moisture content for samples was equally adjusted at 78%.

and marketed in a sealed container, the use of $CaCO_3$ as a whitening agent is not recommended. $CaCO_3$ produces carbon dioxide (CO_2) in the presence of acid (lemon juice) and phosphate (used as a cryoprotectant in frozen surimi). As a result, air bubbles often appear on the surface of the salad.

When TiO_2 is used, careful pH measurement is required. At higher levels of TiO_2 (0.2%), the pH of the meat approaches 6.7 or below. At these conditions, fish proteins do not form an elastic gel during crabstick production and the sheet of surimi paste keeps breaking, resulting in reduced yield.

$CaCO_3$ and TiO_2 showed the ability to make opaque surimi gels (Figure 13.31). TiO_2 was much stronger than $CaCO_3$. In addition, $CaCO_3$ at 1.5% was not as strong as 0.02% TiO_2 for the L* value. Lightness (L*) increased from 75.5 to 78.4 with a level of 1.5% $CaCO_3$ and to 84.9 with 0.08% TiO_2. For this experiment with $CaCO_3$ and TiO_2, shear stress (30.2

Ingredient Technology for Surimi and Surimi Seafood

to 33.8), shear strain (2.45 to 2.63), and pH (7.03 to 7.23) values remained consistent, indicating they did not affect texture at the level used. Other color hues (a^* and b^*) were also affected by these additives but the effect was insignificant.

Vegetable oil is certainly an important coloring ingredient for white chalky body meat. Its effect on color, however, was previously discussed in this chapter.

For the development of more translucent surimi seafood, the use of potato starch or tapioca starch, instead of corn or wheat starch, is recommended. In addition, the use of modified starch will also contribute to the translucent color of surimi gels.

13.3 EVALUATION OF FUNCTIONAL INGREDIENTS

13.3.1 Texture

The gel-forming ability of functional additives can be measured at certain chemical conditions, considering the ionic strength and/or pH of surimi seafood. Additives can be evaluated alone for their individual gel-forming ability or in the presence of surimi. In both cases, the use of an ionic strength and pH (7.0 to 7.2) that is similar to surimi seafood (1.5 to 2.0% NaCl) is recommended.

Depending on the nature of ingredients, physical factors (i.e., hydration, time of salt addition, temperature, and deaeration) have a significant impact on the texture of the surimi seafood gels. For example, various mixing methods significantly affected the function of processed whey protein in surimi seafood (Figure 13.13). Time of whey protein addition in conjunction with salt addition also played a significant role. Processed whey protein added into frozen surimi (30% level) at the very first stage of chopping or applied after prehydration, increased the shear stress value from approximately 47 kPa to between 53 and 56 kPa (Figure 13.14). However, when processed whey protein was mixed after or at the same time with salt addition, shear stress values decreased from approximately 47 kPa to between 34 and 39 kPa.

13.3.2 Color

Color is another important characteristic of surimi seafood. The three color hue values commonly measured in the surimi and surimi seafood industry are CIE L^*, a^*, and b^*. Color quality is often determined by whiteness. Two different indices are used to ascertain whiteness. Whiteness I uses all components of the coloring fraction, while Whiteness II uses L^* and b^* but does not use a^*. According to Park,[8] the a^* values of pollock and whiting gels were consistent, regardless of cooking/setting conditions, moisture content, sample size, or frozen storage.

$$\text{Whiteness I} = 100 - [(100 - L^*)^2 + a^{*2} + b^{*2}]^{1/2}$$

$$\text{Whiteness II} = L^* - 3b^*$$

The whiteness of surimi gels containing 1% dried protein additive was measured using the two different whiteness indices.[24] Whiteness II showed that protein additives affected the whiteness of surimi gels (Figure 13.32). Whiteness I, however, showed only slight differences when compared to the control. The visual differences between gels were observed as well, and no difference occurred in the a^* value of the gels with protein additives.[24]

13.3.3 Formulation Development and Optimization

Similar to other comminuted food products, two major factors are considered for the formula development of surimi seafood. They are production requirements (related to machinery and other infrastructure) and R&D formulation requirements.

For production requirements, the physical conditions of the production facilities must be reviewed when a new product is developed. These include the temperature during the primary cooking and pasteurization, mechanical shear or vacuum during comminution, operation speed, and freezing rate (blast freezer vs. liquid nitrogen). In addition, some chemical conditions that affect formulation development are pH, water quality, available moisture (due to competition from various

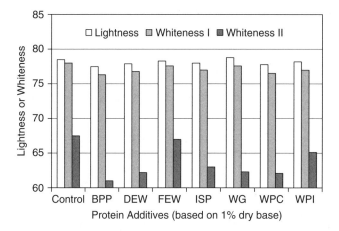

Figure 13.32 Two whiteness indices, as compared with lightness (L*), for surimi gels containing various protein additives. BPP: beef plasma protein; DEW: dried egg white; FEW: frozen liquid egg white; SPI: soy protein isolate; WG: wheat gluten; WPC: whey protein concentrate; WPI: whey protein isolate. (*Source:* Adapted from Ref. 25.)

ingredients), and surimi quality (depending on the presence of proteolytic enzymes).

For formulation requirements, product properties (chemical and physical composition) must be defined, including structure (texture), appearance, and shelf life. The product form, whether filament or formed (molded), must also be determined. In addition, R&D needs to know whether the product is designed to be ready-to-eat or used for additional processing (salad, hot entrée, etc.). Labeling is another key element of formulation requirements. Furthermore, the cost of ingredients must be evaluated based on their functional properties.

ACKNOWLEDGEMENTS

The author would like to acknowledge the leading industrial experts (Dr. Carl Jaundoo of Roquette America, Dr. Akihiro Handa of Henningsen Foods, Dr. David Cantrell of the Solae

Company, and Dr. Rainer Schnee of Budenheim) for their technical advice and contributions to this chapter.

REFERENCES

1. T.C. Lanier. Functional properties of surimi. *Food Technol.*, 40(3), 107–114, 1986.

2. T.C. Lanier, D.D. Hamann, M.C. Wu, and G. Selfridge. Application of functionality measurements to least cost linear programming of surimi-based product formulations. In R.E. Martin and R.L. Collette, Eds. *Engineered Seafood including Surimi*. Washington, D.C.: National Fisheries Institute, 1985, 264–273.

3. G.R. Ziegler and E.A. Foegeding. The gelation of proteins. In *Advances in Food and Nutrition Research, Vol. 34*. New York: Academic Press, 1990, 203–298.

4. A.H. Clark, R.K. Richardson, G. Robinson, S.B. Ross-Murphy, and A.C. Weaver. Structure and mechanical properties of agar/bsa co-gels. *Prog. Food Nut. Sci.*, 6, 149–160, 1982.

5. J. Yongsawatdigul and J.W. Park. Linear heating rate affects gelation of Alaska pollock and Pacific whiting surimi. *J. Food Sci.*, 61, 149–153, 1996.

6. T.C. Lanier. Interactions of muscle and nonmuscle proteins affecting heat-set gel rheology. In N. Parris and R. Barford, Eds. *Interactions of Food Proteins, ACS Symposium Series 454*. Washington, D.C.: American Chemical Society, 1991.

7. D.D. Hamann and G.A. MacDonald, Rheology and texture properties of surimi and surimi-based foods. In T.C. Lanier and C.M. Lee, Eds. *Surimi Technology*. New York: Marcel Dekker, 1992, 429–500.

8. J.W. Park. Surimi gel colors as affected by moisture content and physical conditions. *J. Food Sci.*, 60, 15–18, 1995.

9. C.M. Lee. Countercurrent and continuous washing systems. In R.E. Martin and R.L. Collette, Eds. *Engineered Seafood including Surimi*. Park Ridge, NJ: Noyes Data Corp., 1990, 292–296.

10. M. Okada. Varieties of fish and paste products. In M. Okada, T. Imaki, and M. Yokozeki, Eds. *Kneaded Seafood Product*. Tokyo: Koesisha-Kosikaku Publishing, 1981, 169–176.

11. C.M. Lee, M.C. Wu, and M. Okada. Ingredient and formulation technology for surimi-based products. In T.C. Lanier and C.M. Lee, Eds. *Surimi Technology.* New York: Marcel Dekker, 1992, 273–302.

12. M.C. Wu, D.D. Hamann, and T.C. Lanier. Rheological and calorimetric investigations of starch-fish protein systems during thermal processing. *J. Text. Studies,* 16, 53–74, 1985.

13. V.J. Morris. Starch gelation and retrogradation. *Trends Food Sci. Technol.,* 1, 2–6, 1990.

14. J.W. Park, H. Yang, and S. Patil. Development of temperature-tolerant fish protein gels using starches. In A.M. Spanier, M. Tamura, H. Okai, and O. Mills, Eds. *Chemistry of Novel Foods.* Carol Stream, IL: Allured Publishing Co., 1997, 325–340.

15. M.C. Wu, D.D. Hamann, and T.C. Lanier. Thermal transitions of admixed starch/fish protein systems during heating. *J. Food Sci.,* 50, 20–25, 1985.

16. J.M. Kim and C.M. Lee. Effect of starch of textural properties of surimi gel. *J. Food Sci.,* 52, 722–725, 1987.

17. Y.B. Yoon, B.Y. Kim, and J.W. Park. Linear programming in blending various components of surimi seafood. *J. Food Sci.,* 62, 564–567, 1997.

18. S. Konoo, H. Ogawa, and N. Iso. Effects of amylose and amylopectin on the breaking strength of fish meat gel. *Nippon Suisan Gakkaishi,* 64(1), 69–75, 1998.

19. W.B. Yoon and J.W. Park. Unpublished data. Astoria, OR: Oregon State University Seafood Lab, 1995.

20. P. Pongviratchai and J.W. Park. Control of the stickiness of refrigerated surimi seafood by pre-gelled starch. Abstract #40-5. Presented at the *IFT Annual Meeting,* Chicago, IL, 2003.

21. C.M. Lee. Surimi manufacturing and fabrication of surimi-based products. *Food Technol.,* 40(3), 115–124, 1986.

22. E. Niwa, N. Ogawa, and S. Kanoh. Depression of elasticity of kamaboko induced by pregelatinized starch. *Nippon Suisan Gakkaishi,* 57(1), 157–162, 1991.

23. H. Yang and J.W. Park. Effects of starch properties and thermal processing conditions on surimi-starch gels. *Lebensm.-Wiss U-Tech.,* 31, 344–353, 1998.

24. J.M. Regenstein. Protein-water interactions in muscle foods. *Reciprocal Meat Conference Proceedings,* 37, 44–51, 1984.

25. J.W. Park. Functional protein additives in surimi gels. *J. Food Sci.,* 59, 525–527, 1994.

26. S.L. Taylor. Chemistry and detection of food allergens. *Food Technol.,* 46(5), 146–152, 1992.

27. T. Hamamoto. Japan Food and Agricultural Import Regulations and Standards. New Allergen Labeling Requirements. GAIN Report #JA2001. USDA. U.S. Embassy, Japan, 2002.

28. Y. Kinekawa. Whey proteins — functional ingredients for surimi seafood. Presented at the 3rd OSU Surimi School, Bangkok, Thailand, 2000.

29. A. Handa, K. Hayashi, H. Shidara, and N. Kuroda. Correlation of the protein structure and gelling properties in dried egg white products. *J. Agric. Food Chem.,* 49, 3957–3964, 2001.

30. A. Kato, H.R. Ibrahim, H. Watanabe, K. Honma, and K. Kobayashi. New approach to improve the gelling and surface functional properties of dried egg white by heating in dry state. *J. Agric. Food Chem.,* 37, 433–437, 1989.

31. A. Kato, H.R. Ibrahim, H. Watanabe, K. Honma, and K. Kobayashi. Structural and gelling properties of dry-heating egg white proteins. *J. Agric. Food Chem.,* 38, 32–37, 1990.

32. W.B. Yoon, B.Y. Kim, and J.W. Park. Rheological characteristics of fibrinogen-thrombin solution and its effect on surimi gels. *J. Food Sci.,* 64, 291–294, 1999.

33. Y.J. Choi, M.S. Cho, and J.W. Park. Effects of hydration time and salt addition on gelation properties of major protein additives. *J. Food Sci.,* 65, 1338–1342, 2000.

34. J.W. Park. Effect of salt on rheological properties of soy protein emulsions. Unpublished data. OSU Seafood Lab, Astoria, OR, 2001.

35. K.H. Chung and C.M. Lee. Evaluation of wheat gluten and modified starches for their texture-modifying and freeze-thaw stabilizing effects on surimi based-products. *J. Food Sci. Nutr.,* 1(2), 190–195, 1996.

36. T. Yamashita and N. Seki. Effects of the addition of soybean and wheat proteins on setting of walleye pollock surimi paste. *Nippon Suisan Gakkaishi,* 62(5), 806–812, 1996.

37. M.G. Llanto, C.W. Bullens, J.J. Modliszewski, and A.A. Bushway. Effects of carrageenan on gelling potential of surimi prepared from Atlantic pollock. In M.N. Voigt and J.R. Botta, Eds. *Advances in Fisheries Technology and Biotechnology for Increased Profitability.* Lancaster, PA: Technomic Publishing, 1990, 305–311.

38. C.W. Bullens, M.G. Llanto, C.M. Lee, and J.J. Modliszewski. The function of a carrageenan-based stabilizer to improve quality of fabricated seafood products. In M.N. Voigt and J.R. Botta, Eds. *Advances in Fisheries Technology and Biotechnology for Increased Profitability.* Lancaster, PA: Technomic Publishing, 1990, 313–323.

39. D.J.B. Da Ponte, J.M. Herft, J.P. Roozen, and W. Pilnik. Effects of different types of carrageenans and carboxymethyl cellulose on the stability of frozen stored minced fillets of cod. *J. Food Technol.,* 20, 587–590, 1985.

40. R.J. Tye. Konjac flour: properties and applications. *Food Technol.,* 45(3), 88–92, 1991.

41. Y. Ohta and K. Maekaji. Preparation of konjac mannan gel. *Nippon Nogeikagaku Kaishi,* 54, 741–746, 1980.

42. M.C. Wu and T. Suzuki. Process of Forming Simulated Crustacean Meat. U.S. Patent No. 5028445, 1991.

43. J.W. Park. Temperature-tolerant fish protein gels using konjac flour. *J. Muscle Food,* 7, 165–174, 1996.

44. Takeda Chemical Industries. Curdlan properties and Food Application. Technical information. Takeda Chemical Industries, Tokyo, Japan, 1997.

45. R. Clark. Hydrocolloid applications in fabricated mince fish products. In R.E. Martin, Ed. *3rd National Technical Seminar on Mechanical Recovery and Utilization of Fish Flesh.* Washington, D.C.: National Fisheries Institute, 1980, 284–298.

46. K.S. Yoon and C.M. Lee. Effect of powdered cellulose on the texture and freeze-thaw stability of surimi-based shellfish analog products. *J. Food Sci.,* 55, 87–91, 1990.

47. Y.E. Ueng and C.J. Chow. Effects of cellulose and agar on the quality of surimi products. *Food Sci. Taiwan,* 22(5), 606–614, 1995.

48. C.M. Lee and A. Abdollahi. Effects of hardness of plastic fat on structure and material properties of fish protein gels. *J. Food Sci.,* 46, 1755–1759, 1981.

49. K. Nishimura, M. Ohtsuru, and K. Nigota. Mechanism of improvement effect of ascorbic acid on the thermal gelation of fish meat. *Nippon Suisan Gakkaishi,* 56, 959–964, 1990.

50. R. Yoshinaka, M. Shiraishi, and S. Ikeda. Effect of ascorbic acid on the gel formation of fish meat. *Bull. Jap. Soc. Sci. Fish.,* 38, 511–556, 1972.

51. H.G. Lee, C.M. Lee, K.H. Chung, and S.A. Lavery. Sodium ascorbate affects surimi gel-forming properties. *J. Food Sci.,* 57, 1343–1347, 1992.

52. R. Pacheco-Aguilar and D.L. Crawford. Potassium bromate effects on gel-forming ability of Pacific whiting surimi. *J. Food Sci.,* 59, 786–791, 1994.

53. Y. Yamamoto, T. Okubo, S. Hatayama, M. Naito, and T. Ebisu. Frozen Surimi Product and Process for Preparing. U.S. Patent No. 5,028,444, 1991.

54. D.T. Gordon and G.L. Roberts. Mineral and proximate composition of Pacific coast fish. *J. Agric. Food Chem.,* 25, 1262–1268, 1977.

55. V.D. Sidwell. Chemical and Nutritional Composition of Finfishes, Whales, Crustaceans, Mollusks, and Their Products. NOAA Technical Memorandum NMFS F/SEC-11. Washington, D.C.: U.S. Dept. of Commerce, 1981.

56. N.G. Lee and J.W. Park. Calcium compounds to improve gel functionality of Pacific whiting and Alaska pollock surimi. *J. Food Sci.,* 63, 969–974, 1998.

57. H. Saeki. Gel-forming ability and cryostability of frozen surimi processed with $CaCl_2$-washing. *Nippon Suisan Gakkaishi,* 62(2), 252–256.

58. N. Seki, H. Uno, N.H. Lee, I. Kimura, K. Toyoda, T. Fujita, K. Arai. Transglutaminase activity in Alaska pollock muscle and surimi, and its reaction with myosin B. *Nippon Suisan Gakkaishi,* 56, 125–132, 1990.

59. Ajinomoto USA. Basic Properties of Transglutaminase. Technical bulletin. Ajinomoto USA, Teaneck, NJ, 1998.

60. S.L. Wang and T.C. Lanier. Effects of endogenous fungal and microbial transglutaminase in pollock surimi gelation. Abstract #32-4. Presented at the *IFT Annual Meeting,* Atlanta, GA, 1998.

61. T. Soeda, S. Toiguchi, T. Mumazawa, S. Sakaguchi, and C. Kuhara. Studies on the functionalities of microbial transglutminase for food processing. II. Effects of microbial transglutaminase on the texture of gel prepared from Suketoudara surimi. *J. Jap. Soc. Food Sci. Technol.,* 43(7), 780–786, 1996.

62. T.C. Lanier. Surimi gelation chemistry. In J.W. Park, Ed. *Surimi and Surimi Seafood.* New York: Marcel Dekker, 2000, 237–266.

63. A. Hunt, J.S. Kim, J.W. Park, and R. Schnee. Effect of Various Blends of Phosphate on Fish Proteins during Frozen Storage. Paper # 63-11. Presented at the *Annual IFT Meeting,* Las Vegas, NV, 2004.

14

Surimi Seafood Flavors

CHARLES MANLEY, PH.D. and AMBY MANKOO
Takasago International Corporation, Rockleigh, New Jersey

VÉRONIQUE DUBOSC
Activ International, France

CONTENTS

14.1 Introduction 710
14.2 What Is Flavor? 712
    14.2.1 Creation of a Flavor 712
    14.2.2 Natural Product Chemistry 713
        14.2.2.1 Solvent Extraction 715
        14.2.2.2 Gas Chromatography–
                  Olfactometry (GCO) 716
        14.2.2.3 Headspace Analysis 716
    14.2.3 Building a Flavor 717
14.3 Basic Seafood Flavor Chemistry 720

14.3.1 Sources of Flavor Ingredients 720
    14.3.1.1 Natural Extracts 720
    14.3.1.2 Synthetic Components 721
14.3.2 The Importance of Lipids in Fish Flavors... 721
14.3.3 Important Components Found in
    Seafood Extracts 724
    14.3.3.1 Volatile Compounds 725
    14.3.3.2 Nonvolatile Compounds.................. 726
14.4 Additives and Ingredients Used in Flavors 727
    14.4.1 Glutamate 728
    14.4.2 Ribonucleotides 728
    14.4.3 Hydrolyzed Proteins 728
    14.4.4 Yeast Extracts 729
14.5 The "Off Flavors" of Seafood 730
14.6 Effects of Processing on Seafood 731
14.7 Flavor Release and Interactions 731
14.8 Effects of Ingredients on Flavor 733
    14.8.1 Sorbitol and Sugar 734
    14.8.2 Starch 734
    14.8.3 Surimi (Raw material) 735
    14.8.4 Egg Whites and Soy Proteins 736
    14.8.5 Vegetable Oil 736
    14.8.6 Salt 736
14.9 Processing Factors Affecting Flavors 737
    14.9.1 Adding Additional Flavor/Flavor
    Components 737
    14.9.2 Addition Points 737
    14.9.3 Encapsulation 738
    14.9.4 Storage Conditions and Shelf Life 738
14.10 Flavor Regulations and Labeling 739
14.11 Summary 744
References 745

14.1 INTRODUCTION

The major success of any product lies in the fact that its flavor is acceptable to the consumer. The flavor of any food consists of both the aroma character and taste of the food. The aroma

Surimi Seafood Flavors

can influence taste and the opposite is also true. In addition, compounds that have no flavor can also affect the taste and aroma of a product.

The aroma, or smell, of the product consists of volatile chemicals (low molecular weight compounds) of great variety; whereas the taste components are typically fewer in number, nonvolatile (some may be fairly high molecular weight compounds), and found either naturally in the food or added to food as "flavor." The "flavor" that is added to surimi is the basis of the discussion in this chapter and the business of the flavor industry. This is, of course, an oversimplification of the actual facts that lead one to enjoy a food or beverage.

Surimi, a fabricated food ingredient produced from fish protein, represents a system that is primarily developed to have little or no flavor at all. However, different grades of surimi from different sources and processing methods impart a certain flavor or, in most cases, "off flavor," as noted in previous chapters. Hence, the major objective is to select a process and fish source that will result in the least flavor.

The Japanese taste favors subtle flavors and, therefore, any type of surimi seafood product of great interest to the Japanese must have a subtle flavor. An interesting food product in Japan and one of the oldest uses for surimi seafood products is kamaboko, or basically a bland "steamed" surimi-based seafood. The Japanese also produce other types of surimi seafood that have more flavor than kamaboko, but the flavor is added by the cooking process or by the addition of other foods and seasonings.

Typically, outside of Japan, however, the success of surimi-based foods has been with flavored products. Surimi, being essentially flavorless, is an excellent base for flavoring, and a wide range of flavors have been developed for that use. Many commercial products have included the flavor of crab, salmon, clam, lobster, various sausage-type seasonings, and some other interesting flavors. In addition, the ability to produce surimi with various textures and in different forms with the use of flavors allows for the formation of many analog products (such as salmon, crab, and lobster) that taste like their natural counterparts. Remarkably, many of these products can be made to have very high quality and consistency. The secret is both in the texture and flavor of the surimi used in the analog products.

Flavoring surimi is both an art and a science. It deals with the human senses and with the understanding of both chemistry and physics. This chapter provides insight into the nature of the flavor of various seafood products and the "flavor" created by the flavorist that is added to surimi.

14.2 WHAT IS FLAVOR?

Flavor is both the aroma and taste component of a food or beverage. The components that evoke enjoyment in natural foods can be those that occur in the foods themselves or are the by-products of fermentation, enzyme conversions, or heat processing. Cooking, roasting, or grilling are examples of food preparation that enhance a food's flavor. Certain enzyme conversions, such as the creation of black tea, or the process of microorganisms, such as in the production of pickles, also create desirable flavors.

There are thousands of components that have been found to play a role in what we relate to as flavor. These are either naturally occurring chemicals or those created by their processing as noted above. For well over 100 years, scientists have been identifying these substances and using them in the creation and manufacture of "flavor" that can then be added to foods or used in making newly formulated food and beverage products such as surimi seafood. Since the 1960s, the ability of scientists to identify flavor components has expanded with the use of gas chromatography, mass spectroscopy, and other methods that allow the isolation, separation, and identification of the minor components found in foods, herbs, and spices sometimes at concentrations as low as parts per billion and below.

14.2.1 Creation of a Flavor

The creation of a flavor goes through a series of stages. The flavorist's steps in creating a "flavor" consist of specifying blends of ingredients that include flavors derived from nature and those created by man that meet the taste expectation of the consumer. The flavorist, who is a scientist, artist, and creator of "flavor," has a wide, constantly evolving range of raw materials available. A flavor might be made with ten, twenty, or even hundreds of chemical compounds, of which the relative proportion of each is critical to the flavor's profile and its use in a particular food product.

The known number of different chemical compounds identified as naturally occurring in food products is estimated to be more than 20,000.[1] Table 14.1 indicates the number of flavor components found in some common natural foods, including a number of seafood products. There are also a number of chemicals that have been created by the flavor industry that do not have equivalents in the natural world. Most are similar to those found in nature but have certain attributes, such as heat stability or flavor strength, that are not found in nature. All components are held to a very high degree of review with regard to the safety of their use in food products. The regulatory and safety issues are discussed later in this chapter. All these materials and ingredients then become the "pallet" with which the flavorist works.

A handful of compounds are, typically, considered the major flavor components because they represent about 95% of the total amount of volatile chemicals found in the natural product. However, in many cases, it is the components found at very low concentration that are responsible for the complete character of the food's flavor. It is therefore the complex mixture of components that renders a complete flavor profile. But how is the information obtained that allows the building of a flavor?

14.2.2 Natural Product Chemistry

The development of many analytical tools has contributed to the rapid growth of information about the chemicals that we relate to as the organoleptic properties of the flavor. The most important analytical tool available to the flavor industry has been the gas chromatograph. It is a technique by which a complex mixture of chemicals, such as those found

TABLE 14.1 Major Flavor Components Found in Common Foods, Herbs, and Spices

Item	Major Flavor Component
Lemon	Citral
Star anis	Anethole
Cumin seed	Cuminaldehyde
Bitter almond	Benzaldehyde
Cloves	Eugenol
Basil	Methyl cinnamate
Fennel	Anethole
Mandarin orange	Methyl anthranilate
Orris	Irone
Sage	Cineol, linalool
Thyme	Thymol, carvacrol
Turmeric	Turmerone
Onion	Propyl disulfide, methyl propyl disulfide
Garlic	Allyl propyl disulfide
Cinnamon	Cinnamic acid
Ginger	Zingiberol
Caraway	d-Carvone
Peppermint	l-Menthol
Spearming	l-Carvone
Vanilla	Vanillin
Bell pepper	2-Isobutyl-3-methoxy pyrazine
Potatoes	Methional
Chili pepper	Capscaisin
Black pepper	Piperine

in natural food products, are separated into individual chemical components.

Chromatography was first developed in 1900 by Russian botanist M.S. Tsweet to separate vegetable pigments. His technique separated vegetable "colors" on filter paper with the use of a solvent. The separation of colored (*chroma* in Greek) components gave rise to the term "chromatography" as the science of the separation of complex mixtures.

Gas chromatography is based on the differential migration of organic compounds through a medium (stationary phase) in an inert gas medium (mobile phase) at temperatures that keep the components from being separated into their gaseous or volatile form. By comparing the elution time, which is the time for an individual chemical component to pass through the gas chromatography column, against the elution times of known standards, the individual chemicals can be identified. There are many types of detectors that are used to "see" the components coming off the end of the column. The human nose is one that can be used for the detection of components with smell, and the mass spectrometer is one that allows the chemist to determine the structure of the chemical.

To carry out a good chromatographic analysis, one must have the volatile chemical composition of the material being analyzed. A seafood sample cannot be directly injected into the gas chromatograph easily. Therefore, the chemist must find ways to extract the volatile components of the natural product in its original form, one that represents the actual aroma of the original product. This is no easy task and one that requires the chemist to select a technique from the myriad of well-known isolation methods that will be best suited to the solution of the particular food product. In addition, the analytical chemist must also separate and identify the composition of the resulting complex mixtures. These mixtures comprise a wide range of organic chemicals that possess varying polarities and reactivities, occurring at trace concentrations, and, more than likely, in a complex organic matrix. Fortunately, as previously mentioned, most aroma chemicals are very volatile, and procedures for their isolation from foods, herbs, and spices have been established that take advantage of their volatility.

14.2.2.1 Solvent Extraction

Organic solvents are typically used to dissolve both types of compounds from the natural starting materials. The solvent is subsequently removed, typically via high vacuum, leaving behind compounds, some of which have flavor value. The polarity of the solvent plays a critical role in the type of compounds extracted. Traditional extraction methods use several types of solvents: polar solvents, such as ethanol, which are miscible with water; nonpolar solvents, such as hexane, which are immiscible with water; and organic ester compounds, such as ethyl acetate, which are also immiscible in water. One of the major considerations in choosing a solvent is its boiling point; the lower the boiling point, the less likely highly volatile components will be removed from the extract during its concentration. The extract then can be subjected to gas chromatographic separation. One type of separation using the human nose as a detector is particularly useful in identifying the flavor components in the extract.

14.2.2.2 Gas Chromatography–Olfactometry (GCO)

The use of the human nose as a sensitive detector in gas chromatography was first proposed in 1964. It allows the selection of aroma-active components from a complex mixture. This technique helps detect potent odorants at very low concentrations without knowing their chemical structures. The flavorist will smell the effluent coming off the column and mark the peak of interest with his or her characterization of the aroma of the compound. Serial dilution of the extract will also allow the observer to determine the threshold values for the individual components and to identify those that have the greatest impact on the aroma. As painstaking as this method is, the result of the analysis is extremely useful to the flavorist in recreating the flavor.[2,3]

14.2.2.3 Headspace Analysis

Another method, called headspace analysis, allows for the separation of aroma components without extraction of the food with a solvent. Headspace analysis can be either static (in a closed vessel that has reached equilibrium) or dynamic (sampling in real time by absorption onto a material that can be injected into the gas chromatograph), and provides information about the aroma above the food and additional data about very volatile compounds that are usually lost during conventional sample preparation. These methods and others allow the flavorist to separate and identify components that are important to the aroma of the food and to create a flavor from materials that are part of the flavorist's pallet.

14.2.3 Building a Flavor

Describing a flavor or a taste is not an easy task. It is nearly impossible for everyone to interpret and define identical flavor terms, mainly because each person has a unique sense of taste, smell, and ability to articulate that experience. Flavorists have developed their own language for describing a flavor that allows for communication of flavor information between professionals in the flavor industry. Their skill in describing flavors and their acute sense of taste and smell is a major part of the process in selecting people to become flavorists. Indeed, years of experience under a mentor are needed to learn to be a skilled flavorist. It is imperative to have a standard definition and to develop organoleptic profiles of food flavors to be able to create flavors that relate to natural foods.

When a flavorist tastes crabmeat, for example, he or she must be able to describe the flavor and remember what materials in the flavorist's pallet are important for its recreation. Their first evaluation is that of *smell*, which is perceived only in the nose. After a food product is introduced into the mouth of the evaluator, volatile aroma compounds are released during the chewing process, which adds to the sensory attribute of the flavor. This part can be defined as the *aroma* of the food being eaten. The *taste* is then initiated by the contact of an aqueous solution of a compound with the taste buds on the tongue. For example, the flavor of crab might first appear fruity, lemon-like, floral, and seaweed-like. These are considered the top notes, which are very volatile components that characterize the initial flavor of the product. Top notes or aromas are usually found at very low levels of parts per million, billion, or even trillion of the food.

As one eats food, other flavor notes, such as sulfury, alliaceous, butter-like, grassy, and meat-like, will be noted. These notes characterize what is called the *body* of the flavor of the product. They are usually similar to the aroma of the extract that can be made from the food product. The evaluator can also detect the lower notes such as grilled, shell-like, and meaty. They characterize the long-lasting and less volatile or nonvolatile components of the food. In many cases, this can be described as the *savory* or *seasoning* notes of the flavor. Many of these components are water-soluble materials that have taste rather than aroma.

The final "taste," called *umami,* is very interesting and relates to a *taste-enhancing* effect rather than a true taste itself. It is a Japanese word closely related to the English word meaning delicious.

In 1907, the Japanese scientist Kikunae Ikeda sat before a bowl of tofu in dashi, a kelp-based broth, and pondered dashi's peculiar flavor and its puzzling ability, like salt, to enhance or harmonize the flavor of other foods. Ikeda, a chemist, went to his laboratory bench, where he found that the substance responsible for the umami effect, as he named it, was glutamic acid, one of the 20 common amino acids of which proteins are built.

However, bound glutamate one of the amino acids found in the protein content of all foods, does not produce an umami taste. Only the free form, which is liberated when proteins break down during processes such as aging, fermentation, ripening, cooking, or direct chemical hydrolyses, produces the effect. Many foods, such as tomato, mushroom, and cheese, naturally contain the free glutamate. Chefs for many ages have used these foods to enhance the taste of many fine culinary creations. Glutamic acid, however, tastes sour, so Ikeda experimented with other glutamate compounds, finally settling on its sodium salt: monosodium glutamate, or MSG (Figure 14.1). In 1909, S. Suzuki & Company started to sell the product commercially in Japan as a product called Aji-no-moto, a product so successful that the company subsequently renamed itself Ajinomoto. In 1917, it came to the United States under the brand name "Super Seasoning."

The umami story, however, was complicated by the discovery that another class of molecules, 5'-ribonucleotides, also produces the umami taste. These compounds, commercially available in the form of disodium 5'-inosinate (IMP) and disodium 5'-guanylate (GMP) (Figure 14.2) are effective at far lower concentrations than glutamate. More significantly, they have such a strong synergistic interaction with glutamate that very small quantities of IMP or GMP added to MSG or

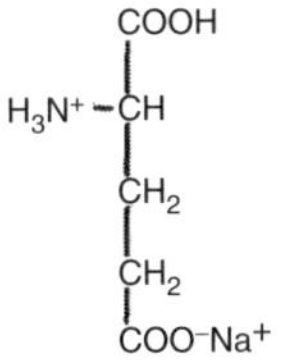
COOH
H3N+ CH
CH2
CH2
COO-Na+

Figure 14.1 Monosodium glutamate (MSG) structure.

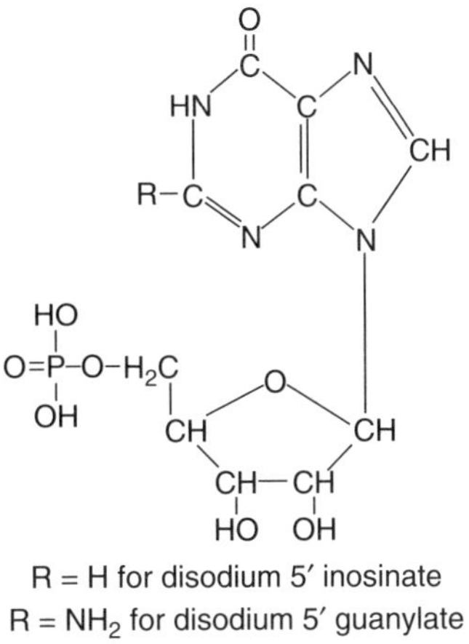
R = H for disodium 5′ inosinate
R = NH2 for disodium 5′ guanylate

Figure 14.2 5′-Ribonucleotide structure.

to foods that naturally contain free glutamate yield an intense umami taste.

Recently, a umami taste receptor has been identified in the human mouth and, therefore, this "taste" can now be considered equal to the four major tastes of sweet, sour, salty, and bitter that are already known. In 2003, Nirupa Chaudhari and Stephen Rope at the University of Miami cloned a specific glutamate receptor from rat tongue that offers definitive proof that the umami sensory property is a taste effect. The cloned receptor was specific only to glutamate, so the search continues for a receptor or scientific proof of the action of the 5′-ribonucleotides.[4]

All of these organoleptic sensations, including umami, can be created from the flavorist's pallet and added to bland surimi to create a crab flavor or any other flavor profile desired by the consumer. The flavorist's challenge is to put together materials that are available, safe, and when blended together give the impression to the consumer that the product does, indeed, taste like a known food or is an acceptable "new" product. Only the imaginations of the flavorist and the product developer are needed to create new, innovative types of surimi seafood.

14.3 BASIC SEAFOOD FLAVOR CHEMISTRY

We have discussed what constitutes a flavor and how the flavorist works to create a flavor, but what are the raw materials on the flavorist's pallet?

14.3.1 Sources of Flavor Ingredients

The flavorist has a broad pallet, starting with natural extracts from many sources and ending with synthetic compounds.

14.3.1.1 Natural Extracts

These are complex and diverse products; they are obtained from animal, plant, or microbial (fermentation) origin or by physical processes (e.g., distillation, extraction, or pressing).

Distilled flavors. Distillation techniques involve using hot water or steam to extract the aromatic materials from foods, herbs, and spices and other natural products. The quality of the flavor depends on the raw materials and the process. The product generally produced from this process is termed a "volatile" or "essential" oil, although further processing can recover other aromatic volatiles. The nature of the raw material dictates the best distillation technique, but water distillation usually produces less heat-sensitive changes.

Extraction. Distillation only removes the volatile portion of a flavor. To obtain characteristic taste attributes provided by the nonvolatiles, extraction techniques involving a polar solvent, including water or alcohol and water mixtures, must be applied.

Concentration. Such extracts must be concentrated to be of commercial value. Again, distillation is used, except in this case it is the concentrate that is used and not the distillate. To concentrate the compounds contributing the flavor, other materials such as sugars, acids, tannins, aromatic compounds, and water are removed.

Supercritical extraction. More recently, extractions have been performed using liquid carbon dioxide. Carbon dioxide liquefies under high-pressure and low-temperature conditions to become a nonpolar to polar solvent, depending on the actual pressure of the carbon dioxide. This method of extraction provides several advantages over traditional extraction methods. Carbon dioxide is colorless, odorless, and easily removed as the release of pressure increases, at which point it returns to it gaseous state. This process does not promote the loss of volatiles. Flavors obtained by super-critical extraction cost significantly more than flavors produced with standard solvents, but their quality is unique and considered superior when compared to other extraction methods.

14.3.1.2 Synthetic Components

The ability of the natural product chemist has allowed the flavor industry to synthesize chemicals found in nature or create new ones with stronger, more stable or more unique flavor profiles. There are more than 1000 flavor ingredients available to the flavorist.

Before the flavorist starts to work, there must be a basic understanding of the flavor chemistry of the material of interest. One of the most important substances related to the flavor in seafood products is the lipid content found in fish and shellfish.

14.3.2 The Importance of Lipids in Fish Flavors

Historically, the major focus in the science of flavor creation was placed on fruits, herbs, spices, and meat products, mainly

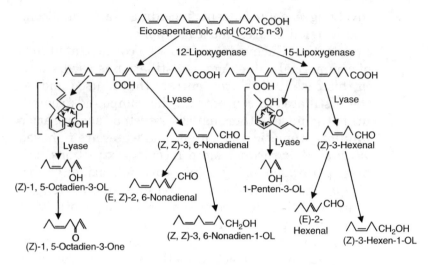

Figure 14.3 Auto-oxidation pathway of fish lipids.

because of the sensory interest in these products, their chemistry, and the economic importance they have to the flavor industry. Seafood flavors, however, have not received the same research attention given to other areas of food. One of the major problems has been that the volatile components in seafood are found in extremely low concentrations, with the exception of components present due to the "spoilage" of the seafood. Many volatile components are the products of lipid degradation from enzymatic or microbiological activity or simple oxidation in the fish or shellfish. Typically, these are the "off-notes" of a spoiled product. A significant amount of the lipid fraction of any fish or shellfish is highly unsaturated lipids and typically of the methyl-interrupted double-bond type structure that is very susceptible to oxidation. These lipids are also prone to oxidation by a number of enzymes. The structure of such a lipid and the proposed mechanisms for the production of aroma compounds are shown in Figure 14.3.[5]

The compounds created from lipid degradation are part of the library of flavor ingredients used by the flavorist in creating flavors for use in surimi seafood products. Many are produced synthetically and have been approved as safe in

TABLE 14.2 Some Important Volatile Flavor Compounds Found in Seafood

Compound	Seafood
Butanol	Cooked shrimp and crayfish
Hexanol	Cooked crayfish and steamed clams
2-Nonanol	Clams, crayfish
Butanal	Squid, crab
2-Pentental	Crab, shrimp, clams
Hexanal	Crayfish, shrimp, lobster
2e,4z-Heptadienal	Oyster
2-Octenal	Clams
Decanal	Shrimp, clams, scallops
2-Butanone	Shrimp, clams
2-Hexanone	Crayfish, crab
Octanone	Crayfish, shrimp, crab
2-Decanone	Crayfish, shrimp, crab
2-Acetyl furan	Shrimp, roasted clam
2-Furfural	Shrimp hydrolysate
Pyrazine	Cooked shrimp, boiled scallop
Pyridine	Roasted clam, broiled scallop
Pyrrole	Roasted clam, cooked crayfish
Trimethylamine	Raw shrimp, crab, boiled scallop
Hexane	Boiled scallop
Decane	Boiled shrimp
Phenol	Cooked crayfish
Ethyl acetate	Raw shrimp, oyster hydrolysate
Methional	Raw, fermented, and cooked shrimp

many countries as ingredients in the fabrication of flavors for the use in foods. Table 14.2 indicates some of the ingredients that are allowed by FEMA (Flavor and Extract Manufacturers Association) and Council of Europe Numbers (CoE). Having these chemicals approved for commercial use indicates that the flavor companies have determined that they play an important role in creating flavors. Table 14.3 indicates the character of each of the flavor ingredients found in some seafood and their flavor threshold.

The *threshold value* is a subjective value given to any substance that has an aroma or taste. It is established using a trained panel to smell or taste various dilutions of the substance until collectively or statistically they cannot taste

TABLE 14.3 Comparison of Some Seafood Volatile Thresholds

Chemical	Carbonyl	Alcohol	Threshold (ppb)
1-Octen-	3-one		0.005
1-Octen-		3-ol	10
1,5-Octadien-	3-one		0.001
1,5-Octadien-		3-ol	10
3,6-Nonadien-	al	1-ol	10
2,6-Nonadien-	al		0.01

or smell the substance. The variability of a given number as a threshold is very broad and is affected by the evaluators, temperature, and nature of the solution in which the media is smelled or tasted.

Spurvey et al.[6] reviewed some of the early research related to the flavor of seafood. For fresh seafood there does not appear to be a single compound that characterizes a particular fish or shellfish aroma or taste. Complex mixtures of autooxidatively derived volatiles, including those shown in Table 14.2, may add characteristic notes to the flavor of certain fish but do not represent the total organoleptic experience. The flavor of fresh seafood is considered very delicate and subtle. The volatile carbonyls found in fish exhibit a stronger aroma than volatile alcohols. Table 14.3 indicates the difference between the threshold values of some carbonyls and alcohols found in fish. Those differences show that the aldehydes with lower threshold values contribute more to the overall fresh fish-like aroma than do the corresponding alcohols.

14.3.3 Important Components Found in Seafood Extracts

The extracts of seafood contain many materials that are used by the flavorist in the creation of flavor for surimi seafood products. These compounds can be categorized into the following groups.

TABLE 14.4 Autooxidation Products of Fish Lipids

z 1,5-Octadiene-3-ol
z 1,5 Octadiene-3-one
z,z 3,6 Nonadienal
e,z, 2,6 Nonadienal
z,z 3,6 Nonadien-1-ol
z 3-Hexenal
1-Penten-3-ol
e 2-Hexenal
z 3 Hexen-1-ol

14.3.3.1 Volatile Compounds

Generally speaking, the volatile aroma compounds can be classified as most of the top notes.

N-containing compounds. Among this category of compounds, the most common chemical is trimethylamine, which results from a degradation of TMAO (trimethylamine oxide) by microbial enzymes. This degradation happens in freshly harvested fish; and when the fish is frozen, degradation of TMAO leads to the formation of formaldehyde and dimethylamine, two other components that contribute to the flavor or "off-flavor" of fish.[7]

Alcohols, aldehydes, and ketones. These components can result either from the oxidative degradation of polyunsaturated fatty acids, as is the case of tetradecatrien-2-one, which yields a cooked shrimp-like note, or from the enzymatic degradation (mainly catalyzed by a lipoxygenase) of the same polyunsaturated fatty acids to produce aldehydes, ketones, and alcohols such as those noted in Table 14.4 and Figure 14.3. The three main precursors for these reactions are arachidonic acid, eicosapentaenoic acid, and docosahexaenoic acid, as shown in Figure 14.3.[8]

Sulfur-containing compounds. The most famous sulfur-containing compound typical of seafood is dimethyl sulfide, resulting from the degradation of sulfury amino acids, such as cysteine and methionine, by endogenous

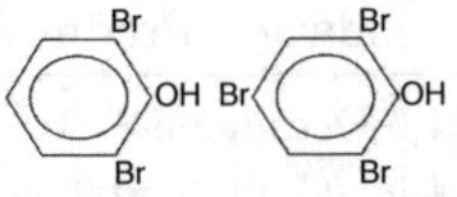

[Structures: 2,6-Dibromophenol and 2,4,6-Tribromophenol]

Figure 14.4 Bromophenols found in seafood.

TABLE 14.5 Some Important Sulfur-Containing Compounds Occurring in Seafood

Compound	Occurrence
Dimethyl sulfide	Shrimp
Carbon disulfide	Cooked shrimp
Dimethyl disulfide	Cooked crayfish, oyster, cooked shrimp, spray-dried shrimp powder
Dimethyl trisulfide	Cooked crayfish, oyster, cooked shrimp
Methyl propyl trisulfide	Cooked shrimp
Dimethyl sulfoxide	Oyster, roasted shrimp
Methanethiol	Cooked shrimp
Methyl trithiomethane	Roasted clam
2-Methylthioethanol	Roasted shrimp
3-Methylthiopropanol	Roasted shrimp
Methional	Raw, fermented, and cooked shrimp
2-(Methylthio)-methyl-2-butenol	Spray-dried shrimp powder
3-Methylthiopropanol	Fermented and cooked shrimp
3-Methylthiobutanal	Cooked shrimp

enzymes, through a Strecker degradation (Figure 14.4). Table 14.5 summarizes the sulfur compounds found in seafood and their characterized flavor.[9]

Bromophenols. These compounds (Figure 14.4) are rather unique to food flavors, but they have been identified in a number of marine species, and are considered to be responsible for the iodine-like flavor of shrimp and fish.[10]

14.3.3.2 Nonvolatile Compounds

The following compounds can be classified as savory or seasonings:

Amino acids. These components are believed to either contribute to the umami taste, like glycine and glutamic acid, or add body to the product, like arginine and alanine. They occur naturally in their free form in crabmeat.

Organic acids. The two acids related to seafood tastes are lactic acid, which is found in the muscle of the fish, and succinic acid, which occurs in mollusks.

Ribonucleotides. As previously mentioned, the 5'-ribonucleotides have a synergistic effect with glutamic acid to enhance the umami impact. The nucleotide 5'-inosinate (IMP) occurs widely in fish and meats, whereas 5'-guanylate (GMP) is found in mushrooms, but not in fish or shellfish.

Peptides. These compounds are chains of five or fewer amino acids, such as creatine, anserine (naturally occurring in salmon), cadaverine, and creatinine. They also have a certain type of umami-enhancing effect in surimi seafood

14.4 ADDITIVES AND INGREDIENTS USED IN FLAVORS

Many additives not found in seafood can also be used by the flavorist to add certain flavor and taste notes that enhance the overall enjoyment or shelf life of the final product. Ingredients such as amino acids and ribonucleotides will help develop lower notes (more taste-type components found in what a chef would call meat or fish stock) like meat or flesh. In addition, these ingredients will add what can be called a well-balanced feeling, mouth feel, and, in many cases, a long-lasting taste effect. White wine or Mirin wine (a Japanese rice wine high in sugar and salt) can be used in the flavor to cover up many of the fishy off-notes coming from the fish used in fabricating the surimi seafood product. These ingredients will also reinforce the flavoring as a whole and bring an additional fruity note. Other ingredients, such as organic salts (Na^+, Ca^{2+}, K^+, Cl^+, PO_4^-, Mg^{2+}) and seasoning blends, are used to complete a major part of the flavor.

Several commercial taste enhancers are also used to deliver the umami effect, including glutamate, ribonucleotides, hydrolyzed proteins, and yeast extracts.

14.4.1 Glutamate

Certain amino acids, in particular, the monosodium salt of glutamic acid, have been found useful taste enhancers. There are a number of excellent reviews that discuss the chemistry and sensory effect of this substance.[11,12] Glutamate has been used by man for many hundreds of years to enhance the savory notes in foods. The flavorist uses it in surimi seafood to help enhance the more subtle notes of the seafood flavors added to the product.

14.4.2 Ribonucleotides

For the past hundred years, ribonucleotides, naturally occurring in most plant and animal cells, have been used for their flavor enhancing impact. The Japanese have been using dried fish, such as bonito, for subtle enhancement of their many seafood products, including soups. The chemical substance involved in this unique flavor effect was the 5′-disodium nucleotide of inosinate (IMP), the structure of which is shown in Figure 14.2. Another ribonucleotide, the 5′-guanylate (GMP), has also been determined to have an enhancing effect and, together in a 50/50 ratio of IMP and GPM, is marketed as a commercial flavor enhancer. The combination is useful in lessening the metallic tastes and off-flavors, as well as masking bitter and acid notes in surimi seafood products.

Although their power is 50 to 100 times greater than MSG, they are typically used in combination with MSG as they have a synergistic effect on each other to decrease the usage levels of both components. More information on this effect can be found in a research paper by Lin.[13]

14.4.3 Hydrolyzed Proteins

The use of hydrolyzed proteins to bring savory notes to foodstuffs is quite old. Adding soy sauce (Japanese) or Southeast

Asian fish sauce (*nampla* or *nuoc-mam*) has been a food practice for centuries. Both seasonings are produced by hydrolyzing vegetable or plant proteins (HVP and HPP) and animal protein (HAP), respectively, using microorganisms. Amino acids and peptides are their main constituents. Beginning in the mid-1800s, the Europeans were producing acid hydrolyzed proteins (also referred to as HVP, HPPP, or HAP) as enhancers and replacers of meat extracts. Their enhancer strength is related to their high content of glutamic acid, other amino acids and their salts, and sodium chloride resulting from the neutralization of the acid in the acid hydrolysis process. Further details about these ingredients can be found in a number of texts.[14,15]

These products are also used in the manufacture of flavors that are commonly called process flavors. The process flavors or reaction flavors are based on Maillard reactions, also referred to as nonenzymatic browning, where there is a complex set of reactions that take place between amines, usually from proteins (free amino groups), and carbonyl compounds, generally simple reducing sugars such as glucose, fructose, maltose, or lactose. The reactions begin to occur at temperatures varying between 100 and 180°C. The results of these reactions include the formation of many flavors that are typically referred to as roasted, cooked, baked, or meaty type notes.[16]

14.4.4 Yeast Extracts

The autolyzed yeast extract (AYE) contains the water-soluble components of the yeast cell, which is mainly composed of amino acids, peptides, and ribonucleotides. Yeast extract is produced by hydrolysis of peptide links by their endogenous, naturally occurring enzymes (autolysis), or by the addition of food-grade enzymes (enzymolysis). Salt can also be added during the process.

AYE is often used to bring body and base taste notes to savory foodstuffs and, once again, enhance the savory taste of many products. As with many of the hydrolyzed proteins, they have traditionally been used in Western countries as substitutes for beef or chicken extracts in food products. Like the other enhancers, their principal taste substances are amino acids, ribonucleotides, and organic acids, such as lactic acid (characteristic of fish) and succinic acid (characteristic of mollusks), the same substances found in beef and chicken extracts.[17]

14.5 THE "OFF FLAVORS" OF SEAFOOD

Although many seafood flavors are very pleasing to humans, other materials can develop many strong "off-flavors" that distract from the value of the food. Some of these "off-flavors," however, have been accepted by the consumer as notes of the particular seafood; but when presented with fresh fish or shellfish, the consumer will appreciate the difference. For products, such as surimi, made from processed fish, it is important to understand these "off-flavors" and eliminate or mask them.

The typical aroma of seafood that has not been stored properly is attributed to trimethylamine and other amines that arise from the microbial reduction of trimethylamine oxide, which is found abundantly only in marine fish.[18] Dimethylamine is also found in marine fish muscle as an endogenous, enzymatically produced product from trimethylamine oxide.[19] Both of these compounds significantly contribute to the fishy aroma notes in most seafood products that have not been properly stored.

Sulfur-containing compounds have also been identified in seafood and are associated with the deterioration of fish through improper storage and handling. However, dimethyl sulfide is a characterizing aroma top note (an aroma that is smelled initially) of cooked or stewed clams and oysters.[20–22] In addition, bis-(methylthio)-methane has been reported in several species of prawns and lobster.[23] Although these compounds are associated with seafood's "off aroma and flavor," they can also play a role in the overall final flavor used in a surimi seafood product. However, it should be noted that these compounds are very powerful, having extremely low thresholds and therefore are very difficult to use in creating good seafood flavors.

14.6 EFFECTS OF PROCESSING ON SEAFOOD

Many types of compounds produced in food products upon cooking or thermal processing are also seen when seafood is subjected to high temperatures. The nonenzymatic browning of food is known to create useful aroma components, such as pyrazines, that are considered very desirable in all types of cooked foods. A full discussion of nonenzymatic browning or Maillard browning is very complex and outside the scope of this discussion; however, for more details on the chemistry of this reaction, there is an excellent review by Hurrell.[24] In addition, a full discussion of the use of the Maillard reaction to create "process flavors" has also been reported by Manley.[25]

The flavor industry uses thermal reactions to develop flavor, employing many types of reactions that occur between amino acid, reducing sugars, and other components found in foods. The resulting flavors from these reactions are commonly referred to as process flavors. These reactions occur when seafood is cooked and therefore the flavor chemist will keep the Maillard reactions in mind when developing flavors for surimi that are related to seafood when processed or cooked.

14.7 FLAVOR RELEASE AND INTERACTIONS

Volatile aroma components are susceptible to losses during thermal treatments and, depending on the degree of volatility, can be severely affected during surimi seafood manufacturing.[8] Flavorings can be coated with a protective material to allow them to withstand degradation by heat and oxidation. Several techniques create encapsulated flavor materials. Spray-drying is widely used by adding encapsulating materials, such as gums and other hydrocolloids, dextrins, waxes, oils, and other food additives, to the slurry prior to the drying operation. The resulting product is a powder with good shelf-life and handling properties while retaining the original character of the flavor through many types of processing conditions. Further details on this type of encapsulation and aspect of flavor release can be found in an ACS Symposium text.[26] There are many other methods to stabilize and present a

TABLE 14.6 Encapsulation Methods

Type	Description	Attributes	Cost
Spray drying	Use of a heated column with various materials to encapsulate	Low cost and many type of food additives commercially available	Low
Dual coat-fluid bed	Use spray-dried product and adds coating in an air suspension	Coating materials can add release or protect properties to the basic spray dried flavor	Moderate
Glass transition/extrusion	Extrusion for sugar alcohols with flavor and color	Protect against oxidation and adds color and size definition	Moderate
Coascervation (scratch and sniff types)	Co-precipitation of flavor with ionic material (i.e., gelatin)	Oxidation protect and release factors	Very high
Inclusion	Trapping inside cyclodextrins	Stable with good release properties	High
Spray cooling	Fat encapsulation	No high heat or chemistry used in the encapsulation Not water soluble	Moderate to High

flavor in a usable form (Table 14.6) and allow it to be released when eaten; however, spray-dried products are one of the most economical and useful of these types of product.

Flavor interactions with food constitute a very complex study. Imagine what has already been discussed in this chapter about how oxidation, thermal effects, and reactions with proteins and other food materials can affect flavor and then the problems that develop when a flavor is put into a food matrix can be understood. This subject is so important and yet so difficult to understand that only by coordination with the flavor manufacturer will problems be recognized prior to

Surimi Seafood Flavors

TABLE 14.7 Flavor Ingredient Interaction in Food Matrix

Type of Reaction	Compounds Affected	Major Volatile Flavor Components Produced
Thermal reaction	Amino acids and sugar	Pyrazines
Oxidation	Lipids	Aldehydes and ketones
Sulfur interaction	Sulfur-containing amino acids	Methional
Dehydration	Sugars	Furaneol
Decomposition	Vitamin B1 (thiamine)	Sulfur-containing compounds

product launch. In addition, a successful commercial product can only develop if a collaboration of companies is established where information regarding the ingredients being used, as well as the processing conditions of the product is shared. Table 14.7 indicates many of the reactions that can have a pronounced effect on flavor.

14.8 EFFECTS OF INGREDIENTS ON FLAVOR

Adding a flavoring ingredient into a complex matrix such as surimi can also lead to a significantly different aromatic perception due to the texture, processing, and composition of the matrix. This partly results from the interactions of the flavor with the chemical reactions in the matrix and the physical nature of the texture on the release of the flavor when the food is eaten.

The addition of flavor to a product is a key part of foodstuff manufacturing. Aromas are typically organic molecules that can be fairly nonpolar or have a higher affinity for fats and oils. Taste molecules are typically water soluble and do not mix well in fats or oil. The affinity that the flavor has for the matrix will have a pronounced effect on how the flavor performs in the mouth. A flavor that works well in a sugar solution may not work well in a high-fat product like mayonnaise. It is the partitioning of the flavor and its components with its food matrix that may cause a major problem in formulating a flavor that works well in the final food product.

Again, compositional knowledge of the food can help the flavorist create a flavor that has the final sensory character preferred by the consumer when eaten.

Surimi may contain some other additives to protect the product during freezing and cold storage. These additives can have a significant influence on how the flavor will perform. Ingredients used in surimi and surimi seafood processing are discussed in detail in Chapter 13.

14.8.1 Sorbitol and Sugar

Sugar and sorbitol are used in surimi manufacturing as cryoprotectants. Their cryoprotective functions are well described in Chapter 5. Like sugar, sorbitol is also naturally present in many fruits and vegetables. It is highly soluble in water and has a relatively low molecular weight. It is twice as efficient as sucrose in lowering the freezing temperature, and it brings a sweetness of around 0.6 of the sweetness of sucrose. Sweetness is generally not an appreciated attribute in surimi seafood in Asia and Europe, while U.S. consumers do prefer slightly sweeter crabsticks. Sorbitol is chemically stable and does not react with proteins as reducing sugars do during Maillard browning reactions. It is typical to find 4 to 5% sorbitol, 4% sugar, and 0.3% sodium polyphosphate in frozen surimi. However, surimi made from warm-water species contains 6% sugar and 0.3% sodium polyphosphate.

14.8.2 Starch

Various forms of food starch and modified starches are used to retain moisture and give the product a certain texture. The moisture content in surimi crabsticks is approximately 72 to 78%. It is essential to bind this water (lower the A_w) and to avoid the release of water during cooking, cooling, and freezing. Starches can also confer a certain type of texture, firmness, and elasticity similar to those of crab legs or lobster. Different native starches can be found on the market, such as potato, maize, tapioca, wheat, or waxy maize starch, and many modified starches are also commercially available. They

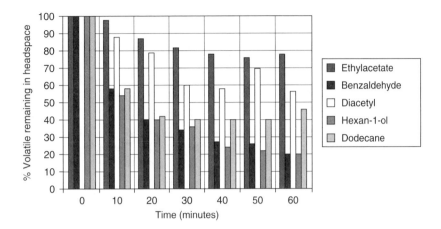

Figure 14.5 Effects of starch on certain flavor compounds.

all have an effect on the flavors released. Figure 14.5 indicates different affinities that different types of starches have toward certain flavor components in solution and therefore how that can release a flavor in use.[27]

Many studies have shown that potato starch gives the highest firmness, gel strength, and elasticity to surimi seafood products. Wheat starch is considered the second choice and is used when there is no local production of potato starch or for cost-saving reasons. It is also more resistant to higher cooking temperatures than potato starch. A blend of 50% potato starch and 50% wheat starch is often considered as giving superior functionality.

14.8.3 Surimi (Raw material)

The inclusion of a large amount of frozen surimi in a surimi seafood product improves the taste of the product. Off-notes can be observed when the level of surimi is reduced or replaced by recovery-grade surimi. But this can be compensated by increasing the flavor level or by adding Mirin wine or some other masking agent, which helps enhance aroma or mask the fishy notes. Furthermore, these fishy notes are more or less perceived, depending on the type of fish used to produce the surimi and its processing.

14.8.4 Egg Whites and Soy Proteins

The purpose of adding egg whites and soy proteins is to make the surimi seafood whiter and glossier, while improving the gelation properties of surimi. The use of egg whites or soy proteins leads to several adjustments in the flavor percentage because they directly affect the texture by increasing gel strength. Proteins must be used carefully because they often generate off-flavors and react with many of the components of the flavor, particularly the aldehydes found in the flavor or extracts used. Hansen and Brooke[28] have shown that the percent binding of benzaldehyde by β-lactoglobulin (a milk protein) averaged 14.1, 26.1, and 60.5%, respectively, for samples of ice cream that contained 2.5, 5.0, and 7.5% of the protein.

14.8.5 Vegetable Oil

Many types of oils are used to improve the texture and increase the whiteness of surimi seafood. Flavor perception is decreased when vegetable oils are used because they decrease the vapor pressure of many of the flavor's components due to their affinity toward the oil. However, flavor release is prolonged, giving the product a more rounded flavor profile. Furthermore, chemicals that tend to evaporate quickly are retained within the oil phase of surimi seafood products and thereby increase the shelf life of the product.

14.8.6 Salt

Salt is essential in extracting myofibrillar proteins, which form a three-dimensional structure upon thermal processing. Salt has historically been used as a flavor enhancer. However, high levels can deteriorate the well-rounded flavor profile as well as mask the sweetness. Salt can also affect the vapor pressure of some the components of the flavor by forcing them out of solution and thereby distorting the flavor profile of the original product.

14.9 PROCESSING FACTORS AFFECTING FLAVORS

Processing ranks equal in importance to formulation from a flavor development standpoint. Because of the volatile nature of many flavor components, exposure to heat, high pressures, or processes that promote oxidation or denaturation negatively impact these flavor components. In addition to the processes that the food undergoes during manufacturing, conditions encountered during shelf life and during consumer use also influence the final flavor.

Heat can be considered the destroyer of flavor quality. The effect can be exacerbated by the presence of water, which can perform a kind of steam distillation of the volatiles into the atmosphere and also facilitate chemical reactions detrimental to the flavor profile. The length of exposure to heat and the degree of the temperature influence these problems. Therefore, short processing times provide an advantage in protecting the flavor.

Several techniques combat this problem and may also alleviate problems due to other severe process conditions

14.9.1 Adding Additional Flavor/Flavor Components

Adding higher levels of the most volatile components can result in a finished product with the desired profile. In principle, the more volatile components may be lost during the manufacture and distribution of the product; therefore, to arrive at a flavor profile that is acceptable to the consumer might warrant starting with a flavor that does not appear balanced prior to its use. Adjusting the profile for the process is not easy but can be done to ensure that the final product has a well-balanced flavor profile.

14.9.2 Addition Points

The best place to add delicate flavors (most seafood is in this category) is toward the end of the process. Flavoring components can be separated into multistage systems in which the stable compounds (taste components) are added in the beginning of the process and those prone to degradation later in the process. However, the product must still receive adequate mixing to ensure proper dispersion. In some cases, a viable alternative is to spray the flavor onto the product after processing is complete.

14.9.3 Encapsulation

Many flavor companies have developed encapsulation techniques to protect volatile flavor components. The goal, currently met with limited success, is to engineer a system that not only prevents heat degradation but also allows flavor release upon consumption, especially when the process includes the presence of moisture.

14.9.4 Storage Conditions and Shelf Life

Flavors have a definite shelf life, varying according to the type of flavor, their form as presented to the customer, and storage conditions. Because of the variability in shelf life due to the complex nature of the flavor, it is best to have the manufacturer specify the expected shelf life and recommended storage conditions. In any case, it is best to store flavors under the best conditions (temperature and relative humidity) available when received and to use them as quickly as possible.

Spray-dried flavors may represent a more stable form of the flavor if stored at reasonable temperatures (max = 90°F) and low RH (max. = 70%), only because the flavor's components are immobile and therefore cannot react with other ingredients in the flavor. Certain combinations of starches and gums can create a spray-dried flavor with superior shelf life; however, the supplier can only help if he has knowledge about the intended use of the flavor, the storage conditions of the flavor, and the environment in which the flavor will be used.

Remember that many of the flavor notes of seafood products are very delicate and can be lost very easily by exposure to oxygen, time, high temperature, and moisture conditions. Once again, the best protection of a flavor is to keep the

Surimi Seafood Flavors 739

package closed until use, store it at low temperature and humidity, and once the container is opened, to use it completely in a short period of time.

14.10 FLAVOR REGULATIONS AND LABELING

All materials added as product flavor are subject to the legal regulations of many governments around the world. However, there is no single, worldwide harmonized listing of flavor ingredients that can be used in food products and, therefore, the use of flavors in any product, including surimi-based products, represents a complex problem. In addition, the three major economies — the United States, Europe, and Japan — regulate flavor in very different ways. Each is summarized below and the view on how the flavor industry collectively tries to work with governments to harmonize the differences in local regulation is also addressed.

There are a number of major aspects of flavor and flavor ingredient use and regulation. They include the safety of the materials being used, the methods of establishing safety, and the label claims that can be made relative to the flavor and its use in specific food products.

The safety of food additives is a major concern of any government and regulatory group. The basic rule of many governments is that a flavor ingredient should be considered safe prior to its use in food. However, the methods used to prove safety in use are not harmonized around the world and, therefore, the following brief summary of the major world economic units and how they approach this problem is presented.

14.10.1 United States

In the United States, the 1958 Food Additive Amendment to the Pure Food Act established a law requiring that all ingredients added to food must be considered safe. The U.S. Congress, in enacting that amendment, recognized that there were different levels of concern for safety related to a material's exposure level, scientific evidence of safety, and history of use. Ingredients that are new, unique, or that will be used at high exposure levels are termed "food additives" and are subject to formal review and approval by the Food and Drug Administration (FDA) prior to food use or marketing. Ingredients that have a long history of use in the human diet (such as sugar), very low exposure potentials (such as flavor ingredients), or are a part of a known food (such as the oil of an orange) are designated "generally recognized as safe" (GRAS).

The GRAS part of the amendment allows experts in the field of food safety or other recognized experts (e.g., toxicologists and medical professionals) to review the ingredient and its potential use in food and determine its safety as a food additive. The Act, however, did not mandate that a government group, such as FDA, review and approve the use of the ingredient. This led the flavor industry in the 1960s, under the leadership of their trade organization, the Flavor and Extract Manufacturers Association (FEMA), to establish an independent "expert panel" to review the safety of flavor ingredients and establish the ingredient's GRAS status. For more information on the GRAS process, refer to the article by Woods and Doull.[29] The result of the process has been a series of so-called "GRAS lists" published over the years in *Food Technology*, a food magazine published by IFT (Chicago, IL). The FDA had started to issue its own GRAS list in the *Code of Federal Regulations* (CFR)[30]; however, it now accepts the ingredients found on the FEMA GRAS list as approved ingredients for use in flavors in the United States and no longer adds to its list. However, the FDA's position does not exempt the manufacturer from liabilities if the ingredient is found to have a safety- or health-related problem. Such food products containing ingredients found not to be safe can be seized by the FDA under the Food Additive Act.

GRAS safety is based on ingredient chemistry, purity, and the level of use in the final food.

14.10.2 European Union (EU)

Prior to 1988, the only pan-European list with any impact on the use of flavor ingredients was the so-called Council of Europe list. This list was developed by a group of experts in

Surimi Seafood Flavors

Europe as a guideline for the safe use of flavor ingredients. Many European countries, however, had exceptions or additions that could be used and an overriding principle of the allowance of "nature-identical" ingredients was firmly in place as a guide to the acceptance of flavor ingredients. In many countries, the use of many "non-nature identical" (those substances not found to occur naturally in food) or, as they are called in Europe, artificial flavor ingredients, are not allowed to be used in food.

The European Union (EU) Parliament in its 1988 Directive established that there is to be a Positive List of ingredients allowed for use in all EU countries and the first official list should be published in 2005.[31] The committee of experts is currently reviewing the safety of ingredients from a list that contains the Council of Europe Ingredients, the FEMA GRAS list (including a list found in the FDA's CFR Title 21), and a number of other ingredients that occur on individual countries' usage lists. Safety data from many sources is being used to ensure that the list reflects the safety of the ingredients for use in EU member states. The current list being reviewed can be found at an EU website (www.flavis.net).

14.10.3 Japan

In Japan, the use of any food additive, including flavors and flavor ingredients, is covered under the Food Sanitation Laws. In the 1950s, a list of acceptable flavor ingredients was issued and that list has been in use ever since. Although the list is considered an open list, it has not been updated with substances that are in use in many other parts of the world. The list contains both chemically identified ingredients and general functional groups that are allowed for use in food. This has created further problems for the flavor industry as flavors created in Europe and the United States may not be legal in Japan because they are not on the Japanese Sanitation List or because companies might misunderstand the meaning of functional group definitions.[32]

The Health Ministry decided in 2003 that a review of the list should be made and a new list be issued that includes ingredients used in other parts of the world, if they are found acceptable to a Japanese expert panel reviewing safety data. The current list of ingredients and functional groups can be found at a Japanese website (www.jetro.go.jp/se/e/standards_regulation/).

14.10.4 A Potential World List — The United Nations

It would be very useful to the world's food industry if the flavor industry had one list of ingredients that was approved for use in all countries around the world. To that end, the flavor industry, through its worldwide trade organization, the International Organization of Flavor Industries (IOFI) and its national trade organization members, have been working to present safety data to a United Nations group under the World Health Organization (WHO) called the Joint Expert Panel of Food Additives (JECFA). The JECFA has been reviewing the safety of ingredients and publishing a list of ingredients it considers safe for use in foods. That list can be found on the JECFA website (www.fao.org/waicent/fao-info/ECOMONICesn/jecfo/index_en.stm).

WHO, with its UN partner organization, the Food and Agriculture Organization (FAO), also issues the Codex Alimentarius, which is the determining rule book for food additives and their specifications and use for many, but not all nations. The Codex website is at www.fao.org/docrep/w9114e/W9114e03.htm#.

The division of flavoring ingredients into categories has been adopted by the FAO/WHO Food Standard Program. The definitions for the classes of flavoring ingredients can be found in the *Codex Alimentarius* and are as follows[33]:

1. *Natural flavors and natural flavoring substances.* The *Codex Alimentarius* defines "natural flavors" and "natural flavoring substances" as preparations and single substances, respectively, acceptable for human consumption, obtained exclusively by physical processes from vegetable and some animal raw materials

either in their natural state or as traditionally processed for human consumption.
2. *Nature-identical flavoring substances.* The *Codex Alimentarius* defines "nature-identical flavoring substances" as substances chemically isolated from aromatic raw materials or obtained synthetically. They should be chemically identical to substances present in natural products intended for human consumption, either processed or not, as defined above.
3. *Artificial flavoring substances.* The *Codex Alimentarius* defines "artificial flavoring substances" as those substances that have not yet been identified in natural products intended for human consumption, either processed or not.

14.10.5 Religious Certification Issues

Religious certification programs add to the problem of developing suitable flavors for various uses around the world. These religious issues can also relate to surimi seafood products as major religious certification programs are in place in certain areas of the world for Kosher (Jewish food laws) and Halal (Moslem food laws) foods.[34,35] To carry a particular religious mark in either of these areas, the ingredients in the products must be proven to be certified for use in the product. Typically, a company will utilize a certifying group that will perform the product reviews and offer a certification that the products meet their religious beliefs and rules. It is, consequently, important that the suppliers know of the need of these certifications prior to offering a flavor.

14.10.6 Worldwide Issues

As one can see, there are many problems associated with the use of a flavor for an international food product. Although the UN has created some rules and an approved ingredient list, they are not universally accepted and that can cause some confusion both on the nature of the ingredients used and their labeling. For example, the titles given to the flavor will have a legal aspect on what food products the flavor can be used in (e.g., flavors cannot be used in Danish dairy products) and what the flavor food can be labeled ("natural flavor" has a different definition in many countries and that definition can have a major impact on the legality of the food product). The allowance for the use of a flavor ingredient can be taken care of by listing the allowed ingredients, as has been done in many areas of the world; however, the labeling issues for food and flavor vary considerably around the world and are not easily addressed in this chapter.

These variations will ultimately have an important effect on the flavor that may be created in a product. There is no one source of information concerning the use and labeling in all parts of the world, including the major countries, but a couple of references can be useful in answering some questions.[36,37] The best position, once again, is to be open and share as much information as possible with the supplier and these issues will be considered when the final flavor is developed for the product. The other organization that can be helpful in answering questions is the local flavor trade organization.

14.11 SUMMARY

Flavor plays an important role in any successful food product. It is, indeed, a critical component of a successful surimi seafood product. This chapter reviewed the many aspects of surimi seafood based on its complicated nature. The complexity of chemistry, processing, and labeling adds to the problem that the flavorist has in creating a flavor that works in the surimi seafood product. Open collaboration between the flavor manufacturer and surimi seafood processor is of great importance for ultimate success. The flavor manufacturer can only produce an outside flavor if total knowledge of its use is known. This knowledge can also be a resource for further creativity in producing new, unique, interesting, and successful products.

REFERENCES

1. G. Reineccius. Flavor chemistry. In *Source Book on Flavors,* 2nd ed. New York: Chapman & Hall, 1994, 61–115.

2. T.A. Acree. Gas chromatography-olfactometry. In C.T. Ho and C.H. Manley, Eds. *Flavor Measurement.* IFT Basic Symposium Series. Chicago: Institute of Food Technologists, 1993, 77–94.

3. F. Ulrich and W. Grosch. Identification of the most intense odor compounds formed during auto-oxidation of linoleic acid. Z. Lebensmitle Unters Forsch., 184(4), 277–282.

4. N. Chaudhari, A.M. Landin, and S.D. Roper. A metabotropic glutamate receptor variant functions as a taste receptor. *Nature Neuroscience,* 3, 113–119, 2000.

5. D.B. Josephson. Mechanisms for the Formation of Volatiles in Seafood Flavors. Ph.D. thesis, University of Wisconsin–Madison, Madison, WI, 1987.

6. S. Spurvey, B.S. Pan, and F. Shahidi. Flavour of shellfish. In F. Shahidi, Ed. *Flavor of Meat, Meat Products and Seafood, 2nd ed.* New York: Blackie A & P, 1998, 159–196.

7. G. Reineccius. Off-flavors in foods. In *Source Book on Flavors, 2nd ed.* New York: Chapman & Hall, 1994, 116–136.

8. R.J. Gordon and D.B. Josephson. Surimi seafood flavors, creation and evaluation. In J.W. Park, Ed. *Surimi Seafood, 1st ed.* New York: Marcel Dekker, 2000, 393–416.

9. H. Iida. Studies of the accumulation of dimethyl-beta-propiothetin and the formation of dimethyl sulfide in aquatic organisms. *Bull.Tokai Regional Fisheries Res. Lab. (Japan),* 1988, 35–111.

10. J.L. Boyle, R.C. Lindsay, and D.A. Stuiber. Occurance and properties of flavor related bromophenols found in the marine environment: a review. *J. Aquat. Food Product Technol.,* 2, 75–82, 1993.

11. G. Mathesis. Flavor modifiers. In P.R. Ashurst, Ed. *Food Flavorings, 3rd ed.* Gaithersburg, MD: Aspen Publications, 1999, 367–406.

12. G. Reineccius. *Flavoring Materials Contributing to Taste,* 2 ed. New York: Chapman & Hall, 1994, 626–654.

13. W. Lin, T. Ogura, and S.C. Kinnamon. Responses to disodium guanosine 5'-monophosphate and monosodium l-glutamate in taste receptor cells of rat fungiorm papillae. *J. Neurophysiol.*, 9, 1434–1439, 2003.

14. C.H. Manley, B.H. Choudhury, and P. Mazeiko. Thermal process flavorings. In P.R. Ashurst, Ed. *Food Flavorings, 3rd ed.* Gaithersburg, MD: Aspen Publications, 1999, 367–406.

15. C.H. Manley. Progress in the science of thermal generation of aroma: a review. In T.H. Parliment, R.J. McGorrin, and C.-T. Ho, Eds. *Thermal Generation of Aromas.* ACS Symposium Series 409. Washington, D.C.: American Chemical Society, 1989, 12–22.

16. C.H. Manley. Process flavors. In G. Reineccius, Ed. *Source Book on Flavors, 2nd ed.* New York: Chapman & Hall, 1994, 139–154.

17. T.W. Nagodawithana. Savory flavors. In A. Gabelman, Ed. *Bioprocess Production and Flavor, Fragrance and Color Ingredients.* New York: John Wiley & Sons, 1994, 135–168.

18. D.B. Josephson and R.C. Lindsay. Enzymatic generation of volatile aroma compounds from fresh fish. In T.H. Parliment and R. Croteau, Eds. *Biogeneration of Aroma.* ACS Symposium Series 317. Washington, D.C.: American Chemical Society, 1986, 201–219.

19. R.C. Lundstrom and L.D. Racicot. Gas chromatographic determination of dimethylamine and trimethylamine in seafood. *J. Assoc. Off. Anal. Chem.*, 66, 1158–1162, 1983.

20. J. Suzuki, N. Ichimura, and T. Etoh. Volatile components of boiled scallop. *Food Rev. Int.*, 6, 537–552, 1990.

21. Y.J. Cha. Volatile compounds in oyster hydrolysate produced by commercial protease. *J Korean Soc Food Nutr* 24: 420-426, 1995.

22. C.T. Ho and Q.Z. Jin. Aroma properties of some alkythiazoles. *Perfumer Flavorist* 9:15-18, 1985.

23. D.B. Josphson. Seafood flavor. In H. Maarse, Ed. *Volatile Compounds in Foods and Beverages.* New York: Marcel Dekker, 1991, 179–202.

24. R.F. Hurrell. Maillard reactions. In I.D. Morton and A.J. MacLeod, Eds. *Food Flavours. Part A: Introduction.* New York: Elsevier, 1982, 399–425.

25. C.H. Manley, B.H. Choudhury, and P. Mazeiko. Thermal process flavorings. In P.R. Ashurst, Ed. *Food Flavorings, 3rd ed.* Gaithersburg, MD: Aspen Publications, 1999, 367–406.

26. F.E. Escher, J. Nuessli, and B. Conde-Petit. Interactions of flavor compounds with starch in food processing. In D.D. Roberts and A.J. Taylor, Eds., *Flavor Release.* ACS Symposium Series 763. Washington, D.C.: American Chemical Society, 200, 230–245.

27. M.Y.M. Hau, D.A. Gray, and A.J. Taylor. Binding of volatiles to starch. In R.J. McGorrin and J.V. Leland, Eds. ACS Symposium Series 633. Washington, D.C., American Chemical Society, 1996, 109–117.

28. A.P. Hansen and D.C. Brooke. Flavor interactions with casein and whey proteins. In R.J. McGorrin and J.V. Leland, Eds. ACS Symposium Series 633. Washington, D.C., American Chemical Society, 1996, 74–97.

29. L.A. Woods and J. Doull. Evaluation of flavoring substances by the Expert Panel of FEMA. *Regulatory Toxicol. Pharmacol.,* 14, 48–58, 1991.

30. *Code of Federal Regulations,* Title 21, Section 184.

31. EEC. Council Directive of 22 June 1988 on the approximation of the laws of the Member States relating to flavouring for use in foodstuffs and to source materials for their production, (88/388/EEC). *Official Journal of the European Communities* No. L 184/61-66.

32. Japanese Minister of Health and Welfare. Guidelines for Designation of Food Additives and for Revision of Standard for Use of Food Additives, 1996.

33. World Health Organization (WHO) and The Food and Agriculture Organization (FAO). Food Standards Programme. *Codex Alimentarius.* Rome.

34. J.M. Regenstein and C.E. Regenstein. Kosher foods and food processing. *Encycl. of Food Sci.,* 2000, 1449–1459.

35. M.M. Chaudry. Islamic food laws: Philosophical basis and practical implication. *Food Technol.,* 1992, 92–93.

36. K. Bauer. Labeling Regulations. In G. Reineccius, Ed. *Source Book on Flavors, 2nd ed.* New York: Chapman & Hall, 1994, 852–875.

37. R.D. Middlekauff and P. Subik. *International Food Regulations Handbook.* New York. Marcel Dekker, 1989.

15

Color Measurement and Colorants for Surimi Seafood

GABRIEL J. LAURO, PH.D
California Polytechnic University, Pomona, California

OSAMU INAMI
T. Hasegawa, Fukaya, Saitama, Japan

CRAIG JOHNSON
Minolta Corporation, Ramsey, New Jersey

CONTENTS

15.1 Introduction 751
15.2 Understanding Color and Measurement 753
    15.2.1 Development of Color Language 753

- 15.2.2 Color Space 754
- 15.2.3 Instrument Development 756
- 15.2.4 Tristimulus Values 756
- 15.2.5 L*a*b* Color Space 757
- 15.2.6 Indices 759
- 15.2.7 Measuring Color 759
  - 15.2.7.1 Tristimulus Measurement 759
  - 15.2.7.2 Spectrophotometric Measurement 762
- 15.3 Coloring Surimi Seafood 765
  - 15.3.1 Preparation of Surimi Paste for Crabsticks 766
  - 15.3.2 Color Application to Crabsticks 766
  - 15.3.3 General Principles 766
- 15.4 Colorants 769
  - 15.4.1 Colorants Requiring Certification 769
  - 15.4.2 Colorants Not Requiring Certification 770
    - 15.4.2.1 Carmine (21 CFR 73.100, EEC No. 120, CI No. 75470, CI Natural Red 4) 771
    - 15.4.2.2 Cochineal Extract (21CFR 73.100, EEC No. E120, CI Number 75470, Natural Red) 773
    - 15.4.2.3 Paprika (21CFR 73.345, EEC No. E 160c) 776
    - 15.4.2.4 Annatto (21 CFR 73.30, EEC No. E 160b, CI 75120, CI Natural Orange 4) 780
    - 15.4.2.5 Turmeric (21CFR 73.600, EEC No. E 100, CI No. 75300, CI Natural Yellow 3) 783
    - 15.4.2.6 Grape Color (21CFR 73.169, EEC No, E163) and Other Anthocyanins 784
    - 15.4.2.7 Beet Juice Concentrate (21CFR 73.260) 785
    - 15.4.2.8 Caramel (21 CFR73.85, EEC No. 150) 786
  - 15.4.3 Monascus Colorants 787

Color Measurement and Colorants for Surimi Seafood

    15.4.4 Nature Identical Colorants 787
            15.4.4.1 Canthaxanthin (21 CFR 73.73.75,
                      EEC No. E161g) 787
            15.4.4.2 β-Carotene (21 CFR 73.95) 788
    15.4.5 Other Naturally Derived Colorants 788
            15.4.5.1 Titanium Dioxide (E171, CI No.
                      77891, CI Pigment White 6) 788
            15.4.5.2 Calcium Carbonate (EE 170,
                      CI No. 77220, CI pigment
                      White 18) 788
            15.4.5.3 Vegetable Oil 789
15.5 Color Quality 789
    15.5.1 Final Product Color 789
    15.5.2 Colorant Quality 789
    15.5.3 Acceptance Criteria 789
    15.5.4 Acceptance Tolerance 792
15.6 Labeling 794
    15.6.1 Requirements in the United States 794
    15.6.2 Religious Requirements 795
15.7 Summary 796
References 796
Additional Reading 798

15.1 INTRODUCTION

In the United States, surimi is a summer product, consumed as a shellfish substitute for crab, shrimp, and lobster in salads, fillings, and soups. The United States, however, has not yet embraced the creativity enjoyed by the Asian, especially the Japanese, market. In the United States, surimi's association with the concept of "imitation" has held back its acceptance and slowed its development. In this chapter, surimi seafood should not be viewed as imitation but as a new food capable of offering highly nutritious protein in a novel form. Surimi seafood, therefore, should also be considered as a new, healthy, and nutritionally satisfying food that can be molded, shaped, flavored, and, of course, colored — limited only by the extent of one's imagination and the present state of color technology.

Surimi seafood, viewed in the eyes of a scientist interested in the application of natural colors, is a food system consisting of comminuted proteinaceous tissue (fish or otherwise), modified by additives (such as flavors, fillers, and colors), that can be shaped/cooked to a desired edible end product. The color additive can color either the surface, the entire piece, or even be co-extruded. The approach is taken that surimi seafood, as a new food, can be made with any appropriate color and therefore the discussion of colorants in this chapter is much broader than just "crab/lobster red." The colorants discussed include both artificial and naturally derived as approved for use in the United States.

The application of colorants to surimi systems requires an understanding of the current practices in the industry, colorant utilization in various types of surimi seafood products in other countries, the chromophores themselves (i.e., chemistry, source, preparation, limitations, and available forms), and the influence of the surimi seafood process/distribution on colorant stability and acceptability. Acceptance always requires a discussion of color, what it is, how it is measured, and how it is evaluated for acceptance. The "how" requires a discussion of the analytical instrumentation utilized to make the valuation more objective than just human observation.

This chapter begins with a discussion of color and the development of a color language, including color measurement. This is followed by a review of today's practices in coloring surimi seafood and how the product is evaluated for appearance and acceptance. Colorants are then defined and individually reviewed. This is important because the properties of each colorant or colorant family differ sufficiently so as to influence specific selections in order to obtain acceptable performance in surimi seafood systems. Color performance is reviewed through a discussion of quality and its measurements. The chapter concludes with a discussion of ingredient color labeling in the United States and to a lesser degree in other parts of the surimi-consuming world.

The intent of this chapter, therefore, is primarily aimed at providing useful information to individuals working with surimi seafood. Building on this information, it is also aimed at those with an entrepreneurial spirit — those who see surimi as the base material for a host of potentially new wholesome foods — foods that may have to be colored for improved acceptance.

15.2 UNDERSTANDING COLOR AND MEASUREMENT

15.2.1 Development of Color Language

The quantification of measured color of a fabricated food such as surimi seafood should always be determined within a tolerance based on a level of visual acceptability. The measurement of color, being a relatively new science concept, has historically been difficult to describe and understand. Scientists and industry experts have struggled with the problem of correlating the visual perception of a human being with the colormetric determination of a sample, as color is perceived differently under differing angles of viewing, illumination, and human physiology. The biggest obstacle is that human vocabulary did not provide an adequate means to consistently differentiate magnitude and direction of color variation even within the same color family. The science of color measurement, however, has evolved quickly over the past century to produce tools that make the determination of the color of objects (and the size and direction of the differences of a standard compared to a sample) rapid and easy to understand. It started, however, with the observation that without light, there is no perception of color.[1]

The basis of color measurement is visible light, which is a form of electromagnetic energy. The electromagnetic spectrum is a continuum of energy waves, some being very short and others being very long. To perceive color, the wavelengths of energy must be such that the detectors in the human eye can receive them. The eye can detect wavelengths of energy between 380 mµ (millimicrons), a violet color, and 780 mµ, a red color. This range of energy is called the visible spectrum. The summation of the wavelengths of energy along the entire range of the visible spectrum is called white light.

When white light is reflected from or transmitted through an object, the energy will be totally or partially absorbed. These wavelengths of energy, which are reflected or transmitted to the eye, are perceived as color. This same energy is also relied on in the instrumental evaluation of color.

15.2.2 Color Space

White light can be divided into seven dominant wavelengths using an optical prism. The wavelengths appear as different colors because of the way the energy stimulates the rods and cones in the human eye. The rods and cones also allow for differentiation between light and dark, as well as the determination of different colors.

The earliest attempt to organize colors occurred in 1915 by an artist named Arthur Munsell. His model ordered color into a system called the Munsell Color Tree. This system described color in three dimensions: value, hue, and saturation (chroma). *Hue* describes the color of the object, that is, red or yellow. *Value* indicates the amount of lightness or darkness in an object compared to a gray scale. Lightness could also be measured separately from hue. The term "saturation" (chroma) addresses the vividness of a color or the intensity of actual color perceived. Pure colors, when mixed, always become reduced in saturation. The amount of color from little to almost dominant can therefore be made into a graduated scale representing saturation.[2]

These three terms were developed to help describe the color of an object based on a specific viewing angle and light source. Consequently, the concept of a color "space" was realized (Figure 15.1). This allowed a color to be positioned or placed in a three-dimensional space relative to its hue, value, and saturation. There were many three-dimensional models developed as color science progressed, each having its own merits and weaknesses, but each contributing to a workable model, such as the Munsell Color System. The latter defined the quality of a color as hue, lightness, and chroma (our original hue, value, and saturation). The work solidified the concept of color space and established that a color having the

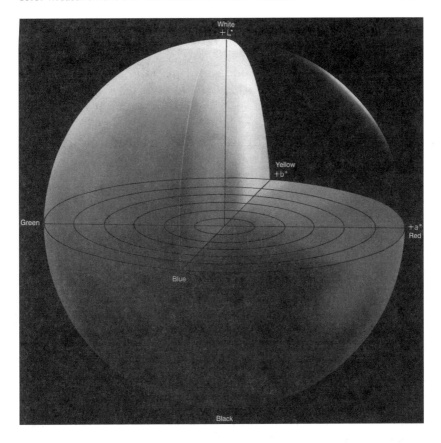

Figure 15.1 Representation of color solid for L*a*b* color space.

same hue, lightness, and chroma under a given illuminating condition would occupy the same point in color space.

The observation that the same point in color space could be perceived differently under different conditions, however, was still a problem. The following five conditions describe situations where perception and measurement might differ:

1. Perceived color was affected by the quality of the light source (sun, candle, fluorescent, incandescent, etc.).
2. Perceived color was affected by the size of the object (the best example is viewing a paint swatch and then seeing the same color on a wall — the wall, or the

larger area, always appears more saturated than the smaller sample).
3. Perceived color was affected by the background (black or white backgrounds appear to change the "lightness" and "saturation" of the object).
4. Perceived color was affected by viewing directional differences (the angle at which an object is observed will affect hue, lightness, and/or saturation).
5. Perceived color was affected by personal differences (humans differ genetically and physiologically in vision and there is an average, but no normal, color vision).

It was therefore concluded that if these conditions could be defined and quantified, then instrumentation could afford a reproducible means of describing and quantifying color.

15.2.3 Instrument Development

Color, under very fixed conditions, could now be fully described by its hue, lightness, and chroma. It was thought that instruments could be developed to incorporate the mathematical and physical replication of these selected conditions. Early instruments frustrated scientists because of the machine's inability to quantify color in a meaningful way. What was needed was a means to define color in the same manner that the human observer interpreted color and color differences. Scientists wanted a numerical value to express color in a way similar to expressing weight or volume. Unfortunately, the Munsell Color System could not be developed further as it used paper color chips viewed under daylight lighting conditions. The Munsell colors, however, were renotated to express color as a letter and number combination using H V C (or hue, value, chroma). While this new system allowed the user to define color in an orderly fashion, the problem with physical observation and interpretation remained an issue.

15.2.4 Tristimulus Values

In 1931, The Commission Internationale de l'Eclairage (referred to as CIE), a scientific body composed of representatives from many countries, set out to establish and recommend constants for both testing and expressing color mathematically. The Commission reported the development of a series of "spectral response curves" corresponding to the human eye's ability to match colored light and thus established the "standard observer" color-matching functions. These spectral response functions utilized three color-matching functions: $x(\lambda)$ as the integration of response of the human eye to red light, $y(\lambda)$ as the integration of the response of the human eye to green light, and $z(\lambda)$ as the integration of response of the human eye to blue light.

The color of an object is a result of the different proportions of red, green, and blue light being reflected or transmitted to the eye. When integrating the spectral power distribution of a light source (a light source spectral curve), the reflectance curve of the object, and the standard observer color-matching functions, the mathematical relationship between color measurement and visual perception could now be established. This mathematical integration created the XYZ, or tristimulus color scale.

The XYZ scale could be used as a tool to determine the amount of red, green, and blue light from a standard and compare those amounts to a sample to determine color difference. The challenge was that color difference was difficult to visualize in numerical terms because of the mathematical calculation of each function. These XYZ values were then converted into a linear transformation of the tristimulus equation and plotted for visualization on a two-dimensional diagram. This second derivation was called the chromaticity scale or Yxy color space. In this scale, the Y value indicated the degree of lightness; and the xy determined the color independent of lightness.

15.2.5 L*a*b* Color Space

A problem still remained, however, when using the Yxy color space. Equal distance on the x,y chromaticity diagram did not equal the perceived visual color difference. In 1976, the CIE introduced an opponent color space based on a nonlinear

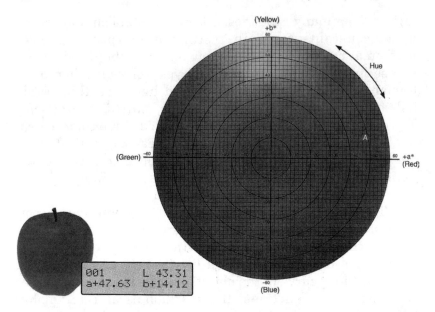

Figure 15.2 Chromicity and lightness. (*Source:* Courtesy of Konica Minolta, Instrument Systems Division.)

transformation of the XYZ values to correlate color difference more closely with the visual perception of color. This scale was called the CIE L*a*b* color space. In this space, L* indicates lightness, while a* and b* are the color coordinates; a* is the red/green axis ("+" being toward the red and "–" being toward the green). Similarly, b* is the yellow/blue axis ("+" being toward the yellow and "–" being toward the blue) (Figure 15.2).

This method of defining color pinpoints the color in a three-dimensional color space. The point on a color axis determines its hue, its distance from the center (achromatic toward purity) establishes saturation, and its position on the vertical axis establishes its lightness. That is, the positive or negative value of the color coordinates defines the hue, while its numerical value establishes the extent of saturation. In addition, the L* value, from 0 to 100, establishes the point's light to dark content. In L* a* b* color space, the intersection of the three values is the perceived color of a sample. This concept of color space is widely used in almost all fields, including surimi seafood.

15.2.6 Indices

As an alternative, one-dimensional scales, called indices, have also been devised in an attempt to simplify and define the color through the use of a single number. These can be very valuable when used under proper conditions and work reasonably well as long as the sample is similar in color to the standard being used to create the scale.

15.2.7 Measuring Color

There are two basic instruments used in the measurement of color: tristimulus colorimeters and spectrophotometric colorimeters. Each is designed to standardize measurement conditions for color determination by fixing the physical setup of the instrument and by integrating the mathematical equations to correlate with visual perception.

15.2.7.1 Tristimulus Measurement

As previously explained, instrumental color measurement is modeled after the physiology of the human eye. The eye utilizes a system of receptors called rods and cones. Rods allow us to see luminosity or light and dark. Cones allow us to detect specific wavelengths of light and are our color receptors. There are several types of cones located in the central region of the retina, called the fovea pit, where color acuity is most sensitive. It was established that the signal generated by these sensors, and transmitted to the brain, determined the many variations of color. By substituting the cones of the eye with red, green, and blue filters attached to photo cells, an output that can provide values specifying the color of the object in the same sensitivity $x\,(\lambda)$, $y\,(\lambda)$, and $z\,(\lambda)$ as the human eye can be obtained. These measurements directly express the tristimulus values X, Y, and Z.

The instrument developed having filters, photocells, and a fixed light source arranged in a standardized geometry, to

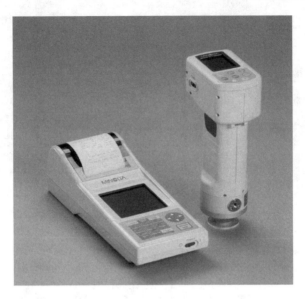

Figure 15.3 Tristimulus colorimeter. (*Source:* Courtesy of Konica Minolta, Instrument Systems Division.)

allow consistent reproduction of tristimulus values for an object's color, is known as a tristimulus colorimeter. Its use is the easiest means of defining color. It does not have the accuracy of a spectrophotometer, but it is more than adequate as a quality control tool to measure differences among samples. In using such an instrument, one must be constantly sensitive to the fact that the tristimulus measurements depend on the measurement geometry, the illuminant, and the observer.

There are two basic measurement geometries incorporated in the design of tristimulus colorimeters: 45/0 and diffused/0 (Figure 15.3). The first function indicates where the light source is situated; the second indicates where the sample is placed and viewed by the detectors. A 45/0 instrument illuminates the sample at a 45° angle and views the sample at 0° from perpendicular of the sample.

In the diffused/0 measurement, the light is "spread out" over the entire surface of the sample by projecting it through a translucent plastic material. The light is then detected at

0° to the light source. In addition, in the 45/0 measurement, the specular component (gloss) is lost to the measurement and only the color is detected. In the second function, the specular component is included with the color. Diffused/0 measurements will also usually indicate a lighter measurement (higher L^* value) than 45/0. However, any change in the angle of illumination or viewing will generate new values. For this reason, it is important to record the conditions of the test when the values are obtained. Consequently, comparison of results, without these details, could result in significant error.

The tristimulus colorimeter is proficient at indicating small color differences, some of which are difficult to be perceived by the human eye. For this reason, a difference may not be a cause for rejection. It is important that an "acceptance tolerance" be established around the product's absolute colormetric values based on where it is located in color space, especially when working with inconsistent materials, such as surimi and surimi seafood. Establishment of acceptance tolerances is reviewed later in this chapter.

The Konica Minolta CR-400 is a typical high-end colorimeter (Figure 15.3), allowing the operator to measure the sample and obtain colorimetric values in any number of color spaces and indices. The instrument is designed with a Xenon light source (offering two standard mathematical illuminants, C and D65), fixed illumination and viewing angles of diffused/0, and a set of photocells filtered to closely match the CIE 1931 standard observer. Different sample area viewing sizes (8 and 50 mm) can be used to reduce the effect of the backgrounds and accommodate different sample sizes. Furthermore, it allows for the determination of absolute color and color differences in a numerically useful form by use of a detachable microprocessor or easy-to-use software program that can be loaded on a personal computer.

Surimi or surimi seafood samples to be measured can range from an uncolored-uncooked paste to a surface-colored or fully colored cooked gel. Each form has its unique presentation requirements that must be standardized for reproducibility within and between labs. The uncolored-uncooked paste form can be presented to the colorimeter by over-filling a small, optically clear plastic Petri dish (60 × 15 mm) and pressing the cover down in such a manner as to eliminate much of the entrapped air. After standardizing the instrument to the black and white plates, at least three measurements of the paste are made through the dish bottom, rotating the dish for each measurement. The values are then averaged for sample-to-sample comparison.

Colored-uncooked paste can be similarly measured by flattening the paste as level as possible in the bottom of a small, optically clear Petri dish. Care must be taken, however, to use the same size sample in the preparation of the dish, as paste thickness affects the colorimeter reading.

The measurement of fully cooked gels from production requires the selection of large enough pieces that have a smooth surface. At least three, but preferably ten, measurements should be made and averaged. Outlying measurements should not be included in the average, but care should be taken not to bias the sampling. Pieces smaller than the aperture or with an irregular surface should not be used because measured values will not be reproducible.

Powdered colorants can also be checked by tristimulus with the use of a sample cell accessory. Colorants high in saturation, such as carmine powder, are measured by mixing the sample with talc (1% dilution) and taking at least ten readings. Any change in the nature of the talc requires that a new standard be prepared with new talc. Colorants, however, can also be measured directly with a Minolta CR-400 series without the dilution step.

15.2.7.2 Spectrophotometric Measurement

The spectrophotometer provides the same results as a colorimeter (Figure 15.4), that is, the determination of absolute colorimetric and color difference values. It differs, however, in the method of light collection and geometry. A spectrophotometer measures and collects the spectral characteristics of light absorbed or reflected from a sample over the entire visible spectrum, called a spectral curve.

Figure 15.4 Spectrophotometric colorimeter. (*Source:* Courtesy of Konica Minolta, Instrument Systems Division.)

The measured value(s) or spectral curve obtained are a fundamental physical property of an object and can be likened to a human fingerprint. The spectral curve of a sample has a unique shape specific to the sample makeup. Just as with the colorimeter, spectrophotometer geometry is of two basic types: 45/0 and diffused/8. The 45/0 geometry is the same as previously discussed for the tristimulus colorimeter. The diffused/8 geometry is also similar to the diffused/0, with the exception that the diffuser is an integrating sphere with three holes cut into it. One hole is where the sample is placed for measurement. The other two holes are placed opposite the sample, each 8° from perpendicular to the sample port. This allows for the inclusion or exclusion of the specular component (gloss). The spectrophotometer can calculate the tristimulus values by integrating the CIE standard observer function with an illuminant table (spectral values for the light source). The light measurement information obtained from the spectrophotometer is then displayed as color data for a specific light source or directly as a spectral reflectance graph and values. Spectrophotometers, such as the Konica

Minolta CM 2600 portable or CM-3600 benchtop, allows for the selection of multiple standard observers, CIE illuminant tables, fixed angle of illumination, and viewing angle. Calibration with a black plate (0% reflection) and a white plate (100% reflection) must also be done before any sample measurement is taken.

Samples of surimi seafood for measurement are presented in the same manner as described earlier for the tristimulus colorimeter. Reflectance measurement is set for the L*a*b* color space. The versatility of the Minolta CM 2002, and now its successor the CM-2600, allows the instrument to be set upside down so that the measurement aperture is aimed upward. The sample Petri dish or surimi gel is then placed directly on the aperture and measured through the bottom of the dish or the gel, respectively.

In addition, the instrument can be used with a simple cuvette attachment (Figure 15.5) for measuring liquid colors. Care must be taken, however, to have the liquid sample high enough in the cuvette to fill the aperture opening.

Figure 15.5 Cuvette attachment for liquid. (*Source:* Courtesy of Konica Minolta, Instrument Systems Division.)

15.3 COLORING SURIMI SEAFOOD

Surimi seafood, as currently colored, mimics shellfish (crab, lobster, shrimp, and scallop) and salmon. Within the family of crabs, however, there are many species such as snow crab, dungeness crab, blue crab, king crab, swimming crab, rock crab, and stone crab, all requiring variation in hue, ranging from dark-red to bright blood-red to dark brown-red. Currently, the industry (worldwide) achieves the desired hue by using or creating blends with such materials as carmine, paprika oleoresin, annatto, caramel, monascus, canthaxanthin, and β-carotene (Figure 15.6). Among these, monascus is not allowed in the United States and carmine is not allowed in Japan. All the colorants are derived from natural sources except canthaxanthin and β-carotene, which are chemically synthesized. These colorants, as well as others that may function for surimi coloration, are reviewed in detail later in this chapter.

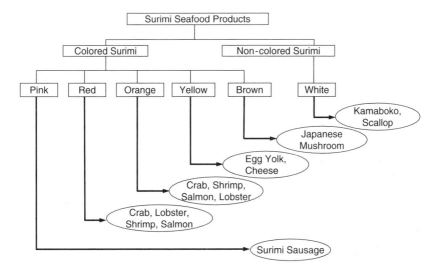

Figure 15.6 Colorants used for surimi seafood.

15.3.1 Preparation of Surimi Paste for Crabsticks

Surimi paste can be prepared by chopping frozen hydro-flaked surimi into a paste as described in Chapter 9. During preparation, salt, ice/water, starch, sodium citrate, and egg white are slowly added. A typical recipe follows:

Frozen surimi	399.2 g (high grade)
Ice water	136.1 g
Wheat starch	51.7
Modified starch	19.0 g
Salt	10.0 g
Sodium citrate	11.8
Egg white	323.9 g
Colorant	X g
TOTAL	951.7 g + X g

It is important that the temperature of the paste is maintained between 0 and 7°C. Colorant is added last and uniformly mixed.

15.3.2 Color Application to Crabsticks

The process involves the application of colored surimi paste onto an uncolored surimi base. There are two industrial methods: (1) direct application onto uncolored, cooked sheet, and (2) application to a plastic film, which then enwraps the bundled rope continuously (Figure 15.7).

After color application, the surimi stick is cut into lengths required for the specific crab being imitated, packed, and heat sterilized. Typical L*a*b* values obtained with common colorants used for surimi-based seafood are listed in Table 15.1 and graphically in L*a*b* color space (Figure 15.8).

15.3.3 General Principles

The general principles for selecting colorants for surimi seafood include:

> 1. The color must be the correct hue and be able to be standardized to meet the customary and traditional coloration of the product.

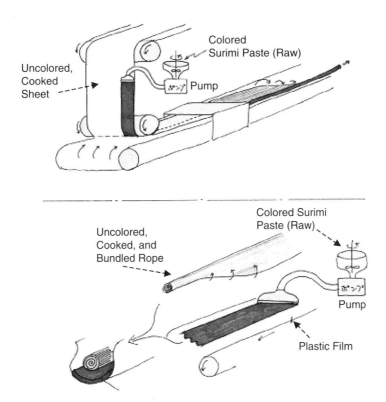

Figure 15.7 Common methods for color application in crabstick processing.

2. The color must not contribute objectionable flavor.
3. The color must be compatible with surimi and other ingredients and not affect gel formation or the texture of the final product.
4. The color must survive steam or radiant heat cooking.
5. The color must not be affected by light during storage, distribution, and display of the final product.
6. The color must not bleed or migrate into the piece during or after thermal processing.
7. The color must be in compliance with 21 CFR 70-82 for sale in the United States.

TABLE 15.1 Color Tone Data of Colored Surimi

Colorant Name	Color Tone		
	L*	a*	b*
Carmine (dosage 0.2%)[a]	38.0	43.9	3.6
Carmine WS (painted on surimi)[b]	30.0	41.6	4.2
Paprika oleoresin (dosage 0.5% 80,000 ICU)	51.6	30.3	26.2
Annatto (painted on surimi)[c]	65.2	10.1	41.0
10% Canthaxanthin (dosage: 0.25%)	50.2	35.7	24.9
1% β-Carotene (dosage: 2.5%)	67.5	7.9	41.1
Monascus (painted on surimi)[d]	31.0	40.0	13.9

[a] Carminic acid content 52%.
[b] Carminic acid content 7.5%.
[c] Norbixin content 7.0%
[d] Color value (E10%) 60.

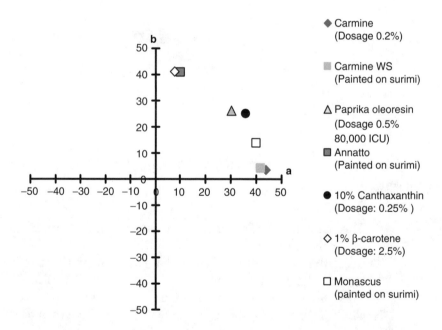

Figure 15.8 Spatial relationship of common surimi colorants: a = redness (+a*) to blueness (–a*); b* = yellowness (+b*) to greenness (–b*). Refer to Table 15.1 for concentration of colorants.

15.4 COLORANTS

Surimi seafood manufacturers have two groups of color additives to choose from when selecting colorants: (1) certified color additives and (2) color additives exempt from certification (exempt colorants). The first group currently contains seven synthetic colorants, referred to in the industry as FD & C (Food, Drug, and Cosmetic). Each batch of these colorants is required by law to be analyzed and certified by the FDA to comply with purity specifications. The second group currently contains 26 colorants that do not require FDA analysis and certification. Within in this group are colorants that are called "natural" in other countries. These "natural" colorants are reviewed in detail later in this chapter as they better fit the consumer-perceived image of "healthy, nutritious, and trendy" surimi seafood.

15.4.1 Colorants Requiring Certification

The FD&C colorants allowed as food additives are listed in Table 15.2. Because the primary colors are represented in the approved additive listing, formulators have the opportunity to blend and create almost any desired hue with FD&C colorants. Although easy to work with, the colorants have been found to behave differently in different food systems. Experience with specific surimi bases will provide confidence in predicting the final hue. Being water soluble, the formulator can expect them to bleed. They may, however, work well in surimi seafood pieces that are colored throughout. Factors specific to the system to be colored that might affect hue stability include pH, the presence of metal contaminants, ingredients (such as sugar), acidulants (such as citric or ascorbic acid), and heat and processing conditions.

The FD&Cs can also be obtained in the form of a "lake" and are usually marketed as a powder. Lakes suspend well in surimi paste but may show some bleeding. The color intensity depends on the conditions of application, including the colorant's sieve size and structure in the surimi's protein matrix. All lakes, with the exception of FD&C Red 40 Lake,

TABLE 15.2 Colorants Requiring Certification

FD&C Name	Common Name	E Number	CI Number	Chemical Class
Red 3	Erythrosine	E127	43430	Xanthine
Red 40	Allure red	E1291	6035	Azo
Yellow 5	Tartrazine	E102	19140	Pyrazolone
Yellow 6	Sunset yellow	E110	15985	Azo
Blue 1	Brilliant blue	E133	42090	Triphenylmethane
Blue 2	Indigotine	E132	73015	Indigoid
Green 3	Fast green	0000	42053	Triphenylmethane

are "provisionally" approved but any one or all can be de-listed or "permanently" listed in the future.

15.4.2 Colorants Not Requiring Certification

The ever-increasing consumer demand for "all natural" products favor natural colorants because they are perceived as being more compatible with the image of surimi seafood. Although there exists a demand for "natural" from both the consumer and the manufacturer, there is confusion as to what is a "natural" color. The industrial world appears to accept, as natural, any colorant that is extracted from agricultural, biological, and at times, mineral sources. The acceptance can be very broad or narrow, depending on the individual agenda. The consumer, on the other hand, believes that natural colors are from fruits and vegetables, and by inference, also healthy. With all of the interest in "natural," one might expect the government regulators to define "natural" but instead, the U.S. Government has opted to *not* recognize the term "natural." Therefore, in a legal sense, "natural colors" do not exist.

In lieu of an artificial/natural classification, the FDA created two categories of colorants: (1) those that require certification and (2) those that do not require certification. The latter category is often incorrectly referred to as the "natural colors," and indeed, includes those colorants that are conventionally called natural (i.e., carmine, paprika, annatto, turmeric, beet juice extract, and grape skin extract).[3]

Within the FDA's uncertified color listing (21CFR 73) there are also colorants that are synthetically derived, much like the certified colors, which behave chemically identical to the similar colorants extracted from natural sources. These are referred to as "nature-identical" and their acceptance as "natural" will depend, again, on one's perspective. Caramel is also on the FDA's uncertified color listing, and some have argued that it should not be accepted as a "natural colorant" because it is a reaction end product. Mineral colorants, such as titanium dioxide and calcium carbonate, have also been characterized as natural by some seeking to satisfy an "all natural" claim. In addition, there are other colorants on the FDA's list of "colorants not requiring certification" that may or may not be accepted as natural, depending again on one's point of view.

One can appreciate how the lack of an agreed-upon definition has led to confusion. To compound this further, the FDA views all colorants (certified or not requiring certification) as "artificial colors" for labeling purposes (unless the colorant is from the characterizing ingredient of the product). The subject area of labeling is reviewed later in this chapter.

For our purpose, the term "natural color" will remain undefined. The traditional six colorants will just be accepted as natural. In addition, other colorants falling under the FDA's generic terms of vegetable juice extracts and fruit juice extracts will be included as natural colorants and brought into the discussion as appropriate. Each colorant will be identified by its FDA citation and also its European Economics Community (EEC) number, Colour Index (CI) number and name, should the colorants be sourced from other than domestic producers.

15.4.2.1 Carmine (21 CFR 73.100, EEC No. 120, CI No. 75470, CI Natural Red 4)

The article of commerce, carmine, is a vivid purple-red powder offered in "lake" form (a lake, by definition, is insoluble in water and functions as a pigment). This colorant is of interest partly because of its current use to color surimi seafood and

Figure 15.9 Photograph of cochineal insect.

partly because of its source, the dried bodies of a small female insect (*Coccus cacti*). (Figure 15.9) The insect, found on cactus growing in the semi-arid areas of Peru, Bolivia, Chili, the Canary Islands, and Mexico, can contain from 10 to 20% of their dry weight as carminic acid, the coloring principle.

Carminic acid is extracted with hot alkalized water containing a small amount of alcohol. Carminic acid can be complexed with metal salts of an aluminum or calcium substrate, forming a precipitate that is then collected and spray-dried. The process results in a bright red powder containing 50 to 53% carminic acid and is totally free of any insect parts. The powder is stable to light, heat, oxidation, and storage.[4] The chemical structure of carmine is shown in Figure 15.10.[5,6] Carmine is used worldwide to color surimi products, except in Japan and for specific religious groups with special dietary requirements.

Carmine is insoluble in a neutral to acidic solution but can be solubilized in an alkaline solution to form a deep

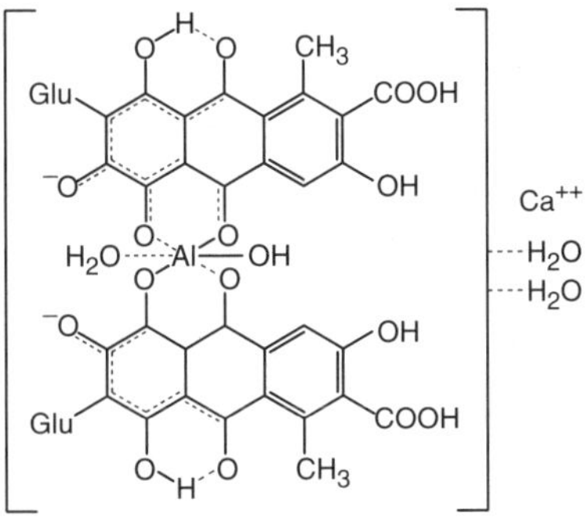

Figure 15.10 Chemical structure of carmine.

purple-red (violet) liquid. Liquid carmine is usually offered with a 3 to 7% carminic acid content, alkalized with either ammonium or potassium hydroxide. The pH can vary from 8 to 11 and therefore could affect, depending on usage level, the pH of the surimi paste. Carmine solution has an affinity for proteinaceous materials and works well in surimi seafood. When applied directly by brushing or by spraying to the surface of the surimi, it adheres well as the solvent(s) evaporates. The red coloration may have a slight blue/purple hue that is often modified by the addition of paprika oleoresin and a stabilizing emulsifier. In addition, when used as a Tentiki color, the hue is modified by the addition of yellow-orange carotenoids in a water/alcohol solution.

15.4.2.2 Cochineal Extract (21CFR 73.100, EEC No. E120, CI Number 75470, Natural Red)

The initial aqueous extract from the cochineal insects, if not processed further into carmine, is offered, per se, as cochineal extract. Tinctorially, it is a weak colorant. It has protein affinity and is therefore used to color surimi seafood. The liquid is typically acidic (pH 5.0 to 5.3) with a solids content of 1 to 5%. It is susceptible, however, to microbial contamination and may be offered with an added preservative such as sodium benzoate (which might compromise the colorant's natural status). The extract can vary in hue from orange to a purple-red, depending on pH. It is water soluble and, like carmine, has good stability to light, oxidation, and heating. The extract is allowed for use in coloring surimi seafood in the United States, Japan, and most parts of the world.

15.4.2.2.1 Application Problems with Carmine

Surimi seafood colored with a carmine colorant (powder, liquid, or cochineal extract), however, has a tendency to bleed. "Bleeding" is an undesirable condition in which a colored piece transfers its color to an uncolored piece that it is in contact with during cooking and storage. The bleed appears as a purple-pink coloration on the white uncolored piece, a condition that the consumer might mistakenly perceive as spoilage. Due to the economic importance of a no-bleed color, finding a solution to this problem is therefore highly important to the industry.

15.4.2.2.2 PGPR (Polyglycerol Polyricinoleate)

The industry has experimented with a novel process for controlling bleed and the process is also a "hot" subject among surimi seafood processors. The process involves the formation of a physical barrier to bleeding rather than chemically altering the colorant or surimi system. The most effective physical barrier between carmine and the wet surimi paste was obtained by the use of water-in-oil emulsifiers. The process required that the emulsion hold until gelling is complete. Once physically trapped, the carmine is not likely to bleed. The most functional emulsifier is the polyglycerol ester of ricinoleic acid (PGPR), a major component (at times, as high as 80 to 90%) of castor oil.

Although this emulsifier has been used on and off by color manufacturers since the inception of the crabstick business (mid-1970s in Japan and early 1980s in the United States), no GRAS (generally recognized as safe) application

TABLE 15.3 Effect of $CaCl_2$ Addition on Controlling Color Bleeding

Addition of $CaCl_2$ (%)	Bleeding Degree	L*	a*	b*	ΔE*
Control	+3	92.3	13.4	−6.8	5.6
0.075	+2	94.2	12.1	−6.0	3.2
0.15	+2	92.8	11.3	−6.1	3.8
0.225	+2	93.7	12.4	−6.2	3.8
0.3	+1	96.1	9.5	−5.6	0

Note: *ΔE: Color difference is compared with 0.3% $CaCl_2$.

of PGPR has been submitted to the U.S. FDA by the seafood manufacturing industry. Strangely, the U.S. chocolate industry is successfully using PGPR at 0.3% based on Quest's GRAS application. The FDA issued no questions on Quest's conclusion that PGPR is GRAS under the intended conditions of use.[7] The agency emphasized that it is the manufacturer's continuing responsibility to ensure that all food ingredients marketed are safe. Although it is allowed to be used for crabstick coloring in Europe, PGPR is not legal in the United States for that purpose.

Other studies indicate that bleeding can be suppressed by the addition of calcium chloride[8] to the surimi paste. The effect of calcium ion addition is shown in Table 15.3. The use of calcium salt, however, can speed up the rate of gel setting.[9] The rate of setting, on the other hand, can be slowed by lowering the processing temperature (Figure 15.11). Satisfactory results can therefore be obtained by balancing setting time against bleed.

15.4.2.2.3 Testing for Bleeding

Testing for bleed is conducted as shown in Figure 15.12. Colored surimi paste is painted onto cooked, uncolored surimi strips (the test sample) and steam-cooked for about 50 min. Each testing laboratory has developed its own test container for the steaming step. For test standardization, it is best that the container be one that can be sealed with a minimum of

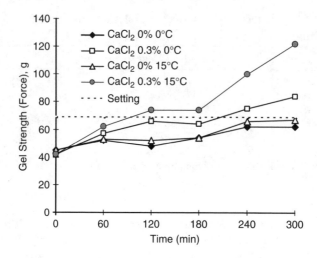

Figure 15.11 Effect of $CaCl_2$ on gel strength of colored surimi gel.

headspace above the test sample, and that no air is trapped between the test strips. After steaming, the two strips are separated and the original uncolored strip is examined for color transfer from its colored mate. The degree of color transfer (bleed) can be quantified with the use of a colorimeter. Quantification is useful when comparing results from different test variables to control bleed. A comparison of bleed test results of processed and unprocessed carmine is shown in Table 15.4.

15.4.2.3 Paprika (21CFR 73.345, EEC No. E 160c)

Paprika, as used in color applications, is an oleoresin. It is made by percolating any one or a combination of approved solvents (such as hexane, acetone, various alcohols, etc.) through dried, ground, or pellets of sweet red pepper (free of seeds) (Figure 15.13). The solvent(s) is removed by vacuum distillation. Oleoresin can also be obtained by supercritical CO_2 extraction.

The oleoresin is a viscous, dark reddish-brown liquid, pourable at room temperature, and contains a mixture of some

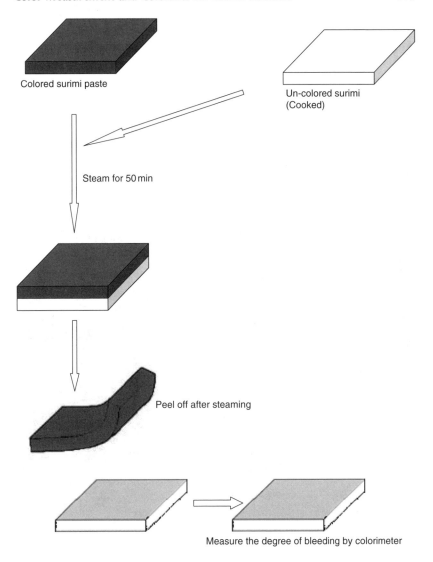

Figure 15.12 Testing diagram for color bleeding.

ten or more carotenoids (such as xanthophyls, capsanthins, and capsorubin).[10,11] Capsorubins (Figure 15.14), the red fraction, can be as high as 40 to 59% of the oleoresin. The exact hue of the oleoresin can vary, depending on growing, harvest-

TABLE 15.4 Results of Bleeding Test

Items	Bleeding Degree	L*	a*	b*	ΔE*
Uncolored surimi	±	85.6	−2.3	2.8	0
0.2% (PGPR + carmine[a,b])	+0.5	82.5	3.4	2.1	6.4
0.2% Carmine[2)]	+5.0	68.9	18.7	−3.7	27.5

Note: *ΔE: Color difference is compared with uncolored surimi.

[a] Colorant contained 52% carminic acid and unknown PGPR.
[b] Colorant contained 52% carminic acid.

Figure 15.13 Sweet pepper, Capsicum annum.

ing, drying, storage, and extraction conditions. It is standardized for color strength (not hue) with vegetable oil to a level referred to as 40,000 and 80,000 CU (although higher concentrations are available). CU values are directly proportional to the optical density of a solution of oleoresin in acetone. Some suppliers prefer to use ASTA values to designate strength. A 1000 ASTA is equivalent to 40,000 CU.

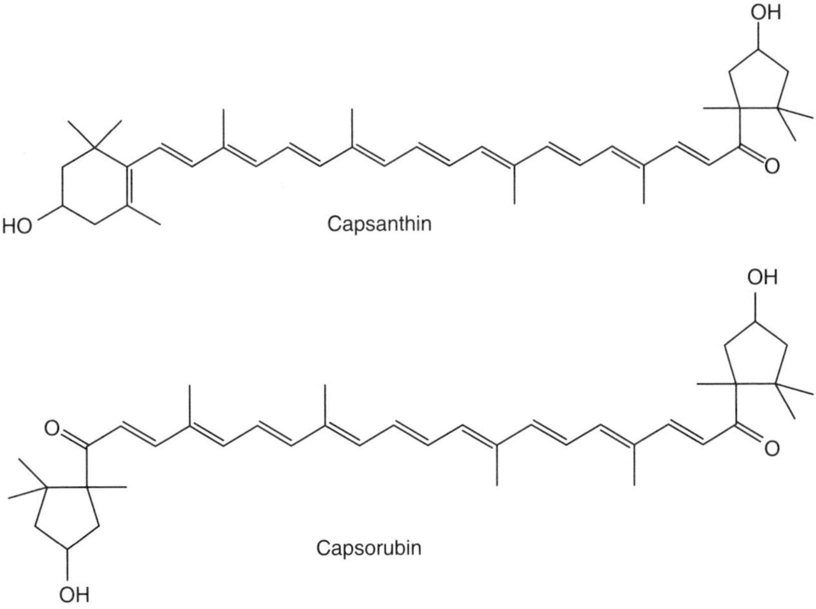

Figure 15.14 Chemical structure of paprika pigment: capsanthin and capsorubin.

15.4.2.3.1 Application of Paprika

Paprika oleoresin is insoluble in water but readily soluble in oil and therefore often diluted with oil. It will disperse in surimi paste, giving a yellow-orange to red-orange color, depending on the hue of the starting material, the level of usage, and the final pH of the surimi seafood. The oleoresin can be susceptible to oxidation and can lose color over time even when stored cold and in the dark. Addition of alpha-tacopherol, however, may help slow the process. One supplier utilizes an extract of the herb rosemary, as a natural antioxidant, to improve the shelf life of paprika.[12]

Paprika oleoresin can be made water dispersible by the addition of polysorbate, usually at a 50% or higher level. In this form, it disperses well in carmine solution to produce the final hue required in surimi for crab seafood. Because paprika can contribute flavor if used at high levels, some suppliers offer a deflavored paprika (reduced in flavor and aroma by a deodorizing distillation process). The more recent CO_2 extraction process produces odor-free oleoresin. At times, unacceptable odor in the final product has been traced to paprika and therefore taste testing of the final product as a QC protocol may be prudent.

Manufacturers in Asia and in Europe have experimented with the substitution of nature-identical carotenoids for paprika — that is, β-carotene (yellow dispersion), β-apo-8-carotenal (orange dispersion), and canthaxanthin (red-orange emulsion) — with varying degrees of success in coloring surimi-based shrimp. Paprika alone is too orange in hue to be accepted for crab, but the colorant has been used in some parts of Europe for imitation crab. Some have also worked with annatto (norbixin), but this colorant is not allowed in Europe for coloring surimi products.

15.4.2.4 Annatto (21 CFR 73.30, EEC No. E 160b, CI 75120, CI Natural Orange 4)

Annatto is pale yellow to golden yellow-orange in color (depending on concentration) when added to surimi paste. It has been used in the United States and Europe for more than 100 years, primarily as a color additive for butter and cheese. Annatto, however, is not allowed in surimi products in EU. The colorant, available as a liquid, powder, or emulsion, and at varying concentrations or mixtures, is essentially of two forms: (1) bixin (oil soluble form) and (2) norbixin (water-soluble form). The colorant is extracted from the seed coat of a fast-growing tropical shrub/tree, *Bixa orellano* (Figure 15.15), typically found in parts of South America, India, and East Africa.[13] A maturing plant produces capsular fruit containing hundreds of seeds with a resinous rust-colored coating, the source material for this colorant. There is no standard method for harvesting or drying the seeds prior to extraction and this has introduced variability.

15.4.2.4.1 Bixin

To produce the bixin form, the colorant is extracted using one or more approved food-grade solvents: ethanol, a chlorinated

Figure 15.15 Photographs of annatto.

hydrocarbon, or a vegetable oil.[14] Bixin is commercially available in three forms: (1) as a liquid (bixin content varying from 0.1 to 0.3% in vegetable oil), (2) a suspension (bixin content of 3.0 to 7.5% in vegetable oil), or (3) as an emulsion of varying concentrations of bixin and norbixin.

When using bixin in surimi paste, keep in mind the potential adverse effect on paste rheology by the varying oil content of the colorant. Incorporation of the oleoresin during

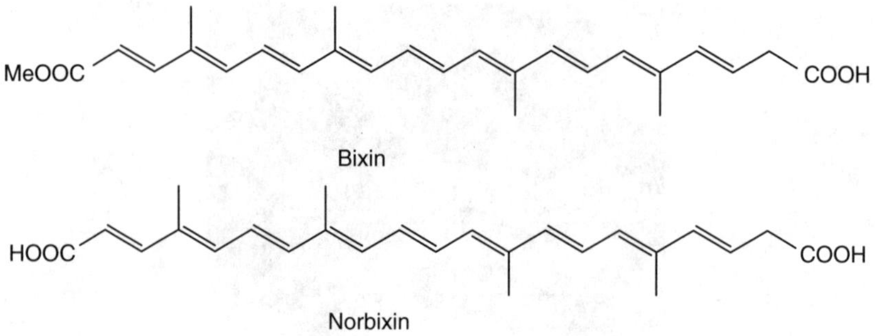

Figure 15.16 Chemical structure of annatto: bixin and norbixin.

the chopping step appears to be best, as a loose emulsion is formed and held during gelation. The colorant is pH sensitive and moves from yellow-orange to a red shade with decreasing pH. The pH shift does not affect color stability, which is very good when the surimi is stored in a dark and frozen state. Light instability, however, should be expected, as bixin is a carotenoid and shows fading upon light exposure, depending on intensity and time. The chemical structure of annatto was determined by Karrer et al.[15] and is shown in Figure 15.16.

15.4.2.4.2 Norbixin

The water-soluble form is produced by extracting the seed coats with alkalized propylene glycol and/or water. The colorant is offered commercially as a liquid (in varying percentages as high as 5% norbixin in KOH solution) or as a powder (varying from 7 to 15% norbixin on a suitable spray-dried carrier).

The norbixin solutions, at concentrations developed for the cheese industry, may be a good vehicle for surimi coloration. These are offered as a single strength (1.25 to 1.4%), double strength (2.5 to 2.8%), and triple strength (3.0 to 3.8%) product and are usually standardized by the supplier for tinctorial strength.

Norbixin solutions have been known to precipitate in products high in calcium ions, high in acidity, or held at frozen temperatures. An acid-stable form, however, is commercially available. Norbixin solution can be spray dried with maltodextrin to form a free-flowing powder, but in this form, the colorant is prone to oxidation with loss in color value over time. Color loss of the powder can be slowed, however, by storage below 35°F.

Norbixin is known to react with protein and form a strong bond. Dilute solutions sprayed onto surimi paste prior to cooking adhere well with minimal penetration, will not fade over a 3-month period in frozen storage, and will not bleed.

Annatto can be blended into carmine solutions to produce a very acceptable color blend for surimi seafood. The blend, however, may bleed due to the presence of carmine.

15.4.2.5 Turmeric (21CFR 73.600, EEC No. E 100, CI No. 75300, CI Natural Yellow 3)

Turmeric produces a bright fluorescent yellow color very close in hue to FD&C Yellow No. 5. The colorant, sometimes called curcuma, is from a tuberous root of a plant *(Curcuma longa)* native to Southern Asia, with India being the major grower. The fingers (extensions of the mother root) are collected, cleaned, boiled in alkali, dried, polished, and subsequently ground (turmeric powder) or solvent extracted (turmeric oleoresin). The principal coloring matter in turmeric powder or its oleoresin is curcumin, and it is this content that forms the basis of purchase specifications.

Turmeric powder is deep orange-yellow, usually of 90 to 95% purity. The powder is insoluble in water but can be made water soluble by the addition of food-grade emulsifiers such as propylene glycol or polysorbates. The emulsified form (usually containing 20 to 25% curcumin) can be used to color surimi paste a bright yellow. The colorant is very strong tinctorially, about four times that of a typical FD&C dye, and usage in the range of 5 to 20 ppm is sufficient to color surimi. Turmeric powder is also soluble in alcohol and in this form can be used in the tentiki method.

Turmeric oleoresin is produced by extraction with ethanol (which is subsequently removed by vacuum distillation).

The oleoresin is commonly standardized to contain 5 to 15% curcumin. It is not water soluble; however, it can be mixed into surimi paste to produce a bright deep yellow to a pale yellow coloration, depending on usage level and the final pH of the paste.

Emulsified turmeric produces an attractive yellow hue but is limited in usage because of its instability toward light. It has good thermal stability, slightly affected by pH, and if protected from light, will have minimal color loss. In surimi seafood, turmeric does not bleed (probably because it complexes with the surimi proteins) and is highly resistant to oxidation. It is therefore a good colorant choice if the end product can be protected from light.

15.4.2.6 Grape Color (21CFR 73.169, EEC No, E163) and Other Anthocyanins

The FDA has made a usage distinction between grape color extract and grape skin extract, the former allowed to color non-beverage foods while the latter is limited to beverages. Therefore, only grape color extract can be legally considered a possible colorant for surimi.

Grape color extract is prepared by extracting pigments from precipitated lees, the sediment found at the bottom of vessels storing Concord grape juice. The aqueous extract is really a "soup" of various anthocyanins, tartrates, sugars, malates, minerals, salts, etc. Grape color extract is available in liquid form (containing about 1.5% anthocyanin) and powder form (containing about 4% anthocyanins). The extract is purchased not on anthocyanin content but on color strength (absorbency/gram). Variability in the commercial colorant can be a function of variety, source, field conditions, and method of extraction.

Grape color extract is soluble in water, and depending on the pH, will be red, blue-red, purple, steel blue, or gray-green to a dull yellow as pH increases from 1 to 13. The pH not only affects color, but also color intensity and stability, each optimized at pH 1 to 3.5. The extract is sensitive to heating and may degrade to a brown precipitant, depending on the temperature, time, and pH of the system. In the laboratory, surimi paste colored with grape color extract survived both pasteurization temperatures and 30 min of steam cooking. The colorant survived 200 hr of light exposure and may have survived longer. Although grape color extract bleeds, it could be successfully applied in block surimi.

It was recently discovered that doubly acylated anthocyanin mixtures exhibited improved stability toward light, heat, and oxidation over the nonacylated or monoacylated anthocyanins. A search for plant material high in doubly acylated anthocyanins consequently led to the identification and commercialization of extracts from new grape varieties,[16] cabbage, carrot, sweet potatoes, radish, and a number of berry fruits. These may function as well as grape or better in surimi if a red/purple/blue hue is sought and bleeding is not considered a problem. Other sources of anthocyanins being researched and maybe approved for use in the future are roselle (hibiscus), cherry plum, cranberry pomace, blue/blackberry, miracle fruit, aronia, various flowers, and purple corn.

15.4.2.7 Beet Juice Concentrate (21CFR 73.260)

This is a bright red-purple color in surimi paste. Beet juice concentrate is available as a liquid or powder. The starting material is the common table beet *(Beta vulgaris ruba)*. Beet root contains both red (betacyanins) and yellow pigments (betaxanthis), collectively known as betalains. The red fraction, called betanin, can be as high as 75 to 90% of the extract. In traditional processing, the beets are blanched, cut, hydraulically pressed (or centrifuged), filtered, and vacuum concentrated to 40 to 65% total solids. The powdered form is made by spray-drying the expressed liquid with maltodextrin. Both forms may contain added stabilizers such as ascorbic or citric acid.

Beet juice extract is soluble in water and provides an intense color, the actual hue depending on the pH of the system. At pH 4 to 7, it is a bright bluish-red and probably the closest match to FD&C Red No. 2. At higher pH, the hue turns to a blue-violet. The extract is most often sold on the basis of betanin content, which can vary from 0.3 to 1% for the liquid, and about half as much for the powder. Beet juice extract easily colors surimi paste, producing a light pink to a bright cherry-red color, depending on the amount of color used (preferably at pH 4). It could substitute for carmine in certain products but it will bleed. Beet juice is known to be sensitive to heat processing, but in surimi it appears to survive steam heating for 30 min. Some color regeneration is seen following heating; therefore, a controlled rest time should be allowed before taking color measurements. It is also known that betanin is susceptible to atmospheric oxidation[17] and the presence of cations, such as sodium, iron, and copper (normal constituents of surimi), will speed up oxidation. Beet colorant can also degrade at increased water activity (a_w), in acid pH and under prolonged exposure to light and air, which limits the use of beet concentrate as a red colorant in surimi seafood.

15.4.2.8 Caramel (21 CFR73.85, EEC No. 150)

Caramel is a water-soluble dark-brown liquid (or powder) made by controlled heating of any number of food-grade carbohydrates, usually with or without acids or alkali, in the presence of sulfite compounds, ammonium compounds, or with no added catalyst. Not much is known about the nature of caramelization; so, simply said, caramel is burnt sugar. In aqueous solution, caramel coloring exhibits colloidal properties, with the particles carrying a small positive or negative charge, depending on the method of manufacture and the pH of the system to be colored. It is important that the particle charge be the same as the system it is to color or mutual particle attraction will occur, causing flocculation. This, however, does not appear to hold true for surimi seafood, as all caramels appear to color the product without adverse effects. In addition, the color has very good stability with minimal hue shift as the pH changes. It is not affected by heat, light, or oxidation. It is tinctorially weak, however, and often requires a high usage level for coloration, raising the potential of contributing to an unwanted burnt flavor. It will bleed, and therefore care should be taken when using it to darken carmine-paprika blends for surimi seafood. Caramel is approved for use in most countries.

15.4.3 Monascus Colorants

Monascus colorant (Angkak or Chinese red rice) has been used for more than a hundred years as a food color in Malaysia and China. It was the color of choice in Japan from the inception of surimi-based crabstick production called *kani kama*. This colorant is produced by solid or liquid mold fermentation of bread crumbs or rice grains. The colorant consists of a mixture of red, yellow, orange, and purple pigments.[18] Like a carmine solution, the monascus colorant has a strong affinity for proteins but its poor light stability limits its usefulness in seafood manufacturing. When monascus-colored surimi products are placed in supermarket showcases, the color fades within a week of exposure to fluorescent lighting. Monascus is not allowed as a food colorant in either the United States or Europe.

15.4.4 Nature Identical Colorants

Within the FDA's uncertified color listing (21CFR 73) there are also colorants that are synthetically derived, much like the certified colors, but behave chemically identical to similar colorants extracted from natural sources. These are referred to as "nature-identical" and their acceptance as "natural" will depend upon one's own definition of natural. Of these, two carotenoids have been used in coloration of surimi seafood (i.e., canthaxanthin and β-carotene).

15.4.4.1 Canthaxanthin (21 CFR 73.73.75, EEC No. E161g)

Canthaxanthin exists widely in nature. It was first discovered in edible mushrooms (*Cantharellus cinnabarinus*) by Haxo in 1950.[19] Today, however, the canthaxanthin commercially available as a colorant is produced by chemical synthesis. Interestingly, modern laboratory analysis techniques cannot distinguish between the synthetic product and that extracted from nature. It is insoluble in water and only very slightly soluble in oil. Several commercial forms are available and have been used as a substitute for paprika in carmine-based blends for surimi crabstick. As with all carotenoids, it is degraded by light, heat, and oxidation.

15.4.4.2 β-Carotene (21 CFR 73.95)

β-Carotene is very similar in functional properties to canthaxanthin. This carotenoid, as a colorant, is also chemically synthesized. Recently, β-carotene was commercially extracted from seaweed (*Dunaliella carotene*) and palm oil (palm carotene) at a significant up-charge. Chemical analysis indicates that naturally derived β-carotene contains *alpha* in addition to the *all-trans* and *cis* forms. As a colorant, β-carotene has been used as a substitute for paprika in carmine blends.

15.4.5 Other Naturally Derived Colorants

At times, some manufacturers have resorted to the use of "whiteners" such as titanium dioxide and calcium carbonate to improve the whiteness of surimi seafood products. Their effects on color values of surimi seafood are shown in Figures 13.24 and 13.30 of Chapter 13.

15.4.5.1 Titanium Dioxide (E171, CI No. 77891, CI Pigment White 6)

Titanium dioxide (TiO_2) is an approved colorant with an end-use restriction of 1% (wt/wt) maximum in the finished product. Although TiO_2 exists in nature, usually in three crystalline forms, only synthetically prepared pigment can be used as a color additive. Some manufacturers have taken the posture that this pigment is nature-identical.

15.4.5.2 Calcium Carbonate (EE 170, CI No. 77220, CI pigment White 18)

Calcium carbonate ($CaCO_3$) is also a synthetically prepared powder composed primarily of precipitated $CaCO_3$. This pigment has considerably less whitening strength compared to TiO_2 and there is always the risk that carbon dioxide gas could be formed.

15.4.5.3 Vegetable Oil

Vegetable oil has been used to improve the perceived whiteness of surimi paste, possibly through the light scattering effect of the paste/oil emulsion.

15.5 COLOR QUALITY

15.5.1 Final Product Color

The methodology of color measurement and color language has been reviewed and can be applied to monitor final product color from production, exiting storage, and into distribution. In addition to the issue of bleed, some chromophores, such as the family of carotenoids, are sensitive to light exposure and can fade to an unacceptable color level if not handled correctly during distribution. Table 15.5 depicts the relative stability to light of the common colorants utilized in surimi.

15.5.2 Colorant Quality

The specifications provided by colorant suppliers are often used to establish internal specifications. Some colorants, for example, are purchased on the basis of concentration, although hue is probably more important. Analytical confirmation of concentration assures purchase value, but confirmation of hue by color difference assures controlled and standardized production. Other factors, such as specific gravity, total solids, lead, arsenic, mercury, and ash, may be specified but these appear to be generally ignored unless a dispute over quality arises.

15.5.3 Acceptance Criteria

Some manufacturers rely solely on visual observation for acceptance of incoming natural color ingredients and the subsequent final product. For visual comparison, incoming liquid

TABLE 15.5 Stability of Colorants against Light over Time

Colorants	Day 0			Day 1				Day 3				Day 7			
	L*	a*	b*	L*	a*	b*	Δ*	L*	a*	b*	Δ*	L*	a*	b*	Δ*
Carmine	37.5	44.7	4.0	38.3	44.5	3.0	1.3	37.9	45.3	3.7	0.8	38.3	42.9	2.9	2.3
Carmine WS	28.9	47.4	4.8	28.7	49.0	4.4	1.7	28.6	49.2	4.2	1.9	27.9	48.4	4.3	1.5
Paprika color	50.6	31.4	26.2	50.6	31.9	25.9	0.6	50.8	32.0	25.9	0.7	50.1	32.5	25.4	1.4
Canthaxanthin	50.9	35.3	25.0	49.6	36.5	24.5	1.8	50.0	36.8	24.4	1.8	49.0	37.7	24.1	3.2
β-Carotene	68.8	6.6	40.7	68.1	8.1	40.7	1.7	67.5	9.7	41.5	3.5	67.9	8.8	40.8	2.4
Annatto WS	66.1	8.6	41.0	66.3	8.8	41.2	0.3	66.3	8.9	41.3	0.5	66.9	7.5	40.6	1.4
Monascus	30.8	40.9	14.3	34.8	39.7	16.6	4.8	36.5	38.8	17.8	7.0	39.1	36.5	19.1	10.5

Note: Refer to Table 15.1 for concentration and dosage of colorants. Light: fluorescent lamp (2500 Lux), storage temp.: 5°C L: lightness, a: redness, b: yellowness, ΔE: color difference color is diluted, or if a powder, dissolved in an appropriate diluent, and placed in one of two similar color comparison tubes (Nessler). A stored standard is identically handled and placed into the second color comparison tube. The two tubes are compared side by side in appropriate lighting. Visual comparison depends on the eyes of skilled personnel to determine whether or not a product is within the acceptance range. This procedure is workable only if the acceptance range is quite broad. In the real world, such evaluation can be affected by the experience and age of the evaluator, time of day, number of samples evaluated, and the viewing conditions. The method also lacks the means to determine the exact concentration difference and limits the ability to adjust for the variance. The visual method is often replaced by tristimulus or spectrophotometric difference measurements, and at times, specific analytical methodology.

Tristimulus difference measurements record differences between the target sample color and the test sample color. The colorimeter expresses the color difference as a single numerical value E*ab. This value indicates the size of the difference, but not how the colors differ. One must therefore examine the specific L*, a*, and b* values of both colors to determine the exact point of each in color space. The spectral reflectance graph and/or the color difference graph displayed by the Minolta CM-2002 spectrophotometer are a considerable aid in this evaluation. The establishment of elliptical tolerance and its display provide the magnitude and direction of the difference. Tolerance is discussed later.

Natural colors are rarely pure. As extractives from natural sources, some contain a "soup" of chromophores (i.e., paprika), while others may contain soluble components extracted along with the chromophore (i.e., carmine). These "tag-along" components can vary along with the chromophore because of environmental and handling conditions and can therefore confuse color measurement. Their presence, however is often ignored.

Manufacturers rely on spectrophotometry to confirm the presence and concentration of chromophores. Concentration is often the basis of sale and a measure of purity in the eyes of

TABLE 15.6 Absorptivity Values of Colorants

Colorant	Sample Size	Solvent System	Maximum Wavelength	Absorption (in L/g-cm)
Bixin	0.1	Chloroform	503 nm	282.6
Norbixin	0.1	2–5% KOH	480	287
Carminic acid	0.1	HCl	494	139
Turmeric	0.1	Acetic acid	540	Compared to Std.

the buyer. Although important, absolute concentration is not the same as applied concentration and these may not be directly related. Two identical colorants of equivalent concentration may present visual color differences in the final product. Applied concentration is important to production as a few percent less or more colorant in a product preparation, to achieve product standard, could mean a significant cost saving or loss.

Absolute concentration, determined by spectrophotometry, is better, faster, and more specific than the classical wet methods. Spectrophotometric measurement first requires the solubilization of the color sample in an appropriate solvent, diluting it to a readable range, and measuring its absorbency at its predetermined maximum absorption wavelength vs. that of a solvent blank. The general formula [% concentration = $A \times 100/(a \times b \times c)$] is then used to establish sample concentration where A is the absorbency reading blank corrected, of the color sample; a is the color's absorptivity (Table 15.6); b is the absorption cell's path length (in centimeters); c is the concentration (in L/g-cm) of the sample solution; and 100 is the percent conversion factor.

The absorptivity values from published articles are considered close approximations of their true values and are used by industry and buyers alike throughout the world. The details of each analytical method are detailed in Marion[20] (pp. 251 to 257) and are not repeated here.

15.5.4 Acceptance Tolerance

Acceptance tolerance establishes how different a color sample can be from the target color and still be acceptable. Typically, acceptable samples differing in lightness are closer to the target, while samples differing in saturation are further away from the target. Said another way, "tolerance" is the limit or boundary around a range of acceptable colors in terms of their difference from the target color.

Tolerance can be expressed in a number of ways in relation to color space, such as ΔE (with various subsets denoting the color space referenced). In the past, and in some companies today, the "box tolerance" system is used, wherein the positive and negative limits are placed along each axis in the color space. The limit points are connected to form a rectangle or box around the target. The assumption is that all sample values falling within the rectangle are acceptable and those outside are not. In the real world, visual assessment of a sample is often not in agreement with the limits of the "box," especially when the sample values approach the corners of the box. It became evident that if the corners were rounded so as to change the box into an ellipse, a family of samples falling within the ellipse (each member differing in some color parameter) matched the visual acceptable sample range selected by consumer/expert panels.

Elliptical tolerance more closely matches visual perception of color difference and thus can be used with confidence for establishing acceptance or rejection of a sample relative to a target sample. In the "box tolerance" system, differences were seen as a straight-line distance from the target point. With elliptical tolerance, the color difference is seen as a combination of differences along each axis. The amount of difference on one axis relates to the amount of difference on the other two axes. Mathematical formulas were then developed to calculate the ellipse around the target point, the shape, size, and direction of the ellipse, depending on the acceptance parameters selected.

The easiest way to establish elliptical tolerance, without having to resort to statistical computations, is to use software provided by Minolta that can automatically set elliptical tolerance parameters. The computer-aided technique starts with defining, by use of a human panel, what is and is not acceptable relative to the target color. At least 30 samples representing color differences in each direction are measured. These can come from production or be made deliberately by varying formula components. The data are analyzed for "best fit" by the program, which then calculates the most suitable tolerance ellipse for the set of sample values presented. This calculation should not be seen as static as it can be updated with new sample measurements, and the size, shape, and direction can be manually modified to fine-tune the tolerance ellipse. The tolerance ellipse, along with sample and target points, can be displayed on the computer monitor for a clear and meaningful visualization of sample point distance from the limits established by the ellipse.

15.6 LABELING

15.6.1 Requirements in the United States

Labeling laws are often amended or changed either by court action, legislative action, or regulatory agency action and therefore this discussion may not be timely. It is recommended that legal counsel be consulted before proceeding with labeling involving the declaration of natural color(s) in an ingredient statement. This discussion will apply to all colorants intended for use in the United States, regardless of the country of origin.

The Guide to U.S. Food Labeling Law (August 1991, Tab 400, Pg. 19) reported that "on January 3, 1978, former FDA Commissioner Donald Kennedy wrote to the 100 largest manufacturers and distributors of packaged foods requesting that they voluntarily label all the specific colors used in their foods. There remains no legal requirement, however, for labeling individual uncertified colors in food." The FDA still encourages firms to voluntarily declare colors, spices, and flavorings when they are added to foods and "supports declaration of ingredients on food labels to the extent that it is practical." (56 FR 28592, Proposed Rule B.2). In this Rule, the FDA indicated it would prefer, as part of the ingredient list, "terms such as 'colored with _____' and '_____(color),'" as an example of appropriate terms for the voluntary declaration of noncertified colors when the blank is completed with the utilized colorants listed in 21 CFR, Part 73.[20]

The Nutrition Labeling and Education Act of 1990 did not change the FDA's position, but it did allow for the natural colorants to be listed generically, that is, without naming the specific colorant (21 CFR, Part 101.22(k)(2)) in place of individual listing. This section states that color additives not subject to certification may be declared as "Artificial Color," "Artificial Color Added," or "Color Added" (or by an equally informative term that makes clear that a color additive has been used in the food). Alternatively, such color additives may be declared as "Colored with _____" or "_____ color," the blanks to be filled in with the color additive listed in the applicable regulation in Part 73.

Some manufacturers want to maximize the value-added positioning obtained by the use of natural colors and have opted to use the word "natural" as a descriptor in the ingredient statement. When this is done, the specific colorant(s) is (are) always declared in descending order, followed by the parenthetical notation (natural colors) or (a natural color). This might highlight the natural colorants to the consumer, but is not an accepted legal posture. Do not assume that the FDA's silence over the past 10 to 15 years of this practice to mean concurrence.

In January 1993 (58 FR 2302-2407), the FDA stated it would not undertake rule-making to establish a definition for "natural" and tabled the activity to some future date. Seven months later (58 FR 44059-44061), the FDA stated that the term "natural" will not be restricted in use except for added color, synthetic substances, and flavors. The agency viewed "natural" as meaning that nothing artificial or synthetic has been added to food.

15.6.2 Religious Requirements

Surimi seafood is an ideal food for groups who have religious reasons for not eating specific meats and require strict adherence to certain dietary practices. Surimi preparation can be controlled to render it fit for their use. Orthodox

Judaism may be the largest group in the United States with a recognizable special requirement (Kosher), but there are many other groups having special requirements. Carmine is designated as "not Kosher" and cannot be used to color surimi for distribution to that market segment. Most grape colors on the market are also "not Kosher," although this colorant can be made Kosher if the market could sustain the added cost of rabbinical supervision during its manufacture. Most other colorants can be obtained Kosher. In this specialized market, respect, concern, and diligence are paramount and must be reflected in the company's quality control program.

15.7 SUMMARY

Surimi (seafood or otherwise) is a versatile system that can be shaped, colored, and flavored into exciting new products. These products can be exploited not only for their nutritional aspects, but also for their controlled texture, flavor, and color. New surimi seafood products may require the use of natural colors beyond the traditional crab/lobster red, colorants that will play a prominent role in acceptance. Hopefully, the information provided in this chapter will do more than act as a guide to natural color application but be a stimulus for creative new surimi-based products.

REFERENCES

1. F.J. Francis and F.M. Clydesdale. *Food Colorimetry Theory and Applications*. Westport, CN: Avi, 1975.
2. Konica Minolta, Precise Color Communication: Color Control from Perception to Instrumentation. *Handbook: The Essentials of Imaging*, Konica Minolta Instrument Division, Japan, 1998.
3. G.J. Lauro. A primer on natural colors. *Cereal Foods World*, 36(11), 949–953, 1991.
4. A.G. Lloyd. Extraction and chemistry of cochineal. *Food Chemistry*, 5, 91–107, 1980.

5. E.G. Kiel and P.M. Heertjies. Metal complexes of alizarian. The structure of the calcium-aluminum lake of alizarin. *J. Soc. Dyers Col.*, 79, 21–27, 1963.

6. E.G. Kiel and P.M. Heertjies. Metal complexes of alizarian. The structure of some metal complexes of aliozarin other than Turkey Red. *J. Soc. Dyers Col.*, 79, 6164, 1963.

7. U.S. FDA. Agency response letter GRAS notice No. GRN 000009. http://vm.cfsan.fda.gov/~rdb/opa-g009.html, 1999.

8. Y. Kanmuri, M. and Hashino, N. Yokota. Japan Patent H06-61237, 1994.

9. H. Saeki, T. Shoji, F. Hirata, M. Nonaka, and K. Arai. Effect of calcium chloride on forming ability and cross-linking reaction of myosin heavy chain in salt-ground meat of skipjack tuna, carp and walleye Pollack. *Nippon Suisan Gakkaishi*, 58, 2137–2146, 1992.

10. M.I. Minguez-Mosquera and D. Hornero-Mendez. Formation and transformation of pigments during the fruit ripening of *Capsicum Annuun* Cv. *Bola* and *Agridulce*. *J. Agric. Chem.*, 42, 38–44, 1994.

11. M.I. Minguez-Mosquera, M. Jaren-Galan, and J. Garrido-Fernandez. Color quality in paprika. *J. Agric. Chem.*, 40, 2384–2388, 1994.

12. Kalsec Inc., P.O. Box 50511, Kalamazoo, MI (www.kalsec.com).

13. H.D. Preston and M.D. Richard. Extraction and chemistry of annatto. *Food Chem.*, 5, 47–56, 1980.

14. T.N. Murthi, V.D. Devdhara, J.S. Punjrath, and R.P. Aneja. Extraction of annatto colors from the seeds of *bixa orellana* using edible oils. *Ind. J. Dairy Sci.*, 42(4), 750–756, 1989.

15. P. Karrer, A. Helfenstein, R. Widmer, and T.B.V Itallie. Ueber bixin. *Helv. Chim. Acta*, 12, 741–756, 1929.

16. Mega Natural Grape Color, Canandaigua Wine Co., P.O. Box 99, Madera, CA.

17. E. Attoe and J.H. von Elbe. Oxygen involvement in betanin degradation. *Z.Lebensmittee-Untersuchung and Forsch.*, 179(3), 232–236, 1984.

18. P.J. Black, MO. Loret, A.L. Santerre, A. Pareilleux, D. Prome, J.C. Prome, J.P. Lausac, and G. Goma. Pigments of monascus. *J. Food Sci.*, 59, 862–865, 1994.

19. F. Haxo. Carotenoids in the mushroom *Cantharellus cinnabarinus*. *Botan. Gazz.*, 112, 228–232, 1950.

20. D.M. Marmion. *Handbook of U.S. Colorants: Food, Drugs, Cosmetics, and Medical Devices, 3rd ed.* New York: John Wiley & Sons, 1991, 251–257.

Additional Reading

Should anyone want to read further, the following listing is provided.

Color Measurement

M.V. Halsted. Colour rendering — past, present and future. *Proceedings of the 3rd Congress of the International Color Association, Renssalaer.* Bristol: Hilger, 1997, 97–127.

J.B. Hutchings. *Food Colour and Appearance.* London: Blackie Academic and Professional, 1994.

K. Loughrey. The Measurement of Color. In G.J. Lauro and F.J. Francis, Eds. *Natural Food Colorants, Science and Technology.* New York: Marcel Dekker, 2000, 273–288.

Analytical

Food Additive Analytical Manual, Vol 1. Washington, D.C.: U.S. Dept of Health, Education and Welfare, Food and Drug Administration, 1967.

Food Chemical Codex, 3rd ed. Washington, D.C.: National Academy Press, 1981.

T. Maitani, H. Kubota, and T. Yamada. *Food Additives and Contaminants,* 13(8), 1001–1008, 1996.

D.M. Marmion. *Handbook of U.S. Colorants: Food, Drugs, Cosmetics, and Medical Devices, 3rd ed.* New York: John Wiley & Sons, 1991.

JECFA, *Compendium of Food Additives Specification,* 1, 345–355, 1992.

Natural Colors

J.M. Dalzell. *LFRA Ingredients Handbook. Food Colours.* Surrey, U.K.: Letherhead Food RA, 1997.

P.R. Freund, C.J. Washam, and M. Maggion. Natural colors for use in foods. *Cereal Foods World,* 33(7), 171–173, 1988.

Natural Food Colourants 2nd ed.. G.A.F. Hendry, J.D. Houghton, Eds. Glasgow: Blackie Academic and Professional, 1996.

J. Noonan. Color additives in foods. In T.E. Furia, Ed. *Handbook of Food Additives.* Cleveland, OH: CRC Press, 1972, 587–615.

Carmine

F.L.C. Barenyovits. Cochineal carmine: an ancient dye with a modern role. *Endeavour,* 2(2), 85–92, 1978.

V. Flores-Flores and A. Tekelenburg. Dacti Dye Production, FAO Plant Production and Protection Paper (No. 132). Ago-ecology, cultivation and use of cactus pear, 1995, 167–185.

J. Schul. Carmine. In G.J. Lauro and F.J. Francis, Eds. *Natural Food Colorants, Science and Technology.* New York: Marcel Dekker, 2000, 1–10.

Paprika

J.G. Lease and E.J. Lease. Factors affecting the retention of red color in pepper. *Food Technol.,* 10(8), 368, 1956.

J.B. Moster, A.N. Trater. Color of capsicum species. IV Oleoresin paprika. *Food Technol.,* 11(4), 226, 1957.

C. Locey, J.A. Guzinski. Paprika. In G.J. Lauro and F.J. Francis, Eds. *Natural Food Colorants, Science and Technology.* New York: Marcel Dekker, 2000, 97–114.

Annatto

K.D. Aparnathi, R. Lata, and R.S. Sharma. Annatto — its cultivation, preparation and usage. *Int. J. Tropical Agric.,* 8(1), 80–86, 1990.

P. Collin. The role of annatto in food coloring. *Food Ingredients Processing, Int.,* February, 23–27, 1992.

L.W. Levy and D.M. Rivadeneiro. Annotto. In G.J. Lauro and F.J. Francis, Eds. *Natural Food Colorants, Science and Technology.* New York: Marcel Dekker, 2000, 115–152.

Turmeric

V.S. Govindarajan. Tumeric — chemistry, technology and quality. *Crit. Rev. Food Sci. Nutr.,* 12(3), 199–301, 1980.

H.H. Tonnersen. Evaluation of curcuma product by the use of standardized reference color values. *Z. Lebensmittee-Untersuchung Forsch.,* 194(2), 129–130, 1992.

Natural Food Colourants, 2nd ed. G.A.F. Hendry, JD. Houghton, Eds. Glasgow: Blackie Academic and Professional, 1996, 68–71

R. Buescher and L. Yang. Turmeric. In G.J. Lauro, F.J. Francis, Eds. *Natural Food Colorants, Science and Technology.* New York: Marcel Dekker, 2000, 205–226.

Grape Color

F.J. Francis. Anthocyanins as Food Colorants. *The World of Ingredients,* August 16–18, 1996.

R.L. Jackman. Anthocyanins as food colorants: A review. *J. Food Biochem.,* 11, 201–47, 1987.

cf. Timberlake. Anthocyanins: occurrence, extraction and chemistry. *Food Chemistry (UK),* 5, 69–80, 1980.

R. Wrolstad. Anthocyanins. In G.J. Lauro and F.J. Francis, Eds. *Natural Food Colorants, Science and Technology.* New York: Marcel Dekker, 2000, 237–252.

Beet Juice Extract

M. Drdak, M. Vallova, G. Greig, P. Sinko, and P. Kusy. Influence of water activity on the stability of betanin. *Z. Lebensmittee-Untersuchung Forsch.,* 190(2), 121–122, 1990.

T.J. Mabry. Betalains. *Encycl. Plant Physiol.,* 8, 513–533, 1979.

D. Megard. Stability of red beet pigments for use as food colorant: a review. *Food and Food Ingredients J.,* 158, 130–150, 1993.

G. Patkai and J. Borta. Decomposition of betacyanins and betaxanthins by heat and pH changes. *Nahrung,* 40(5), 267–270, 1996.

I. Saguy, I.J. Kopelman, and S.J. Mizraki. Computer aided determination of beet pigments. *Food Sci.,* 43, 124–127, 1978.

J.H. von Elbe, I.Y. Maing, and C.H. Amundson. Color stability of betanin. *J. Food Sci.,* 39(2), 334–337, 1974.

J.H. von Elbe and I.L. Goldman, The betanins. In G.J. Lauro and F.J. Francis, Eds. *Natural Food Colorants, Science and Technology.* New York: Marcel Dekker, 2000, 11–30.

Caramel Color

W. Kamuf, A. Nixon, and O. Parker. Caramel color. In G.J. Lauro and F.J. Francis, Eds. *Natural Food Colorants, Science and Technology.* New York: Marcel Dekker, 2000, 97–114.

Monascus

R.E. Mudgett. Monascus. In G.J. Lauro and F.J. Francis, Eds. *Natural Food Colorants, Science and Technology.* New York: Marcel Dekker, 2000, 97–114.

Labeling

J.B. Hollagan. Safety assessment and regulatory issues for natural colors and flavors: global challenges. Speech No. 13. *Intertech Conference on Natural Colors and Flavors,* Washington, D.C., 1998.

U.S. Code of Federal Regulations, 21 CFR 73, Listing of Color Additives Exempt from Certification, Subpart A: Foods, 1987.

B. Henry. Regulations in Europe and Japan. In G.J. Lauro and F.J. Francis, Eds. *Natural Food Colorants, Science and Technology.* New York: Marcel Dekker, 2000, 315–327.

16

Application of Sensory Science to Surimi Seafood

JEAN-MARC SIEFFERMANN, PH.D.

Laboratoire de Perception Sensorielle et Sensométrie,
Massy, France

CONTENTS

16.1 Introduction 805
    16.1.1 What Is Sensory Evaluation? 805
    16.1.2 Why Should We Care about Sensory
           Evaluation? 805
    16.1.3 Brief History 806
    16.1.4 Fundamentals of Sensory Evaluation 807
           16.1.4.1 Complex Nature of Sensory
                    Measurement 807

16.1.4.2 Sensory Perception Is More than Product Measurement.......... 807
16.1.4.3 Multidimensional Sensory Answer............................. 809
16.2 Who Is Sensory Evaluation Working For? 811
  16.2.1 Research and Development (R&D) 811
  16.2.2 Production 812
  16.2.3 Marketing... 812
16.3 Developing a Sensory Approach............................... 813
  16.3.1 What Is the Problem? 813
  16.3.2 Sensory Human Resources 817
    16.3.2.1 Experimenter................................. 817
    16.3.2.2 Panel 818
  16.3.3 Sensory Laboratory and Sensory Test Conditions 820
    16.3.3.1 Analytical Studies.......................... 820
    16.3.3.2 Hedonic Studies 822
  16.3.4 Sensory Tests 823
    16.3.4.1 Difference Tests............................. 823
    16.3.4.2 Threshold Tests............................. 823
    16.3.4.3 Descriptive Tests........................... 824
  16.3.5 Statistics for Descriptive Analysis 826
    16.3.5.1 Basic Statistical Analysis.............. 828
    16.3.5.2 Variance Analysis.......................... 828
    16.3.5.3 Multivariate Analysis 828
  16.3.6 Consumer Tests 828
    16.3.6.1 Declarative Methodologies 829
    16.3.6.2 Behavioral Methodologies 829
  16.3.7 Summary: Which Tests for Which Panelists? 831
16.4 Correlating Sensory Evaluation with Instrumental and Consumer Measures 832
  16.4.1 Why Is Instrumental Not Enough?............... 832
  16.4.2 Linking Consumer Data with Analytical Data: Preference Mapping Techniques 835
16.5 Conclusion: Sensory Evaluation from the Lab to the Consumers 837
References 844

16.1 INTRODUCTION

The object of this chapter is to provide a global and general introduction to sensory analysis and its potential applications for surimi seafood. Many good reference books exist on sensory evaluation.[1-6] The purpose of this chapter, therefore, is not to replace them or provide a complete practical sensory evaluation manual, but rather to give readers a quick, broad overview of what sensory analysis is, what are its possibilities, and the ways it can be used in an industrial environment.

16.1.1 What Is Sensory Evaluation?

A classic definition of sensory evaluation is a scientific discipline using humans as measuring tools to interpret product characteristics as they are perceived by the human senses. A slightly different definition introduces it as a scientific field taking into account human judgment based on the perceived sensory characteristics and their consequences. Either way, it incorporates a full set of methods designed to measure, study, and interpret the food characteristics as they are perceived by the human senses (vision, touch, hearing, smell, and taste). As such, it can also be seen as a complete methodological "human toolbox" with different analytical tools based on the human senses.

16.1.2 Why Should We Care about Sensory Evaluation?

Because sensory analysis uses human subjects as instruments, it is obviously similar to what actual human consumers perceive. As such, it should be considered an almost necessary measurement when dealing with human consumers and everyday products. In fact, sensory characteristics are often put forward in the food industry.

At the same time, however, sensory evaluation itself is rarely developed or even seen as a strong and serious analytical tool. Apart from the intrinsic difficulties associated with the development of sensory activities, sensory is generally not the first or most important characteristic to be considered by the consumer. Instead, price, nutrition, toxicological characteristics, retailers' availability, brand image, and packaging characteristics are the predominant purchase drivers. This is especially true in the case of a first-try purchase. However, from a food manufacturer's point of view, sensory evaluation is paramount when competing for repeat purchases or developing products for human consumers in a highly competitive market, because sensory analysis is a powerful way to identify and differentiate competitors' products.

16.1.3 Brief History

Sensory analysis originated during the 1950s with the food industry. It developed with the increased competition between manufacturers and the increasing variety of products offered to consumers. The food industry has naturally been inclined to take into account the sensory "in-mouth" judgment and, consequently, sensory analysis is sometimes wrongly associated to an "in-mouth" and nose judgment only, restricted to taste and smell. Sensory analysis, however, concerns any information going through the sensory organs, and that means visual (shape, color, brittleness), tactile (warmth, mastication, touch, irritation), and auditory information (biting through a product, stirring it), as well as smelling the product and placing it into the mouth for taste, retroolfactive, and tactile information.

Although originally limited to the food industry only, sensory analysis is now rapidly developing in industries facing increased competition and higher consumer expectations, such as the cosmetic and car industries. Still, despite the long history with the food industry, very few documents are available specifically regarding surimi and sensory. This will surely change in the near future. The international development of the surimi market, development of new fish species involved in fabrication, increased competition among producers, development of new products beyond simple "crab imitation" are setting the stage for the development of sensory evaluation techniques relating to surimi and surimi seafood products.

16.1.4 Fundamentals of Sensory Evaluation

Despite its initial simplicity (asking humans how they feel and what they think about products they are consuming), sensory evaluation is complex in nature and sensory characteristics are often difficult to apprehend.

16.1.4.1 Complex Nature of Sensory Measurement

Depending on the human senses requires strong knowledge of neurophysiology, such as the capabilities, limits, and uses of the human body as an analytical tool. Knowledge of how the sensory system works, how the information is processed to the brain, and how that information is used in a given context is also necessary. Some human senses are well documented (e.g., vision and hearing); other senses, however, are still poorly understood (this is especially true for taste and smell [i.e., chemoreception senses]). In addition, brain research is only seriously beginning, and little is known about the links between sensory information processing and memory or emotions systems.

One thing is sure, however; the huge diversity in body shapes and appearances among human beings is also true for sensory systems, specifically for the olfactory and gustatory systems where huge inter-individual differences in sensitivity and interpretation abound. Taking into account the huge genetic and physiological human variability is the first difficulty in performing sensory analyses.

16.1.4.2 Sensory Perception Is More than Product Measurement

Sensory evaluation involves more than "product-only" characteristics. Instead, it relates to the interaction between a product and a human organism. In addition to the many product differences that exist, a specific product will provide a stimulus that will be potentially received by different human beings (with different bodies and sensory systems) under different contextual environments, which could result in dramatically different responses among consumers.

For example, the color of a product does not exist in itself. Color depends on product composition and the way it is perceived by a specific human. The same object will not have the same color under different illumination conditions or when seen by different animal species or even different human beings with different color visual abilities. People with specific color blindness will not perceive a product the same way standard people will, although the product or lighting conditions remained the same. What is true for color can be generalized for every human sense and body characteristic. The genetic variability among humans will lead to significantly different sensory potentialities and perceptions (different visual acuity and sensibilities, smell potentialities, teeth and in-mouth sensations, etc.).

In addition to different people's perception levels, their personal and cultural history and the context of product consumption will be used to differentiate consumers' behaviors. The context environment, in which a particular object will be perceived, might dramatically change the comportment regarding this product even if it does not modify the way the senses perceive the product. Trying a food with strong taste characteristics you dislike might give dramatically different results, depending on whether it is consumed alone (you might spit it), with friends as an initiation experience (first glass of alcohol), with family, or at a fancy official reception (maybe you will force yourself to finish the plate and maybe even compliment it).

Products are different, people are different, the way people eat food is different, and the history and product cultural experiences of people are different. All this makes studying sensory characteristics an exciting but difficult task. Figure 16.1 summarizes complex product–human interaction sensory information.

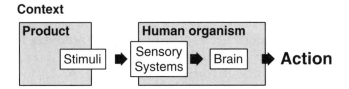

Figure 16.1 Sensory science based on contextual product and human interaction.

16.1.4.3 Multidimensional Sensory Answer

Another fundamental characteristic of sensory evaluation is that the sensory answer is multidimensional and integrates both objective and subjective measurements. Following product consumption with the question "how was it?," the spontaneous and natural answer will always concern a preference or a hedonic point of view: "It was great!," "I did not like it," or "it is better than the former one." Although these preferences and hedonic answers are of great interest for that particular individual, they can not be easily generalized to anyone else. The same question, asked of a different person with the same product, could yield a dramatically different answer. Some people will prefer a surimi stick with soft textural characteristics; others will prefer firmer ones. Some people will like a strong crab flavor; others will only like them with a low odor impression. This part of sensory analysis, which deals with preference or hedonic answers, is intrinsically subjective, strongly depending on the individuals answering and their personal history.

Moreover, this hedonic answer, although of great interest concerning final consumer preferences and choices, does not provide any information regarding the product's characteristics. Answering "I don't like it very much" to the question "how do you find the salt level in that surimi stick?" does not provide much information about the specific characteristics of the product or how to improve them. To be able to interpret and react to consumer preferences, the hedonic aspects should be accompanied by more analytical descriptive results, such as "is it because it was too salty or not salty enough?"

The more informative part of sensory analysis is called "analytical sensory" and is intended to describe the product's characteristics without initially concerning the hedonic consequences. It includes both qualitative information (how to describe the perceptions) and quantitative information (how much of the described characteristic is present). Although it still has to overcome inter-individual genetic differences, it yields more objective information that will be supported by precisely defined descriptions and intensity quantification.

These two fundamental aspects of sensory analysis (Figure 16.2) are both relevant and complementary to each other but are not of the same level of difficulty (Table 16.1). In addition, they strongly differ in the questions they can answer and the methods they will use. The analytical part of sensory analysis will concentrate on precisely describing and measuring sensory perceptions with methods similar to the ones used in instrumental analysis, while the hedonic part of sensory analysis will rely more on consumers' appreciation and diversity.

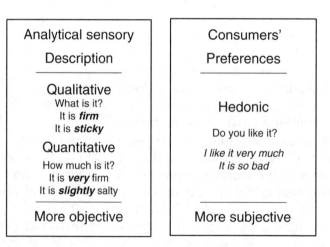

Figure 16.2 Sensory evaluation: objective (left) and subjective (right) evaluation.

TABLE 16.1 Difficulty Levels in Asking Humans

Type of Information	Task Easiness
I sense something or not	Easy
I like it or not	Easy
It is weak/strong	Difficult
I will describe it as	Very difficult

16.2 WHO IS SENSORY EVALUATION WORKING FOR?

As an ideal interface tool, the industrial potentialities of sensory evaluations are many and will generally cover the three following domains.

16.2.1 Research and Development (R&D)

Most of the time, sensory analysis will be used in Research and Development (R&D) on the analytical side. Sensory evaluation can be extremely useful in product development, helping product design and formulation. More specifically, sensory analysis will be used as one of the characterization tools, allowing full and precise product evaluations and comparisons of different formulations. This will usually be done in parallel with analytical instrumental measurements to further conduct instrumental-sensory correlation studies. The following problems are the most frequently addressed:

- Working on the marketing brief and selecting a formula according to it
- Mimicking existing competitive products
- Identifying the right molecular mix to give a product specific sensory characteristics
- Identifying key physical properties affecting sensory characteristics
- Replacing one ingredient by another one
- Looking for new sensory concepts and products
- Identifying pertinent analytical instruments to replace or compensate for human panelists

- Providing sensory markers to identify a company's full product range

16.2.2 Production

Production people will be more interested in assurance and quality control sensory applications. Sensory measurements will be done to control sensory specifications of either incoming raw materials or finished products. They can also be used as quality control calibration reference measurements to instrumental measurements. Sensory measures will ensure that the finished product is what was expected and that its sensory specifications are well conserved. The following problems are the most frequently addressed:

- Checking the sensory quality of incoming raw materials
- Controlling the sensory quality during product processing
- Comparing different processing units according to the sensory results
- Looking for off-flavors
- Checking the sensory quality of finished products
- Looking at the sensory consequences of changing processing units
- Looking at the sensory consequences of different storage conditions

16.2.3 Marketing

Marketing people will be interested in comparing products' sensory attractiveness and evaluating the impact of sensory characteristics on consumer choices. They will use sensory analysis to both develop and compare product performances in the market, in addition to price, image, and concept studies. Most of the time, when it comes to new product development, sensory will play a key interface role and should be done in close cooperation between marketing and development and formulation people. The following problems are the most frequently addressed:

Application of Sensory Science to Surimi Seafood

- Comparing different competitive products in terms of sensory preferences
- Understanding and identifying consumer sensory expectations
- Reacting to a market change
- Identifying key sensory properties for increasing consumer appreciation
- Choosing the best possible packaging regarding the product's sensory characteristics
- Identifying possible development of new products
- Adapting a product to another country

16.3 DEVELOPING A SENSORY APPROACH

In a food business, sensory evaluation is fundamentally an interface measurement, with potential applications and interests in many company levels. Table 16.2 lists some of the more frequent sensory-related questions and their potential applications to the surimi business.

To develop a sensory evaluation, a few systematic decisions should be taken:

- Understand the exact nature of the sensory problem.
- Choose the right human beings for the task.
- Choose the test conditions.
- Choose the sensory tests.
- Choose the statistic data analysis.

16.3.1 What Is the Problem?

Most of the questions that will be addressed by sensory people will enter one the following four categories:

1. *Is there any effect or difference?* This relates to a change of ingredients, the development of a "me-too product," or a change in processing equipment. These questions will generally be addressed through difference tests.
2. *When does the difference or effect start?* This concerns formulation questions relative to finding the mini-

TABLE 16.2 Examples of Surimi Industry Sensory-Related Questions

Typical Frequently Asked Surimi Sensory Questions	Comments
How do all of the market products differ from a sensory point of view?	Looking at the market
How does my surimi stick compare to the ones sold on the French or the Japanese market?	
All of those Japanese and French surimi sticks are different. How can I describe and summarize this?	
Among all those products, which are the preferred ones?	
How do I quantify the sensory attractiveness of a surimi product?	
Can I imagine a better product than the existing ones to match consumer expectations?	Imagining the future
How do I know what the consumers want?	
What are the main liking drivers for consumers? Aroma, flavor, texture?	
How will the consumer react to new products?	
I want to market a low-calorie and healthy surimi stick and another one as a kid snack. Which sensory characteristics should I emphasize?	Let's go to development
The marketing people want a "tropical vacation" flavor to add to their new surimi sticks. How do I get this?	
What is our best test product to match with this specific marketing brief?	

Application of Sensory Science to Surimi Seafood

Formulation work

Is it interesting to manufacture surimi stick with lemon or basil in addition to this crab flavor? Or will consumers prefer a scallop flavor?
Would a firmer texture be preferred by consumers? Or maybe a more elastic one?
Which one of the surimi stick textures do consumers prefer?
I think this formula is ok regarding the citrus flavor. What does the sensory panel think of it?
I have put in that amount of aroma. Is that enough?
Is my aroma well balanced?
Our senior flavorist suggests this aroma. But is he really on the same wavelength as our customers?
Do all our consumers perceive this newly added 2% lemon flavor?

Keeping in touch with the consumers

I think all those formula attempts are interesting. How do they perform compared with the actual market products?
What do the consumers think of it?
Does one of our products match the market leader ?

Going for the market leader.

One of the objectives can be to manufacture a surimi stick with sensory characteristics similar to an existing product.
Sensory will help determine which product, among the different test formulas is the closest to the target and if yes or no, the developed product is indistinguishable from the target product, without any price or packaging information.

Adjusting the process.

One might study the consequences of using different adjustments levels (temperature process, length of process, etc.) on the sensory quality of the finished product.

How sensitive is the sensory quality to the specification levels of the manufacturing equipment?
What level should I put as an acceptable level for my control texturometer?

TABLE 16.3 Examples of Surimi Industry Sensory-Related Questions (Continued)

Typical Frequently Asked Surimi Sensory Questions	Comments
What are the sensory consequences of a change in manufacturing equipment?	Modifying the process.
I have to rebuild my factory with new processing equipment. Are you sure my surimi stick taste will be the same?	Replacing previous machinery or transferring a production line can lead to a change in the finished product sensory characteristics.
To which adjustment level should my equipment be reset to control my process?	Manufacturing surimi seafood with new fish species might lead to very different sensory characteristics.
How to grade a food ("excellent", "prime", "good", "fair", "reject") using a sensory judgment?	Controlling the process.
Could I set an instrument to do the same job?	
What are the consequences of this new raw material?	Modifying ingredients.
Is that new crab flavor identical to the previous, more expensive one?	Aroma from a competitor might lead to unexpected changes.
If the MSG is removed, will my surimi seafood profile be changed?	Will it be perceived the same or not?
Would slightly decreasing this aroma level be perceptible?	The finished product sensory properties might evolve differently from toxicology or microbiology changes.
How is product storage length and conditions affecting the sensory characteristics?	Different plastic polymers used to wrap my products might affect product–packaging interactions negatively.
How does packaging interact with my product sensory properties?	

mum quantity of a specific ingredient to be perceived by part of a population. These questions will generally be answered through threshold tests.
3. *What are the differences?* This addresses questions relative to the sensory description of a specific product set. This will be answered using sensory profile descriptive methods.
4. *How important is this to the consumers?* This concerns all consumer appreciation relative to the product's sensory characteristics and will be addressed by consumer tests.

16.3.2 Sensory Human Resources

16.3.2.1 Experimenter

To be effective, a sensory measurement should rely on people having different levels of expertise. This comes as a consequence of the fundamental product–human interaction in sensory studies. Four scientific competencies are usually required:

1. A good understanding of basic neurobiology and neurophysiology (understanding how the human senses function) associated with basic psychological behavior knowledge (understanding how a human being will behave in a group)
2. A strong knowledge on the studied product diversity: physical and chemical composition, formulation, process, and market competition
3. A complete knowledge of the different and various sensory tests and methodologies
4. A thorough understanding of the different statistical methods that can be applied to the obtained data

Apart from finding the perfect, omniscient individual, this will generally be accomplished by teamwork. Sometimes this makes sensory evaluation a difficult task to handle for smaller companies with limited human resources. This constraint should, however, never be underestimated: poor sensory analysis too often relies on simple sensory test methodology and product expertise only, without taking into account a necessary reflection on the real human physiological and psychological possibilities. Mastering the diversity and subtleness of statistical methods to deal with the huge products and human diversity is often a case for limitation.

At a minimum, the person in charge of the sensory tests should be highly trained in the following ways:

- Prepare products and tests
- Motivate panelists
- Able to animate tests and meetings
- Process and interpret data
- Present results according to the objectives

16.3.2.2 Panel

Sensory panelists should be carefully selected according to the objectives of the study. A strong difference will be made between analytical sensory evaluation and consumer studies.

16.3.2.2.1 Analytical Studies

In analytical studies, the panel should be chosen to be the most effective instrument for measuring the different attributes of the tested products. Human panelists will be chosen not only according to their motivation and availability, but also because of their product experience and communication abilities. Panelist selection will, above all, be made according to their physiological sensory abilities. Selecting the right panelist is in no way a random sampling of the population but rather a careful selection to a specific task. If one is interested in odor modification and perceptions, then one should test the panelists according to that olfactory sense and the molecules one might encounter. On the contrary, if one is more interested in the visual properties of the products, one will make sure that the selected panelists do not have any visual incapability.

Contrary to popular belief, the number of panelists is not a quality criterion in analytical sensory studies. As for any instrumental evaluation, one sensitive, pertinent, reliable sensory panelist might be enough to analyze the products.

But this is true if, and only if, one is able to select that best panelist. Generally, the number of human subjects in a sensory panel will range between 10 and 20. This is done, first, as a precaution so that one does not have to depend on the availability of a too-small group of panelists, and second, because it is often difficult to clearly identify the different physiological sensitivities to be tested, especially in the case of taste and smell evaluations and/or large product sets.

In any case, one should never consider the panel population as representative of the consumer population. The sensory panelists are not a sample of consumers; they should be dedicated to the analytical examination of the products only. It is useful to train the sensory judges to specific tests and sensory methodologies. Training can be done for a specific test methodology, for providing panelists with a full range of product experiences, for training them to quantify perceived intensities, or to train them in the use of specific vocabulary terms.

16.3.2.2.2 Hedonic Studies

In consumer testing, panelists are chosen according to the target population. Concerning the number of consumers: the more consumers, the better the precision will be and one usually considers the number of 60 consumers as the minimum.[6] Consumer panelists should not only be representative of the actual customer population, but also the potential buying population. They can be chosen at random but should not have any specific or particular knowledge of the products. This means that they should neither be recruited from the company manufacturing nor selling the products. In addition, they should not have any particular training regarding the products.

16.3.2.2.3 What about the Individual Experts?

Sensory analysis does not necessarily mean using a sensory panel. A personality having excellent product background knowledge, such as a chef, an oenologist, or a flavorist, will be able to provide some kind of pertinent sensory judgment regarding the products. Their appreciation will be based on strong product knowledge, but they will however usually undermine the value of their description by using a strong personal vocabulary (not necessarily understandable by another person) and by mixing it with purely personal recommendations of what a "good" product should be. Table 16.3 summarizes the pros and cons of the different human resources that can be used in sensory evaluation.

16.3.3 Sensory Laboratory and Sensory Test Conditions

As for the sensory panel, the conditions under which the sensory tests will occur will greatly vary between analytical sensory evaluation and consumer studies

16.3.3.1 Analytical Studies

Sensory labs can vary in design, depending on the type of products or on the goals of testing. Three general principles should always be kept in mind:

1. *Allow for individual evaluations.* Panelists should be able to concentrate and work quietly without being disturbed by either the experimenter or the other panelists.
2. *Allow for standardized test and preparation procedures.* The experimenter should be able to prepare and present the tests samples in the exact same conditions over and over.
3. *Allow for product "sensory-only" evaluations,* without any context, brand image, or price interference. As little information as possible should be given to the panelist in order to avoid any unwanted preconceived intrusion with the perceived sensations.

A meeting room will prove to be useful before and after testing to allow panel discussion or methodological explanations.

TABLE 16.3 Pros and Cons of Different Human Resources for Sensory Evaluation

Human Sensory Tool	Who Are They?	Pros	Cons
Individual expertise (flavorist, chef, oenologist, perfumer, etc.)	A personality having an excellent product background knowledge	Rich description Very strong product expertise	Very low consumer representativeness No consumer predictability Very personal methodology Very personal vocabulary
Sensory panel	A limited number of trained and experienced subjects	Strong analytical description Consensual vocabulary Detailed procedures and methodologies	Low consumer representativeness Not necessarily representative of the consumers' sensitivity Need product expertise training
Consumer panel	A large and representative number of naive subjects	The final customers	Huge diversity Poor explanations or characterizations of the products

The most common sensory lab will include a few individual sensory booths adjacent to a separate preparation room. Each booth is separated from one another using dividers, thus allowing panelist to assess the products without being distracted by others. Each booth should have standardized light, temperature, and possibly humidity control. Air extraction should be available with an efficient filter to permit odor evaluation. Most sensory booths in the food industry incorporate a sink to allow the panelist to rinse between samples. Sensory booths should be large enough to allow the correct display of the products, plus allow the panelist to fill in the evaluation form.

Most of the time, the sensory booths will directly communicate with a preparation room with the help of serving hatches. In the food industry, the preparation room will generally be very similar to a professional kitchen, incorporating washing, cooking, and freezing apparati. It should be large enough to store product display materials and to allow for proper product preparation. The samples to be tested will be prepared in neutral coded and anonymous presentation kits.

Judges' responses are generally written but computerized sensory systems are now widely available that allow faster panelist data input and better judge and product database controls. They also provide most of the necessary statistical data treatments.

16.3.3.2 Hedonic Studies

In the case of consumer studies, using the sensory booths should always be considered as an imperfect solution. They, of course, allow for better control of the test conditions but they are poorly representative of a normal consumer environment. The preference results obtained under such conditions should always be considered with a critical eye and they might mean nothing more than trend indications.

More satisfying test conditions should try to study the consumers' behavior without directly interfering with them. Giving products to the panelists to be tested at home could do this, for example, or using central location tests during a supermarket visit or in semi-controlled cafeteria conditions.

16.3.4 Sensory Tests

The four main test categories will correspond to the most frequent sensory questions.

16.3.4.1 Difference Tests

Discrimination or difference tests are used to determine if a difference (or similarity) in perception exists between two products. Many difference tests exist, including: the triangle test, pair comparison test, duo-trio test, 2 out of 5 test, 2-AFC test, 3-AFC test, etc. Among these, the triangle test is the most widely used sensory difference test. It consists of presenting three coded samples to the panelist; two of them filled with one of the two tested products, the third sample being filled with the second product. The panelist must indicate which sample is different from the other two (forced choice). Data analysis will be done using statistical tables that allow for a comparison of the number of correct answers compared to the total number of answers and the probability of giving the correct answer at random (1/3 in the case of the triangle test).

In general, difference tests will be done on large numbers of panelists to estimate the proportion of the population that is able to differentiate between the two products. It is also possible to focus the test on a smaller number of panelists: either very sensitive ones to make sure that no difference is noticeable, or, on the contrary, to poorly sensitive panelists to confirm the existence of a surely perceptible modification.

These discriminating tests will be particularly useful when performing cost reduction studies, ingredient modifications, or "product match" formulations.

16.3.4.2 Threshold Tests

Threshold tests are used to evaluate the lowest amount of a given stimulus necessary to be noticeable to a human being. Absolute thresholds will concentrate the minimum level to be noticeable, while relative thresholds will give information on the minimum level to be added (or removed) from a previous concentration to be perceived as different. Threshold tests rely either on testing the possible perception of a succession of different samples at different concentrations, or on ranking procedures that ask the panelist to rearrange the different samples in order of perception intensity.

Threshold tests are generally done at an individual level but must be applied to population samples to be interesting to interpret. They will give information about the smallest difference in physical or chemical intensity that can be perceived by different population segments. They are invaluable for appreciating the population heterogeneity in sensitivity. Threshold tests are primarily used in neuro-physiological studies but they still have interesting potential applications in food formulation, especially when having to decide the level of a specific ingredient (aroma, for example) to add to be perceptible by a predetermined minimum percentage of the consumer population.

16.3.4.3 Descriptive Tests

Descriptive tests, also called sensory profiling techniques, are used to describe predetermined subsets of products using quantified consensual and well-defined descriptive terms. There are many descriptive methods: the flavor profile,[7] QDA®,[8] texture profile method,[9,10] quantitative flavor profiling,[11] spectrum method™,[12] free choice profiling,[13,14] and flash profile.[15–17] They differ with respect to the training of the panelists and the construction of the vocabulary, but most of them are based on the following assumptions:

- Use of a small, limited number of trained panelists
- Use of a descriptive evaluation without any hedonic connotation
- Use of a consensual and well-defined vocabulary. (This is not the case for the free choice profile and flash profile, which do not rely on a consensual semantic description but permit more semantic flexibility in an effort to gain development time.)
- Use of a quantitative evaluation

Firmness (tactile): This attribute concerns the stick's resistance to an applied pressure. ⬅ *Definition*

Procedure: Hold the stick between the thumb and the forefinger (one on the white part, the other one on the colored part), apply a pressure until strong resistance is felt. Evaluate the overall resistance of the stick to the pressure. ⬅ *Standardized procedure*   Reference: Sur27

0 ————————————————————————→ +

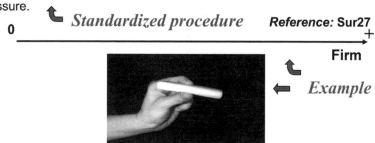

Firm ⬅ *Example*

Figure 16.3 Example of a descriptive sensory attribute used in a surimi seafood sensory glossary.

The common objective among these methods is to develop a strong consensual vocabulary. This vocabulary should be both efficient and adapted to the products being studied. Each sensory attribute will be backed by a definition (i.e., what is the meaning of the used semantic term); a standardized procedure (how to proceed to best experience the sensation); and reference products (strong characteristic examples of the semantic attribute) (Figure 16.3). This is done to leave no room for individual interpretations of the semantic terms and also to be able to use this vocabulary as a reference language when discussing product characteristics.

Table 16.4 lists different possible semantic attributes that are or could be used for surimi products. The table does not, however, give the corresponding definitions, procedures, or examples to render the terms fully understandable. This descriptive work must be done by and for the panel that will use the terms. Finally, the number of sensory attributes that should be selected for a fully operational sensory vocabulary should be kept small enough to keep the sensory glossary manageable and to avoid using redundant and correlated

TABLE 16.4 Potential Surimi Attributes Classified According to Major Sensory Perceptions

Visual	Visiotactile	Olfactive	Flavor (in-mouth taste and olfaction)	In-Mouth Texture
Artificial	Compactness	Artificial	Bitter	Compactness
Color intensity	Crumbliness	Crab	Metallic	Crumbliness
Colored	Crumbly	Fishy	Salty	Crumbly
Clear	Elastic	Fresh	Sour	Elastic
Flesh lightness	Firm	Grilled	Sweet	Fibrous
From white to gray	Fibrous	Odor intensity	Umami	Firmness
	Firmness			Moistness
Grilled	Moistness	Pungent	Astringent	Smoothness
Orange	Smoothness	Seaweed		Soft
Uniform (color)	Stringy	Shrimp	Spicy	Stringy
Yellow		Smoked		
			Aroma intensity	
Dull			Artificial	
Pearl-shiny			Crab	
Matt to shiny			Fishy	
Opalescent			Fresh	
Opaque			Grilled	
Shiny			Seaweed	
Transluscent			Shrimp	
			Smoked	

terms. Between 15 and 20 attributes is generally considered a reasonable maximum number.

Products will then be evaluated according to the intensity perceived on each of those sensory attributes. The quantitative evaluation will be done using scoring or ranking techniques, asking panelists to estimate the level of intensity on each of the sensory attributes. This will permit the building of a complete product ID (Figure 16.4). This will also allow product comparisons using a full range of statistic methods.

16.3.5 Statistics for Descriptive Analysis

The number of statistics methods available for sensory analysis is enormous.[18] Three statistical techniques will be regularly used in conjunction with sensory descriptive analysis.

Application of Sensory Science to Surimi Seafood

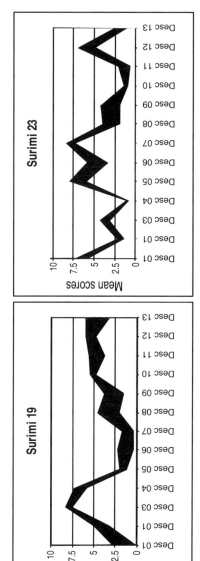

Figure 16.4 Example sensory profiles of two different surimi sticks. Means and standard deviations of the sensory panel scores are represented on a 10-point scale for 13 different sensory attributes.

16.3.5.1 Basic Statistical Analysis

Means, standard deviations, and histograms will be used to represent the individual results of sensory profiles. They will allow for a quick understanding of the major sensory characteristics of specific products (Figure 16.4).

16.3.5.2 Variance Analysis

Variance analysis will permit a comparison of different products on each of the sensory attributes, taking into account the different panelists' scores. It will emphasize the most important attributes to discriminate between products. It will usually be completed by post hoc tests enabling two-by-two product comparisons.

16.3.5.3 Multivariate Analysis

Multivariate analysis will simultaneously use all of the information coming from the different scores obtained by the different products on all the attributes to represent the most efficient product maps, sorting products into groups according to the main sensory characteristics that differentiate them. Many techniques are available, but Principal Component Analysis (PCA) (Figure 16.5) and Cluster Analysis are the most frequently used among them. The products' relative positions will be interpreted according to the correlated terms associated to the products. In the case of Figure 16.5, the products will be essentially differentiated based on aromatic and visual properties along one main direction and on textural characteristics in another direction (Figure 16.6)

16.3.6 Consumer Tests

Two major sets of methods will be used in consumer testing: declarative and behavioral.

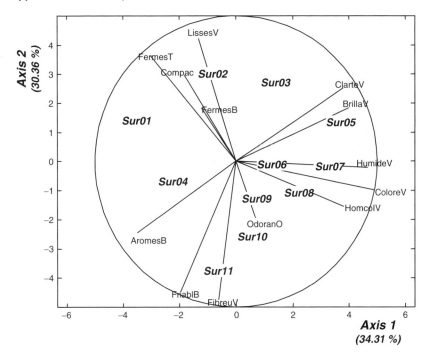

Figure 16.5 Example PCA performed on 11 surimi seafood sticks using 13 sensory attributes.

16.3.6.1 Declarative Methodologies

Declarative methodologies rely on asking consumers to verbally answer (or in writing) questions regarding the hedonic aspects of tested products. The consumer reacts by declaring opinions after having been encouraged to do so. These methods heavily depend on the use of questionnaires.

16.3.6.2 Behavioral Methodologies

These tests should be as close as possible to a normal consumer behavior, without asking questions, but looking at the reactions of panelists when confronted with different products in real-life situations. A few methods (mostly confidential)

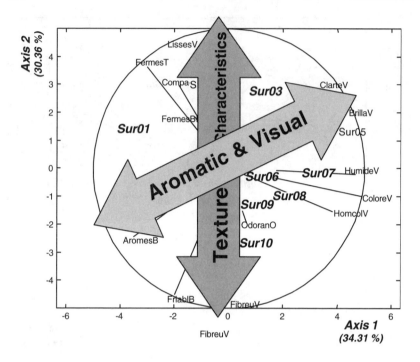

Figure 16.6 Example potential interpretation of the major sensory differences among the 11 previous (Figure 16.6) surimi seafood sticks.

have been used by large food and body care companies. They usually rely on observing consumers in real, yet controlled situations (a supermarket, coffee shop, restaurant, game prices, fidelity cards, etc.). These methods usually require a lot of preparation.

In theory, consumers should be used in their purest form, without asking them to score or describe their sensations (i.e., what they never do in regular consumer situations). One should, on the contrary, concentrate on their real purchases and consumption (i.e., what they really choose, how much they eat, how long it takes, etc.). In practice, declarative procedures, sometimes done in sensory booths, are most frequently used. This is because they are cheap to set up and process compared to behavioral methodologies.

One should, however, be very careful when interpreting and generalizing the results obtained from questionnaires. Answering a questionnaire does not deeply involve the consumer; real consumer behaviors could well be very distinct from what they declare when evaluating the same products.

If questionnaires are still to be used, the following recommendations should be kept in mind:

- It is easy to ask consumers for their preference on products.
- It is not easy to observe their natural behavior.
- It is difficult to ask consumers about the reasons for their preferences.
- It is illusory to think that one will be able to extract any clear and pertinent information in the pseudo-descriptive vocabulary generated to answer open questions in a questionnaire.

The actual recommendations for consumer questionnaires are as follows:

- Present the product in situations as close as possible to normal consumer behaviors.
- Asking one question is best: a preference or overall acceptance question on the product.
- If many questions must be asked, reduce them to a minimum. The less questions, the better.
- If many questions must be asked, the overall score should always be asked first.
- If many questions must be asked concerning different aspects of the product, it is better to ask specific questions after having presented the products to the panelist a second time, to avoid the panelist adjusting the overall score previously given.

16.3.7 Summary: Which Tests for Which Panelists?

One should never forget that it is a fundamental mistake to ask sensory experts from an analytical sensory panel to provide preference information. And it is also a mistake to ask consumers to explain or describe the reasons for their preference. One should also never forget that the panelists used for analytical sensory evaluation are instruments: they are here to test and give information about the products. As such, one should always primarily focus on the products and not on the panelists. The food samples are the ones to be intensely tested — not the panelists. Table 16.5 summarizes the major sensory tests and their corresponding human panels.

16.4 CORRELATING SENSORY EVALUATION WITH INSTRUMENTAL AND CONSUMER MEASURES

In a food company, sensory evaluation, informative and interesting as it might be, is worthless if it cannot be linked to consumer purchases. Those consumer preferences and purchases depend on many factors, some of them being potentially explained or linked to sensory characteristics. Some consumer preferences might even be directly linked to formulation or ingredient composition. So there is a natural tendency to question the link between consumer preferences and the sensory characteristics and/or product composition.

Figure 16.7 summarizes the way sensory studies should be correlated to instrumental studies and preference data. On the one hand, one tries to explain and predict consumer preferences based on either analytical sensory or instrumental studies, thus helping R&D people identify key ingredients to formulate more interesting products. On the other hand, one will try to correlate instrumental analysis and sensory analytical data. At this point, the very evident question — "Could correlating preference data with instrumental data be sufficient?" — must be asked.

16.4.1 Why Is Instrumental Not Enough?

One of the frequently asked questions about sensory analysis in the food industry is: "Can an instrument provide the same information?" This, of course, is a most relevant issue for in-

TABLE 16.5 Major Sensory Tests and the Corresponding Sensory Panels

Tests to Study What ?	Which Tests ?	Which Panel ?	Additional Comments
Sensory characteristics	Descriptive analysis	A small number of trained panelists	Few panelists The more sensitive, precise, and understandable ones Consumer-relevant standardized procedures
Sensory effects	Difference tests Threshold tests	Either a small number of trained panelists or a large and representative number of the population	
Consumer preferences	Declarative tests Behavior tests	A large and representative number of consumers	The more, the better, and never less than 60 Should be representative of consumer target population The less questions, the better Close to a normal consumer behavior

line quality control measurements, but not exclusively. In general, instruments allow faster and simpler measurements than sensory evaluation with humans and, therefore, replacing a sensory tool with an instrument must be taken into account whenever possible. However, instrumental measurements are often not sufficient to render the information provided by sensory evaluation. There are several reasons for this.

First of all, in some cases, the available instruments are less sensitive than the human senses. This is true for aroma perception because the human detection threshold of some aroma compounds is lower than that of most sensitive instrumental detectors, such as single ion monitoring mass spectrometers and much lower than the commonly available flame

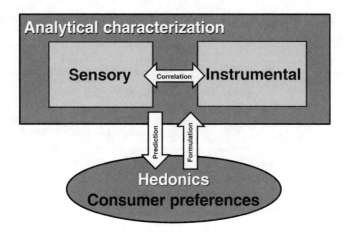

Figure 16.7 Main connections between sensory analysis, instrumental analysis, and consumer preferences in industry.

ionization detector associated with gas chromatography. On the other hand, some volatile compounds that can be detected instrumentally, such as carbon dioxide, are odorless. This means that in an attempt to link instrumental and sensory data, one must make sure that the instrumental data is truly relevant in terms of sensory perception. Unfortunately, in most cases, the relationship between instrumental and sensory data is not straightforward. This is because of the way we perceive things. Indeed, the intensity of a stimulus perceived with our senses is never a simple linear function of the physical intensity of that stimulus (e.g., the concentration of a flavor compound, the wavelength of a color, the viscosity of a cream, etc.).

Second, when using an analytical instrument, only one thing is measured at a time (i.e., color brightness, gel strength, pH, or salt concentration, etc,). However, this is not the way the brain perceives things. Human perception of objects is much more integrated and we tend to perceive, say the flavor of a surimi stick, as a unitary perception, not a combination of taste, aroma, odor, texture, etc. This is especially true for consumers who are not trained to analyze their sensations.

And indeed, the most accurate and elaborate chromatogram does not provide any idea of what a lobster flavor tastes like.

This is probably the most important reason why replacing a sensory panel with an instrument is not an easy thing to do, except maybe in some simpler textural measurements.[19–21] Accordingly, instrumental measurements will never provide information about consumer preferences. An exception to this is quality control, when a specific sensory defect is identified (e.g., an odorous volatile compound possibly migrating from the packaging material to the food) or when the acceptable range of a specific characteristic, such as firmness or color intensity, is determined in a consumer test and has shown correlation with human perception.

16.4.2 Linking Consumer Data with Analytical Data: Preference Mapping Techniques

Internal and external preference mapping studies have been available to the food industry since 1990.[22] They enable the scientist to link and interpret consumer preferences with analytical sensory and instrumental expertise. To do this, it is important to collect hedonic consumer judgments separately from the analytical work before linking them back together through sophisticated statistical models. Thus, they provide an effective way of identifying potentially successful market products with clear sensory and/or instrumentally descriptive information.

The more powerful technique is the quadratic preference mapping model.[23] It involves three steps. The first step is to establish a sensory map using the Principal Component Analysis (PCA) method. A PCA is performed on the expert sensory data to obtain a sensory map of the products. On the map, products lying close together are quite similar from a sensory point of view, and products lying farther apart strongly differ on some sensory attributes. The location of the products on the map is fully explained by the most influential sensory attributes, and corresponds to the consensual judgment of the expert panel.

The second step is to perform an individual modeling (consumer by consumer) of all the consumers' preferences. Considering the preference scores of a given panelist, a quadratic model is calculated to predict the preference score for a given product by the location of this product on the sensory map. This quadratic model gives three kinds of shapes for the response surfaces corresponding to three types of consumer behaviors:

1. Preferers: consumers who have an optimal product.
2. Rejecters: consumers who have an anti-optimal product.
3. Eclectics: consumers who have two opposing preferences.

Because the optimal product is almost never an extreme product but rather a balanced one, this quadratic model better fits real consumer preferences compared to the older and simpler linear models used in many previous studies.

The third step is to determine a global model of the preferences. The aim of this step is to find the best compromise between the different locations of the optimal products for the different consumers. This means that rather than obtaining a single best product for everybody, which in reality does not exist, the obtained product will be considered by the maximum number of consumers as a good product. The final result is a three-dimensional surface with lows and highs (Figure 16.8). Considering all the consumer opinions, the summits correspond to virtual "optimal" products and the lowest points correspond to the worst products. The results are expressed in percentage instead of an absolute number of consumers. To locate the optimum product, one can also take a look at the easier to read and interpret corresponding level curves (Figure 16.9).

A further and more recent improvement in the method is the clustering of the consumer population. Consumers rarely agree on preferences. Consequently, unanimous opinions on the same product are almost nonexistent (Figure 16.10). So, instead of working with the consumer group taken as a whole, it is far better to divide it into homogeneous subgroups before applying the appropriate preference mapping

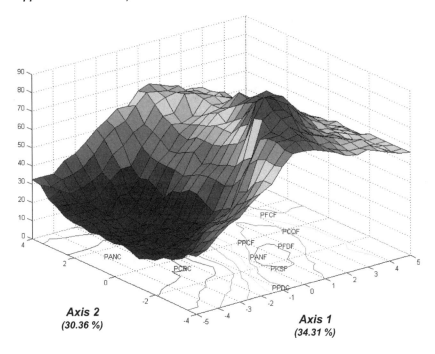

Figure 16.8 Example of a typical preference mapping result. The consumer preferences levels are on top of the sensory map of the surimi seafood products.

technique. This will allow for a more specific and accurate definition of the optimal product for specific consumer targets. Consumer groups should not be based on the classical socio-professional groups but on the declared preferences of the products. This will be done by cluster analysis on the panelists × products data (Figure 16.11). The consumer clusters will then be used to calculate specific preference mapping studies (Figure 16.12).

16.5 CONCLUSION: SENSORY EVALUATION FROM THE LAB TO THE CONSUMERS

Table 16.6 was developed based on Table 16.2 and they complement each other. Table 16.6 summarizes all the major

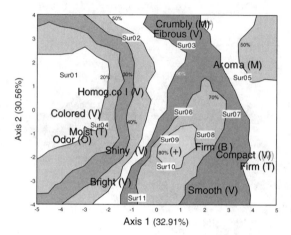

Figure 16.9 Curve levels representation of the same preference mapping study. The sensory attributes generated by the sensory panel illustrate and clarify the sensory reasons for the differences in consumer preferences.

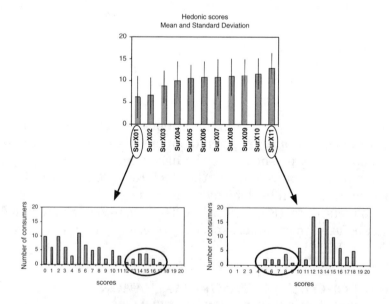

Figure 16.10 Example of the diversity of the consumers' hedonics scores on two surimi seafood sticks. Although surimi seafood sticks X01 and X11 have very different mean hedonic scores, they are either liked or disliked by some of the consumers.

Application of Sensory Science to Surimi Seafood 839

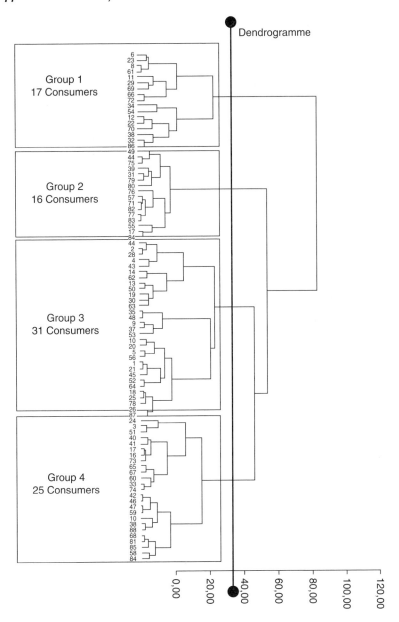

Figure 16.11 Example cluster analysis performed on 89 consumers and based on their preferences scores on the same 11 surimi seafood sticks.

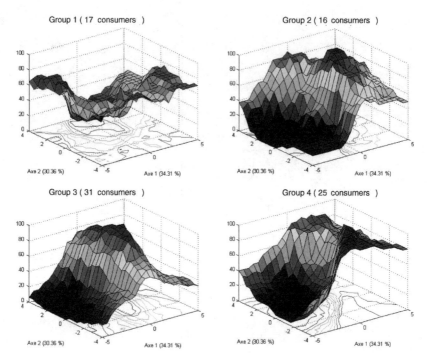

Figure 16.12 Example of the different preferences mapping results obtained on the four consumer groups after the previous cluster analysis.

concerns and the potential applications of sensory studies to answer specific surimi-related problems. However exhaustive and wide-ranging it might be, one should nevertheless not forget that people like a specific product not only for its sensory characteristics, but also because of price, availability, personal culinary reference history, consumption patterns, packaging, nutritional aspects, convenience, product image, and the body potentialities it permits. Consequently, food companies should definitively not forget the importance of sensory results, yet they should not, on the contrary, consider them as the key "silver bullet" that will provide all answers to their marketing and development problems.

TABLE 16.6 Surimi Sensory Related Questions Classified According to Their Main Application and Sensory Categories

Main Application Category	Frequently Asked Question	Nature of the Question	Example of Possible Sensory Tests
Market research	All those Japanese and French surimi sticks are so different. How can I describe and summarize this?	Analytical	Descriptive analysis
Market research	Does one of our products match the market leader?	Analytical	Flash profile Difference test
Market research	How do all the market products differ from a sensory point of view?	Analytical	Descriptive analysis
Market research	How do our surimi sticks compare to those on the French and Japanese markets?	Analytical	Descriptive analysis
Market research	How do we know what the consumers want?	Hedonic	Preference mapping
Market research	How do I quantify the attractiveness of a surimi product?	Hedonic	Hedonic test
Market research	How will the consumers react to new products?	Hedonic	Consumer test
Market research	I think all those formula tries are interesting; how do they compare with the actual market products?	Analytical Hedonic	Flash profile
Market research	Is it interesting to manufacture surimi sticks with lemon or basil in addition to this crab flavor? Or, will consumers prefer a scallop flavor?	Analytical Hedonic	Consumer test
Market research	Is our senior flavorist's suggestion fitting what our customers really like?	Hedonic	Consumer test
Market research	What do the consumers think of it?	Hedonic	Consumer test

TABLE 16.6 Surimi Sensory Related Questions Classified According to Their Main Application and Sensory Categories (Continued)

Main Application Category	Frequently Asked Question	Nature of the Question	Example of Possible Sensory Tests
New product design	Applying a marketing brief (i.e., what is a "tropical vacation" surimi flavor?).	Analytical	Preference mapping
Formula/Process optimization	Can I create a product better than existing products in order to fit consumer expectations?	Analytical Hedonic	Preference mapping
Formula/Process optimization	I think this formula is OK regarding that citrus flavor. What does the sensory panel think of it?	Analytical Hedonic	Difference test Threshold test Descriptive analysis
Formula/Process optimization	Is that amount of flavor enough?	Hedonic	Threshold test
Formula/Process optimization	What are the main liking drivers for consumers? Aroma, flavor, texture?	Analytical Hedonic	Descriptive analysis Hedonic test Preference mapping
Formula/Process optimization	What are the optimal process settings with that new fish species?	Analytical	Correlation study
Formula/Process optimization	Would a firmer texture be preferred by consumers? Or perhaps a more spongy one?	Analytical Hedonic	Descriptive analysis Hedonic test Preference mapping
Product optimization	Among all those prototype products, which are the preferred ones?	Hedonic	Consumer test
Product optimization	Which of the surimi stick textures do the consumers prefer?	Hedonic	Consumer test
Cost reduction	Do all our consumers perceive this newly added 2% lemon flavor?	Analytical	Threshold tests Difference tests

Cost reduction	Is that new crab flavor identical to the former but more expensive one?	Analytical	Difference tests
Cost reduction	What are the sensory consequences of replacing ingredients?	Analytical	Difference tests
Cost reduction	Would slightly decreasing the flavor level be perceptible?	Analytical	Difference tests
Quality control	How does the packaging interact with the sensory properties of our product?	Analytical	Descriptive analysis Difference test
Quality control	How do product storage times and conditions affect the sensory characteristics?	Analytical	Descriptive analysis
Quality control	How sensitive is the sensory quality to the manufacturing process specifications?	Analytical	Descriptive analysis
Quality control	How do I grade a food ("excellent," "prime," "good," "fair," "reject") using sensory judgment?	Analytical Hedonic	Sensory profile Correlation with consumer acceptation levels
Quality control	I must build a new factory with a new processing equipment. Are you sure my surimi will taste the same?	Analytical	Sensory profile Differnece test
Quality control	What are the sensory consequences of a change in manufacturing equipment?	Analytical	Descriptive analysis
Quality control	What level should I define as "acceptable" for my control texturometer?	Analytical	Correlation study
Quality control	Could I set an instrument to do the same job?	Analytical	Correlation study

REFERENCES

1. H.T. Lawless, H. Heymann. *Sensory evaluation of food. Principles and Practices.* New York: Kluwer Academic/Plenum Publishers, 1998.

2. M. Meilgaard, G.V. Civille, and T.B. Carr. *Sensory evaluation techniques, 3rd ed.* Boca Raton, FL: CRC Press, 1999.

3. H. Stone and J.L. Sidel. *Sensory evaluation practices, 2nd ed.* London, U.K.: Academic Press, 1993, 337.

4. ACTIA. *Evaluation sensorielle — Guide de bonnes pratiques.* Paris: ACTIA, 1999.

5. SSHA. Evaluation sensorielle. *Manuel Méthodologique, 2nd ed.* Paris: Lavoisier, 1998.

6. AFNOR. *Analyse Sensorielle, 6eme ed.* Paris, AFNOR, 2002.

7. S.E. Cairncross and L.B. Sjostrom. Flavor profiles — a new approach to flavor problems. *Food Technol.,* 4, 308–311, 1950.

8. H. Stone, J. Sidel, S. Oliver, A. Woosley, and R.C. Singleton. Sensory evaluation — quantitative descriptive analysis. *Food Technol.,* 28, 24 34, 1974.

9. G.V. Civille and A. Szczesniak. Guidelines to training a texture profile panel. *J. Texture Studies,* 4, 204–223, 1973.

10. M.A. Brandt, E. Skinner, J. Coleman. Texture profile method. *J. Food Sci.,* 28, 404–410, 1963.

11. C.R. Stampanoni. The "quantitative flavor profiling" technique. *Perfumer and Flavorist,* 18, 19–24, 1993.

12. A.M. Muñoz and G.V. Civille. The spectrum descriptive analysis method. In ASTM, Ed. *Manual on Descriptive Analysis Testing for Senory Evaluation.* Baltimore: R.C. Hootman, 1992, 22–34.

13. A.A. Williams and S.P. Langron. The use of free-choice profiling for the evaluation of commercial ports. *J. Sci. Food Agric.,* 35, 558–568, 1984.

14. A.A. Williams and G.M. Arnold. A comparison of the aromas of six coffees characterized by conventional profiling, free-choice profiling, and similarity methods. *J. Sci. Food Agric.,* 36, 204–214, 1985.

15. J.M. Sieffermann. Le profil flash — un outil rapide et innovant d'évaluation sensorielle descriptive. *Proceedings of AGORAL 2000, XIIèmes rencontres "L'innovation: de l'idée au succès."* Montpellier, France, 2000, 335–340.

16. J.M. Sieffermann. Flash profiling: a new method of sensory descriptive analysis. In *AIFST 35th Convention,* Sidney, Australia, 2002.

17. J. Delarue and J.M. Sieffermann. Sensory mapping using flash profile. Comparison with a conventional descriptive method for the evaluation of the flavour of fruit dairy products. *Food Quality and Preference,* 15, 383–392, 2004.

18. M. O'Mahony. *Sensory Evaluation of Food — Statistical Methods and Procedures.* New York: Marcel Dekker, 1986.

19. A.S. Szczesniak, M.A. Brandt, and H.H. Friedman. Development of standard rating scales for mechanical parameters of texture and correlation between the objective and the sensory methods of sensory evaluation. *J. Food Sci.,* 28,397–403, 1963.

20. A. Henry-Bressolette. Texture of small toasted bread: a case study incorrelating sensorial and instrumental data beteen weakly differentiated crisp products. In *Pangborn 2nd Sensory Science Symposium,* Davis, CA, 1995.

21. A. Henry-Bressolette, B. Launay, and M. Danzart. Mode d'établissement de corrélations entre données sensorielles et instrumentales pour des produits croustillants présentant de faibles différences de texture. *Sciences des Aliments,* 16, 3–22, 1996.

22. P. Schlich and J.A. McEwan. Cartographie des préférences — un outil statistique pour l'industrie alimentaire. *Sciences des Aliments,* 12, 339–355, 1992.

23. M. Danzart. Cartographie des préférences. In SSHA, Ed. *Évaluation Sensorielle. Manuel Méthodologique. 2nd ed.* Paris: Lavoisier, 1998, 290–297.

17

New Developments and Trends in Kamaboko and Related Research in Japan

KUNIHIKO KONNO, PH.D.

Hokkaido University, Minato, Hakodate, Japan

CONTENTS

17.1 History of Kamaboko 848
17.2 Variations in Kamaboko Products in Japan 849
    17.2.1 Steamed Kamaboko on a Wooden Board:
            Itatsuke Kamaboko 850
    17.2.2 Grilled Kamaboko on Wooden Board:
            Yakinuki Kamaboko....................................... 851
    17.2.3 Grilled Kamaboko on Bamboo Stick:
            Chikuwa 851

17.2.4 Deep-Fried Kamaboko: Age-Kamaboko 851
17.2.5 Boiled Kamaboko: Hanpen and Tsumire 852
17.2.6 Crab Leg Meat Analog, Crabstick: Kani-ashi Kamaboko or Kanikama 853
17.2.7 Fish Sausage and Ham 853
17.2.8 Other Kamaboko 853
17.3 Change in Fish Species Used for Kamaboko Production 855
17.4 Trends of Kamaboko Products: Quality, Variety, and Nutrition 857
17.5 Scientific and Technological Enhancement in Kamaboko in Japan during the Past 10 to 15 Years 859
17.5.1 Recent Progress in Gelation Mechanism 859
17.5.2 Recent Progress in the Understanding of Setting 860
17.5.3 Recent Progress in the Myosin Denaturation Study in Relation to Gelation 862
17.5.4 Myosin Rod Aggregation at High Temperature in Relation to Gelation 863
17.5.5 Biochemical Index for the Quality Evaluation of Frozen Surimi 865
References 866

17.1 HISTORY OF KAMABOKO

Kamaboko is a fish meat-based, elastic gel product developed in Japan. A long history of kamaboko production in Japan illustrates that the Japanese have enjoyed fish meat as a food in various ways from the early days. According to the literature, kamaboko was first developed as early as 1115. The development of kamaboko gave new value to fish meat by providing a unique texture. The name "kamaboko" comes from its original shape. The original shape of kamaboko, as described in the literature, was very similar to *chikuwa*, the modern product of grilled fish meat paste on a bamboo stick. The shape resembles the ear of cattail (*gama no ho*). The Japanese term *gama no ho* was then changed to kamaboko.

The fish species used in the original production of kamaboko were freshwater species such as carp or catfish. Carp was a precious fish during that time period, so kamaboko was also considered a precious food exclusively for special people in the Imperial Court. In the 15th and 16th centuries, kamaboko was an auspicious food for the Samurai society because the "boko" in kamaboko means the pike, a symbol of Samurai spirit. During the Edo (Tokugawa) period (17th to 19th centuries), kamaboko gradually became popular among ordinary people.

Once its elastic texture was accepted by the people as a delicious food, kamaboko became a commercial product at an affordable price. The popularity of kamaboko was achieved by the introduction of new cooking methods for production, such as boiling and steaming. These methods increased the productivity as well as provided variations in the products. Availability of vegetable oil from rapeseed also added variations to kamaboko and, during the Edo period, resulted in deep-fried kamaboko.

For the mass production of kamaboko, a constant supply of fish meat was required to meet the people's demand. Manufacturers searched for fish meat suitable for kamaboko production among the species captured nearby. They also modified the processing method so that it was suitable for the species used. Consequently, region-based characteristics of local kamaboko products were developed. Even now, such traditional kamaboko can be seen throughout Japan as that described below.

17.2 VARIATIONS IN KAMABOKO PRODUCTS IN JAPAN

The Japanese developed various types of kamaboko, depending upon the method of cooking, ingredients, and shape. Kamaboko products are grouped according to heating method, shape, and ingredients. Kamaboko products in Japan include the following.

17.2.1 Steamed Kamaboko on a Wooden Board: Itatsuke Kamaboko

Steamed kamaboko is the most popular type of kamaboko in Japan. When the term "kamaboko" is used without specification, it means steamed kamaboko. Meat paste is piled up semi-cylindrically onto a wooden board and heated by steaming in a chamber. The typical size is approximately 50×150 mm with a height of 40 mm (~260 g). Products of various sizes were subsequently produced in response to consumer demand. For mass production, a slightly modified method was developed in Niigata, in which meat paste on a board was wrapped with plastic film to prevent the deformation of salted meat paste during processing. It also allowed the manufacturer to use low-quality fish meat as a raw material. This type of kamaboko is produced throughout Japan from Hokkaido to Kyushu. The most well-known place is Odawara (Kanagawa). Other well-known areas include Niigata, Toyama, Osaka, Koh-chi, and Yamaguchi.

Although the shape is the same, the gel properties are area specific. Kamaboko produced in different areas are therefore clearly distinguishable from each other by their gel properties. For example, kamaboko in the Odawara area is characterized by its high elasticity, which is achieved by the introduction of a setting process. People in Osaka and Toyama, however, prefer less elastic and soft Kamaboko, so manufacturers in these regions skip the setting process during kamaboko production. In addition to local production, large companies also produce this type. Traditionally, the gel properties of Odawara-style kamaboko are regarded as the standard in Japan.

Steamed kamaboko is the essential food for the New Year Traditional Dish in Japan. People buy it for a special dish at this holiday although they usually do not buy it during other times of the year. Consequently, it is no exaggeration to say that half of the steamed kamaboko produced in Japan is consumed during this special occasion.

17.2.2 Grilled Kamaboko on Wooden Board: Yakinuki Kamaboko

Salted fish meat placed on a wooden board, as above, is heated by grilling through the wooden board with charcoal or gas placed beneath it. Lizardfish meat is believed to be the best for this type of kamaboko. The main places for this product are Senzaki (Yamaguchi), Osaka, and Uwajima (Ehime).

17.2.3 Grilled Kamaboko on Bamboo Stick: Chikuwa

Chikuwa inherits its shape from the original kamaboko. The name *chikuwa* also comes from its shape resembling a bamboo stick. Meat paste on a wooden, originally bamboo, stick or steel skewer (recently) with a length of about 150 to 200 mm and a diameter of about 50 mm was grilled directly. Grilling gives a different flavor from the steamed product. In the modern process, meat paste is placed on a steel skewer with constant rotation and grilled by passing it through a linearly aligned gas heater. The sticks for holding meat are usually removed before shipping, which results in a hole and the formation of a bamboo stick-like shape This is one of the most popularly consumed forms of kamaboko in Japan. Many areas, such as Toyohashi (Aichi), Shimane, Nagasaki, Tokushima, Tottori, and Sendai (Miyagi), are well known for this type of production.

17.2.4 Deep-Fried Kamaboko: Age-Kamaboko

The whiteness of steamed kamaboko is one of the most important factors in determining kamaboko quality. Consequently, white-meat fish have been used as material for kamaboko production. When the deep-frying method was introduced, however, the situation changed. As deep-frying results in a dark color on the surface of the products, whiteness is not the primary factor for choosing the fish meat. Fish species having dark meat, such as sardine and mackerel, could therefore be used as material for deep-fried production. Furthermore, deep-frying is an excellent heating method to form the shape. In addition, low-quality fish meat, which was not used for other types of kamaboko production because of low gel-forming ability, can also be used.

Ironically, production of deep-fried kamaboko is the most abundant and most popularly consumed in Japan. It is very likely that its taste is well-matched to modern life. Original age-kamaboko contained no additional ingredients. However, the products changed gradually, and recent products contain peas, burdock, carrot, squid, shrimp, octopus, etc. to give more variation. This type of kamaboko is produced throughout Japan. The areas especially well known for the products included Kagoshima, Ehime, and Osaka.

17.2.5 Boiled Kamaboko: Hanpen and Tsumire

Boiled kamaboko is quite different from other kamaboko with regard to texture. Hanpen is quite an old form of kamaboko. The uniqueness of hanpen is due to its materials; shark meat and ground yam potato are essential. These two ingredients give a marshmallow-like, soft texture to the product by trapping large amounts of fine foam inside. Whipped salted shark meat with yam potato prepared as above is cooked by floating on boiling water. The usual product size is $120 \times 120 \times 20$ mm.

Tsumire is another boiled kamaboko, which is usually produced from unwashed dark fish meat, such as sardine. Ground meat, in the shape of a small dumpling with a size of 30 to 40 mm, is directly cooked by boiling. The product, as well as hanpen, is less elastic and softer than other kamaboko. Tsumire is characterized by its strong fishy smell and taste compared to other kamaboko products.

These two boiled kamaboko products, as well as deep-fried kamaboko, are essential components for the Japanese hodge-podge dish during the winter season.

17.2.6 Crab Leg Meat Analog, Crabstick: Kani-ashi Kamaboko or Kanikama

An innovative kamaboko, crab leg meat analog, was developed in 1970 using Alaska pollock surimi. Two different methods for the production were proposed. The first method involved cooked kamaboko that is finely sliced and assembled into a crabmeat fiber-like shape using meat paste as a binder of the slices. The other method consists of a sheet-like kamaboko (e.g., 1.5 mm thick and several 100 mm wide) that was first prepared, and then narrow grooves were added to it by passing it through a machine similar to a noodle-making machine. The kamaboko sheet with grooves was rolled into a crab-meat-like shape. The surface of the assembled kamaboko was then colored with red colorant to imitate the appearance of natural crab. Crab flavor was also added. The same method can be used for scallop adductor meat analog as well by altering the size.

17.2.7 Fish Sausage and Ham

This is the kamaboko that imitates pork sausage and ham. However, it is recognized as a completely different food from pork sausage in Japan. The development of this type of product increased the total consumption of kamaboko in the '60s and '70s. Originally, tuna meat was used for the production. Recent products, however, are mainly made from frozen Alaska pollock surimi. To convey the sausage-like characteristics to the product, solid pork fat, chunk meat, and spices similar to those used for pork sausage were added. Meat paste combined with the above ingredients were then stuffed into a plastic casing and cooked by steaming. The process usually does not include a smoking step. The typical size is 30 (diameter) × 200 mm. Fish ham contains more chunk meat. It the larger companies that mainly provide this product.

17.2.8 Other Kamaboko

There are several varieties of kamaboko available in supermarkets as daily foods. Manufacturers offer new products such as smoked kamaboko, kamaboko containing soy protein or tofu giving a soft texture, and one containing cheese to give a "Western" taste.

Historically well-known kamaboko-producing areas in Japan with their respective types of kamaboko are highlighted in Figure 17.1. Changes in the production of kamaboko in Japan over the past 25 years are shown in Table 17.1. Total production of Kamaboko decreased gradually and production in 2000 was about 63% of that in 1975. Crabstick also decreased in quantity by about 22% during these years. Among the items, the largest decrease was in the traditional steamed kamaboko, with a 58% reduction. Deep-fried kamaboko, however, showed the least decrease at 27%; and among all products in 2000, its production was the largest. This fact demonstrated that deep-fried kamaboko is most accepted by modern Japanese.

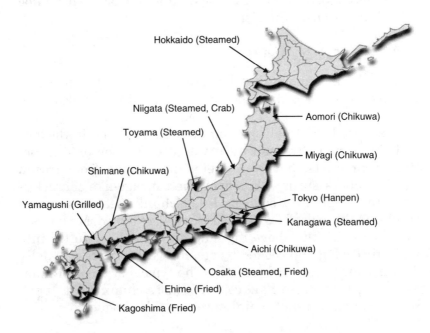

Figure 17.1 Geographical distribution of various types of kamaboko produced in Japan.

TABLE 17.1 Change in Kamaboko Production in Japan in 25 Years

Year	Total	Grilled	Steamed	Fried	Boiled	Crabstick	Others
1975	1034	259	362	327	85	—	1
1985	891	200	242	291	86	—	73
1990	829	182	223	280	54	65	25
1995	735	170	178	259	45	59	24
2000	654	160	153	236	38	51	16

Note: Unit: × 1000 metric tons.

17.3 CHANGE IN FISH SPECIES USED FOR KAMABOKO PRODUCTION

It is generally believed that fish meat easily loses its quality due to its low thermal stability.[1] Thus, the storage and transportation of fish as a raw material for kamaboko production without losing its quality was difficult in the early periods and, consequently, only local fish were used for production. To achieve a consistent quality of kamaboko, a constant supply of raw material with the same quality is also essential. Conger eel, cutlass fish, flying fish, croaker, and lizard fish were the species used in the western areas of Japan (up to the late 19th century). Modern research determined that these species produce high-quality thermal gels. A seasonal change of the catch, however, made the supply of these fish unstable.

The next stage in kamaboko production occurred when the fish caught in the Yellow Sea, such as croaker and lizard fish, became available as a material (in the early 20th century). These species were valuable species as fresh fish; however, the fish used for kamaboko were small and had low value as fresh fish.

The third stage would be the period of frozen Alaska pollock surimi. In 1960, frozen surimi was developed in Hokkaido, Japan, from the low-value fish, Alaska pollock, which is, for various reasons, used as walleye pollock in Japan. Alaska pollock had been caught in Japan mainly for its roe. The meat portion had minimal value. The development of frozen surimi, however, solved the problem of an unstable supply of raw material.

The original surimi industry in Hokkaido relied on fish caught near the coast. However, once frozen surimi proved to be an excellent material for kamaboko production, major companies sent their factory ships to the North Pacific Ocean for the production of high-quality frozen surimi using fresh fish immediately after catch. In 1970, Japan produced 0.4 million tons (t) of frozen surimi from 2.8 million t of Alaska pollock. In 1977, Russia and the United States declared a 200-mile economic restriction area, and Japanese factory ships were subsequently excluded from this area. The United States then began producing frozen surimi from Alaska pollock by introducing technology established in Japan. The surimi industry in Hokkaido was consequently beaten out by the frozen surimi produced in various other countries, especially the United States. At present, almost all of the frozen surimi consumed in Japan is imported. A main exporter is the United States.

The merits of frozen surimi are the stable supply of the raw material that enables the manufacturers to use material with constant quality and no waste processing requirement, which is necessary if whole fish are used. The unstable nature of Alaska pollock myofibrillar proteins was subsequently overcome using a cryoprotectant, such as sucrose and sorbitol. The spread of frozen Alaska pollock surimi, however, diminished the locally distinguishing characteristics of kamaboko. The sweetness derived from these additives and the loss of taste caused by the thorough washing of fish mince are some of the disadvantages of frozen surimi. Loss of character, however, was good for Western people who disliked a fishy smell. Frozen surimi therefore greatly contributed to the popularization of the various surimi seafoods in the Western world.

The next stage came about when frozen surimi was produced from other fish species. Underutilized fish species with good gelling properties have been explored. Consequently, grenadier, atka mackerel, salmon, Pacific whiting, sardine, Southern blue whiting, jurel, threadfin bream, and various other warm-water species were proven useful for surimi utilization.

17.4 TRENDS OF KAMABOKO PRODUCTS: QUALITY, VARIETY, AND NUTRITION

During surimi production, minced fish meat is washed with water to remove unfavorable flavors, blood, and fat, resulting in a concentration of myofibrillar proteins. Kamaboko might therefore be referred to as low-fat, low-calorie, low-cholesterol, as well as a high-protein food. These properties are favorable for people worrying about obesity and lifestyle diseases. In addition, amino acid composition analysis of fish myosin, a major protein component of muscle, revealed that the nutritional value, as evaluated by amino acid score, is comparable to that of land-animal myosin. Despite these advantages, however, consumption of fish itself by Japanese people has shown a decreasing trend. This is also true for kamaboko, as shown by the decreased production of kamaboko. This is partly due to the availability of cheap chicken, pork, and beef, mainly imported, compared to fish.

The washing process of fish mince, an essential step in surimi production, removes several components favorable to human health, such as fish lipids, vitamins, and minerals. These health-promoting components, however, can be returned to fish meat by adding purified compounds to the kamaboko product. Surimi is an excellent material for incorporation of these additives. The gelling ability of salted surimi makes it possible to incorporate water-insoluble fish lipids. Because the purified fish lipid has no fishy smell, the health-promoting fish lipid can be returned to the fish meat without imparting a bad smell to the product.

During the development of kamaboko containing fish lipid, the effect of fish lipid on the thermal gelation of surimi was studied.[2] Fish lipid is rich in polyunsaturated fatty acids, such as docosahexaenoic acid (DHA) and eicosapentaenoic acid (EPA). Triglyceride and phospholipids, however, give different effects (Figure 17.2). Triglycerides from sardine can be mixed homogeneously with the meat paste, up to 10%, without affecting the gelation profile and resulting in the same breaking force and strain values of the gel. This was also true for vegetable oil. On the other hand, phospholipids, both from krill

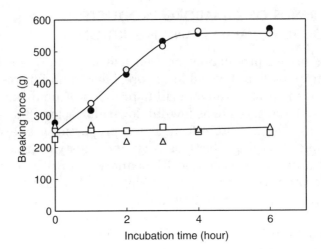

Figure 17.2 Thermal gelation of surimi in the presence of lipid. Salted surimi of Alaska pollock was incubated at 25°C for setting. The meat contained no added lipid (closed circles), 10% sardine oil (open circles), 5% phospholipid from salmon egg (squares), and one from krill (triangles). Set samples were cooked at 90°C for 30 min for thermal gel formation.

and salmon egg (and chicken egg), severely impaired the setting effect at 5% addition. Although the breaking force and strain for the directly heated gel at 90°C were unaffected by the addition, the effect was observed during the setting process. Although the myosin cross-linking reaction is believed an essential event for achieving setting,[3–5] according to SDS-PAGE pattern (Figure 17.3), the formation of cross-links (HC2, HC3, and HC4) was not affected by the presence of phospholipids (PL). This clearly demonstrated that cross-linking of myosin is not always accompanied by the setting effect.

To clarify what component is essential for the inhibition of setting by the addition of phospholipids, the effect of phosphatidylcholine and its components (lisophosphatidylcholine, phosphorylcholine, and choline) on setting was studied. It was concluded that the liposome structure in an aquatic medium was essential for causing the loss of the setting effect because only phosphatidylcholine and lisophosphatidylcholine sup-

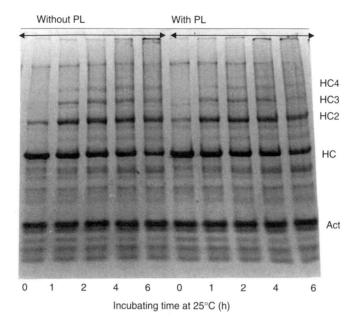

Figure 17.3 Myosin cross-linking in the presence of phospholipids. Cross-linking with or without salmon phospholipids (PL) (5%) was analyzed on SDS-PAGE with 2.5% polyacrylamide/0.5% agarose gel. HC and Act are myosin heavy chain and actin bands. Numbers with HC denote the extent of myosin heavy chain cross-linking.

pressed setting. The binding experiment of triglyceride and phospholipids to heated myofibrils showed that the inhibition of setting by phospholipids was caused by their inability to bind the myofibrils. This strongly supports the idea that hydrophobic interactions are important in the setting process, in addition to cross-linking.[6]

17.5 SCIENTIFIC AND TECHNOLOGICAL ENHANCEMENT IN KAMABOKO IN JAPAN DURING THE PAST 10 TO 15 YEARS

17.5.1 Recent Progress in Gelation Mechanism

Myosin, especially the rod portion, plays the most important role in kamaboko gel formation because the head portion of myosin alone does not form an elastic, thermal gel, unlike the rod. Gel formation requires the presence of salt to dissolve the myofibrillar proteins. These factors are essentially the same for pork sausage processing. However, there are several differences between the two products. For example, the optimal pH for sausage production is slightly acidic[7] while it is neutral for kamaboko, and kamaboko production often introduces a setting process, preheating, before cocking at high temperature, which is never employed for pork sausage.

There have been many reports on events happening during the setting and heating processes. The prominent finding was myosin cross-linking during the setting process. Analysis of the set gel on low-density polyacrylamide gel or gel containing agarose revealed that myosin heavy chain was cross-linked to produce higher molecular weight polymers (Figure 17.3). The enzyme involved in the reaction is transglutaminase.[3] The enzyme was found in Alaska pollock surimi and later in many species of fish meat, but not in land animal meat.

The reaction is certainly important to form elastic, thermal kamaboko gels. However, as described above, in the presence of phospholipid, the setting effect disappeared completely although cross-linking equally occurred (Figure 17.3). Moreover, direct heating of surimi paste without cross-linking of the myosin can produce kamaboko gel. The latter fact clearly shows that the cross-linking reaction is not essential for kamaboko gel formation and noncovalent bonds, especially hydrophobic interaction, can form kamaboko gel.

Myosin cross-linking is easily detectable on SDS-PAGE. However, myosin association by noncovalent bonds is not easy to detect. Thus, the importance of the cross-linking mechanism to the gelling mechanism was overestimated. Events occurring in setting and in heating should therefore be considered separately.

17.5.2 Recent Progress in the Understanding of Setting

Once the myosin cross-linking enzyme, transglutaminase, was found in Alaska pollock surimi, the distribution of the enzyme was surveyed, and it was concluded that all fish meat contained it although with different amounts of activity. However, there was no relationship between the activity detected and the setting effect observed for the species. For example, freshwater species, such as carp, contain a comparable or much higher activity than pollock. Nevertheless, neither the setting effect nor cross-linking occurs with this species. This result clearly showed that the properties of myosin, as the substrate of the enzyme, rather than the enzyme activity itself determined the cross-linking reaction.[8]

A similar enzyme was extracted from bacteria, *Streptoverticillum* sp.[9,10] The most striking difference of this enzyme from the endogenous one was no calcium ion requirement for activity, as well as high stability. The stable nature of the enzyme is also favorable when considering its storage as an additive, but it is not always favorable. The bacterial enzyme is more stable than the endogenous enzyme and it catalyzes the reaction through higher temperatures. But the reaction terminates as the temperature reaches more than 60°C. This would cause over-cross-linking of myosin and result in the formation of gel with much more brittle and less elastic properties than the gel produced with endogenous enzyme.[11] That is, the unstable nature of the endogenous enzyme is rather favorable to achieve the suitable extent of cross-linking; a gradual progress of the inactivation together with cross-linking occurs during the preheating process.

The importance of myosin denaturation in the setting process was proposed by studying the events happening during the preheating of salted Alaska pollock surimi.[12] The indicators used are myosin ATPase inactivation, myosin aggregation, loss of solubility into salt and into urea solutions, and the cross-linking reaction (Figure 17.4). The earliest events are loss of myosin solubility into salt and urea. The ATPase inactivation and myosin cross-linking reactions then follow and both occur at almost the same time. This order of events clearly demonstrated that cross-linking took place with the aggregated myosin in salted surimi, and the cross-linking is the event following myosin denaturation. It is said that denatured myosin is a better substrate for transglutaminase,[8]

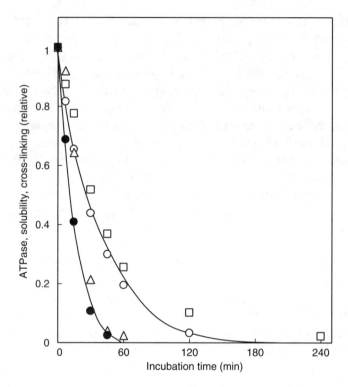

Figure 17.4 Myosin denaturation and cross-linking during the setting process. Salted Alaska pollock surimi was incubated at 25°C. Changes in Ca-ATPase activity (open circleS), solubility in 0.5 M KCl (closed circles), solubility in 8 M urea (open triangles), and the decrease in myosin heavy chain as a result of cross-linking (open squares) were evaluated.

although its gel-forming ability is low. These findings suggest that well-balanced myosin denaturation and cross-linking is therefore important for the setting effect.

17.5.3 Recent Progress in the Myosin Denaturation Study in Relation to Gelation

The flexible structure of the myosin molecule is important for biological function. Myosin conformation, like other proteins, is maintained by several types of noncovalent weak bonds, such as hydrogen, electrostatic, and hydrophobic bonds. These bonds are readily cleaved upon application of heat, but the same bonding is never formed upon cooling. The phenomenon is termed "denaturation." Thermal gelation of surimi is a kind of controlled myosin denaturation process. Thus, understanding the conformational change in myosin upon heating has been very important in understanding the gelation mechanism. Circular dichroism (CD) spectroscopy allows the the determination of the conformational change of myosin, especially the tail portion, because α-helix unfolding is the major change detectable with myosin upon heating.[13] Differential scanning calorimetry also gives similar information regarding the destruction of the myosin molecule.[14]

The chymotryptic digestion technique was introduced to study the conformational changes occurring in the head and tail portion of the myosin molecule.[15,16] Chymotrypsin cleaves the carboxyl end of the hydrophobic residues, which are usually buried inside the protein molecule and exposed only when irreversible conformational changes occur. Cleavage at the exposed sites is then easily detectable by analysis of the produced fragments on SDS-PAGE. The amino acid sequences of the myosin heavy chain for several useful species of fish have been determined;[17,18] cleavage sites can be easily identified by the determination of several amino acids of the new fragment from its amino end.

17.5.4 Myosin Rod Aggregation at High Temperature in Relation to Gelation

It is well established that myosin rod is directly involved in thermal gelation, but thermal gelation is fish species specific. Species-specific aggregation by rod as an indicator of gelation and its unfolding upon heating was studied using carp and pollock rod (Figure 17.5). Carp does not form thermal gels very easily, whereas pollock readily does so. In the unfolding profile, carp myosin rod was clearly distinguishable from pollock: two-step unfolding for carp and one-step unfolding for pollock. About two thirds of the unfolding occurred by around

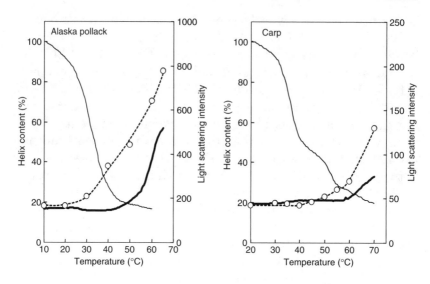

Figure 17.5 Comparison of thermal aggregation and unfolding patterns between carp and Alaska pollock myosin rod. Rod solution in 0.5 M KCl, pH 7.5, was gradually heated from 20 to 70°C at a rate of 1°C/min. Aggregation at the temperature (bold solid line), and after cooling (circles with dotted line) were both measured. The decrease in helix content (thin solid line) under identical conditions was also evaluated.

40°C, and the rest of the unfolding required 54°C, as shown in the pattern of reduction in helix content (Figure 17.5). Two-step unfolding was clearly detectable when the derivative of the unfolding profile of carp myosin rod was taken. The analysis revealed that carp rod had an unfolding peak around 36°C, corresponding to the unfolding of the light meromyosin (LMM) portion and at 52°C corresponding to that of the amino terminal 40-kDa region of subframgent-2 (S-2).

The aggregation profile was also different for the two species. Rod aggregation is the event occurring when the unfolded rod refolds upon cooling. Pollock rod forms aggregates in an early phase of unfolding with a starting temperature of 25°C, where more than 80% of the helix structure remains. On the other hand, carp rod started to aggregate at 50°C, where about 30% of the helix remains. The α-helix content recovered upon cooling the heated rod was more than 80%, although almost complete unfolding was achieved at high temperature. Comparing rod aggregation at the heating temperature with one after cooling, it was demonstrated that the cooling process is very important for aggregate formation with both fish species.

Studies with other species of fish rod revealed that the difference in the aggregation profile originates from whether or not the unfolding profile contained a high-temperature peak of the stable S-2 portion. Species that contain a stable S-2 region require higher temperatures for aggregation compared with those lacking the peak. These results suggest that the aggregation properties of myosin are determined by its myosin rod structure.

17.5.5 Biochemical Index for the Quality Evaluation of Frozen Surimi

There are many proposals to evaluate the quality of frozen surimi. As surimi is a raw material for kamaboko gel formation, indicators commonly used are rheological parameters of the final thermal gel obtained by the puncture test or torsion test. Indeed, these are a good index for the quality of kamaboko gel but are indirect in their approach to surimi. Because surimi is referred to as concentrated myofibrils and myosin is the key protein in gelation, it is reasonable to assume that myosin content and its extent of denaturation in surimi would be the index for its "true" quality.

The first proposal was made as early as 1970 by Arai's group at Hokkaido University.[19,20] They extracted actomyosin from surimi and its specific Ca-ATPase activity was measured as the index of native myosin content. As the extraction of actomyosin was too complicated to perform, they modified the method in 1979, using myofibrils instead of actomyosin for analysis.[21] Total activity calculated from the specific Ca-ATPase activity and total protein content of the surimi was then employed as the index for surimi quality. Arai et al. showed that there was a positive correlation between the total Ca-ATPase activity and the resulting gel strength of the kamaboko. However, the index has been rarely used since then, probably because the measurement process, especially ATPase measurement, is too complicated to employ.

For the people who handle myofibrillar protein, ATPase activity is one of the most useful and sensitive indicators in studying myosin denaturation, not only in surimi but also in many other situations. The ATPase activity is usually assayed by measuring inorganic phosphate (Pi) liberated from ATP catalyzed by myosin. Recently, an innovative method of pH-stat titration was introduced as an ATPase assay system. In the ATPase assay system, the hydrolysis of ATP liberates Pi, and subsequently Pi liberates H^+. The apparatus detects the liberated H^+ just like a pH meter, and pumps out alkaline solution to compensate. Consequently, the ATP hydrolysis reaction can be monitored by reading the amount of NaOH solution pumped to neutralize the liberated H^+. The ease of this method could therefore expand the application of ATPase activity measurement for the study of fish muscle protein.

REFERENCES

1. J.J. Connell. Studies on the proteins of fish skeletal muscle. 7. Denaturation and aggregation of cod myosin. *Biochem. J.*, 75, 530–538, 1960.

2. K. Konno, H. Naraoka, and K. Akamatsu. Suppressive effect of phosphatidylcholine on the thermal gelation of Alaska pollock surimi. *J. Agric. Food Chem.*, 46, 1262–1267, 1998.

3. N. Seki, H. Uno, N.-H. Lee, I. Kimura, K. Toyoda, T. Fujita, and K Arai. Transglutaminase activity in Alaska pollock muscle and surimi, and its reaction with myosin B. *Nippon Suisan Gakkaishi*, 56, 125–132, 1990.

4. T. Numakura, N. Seki, I. Kimura, K. Toyoda, T. Fujita, K. Takama, and K. Arai. Cross-linking reaction of myosin in the fish paste during setting (Suwari). *Nippon Suisan Gakkaishi*, 51, 1559–1565, 1985.

5. I. Kimura, M. Sugimoto, K. Toyoda, N. Seki, K. Arai, and T. Fujita. A study on the cross-linking reaction of myosin in kamaboko "Suwari" gels. *Nippon Suisan Gakkaishi*, 57, 1386–1396, 1991.

6. E. Niwa, T. Nakayama, and I. Hamada. The third evidence for the participation of hydrophobic interactions in fish flesh gel formation. *Nippon Suisan Gakkaishi,* 49, 1763, 1983.

7. K. Samejima, M. Ishioroshi, and T. Yasui. Relative roles of the head and tail portions of the molecule in heat-induced gelation of myosin. *J. Food Sci.,* 46, 1412–1418, 1981.

8. H. Araki and N. Seki. Comparison of reactivity of transglutaminase to various fish actomyosin. *Nippon Suisan Gakkaishi,* 59, 711–716, 1993.

9. K. Seguro, Y. Kumazawa, T. Ohtsuka, S. Toiguchi, and M. Motoki. Microbial transglutaminase and ε-(γ-glutamyl)lysine cross-link effects on elastic properties of kamaboko gels. *J. Food Sci.,* 60, 305–311, 1995.

10. H. Ando, M. Adachi, K. Umeda, A. Matsuura, M. Nonaka, R. Uchio, H. Tanaka, and M. Motoki. Purification and characteristics of a novel transglutaminase derived from microorganism. *Agric. Biol. Chem.,* 53, 2613–2617, 1989.

11. Y. Abe. Quality of Kamaboko gel prepared from walleye pollock surimi with an additive containing transglutaminase. *Nippon Suisan Gakkaishi,* 60, 381–387, 1994.

12. K. Konno and K. Imamura. Identification of the 150 and 70 kDa fragments generated during the incubation of salted surimi paste of walleye pollock. *Nippon Suisan Gakkaishi,* 66, 869–875, 2000.

13. M. Nakaya, M. Kakinuma, S. Watabe, and T. Ooi. Differential scanning calorimetry and CD spectrometry of acclimation temperature-associated types of carp light meromyosin. *Biochemistry,* 36, 9179–9184, 1997.

14. M. Nakaya, S. Watabe, and T. Ooi. Differences in the thermal stability of acclimation temperature-associated types of carp myosin and its rod on differential scanning calorimetry. *Biochemistry,* 34, 3114–3120, 1995.

15. M. Nakaya, M. Kakinuma, S. Watabe, and T. Ooi. Differential scanning calorimetry and CD Spectrometry of acclimation temperature-associated types of carp light meromyosin. *Biochemistry,* 36, 9179–9184, 1997.

16. K. Konno, T. Yamamoto, M. Takahashi, and S. Kato. Early structural changes in myosin rod upon heating of carp myofibrils. *J. Agric. Food Chem.*, 48, 4905–4909, 2000.

17. M. Togashi, M. Kakimuma, Y. Hirayama, H. Fukushima, S. Watabe, T. Ojima, and K. Nishita. cDNA cloning of myosin rod and the complete primary structure of myosin heavy chain of walleye pollock fast skeletal muscle. *Fisheries Sci.*, 66, 349–357, 2000.

18. S.H. Yoon, M. Kakinuma, Y. Hirayama, T. Yamamoto, and S. Watabe. cDNA cloning of myosin heavy chain from white croaker fast skeletal muscle and characterization of its complete primary structure. *Fisheries Sci.*, 66, 1163–1171, 2000.

19. K. Kawashima, K. Arai, and T. Saito. Studies on muscular proteins of fish. X. The amount of actomyosin in frozen "Surimi" from Alaska pollock. *Nippon Suisan Gakkaishi*, 39, 525–532, 1973.

20. K. Kawashima, K. Arai, and T. Saito. Studies on muscular proteins of fish. XIII. Relationship between the amount of actomyosin in frozen surimi and the quality of kamaboko from the same material in Alaska pollock. *Nippon Suisan Gakkaishi*, 39, 1201–1209, 1973.

21. N. Katoh, H. Nozaki, K. Komatsu, K. Arai. A new method for evaluation of the quality of frozen surimi from Alaska pollock. Relationship between myofibrillar ATPase activity and Kamaboko forming ability of frozen surimi. *Nippon Suisan Gakkaishi*, 45, 1027–1032, 1979.

Appendix

Code of Practice for Frozen Surimi

Joint FAO/WHO Food Standards Program
Codex Alimentarius Commission

This Code of Practice has been written for the use of those engaged in the frozen surimi production industry.

Frozen surimi, which is raw material and not intended for direct human consumption, in brief terms, is myofibrillar protein isolated from fish meat protein by washing that is further heat-treated and consumed in the form of surimi-based products. It should be kept in mind that frozen surimi was originally developed as raw material for surimi-gel that is produced by taking advantage of the gel-forming ability possessed by myofibrillar proteins. Therefore, certain properties specifically required for surimi-based products should be taken into consideration, and it should be fully understood that it is in this point that code of practice for frozen surimi is different from the codes of practice for all the other fish products.

This Code of Practice provides for technological, essential hygienic and quality inspection requirements for the production of frozen surimi that can be used for manufacturing high-quality surimi-based products, and is based on established and recognized good commercial practices.

In addition, this Code is intended for guidelines for the elaboration of quality standards and quality control inspection programmes in countries where these have not yet been established.

However, since most of the practical information pertaining to the technology and hygiene of the production of frozen surimi has been based upon experiences gained in Japan and the United States of America, this Code is not intended to be strictly applied in all countries producing frozen surimi. The establishment of a code of any country, in accordance with this Code, will probably require the consideration of various conditions and consumers' tastes in the country concerned. In other words, a national code of practice of any country could be developed from the information contained in this Code supplemented by taking into consideration the species of fish and the various conditions of the country in question.

Moreover, this Code has been prepared based on Alaska pollock (*Theragra chalcogramma*), which constitutes the great majority of frozen surimi production in the world. This Code, though, will require periodic revision, since the increase of surimi made from other fish species, as well as further technological development, can be foreseen.

I. ESSENTIAL FINAL PRODUCT REQUIREMENTS

These end-product specifications describe the essential requirements for frozen surimi. These essential requirements are factors describing the minimal health and hygiene provisions, which must be met in order to comply with the requirements contained in Codex standards.

Frozen surimi is myofibrillar protein concentrate prepared from fish meat without retaining the original shape of fish, so that it is difficult to determine its quality from its appearance. Moreover, it is generally not consumed directly, but further processed. This means, therefore, that the quality of frozen surimi is measured by the compositional properties and the functional properties of surimi-based products. Therefore, it is strongly recommended to inspect functional properties, such as those outlined in Appendix II, which are different from those for other fishery products.

A. Essential Health and Hygiene Requirements

When tested by appropriate methods of sampling and examination prescribed by Codex Alimentarius Commission (CAC), the product:

1. Shall be free from microorganisms or substances originating from microorganisms in amounts that may represent a hazard to health in accordance with standards established by the CAC
2. Shall not contain any other substance in amounts that may represent a hazard to health in accordance with standards established by the CAC

The final product shall be free from any foreign material that poses a threat to human health.

B. Essential Final Product Quality Requirements

When tested by appropriate sampling and acceptance procedures prescribed by the CAC, the presence of the following defects in sample units of final product will render the sample unit in noncompliance.

1. Foreign Matter

The presence in a sample unit of any matter that has not been derived from fish (excluding packaging material), does not pose a threat to human health, and is readily recognized without magnification or is present at a level determined by any method, including magnification, that indicates noncompliance with good manufacturing and sanitation practices.

2. Odor and Flavor

Surimi affected by persistent and distinct objectionable odors or flavors indicative of decomposition or rancidity.

II. OPTIONAL FINAL PRODUCT REQUIREMENTS

These end-product specifications describe the optional defects for frozen surimi. The descriptions of optional defects will assist buyers and sellers in describing the defect provisions, which are often used in commercial transactions or in designing specifications for final products.

Frozen surimi is myofibrillar protein concentrate prepared from fish meat without retaining the original shape of fish, so that it is difficult to determine its quality from its appearance. Moreover, it is generally not consumed directly, but further processed. This means that the quality of frozen surimi is measured by the compositional properties and the functional properties of surimi-based products. Therefore, it is strongly recommended to inspect functional properties, such as the following quality attributes, which are different from those for other fishery products.

It is most important to evaluate the following primary test attributes: moisture content, pH, and objectionable matter of raw surimi and gel strength, deformability, and color of cooked surimi gel. Other secondary attributes may be measured as desired.

1. Primary Quality Attribute

1.1 Raw Surimi Tests

Preparation of test sample:

Put 2–10 kg frozen surimi in a polyethylene bag, seal the bag, and temper the surimi at room temperature (20°C) or below so that the temperature of the surimi rises to approximately –5°C. Do not soften the surface of the test sample.

1.1.1 Moisture

Samples for moisture content should be taken from the interior of a surimi block to ensure that no freezer burn (surface dehydration) of the sample has occurred. Put the test sample in a polyethylene bag or polyethylene bottle, seal the bag or bottle, and let the test sample thaw so the temperature of the sealed article rises to room temperature. Then measure the moisture using any of the following methods:

> In case of using a drying oven method, see AOAC Method.
> In case of using an infrared lamp moisture tester, take out 5 g of the test sample precisely weighed with a sample tray, and dry it immediately;
> In case of using a microwave drying moisture tester, see AOAC Method.

Calculate the moisture according to the following formula to the first decimal place.

In using any of the measurement methods, test two or more pieces of the test sample and indicate the average value obtained therefrom.

When measuring a fatty test sample with a microwave drying moisture tester, cover the top of the sample tray with glass fiber paper to prevent fat from splashing, while drying.

$$\text{Moisture } (\%) = \frac{\text{Pre-dry weight (g)} - \text{After-dry weight (g)}}{\text{Pre-dry weight}} \times 100$$

1.1.2 pH

Add 90 or 190 ml distilled water to 10 g test sample as needed to disperse. Homogenize the mixture and then measure pH of the suspension with a glass electrode pH meter to the second decimal place. Indicate the value obtained thereby.

1.1.3 Objectionable Matter

The term "objectionable matter" as used in this item shall mean skin, small bone, and any objectionable matter other than fish meat.

Spread 10 g test sample to the thickness of 1 mm or less and count the number of visible objectionable matter in it. Indicate the value obtained there by, provided an objectionable matter of 2 mm or larger shall be counted as one and an objectionable matter smaller than 2 mm shall be counted as one half and any unnoticeable matter smaller than 1 mm shall be disregarded.
The inspection method for distinguishing scales visibly unnoticeable is specified in Appendix 2.1.1.

1.2 Cooked Surimi Gel Tests

1.2.1 Gel Gtrength and Deformability

Two methods are presented here. The test to use should be decided upon between buyer and seller.

1.2.1.1 Puncture Test. Preparation of test sample:
Put 2-10 kg of frozen surimi in a polyethylene bag, seal the bag, and temper the surimi at room temperature (20°C) or below so the temperature of the surimi rises to approximately −5°C. Do not soften the surface of the test sample.

Preparation of surimi gel for testing: surimi gel not containing added starch

A. Comminution

Sample volume necessary for surimi paste preparation depends on the capacity of mixing instrument used. Use of 1.5 kg or more is necessary to represent the property of a 10-kg block. Regarding that enough amount of surimi is necessary for consistency of testing, equipment of large capacity, which can mix surimi of 1.5 kg or more, must be installed in laboratory. When you use larger size of the equipment, you also need to put in adequate amount of surimi in accordance with equipment to secure enough texture of surimi paste. Crush 1.5 kg or more of the test sample with a silent cutter, then add 3% of salt. Further grind and mash the sample for 10 minutes or more into homogenized meat paste. Remember to keep the temperature of the material to be tested at 10°C or less.

Desirable timing for adding salt is at −1.5°C.
Desirable temperature of the test material is 5–8°C.

B.   Stuffing

Stuff approximately 150 g of the meat paste into a polyvinylidene chloride tube of 48 mm width (30 mm in diameter) and then flatten (resulting in approximately 20 cm in length). Stuffing can be done with an 18 mm diameter stuffing tube. After stuffing, tie both ends of the tube.

C.   Heating

Heat the test material in hot water of 87 ± 3°C for 30 minutes. At the time the test material is put in, the temperature drop should not exceed 3°C.

D.   Cooling

Immediately after finishing the heating treatment, put the test material in cold water and fully cool it, and then leave it at the room temperature for 3 hours or longer.

Test Method

Perform between 24 and 48 hours after cooking the following measurements of the prepared inspection sample of surimi gel. The temperature of gel should equilibrate to the room temperature and record the temperature of the sample at the time of measurement.

Measure gel strength and deformability of the inspection sample of surimi gel with a squeeze stress tester (rheometer). Use a spherical plunger, of which diameter shall be 5 mm and speed shall be 60 mm/minute.

Remove film off the inspection sample of surimi gel, cut into 25 mm long test specimen, and place test specimen on the sample deck of the tester so the center of the test specimen will come just under the plunger. Apply load to the plunger and measure the penetration force in g and the deformation in mm at breakage.

Record the obtained value of the penetration and deformation in g by integral number. Record the obtained value of the deformation in mm to the first decimal place.

Prepare six or more test specimens from the same inspection sample of surimi gel and test each of them. Record the average values obtained thereby.

1.2.1.2 Torsion Test

Preparation of the surimi gel test specimen

A. Comminution

Temper frozen surimi at room temperature (near 25°C) for 1 hour or in a refrigerated tempering room to approximately −5°C. Cut the tempered surimi blocks into slices or chunks and load into the bowl of a silent cutter or cutter/mixer equipped for vacuum use. First, reduce the frozen surimi to a powder by comminution at low speed without vacuum. Add sodium chloride (2% on total batch weight basis) and ice/water (sufficient to obtain 78% final moisture content on total batch weight basis). Secure the lid and begin chopping again at low speed with no vacuum, gradually (if possible) increasing to high speed (about 2000 rpm). At the point that the mixture becomes a single mass, turn on the vacuum pump and allow approximately 70–80% of a full vacuum (approximately 20–25 inch Hg or 500–650 mm Hg) to be obtained. During comminution, ensure that paste is scraped from the walls and balls of paste are forced down into the blades of a cutter/mixer. Discontinue chopping when a temperature of 5–8°C is obtained. A minimum 6 min chopping time is recommended.

B. Stuffing

Transfer the paste to the sausage stuffer with a minimum of air incorporation. Maintain paste temperature below 10°C at all times. Stuff into polycarbonate or stainless steel tubes 1.9 cm (i.d.) of an appropriate length, typically about 20 cm. Tubes should be sprayed with lecithin release agent before filling. Stuff the paste uniformly and without air pockets into tubes. Cap or seal both ends and place in ice bath until ready to heat process (within one hour).

C. Heating

Heat process by immersing filled tubes in a water bath previously equilibrated to the proper temperature. Time–temperature relationships for thermal processing are: low temperature setting ability: 0–4°C for 12–18 hours, followed by 90°C for 15 min; median temperature setting ability: 25°C for 3 hours, followed immediately by 90°C for 15 min; high temperature setting ability: 40°C for 30 min, followed immediately by 90°C for 15 min; evaluation of protease activity: 60°C for 30 min, followed immediately by 90°C for 15 min; rapid cooking effect: 90°C for 15 min. It is recommended that water baths be heated to about 5°C higher than the intended treatment temperature, to account for the heat loss experienced upon loading, and the temperature be adjusted approximately within 2 min, possibly requiring ice addition.

Only cold water species will demonstrate good setting ability at lower temperatures. The heat process used to prepare the sample should be specified; if not, it is assumed that only the rapid cooking effect is being assessed. Relative proteolytic activity is assessed by comparing tests conducted on gels prepared at 60°/90°C with those processed only at 90°C.

Ohmic heating can be used as a means of heating method. Heat is uniformly generated through electrical resistance. Paste placed in a chlorinated PVC tube is heated between two electrodes. Internal temperature of 90°C can be reached within 1 min. Heating rate (fast and slow) can be controlled linearly. This method provides another advantage: Pacific whiting surimi or others with proteolytic enzymes can be successfully gelled (without enzyme inhibitors) under ohmic heating because fast heating can inactivate the enzyme.

D. Cooling

After heat processing, quickly transfer tubes to an ice water bath and equilibrate to 0°C. Remove gels from tubes with a plunger and seal in plastic bags. Keep samples refrigerated until tested (within 48 hours).

Test Method

Perform within 24 hours the following measurements of the prepared inspection sample of surimi gel, whose temperature should be equilibrated to the room temperature (20–25°C).

Measurement of Stress and Strain

The gel-forming ability of surimi is evidenced by the fundamental rheological properties of the test product when strained to failure (breakage). Allow refrigerated samples to reach room temperature (near 25°C) before testing. Cut test specimens to length of about 30 mm. Attach specimens to mounting discs at each flat end with cyanoacrylate glue, being careful to place samples in center of mounting discs. Mill center of test specimens to a capstan shape, the milled portion being 1 cm. in diameter. Mount the milled test specimen in the torsion rheometer. Rotate top of sample to the point of sample failure (breakage), record torque and rotational distance at this point. Calculate and report stress and strain, respectively, at sample failure as: Stress = τ = 1581 × (torque units); Strain = $\ln[1+(\gamma^2/2) + \gamma(1+\gamma^2/4)^{0.5}]$, where γ = 0.150 × (rotational distance, mm) − 0.00847 × (torque units). In practice these equations are normally programmed onto a computer linked to the torsion rheometer for data acquisition and analysis, thus yielding directly the stress and strain measurements.

1.2.2 Color

Cut the inspection sample of surimi gel into flat and smooth slices of 15 mm or more thickness, and immediately measure, with a color-difference meter, the cross section of the slice pieces for the values L^* (lightness), a^* (red-green), and b^* (yellow-blue) to the first decimal place. Test three or more slice pieces and indicate the averages of the values obtained thereby.

2. Secondary Quality Attributes

2.1 Raw Surimi Tests

Preparation of test sample:
Put 2–10 kg of frozen surimi in a polyethylene bag, seal the bag, and defrost the surimi at room temperature (20°C) or below so the temperature of the surimi rises to approximately −5°C. Do not soften the surface of the test sample.

2.1.1 Objectionable Matter (Scales)

After measurement, according to Appendix 1.1.3, add 100 ml of water to the same test sample, homogenize, add 100 ml of 0.2 M NaOH solution to it and dissolve with a stirrer. Filter the dissolved solution with filter paper (No. 2), wash the residue with water, and then dry it at 105°C for 2 hours. Count the number of scales obtained thereby and indicate that number in brackets appearing subsequent to the number of the objectionable matter according to Appendix 1.1.3.

After having dissolved, leave the dissolved solution still to ensure precipitation, and scoop up as much skim as possible before filtration.

2.1.2 Crude Protein Content

AOAC Kjeldahl Method

2.1.3 Sugar Content

Precisely weigh 10 g of the test sample, put it in a 50 ml beaker, add 10 ml of 2% trichloroacetic acid (TCA) solution, and fully stir the material. Leave it still for approximately 10 min, stir again, and leave still for 10 min. Filter with filter paper (No. 2), drop some part of the filtered liquid on a refractometer (for Brix 0-10% use), and read the graduation on the refractometer. Apply the reading to the following formula and calculate a value to the first decimal place. Indicate the value obtained thereby.

Calibrate in advance the refractometer at a specified temperature with distilled water.

$$\text{Sugar (\%)} = 2.04 \times \text{Brix (\%)} - 2.98$$

2.1.4 Crude Fat Content

Put in a mortar, a precisely weighed 5–10 g of test sample with approximately the same quantity of anhydrous sodium sulfate and a small amount of refined sea sand. Mash the material uniformly into dry powder and put it in a cylindrical filter paper. Do not fail to take out and put in the cylindrical filter paper the powder remaining in the mortar by the use of a small amount of ethyl ether and absorbent cotton. Extract and determine the fat according to Soxhlet method and calculate a value according to the following formula to the first decimal place. Indicate the value obtained thereby.

Fill the ends of the cylindrical filter paper with a slight amount of absorbent cotton so the material to be tested will not fall out.

Dry the extraction receptacle in advance at 100–106°C, and weigh it. Extraction speed shall be 20 times per hour.

$$\text{Crude Fat (\%)} = \frac{(W_1 - W_0)}{S} \times 100$$

where
S = Quantity of test sample taken (g)
W_0 = Weight of receptacle (g)
W_1 = Weight of receptacle after fat has been extracted (g)

2.1.5 Color and Whiteness

Color: Temper frozen surimi completely to room temperature (near 25°C). Fill into a 50 ml glass beaker (4 cm diameter, 5.5 cm height) and measure color values of L^*, a^*, and b^* (CIE Lab system) to the first decimal point. Complete contact between the test specimen and the colorimeter measurement port, as well as filling of the beaker with no voids, is recommended for consistent results. Measure three or more samples and record the average value.

Whiteness: Whiteness can be calculated as:

$$\text{Whiteness} = L^* - 3b^*$$

or

$$\text{Whiteness} = 100 - [(100 - L^*)^2 + a^{*2} + b^{*2}]^{0.5}.$$

2.1.6 Pressure Induced Drip

Defrost 50 g of the test sample and put it in a circular cylinder of 35 mm inner diameter and 120–150 mm long made of stainless steel or synthetic resin, which has 21 holes of 1.5 mm diameter, 3 mm from each other, opened in the bottom. Immediately apply 1 kg of load with a pressurizing cylindrical rod of 34 mm diameter, of which weight shall be included in the load. Leave as it is for 20 minutes and then measure the weight of the dripped liquid. Calculate its percentage to the weight of the test sample to the first decimal place. Indicate the value obtained thereby.

2.2 Cooked Surimi Tests

2.2.1 Preparation of Test Sample

2.2.1.1 Water-Added Surimi Gel:

A. Comminution

Sample volume necessary for surimi paste preparation depends on the capacity of mixing instrument used. Use of

1.5 kg or more is necessary to represent the property of 10 kg of block. Regarding that enough amount of surimi is necessary for consistency of testing, equipment of large capacity, which can mix surimi of 1.5 kg or more, must be installed in laboratory. When you use larger size of the equipment, you also need to put in adequate amount of surimi in accordance with equipment to secure enough texture of surimi paste. Crush 1.5 kg or more of the test sample with a silent cutter, then add 3% of salt and 20% of 3% cooled salt water, and further grind and mash, for 10 min or more, into a homogenized meat paste. However, if using the remaining water-unadded, starch-unadded test material under Appendix 1.2.1.1.A, add 20% of 3% cooled salt water only and further grind and mash it for 5 min into homogenized meat paste, while keeping the temperature at 10°C or less for cold water species, such as Alaska pollock (*Theragra chalcogramma*). Warm water species may be processed at a slightly higher temperature (not to exceed 15°C). However, better quality will be achieved at a lower temperature.

B.   Casing

Same as Appendix 1.2.1.1.B.

C.   Heating

Same as Appendix 1.2.1.1.C.

D.   Cooling

Same as Appendix 1.2.1.1.D.

2.2.1.2   Starch-Added Surimi Gel

A.   Comminution

Add 5% of potato starch to the meat paste prepared according to the method under Appendix 1.2.1.2.A and mix (homogenize) within 5 min. Remember to keep the temperature of the test material at 10°C or below all the while. Desirable temperature of the test material is 7–8°C.

B. Stuffing

Same as Appendix 1.2.1.2.B.

C. Heating

Same as Appendix 1.2.1.2.C. However, if performing treatment to secure suwari (setting), same as Appendix 2.2.4.1.C suwari-treated surimi gel c.

D. Cooling

Same as Appendix 1.2.1.1.D.

2.2.1.3   Suwari (Setting)-Treated Surimi Gel

A. Grinding and Mashing

Same as Appendix 1.2.1.1.A.

B. Casing

Same as Appendix 1.2.1.1.B.

C. Heating

After treatment to secure Suwari (setting) in warm water of 30 ± 2°C for 60 minutes and perform the same heating as Appendix 1.2.1.1.C.

D. Cooling

Same as Appendix 1.2.1.1.D.

2.2.2   Test Method

Perform between 24 and 48 hours after cooking the following measurements of the prepared inspection sample of surimi gel whose temperature should equilibrate to the room temperature and record the temperature of the sample at the time of measurement.

2.2.2.1 Whiteness

Whiteness, as an index for the general appearance of a surimi gel, can be calculated as:

$$\text{Whiteness} = L^* - 3b^*.$$

or:

$$\text{Whiteness} = 100 - [(100 - L^*)^2 + a^{*2} + b^{*2}]^{0.5}.$$

2.2.2.2 Expressible Moisture

Place a slice of surimi gel (2 cm diameter, 0.3 cm thick, and about 1 g in weight) between two filter papers and press them by an oil pressure equipment under a fixed pressure (10 kg/cm^2) for 20 sec.

Calculate the expressible water according to the following formula to the first decimal place. Test three or more pieces of the test sample, and indicate the average value obtained thereby.

$$\text{Expressible water (\%)} = \frac{\text{Pre-pressed weight (g)-after-pressed weight (g)}}{\text{Pre-pressed weight (g)}}$$

Water holding capacity is also used as an index of surimi gel as well as the expressible water. Water holding capacity (%) is calculated as follows.

$$\text{Water holding capacity (\%)} = \frac{\text{Expressible water content (g)}}{\text{Total moisture content of pre-pressed sample (g)}} \times 100$$

2.2.2.3 Folding Test

The folding test is conducted by folding a 5-millimeter thick slice of gel slowly in half and in half again while examining it for signs of structural failure (cracks). Make sure the sample is folded completely in half. Keep the folded state for five sec, then evaluate the change in the shape by 5-stage merit marks. The minimum amount of folding required to produce a crack in the gel determines the score for this test. Test three or more slice pieces of the same inspection sample and indicate the average mark obtained. In case of folding by hand, apply constant power throughout the folding surface.

Merit Mark	Property
5	No crack occurs even if folded in four.
4	No crack occurs if folded in two, but a crack(s) occur(s) if folded in four.
3	No crack occurs if folded in two, but splits if folded in four.
2	Cracks if folded in two.
1	Splits into two if folded in two.

2.2.2.4 Sensory (Biting) Test

Bite a 5 mm thick slice piece of the gel sample and evaluate its resilience upon touch to teeth and cohesiveness upon bite by 10-stage merit marks. Test three or more slice pieces of the same inspection sample by a panel consisting of three or more experts and indicate the average mark obtained thereby. Merit marks 2, 3, 4, 5, and 6 corresponds to the folding merit marks 1, 2, 3, 4, and 5 under (2), respectively.

Merit Mark	"Ashi" (footing) Strength	Merit Mark	"Ashi" (footing) Strength
10	Extremely strong	5	Slightly weak
9	Very strong	4	Weak
8	Strong	3	Very weak
7	Slightly strong	2	Extremely weak
6	Fair	1	Incapable to form gel

Index

A

ABC, *see* Allowable biological catch
Abdominal cavity contents, removal of, 123
Acid detergents, 155
Acid proteases, 232
Actin, 441, 444
Activated sludge, 297, 298
Activation energy, species habitat temperature and, 171, 172
Actomyosin, 35, 437
   cross-linking between myofibrillar protein constituents and, 463
   pyrophosphate and, 120
   sodium lactate and, 188
Adenosine triphosphate (ATP), 161, 472
   ability of myosin to split, 473
   bioluminometric assay, 161
   breakdown, 472
   nucleotide of, 196
ADISUR, *see* Association for the Development of Surimi Industry
ADP, nucleotide of, 196
Aeromonas hydrophila, 587, 589
AFA, *see* American Fisheries Act
AFGPs, *see* Antifreeze glycoproteins
AFPs, *see* Antifreeze proteins
Age-kamaboko, 851
Aggregation, 446
Agricultural fertilizers, fish by-products as, 289
Air blast freezer, 335
Airflow freezers, 331
Air-included packaging, 637
Alaska pollock
   Code of Practice based on, 870
   compositional properties, 49, 50, 52
   effectiveness of BPP on gels from, 253
   flow behavior of sols from, 503
   freeze denaturation of proteins in, 35
   gels, fracture shear stress of, 453
   harvesting of, 4, 54, 376
   infestation of with microsporia, 49
   microbial populations associated with, 590
   processing, increased recovery of surimi in, 11
   surimi
      development of, 855
      electron penetration experiments, 631
      electrophoresis of, 635, 636

production of by main
producers, 9
sol, dynamic properties of,
513, 515
transglutaminase, 860–861
surimi gels
properties, 91
stress values, 527
surimi paste
changes in dynamic
properties of, 517
viscosity of, 552, 553, 554
thermal treatments in forming
gels from, 544–545
use of in United States, 388
utilization, U.S., 10
washwater, recovery of proteins
from, 305
Alcaligenes faecalis var. myogenes,
684
Alginate, uses, 685
Alkaline processing, 119, 120
Alkaline proteinases, 245
Allowable biological catch (ABC),
17
Ambient-air cooling, 623
Ambient temperature, 350
American Fisheries Act (AFA), 12,
292
Amino acids, 127, 727
Ammonia refrigerant, 329, 330
Amylase, 657
Amylopectin, 657, 662
Amylose, 662, 659
Anhydrobiosis, 206
Animal body temperature, 170
ANN, *see* Artificial neural network
Annatto
achievement of desired hue
using, 756, 780
bixin form, 780
chemical structure of, 782
norbixin form, 782
photographs of, 780
Anserine, 309

Antarctic fish, myosin rod purified
from, 171
Anthocyanins, 784
Antifreeze
glycoproteins (AFGPs), 176
proteins (AFPs), 165, 175, 176,
211
Antioxidant(s), 189, 310
mixture, 121
surimi shelf life and, 211
Apparent viscosity, 552
Arginine, 452
Argyrosomus argenteus, 23
Aroma components, volatile, 731
Arrhenius relation, 562
Arrowtooth flounder, 16, 49
effectiveness of BPP on gels
from, 253
gel softening in, 308
protease inhibitors, 17
proteolytic activity, 229
proteolytic enzyme levels in,
449
Artificial flavoring substances,
definition of, 743
Artificial neural network (ANN), 77
automatic learning by, 79
basic architecture of, 78
Ascaris lumbricoides, 237
Aseptic packaging, 637
Asperigillus oryzae, 251
Association for the Development of
Surimi Industry
(ADISUR), 384, 386
Atheresthes stomias, 16, 49
Atlantic croaker
alkaline proteinase in, 246–247
gel texture, 417
proteinases, potato extracts and,
253
strongest gels obtained, 467
thermal treatments in forming
gels from, 544–545
Atlantic herring, 120
Atlantic mackerel, hemoglobin
present in washed, 121

Atlantic menhaden, 49
  alkaline proteinase in, 247
  effectiveness of BPP on gels from, 253
  proteolytic activity, 229
  proteolytic enzyme levels in, 449
Atlantic salmon, alkaline proteinase in, 246–247
ATP, *see* Adenosine triphosphate
ATPase activity, residual, 437
Atrobucca nibe, 23
At-sea processors, 40
Australia, crabstick industry in, 387
Autolyzed yeast extract (AYE), 729
Axial compression, 519
Axial tensile test, 499
AYE, *see* Autolyzed yeast extract

B

Bacillus subtilis, 638
Bacteria
  -eating protozoa, 297
  Gram-positive, 629
  inhibited growth, 586
Barracuda, 19, 23, 24
Batch blast freezer, 335
Batch-continuous systems, 334
Beam theory, 521
Beef plasma protein (BPP), 253, 254, 676, 677
Beet juice concentrate, 785
Behavioral methodologies, consumer testing, 829
Bench-top microwave testing unit, 96
Bereche, 25, 36
Beta vulgaris ruba, 785
Bigeye snapper, 19, 22, 246–247
Bioactive compounds, recovery of, 306, 310
Biological oxygen demand (BOD), 286, 294, 301, 315
Biological treatment
  aerobic process, 297
  anaerobic process, 298
Biting test, 885
Bixa orellano, 780
Blast freezer, 334, 335, 336, 358, 367
Bleeding
  description of, 774
  testing for, 775, 777, 778
Block(s)
  formation, 417
  freezing rate, 342
  freezing time, 357
  over-filled, 361
  simulated temperatures in, 362
  temperature, simulated, 365
  thickness, predicted freezing time vs., 360
BOD, *see* Biological oxygen demand
Boltzmann constant, 562
Botulism, outbreaks of human, 596
Bovine cathepsin E activity, 237
Bovine serum albumin (BSA), 211, 653
Bovine spongiform encephalopathy (BSE), 43, 256
Box tolerance system, color, 793
BPP, *see* Beef plasma protein
Brevoorti tyrannus, 49
Brine
  calcium chloride, 337
  freezers, 336
  sodium chloride, 337
Brookfield viscometer, 526
Brown refiner, 41
BSA, *see* Bovine serum albumin
BSE, *see* Bovine spongiform encephalopathy
Bulk modulus, 494
Bundling process, description of, 405, 406
By-product(s)
  categorizing of, 283
  non-food-grade, 287
  recovery, 316

surimi processing, 313
utilization, as profit center for seafood plants, 290

C

CAC, *see* Codex Alimentarius Commission
Caesio erythrogaster, 24
Calcium
   carbonate, 788
   chelating agents, 466
   chloride brine, 337
   effects of on frozen surimi, 691
Calmodulin, 307
Calpain, 242, 243, 246, 253
Calpastatin, 243, 253
Canola oil, 686
Cantharellus cinnabarinus, 787
Canthaxanthin, achievement of desired hue using, 756, 787
Capillary extrusion viscometer, 551
Capsicum annum, 777
Capsorubins, 777
Caramel, achievement of desired hue using, 756, 771, 786
Carbohydrate(s)
   cryoprotective additives, 186
   freezing point depression and, 190
   gelling, 436
Carmine
   achievement of desired hue using, 756, 771
   application problems with, 774
   chemical structure of, 773
Carminic acid, 772
Carnosine, 309
β-Carotene, achievement of desired hue using, 756, 788
Carp
   actomyosin from, 461
   alkaline proteinase in, 246–247
   cathepsin B purified from, 235

muscle
   calpain purified from, 242
   preparation of lyophilized myofibrillar powder from, 203
   trypsin inhibitor from, 244
   thermal gels from, 863
Carrageenan
   common types of, 683
   interaction, 557
   uses, 682
Carreau model parameters, 574
Cathepsin(s), 48, 229
   action of on hemoglobin, 234
   activities of in fish mince, 239
   lysosomal, 232
   properties of, 233
Cathepsin A, 233
Cathepsin B, 234, 252
Cathepsin C, 235
Cathepsin D, 236
Cathepsin E, 237
Cathepsin H, 237, 252
Cathepsin L, 238, 252, 308
CCPs, *see* Critical Control Points
CDQ, *see* Community development quota
CD spectroscopy, *see* Circular dichroism spectroscopy
Cellular components, hydroxyl radical damage to, 112
Cellulose, 686
Centrifugation, 301
Certificates of Acceptance (COAs), 153
Certification, colorants requiring, 769, 770
CFRs, *see* Code of Federal Regulations
Champagne seawater (CSW) system, 56, 57
Cheese manufacturing, 253–254, 303
Chelating agent, 468
Chemical sanitizer, properties of, 158

Index

Chikuwa, 376, 378, 390, 391, 851
Chilean jack mackerel, 36
Chilean whiting, 19
Chilling, rapid, 411
Chirocentrus dorab, 24
Chitosan, 296
Chopping
  principle, 401
  procedures, U.S. surimi seafood industry, 673
  temperature, 402
Chromatography, development of, 714
Chymotrypsin inhibitors, 251, 252
Chymotryptic digestion, 863
Chymotryptic enzymes, contamination from, 248
CI number, *see* Colour Index number
CIP system, *see* Clean–in-place system
Circular dichroism (CD) spectroscopy, 863
Cleaning
  achievement of, 155
  definition of, 154
  verification of, 161
Clean–in-place (CIP) system, 155
Clostridium
  botulinum, 149, 150, 152, 413, 590, 593, 595
    categories, 596
    FDA guidance for inactivation of, 616
    lethality calculation, 620–621
    production of toxins by, 599
    spores, resistance of to heat processing, 597
  perfringens, 589
  spp., 586
CLs, *see* Critical limits
Clupea harengus, 120
COAs, *see* Certificates of Acceptance
Coastal trawlers, 54, 55
Coccus cacti, 772

Cochineal extract, 773
Cod
  alkaline proteinase in, 246–247
  muscle, protein solubility of, 197
Code of Federal Regulations (CFRs), 144
Code of Practice, 869–885
  essential final product requirements, 870–872
  essential final product quality requirements, 871–872
  essential health and hygiene requirements, 871
  optional final product requirements, 872–885
  primary quality attribute, 872–879
  secondary quality attributes, 879–885
Codex Alimentarius Commission (CAC), 871
Codex Code, frozen surimi, 91
Co-extruded color, 408
Cold temperature sink, 357
Cold-water species, cryoprotectants for, 211
Collagen, 168–169, 176, 292, 309
Collagenase, isolation of, 308
Color
  acceptance criteria, 789
  acceptance tolerance, 792
  application, 407
  body meat, 696
  box tolerance system, 793
  co-extruded, 408
  enhancers, 43
  hue values, 700
  surimi seafood surface, 696
  tone data, 768
Colorimeter
  high-end, 761
  spectrophotometric, 763
  tristimulus, 761
Coloring agents, 696
Color measurement and colorants, 749–801

colorants, 769–789
    colorants not requiring
        certification, 770–787
    colorants requiring
        certification, 769–770
    monascus colorants, 787
    nature identical colorants,
        787–788
    other naturally derived
        colorants, 788–789
coloring surimi seafood, 765–768
    color application to
        crabsticks, 766
    general principles, 766–768
    preparation of surimi paste
        for crabsticks, 766
color quality, 789–794
    acceptance criteria, 789–792
    acceptance tolerance,
        792–794
    colorant quality, 789
    final product color, 789
    labeling, 794–796
        religious requirements,
            795–796
        requirements in United
            States, 794–795
    understanding color and
        measurement, 753–764
        color space, 754–756
        development of color
            language, 753–754
        indices, 759
        instrument development, 756
        L*a*b* color space, 757–759
        measuring color, 759–764
        tristimulus values, 756–757
Colour Index (CI) number, 771
Commercial freezer, role of, 326
Community development quota
    (CDQ), 7
Compost piles, 289
Compressive test, 501
    convex shape samples, 520
    rod-type samples, 521

Compressor selection, diversified,
    367
Concentric cylinder geometry,
    schematic diagram of,
    555
Conger eel, 19
    harvesting of in sea around
        Japan, 376
    use of in kamaboko, 855
Congresox
    spp., 19
    talabon, 24
    talabonoides, 23
Connective tissue proteins, 444
Consumer
    preference(s)
        scores, 839
        sensory analysis and, 834
    purchases, sensory evaluation
        and, 832
    questionnaires,
        recommendations, 831
    tests, 828
        behavioral methodologies,
            829
        declarative methodologies,
            829
Contaminants, tolerance levels for,
    144
Contamination
    fish organ, 248
    microbial, 593
    production machinery and, 148
Cooked surimi tests, 874, 881
Cooking method, 93
Cooling, ambient-air, 623
Coral fish, 24
Coronary heart disease, 290
Cost reduction, 823, 842
Cox-Merz rule, 556, 568, 569, 570
Crab-flavored surimi seafood, types
    of cut in, 381
Crabmeat style, categories of, 395
Crabstick(s), 21, 92, 376
    cheap, 661
    color application to, 766, 767

industry, Australia, 387
invention of, 382
line, LIBETTI, 404
manufacturer, world's largest, 386
manufacturing
　flow diagram of, 397
　Thailand, 387
　pasteurization studies using, 413
　production, South Korea, 385
　stickiness of, 664
Crab-style products, 396
Critical Control Points (CCPs), 146
　corrective actions, 147
　critical limits for, 147
　identification of, 147
　monitoring procedures, 147
Critical limits (CLs), 607
Croaker(s), 19, 23, 27
　harvesting of in sea around Japan, 376
　use of in kamaboko, 855
Cross-flow filtration, 303, 304
Cross-linking, definition of, 658
Cryogenic freezers, 338
Cryogenics, advantages of, 339
Cryomechanical freezer, 339
Cryoprotectant(s), 6, 35, 91, 128
　cold-water species, 211
　commercial practices for mixing, 42
　concentration of on in surimi seafood, 343
　definition of, 476
　diversity of, 212
　DMA production and, 200
　freeze denaturation and, 84
　phosphates as, 695
　polyphosphate as, 470
　stabilizing surimi with, 42
　sugar as, 184
Cryoprotection, tilapia actomyosin, 187, 188
Cryoprotective additives, carbohydrate, 186

Cryostabilization, 191, 193
CSW system, *see* Champagne seawater system
Culpea harengus, 36
Curcuma longa, 783
Curdlan, 684
　gels, formation of, 685
　solubility of, 684
Cutlass fish, use of in kamaboko, 855
Cylinder type gels, axial compression for, 519
Cysteine proteinase(s)
　gel degradation and, 255
　inhibitor, 252, 255
Cytochromes, 114

D

DAF, *see* Dissolved air flotation
Dark muscle, 447
　mechanical strength of, 119
　myoglobin in, 122
　proteolytic activity of, 117
　tissue, oxidation of, 113
Dark-muscled fish, 108
　characteristics, 109
　lipids, 113
　problems with processing of, 120
　processing of pelagic fish, 117
　proteins, 115, 123
Dead-end filtration, 303, 304
Deboning, methods to prepare fish for, 37
Decanter
　centrifuge(s), 86, 287, 302
　　bowl, cross-sectional view of, 88
　　main component of, 87
　　sedimenting of cellular membranes using, 126
　　separation of mince and water in, 88
　technology, 4, 88

Decimal reduction time (D-value), 602, 603
Declarative methodologies, consumer testing, 829
Deformation
definition of, 498
gel resistance to, 548
rate, 499
Defrost schedule, 366
Degreasing detergents, 155
Denaturants, generation of, 474
Denaturation, 446, 863
Descriptive analysis, statistics for, 826
Descriptive tests, 824
Detergents, 155
DEW, see Dried egg whites
DHA, see Docosahexaenoic acid
Difference tests, 823
Digital image analysis, 81, 82
Dilute extract measurement, 550
Dimethylamine (DMA), 174, 199, 201
cryoprotectants and, 200
generation, H&G fish, 199
Dimethyl sulfide, aroma of, 720
Dipeptidyl peptidase, 252
Dipeptidyl transferase, 235
Direct surface contact, 160
Disaccharide(s)
use of as protectants during drying, 206
water replacement mechanism, 208
Disinfecting, verification of, 161
Dissolved air flotation (DAF), 299, 300
Dissolved solids, recovery of, 287
Distilled flavors, 720
Disulfide bonding, 455, 456
formation, 458
gel network and, 460
role of in myosin/actomyosin gelation, 459
Dithiothreitol (DTT), 460
DMA, see Dimethylamine

DNA, 309
Docosahexaenoic acid (DHA), 290, 857
Dosidiscus gigas, 36
Dried egg whites (DEW), 674–675
Drum cooker, 403
Drum freezing, 44
Dry expansion system, disadvantage of, 330
Dry ice, 340
DTT, see Dithiothreitol
Duck ovostatin, 253
Dunaliella carotene, 788
Dungeness crabmeat, pasteurization process for, 597
D-value, see Decimal reduction time

E

EDTA, 235, 464, 466
Edwardsiella tarda, 587
EEZ, see Exclusive Economic Zone
Egg white(s)
proteins, 673
purpose of adding, 736
use of in surimi production, 14
Eicosapentaenoic acid (EPA), 290, 857
Electrocoagulation, 301
Electromagnetic spectrum, color measurement and, 753
Electron beam radiation, 630
Electron penetration, dose map, 632
Electron reduction, changes of molecular oxygen with, 112
Empirical measurement, comparison of accuracy in, 536
Empirical testing, 494, 528
Energy
conservation, freezing and, 362
red muscle and, 110

Engraulis ringens, 25, 36
Enzyme(s), 447
  freeze-thaw experiments on, 184
  inhibition, 43, 251, 424, 690
  myosin, 473
  proteolytic, 40, 449
  TMAO demethylase, 175, 448, 449
  transglutaminase, 461, 692
  types in fish, 307
EPIXDMA, 295
Equilibrium stress, 505
Equipment manufacturers, major, 402
Equivalent lethality, 606
Escherichia coli, 151, 152, 157, 257, 258, 630, 638
Eso, 19, 27
Essential fatty acids, 291
Ethylene glycol, 338
Ethylenevinyl alcohol (EVOH), 408–409, 410
Euchema
  cottonii, 683
  spinosum, 683
European market, surimi processing for, 28
European Union, flavor regulations, 740
Evaporator, oversized, 364
EVOH, *see* Ethylenevinyl alcohol
Exclusive Economic Zone (EEZ), 6–7
Experience feedback, 78
Extraction methods, 715
Extruded meat, 419

F

Factory trawlers, 38, 54, 57
Failure test, 518
FAO/WHO
  Codex Code, 91, 538, 742
  Food Standard Program, 742
Fast freezing, 45, 418

Fat
  oxidation, 49
  replacer, 686, 688
Fatty acid composition, 113
Fatty fish, 113, 291
FDA, *see* U.S. Food and Drug Administration
FD&C Act, *see* Food, Drug, and Cosmetic Act
Feeding period, harvesting of fish during, 69
Fertilizers
  environmental limitations in producing, 312
  surimi solid waste made into, 289
Fiberization, 403
Fibrinogen, 517, 676
Filament meat, 395
Filament products
  cutting of, 407
  fiberization for, 403
  ohmic cooking, 422
Final product quality requirements
  foreign matter, 871
  odor and flavor, 872
Fish
  Antarctic, myosin rod purified from, 171
  dark-muscled, 108
  deterioration, 720
  endopeptidases, subgroups, 231
  fatty, 113, 291
  fiber types in, 447
  flesh, myofibrillar proteins in, 294
  freshness, 196, 540
  gelatin, 309
  handling practices, on-board, 588
  lobster-flavored, 381
  meat separators, 37
  milt, 309
  muscle
    gel-forming ability of, 182
    lipid components, 450

microstructure, 438
moisture content of, 50–51
softening of, 229
striated, 437
oil recovery, 290
organs, contamination from, 248
processing discards, enzymes in, 307
protein(s)
  demand for, 108
  effect of thermal sensitivity on, 401
  extracts, viscosity of, 549
  hydrolysates (FPH), 288
  intrinsic stability of, 170
  paste, steady shear viscosity of, 567
saltwater, muscle of, 70
sauce fermentation, 290
sausage, 378, 853
shrimp-flavored, 381
silage, 289
skin, collagen extracted from, 292
solids recovery, limitations of, 311
species, underutilized, 856
Fish ball(s)
  ingredients used for, 393
  most common display of, 394
  popularity of in Singapore, 388
  processing flow of, 395
  raw, 394
  setting/cooking of, 396
Fishery(ies)
  Americanization of, 6
  foreign, 7
  management, 5
  Olympic, 7, 12, 315
  U.S. pollock, 8, 11
Fishing operations, regulation of, 77
Flash profile, 824
Flatfish, harvesting of in sea around Japan, 376

Flavor(s), 709–748
  additives and ingredients used in flavors, 727–730
    glutamate, 728
    hydrolyzed proteins, 728–729
    ribonucleotides, 728
    yeast extracts, 729–730
  basic seafood flavor chemistry, 720–727
    importance of lipids in fish flavors, 721–724
    important components found in seafood extracts, 724–727
    sources of flavor ingredients, 720–721
  components, major, 714
  definition of flavor, 712–720
    building of flavor, 717–720
    creation of flavor, 712–713
    natural product chemistry, 713–716
  description of, 717
  effects of ingredients on flavor, 733–736
    egg whites and soy proteins, 736
    salt, 736
    sorbitol and sugar, 734
    starch, 734–735
    surimi (raw material), 735–736
    vegetable oil, 736
  effects of processing on seafood, 731
  encapsulation, 732, 738
  off flavors of seafood, 730
  organoleptic properties of, 713
  processing factors affecting flavors, 737–739
    adding additional flavor/flavor components, 737
    addition points, 737–738
    encapsulation, 738

storage conditions and shelf life, 738–739
quality, destroyer of, 737
regulations and labeling, 739–744
  European Union, 740–741
  Japan, 741–742
  religious certification issues, 743
  United Nations, 742–743
  United States, 739–740
  worldwide issues, 743–744
release and interactions, 731–733
spray-dried, 731, 738
synthetic components, 721
Flavor and Extract Manufacturers Association, 723, 740
Flounder, harvesting of in sea around Japan, 376
Flying fish, use of in kamaboko, 855
Folding test, 885
Food(s)
  additives, safety of, 739
  aroma of, 710
  -borne disease, 142, 601
  components, chemical interactions of, 495
  flavor interactions with, 732, 733
  fracture property measurements of, 531
  Halal, 743
  industry, preference mapping studies, 835
  irradiation, 629
  Kosher, 743, 796
  nonenzymatic browning of, 729, 731
  preservation
    methods, 587
    vitrification and, 195
  product
    illness associated with, 142
    pH value of, 586
  safety, 589
    regulatory agencies and, 142
  sanitation information and, 143
  threat, 155
  sanitizing agents, 589
  taste of, 710
  testing machines used in, 499
Food, Drug, and Cosmetic (FD&C) Act, 143, 769
Food-grade chemical compounds, 689–699
  calcium compounds, 689–692
  coloring agents, 696–699
  oxidizing agents, 689
  phosphate, 695–696
  transglutaminase, 692–694
Force(s)
  definition of, 496
  relationship between shear stress and, 539
  resistance to, 510
Formaldehyde, 184, 199
FPH, *see* Fish protein hydrolysates
Fracture property measurements, food, 531
France
  change of surimi seafood consumption in, 384
  industry-driven quality assurance in, 386
  success of surimi seafood in, 384
  surimi seafood market, 383
Free choice profiling, 824
Freeze-drying, 202, 204
Freezer(s)
  airflow, 331
  blast, 334, 335, 336, 358, 367
  brine, 336
  cabinet size, freeze time affecting, 349
  cryogenic, 338
  cryomechanical, 339
  design and operation, 363
  horizontal plate, 327, 328
  liquid nitrogen, 340, 341, 342
  plate, 346, 356, 361
  spiral, 331, 332, 340

stationary, thermocouple meter used for, 356
tunnel, 332, 333
Freeze resistance, plant, 195
Freeze-thawing
  damage, 546, 674
  denaturation during, 187
  LDH, 207, 209
  trehalose, 207
Freezing, 44
  capacity, definition of, 344
  drum, 44
  effect of on surface water, 182
  fast, 45
  -point depression, 189, 190
  predicted influence of air velocity on, 355
  rapid, 415
  rate, 46, 197, 342
  slow, 45
  time(s)
    datalogger measurement of, 355
    excessive, 360
    predicted, 360, 361
    salmon, 353
    simulated, 359, 363
    ways to determine, 354
Freezing technology, 325–371
  airflow freezers, 331–336
    blast freezer, 334–336
    spiral freezer, 331
    tunnel freezer, 332–334
  brine freezers, 336–338
    calcium chloride, 337–338
    glycols, 338
    other, 338
    sodium chloride, 337
  cryogenic freezers, 338–342
    carbon dioxide, 340–342
    liquid nitrogen, 339–340
  energy conservation, 362–368
    blast freezer case, 367–368
    freezer design and operation, 363–366
    refrigeration machinery options, 366–367
  freezing capacity, 344–346
  freezing of product, 342–344
  freezing time, 347–356
  horizontal plate freezers, 327–331
  "what-if" effects on freezing time, 356–362
    block thickness, 357
    cold temperature sink, 357–359
    heat transfer coefficient, 359–362
Freon, 330
Fresh water pollution, 589
Frozen fish, production of surimi from, 121
Frozen meat products, quality losses of, 178
Frozen surimi
  biochemical index for quality evaluation of, 865
  Codex Code for, 91
  essential requirements for, 870
  fresh surimi vs., 85
  merits of, 856
  proteins, stability of, 176
  water content of, 201
FSPC, *see* Functional soy protein concentrates
Fukoku refiner, 41
Functional additives, effect of, 546
Functional proteins, recovery of by pH shifts, 107–139
  characteristics of dark muscle fish crucial to surimi processing, 109–123
    dark muscle, 109–113
    lipids, 113–115
    muscle proteins, 115–117
    processing of pelagic fish, 117–123
  new approach, 123–132
Functional soy protein concentrates (FSPC), 678, 679

Fundamental test
  deformation and strain, 498
  flow and rate of strain, 499
  force and stress, 496
  rheological testing using large strain, 518
  rheological testing using small strain, 501
Fuzzy logic, 77
F-value, 605, 615, 617

G

Gas chromatography
  basis of, 714
  flavor components identified by, 712
  –olfactometry (GCO), 716
Gas purgers, automatic, 367
Gastric protease, 307
Gastroenteritis, 587
GCO, see Gas chromatography–olfactometry
Gel(s)
  -breaking force values, 89
  cohesiveness, 46, 76, 80, 654
    change in, 200
    loss of, 196
    mince stability and, 198
  color hues of, 655
  compressive test for, 501
  cooking methods, 93, 94
  curdlan, 685
  cylinder type, 519
  -degradation inducing factor (GIF), 248
  effect of refrigerated storage of, 540
  elastic behavior of, 526
  enhancers, 43, 689
  formation, myofibrillar proteins and, 115
  -forming properties, 493
  fracture, planes of by torsion, 528
  gelation kinetics of, 559
  hardness, 654
  matrix development, 653
  modori phenomenon in, 307, 674
  point, definition of, 563
  preparation, vacuum silent cutter for, 94
  pressure-induced, 464
  properties
    Alaska pollock surimi, 91
    kamaboko, 850
    use of rheological data to characterize, 558
  setting, calcium salt use and, 775
  shear stress of, 44, 45, 634
  shear type test, 501
  softening, 49, 308
  strength, 76, 229, 230, 874
    different gels with same, 533
    effects of HHP on, 424
    food scientist view of, 77
    oxidants and, 458
    surimi quality and, 531–532
    technique for evaluating, 529
  temperature dependency of pollock, 541
  texture
    effects of chopping temperature on, 401, 402
    effects of ingredients on, 651
    influence of processing factors on, 575
  thermo-irreversible, 495
  values, effect of chopping temperature on, 92
  viscoelasticity of, 459
  weakening, 49
Gelation
  activation energy during, 565
  kinetic models, 566
  mechanism, recent progress in, 859
  thermal, 863

Gelation chemistry, 435–489
  bonding mechanisms, 451–471
    covalent bonds, 456–471
    hydrogen bonds, 451–452
    hydrophobic interactions, 455–456
    ionic linkages, 452–455
  factors affecting fish protein denaturation and aggregation, 471–476
    frozen storage stability of surimi, 476
    importance of muscle pH, 475
  factors affecting heat-induced gelling properties of surimi, 476–477
  lipid components of fish muscle, 450–451
  protein components of surimi, 437–450
    myofibrillar proteins, 437–444
    sarcoplasmic proteins, 444–450
    stroma proteins, 444
Gene cloning technology, food-grade proteinase inhibitors and, 257
General Foods Corporation Technical Center, 534
Generally recognized as safe (GRAS), 131, 407, 740, 774
General Method calculation, 618, 619
Giant squid, 36
Gibbs free energy, 185
GIF, *see* Gel-degradation inducing factor
Gigartina acicularis, 683
Glass transition temperature, 190, 191, 192
Gliadin, 681
Globular actin, 441
Globular proteins, lowered solubility of, 185
Glutamate, 728
Glycols, 338
GMPs, *see* Good Manufacturing Practices
Goat fish, 19, 23
Good Manufacturing Practices (GMPs), 143, 144
Gordon and Taylor equation, 193
Government regulations, discharge and waste disposal, 283
Gram-positive bacteria, 629
Grape color, 784
GRAS, *see* Generally recognized as safe
Grass prawn, 253
Green production, 283
Grey mullet, cathepsin B purified from, 235
Guchi, 19, 27

H

HACCP, *see* Hazard Analysis Critical Control Point
Hairtail, 19, 23
Halal foods, 743
Halocarbon refrigerant, 329
Hamann torsion gelometer, 524, 525
Hamo, 23
Hanpen, 378, 392, 852
Hard water, softening of, 59
Hazard Analysis Critical Control Point (HACCP), 143, 610, *see also* Sanitation and HACCP
  -based system, integration of microbial criteria into, 151
  monitoring procedures, design of, 622, 624
  program, metal detection and, 47
  record keeping, 148
  verification procedures, 148

HDPE, *see* High-density polyethylene
Headed and gutted (H&G) products, 8, 9, 10, 132, 199
Headspace analysis, 716
Heat
   capacity change, 170
   coagulation, 300
   -induced gelation, 260
   leakage, 364
   pasteurization, 598
   penetration (HP), 607, 610
   process establishment, 606
   resistance, 608
      data, 601
      target microorganism, 614
   sanitizing, 157
   transfer
      coefficient, 337, 351, 353, 359
      fluids, 338
   waste, 367
Heating
   methods, rapid, 259
   microwave, 259
   ohmic, 259, 260, 419, 877
   radio frequency, 259
Heavy meromyosin (HMM), 460
Hedonic studies, 819, 822
Heme pigments, decoloring of, 458
Heme proteins, 49, 110, 446
Hemoglobin, 130, 234, 446
Herring, alkaline proteinase in, 246–247
Hertz theory, 521
H&G products, *see* Headed and gutted products
HHP, *see* High hydrostatic pressure
High-density polyethylene (HDPE), 407, 409
High hydrostatic pressure (HHP), 424, 425
High-pressure processing, 628
Himeji, 19, 23, 27
HMM, *see* Heavy meromyosin

Hoki, 17, 36
   loss of gel-forming ability, 540
   mechanical deboning of, 198
   surimi, gel strength, 18
Hokkaido Fisheries Research Station, 35
Hooke's law, 521
Horizontal plate freezers, 327, 328
HP, *see* Heat penetration
HPP, *see* Hydrolyzed plant proteins
Human variability, sensory measurement and, 807
HVP, *see* Hydrolyzed vegetable proteins
Hydration forces, 179, 186
Hydrocolloids, 681
Hydrogels, 436
Hydrogen bonding, 557
Hydrolyzed plant proteins (HPP), 729
Hydrolyzed proteins, 728, 729
Hydrolyzed vegetable proteins (HVP), 729
Hydrophobic interaction, 457

I

Ice
   crystal growth, 176
   crystallization, 177
   formation, reactants and, 189
   –protein interface, polarization of water in, 176
   recrystallization, 195
Iceboats, 55
Ichithyophonus hoferi, 49
Imitation crabmeat, consumers' attitudes toward, 380–381
Imitative tests, 495
IMP, nucleotide of, 196
Impurity measurement, digital image analysis for, 81–82
Individually quick-frozen (IQF) product, 334, 408, 418

Ingredient(s)
  blending, 398
  effects of on flavor, 733
  modifications, discriminating tests and, 823
Ingredient technology, 649–707
  cellulose, 686
  evaluation of functional ingredients, 699–701
    color, 700
    formulation development and optimization, 700–701
    texture, 699
  food-grade chemical compounds, 689–699
    calcium compounds, 689–692
    coloring agents, 696–699
    oxidizing agents, 689
    phosphate, 695–696
    transglutaminase, 692–694
  hydrocolloids, 681–686
    alginate, 685–686
    carrageenan, 682–683
    curdlan, 684–685
    konjac, 683–684
  protein additives, 668–681
    egg white proteins, 673–675
    plasma proteins, 676–677
    soy proteins, 677–681
    wheat gluten and wheat flour, 681
    whey proteins, 669–672
  starch, 657–667
    composition of starch, 657
    modification of starch, 657–659
    starch as functional ingredient for surimi seafood, 659–667
  vegetable oil and fat replacer, 686–688
  water, 653–656
Inhibitor activity staining, BPP and, 254
Initial temperature (IT), 611
In-mouth judgment, 806

Instrumental analysis, classes of, 494
International Committee on Microbiological Specifications for Food, 151
International Organization of Flavor Industries (IOFI), 742
International Union of Biochemistry (IUB) Nomenclature Committee, 231
IOFI, *see* International Organization of Flavor Industries
Ionizing radiation, 629
Iota-carrageenan, 683
IQF product, *see* Individually quick-frozen product
Isolated soy proteins (ISPs), 678, 679
ISPs, *see* Isolated soy proteins
IT, *see* Initial temperature
Itatsuke kamaboko, 850
Itoyori, 19
IUB Nomenclature Committee, *see* International Union of Biochemistry Nomenclature Committee

J

Jack mackerel, 25
Japan
  consumption of surimi seafood in, 5
  flavor preference in, 711
  flavor regulations, 741
  Hokkaido Fisheries Research Station, 35
  kamaboko products of, 23, 35, 84, 854

Index

kinds of fish harvested in sea around, 376
New Year Traditional Dish in, 850
overfishing by, 8
pelagic fish used in, 25
shore-side surimi processing in, 85
surimi-based products in, 377, 378
trend of surimi usage in, 379
Japanese sardine, 36
Jewish food laws, 743
Johnius spp., 19, 23
Joint Expert Committee on Food Irradiation, 631
Joint-venture (JV) operations, 7, 14
JV operations *see* Joint-venture operations

K

Kamaboko, 21, 92, 376, 378, 388, 389, 393, 637, 711
Kamaboko, new development and trends in, 847–868
   change in fish species used for kamaboko production, 855–856
   history of kamaboko, 848–849
   scientific and technological enhancement, 859–866
     biochemical index for quality evaluation of frozen surimi, 865–866
     myosin rod aggregation at high temperature in relation to gelation, 863–865
     recent progress in gelation mechanism, 859–860
     recent progress in myosin denaturation study in relation to gelation, 862–863
     recent progress in understanding of setting, 860–862
   trends of kamaboko products, 857–859
   variations in kamaboko products in Japan, 849–854
     boiled kamaboko, 852
     crab leg meat analog, crabstick, 853
     deep-fried kamaboko, 851–852
     fish sausage and ham, 853
     grilled kamaboko on bamboo stick, 851
     grilled kamaboko on wooden board, 851
     other kamaboko, 853–854
     steamed kamaboko on wooden board, 850–851
Kamasu, 22
Kani-ashi Kamaboko, 853
Kani-kama, 21, 787
Kappa-carrageenan, 683
Keratin, 168–169
Kinetic model
   nonisothermal, 561
   thermo-rheological properties and, 559
Kininogens, proteinaceous, 252
Kinme-dai, 22, 27
Klathi, 23
Klebsiella spp., 587
Konjac, 682
   flour, solubility of, 683
   gel, 684
   patented use of, 684
Korea
   overfishing by, 8
   shore-side surimi processing in, 85
Kosher foods, 743, 796
Kramer shear cell, 529
Kudoa, 240

L

L*a*b* color space, 755, 757, 758, 764
Labeling laws, 794
Lambda-carrageenan, 683
Larimus pacificus, 25, 36
LCs, *see* Light chains
LDH, freeze-thawing of, 207, 209
LDL, *see* Low-density lipoproteins
LDPE, *see* Low-density polyethylene
Leaching, surimi stability and, 197
Least cost formulation, quality control and, 425
Leather jacket, 19, 23, 24, 25
Lethality value, 605, 617
LIBETTI crabstick line, 404
Ligand
   binding, 188
   –protein interaction, 189
Light chains (LCs), 439
Light meromyosin (LMM), 439, 460, 864
Linear viscoelastic region (LVER), 557, 559
Lipid(s), 113
   antioxidants of, 189
   auto-oxidation pathway of, 722
   importance of in fish flavors, 721
   oxidation
      lag phases of, 122
      prime catalyst of, 130
      rancid odors from, 121
      stability, 131
   pro-oxidants and, 450
Liquid nitrogen (LN), 338, 339, 340, 341, 342
Listeria monocytogenes, 149, 152, 156, 585, 590, 591, 594, 615
Listeriosis, 591, 592
Lizardfish, 19, 21, 27, 49
   harvesting of in sea around Japan, 376
   use of in kamaboko, 851, 855
LMM, *see* Light meromyosin

LN, *see* Liquid nitrogen
Lobster-flavored fish, 381
Low-density lipoproteins (LDL), 290
Low-density polyethylene (LDPE), 407, 409
Lumptail sea robin, 25, 36
LVER, *see* Linear viscoelastic region
Lyoprotection, 206
Lysine, 452
Lysosomal cathepsins, 232

M

Mackerel
   actomyosin, denaturation of, 182, 183
   alkaline proteinase in, 246–247
   cathepsin B purified from, 235
   fish oil production from, 291
   harvesting of in sea around Japan, 376
   light muscle surimi, 109
   myosin heavy chain, proteolysis of, 252
   poor-quality, 122
Macruronus
   magellanicus, 17
   navaezelandiae, 36
Mad cow disease, 14, 43, 256
Magnuson Fisheries Act, 10
Maillard browning, 211, 729, 731
Maltodextrins, cryoprotective properties of, 194
Manufacturing and evaluation of surimi, 33–106
   biological (intrinsic) factors affecting surimi quality, 48–53
      freshness or rigor, 52–53
      seasonality and sexual maturity, 49–52
      species, 48–49
   decanter technology, 86–91

processing (extrinsic) factors
     affecting surimi quality,
     53–75
  harvesting, 53–55
  on-board handling, 55–58
  salinity and pH, 69–75
  solubilization of myofibrillar
     proteins during
     processing, 62–66
  time/temperature of
     processing, 60–62
  washing cycle and wash
     water ratio, 66–69
  water, 58–60
processing technologies that
     enhance efficiency and
     profitability, 76–85
  digital image analysis for
     impurity measurement,
     81–82
  fresh surimi, 84–85
  innovative technology for
     wastewater, 82–84
  neural network, 76–79
  processing automation,
     80–81
processing technology and
     sequence, 37–48
  freezing, 44–46
  heading, gutting, and
     deboning, 37–38
  metal detection, 47–48
  mincing, 38–39
  refining, 40–41
  screw press, 41–42
  stabilizing surimi with
     cryoprotectants, 42–44
  washing and dewatering,
     39–40
surimi gel preparation for better
     quality control, 91–95
  chopping, 91–93
  cooking, 93–95
MAP, *see* Modified atmosphere
     packaging
Marine Stewardship Council, 13

Market
  acceptance, 312
  research, 841
Marketing, sensory analysis and,
     812
Mass spectroscopy, flavor
     components identified by,
     712
Maxwell model, 507
MBSP, *see* Myofibril-bound serine
     proteinase
Meat
  particles, recovery strategy of, 87
  separation technique, roll-type,
     38
Mechanical chopping, 93
Mechanical relaxation
     phenomenon, 504
Membrane(s)
  filtration, types of, 303
  fouling, 306
  lipids, 114
  separation of from proteins, 126
Merluccius
  capensis, 13
  gayi, 13, 19
  gayi peruanus, 13, 25
  hubbsi, 13
  productus, 13, 36
Metal
  chelating agents, 111
  contamination, most common
     types of, 47
  detection, 47, 411
Metalloproteases, inhibition of, 236
Methyl cellulose, manufacture of,
     686
MHC, *see* Myosin heavy chain
Microbial contamination, 585, 592
  air as source of, 160
  reduced likelihood of, 153
Microbial inactivation, predictive
     model for, 626
Microbial transglutaminase
     (MTGase), 469, 470, 693
Microbiological criteria, 151

Microbiology and pasteurization, 583–648
  growth of microorganisms in foods, 585–587
  microbial safety of surimi seafood, 590–596
    Clostridium botulinum, 593–596
    Listeria monocytogenes, 591–593
  new technologies for pasteurization, 628–637
    effect of e-beam on other functional properties, 634–637
    electron beam, 630–631
    electron penetration in surimi seafood, 631–633
    food irradiation, 629–630
    high-pressure processing, 628–629
    microbial inactivation in surimi seafood, 633–634
  packaging considerations, 637–638
  pasteurization of surimi seafood, 596–600
  predictive model for microbial inactivation, 626–628
  process considerations and pasteurization verification, 600–624
    analyzing pasteurization penetration data, 617–624
    D-value, 602
    F-value, 605–606
    general considerations for heat process establishment, 606–607
    heat penetration test design, 610–611
    heat resistance of selected target microorganism, 614–617
    initial temperature and product size, 611
    principles of thermal processing to surimi seafood pasteurization, 601–602
    product preparation/formulation, 611–614
    study design and factors affecting pasteurization process, 607–608
    temperature distribution test design, 608–610
    z-value, 602–605
  surimi microbiology, 587–590
  temperature prediction model for thermal processing of surimi seafood, 624–626
Micromesistius
  australis, 17, 36
  australis australis, 18
  australis pallidus, 18
  poutassou, 19, 36
Microorganism(s)
  categories of, 586
  detection of after disinfecting, 158
  growth of in foods, 585
  heat resistance of target, 614
Micropoga undulatus, 417
Micropogon opercularis, 49
Microwave heating, 259
Milk coagulation, 511
Milling process, 527–528
Mince
  activities of cathepsins in, 239
  gel-forming ability, sorbitol and, 202, 204
  moisture removal, 59
  process for making stabilized, 198
Mincing, 38
Mirin wine, 727, 735
Model(s)
  bio-economic, 50

Index

electron penetration, 633
gelation kinetic, 566
kinetic
   nonisothermal, 561
   thermo-rheological properties and, 559
Maxwell, 507
spatial partitioning of gelling protein, 651, 652
surface response, 626
temperature prediction, 415, 624, 625
Modified atmosphere packaging (MAP), 637, 638
Modulus of rigidity, 502
Moisture content, dependence of gelation temperature on, 563
Moisture levels, maintenance of during surimi production, 80
Molded products, 420
Molds, thermo-tolerant, 622
Molecular oxygen, 111
Molecular weight cut-off (MWCO), 305
Monascus, achievement of desired hue using, 756, 787
Monosodium glutamate (MSG), 393, 465–466, 718, 719
Moslem food laws, 743
Motherships, 54
Mouth-feel, fat replacers and, 688
MSG, see Monosodium glutamate
MTGase, see Microbial transglutaminase
Multivariate analysis, 828
Munsell Color System, 754, 756
Muraenesocidae, 23
Muraenosox cinereus, 24
Muscle
   alkaline proteinase in, 245
   carp
      calpain purified from, 242
      trypsin inhibitor from, 244
   catheptic activity of, 232–233

dark, 447
dehydrated, 205
extraction of myosin from, 165
fibers, types of, 447
freshness, 472
lysosomal proteinases found in, 233
pH, 475
polar lipids in, 113
protein(s), 115
   extreme pH values and, 128
   polymers, integrity of, 228
   solubility of, 72, 75, 116, 127
proteinases, 308
proteolysis, species classification based on, 229
softening of, 229
tissue
   homogenized, 126
   major contractile proteins of, 124
Z-disk in pre-rigor, 241
Mushi kamaboko, 390
MWCO, see Molecular weight cut-off
Myofibril-bound serine proteinase (MBSP), 250
Myofibrillar protein(s), 39, 49, 437
   actin, 441
   classification of, 70
   cold destabilization of, 177, 194–195
   covalent bonds, 456
   dewatered, 35
   extraction, use of salt in, 736
   gel formation and, 115, 117
   gel hardness and, 540
   heat-induced gelation of, 451
   hydrogen bonds, 451
   hydrophobic interactions, 455
   ionic linkages, 452
   liquid movement, 501–502
   loss of in washwater, 83
   loss of in wastewater, 63
   minimum concentration of needed for gelation, 653

most abundant, 165
myosin, 439
net negative charge of, 454
percentage of in fish flesh, 294
recovery of, 302
solubilization, 62, 400
thick filament assembly, 442
Myoglobin, 110, 446
Myosin, 439
  amino acid composition analysis, 857
  ATPase inactivation, 861
  cross-linking reaction, 858, 859
  denaturation, 862
    freezing rate and, 197
    mechanisms for, 195
  extraction of, 165
  heavy chain (MHC), 61, 62, 68, 469, 635
    degradation of, 61, 62, 421
    mackerel, 252
    washing cycles and, 69
  postmortem, 437
  rod aggregation, 863
  solubility, 444
  stability, 166
    globular head region, 166
    helix-coil transition in myosin rod, 168
Myxosporidean parasite, 14
Myxosporidia, 450

N

National Advisory Committee on Microbiological Criteria of Foods, 146
National Food Processors Association, 413, 608
National Sanitation Foundation, 143
Natural colorant, 771
Natural extracts
  concentration, 721
  distilled flavors, 720
  extraction, 720
  supercritical extraction, 721
Natural flavors, definition of, 742
Nature identical colorants, 787
Navodon modestus, 20, 24, 25, 49
Near-infrared (NIR) energy, 80, 81
Near-real-time analysis, 161
Nemipterus
  bathybius, 49
  japonicus, 20, 36
  marginatus, 20
  mesoprion, 20
  nematophorus, 20
  peronii, 20
  spp., 19, 20
  tolu, 473
Neural network, 76, 78
Neutraceuticals, recovery of, 306
NIR energy, see Near-infrared energy
Nobbing, 123
Nonenzymatic browning, 729, 731
Nonvolatile compounds, 726
  amino acids, 727
  organic acids, 727
  peptides, 727
  ribonucleotides, 727
Northern blue whiting, 19, 36
North Pacific Fisheries Management Council, 12
Nose judgment, 806
Nucleotides, 309
Nutrition Labeling and Education Act of 1990, 795
Nylon, 410

O

Objectionable matter, 874, 879
Oden, 379, 385
Odor modification, 818
Off flavors, 730
Ohmic cooker, 422, 423
Ohmic heating, 259, 260, 419, 877
Ohmic test unit, 96

Okada gelometer, 530
Oleoresin, achievement of desired hue using, 756, 773, 776
Olympic fishery, 7, 12, 315
Omega-3 fatty acids, 290, 310
Oncorhynchus gorbuscha, 36
On-line sensors, 80
Oreochromic niloticus, 448
Organic acids, 727
Oscillatory dynamic test, 508
Osmolytes, naturally occurring, 172
Ovalbumin, 653
Oval filefish, 49
Over-thawing, 398
Oxidants, 476
Oxidations, dark muscle tissue undergoing, 113
Oxidizing agents, 689
Oxyhemoglobin, autoxidation of, 112
Ozone, 60, 589

P

PA, *see* Polyamide
Pacific herring, 36
Pacific mackerel
    myofibrils, denaturation rate of, 475
    sarcoplasmic protein in surimi made from, 448
Pacific whiting, 13, 36
    beef plasma addition to, 253
    bio-economic models for, 50
    cathepsin B activity in, 235
    cathepsin L from, 238
    compositional properties of, 51
    cysteine proteinase inhibitor in, 255
    enzyme inhibitors, 43, 690
    factors affecting quality, 79
    gel(s)
        formation, 421
        softening in, 308
    strain and stress values of, 256
    strength, 470
    harvesting of, 54
    heat-stable cathepsins, 232
    myosin
        dynamic test curves of, 512
        elastic behavior of, 511
    protease enzyme in, 424
    proteins, 65, 72, 75
    proteolytic enzymes, 250, 449
    spawning of, 15
    surimi
        manufacturing, removal of fillets in, 38
        paste, storage modulus values of, 514
        production, 16
    use of in United States, 388
Package thickness(es)
    heat penetration curves for surimi seafood with various, 612
    maximum, 613
Packaging, 408
    machine, 413
    types of, 6537
    typical films used in, 412
Paprika
    achievement of desired hue using, 756, 773, 776
    application of, 779
    chemical structure of, 779
Parasite contamination, 81
Parupeneus spp., 19, 23
Parvalbumins, 307
Paste state, 557
Pasteurization, 149, 411, 588, 601
    crabstick studies, 413
    definition of, 596
    heat, 598
    penetration data, 617
    process
        adequacy of, 607, 613
        factors affecting, 607
    sulfur odor detected, 673

system
    cold spots, 608, 609
    continuous conveyance, 609
    temperature during, 700
    time/temperature, optimization of, 621
PBO fillets, see Pinbone out fillets
PCA, see Principal Component Analysis
PCC, see Pollock Conservation Cooperative
PDEase software, 356
PE, see Polyethylene
PEG, see Polyethylene glycol
Pelagic fish, 25
    acid content of, 118
    importance of as human food source, 117
    problems encountered with small, 109
    seasonality of catch, 118
Penaeus monodon, 253
Pennahia, 19
Pepstatin, cathepsin D and, 236
Peptides, 727
PER, see Protein efficiency ratio
Perceived color, 755, 756
Perception levels, sensory evaluation and, 808
Peruvian anchovy, 25, 36
Peruvian whiting, 25
PET, see Polyester
PGPR, see Polyglycerol polyricinoleate
Pharmaceutical industry, 308
Phosphates, 695, 696
Phosphatidylcholine, effect of on setting, 858
Phosphofructokinase, 189
Phospholipids (PL), 858
pH shift process, 29, 131
Pike-conger eel, 23
Pinbone out (PBO) fillets, 10
Pineapple stem acetone powder, 252
Pink salmon, 36

PL, see Phospholipids
Plank's equation, freezing time and, 349, 354
Plant locations, microbial biofilms in, 155
Plate freezers, 346
    common malady for, 361
    thermocouple meter used for, 356
Plesiomonas shigelloides, 587
Pleurogrammus
    azonus, 36
    monopterygius, 25
Poisson's ratio, 494, 519
Pollachius virens, 683
Pollock
    blending grades of, 426
    gels, temperature dependency of, 541
    harvest, decrease in, 26
    milt, 292
    recovery-grade, 28
    -starch gels, 546
    surimi
        gels, 424, 465, 546, 694
        pastes, cystine addition to, 458
        shear strain values of, 654
Pollock Conservation Cooperative (PCC), 7, 11–12
Pollutant discharge, 283
Polyacrylamide, 295
Polyamide (PA), 410
Polyclonal antibodies, 243
PolyDADMAC, 295
Polyester (PET), 409, 410
Polyethylene (PE), 409
Polyethylene glycol (PEG), 211
Polyglycerol polyricinoleate (PGPR), 774
Polymeric coagulants, 296
Polymers, high charge density synthetic, 295
Polypropylene (PP), 410
Polystyrene (PS), 410
Polyunsaturated fatty acids, 857

Polyvinylidene chloride (PVDC), 408, 410
Polyvinylpyrrolidone (PVP), 211
Pork plasma protein (PPP), 676, 677
Potassium bromate, 689
Potato
   extract, 253, 255
   starch, 659, 661, 663, 699
PP, see Polypropylene
PPP, see Pork plasma protein
Prechilling, 347
Preference(s)
   global model of, 836
   mapping
      result, example, 837
      techniques, 835
Priacanthus
   maracanthus, 22
   tayenus, 22
Principal Component Analysis (PCA), 828, 829, 835
Prionotus stephanophyrys, 25, 36
Processed whey proteins (PWP), 672
Processing water, treatment of, 60
Process optimization, 842
Product(s)
   characteristics, description of, 810
   competitive
      mimicking of, 811
      sensory preferences of, 813
   composition, color and, 808
   heating, factors affecting rate of, 611
   heat penetration tests, 614
   –human interaction, 817
   preference
      reasons for, 840
      score, quadratic model to predict, 836
   size, initial temperature and, 611
   thickness, 350
   virtual optimal, 836
Products, market, and manufacturing, 375–433
   France, 383–385
   Japan and United States, 377–383
   manufacture of surimi-based products, 388–419
      chikuwa, 390
      fish ball, 393–395
      hanpen, 392–393
      kamaboko, 388–390
      satsuma-age/tenpura, 390–391
      surimi seafood, 395–419
   other countries, 385–388
   other processing technology, 419–428
      high hydrostatic pressure, 424–425
      least-cost linear programming, 425–428
      ohmic heating, 419–424
Pro-oxidant, hemoglobin as, 121
Propylene glycol, 338
Protease(s)
   acid, 232
   inhibitors, use of in surimi production, 14, 17
Protein(s), see also Myofibrillar proteins; Proteins, stabilization of
   additives, 547, 668
   antifreeze, 165, 175, 176, 211
   autolysis, 228–229
   beef plasma, 676, 677
   chemical stability of solid-state, 202
   coagulation, 301
   concentration of in wastewater, 293
   content, moisture and, 92
   contractile, 124
   dark-muscled fish, 115
   degradation, post-mortem, 229
   denaturation, 35, 165, 471, 474
      gel-forming ability and, 228

salt and, 42
dried, 202
efficiency ratio (PER), 668
egg white, 673
electrophoretic patterns of, 64, 73, 74
extractability of, 116
fish muscle, intrinsic stability of, 170
flexibility, 172
formation of calcium cross-links between, 454
frozen surimi, 176
gels, rheological behavior of cross-linked, 459
globular, lowered solubility of, 185
heme, 49, 110, 446
hydrolysates, 288
hydrolyzed, 728, 729
insoluble, 311
instability, 187
isoelectric point of, 295
isolates, pH shift processes and, 132
–lipid–water interactions, 668
loss of during washing process, 307
muscle, 72, 75, 115, 127
myofibrillar, most abundant, 165
non-heme iron, 112, 114
oxidants, 476
Pacific whiting, solubility of, 65
partially unfolded globular, 167
paste, fish muscle, 567
pork plasma, 676
potent denaturant of, 199
–protein interaction, 71, 445, 446, 516, 547
recovery, 131, 239, 240
rockfish, 72
sarcoplasmic, 39, 63, 444
  removal of, 64, 67
  solubility, 70–71, 82–83, 116, 119
separation of membranes from, 126
solubility
  cod muscle, 197
  effects of different salts on, 75
soy, 424, 475, 547, 677
  major proteins in, 678
  purpose of adding, 736
stabilization, trehalose and, 210
stroma, 444
structure, spray-drying and, 206
–sucrose hydrogen bonding network, 206, 207
surface(s)
  hydrophobicity, 463
  solute exclusion from, 184
unfolding, 170, 456
washing conditions and total extractable, 68
water-holding capacity of, 128
–water interface, 186
whey, 547, 669
  gelation of, 671
  processing diagram of, 670
  timing of addition of, 699
Proteinase(s), 228–229
  activity, heat processing and, 446
  alkaline, 245
  Ca^{2+}-activated, 241
  cysteine, 252, 255
  histidine-linked, 231
  inhibitors, 250
    food-grade, 253
    gene cloning technology and, 257
    soy, 258
  muscle, 308
  myofibril-bound serine, 250
  neutral, 241
  sarcoplasmic vs. myofibrillar, 249
  serine, 251
  texture softening and, 307
  viscera, 307

Index

Proteins, stabilization of, 163–225, *see also* Protein
- future developments in fish protein stabilization, 210–212
- intrinsic stability of fish muscle proteins, 170–176
  - antifreeze proteins, 175–176
  - influence of animal body temperature, 170–172
  - naturally occurring osmolytes, 172–175
- mechanisms for cryoprotection and cryostabilization, 184–195
  - antioxidants, 189
  - cryostabilization by high molecular weight additives, 191–195
  - freezing-point depression, 189–191
  - ligand binding, 188–189
  - solute exclusion from protein surfaces, 184–188
  - vitrification, 195
- myosin and fish proteins, 165–166
- processing effects on surimi stability, 196–198
  - fish freshness, 196
  - freezing rate, 197–198
  - leaching, 197
- stability of frozen surimi proteins, 176–184
  - cold destabilization, 177
  - hydration and hydration forces, 179–181
  - ice crystallization, 177–179
  - other destabilizing factors during frozen storage, 181–184
- stability of myosin, 166–170
  - helix-coil transition in myosin rod, 168–170
  - stabilization of globular head region, 166–168
- stabilization of fish proteins to drying, 201–210
  - potential additives and mechanisms of lyoprotection, 206–210
  - processes for drying of surimi, 202–206
- stabilized fish mince, 198–201

Proteolysis
- low temperatures and, 61
- minimization of by process control, 259

Proteolytic enzymes and control in surimi, 227–277
- classification of proteolytic enzymes, 231–249
  - acid proteases (lysosomal cathepsins), 232–241
  - alkaline proteinases, 245–249
  - neutral and Ca^{2+}-activated proteinases, 241–244
- control of heat-stable fish proteinases, 250–260
  - food-grade proteinase inhibitors, 253–258
  - minimization of proteolysis by process control, 259–260
  - proteinase inhibitors, 250–253
  - sarcoplasmic vs. myofibrillar proteinases, 249–250

PS, *see* Polystyrene
Pseudosciaena
  manchurica, 36
  polytis, 23
Psychological behavior knowledge, 817
Punch test, 92, 529, 533, 537
Purchase drivers, 806
Pure Food Act, Food Additive Amendment to, 739
PVDC, *see* Polyvinylidene chloride
PVP, *see* Polyvinylpyrrolidone
PWP, *see* Processed whey proteins

Q

Quality control, 425, 843
Quinols, 112

R

Radiation processing, 630
Radio frequency (RF) heating, 259
Rainbow trout, alkaline proteinase in, 246–247
Raman spectroscopy, 463
Rancidity, development of, 112
Random fiber products, 418
Raoult's law, 190
Rapid cooling, 149, 411
Rapid freezing, 415
Raw material(s)
   fluid properties of, 495
   inspection of, 396, 398
   microbial load of, 148
   minimized cost of, 543
Raw surimi tests, 872, 879
R&D, *see* Research and Development
Reactive oxygen species
   destructive effects of, 111
   formation, prevention of, 112
Ready-to-eat (RTE) products, 152, 593
Recombinant technology, 257, 270
Recovered protein, cathepsin activities of, 239, 240
Recovery-grade surimi, 89, 302
Red mullet, 19, 23, 27
Refrigerant
   ammonia as, 329
   condensing temperature, 366
   evaporating temperature, 366
   flow rate of boiling, 351
   halocarbon as, 329, 330
   secondary, 330
   vaporizing, 336
Refrigerated seawater (RSW) system, 56, 57

Refrigeration
   machinery options, 366
   system(s)
      temperature profiles, 57
      types of, 56
Regulatory agencies, surveillance programs of, 142
Reiner-Riwlin equation, 554
Relaxation modulus, 504
Religious requirements, 795
Research and Development (R&D), 811, 832
Resources, surimi, 3–32
   changes in surimi supply and demand, 26–29
   cold-water whitefish used for surimi, 5–19
      Alaska pollock, 5–13
      arrowtooth flounder, 16–17
      Northern blue whiting, 19
      other whitefish resources (South America), 19
      Pacific whiting, 13–16
      Southern blue whiting and hoki, 17–19
   pelagic fish used for surimi, 25–26
   tropical fish used for surimi, 19–25
      bigeye snapper, 22–23
      croaker, 23
      lizardfish, 21–22
      other species, 23–25
      threadfin bream, 20–21
Retort packaging, 637
Reverse osmosis (RO), 59
RF heating, *see* Radio frequency heating
Rheology and texture properties of surimi gels, 491–581
   effects of processing parameters, 540–549
      fish freshness/rigor condition, 540
      freeze-thaw abuse, 546
      functional additives, 546–548

low temperature setting,
544–545
moisture content, 543
refrigerated storage of gels,
540–541
sample temperature at
measurement, 541–543
texture map, 548–549
empirical tests, 528–539
punch test, 529–534
relationship between torsion
and punch test data,
537–539
texture profile analysis,
534–537
fundamental test, 496–528
deformation and strain,
498–499
flow and rate of strain,
499–500
force and stress, 496–498
rheological testing using
large strain, 518–528
rheological testing using
small strain, 501–518
practical application of dynamic
rheological
measurements, 558–574
estimation of steady shear
viscosity of fish muscle
protein paste, 567–574
gelation kinetics of surimi
gels, 559–567
viscosity measurements,
549–558
measurement of dilute
extract, 550
measurement of surimi
seafood pastes, 550–556
rheological behavior of surimi
paste, 556–558
Ribonucleotides, 727, 728
Rigor mortis, 51
Rinse water, 160
RO, *see* Reverse osmosis
Rockfish proteins, 72

Rod shape, application of in
bending, 522
RSW system, *see* Refrigerated
seawater system
RTE products, *see* Ready-to-eat
products
Rubber elastic theory, 562

S

SA, *see* Sodium-L-ascorbate
Salmon
freezing time, measured
variation of, 353
spawning migration, cathepsin
activity during, 233
Salmonella
enteritidis, 585
species, 152
spp., 589
Salt
bridges, 452
myofibrillar protein extraction
and, 736
Salted-dried fish, 24
Saltwater fish, muscle of, 70
Sanitation
chemical technical
representative, 154
definition of, 143
standard operating procedures
(SSOPs), 145, 153
Sanitation and HACCP, 141–162
cleaners and sanitizers, 154–158
good manufacturing practices,
144–15
HACCP for surimi production,
148–149
HACCP for surimi seafood
production, 149–151
Hazard Analysis Critical Control
Point, 145–146
microbiological standards and
specifications for surimi
seafood, 151–153

principles of HACCP system, 146–148
sanitation standard operating procedures, 153
verification, 158–161
Sanitizers
commonly used, 159
objective of, 156
Sanitizing, definition of, 154
SAOS measurements, see Small amplitude oscillatory shear measurements
Sarcoplasmic proteins, 39, 63, 444
enzymes, 447
heme proteins, 446
removal of, 64, 67
solubility, 70–71, 82–83, 116, 119
Sardine
alkaline proteinase in, 246–247
harvesting of in sea around Japan, 376
ordinary muscle, 109
proteolytic enzyme levels in, 449
triglycerides from, 857
Sardinops melanostrictus, 36
Satsuma-age, 376, 378, 390
Saurida
spp., 19, 49
tumbil, 22
undosquamis, 22
waniese, 22
SBR system, see Sequencing batch reactor system
Scallop-style products, 396
Sciaenidae, 23
Scombroid poisoning, 588
Screw press, 41
SDR, see Structure development rate
SDS, see Sodium dodecyl sulfate
SDS-PAGE, 863
Alaska pollock proteins, 636
MHC degradation and, 635
myosin cross-linking reaction, 858, 859, 860

Seafood
extracts, components found in, 724
off flavors of, 730
plants, profit center for, 290
spoilage, 722
volatile flavor compounds found in, 723
SECODIP, 385
Secondary refrigerant, 330
Semiquinones, 112
Sensory evaluation, human resources for, 821
Sensory lab(s)
most common, 822
principles of, 820
Sensory science, application of to surimi seafood, 803–845
brief history, 806–807
correlating sensory evaluation with instrumental and consumer measures, 832–837
linking consumer data with analytical data, 835–837
why instrumental measures are not enough, 832–835
definition of sensory evaluation, 805
development of sensory approach, 813–832
consumer tests, 828–831
problem, 813–817
sensory human resources, 817–820
sensory laboratory and sensory test conditions, 820–822
sensory tests, 823–826
statistics for descriptive analysis, 826–828
tests and panelists, 831–832
fundamentals of sensory evaluation, 807–810
multidimensional sensory answer, 809–810

Index

sensory measurement, 807
sensory perception, 807–808
industrial potentialities, 811–813
  marketing, 812–813
  production, 812
  research and development, 811–812
sensory evaluation from lab to consumers, 837–843
why we should care about sensory evaluation, 805–806
Sepsis, 587
Sequencing batch reactor (SBR) system, 298
Serine proteases, inhibition of, 236
Setting, protein stability effects on, 467
Shear
  modulus, 494
  oscillatory, 510
  pure, 524
  strain, 538
    gel cohesiveness and, 46
    relationship between shear stress and, 539
  stress, 498, 538
  type test, 501
Shelf life, 738
Shore-side surimi operation, water usage in, 285
Shrimp-flavored fish, 381
SI system, *see* Slush ice system
Skin-on fillet, 39
Slow freezing, 45
Slow setting gels, 24
Slush ice (SI) system, 56
Small amplitude oscillatory shear (SAOS) measurements, 561
Small strain testing, 495
Sodium-L-ascorbate (SA), 689
Sodium chloride brine, 337
Sodium dodecyl sulfate (SDS), 635

Sodium lactate, actomyosin and, 188
Software of the mind, 77
Soft water, 59
Solid–liquid separation, 301
Solid meat, 395
Solid waste, 288–293
  diluted, 294
  fish meal and fish protein hydrolysates, 288–290
  fish oil recovery, 290–292
  flow chart of surimi processing, 289
  separation of from wastewater, 282
  specialty products, 292–293
Solute exclusion principle, 187
Solution viscosity, 191
Solvent extraction, 715
Sorbitol, 202, 211, 465–466, 734
South America, surimi production in, 27
Southern blue whiting, 17, 18, 36
South Korea, surimi-based products manufactured in, 385
Soy
  cystatin inhibitor, 14
  protein(s), 475, 677
    isolate, 547
    major proteins in, 678
    purpose of adding, 736
  proteinase inhibitors, 258
Soybean oil, 686
Species habitat temperature, activation energy and, 171, 172
Spectrophotometer, 762
Sphyraena
  langsar, 24
  obtusata, 24
  spp., 19
Spiral freezers, 331, 332, 340
Spray-drying, 205, 731, 738
SSOPs, *see* Sanitation standard operating procedures

Stabilization, 186
Standard Sanitation Operating
    Procedures (SSOPs), 143
Staphylococcal enterotoxins, 585
Staphylococcus aureus, 151, 152,
    586, 589, 633
Starch(es)
  amylose, 659
  composition of, 657, 660
  effects of on flavor compounds,
    735
  gelatinization of, 661
  granules, expansion of, 661
  modification of, 657
  native, 664
  potato, 659, 661, 663, 699
  pre-gelatinized, 664
  uses, 734
  wheat, 681
Stationary freezers, thermocouple
    meter used for, 356
Stephanoleptis cirrhifer, 20, 24, 25
Storage
  conditions, 738
  temperature, 417
Strain
  rate of, 499
  true, 519
Streptococcus iniae, 587
Streptoverticillum sp., 861
Stress(es)
  decay plots, 505
  definition of, 496
  equilibrium, 505
  measurement, low-temperature
    setting and, 544
  relaxation
    curve, 505
    relationships between
      rupture properties and,
      508
    rheological model for, 506
    test, 503
  shape-changing, 523
  shear, 498
Stroma proteins, 444

Structure development rate (SDR),
    566
Substitution, purpose of, 658
Sucrose addition, 465
Sugar, 194, 734
Sulfur-containing compounds, 720,
    725–726
Supercritical extraction, 721
Superoxide, source of, 112
Surface tension measurements, 187
Surimi
  blended, 428
  block, heat capacity of, 345
  flavor of, 711
  freeze-dried, 46
  frozen shelf life of, 690, 696
  frozen storage stability, 476
  gel-forming ability, 39, 878
  gel strength, 437
  price, gelling ability and, 230
  recovery-grade, 89, 302
  –starch gels, color of, 665
  test gel, heating pattern, 95
Surimi-based products
  8th century, 376
  original model of, 390
  value-added, 382
Surimi industry
  biggest problem for, 315
  sensory-related questions,
    814–816
Surimi manufacturing, 37–48
  flow chart of, 37
  freezing, 44–46
  heading, gutting, and deboning,
    37–38
  metal detection, 47–48
  mincing, 38–39
  refining, 40–41
  screw press, 41–42
  stabilizing surimi with
    cryoprotectants, 42–44
  washing and dewatering, 39–40
Surimi paste
  changes in viscosity of, 503, 504
  color-uncooked, 762

Index   919

flow properties of, 502, 549
rheological behavior of, 556
storage modulus of, 560
zero viscosity, 572
Surimi processing
    automated, 77
    major stages of, 281
    waste water, flow chart of, 282
Surimi production
    history of world's, 36
    machinery, contamination and, 148
    rate, 84
    world trends, 26
Surimi products, semantic attributes, 825, 826
Surimi quality
    factors affecting, 76
    harvesting conditions and, 53
    intrinsic factors affecting, 48
Surimi resources, world map showing major, 6
Surimi seafood
    color and acceptance of, 752
    concentration of cryoprotectants in, 343
    crab-flavored, 381
    e-beam processing of commercial, 634
    electron penetration in, 631
    filament-style, 403
    formula development, 700
    major functional properties of, 650
    major ingredient of, 148
    manufacturers, color additive groups of, 769
    microbial inactivation in, 633
    mosaic shape, 409
    packaging, 412, 637
    pastes, measurement of, 550
    pasteurized, 411
    pathogenic microorganisms of concern in, 600
    potential hazards for, 146
    production, flow chart for, 150
    religious requirements, 795
    sensory glossary, 825
    success of in France, 384
    textural properties of, 559
    vacuum-packed, 150, 617
Sustainable fisheries, 5
Swab test, 160

T

TAC, see Total allowable catch
Tachiuo, 23
Taste
    molecules, 733
    receptor, umani, 719
TA XT plus Texture Analyzer, 534, 535
TCs, see Thermocouples
TDT, see Thermal death time
Temperature
    distribution test design, 608
    prediction model, 415, 624, 625
Tenpura, 390, 392
Tentiki color, 773
Test(s)
    axial tensile, 499
    biting, 885
    compressive, 501
        convex shape samples, 520
        rod-type samples, 521
    consumer, 828
    cooked surimi, 874, 881
    deformation, 499
    descriptive, 824
    difference, 823
    empirical, 528
    failure, 518
    folding, 885
    imitative, 495
    oscillatory dynamic, 508
    punch, 529, 533
    puncture, 874
    raw surimi, 872, 879
    rheological, 501
    sensory, 833

shear type, 501
stress relaxation, 503
threshold, 823
torsion, 523, 528, 876
Testing
　empirical, 494
　machines, 499
　rheological, 518
　small strain, 495
Textural deterioration, 49
Texture
　degradation, heat-induced, 259–260
　development, 403
　map, 548, 651
　modifier, soybean oil as, 686
　profile analysis (TPA), 534
　　curves, 536
　　problem using, 535
　random fibers, 418
　slitting degree and, 405
　softening, 307
TGase, see Transglutaminase
Thailand
　crabstick manufacturing in, 387
　current practice of surimi from, 53
　harvesting of threadfin bream in, 55
Theragra chalcogramma, 5, 35, 49, 870
Thermal arrest zone, 361–362
Thermal death time (TDT), 601, 602, 604, 608
Thermal gelation, 863
Thermal hysteresis proteins (THPs), 211
Thermal stability, 467
Thermister probes, 356
Thermocouples (TCs), 608
THPs, see Thermal hysteresis proteins
Threadfin bream, 19, 20, 36, 49
　current practice of surimi from, 53
　frozen-on-board, 21

harvesting, 55, 376
surimi
　processing of, 28
　quality of, 28
Threshold
　tests, 823
　value, definition of, 723
Thrombin, 676
Tilapia, 448
　actomyosin, cryoprotection of, 187, 188
　cathepsin B purified from, 235
　proteolysis of, 250
Titanium dioxide, 788
TMA, see Trimethylamines
TMAO, see Trimethylamine oxide
TMAOase, see TMAO demethylase
Torsion
　mold, 527
　test, 523, 528, 876
Total allowable catch (TAC), 7
Total lethality, 606
Total solids (TS), 286, 294, 315
Total suspended solids (TSS), 286, 296–297, 301, 315
TPA, see Texture profile analysis
Trachurus murphyi, 25, 36
Transglutaminase (TGase), 308, 445, 461, 692
　activity, variability in, 468
　Alaska pollock surimi, 860–861
　mass production of, 693
　role of in setting of surimi, 464
　thermal stability of, 694, 695
Trawlers
　coastal, 54, 55
　factory, 54, 57
　shore-side pollock, 55
Trehalose, 43, 45
　–borate complexation, 209
　crystallization, 207
　freeze-thawing and, 207
　glass-forming characteristics, 208
　protein stabilization and, 210
Triacylglycerides, 450

Triacylglycerols, 113, 114
Trichiurus
   lepturus, 24
   spp., 19
Triglycerides, 857
Trimethylamine oxide (TMAO), 173, 174
   breakdown, 474, 476
   demethylase (TMAOase), 175, 449
      inhibition of activity, 201
      slowing of activity, 476
Trimethylamines (TMA), 311
Tristimulus
   colorimeter, 760, 761
   difference measurements, 791
Tropical fish, 19
Trout fillets, hemoglobin present in washed, 121
True strain, 519
Trypsin, 251, 307
TS, *see* Total solids
TSS, *see* Total suspended solids
Tsuke-age, 390
Tsumire, 852
Tunnel freezer, 332, 333
Turmeric, color produced by, 783

U

Ultraviolet (UV) sterilization, 638
Umami, 718, 719, 727
United Nations, flavor regulations, 742
United States
   Alaska pollock utilization, 10
   changes of surimi seafood consumption in, 377
   common refrigeration systems in, 56
   crabstick preference in, 734
   Department of Agriculture, listeriosis cases estimated by, 591
   flavor regulations, 739
   Food and Drug Administration (FDA), 592
      categories of colorants, 770
      chocolate industry GRAS application to, 407
      food additives subject to approval by, 739–740
      guidance, adequacy of pasteurization process, 607
      health claim for soy protein, 677
      Health Hazard Evaluation Board, 47, 151
      pasteurization process, 601
      request for labeling, 794
      uncertified color listing, 787
   Food Labeling Law, 794
   Government, fishery management by, 27
      name for surimi-based seafood in, 380
   PGPR use in, 775
   pollock surimi, 26
   surimi industry
      chopping procedures, 673
      criticism of, 10
      use of frozen surimi in, 388
Universal Testing Machine, 499, 521
Unsaturated fatty acids, 111
Upeneus spp., 19, 23
Urophycis
   chuss, 683
   tenuis, 473
UV sterilization, *see* Ultraviolet sterilization

V

Vacuum drying, 202
Vacuum packaging, 637
Vacuum silent cutter, 399, 672
Value-added surimi-based products, 382

Variable frequency drive (VFD), 368
Variance analysis, 828
Vegetable oil, 686, 687
  perceived whiteness and, 789
  uses, 736
VFD, *see* Variable frequency drive
Vibrio
  cholerae, 152
  parahaemolyticus, 152, 585, 590
  spp., 587
  vulnificus, 152
Viscera proteinases, 307
Viscoelastic properties, material, 509
Viscoelastic solids, compressed stress decay, 505
Viscosity measurements, uses of, 549
Vitamins, fat-soluble, 310
Vitrification, 195
Volatile compounds, 723
  alcohols, 725
  aldehydes, 725
  bromophenols, 726
  ketones, 725
  N-containing compounds, 725
  sulfur-containing compounds, 725–726

W

Wash cycles (WC), 66, 67
Wash time (WT), 67
Washwater
  loss of myofibrillar proteins in, 83
  ratio, 66
  recycled, 83
Waste
  heat, recovery of, 367
  solids recovery, 287
  -soluble proteins, recovery of, 304
  streams, classification of, 286
  utilization, options for, 288
Waste management and by-product utilization, 279–323
  opportunities and challenges, 311–315
    current and future potential, 313–315
    limitations of fish solids recovery, 311–313
  recovery of bioactive components and neutraceuticals, 306–311
    bioactive compounds, 306–310
    recovery of bioactive compounds, 310–311
  solid waste, 288–293
    fish meal and fish protein hydrolysates, 288–290
    fish oil recovery, 290–292
    specialty products, 292–293
  surimi waste management and compliance, 283–288
    implementation of waste management program, 286–288
    measurements needed for compliance, 284–286
  surimi wastewater, 293–306
    biological methods, 297–299
    chemical methods, 295–297
    physical methods, 299–306
Wastewater
  application of polyacrylamide to, 295–296
  biological treatment of, 297
  chemical treatment of, 295
  composition of, 293
  disposal, 66
  flow chart of, 282
  flow meters, 284
  innovative technology for, 82
  pH, 295
  physical treatment of, 299

streams, recovery of soluble proteins in, 314
treatment methods, 295
Water
  freezing point of, 190
  hard, 155
  leaching process, 184
  /meat (W/M) ratio, 67, 68
  -in-oil emulsifiers, 774
  -phase salt (WPS), 596, 598
  polar nature of, 653
  quality, 655
  recycling, 83
  reduction/reuse, 287
  replacement mechanism, 208
  restructuring of on hydrophobic surfaces, 180
  rinse, 154
  solidification of, 179
  usage, shore-side surimi operation, 285
  well, 59
WC, see wash cycles
Wheat gluten, 437, 547, 668, 681
Whey protein(s), 547, 669
  addition, timing of, 699
  concentrate, 253–254
  gelation of, 671
  processing diagram of, 670
  use of in surimi production, 14
White croaker, 49
  alkaline proteinase in, 246–247
  proteolytic activity, 229
Whitefish
  cold-water, 5–19
    Alaska pollock, 5–13
    arrowtooth flounder, 16–17
    Northern blue whiting, 19
    other sources, 19
    Pacific whiting, 13–16
    Southern blue whiting and hoki, 17–19
  supply, global decrease in, 4
  total harvests of in main fishing areas, 9
Whiteness indices, 700, 701

Whiting, see also Pacific whiting
  blending grades of, 426
  Chilean, 19
  nonrefrigerated, 79
  Northern blue, 19, 36
  Peruvian, 25
  Southern blue, 17, 36
  surimi, quality of freeze-dried vs. frozen, 204–205
Whiting Conservation Cooperative, 16
Williams-Landel-Ferry kinetics, 192
W/M ratio, see Water/meat ratio
Wolf herring, 24
World Food Logistics Organization, 354
WPS, see Water-phase salt
WT, see Wash time

X

XYZ scale, 757

Y

Yaki-ida, 390
Yakinuki kamaboko, 851
Yeast
  extracts, 729
  thermo-tolerant, 622
Yellow croaker, 36
Yersina enterocolitica, 589
Young's modulus, 494, 521

Z

Z-disk, 116
  degradation, 243, 244
  disappearance of, 241
  pre-rigor muscle, 241
Zero viscosity, 572, 573
z-value, 602, 615